Audio Power Amplifier Design

Audio Power Amplifier Design

Sixth Edition

Douglas Self

Focal Press
Taylor & Francis Group

NEW YORK AND LONDON

This edition published 2013
by Focal Press
70 Blanchard Rd Suite 402, Burlington, MA 01803

Simultaneously published in the UK
by Focal Press
2 Park Square, Milton Park, Abingdon, Oxon OX14 4RN

Focal Press is an imprint of the Taylor & Francis Group, an informa business

Notices

Knowledge and best practice in this field are constantly changing. As new research and experience broaden our understanding, changes in research methods, professional practices, or medical treatment may become necessary.

Practitioners and researchers must always rely on their own experience and knowledge in evaluating and using any information, methods, compounds, or experiments described herein. In using such information or methods they should be mindful of their own safety and the safety of others, including parties for whom they have a professional responsibility.

Product or corporate names may be trademarks or registered trademarks, and are used only for identification and explanation without intent to infringe.

Library of Congress Cataloging in Publication Data
Self, Douglas.
Audio power amplifier design / Douglas Self. – Sixth edition.
pages cm
1. Audio amplifiers–Design and construction. 2. Power amplifiers–Design and construction. I. Title.
TK7871.58.A9S45 2013
621.389'3–dc23
2012048301

ISBN: 978-0-240-52613-3
ISBN: 978-0-240-52614-0

Typeset in Times New Roman
By TNQ Books and Journals, Chennai, India

Bound to Create

You are a creator.

Whatever your form of expression — photography, filmmaking, animation, games, audio, media communication, web design, or theatre — you simply want to create without limitation. Bound by nothing except your own creativity and determination.

Focal Press can help.

For over 75 years Focal has published books that support your creative goals. Our founder, Andor Kraszna-Krausz, established Focal in 1938 so you could have access to leading-edge expert knowledge, techniques, and tools that allow you to create without constraint. We strive to create exceptional, engaging, and practical content that helps you master your passion.

Focal Press and you.

Bound to create.

We'd love to hear how we've helped you create. Share your experience:
www.focalpress.com/boundtocreate

This book is dedicated to Julie,
with all my love, and with gratitude
for all her help and support in the endeavour

Contents

Preface to Sixth Edition

This book appeared in its Fifth edition in 2009. Since then I have done a good deal of research to improve our knowledge of solid-state power amplifiers and their design issues, partly in the course of my consultancy work, and partly in the spirit of scientific enquiry. As a result this edition has been greatly extended, the number of chapters increasing from 21 to 30.

The many new topics include: the characteristics of the audio signal, the principles of distortion, error-correction, non-switching output stages, the details of VAS distortion, push-pull VAS configurations, opamp-array power amplifiers, series output stages, output-inclusive compensation, the optimisation of output coils, the power dissipation in various kinds of amplifiers, and a brief history of solid-state amplification. A complete procedure for designing the heatsinking and power supply is given, going down to the detail of fuse ratings and reservoir ripple-current. Five amplifier design examples that illustrate important design principles are closely examined.

The only material that has been removed is that dealing with some of the more specialised types of balanced line input stages. This information can now be found, in much expanded form, in my book, *Small Signal Audio Design*. However, the line input chapter has also had some completely new material added on instrumentation amplifiers which is of special importance to power amplifiers.

A criticism sometimes voiced of previous editions of this book was that it focused mainly on one configuration, the three-stage amplifier consisting of an input differential pair with current-mirror load, driving a constant-current Voltage-Amplifier Stage (VAS), which is enhanced by adding either an emitter-follower inside the Cdom loop, or a cascode transistor to the VAS collector. The VAS then drives a unity-gain output stage. This may be the most conventional configuration of the lot, but it has become clear that it also the most effective and economical, so it still forms the basis of this book. My recent researches underlined how free of vices and complications it is, compared with other configurations that on the surface appear more sophisticated. In particular the push-pull VAS has now been thoroughly investigated, and two well-known forms of it have been found to be unsatisfactory, though a third is workable. There seems no need to ever go to the extra complication of a four-stage amplifier.

I believe that this edition goes a level deeper than ever before in describing power amplifier behaviour. A major contribution to this has been made by Samuel Groner in his commentary on the Fifth edition of this book. The commentary should be read by anyone with an interest in power amplifier design; it can be found on his website at http://www.sg-acoustics.ch/analogue_audio/index.html. It was this that led me to study VAS distortion more closely, and I can say that in every topic that I have explored, I have found Samuel's findings to be absolutely correct.

Perhaps the most important new information in this edition is that relating to methods of compensation that are more advanced than the near-universal Miller dominant pole. I describe how to make a stable amplifier in which the output stage is enclosed in the Miller loop. This allows much more negative feedback to be applied to the stage which generates the troublesome crossover distortion, and the improvement in linearity is dramatic. Chapter 12 describes the performance of two amplifiers with output-inclusive Miller compensation. One version gives 0.00078 % (7.8 ppm) at 10 kHz and 0.002% (20 ppm) at 20 kHz. At the time of writing this is something of a Personal Best.

The distortion performance is considerably better than that obtainable from a straightforward Blameless amplifier, and I consider it needs a name of its own. In short, it is an Inclusive Amplifier. Beyond Blameless!

Chapter 16 demonstrates that Class-A amplifiers are not just inefficient; with musical signals they are hopelessly inefficient — somewhere around 1%. There is now really no need to resort to Class-A just to get low distortion, so their attraction is as much philosophical as anything else. I believe you should think hard before planning to build one with more output than about 20 W/8 Ω.

Interest in vintage hi-fi, especially pre and power amplifiers of the 1970s and 1980s, has increased greatly in the last few years; in response to this I have added many references to significant models and their technologies. (The date of introduction is in brackets after the model name.) I have also added a chapter that gives a brief overview of the development of solid-state amplifiers; how we got from there to here. One strand of amplifier history is the intriguing topic of amplifiers with non-switching output stages; this topic seems to be surfacing on the audio bulletin boards with increasing frequency at present. It is therefore extensively described in Chapter 4.

My research into power amplifier design is an ongoing activity. It is has not been possible to tie everything up neatly, and at a few points I have simply had to say 'This is as far as I have got.'

This is the only power amplifier book that is filled with measurements conducted on real amplifiers rather than with simulations of dubious accuracy. The new measurements for this edition were done using the state-of-the-art Audio Precision SYS-2702 THD analyser, and show lower noise and lower distortion residuals in the LF regions of the distortion plots. I have not yet tackled the Herculean task of repeating every measurement in the Fifth edition, so please bear in mind that many of these were done with an AP System 1, and this accounts for some apparent differences in performance.

The ultimate aim of this book is to enable the design of the perfect power amplifier. The major parameters by which a piece of audio equipment is judged are frequency response, noise, slew-rate and distortion. Designing power amplifiers with any reasonable frequency response presents no problems at all. The noise level of existing Blameless designs is already less than that of almost any small-signal stage that you can put in front of it. There are few problems in achieving a slew-rate that is many times greater than that required by any credible audio signal.

That leaves distortion as the great unsolved power amplifier problem; in particular, crossover distortion in a Class-B output stage is still very much with us, caused by a small wobble in the output stage gain that looks, but is not, insignificant. My introduction of Class XD (crossover displacement) moves the wobble so it only occurs at medium output levels, and in its push-pull form reduces its size, but it does not abolish it. We have been designing solid-state amplifiers for more than forty years but the distortionless amplifier remains as a challenge to humanity, and that is why much of this book is concerned with distortion. One aspect of this is the difficulty in defining levels at which distortion is definitely inaudible. My answer to this has always been the same; reduce the distortion until it is beneath the noise floor of the best testgear available; if you can't do that, then make it so low that no one could rationally argue it was perceptible. Something like 0.002% might be a possible criterion for that, but the issue is much complicated by the fear of what effects crossover distortion might have at low levels. Class XD at least removes that worry.

Sinking distortion beneath the noise floor assumes THD techniques and visual assessment of the residual on an oscilloscope. A criterion of 'visually lost in the noise' does not definitively mean 'always inaudible' but I suggest it is a pretty good start. This may give slightly different results for analogue and digital scopes. Obviously FFT methods can winkle out harmonics below the noise floor, and make a more demanding distortion criterion possible, but the relevance of this to human perception is unclear. Unlike noise, distortion can in theory be reduced to zero.

Another area in which we could do with a bit more progress is amplifier efficiency. Class-D amplifiers are certainly efficient, but in every other area their performance falls well below that of Class-B and other analogue techniques. Class-G, which I think in many cases is a better solution, especially if combined with advanced compensation techniques, is dealt with in detail in Chapter 19.

As you will have gathered, I am still fascinated by the apparently simple but actually complex business of making small signals big enough in voltage and current to drive a loudspeaker. A significant part of the lure of electronics as a pursuit is the speed with which ideas can be turned into physical reality. In audio amplifier design, you very often just need just a handful of components, a piece of prototype board and a few minutes to see if the latest notion really is correct. If you come up with a brilliant new way of designing large concrete dams, then it is going to take more than an afternoon to prove that it works.

You will also see, in Chapter 1, that in the past few years I have found no reason to alter my views on the pernicious irrationality of Subjectivism. In that period I have repeatedly been involved in double-blind listening tests using experienced subjects and proper statistical analysis, which confirmed every time that if you can't measure it, it's not there, and if it's not there, you can't hear it. Nevertheless Subjectivism limps on, and very tedious it is too. In the words of Elvis Costello, I used to be disgusted; now I try to be amused.

There is in this book a certain emphasis on commercial manufacture, which I hope does not offend those mainly interested in amateur construction or pure intellectual enquiry. This is based on my experience designing for a number of well-known audio companies, and on my current consultancy activities. Commercial equipment has to work. And keep working. This is still a valuable discipline even if you are making a one-off design to test some new ideas; if the design is not reliable, then it may be unsound in some way that has more impact on its operation than you think.

The ideas in this book are embodied in the designs sold by The Signal Transfer Company, which supplies them as bare PCBs, kits, or built and tested units. These designs are all approved by me, and are the best starting point if you want to experiment, as potentially tricky problems like inductive distortion are eliminated by the PCB layout. The Signal Transfer Company can be found at http://www.signaltransfer.freeuk.com/.

To the best of my knowledge, this book has been completed without supernatural assistance.

Writers in the eighteenth century were wont to boast that new editions of their books had been 'Free'd of all Errors and Obscurities'. I don't know if I can claim that, but I have done my best. I hope you enjoy it. Any suggestions for the improvement of this book that do not involve its public burning will be gratefully received. You will find my email address on the front page of my website at douglas-self.com.

Acknowledgements

My heartfelt thanks to:

Samuel Groner for inspiring and supporting the investigations into VAS distortion and current-mirror noise, and for proof-reading the relevant chapters.

Gareth Connor of The Signal Transfer Company for unfailing encouragement, and providing the facilities with which some of the experiments in this book were done.

Averil Donohoe for her help with the trickier bits of mathematics.

List of Abbreviations

I have kept the number of abbreviations used to a minimum. However, those few are used extensively, so a list is given in case they are not all blindingly obvious:

BJT	Bipolar junction transistor
CFP	Complementary feedback pair
C/L	Closed loop
CM	Common mode
CMOS	Complementary metal oxide semiconductor
CMRR	Common-mode rejection ratio
CTF	Current timing factor
DF	Damping factor
DIS-PP-VAS	Double Input Stage Push-Pull VAS
DSP	Digital signal processing
EF	Emitter-follower
EFA	Emitter-follower added
EIN	Equivalent input noise
ESR	Equivalent series resistance
FEA	Finite element analysis
FET	Field-effect transistor
HF	Amplifier behaviour above the dominant pole frequency, where the open-loop gain is usually falling at 6 dB/octave
IAE	Integrated absolute error
IC	Integrated circuit
IGBT	Insulated-gate bipolar transistor
I/P	Input
ISE	Integrated square error
LED	Light-emitting diode
LF	Relating to amplifier behaviour below the dominant pole, where the open-loop gain is assumed to be essentially flat with frequency
LSN	Large-signal non-linearity

MOSFET	Metal oxide semiconductor field-effect transistor
NF	Noise figure
NFB	Negative feedback
O/L	Open loop
O/P	Output
P1	The first O/L response pole, and its frequency in Hz (i.e. the 23 dB point of a 6 dB/octave roll-off)
P2	The second response pole, at a higher frequency
PA	Public address
PCB	Printed-circuit board
PDF	Probability density function
PPD	Power partition diagram
PSRR	Power-supply rejection ratio
PSU	Power-supply unit
PWM	Pulse width modulation
RF	Radio frequency
SID	Slew-induced distortion
SOA, SOAR	Safe operating area
SPL	Sound pressure level
Tempco	Temperature coefficient
THD	Total harmonic distortion
TID	Transient intermodulation distortion
TIM	Transient intermodulation
VAS	Voltage-amplifier stage
VCIS	Voltage-controlled current source
VCVS	Voltage-controlled voltage source
VI	Voltage/current

Amplifiers and The Audio Signal

The microphone-amplifier-loudspeaker combination is having an enormous effect on our civilization. Not all of it is good!
Lee De Forest, inventor of the triode valve

The Economic Importance of Power Amplifiers

Audio power amplifiers are of considerable economic importance. They are built in their hundreds of thousands every year, and have a history extending back to the 1920s. It is therefore surprising there have been so few books dealing in any depth with solid-state power amplifier design.

The first aim of this text is to fill that need, by providing a detailed guide to the many design decisions that must be taken when a power amplifier is designed.

The second aim is to disseminate the original work I have done on amplifier design in the last few years. The result of these investigations was to show that power amplifiers of extraordinarily low distortion could be designed as a matter of routine, without any unwelcome side-effects, so long as a relatively simple design methodology was followed. I have called these Blameless amplifiers, to emphasise that their excellent performance is obtained more by avoiding mistakes which are fairly obvious when pointed out, rather than by using radically new circuitry. The Blameless methodology is explained in detail in Chapters 4 and 5 in this book. My latest studies on compensation techniques have moved things on a step beyond Blameless.

I hope that the techniques explained in this book have a relevance beyond power amplifiers. Applications obviously include discrete opamp-based pre-amplifiers,[1] and extend to any amplifier aiming at static or dynamic precision.

Assumptions

To keep its length reasonable, a book such as this must assume a basic knowledge of audio electronics. I do not propose to plough through the definitions of frequency response, THD and signal-to-noise ratio; this can be found anywhere. Commonplace facts have been ruthlessly omitted where their absence makes room for something new or unusual, so this is not the place to start learning electronics from scratch. Mathematics has been confined to a few simple equations determining vital parameters such as open-loop gain; anything more complex is best left to a circuit simulator you trust. Your assumptions, and hence the output, may be wrong, but at least the calculations in-between will be correct …

The principles of negative feedback as applied to power amplifiers are explained in detail, as there is still widespread confusion as to exactly how it works.

Origins and Aims

The original core of this book was a series of eight articles originally published in *Electronics World* as 'Distortion in Power Amplifiers'. This series was primarily concerned with distortion as the most variable feature of power amplifier performance. You may have two units placed side by side, one giving 2% THD and the other 0.0005% at full power, and both claiming to provide the ultimate audio experience. The ratio between the two figures is a staggering 4000:1, and this is clearly a remarkable state of affairs. One might be forgiven for concluding that distortion was not a very important parameter. What is even more surprising to those who have not followed the evolution of audio over the past two decades is that the more distortive amplifier will almost certainly be the more expensive. I shall deal in detail with the reasons for this astonishing range of variation.

The original series was inspired by the desire to invent a new output stage that would be as linear as Class-A, without the daunting heat problems. In the course of this work it emerged that output stage distortion was completely obscured by non-linearities in the small-signal stages, and it was clear that these distortions would need to be eliminated before any progress could be made. The small-signal stages were therefore studied in isolation, using *model* amplifiers with low-power and very linear Class-A output stages, until the various overlapping distortion mechanisms had been separated out. It has to be said this was not an easy process. In each case there proved to be a simple, and sometimes well-known cure, and perhaps the most novel part of my approach is that all these mechanisms are dealt with, rather than one or two, and the final result is an amplifier with unusually low distortion, using only modest and safe amounts of global negative feedback.

Much of this book concentrates on the distortion performance of amplifiers. One reason is that this varies more than any other parameter — by up to a factor of a thousand. Amplifier distortion was until recently an enigmatic field — it was clear that there were several overlapping distortion mechanisms in the typical amplifier, but it is the work reported here that shows how to disentangle them, so they may be separately studied and then with the knowledge thus gained, minimised.

I assume here that distortion is a bad thing, and should be minimised; I make no apology for putting it as plainly as that. Alternative philosophies hold that some forms of non-linearity are considered harmless or even euphonic, and thus should be encouraged, or at any rate not positively discouraged. I state plainly that I have no sympathy with the latter view; to my mind, the goal is to make the audio path as transparent as possible. If some sort of distortion is considered desirable, then surely the logical way to introduce it is by an outboard processor, working at line level. This is not only more cost-effective than generating distortion with directly heated triodes, but has the important attribute that *it can be switched off*. Those who have brought into being our current signal-delivery chain, i.e., mixing consoles, multi-track recorders, CDs, etc., have done us proud in the matter of low distortion, and to wilfully throw away this achievement at the very last stage strikes me as curious at best.

In this book I hope to provide information that is useful to all those interested in power amplifiers. Britain has a long tradition of small and very small audio companies, whose technical and production resources may not differ very greatly from those available to the committed amateur. I have tried to make this volume of service to both. I also hope that the techniques explained in this book have a relevance beyond power amplifiers. Applications obviously include discrete opamp-based pre-amplifiers,[2] and extend to any amplifier aiming at static or dynamic precision.

I have endeavoured to address both the quest for technical perfection — which is certainly not over, as far as I am concerned — and also the commercial necessity of achieving good specifications at minimum cost.

The field of audio is full of statements that appear plausible but in fact have never been tested and often turn out to be quite untrue. For this reason, I have confined myself as closely as possible to facts that I have verified myself. This volume may therefore appear somewhat idiosyncratic in places; for example, FET output stages receive much less coverage than bipolar ones because the conclusion appears to be inescapable that FETs are both more expensive and less linear; I have therefore not pursued the FET route very far. Similarly, most of my practical design experience has been on amplifiers of less than 300 Watts power output, and so heavy-duty designs for large-scale PA work are also under-represented. I think this is preferable to setting down untested speculation.

The Study of Amplifier Design

Although solid-state amplifiers have been around for some 40 years, it would be a great mistake to assume that everything possible is known about them. In the course of my investigations, I discovered several matters which, not appearing in the technical literature, appear to be novel, at least in their combined application:

- The need to precisely balance the input pair to prevent second-harmonic generation.
- The demonstration of how a beta-enhancement transistor increases the linearity and reduces the collector impedance of the Voltage-Amplifier Stage (VAS).
- An explanation of why BJT output stages always distort more into 4 Ω than 8 Ω.
- In a conventional BJT output stage, quiescent current as such is of little importance. What is crucial is the voltage between the transistor emitters.
- Power FETs, though for many years touted as superior in linearity, are actually far less linear than bipolar output devices.
- In most amplifiers, the major source of distortion is not inherent in the amplifying stages, but results from avoidable problems such as induction of supply-rail currents and poor power-supply rejection.
- Any number of oscillograms of square-waves with ringing have been published that claim to be the transient response of an amplifier into a capacitive load. In actual fact this ringing is due to the output inductor resonating with the load, and tells you precisely nothing about amplifier stability.

The above list is by no means complete.

As in any developing field, this book cannot claim to be the last word on the subject; rather it hopes to be a snapshot of the state of understanding at this time. Similarly, I certainly do not claim that this book is fully comprehensive; a work that covered every possible aspect of every conceivable power amplifier would run to thousands of pages. On many occasions I have found myself about to write: '*It would take a whole book to deal properly with* ...' Within a limited compass I have tried to be innovative as well as comprehensive, but in many cases the best I can do is to give a good selection of references that will enable the interested to pursue matters further. The appearance of a reference means that I consider it worth reading, and not that I think it to be correct in every respect.

Sometimes it is said that discrete power amplifier design is rather unenterprising, given the enormous outpouring of ingenuity in the design of analogue ICs.

Advances in opamp design would appear to be particularly relevant. I have therefore spent some considerable time studying this massive body of material and I have had to regretfully conclude that it is actually a very sparse source of inspiration for new audio power amplifier techniques; there are several reasons for this, and it may spare the time of others if I quickly enumerate them here:

- A large part of the existing data refers only to small-signal MOSFETs, such as those used in CMOS opamps, and is dominated by the ways in which they differ from BJTs, for example, in their low transconductance. CMOS devices can have their characteristics customised to a certain extent by manipulating the width/length ratio of the channel.
- In general, only the earlier material refers to BJT circuitry, and then it is often mainly concerned with the difficulties of making complementary circuitry when the only PNP transistors available are the slow lateral kind with limited beta and poor frequency response. Modern processes eliminated this problem a long time ago.
- Many of the CMOS opamps studied are transconductance amplifiers, i.e., voltage-difference-in, current out. Compensation is usually based on putting a specified load capacitance across the high-impedance output. This does not appear to be a promising approach to making audio power amplifiers.
- Much of the opamp material is concerned with the common-mode performance of the input stage. This is pretty much irrelevant to power amplifier design.
- Many circuit techniques rely heavily on the matching of device characteristics possible in IC fabrication, and there is also an emphasis on minimising chip area to reduce cost.
- A good many IC techniques are only necessary because it is (or was) difficult to make precise and linear IC resistors. Circuit design is also influenced by the need to keep compensation capacitors as small as possible, as they take up a disproportionately large amount of chip area for their function.

The material here is aimed at all audio power amplifiers that are still primarily built from discrete components, which can include anything from 10W mid-fi systems to the most rarefied reaches of what is sometimes called the 'high end', though the 'expensive end' might be a more accurate term. There are of course a large number of IC and hybrid amplifiers, but since their design details are fixed and inaccessible, they are not dealt with here. Their use is (or at any rate should be) simply a matter of

following the relevant application note. The quality and reliability of IC power amps have improved noticeably over the past decade, but low distortion and high power still remain the province of discrete circuitry, and this situation seems likely to persist for the foreseeable future.

Power amplifier design has often been treated as something of a black art, with the implication that the design process is extremely complex and its outcome not very predictable. I hope to show that this need no longer be the case, and that power amplifiers are now designable — in other words, it is possible to predict reasonably accurately the practical performance of a purely theoretical design. I have done a considerable amount of research work on amplifier design, much of which appears to have been done for the first time, and it is now possible for me to put forward a design methodology that allows an amplifier to be designed for a specific negative-feedback factor at a given frequency, and to a large extent allows the distortion performance to be predicted. I shall show that this methodology allows amplifiers of extremely low distortion (sub 0.001% at 1 kHz) to be designed and built as a matter of routine, using only modest amounts of global negative feedback.

The Characteristics of the Audio Signal

If we are designing a device to handle audio, it is useful to know the characteristics of the typical audio signal. Two of the most important parameters are the distribution of signal levels with time and with frequency.

Amplitude Distribution with Time

It is well known that normal audio signals spend most of their time at relatively low levels, with peaks that exploit most or all of the dynamic range being relatively rare. This is still true of music that has been compressed, but may not hold for the very heavy degrees of multi-band processing often used by radio stations, and regrettably sometimes for 'gain wars' CD mastering. It is of vital importance in the design of economical audio power amplifiers; for example, it is quite unnecessary to specify a mains transformer that can sustain the maximum sinewave output into the minimum load impedance indefinitely. For domestic hi-fi, de-rating the transformer so it can supply only 70% of the maximum sinewave current indefinitely is quite usual, and saves a very significant amount of money — and also weight, which reduces shipping costs. The only time this goes wrong is when someone who is not acquainted with this state of affairs subjects an amplifier to a long-term test at full power; as a very rough guide the transformer will overheat and fail in about an hour, unless it is fitted with a resettable thermal cutout. There is more on this in Chapter 26 on power supplies.

The amplitude distribution also has a major influence on the design of heat sinking, rating of bridge rectifiers, and so on.

One way to plot the distribution of signal levels over time is the Cumulative Distribution Function, or CDF. This sounds formidable, but is just a plot of level on the X-axis, with the probability that at any instant the signal exceeds that level on the Y-axis. Figure 1.1 shows the results I measured for three rock tracks and one classical track. You will note that the probability reaches 100% at the bottom limit of zero signal, and falls in a smooth curve to 0%, as obviously the probability that the signal will exceed the maximum possible is zero. There was no significant difference in the curves for the two genres, and other measurements on rock and classical music have given very similar results. The amplitude-time distribution, and its important implications for the power dissipated in different kinds of amplifier, are dealt with in much more detail in Chapter 16.

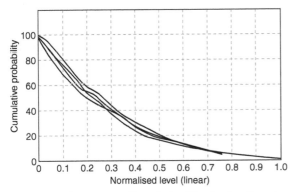

Figure 1.1. The measure Cumulative Distribution Function of instantaneous levels in musical signals from CDs (3 rock tracks and 1 classical).

Amplitude Distribution with Frequency

Once again, relatively little has been written on this important property of audio signals. While it is generally accepted that signal levels are significantly lower at high frequencies, actual figures are rare. I derived the distribution shown in Figure 1.2 from data published by

Greiner and Eggars[3] who analysed 30 CDs of widely varying musical genres. Note that they did this in 1989, before the so-called 'loudness wars' in CD mastering broke out.

Figure 1.2 shows that levels in the top octaves are some 15–20 dB lower than the maximum levels occurring between 100 Hz and 1 kHz, and this appears to be quite dependable across various styles of music. The levels at the bass end are more variable – obviously chamber music will have much lower levels than heavy rock, or organ music with determined use of the 32-foot stop (Bottom C = 16 Hz). It is therefore risky to try and economise on, say, reservoir capacitor size. However, other small economies suggest themselves. It is not necessary to design the Zobel network at an amplifier's output to withstand a sustained maximum output at 20 kHz, as this will never occur in real life. The downside to this is that a relatively modest amount of HF instability may fry the Zobel resistor, but then that won't be happening after the development and testing phases are completed …, will it?

The Performance Requirements for Amplifiers

This section is not a recapitulation of international standards, which are intended to provide a minimum level of quality rather than extend the art. It is rather my own view of what you should be worrying about at the start of the design process, and the first items to consider are the brutally pragmatic ones related to keeping you in business and out of prison.

Safety

In the drive to produce the finest amplifier ever made, do not forget that the Prime Directive of audio design is – Thou Shalt Not Kill. Every other consideration comes a poor second, not only for ethical reasons, but also because one serious lawsuit will close down most audio companies forever.

Reliability

If you are in the business of manufacturing, you had better make sure that your equipment keeps working, so that you too can keep working. It has to be admitted that power amplifiers – especially the more powerful ones – have a reputation for reliability that is poor compared with most branches of electronics. The 'high end' in particular has gathered unto itself a bad reputation for dependability.[4]

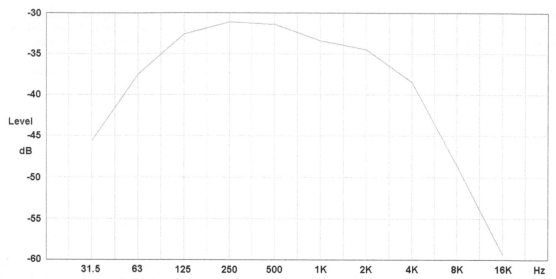

Figure 1.2. The average spectral levels versus frequency for musical signals from thirty CDs of varying genres (after Greiner and Eggars).

Power Output

In commercial practice, this is decided for you by the marketing department. Even if you can please yourself, the power output capability needs to be carefully thought out as it has a powerful and non-linear effect on the cost.

The last statement requires explanation. As the output power increases, a point is reached when single output devices are incapable of sustaining the thermal dissipation, parallel pairs are required, and the price jumps up. More devices, more mounting area on the heatsink, and more mounting hardware. Similarly, transformer laminations come in standard sizes, so the transformer size and cost will also increase in discrete steps.

Domestic hi-fi amplifiers usually range from 20W to 150W into 8Ω though with a scattering of much higher powers. PA units will range from 50W, for foldback purposes (i.e., the sound the musician actually hears, to monitor his/her playing, as opposed to that thrown out forwards by the main PA stacks; also called stage monitoring) to 1 kW or more. Amplifiers of extreme high power are not popular, partly because the economies of scale are small, but mainly because it means putting all your eggs in one basket, and a failure becomes disastrous. This is accentuated by the statistically unproven but almost universally held opinion that high-power solid-state amplifiers are inherently less reliable than those of lower capability.

If an amplifier gives a certain output into 8 Ω, it will not give exactly twice as much into 4 Ω loads; in fact it

will probably be much less than this, due to the increased resistive losses in 4 Ω operation, and the way that power alters as the square of voltage. Typically, an amplifier giving 180 W into 8 Ω might be expected to yield 260 W into 4 Ω and 350 W into 2 Ω, if it can drive so low a load at all. These figures are approximate, depending very much on power supply design.

Nominally 8 Ω loudspeakers are the most common in hi-fi applications. The *nominal* title accommodates the fact that all loudspeakers, especially multi-element types, have marked changes in input impedance with frequency, and are only resistive at a few spot frequencies. Nominal 8 Ω loudspeakers may be expected to drop to at least 6 Ω in some part of the audio spectrum. To allow for this, almost all amplifiers are rated as capable of 4 Ω as well as 8 Ω loads. This takes care of almost any nominal 8 Ω speaker, but leaves no safety margin for nominal 4 Ω designs, which are likely to dip to 3 Ω or less. Extending amplifier capability to deal with lower load impedances for anything other than very short periods has serious cost implications for the power-supply transformer and heatsinking; these already represent the bulk of the cost.

The most important thing to remember in specifying output power is that you have to increase it by an awful lot to make the amplifier significantly louder. We do not perceive acoustic power as such — there is no way we could possibly integrate the energy liberated in a room, and it would be a singularly useless thing to perceive if we could. What we actually perceive is the sound pressure on our eardrums. It is well known that power in watts must be quadrupled to double sound pressure level (SPL) but this is not the same as doubling subjective loudness; this is measured in Sones rather than dB above threshold, and some psychoacousticians have reported that doubling subjective loudness requires a 10 dB rather than a 6 dB rise in SPL, implying that amplifier power must be increased tenfold, rather than merely quadrupled.[5] It is any rate clear that changing from a 25 W to a 30 W amplifier will not give an audible increase in level.

This does not mean that fractions of a watt are never of interest. They can matter either in pursuit of maximum efficiency for its own sake, or because a design is only just capable of meeting its output specification.

Amplifier output power is normally specified as that obtained when clipping has just begun, and a given amount of extra distortion is therefore being generated. Both 0.1% THD and 1% THD are used as criteria; 10% THD is commonly used for Class-D amplifiers as it squeezes out a little more power at the expense of gross distortion. With some eccentric designs of poor linearity, these distortion levels may be reached well before actual clipping occurs.

This seems like a very straightforward measurement to make, but it is actually quite hard with an amplifier with an unregulated power supply, because the clipping point is very sensitive to mains voltage variations, and the THD reading fluctuates significantly. The general prevalence of distorted mains waveforms further complicates things, and it is best to make sure that the amplifier gives its advertised power with a safety margin in hand.

Some hi-fi reviewers set great value on very high peak current capability for short periods. While it is possible to think up special test waveforms that demand unusually large peak currents, any evidence that this effect is important in use is so far lacking.

Frequency Response

This can be dealt with crisply; the minimum is 20 Hz to 20 kHz, ±0.5 dB, though there should never be any *plus* about it when solid-state amplifiers are concerned. Any hint of a peak before the roll-off should be looked at with extreme suspicion, as it probably means doubtful HF stability. This is less true of valve amplifiers, where the bandwidth limits of the output transformer mean that even modest NFB factors tend to cause peaking at both high and low ends of the spectrum.

Having dealt with the issue succinctly, there is no hope that everyone will agree that this is adequate. CDs do not have the built-in LF limitations of vinyl and could presumably encode the barometric pressure in the recording studio if this was felt to be desirable, and so an extension to −0.5 dB at 5 or 10 Hz is perfectly feasible. However, if infrabass information does exist down at these frequencies, no domestic loudspeaker will reproduce them.

Noise

There should be as little as possible without compromising other parameters. The noise performance of a power amplifier is not an irrelevance,[6] especially in a domestic setting.

Distortion

While things happening to an audio signal such as frequency response alteration or phase shifts are sometimes called 'linear distortions' as they do not introduce any new frequency components into the signal, I find the phrase gratingly oxymoronic, and I firmly hold that distortion means non-linear distortion. The non-linearity means that new frequencies are created in the signal. My philosophy is the simple one that distortion is bad, and high-order distortion is worse. The first part of this statement is, I suggest, beyond argument; more on the second part shortly.

Non-linear distortion can be divided into harmonic distortion and intermodulation distortion:

Harmonic Distortion

Harmonic distortion only occurs alone, without intermodulation distortion, when a single note is being handled. It adds harmonics that are always integer multiples of the lowest or fundamental frequency, or if harmonics are already present (as they will be for any input signal except a sine wave), their level is modified by either cancellation or reinforcement. The classic example of harmonic distortion is the use of highly non-linear guitar amplifiers or fuzz boxes to create distinctive timbres for lead guitars. Here the amplifier is a part of a composite guitar-amplifier speaker instrument. We will here ignore the views of those who feel a power amplifier in a hi-fi system is an appropriate device for adding distortion to a complete musical performance.

It is generally accepted that the higher the order of a harmonic, the more unpleasant it sounds, so there is

Table 1.1. The RMAA and Shorter harmonic weighting factors

Harmonic	RMAA factor	RMAA dB	Shorter factor	Shorter dB
2nd	1.0	0	1.00	0
3rd	1.5	3.52	2.25	7.04
4th	2.0	6.02	4.00	12.04
5th	2.5	7.96	6.25	15.92
6th	3.0	9.54	9.00	19.08
7th	3.5	10.88	12.25	21.76
8th	4.0	12.04	16.00	24.08
9th	4.5	13.06	20.25	26.13

a good reason to strive to keep high harmonics at the lowest possible level. From this flows the idea of weighting the high-order harmonics by multiplying their level by a factor that increases with order, to get a better correlation between a single THD figure and the subjective deterioration. As is so often the case, the history of this goes back further than you might think. The Radio Manufacturer's Association of America (RMAA) proposed in 1937[7] that the level of the nth harmonic should be multiplied by n/2, which leaves the second harmonic unchanged, but scales up the third by 3/2 = 1.5 times, the fourth by 4/2 = 2 times, and so on. This is summarised in Table 1.1 up to the ninth harmonic.

Later work by Shorter in 1950[8] showed that better correlation with subjective impairment was given by weighting the *n*th harmonic by $n^2/4$. This again leaves the second harmonic unchanged, but raises the third by 9/4 = 2.25 times, the fourth by 16/4 = 4 times, and so on; see Table 1.1. Using this weighting function, the higher harmonics receive a considerable boost; the ninth harmonic is multiplied by 19.08 times, a massive increase of more than 26 dB, and 13 dB more than the RMAA weighting. It is clear that even relatively low levels of high-order harmonics need to be taken very seriously, and this is one reason why even small amounts of crossover distortion are of such concern.

Shorter pointed out that a weighting proportional to the square of frequency, i.e., by adding a factor of 12 dB per octave, is equivalent to differentiating twice, and so gives a measure of the radius of curvature of the waveform; in other words, the sharpness of its corners. The Shorter weighting is also considered to give a much better measure of the amount of intermodulation distortion created by complex signals such as multi-instrument musical pieces.

Neither of these harmonic weighting methods caught on. It has been suggested that this was because of the difficulty of measuring individual harmonics when THD testgear was the dominant apparatus. It is now of course easy to do, so long as you can afford a measuring system with a digital FFT capability. However, I am not wholly convinced by this argument. Back in 1950, individual harmonics could be measured on a wave analyser, which is basically just a precisely tuneable bandpass filter plus suitable amplification and measuring facilities. The first distortion measurements I ever made were with a Marconi wave analyser, the harmonic levels being RMS-summed (yes, with a slide-rule) to obtain the THD figure. Wave analysers were, however, precision instruments, and correspondingly expensive, and they seem to have been rare in the audio community at the time.

A more sophisticated weighting scheme was introduced by Lidia Lee and Earl Geddes in 2003[9] The method requires an FFT process that gives the phase as well as the magnitude of the harmonics, so the actual waveform can be reconstructed, and from this a fairly complex equation is used to generate a series of weighting factors. The GedLee equation takes into account the amplitude/time distribution of musical signals (see above), which always spend most of the time fairly near zero, with only rare peak excursions, allowing for the proper assessment of crossover distortion.

Intermodulation Distortion

Intermodulation distortion occurs when signals containing two or more different frequencies pass through a non-linear system. Harmonics that are integer multiples of the fundamental frequencies are generated as in harmonic distortion, but, in addition, intermodulation between any pair of frequency component creates new components at the sum and difference frequencies of that pair, and at multiples of those sum and difference frequencies. Thus, if the input signal is composed of two sine waves at f1 and f2, harmonics will be generated at 2f1, 3f1, 4f1, etc., and at 2f2, 3f2, 4f2, and so on. In addition, inharmonic components are generated at f1+f2, f1 -f2, 2fi+f2, and so on. The number of frequencies generated is already large, and increases significantly if three or four signals with non-commensurate frequencies are present in the input. Very soon the total power in the inharmonic sum-and-difference frequencies completely dominates the output and gives the unpleasant muddled and crunchy sound associated with the distortion of music.

Orchestral music can easily be composed of a hundred or more signals from individual instruments, so the sensation of intermodulation distortion is often not unlike the addition of white noise.

Historically, intermodulation distortion was used as a test for linearity as it used relatively simple fixed filters rather than a notch filter that had to be kept tuned with great precision to reject the fundamental. As described above, the majority of the degradation due to non-linear distortion is due to the intermodulation products, so such tests are useful for systems of doubtful acceptability like radio links. There are several standard intermodulation tests; one example is the Difference-Tone IMD test, which employs two equal-amplitude closely spaced high frequency signals; 19 kHz and 20 kHz are normally used when testing in the full audio bandwidth. This test usually measures only the second-order product generated at f2-f1, at 1 kHz, the high frequency signals being easily removed by a fixed lowpass filter. The main standard for this test is IEC 60268-3.

The measurement of intermodulation distortion is of little or no use in checking amplifier design decisions or in diagnosing problems. The reduction of intermodulation distortion is not usually a design goal in itself; if an amplifier has low harmonic distortion, then it must have low intermodulation distortion.

How Much Distortion is Perceptible?

Great technical resources are available in electronics. It should be straightforward to design a power amplifier with distortion that could not possibly be perceived under any circumstances. All we need to do is determine how much distortion, of what sort, represents the minimum that is perceptible. Unfortunately that is not a simple question to answer. We have just seen that the apparently simple task of weighting harmonics to reflect their unpleasantness gets quite complicated. Historically, you could say roughly that the maximum for a good amplifier was regarded as 1% THD in the 1950s, 0.1% THD in the 1960s, and 0.01% THD in the early 1970s. After 1976, the regrettable rise of Subjectivism split the market, with some of us pressing on to 0.001% THD while others convinced themselves that 5% THD from an ancient triode was the only route to audio happiness.

The rapid reduction in what was regarded as acceptable distortion from 1950 to 1976 was not based on an amazing series of discoveries in psychoacoustics. It was driven by the available technology of the time. The 1950s' amplifiers were of course all valve, with output transformers, and achieving low distortion

presented formidable difficulties. In the 1960s solid-state amplifiers appeared, and output transformers disappeared, allowing much more negative feedback to be used to reduce distortion, and 0.1% THD could be achieved reliably. As the 1970s began, the design of solid-state amplifiers had become more sophisticated, including differential-pair inputs, output triples, and so on. Designs by Quad, Radford, and others easily reached 0.01% THD and less. The process was also driven by amplifier designers publishing in serious journals like *Wireless World*; if you've just designed an amplifier that gives 0.05% rather than 0.1%, there is an obvious temptation to argue that 0.1% is really just a bit too high.

Psychoacoustic testing to find the just-perceptible levels of distortion has not been as helpful as one might have hoped. An intractable problem is that you cannot simply take a piece of music and say, 'I've added 0.1% THD. Can you hear it?' (You would do it double-blind in reality, of course.) You can set the THD to 0.1% on a steady sinewave with a certain nominal level, perhaps using one of the distortion models described in Chapter 2, but since music is constantly changing in level, the amount of distortion is constantly changing too. You can of course investigate how much distortion is perceptible on steady test tones, but that is a long way from the reality of amplifier use. The sort of distortion model used naturally has a great effect on the results. Alex Voishvillo of JBL has demonstrated that hard clipping giving 22.6% THD was less objectionable than crossover distortion giving 2.8% THD.[10] This paper also gives a lot of interesting information about the way that masking mitigates the irritation caused by distortion products, and also shows how these are more objectionable when they fall outside the spectrum of the signal from which they are derived.

Another problem is that loudspeakers have significant and complicated non-linearities themselves, and this may be why the relatively few studies on the perception of distortion tend to come up with worryingly high figures. James Moir reported that Just Detectable Distortion (JDD) levels could be no lower than 1% in 1981.[11] The JDD levels fell as listeners learned to recognise distortion. The tests already referred to by D.E.L Shorter in 1950[8] gave just perceivable distortion values of 0.8% to 1.3%. Jacobs and Whitman claimed in 1964[12] that harmonic distortion at 5% was not detectable, but this appears to have derived from single-tone tests. A study by P.A. Fryer in 1979 gave levels of just detectable intermodulation distortion of 2−4% for piano music, and 5% in other types of test signals; the test material used first-order intermodulation products. Some more research has been done recently (2007) by Eric de Santis and Simon Henin.[13]

All these figures quoted are so high that you might wonder why amplifier design did not simply grind to a halt many years ago. One reason is that the figures are so high, and so variable, that in many quarters they are regarded as implausible. There is of course the commercial impetus to come up with better specifications whether or not they are likely to have an audible effect.

So where does that leave us? My answer to this is the same as it has always been. If there is ever a scintilla of doubt as to what level of distortion is perceptible, it should be so reduced that there can be no rational argument that it is audible. 0.001% should do it, but bear in mind that while 0.001% at 1 kHz can be routinely achieved, and has been possible for at least 15 years, 0.001% at 20 kHz is quite another matter. In fact, a Blameless amplifier should have its distortion products well below the noise floor at 1 kHz, i.e., less than about 0.0005%, and I think it would be very hard indeed to argue that it could ever be audible to anyone. In distortion, as in other matters, the only way to be sure is to nuke it from orbit.

This philosophy would not be very attractive if reducing amplifier distortion required lots of expensive components or horribly complex circuitry. Fortunately that is not the case. Most of the linearising techniques I present in this book are of trivial extra cost. For example, adding a small-signal transistor and a resistor to a Voltage-Amplifier Stage completely transforms its performance, and the extra expense is negligible compared with the cost of power devices, heatsinks, and mains transformers. Adding more power devices to reduce distortion into low-impedance loads is somewhat more costly, but still very straightforward. The only distortion-reducing strategy that really costs money is going from Class-B to Class-A. The cost of heatsinking and power supplies is seriously increased. However, such a move is really not necessary for excellent performance, and in this edition of *Audio Power Amplifier Design* I argue that Class-A power amplifiers are so hopelessly inefficient when handling real signals (as opposed to sine waves) that they should not be considered for power outputs above 20 Watts or so.

I am painfully aware that there is a school of thought that regards low THD as inherently immoral, but this is to confuse electronics with religion. The implication is, I suppose, that very low THD can only be obtained by huge global NFB factors that require heavy dominant-pole compensation that severely degrades slew-rate; the obvious flaw in this argument is that once the compensation is applied, the amplifier no longer has a large global NFB factor, and so its distortion performance presumably reverts to mediocrity, further burdened with a slew-rate of 4 Volts per fortnight. Digital audio now routinely delivers the signal with very low distortion, and I can earnestly vouch for the fact that analogue console designers work furiously to keep the distortion in long complex signal paths down to similar levels. I think it an insult to allow the very last piece of electronics in the chain to make nonsense of these efforts.

To me, low distortion has its own aesthetic and philosophical appeal; it is satisfying to know that the amplifier you have just designed and built is so linear that there simply is no realistic possibility of it distorting your favourite material. Most of the linearity-enhancing strategies examined in this book are of minimal cost (the notable exception being resort to Class-A) compared with the essential heatsinks, transformer, etc., and so why not have ultra-low distortion? Why put up with more than you must?

"Damping Factor"

Audio amplifiers, with a few very special exceptions,[14] approximate to perfect voltage sources, i.e., they aspire to a zero output impedance across the audio band. The result is that amplifier output is unaffected by loading, so that the frequency-variable impedance of loudspeakers does not give an equally variable frequency response, and there is some control of speaker cone resonances.

While an actual zero impedance is impossible simply by using negative voltage feedback, a very close approximation is possible if large negative-feedback factors are used. (A judicious mixture of voltage and current feedback will make the output impedance zero, or even negative, i.e., increasing the loading makes the output voltage increase. This is clever, but usually pointless, as will be seen.) Solid-state amplifiers are quite happy with lots of feedback, but it is usually impractical in valve designs.

The so-called "damping factor" is defined as the ratio of the load impedance R_{load} to the amplifier output resistance R_{out}:

$$\text{Damping factor} = \frac{R_{load}}{R_{out}} \qquad \text{Equation 1.1}$$

A solid-state amplifier typically has output resistance of the order of 0.05 Ω, so if it drives an 8 Ω speaker, we get a damping factor of 160 times. This simple definition ignores the fact that amplifier output impedance usually varies considerably across the audio band, increasing with frequency as the negative feedback factor falls; this indicates that the output resistance is actually more like an inductive reactance. The presence of an output inductor to give stability with capacitive loads further complicates the issue.

Mercifully, the damping factor as such has very little effect on loudspeaker performance. A damping factor of 160 times, as derived above, seems to imply a truly radical effect on cone response — it implies that resonances and such have been reduced by 160 times as the amplifier output takes an iron grip on the cone movement. Nothing could be further from the truth.

The resonance of a loudspeaker unit depends on the total resistance in the circuit. Ignoring the complexities of crossover circuitry in multi-element speakers, the total series resistance is the sum of the speaker coil resistance, the speaker cabling, and, last of all, the amplifier output impedance. The values will be typically 7, 0.5 and 0.05 Ω respectively, so the amplifier only contributes 0.67% to the total, and its contribution to speaker dynamics must be negligible.

The highest output impedances are usually found in valve equipment, where global feedback, including the output transformer, is low or nonexistent; values around 0.5 Ω are usual. However, idiosyncratic semiconductor designs sometimes also have high output resistances; see Olsher[15] for a design with Rout = 0.6 Ω, which I feel is far too high.

The fact is that 'damping factor' is a thoroughly bad and confusing term. It was invented around 1946 by F. Langford-Smith, the editor of the famous *Radio Designers Handbook*, but in August 1947 he wrote a letter to *Wireless World*[16] admitting that loudspeaker damping was essentially determined by the series resistance of the loudspeaker, and concluding: '… there is very little gained by attempting to achieve excessively low (amplifier) output impedances'. This was of course in the days of valve amplifiers, when the output impedance of a good design would be of the order of 0.5 Ω.

This view of the matter was practically investigated and fully confirmed by James Moir in 1950,[17] though this has not prevented enduring confusion and periodic resurgences of controversy. The only reason to strive for a high damping factor — which can, after all, do no harm — is the usual numbers game of impressing potential customers with specification figures. It is as certain as anything can be that the subjective difference between two amplifiers, one with a DF of 100, and the other boasting 2000, is undetectable by human perception. Nonetheless, the specifications look very different in the brochure, so means of maximising the DF may be of some interest. This is examined further in Chapters 14 and 25.

Absolute Phase

Concern for absolute phase has for a long time hovered ambiguously between real audio concerns like noise and distortion, and the Subjectivist realm where solid copper is allegedly audible. Absolute phase means the preservation of signal phase all the way from microphone to loudspeaker, so that a drum impact that sends an initial wave of positive pressure towards the live audience is reproduced as a similar positive pressure wave from the loudspeaker. Since it is known that the neural impulses from the ear retain the periodicity of the waveform at low frequencies, and distinguish between compression and rarefaction, there is a prima facie case for the audibility of absolute phase.

It is unclear how this applies to instruments less physical than a kick drum. For the drum, the situation is simple — you kick it, the diaphragm moves outwards and the start of the transient must be a wave of compression in the air. (Followed almost at once by a wave of rarefaction.) But what about an electric guitar? A similar line of reasoning — plucking the string moves it in a given direction, which gives such-and-such a signal polarity, which leads to whatever movement of the cone in the guitar amp speaker cabinet — breaks down at every point in the chain. There is no way to know how the pickups are wound, and indeed the guitar will almost certainly have a switch for reversing the phase of one of them. I also suggest that the preservation of absolute phase is not the prime concern of those who design and build guitar amplifiers.

The situation is even less clear if more than one instrument is concerned, which is of course almost all the time. It is very difficult to see how two electric guitars played together could have a 'correct' phase in which to listen to them.

Recent work on the audibility of absolute phase[18,19] shows it is sometimes detectable. A single tone flipped back and forth in phase, providing it has a spiky asymmetrical waveform and an associated harsh sound, will show a change in perceived timbre and, according to some experimenters, a perceived change in pitch. A monaural presentation has to be used to yield a clear effect. A complex sound, however, such as that produced by a musical ensemble, does not in general show a detectable difference. The most common asymmetrical waveforms encountered in real listening are of speech or a solo unaccompanied singing voice.

Proposed standards for the maintenance of absolute phase have appeared,[20] and the implication for amplifier designers is clear; whether absolute phase really matters or not, it is simple to maintain phase in a power amplifier and so it should be done. (Compare a complex mixing console, where correct phase is absolutely vital, and there are hundreds of inputs and outputs, all of which must be in phase in every possible configuration of every control.) In fact, it probably already has been done, even if the designer has not given absolute phase a thought, because almost all power amplifiers use series negative feedback, and this is inherently non-inverting. Care is, however, required if there are stages such as balanced line input amplifiers before the power amplifier itself; if the hot and cold inputs get swapped by mistake, then the amplifier output will be phase-inverted.

Concern about the need for absolute phase to be preserved is not exactly new. It was discussed at length in the Letters to The Editor section of *Wireless World* in 1964.[21] One correspondent asserted that 'male speech is markedly asymmetrical' though without saying how he thought female speech compared. Different times.

Amplifier Formats

When the first edition of this book appeared in 1996, the vast majority of domestic amplifiers were two-channel stereo units. Since then there has been a great increase in other formats, particularly in multichannel units having seven or more channels for audio-visual use, and in single-channel amplifiers built into sub-woofer loudspeakers.

Multichannel amplifiers come in two kinds. The most cost-effective way to build a multichannel amplifier is to put as many power amplifier channels as convenient on each PCB, and group them around a large toroidal transformer that provides a common power supply for all of them. While this keeps the costs down, there are inevitable compromises on inter-channel crosstalk and rejection of the transformer's stray magnetic fields. The other method is to make each channel (or in some cases, each pair of channels) into a separate amplifier module with its own transformer, power supply, heatsinks, and separate input and output connections; a sort of multiple-monobloc format. The modules usually share a microcontroller housekeeping system but nothing else. This form of construction gives much superior inter-channel crosstalk, as the various audio circuits need have no connection with each other, and much less trouble with transformer hum as

the modules are relatively long and thin so that a row of them can be fitted into a chassis, and thus the mains transformer can be put right at one end and the sensitive input circuitry right at the other. Inevitably this is a more expensive form of construction.

Sub-woofer amplifiers are single channel and of high power. There seems to be a general consensus that the quality of sub-woofer amplifiers is less critical than that of other amplifiers, and this has meant that both Class-G and Class-D designs have found homes in sub-woofer enclosures. Sub-woofer amplifiers differ from others in that they often incorporate their own specialised filtering (typically at 200 Hz) and equalisation circuitry.

Misinformation in Audio

Probably no field of technical endeavour is more plagued with errors, mis-statements, confusion and downright lying than audio. In the past 30 years or so, the rise of controversial and non-rational audio hypotheses, gathered under the title *Subjectivism* has created these difficulties. It is common for hi-fi reviewers to claim that they have perceived subtle audio differences that cannot be related to electrical performance measurements. These claims include the alleged production of a 'three-dimensional sound-stage and protests that the rhythm of the music has been altered'; these statements are typically produced in isolation, with no attempt made to correlate them to objective test results. The latter in particular appears to be a quite impossible claim.

This volume does not address the implementation of Subjectivist notions, but confines itself to the measurable, the rational, and the repeatable. This is not as restrictive as it may appear; there is nothing to prevent you using the methodology presented here to design an amplifier that is technically excellent, and then gilding the lily by using whatever brands of expensive resistor or capacitor are currently fashionable, and doing the internal wiring with cable that costs more per metre than the rest of the unit put together. Such nods to Subjectivist convention are unlikely to damage the real performance; this, however, is not the case with some of the more damaging hypotheses, such as the claim that negative feedback is inherently harmful. Reduce the feedback factor and you will degrade the real-life operation of almost any design.

Such problems arise because audio electronics is a more technically complex subject than it at first appears. It is easy to cobble together some sort of power amplifier that works, and this can give people an altogether exaggerated view of how deeply they understand what they have created. In contrast, no one is likely to take a 'subjective' approach to the design of an aeroplane wing or a rocket engine; the margins for error are rather smaller, and the consequences of malfunction somewhat more serious.

The Subjectivist position is of no help to anyone hoping to design a good power amplifier. However, it promises to be with us for some further time yet, and it is appropriate to review it here and show why it need not be considered at the design stage. The marketing stage is of course another matter.

Science and Subjectivism

Audio engineering is in a singular position. There can be few branches of engineering science rent from top to bottom by such a basic division as the Subjectivist/rationalist dichotomy. Subjectivism is still a significant issue in the hi-fi section of the industry, but mercifully has made little headway in professional audio, where an intimate acquaintance with the original sound, and the need to earn a living with reliable and affordable equipment, provide an effective barrier against most of the irrational influences. (Note that the opposite of Subjectivist is not 'Objectivist'. This term refers to the followers of the philosophy of Ayn Rand).

Most fields of technology have defined and accepted measures of excellence; car makers compete to improve MPH and MPG; computer manufacturers boast of MIPs (millions of instructions per second), and so on. Improvement in these real quantities is regarded as unequivocally a step forward. In the field of hi-fi, many people seem to have difficulty in deciding which direction forward is.

Working as a professional audio designer, I often encounter opinions which, while an integral part of the Subjectivist offshoot of hi-fi, are treated with ridicule by practitioners of other branches of electrical engineering. The would-be designer is not likely to be encouraged by being told that audio is not far removed from witchcraft, and that no one truly knows what they are doing. I have been told by a Subjectivist that the operation of the human ear is so complex that its interaction with measurable parameters lies forever beyond human comprehension. I hope this is an extreme position; it was, I may add, proffered as a flat statement rather than a basis for discussion.

I have studied audio design from the viewpoints of electronic design, psychoacoustics, and my own humble efforts at musical creativity. I have found complete scepticism towards Subjectivism to be the only tenable position. Nonetheless, if hitherto unsuspected dimensions of audio quality are ever shown to exist, then I look forward keenly to exploiting them. At this point I should say that no doubt most of the esoteric opinions are held in complete sincerity.

The Subjectivist Position

A short definition of the Subjectivist position on power amplifiers might read as follows:

- Objective measurements of an amplifier's performance are unimportant compared with the subjective impressions received in informal listening tests. Should the two contradict, the objective results may be dismissed.
- Degradation effects exist in amplifiers that are unknown to orthodox engineering science, and are not revealed by the usual objective tests.
- Considerable latitude may be employed in suggesting hypothetical mechanisms of audio impairment, such as mysterious capacitor shortcomings and subtle cable defects, without reference to the plausibility of the concept, or the gathering of objective evidence of any kind.

I hope that this is considered a reasonable statement of the situation; meanwhile the great majority of the paying public continue to buy conventional hi-fi systems, ignoring the expensive and esoteric high-end sector where the debate is fiercest.

It may appear unlikely that a sizeable part of an industry could have set off in a direction that is quite counter to the facts; it could be objected that such a loss of direction in a scientific subject would be unprecedented. This is not so.

Parallel events that suggest themselves include the destruction of the study of genetics under Lysenko in the USSR.[22] Another possibility is the study of parapsychology, now in deep trouble because after more than 100 years of investigation it has not uncovered the ghost (sorry) of a repeatable phenomenon.[23] This sounds all too familiar. It could be argued that parapsychology is a poor analogy because most people would accept that there was nothing there to study in the first place, whereas nobody would assert that objective measurements and subjective sound quality have no correlation at all; one need only pick up the telephone to remind oneself what a 4 kHz bandwidth and 10% or so THD sound like.

The most startling parallel I have found in the history of science is the almost-forgotten affair of Blondlot and the N-rays.[24] In 1903, René Blondlot, a respected French physicist, claimed to have discovered a new form of radiation he called 'N-rays'. (This was shortly after the discovery of X-rays by Roentgen, so rays were in the air, as it were.) This invisible radiation was apparently mysteriously refracted by aluminium prisms; but the crucial factor was that its presence could only be shown by subjective assessment of the brightness of an electric arc allegedly affected by N-rays. No objective measurement appeared to be possible. To Blondlot, and at least fourteen of his professional colleagues, the subtle changes in brightness were real, and the French Academy published more than a hundred papers on the subject.

Unfortunately N-rays were completely imaginary, a product of the 'experimenter-expectancy' effect. This was demonstrated by American scientist Robert Wood, who quietly pocketed the aluminium prism during a demonstration, without affecting Blondlot's recital of the results. After this the N-ray industry collapsed very quickly, and while it was a major embarrassment at the time, it is now almost forgotten.

The conclusion is inescapable that it is quite possible for large numbers of sincere people to deceive themselves when dealing with subjective assessments of phenomena.

A Short History of Subjectivism

The early history of sound reproduction is notable for the number of times that observers reported that an acoustic gramophone gave results indistinguishable from reality. The mere existence of such statements throws light on how powerfully mind-set affects subjective impressions. Interest in sound reproduction intensified in the post-war period, and technical standards such as DIN 45-500 were set, though they were soon criticised as too permissive. By the late 1960s it was widely accepted that the requirements for hi-fi would be satisfied by 'THD less than 0.1%, with no significant crossover distortion, frequency response 20–20 kHz, and as little noise as possible, please'. The early 1970s saw this expanded to include slew-rates and properly behaved overload protection, but the approach was always scientific and it was normal to read amplifier reviews in which measurements were dissected but no mention made of listening tests.

Following the growth of Subjectivism through the pages of one of the leading Subjectivist magazines (*Hi-Fi News*), the first intimation of what was to come was the commencement of Paul Messenger's column *Subjective Sounds* in September 1976, in which he said: 'The assessment will be (almost) purely subjective, which has both strengths and weaknesses, as the inclusion of laboratory data would involve too much time and space, and although the ear may be the most fallible, it is also the most sensitive evaluation instrument.' Subjectivism as expedient rather than policy. Significantly, none of the early instalments contained references to amplifier sound. In March 1977, an article by Jean Hiraga was published vilifying high levels of negative feedback and praising the sound of an amplifier with 2% THD. In the same issue, Paul Messenger stated that a Radford valve amplifier sounded better than a transistor one, and by the end of the year the amplifier-sound bandwagon was rolling. Hiraga returned in August 1977 with a highly contentious set of claims about audible speaker cables, and after that no hypothesis was too unlikely to receive attention.

The fallibility of informal listening tests was appreciated a long way back. The famous writer Free Grid (Norman Preston Vincer-Minter) said in *Wireless World* in 1950 that the results 'Depend more on what is being listened *for* than what is being listened *to*' and it is hard to improve on that as a succinct judgement.

The Limits of Hearing

In evaluating the Subjectivist position, it is essential to consider the known abilities of the human ear. Contrary to the impression given by some commentators, who call constantly for more psychoacoustical research, a vast amount of hard scientific information already exists on this subject, and some of it may be briefly summarised thus:

- The smallest step-change in amplitude that can be detected is about 0.3 dB for a pure tone. In more realistic situations it is 0.5 to 1.0 dB. This is about a 10% change.[25]
- The smallest detectable change in frequency of a tone is about 0.2% in the band 500 Hz–2 kHz. In percentage terms, this is the parameter for which the ear is most sensitive.[26]
- The least detectable amount of harmonic distortion is not an easy figure to determine, as there is a multitude of variables involved, and in particular the continuously varying level of programme means that the level of THD introduced is also dynamically changing. With mostly low-order harmonics present, the just-detectable amount is about 1%, though crossover effects can be picked up at 0.3%, and probably lower.[17] This issue is dealt with in detail earlier in this chapter.

It is acknowledged that THD measurements, taken with the usual notch-type analyser, are of limited use in predicting the subjective impairment produced by an imperfect audio path. With music, etc. intermodulation effects are demonstrably more important than harmonics. However, THD tests have the unique advantage that visual inspection of the distortion residual gives an experienced observer a great deal of information about the root cause of the non-linearity. Many other distortion tests exist, which, while yielding very little information to the designer, exercise the whole audio bandwidth at once and correlate well with properly conducted tests for subjective impairment by distortion. The Belcher intermodulation test (the principle is shown in Figure 1.3) deserves more attention than it has received, and may become more popular now that DSP chips are cheaper.

One of the objections often made to THD tests is that their resolution does not allow verification that no non-linearities exist at very low level; a sort of micro-crossover distortion. Hawksford, for example, has stated 'Low-level threshold phenomena ... set bounds upon the ultimate transparency of an audio system'[27] and several commentators have stated their belief that some metallic contacts consist of a morass of so-called 'micro-diodes'. In fact, this kind of mischievous hypothesis can be disposed of simply by using THD techniques.

I had the gravest doubts about this. The physics of conduction in metals is well established and absolutely forbids the existence of 'micro-diodes' or any analogous nonsense. To test it for myself I evolved a method of measuring THD down to 0.01% at 200 μV rms, and applied it to large electrolytics, connectors of varying provenance, and lengths of copper cable with and without alleged magical properties. The method

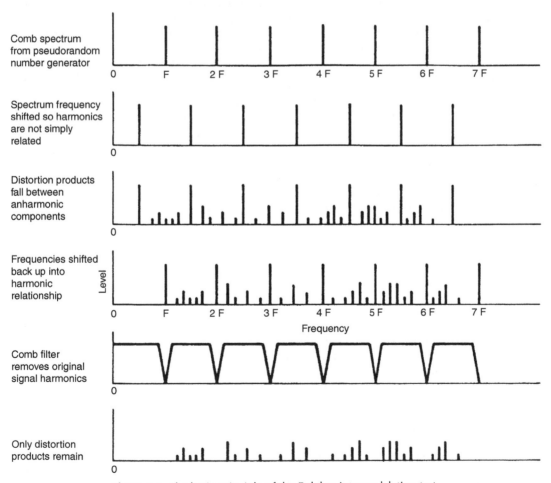

Figure 1.3. The basic principle of the Belcher intermodulation test.

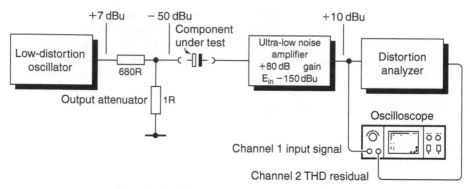

Figure 1.4. THD measurements at very low levels.

required the design of an ultra-low noise (EIN = -150 dBu for a 10 Ω source resistance) and very low THD.[28] The measurement method is shown in Figure 1.4; using an attenuator with a very low value of resistance to reduce the incoming signal keeps the Johnson noise to a minimum. In no case was any unusual distortion detected, and it would be nice to think that this red herring has been laid to rest.

- Interchannel crosstalk can obviously degrade stereo separation, but the effect is not detectable until it is worse than 20 dB, which would be a very bad amplifier indeed.[29]

The Limits of Hearing: Phase Perception

The audibility of phase-shift and group delay have been an area of dispute in the audio community for a long time, so here it gets its own separate section. As Stanley Lipshitz et al. have pointed out,[32] these effects are obviously perceptible if they are gross enough; if an amplifier was so heroically misconceived as to produce the top half of the audio spectrum three hours after the bottom, there would be no room for argument.

If the phase-shift is proportional to frequency, then the group delay is constant with frequency and we have a linear-phase system; we just get a pure time-delay with no audible consequences. However, in most cases the phase-shift is not remotely proportional to frequency, and so the group delay varies with frequency. This is sometimes called group delay distortion, which is perhaps not ideal as 'distortion' implies non-linearity to most people, while here we are talking about a linear process.

Most of the components in the microphone—recording—loudspeaker chain are minimum-phase; they impose only the phase-shift that would be expected

and can be predicted from their amplitude/frequency response. The great exception to this is … the multi-way loudspeaker. The other great exception was the analogue magnetic tape-recorder, which showed startlingly rapid phase-changes at the bottom of the audio spectrum, usually going several times round the clock.[30] Nobody ever complained about it.

Concern about phase problems has centred on loudspeakers and their crossovers, as they are known to have dramatic phase changes in their summed output around the crossover points. A second-order crossover acts like a first-order all-pass filter, with one extreme of the frequency spectrum shifted by 180° with respect to the other. Fourth-order crossovers, like the popular Linkwitz—Riley configuration, act like a second-order all-pass filter, with one extreme of the frequency spectrum shifted by 360° from the other. These radical phase-shifts are accepted every day for the simple reason they are not audible.

We are going to have multi-way loudspeaker systems around for the foreseeable future, and most of them have allpass crossovers. Clearly an understanding of what degradation — if any — this allpass behaviour causes is vital. Much experimentation has been done and there is only space for a summary here.

One of the earliest findings on phase perception was Ohm's Law. No, not that one, Ohm's *Other* Law, proposed in 1843.[31] In its original form it simply said that a musical sound is perceived by the ear as a set of sinusoidal harmonics. Hermann von Helmholtz extended it in the 1860s into what today is known as Ohm's Acoustic Law, by stating that the timbre of musical tone depends solely on the number and relative level of its harmonics, and *not* on their relative phases.

An important paper on the audibility of midrange phase distortion was published by Lipshitz, Pocock and Vanderkooy in 1982.[32] They summarised their conclusions as follows:

1. Quite small phase non-linearities can be audible using suitable test signals (usually isolated clicks).
2. Phase audibility is far more pronounced when using headphones instead of loudspeakers.
3. Simple acoustic signals generated in an anechoic environment show clear phase audibility when headphones are used.
4. On normal music or speech signals phase distortion is not generally audible.

At the end of their paper the authors say: 'It is stressed that none of these experiments thus far has indicated a present requirement for phase linearity in loudspeakers for the reproduction of music and speech.' James Moir reached the same conclusion,[33] and so did Harwood.[34]

Siegfried Linkwitz has done listening tests where either a first-order allpass filter, a second-order allpass filter (both at 100 Hz) or a direct connection could be switched into the audio path.[35] These filters have similar phase characteristics to allpass crossovers and cause gross visible distortions of a square waveform, but are in practice inaudible. He reports: 'I have not found a signal for which I can hear a difference. This seems to confirm Ohm's Acoustic Law that we do not hear waveform distortion.'

If we now consider the findings of neurophysiologists, we note that the auditory nerves do not fire in synchrony with the sound waveform above 2 kHz, so unless some truly subtle encoding is going on (and there is no reason to suppose that there is), then perception of phase above this frequency would appear to be inherently impossible.

Having said this, it should not be supposed that the ear operates simply as a spectrum analyser. This is known not to be the case. A classic demonstration of this is the phenomenon of 'beats'. If a 1000 Hz tone and a 1005 Hz tone are applied to the ear together, it is common knowledge that a pulsation at 5 Hz is heard. There is no actual physical component at 5 Hz, as summing the two tones is a linear process. (If instead the two tones were multiplied, as in a radio mixer stage, new components *would* be generated.) Likewise non-linearity in the ear itself can be ruled out if appropriate levels are used.

What the brain is actually responding to is the envelope or peak amplitude of the combined tones, which does indeed go up and down at 5 Hz as the phase relationship between the two waveforms continuously changes. Thus the ear is in this case acting more like

an oscilloscope than a spectrum analyser. It does not, however, seem to work as a phase-sensitive detector.

The issue of phase perception is of limited importance to amplifier designers, as it would take spectacular incompetence to produce a circuit that included an accidental all-pass filter. So long as you don't do that, the phase response of an amplifier is completely defined by its frequency response, and vice versa; in Control Theory this is Bode's Second Law,[36] and it should be much more widely known in the hi-fi world than it is. A properly designed amplifier has its response roll-off points not too far outside the audio band, and these will have accompanying phase-shifts; once again there is no evidence that these are perceptible.

The picture of the ear that emerges from psychoacoustics and related fields is not that of a precision instrument. Its ultimate sensitivity, directional capabilities and dynamic range are far more impressive than its ability to measure small level changes or detect correlated low-level signals like distortion harmonics. This is unsurprising; from an evolutionary viewpoint the functions of the ear are to warn of approaching danger (sensitivity and direction-finding being paramount) and for speech. In speech perception, the identification of formants (the bands of harmonics from vocal-chord pulse excitation, selectively emphasised by vocal-tract resonances) and vowel/consonant discriminations, are infinitely more important than any hi-fi parameter. Presumably the whole existence of music as a source of pleasure is an accidental side-effect of our remarkable powers of speech perception: how it acts as a direct route to the emotions remains profoundly mysterious.

Articles of Faith: The Tenets of Subjectivism

All of the alleged effects listed below have received considerable affirmation in the audio press, to the point where some are treated as facts. The reality is that none of them has in the past fifteen years proved susceptible to objective confirmation. This sad record is perhaps equalled only by students of parapsychology. I hope that the brief statements below are considered fair by their proponents. If not, I have no doubt I shall soon hear about it:

- *Sinewaves are steady-state signals that represent too easy a test for amplifiers, compared with the complexities of music.*

This is presumably meant to imply that sinewaves are in some way particularly easy for an amplifier to deal with, the implication being that anyone using a THD analyser must be hopelessly naïve. Since sines and cosines have an unending series of non-zero differentials, steady hardly comes into it. I know of no evidence that sinewaves of randomly varying amplitude (for example) would provide a more searching test of amplifier competence.

I hold this sort of view to be the result of anthropomorphic thinking about amplifiers; treating them as though they think about what they amplify. Twenty sinewaves of different frequencies may be conceptually complex to us, and the output of a symphony orchestra even more so, but to an amplifier both composite signals resolve to a single instantaneous voltage that must be increased in amplitude and presented at low impedance. An amplifier has no perspective on the signal arriving at its input, but must literally take it as it comes.

- *Capacitors affect the signal passing through them in a way invisible to distortion measurements.*

Several writers have praised the technique of subtracting pulse signals passed through two different sorts of capacitor, claiming that the non-zero residue proves that capacitors can introduce audible errors. My view is that these tests expose only well-known capacitor shortcomings such as dielectric absorption and series resistance, plus perhaps the vulnerability of the dielectric film in electrolytics to reverse-biasing. No one has yet shown how these relate to capacitor audibility in properly designed equipment.

- *Passing an audio signal through cables, PCB tracks or switch contacts causes a cumulative deterioration. Precious metal contact surfaces alleviate but do not eliminate the problem. This too is undetectable by tests for non-linearity.*

Concern over cables is widespread, but it can be said with confidence that there is as yet not a shred of evidence to support it. Any piece of wire passes a sinewave with unmeasurable distortion, and so simple notions of inter-crystal rectification or 'micro-diodes' can be discounted, quite apart from the fact that such behaviour is absolutely ruled out by established materials science. No plausible means of detecting, let alone measuring, cable degradation has ever been proposed.

The most significant parameter of a loudspeaker cable is probably its lumped inductance. This can cause minor variations in frequency response at the very top of the audio band, given a demanding load impedance. These deviations are unlikely to exceed 0.1 dB for reasonable cable constructions (say, inductance less than 4 μH). The resistance of a typical cable (say, 0.1 Ω) causes response variations across the band, following the speaker impedance curve, but these are usually even smaller at around 0.05 dB. This is not audible.

Corrosion is often blamed for subtle signal degradation at switch and connector contacts; this is unlikely. By far the most common form of contact degradation is the formation of an insulating sulphide layer on silver contacts, derived from hydrogen sulphide air pollution. This typically cuts the signal altogether, except when signal peaks temporarily punch through the sulphide layer. The effect is gross and seems inapplicable to theories of subtle degradation. Gold-plating is the only certain cure. It costs money.

- *Cables are directional, and pass audio better in one direction than the other.*

Audio signals are AC. Cables cannot be directional any more than 2 + 2 can equal 5. Anyone prepared to believe this nonsense will not be capable of designing amplifiers, so there seems no point in further comment.

- *The sound of valves is inherently superior to that of any kind of semiconductor.*

The 'valve sound' is one phenomenon that may have a real existence; it has been known for a long time that listeners sometimes prefer to have a certain amount of second-harmonic distortion added in,[37] and most valve amplifiers provide just that, due to grave difficulties in providing good linearity with modest feedback factors. While this may well sound nice, hi-fi is supposedly about accuracy, and if the sound is to be thus modified, it should be controllable from the front panel by a 'niceness' knob.

The use of valves leads to some intractable problems of linearity, reliability and the need for intimidatingly expensive (and once more, non-linear) iron-cored transformers. The current fashion is for exposed valves, and it is not at all clear to me that a fragile glass bottle, containing a red-hot anode with hundreds of volts DC on it, is wholly satisfactory for domestic safety.

A recent development in Subjectivism is enthusiasm for single-ended directly-heated triodes, usually in extremely expensive monobloc systems. Such an amplifier generates large amounts of second-harmonic distortion, due to the asymmetry of single-ended operation, and requires a very large output transformer as its primary carries the full DC anode current, and core saturation must be avoided. Power outputs are inevitably very limited at 10W or less. In a recent review, the Cary CAD-300SEI triode amplifier yielded 3% THD at 9W, at a cost of £3400.[38] And you still need to buy a pre-amp.

- *Negative feedback is inherently a bad thing; the less it is used, the better the amplifier sounds, without qualification.*

Negative feedback is not inherently a bad thing; it is an absolutely indispensable principle of electronic design, and if used properly has the remarkable ability to make just about every parameter better. It is usually global feedback that the critic has in mind. Local negative feedback is grudgingly regarded as acceptable, probably because making a circuit with no feedback of any kind is near-impossible. It is often said that high levels of NFB enforce a low slew-rate. This is quite untrue; and this thorny issue is dealt with in detail later in this book. For more on slew-rate, see also.[39]

- *Tone-controls cause an audible deterioration even when set to the flat position.*

This is usually blamed on *phase-shift*. At the time of writing, tone controls on a pre-amp badly damage its chances of street (or rather sitting-room) credibility, for no good reason. Tone-controls set to *flat* cannot possibly contribute any extra phase-shift and must be inaudible. My view is that they are absolutely indispensable for correcting room acoustics, loudspeaker shortcomings, or tonal balance of the source material, and that a lot of people are suffering sub-optimal sound as a result of this fashion. It is now commonplace for audio critics to suggest that frequency-response inadequacies should be corrected by changing loudspeakers. This is an extraordinarily expensive way of avoiding tone-controls.

- *The design of the power supply has subtle effects on the sound, quite apart from ordinary dangers like ripple injection.*

All good amplifier stages ignore imperfections in their power supplies, opamps in particular excelling at power-supply rejection-ratio. More nonsense has been written on the subject of subtle PSU failings than on most audio topics; recommendations of hard-wiring the mains or using gold-plated 13A plugs would seem to hold no residual shred of rationality, in view of the usual processes of rectification and smoothing that the raw AC undergoes. And where do you stop? At the local sub-station? Should we gold-plate the pylons?

- *Monobloc construction (i.e., two separate power amplifier boxes) is always audibly superior, due to the reduction in crosstalk.*

There is no need to go to the expense of monobloc power amplifiers in order to keep crosstalk under control, even when making it substantially better than the −20 dB that is actually necessary. The techniques are conventional; the last stereo power amplifier I designed managed an easy −90 dB at 10 kHz without anything other than the usual precautions. In this area

dedicated followers of fashion pay dearly for the privilege, as the cost of the mechanical parts will be nearly doubled.

- *Microphony is an important factor in the sound of an amplifier, so any attempt at vibration-damping is a good idea.*

Microphony is essentially something that happens in sensitive valve preamplifiers. If it happens in solid-state power amplifiers, the level is so far below the noise it is effectively non-existent.

Experiments on this sort of thing are rare (if not unheard of) and so I offer the only scrap of evidence I have. Take a microphone pre-amp operating at a gain of +70 dB, and tap the input capacitors (assumed electrolytic) sharply with a screwdriver; the pre-amp output will be a dull thump, at low level. The physical impact on the electrolytics (the only components that show this effect) is hugely greater than that of any acoustic vibration; and I think the effect in power amps, if any, must be so vanishingly small that it could never be found under the inherent circuit noise.

- *We can invent new words for imaginary impairments of audio whenever we like.*

A good example of this is 'slam'. This quality has become frequently used in the last ten years or so. It is hard to define exactly what it means, as it has no objective reality, but it appears to be something to do with the effective reproduction of loud low-frequency sounds. 'Slam' naturally cannot be measured, and so does not have units, but I suggest it's time that changed. My proposed unit of Slam is of course the Door. Thus a rousing performance of 'Light My Fire' would be rated at 4 Doors.

Let us for a moment assume that some or all of the above hypotheses are true, and explore the implications. The effects are not detectable by conventional measurement, but are assumed to be audible. First, it can presumably be taken as axiomatic that for each audible defect some change occurs in the pattern of pressure fluctuations reaching the ears, and therefore a corresponding modification has occurred to the electrical signal passing through the amplifier. Any other starting point supposes that there is some other route conveying information apart from the electrical signals, and we are faced with magic or forces-unknown-to-science. Mercifully no commentator has (so far) suggested this. Hence there must be defects in the audio signals, but they are not revealed by the usual test methods. How could this situation exist? There seem two possible explanations for this failure

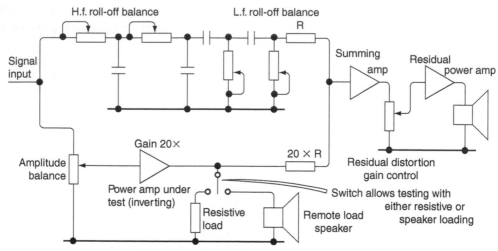

Figure 1.5. Baxandall cancellation technique.

of detection: one is that the standard measurements are relevant, but of insufficient resolution, and we should be measuring frequency response, etc., to thousandths of a dB. There is no evidence whatsoever that such micro-deviations are audible under any circumstances.

An alternative (and more popular) explanation is that standard sinewave THD measurements miss the point by failing to excite subtle distortion mechanisms that are triggered only by music, the spoken word, or whatever. This assumes that these music-only distortions are also left undisturbed by multi-tone intermodulation tests, and even the complex pseudorandom signals used in the Belcher distortion test.[40] The Belcher method effectively tests the audio path at all frequencies at once, and it is hard to conceive of a real defect that could escape it.

The most positive proof that Subjectivism is fallacious is given by subtraction testing. This is the devastatingly simple technique of subtracting before-and-after amplifier signals and demonstrating that nothing audibly detectable remains. It transpires that these alleged music-only mechanisms are not even revealed by music, or indeed anything else, and it appears the subtraction test has finally shown as non-existent these elusive degradation mechanisms.

The subtraction technique was proposed by Baxandall in 1977.[41] The principle is shown in Figure 1.5; careful adjustment of the roll-off balance network prevents minor bandwidth variations from swamping the true distortion residual. In the intervening years the Subjectivist camp has made no effective reply.

A simplified version of the test was introduced by Hafler.[42] This method is less sensitive, but has the advantage that there is less electronics in the signal path for anyone to argue about. See Figure 1.6.

A prominent Subjectivist reviewer, on trying this demonstration, was reduced to claiming that the passive switchbox used to implement the Hafler test was causing so much sonic degradation that all amplifier performance was swamped.[43] I do not feel that this is a tenable position. So far all experiments such as these have been ignored or brushed aside by the Subjectivist camp; no attempt has been made to answer the extremely serious objections that this demonstration raises.

In the twenty or so years that have elapsed since the emergence of the Subjectivist Tendency, no hitherto unsuspected parameters of audio quality have emerged.

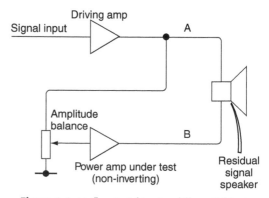

Figure 1.6. Hafler straight-wire differential test.

The Length of the Audio Chain

An apparently insurmountable objection to the existence of non-measurable amplifier quirks is that recorded sound of almost any pedigree has passed through a complex mixing console at least once; prominent parts like vocals or lead guitar will almost certainly have passed through at least twice, once for recording and once at mix-down. More significantly, it must have passed through the potential quality-bottleneck of an analogue tape machine or more likely the A-D converters of digital equipment. In its long path from here to ear the audio passes through at least a hundred opamps, dozens of connectors and several hundred metres of ordinary screened cable. If mystical degradations can occur, it defies reason to insist that those introduced by the last 1% of the path are the critical ones.

The Implications

This confused state of amplifier criticism has negative consequences. First, if equipment is reviewed with results that appear arbitrary, and which are in particular incapable of replication or confirmation, this can be grossly unfair to manufacturers who lose out in the lottery. Since subjective assessments cannot be replicated, the commercial success of a given make can depend entirely on the vagaries of fashion. While this is fine in the realm of clothing or soft furnishings, the hi-fi business is still claiming accuracy of reproduction as its *raison d'être*, and therefore you would expect the technical element to be dominant.

A second consequence of placing Subjectivism above measurements is that it places designers in a most unenviable position. No degree of ingenuity or attention to technical detail can ensure a good review, and the pressure to adopt fashionable and expensive expedients (such as linear-crystal internal wiring) is great, even if the designer is certain that they have no audible effect for good or evil. Designers are faced with a choice between swallowing the Subjectivist credo whole or keeping very quiet and leaving the talking to the marketing department.

If objective measurements are disregarded, it is inevitable that poor amplifiers will be produced, some so bad that their defects are unquestionably audible. In recent reviews,[44] it was easy to find a £795 pre-amplifier (Counterpoint SA7) that boasted a feeble 12 dB disc overload margin (another pre-amp costing £2040 struggled up to 15 dB — Burmester 838/846) and another, costing £1550 that could only manage a 1 kHz distortion performance of 1%; a lack of linearity that would have

caused consternation ten years ago (Quicksilver). However, by paying £5700 one could inch this down to 0.3% (Audio Research M100–2 monoblocs). This does not of course mean that it is impossible to buy an *audiophile* amplifier that does measure well; another example would be the pre-amplifier/power amplifier combination that provides a very respectable disc overload margin of 31 dB and 1 kHz rated-power distortion below 0.003%; the total cost being £725 (Audiolab 8000C/8000P). I believe this to be a representative sample, and we appear to be in the paradoxical situation that the most expensive equipment provides the worst objective performance. Whatever the rights and wrongs of subjective assessment, I think that most people would agree that this is a strange state of affairs. Finally, it is surely a morally ambiguous position to persuade non-technical people that to get a really good sound they have to buy £2000 pre-amps, and so on, when both technical orthodoxy and common sense indicate that this is quite unnecessary.

The Reasons Why

Some tentative conclusions are possible as to why hi-fi engineering has reached the pass that it has. I believe one basic reason is the difficulty of defining the quality of an audio experience; you cannot draw a diagram to communicate what something sounded like. In the same way, acoustical memory is more evanescent than visual memory. It is far easier to visualise what a London bus looks like than to recall the details of a musical performance. Similarly, it is difficult to 'look more closely'; turning up the volume is more like turning up the brightness of a TV picture; once an optimal level is reached, any further increase becomes annoying, then painful.

It has been universally recognised for many years in experimental psychology, particularly in experiments about perception, that people tend to perceive what they want to perceive. This is often called the *experimenter-expectancy* effect; it is more subtle and insidious than it sounds, and the history of science is littered with the wrecked careers of those who failed to guard against it. Such self-deception has most often occurred in fields like biology, where although the raw data may be numerical, there is no real mathematical theory to check it against. When the only 'results' are vague subjective impressions, the danger is clearly much greater, no matter how absolute the integrity of the experimenter. Thus, in psychological work great care is necessary in the use of impartial observers, double-blind techniques, and rigorous statistical tests for significance. The vast majority of Subjectivist writings wholly ignore these precautions, with predictable results. In a few cases properly controlled listening tests have been done, and at the time of writing all have resulted in different amplifiers sounding indistinguishable. I believe the conclusion is inescapable that experimenter-expectancy has played a dominant role in the growth of Subjectivism.

It is notable that in Subjectivist audio the 'correct' answer is always the more expensive or inconvenient one. Electronics is rarely as simple as that. A major improvement is more likely to be linked with a new circuit topology or new type of semiconductor, than with mindlessly specifying more expensive components of the same type; cars do not go faster with platinum pistons.

It might be difficult to produce a rigorous statistical analysis, but it is my view that the reported subjective quality of a piece of equipment correlates far more with the price than with anything else. There is perhaps here an echo of the Protestant Work Ethic; you must suffer now to enjoy yourself later. Another reason for the relatively effortless rise of Subjectivism is the *me-too* effect; many people are reluctant to admit that they cannot detect acoustic subtleties as nobody wants to be labelled as insensitive, outmoded, or just plain deaf. It is also virtually impossible to absolutely disprove any claims, as the claimant can always retreat a fraction and say that there was something special about the combination of hardware in use during the disputed tests, or complain that the phenomena are too delicate for brutal logic to be used on them. In any case, most competent engineers with a taste for rationality probably have better things to do than dispute every controversial report.

Under these conditions, vague claims tend, by a kind of intellectual inflation, to gradually become regarded as facts. Manufacturers have some incentive to support the Subjectivist camp as they can claim that only they understand a particular non-measurable effect, but this is no guarantee that the dice may not fall badly in a subjective review.

The Outlook

It seems unlikely that Subjectivism will disappear for a long time, if ever, given the momentum that it has gained, the entrenched positions that some people have taken up, and the sadly uncritical way in which people accept an unsupported assertion as the truth simply because it is asserted with frequency and conviction. In an ideal world every such statement would be greeted by loud demands for evidence. However, the history of the world sometimes leads one to suppose pessimistically that people will believe anything. By analogy, one might suppose that Subjectivism would persist for the same reason that parapsychology has; there will always be people who will believe what they want to believe rather than what the hard facts indicate.

More than ten years have passed since some of the above material on Subjectivism was written, but there seems to be no reason to change a word of it. Amplifier reviews continue to make completely unsupportable assertions, of which the most obtrusive these days is the notion that an amplifier can in some way alter the 'timing' of music. This would be a remarkable feat to accomplish with a handful of transistors, were it not wholly imaginary.

During my sojourn at TAG-McLaren Audio, we conducted an extensive set of double-blind listening tests, using a lot of experienced people from various quarters of the hi-fi industry. An amplifier loosely based on the Otala four-stage architecture was compared with a Blameless three-stage architecture perpetrated by myself. (These terms are fully explained in Chapter 4.) The two amplifiers could not have been more different — the four-stage had complex lead-lag compensation and a buffered CFP output, while my three-stage had conventional Miller dominant pole compensation and an EF output. There were too many other detail differences to list here. After a rigorous statistical analysis the result — as you may have guessed — was that nobody could tell the two amplifiers apart.

Technical Errors

Misinformation also arises in the purely technical domain; I have also found some of the most enduring and widely held technical beliefs to be unfounded. For example, if you take a Class-B amplifier and increase its quiescent current so that it runs in Class-A at low levels, i.e., in Class-AB, most people will tell you that the distortion will be reduced as you have moved nearer to the full Class-A condition. This is untrue. A correctly configured amplifier gives more distortion in Class-AB, not less, because of the abrupt gain changes inherent in switching from A to B every cycle.

Discoveries like this can only be made because it is now straightforward to make testbed amplifiers with ultra-low distortion — lower than that which used to be thought possible. The reduction of distortion to the basic or inherent level that a circuit configuration is capable of is a fundamental requirement for serious design work in this field; in Class-B at least this gives a defined and repeatable standard of performance that in later chapters I name a Blameless amplifier, so-called because it avoids error rather than claiming new virtues.

It has proved possible to take the standard Class-B power amplifier configuration, and by minor modifications, reduce the distortion to below the noise floor at low frequencies. This represents approximately 0.0005 to 0.0008% THD, depending on the exact design of the circuitry, and the actual distortion can be shown to be substantially below this if spectrum-analysis techniques are used to separate the harmonics from the noise.

References

1. Self, D, Advanced Preamplifier Design, *Wireless World*, Nov. 1976, p. 41.

2. Self, D, *Small Signal Audio Design*, Burlington, MA: Focal Press, 2010. ISBN 978-0-240-52177-0.

3. Greiner, R and Eggars, J, The Spectral Amplitude Distribution of Selected Compact Discs, *Journal of the Audio Engineering Society*, 37, April 1989, pp. 246–275.

4. Lawry, R H, High End Difficulties, *Stereophile*, May 1995, p. 23.

5. Moore, B J, *An Introduction to the Psychology of Hearing*, London: Academic Press, 1982, pp. 48–50.

6. Fielder, L, Dynamic Range Issues in the Modern Digital Audio Environment, *JAES*, 43(5), May 1995, pp. 322–399.

7. Radio Manufacturers' Association of America, 1937.

8. Shorter, D E L, The Influence of High-Order Products in Non-Linear Distortion, *Electronic Engineering*, 22(266), April 1950, pp. 152–153.

9. Geddes, E and Lee, L, Auditory Perception of Nonlinear Distortion, paper presented at 115th AES Convention, Oct. 2003.

10. Voishvillo, A, Assessment of Nonlinearity in Transducers and Sound Systems: From THD to Perceptual Models, paper presented at AES Convention 121, Oct. 2006, Paper 6910.

11. Moir, J, Just Detectable Distortion Levels, *Wireless World*, Feb. 1981, pp. 32–34.

12. Jacobs, J E and Whitman, P, Psychoacoustic: The Determining Factor in Disc Recording, *JAES*, 12, 1964, pp. 115–123.

13. de Santis, E and Henin, S, Perception and Thresholds of Nonlinear Distortion using Complex Signals, thesis, Institute of Electronic Systems, Aalborg University, Denmark, 2007.

14. Mills, P G and Hawksford, M, Transconductance Power Amplifier Systems for Current-Driven Loudspeakers, *JAES*, 37(10), Oct. 1989, pp. 809–822.

15. Olsher, D, Times One RFS400 Power Amplifier Review, *Stereophile*, Aug. 1995, p. 187.

16. Langford-Smith, F. Loudspeaker Damping, *Wireless World*, Aug. 1947, p. 309.

17. Moir, J, Transients and Loudspeaker Damping, *Wireless World*, May 1950, p. 166.

18. Greiner, R and Melton, D A, Quest for the Audibility of Polarity, *Audio*, Dec. 1993, p. 40.

19. Greiner, R and Melton, D A, Observations on the Audibility of Acoustic Polarity, *JAES*, 42(4) April 1994, pp. 245–253.

20. AES, *Draft AES Recommended Practice Standard for Professional Audio – Conservation of the Polarity of Audio Signals*. Inserted in: *JAES*, 42 (1994).

21. Johnson, K C, and Oliver, R, Letters to The Editor, *Wireless World*, Aug. 1964, p. 399.

22. Gardner, M, *Fads & Fallacies in the Name of Science*, Ch. 12, New York: Dover, 1957, pp. 140–151.

23. David, F, Investigating the Paranormal, *Nature*, 320, 13 March 1986.

24. Randi, J, *Flim-Flam! Psychics, ESP Unicorns and Other Delusions*, Buffalo, NY: Prometheus Books, 1982, pp. 196–198.

25. Harris, J D, Loudness Discrimination, *J. Speech Hear.*, Dis. Monogr. Suppl. 11, pp. 1–63.

26. Moore, B C J, Relation Between the Critical Bandwidth and the Frequency-Difference Limen, *J. Acoust. Soc. Am.*, 55, 1974, p. 359.

27. Hawksford, M, The Essex Echo, *Hi-fi News & RR*, May 1986, p. 53.

28. Self, D, Ultra-Low-Noise Amplifiers & Granularity Distortion, *JAES*, Nov. 1987, pp. 907–915.

29. Harwood, H D and Shorter, D E L, *Stereophony and the Effect of Crosstalk Between Left and Right Channels*, BBC Engineering Monograph, No. 52, 1964.

30. Fincham, L, The Subjective Importance of Uniform Group Delay at Low Frequencies, *JAES*, 33(6), June 1985, pp. 436–439.

31. http://en.wikipedia.org/wiki/Ohm%27s_acoustic_law (accessed Sept. 2012).

32. Lipshitz, S P, Pocock, M and Vanderkooy, J, On the Audibility of Midrange Phase Distortion in Audio Systems, *JAES*, Sept. 1982, pp. 580–595.

33. Moir, J, Comment on 'Audibility of Midrange Phase Distortion in Audio Systems, *JAES*, Dec. 1983, p. 939.

34. Harwood, H, Audibility of Phase Effects in Loudspeakers, *Wireless World*, Jan. 1976, pp. 30–32.

35. http://www.linkwitzlab.com/phs-dist.htm (allpass audibility) (accessed Sept. 2012).

36. Shinners, S, *Modern Control System Theory and Application*, Reading, MA: Addison-Wesley, 1978, p. 310.

37. King, G, Hi-fi reviewing, *Hi-fi News & RR*, May 1978, p. 77.

38. Harley, R, Review of Cary CAD-300SEI Single-Ended Triode Amplifier, *Stereophile*, Sept. 1995, p. 141.

39. Baxandall, P, Audio Power Amplifier Design, *Wireless World*, Jan. 1978, p. 56.

40. Belcher, R A, A New Distortion Measurement, *Wireless World*, May 1978, pp. 36–41.

41. Baxandall, P, Audible Amplifier Distortion Is Not A Mystery, *Wireless World*, Nov. 1977, pp. 63–66.

42. Hafler, D A, Listening Test for Amplifier Distortion, *Hi-fi News & RR*, Nov. 1986, pp. 25–29.

43. Colloms, M, Hafler XL-280 Test, *Hi-Fi News & RR*, June 1987, pp. 65–67.

44. Hi-fi Choice, The Selection, *Sportscene*, 1986.

The Basics of Distortion

Most of western culture is a distortion of reality. But reality should be distorted; that is, imaginatively amended.

Camille Paglia, 2012

Models of Non-linearity

A lot of this book is about distortion and ways of reducing it. It is therefore very useful to have a basic knowledge of how distortion works – how various deficiencies in the transfer characteristics of a stage, or even in just one component, create different harmonics from a pure input.

A system that has a perfectly linear relationship between input and output generates no distortion. The input/output law is a straight line passing through zero, as in Figure 2.1. The slope of the line determines the gain; here it is unity.

Whenever the input/output law deviates from a straight line, the amplifier becomes non-linear and distortion is produced, in the form of extra harmonics at integer multiples of the input frequency. This deviation from linearity can occur in many different ways, and we shall look at a few. Some of them are useless as models for real

amplifiers, but are mathematically simple and show how the distortion business works.

SPICE provides an extremely useful tool for investigating distortion models called Analogue Behavioural Modelling. This allows a non-linearity to be defined by a mathematical equation, as opposed to trying to get real components to act in a mathematically simple way (which they don't, not even allegedly square law FETs). For our purposes, the handiest approach is to use the SPICE part E, which denotes a voltage-controlled voltage source (VCVS). Here is the code for a completely linear amplifier with a gain of 100 times:

```
E1 7 0 VALUE = {100*V(3)}          ; 100x amplifier
Rdummmy1 7 0 10G
```

Here the output terminals of the VCVS are numbered 7 and 0, the latter being ground by definition in SPICE. The voltage across those terminals is set by the VALUE statement, where a mathematical expression, in SPICE's BASIC-like language, is enclosed in a tasteful pair of curly brackets. Here the output is the voltage at node 3 (the 'input' of our amplifier) multiplied by a factor of one hundred. The mathematical expression can be very much more complicated than this, and can

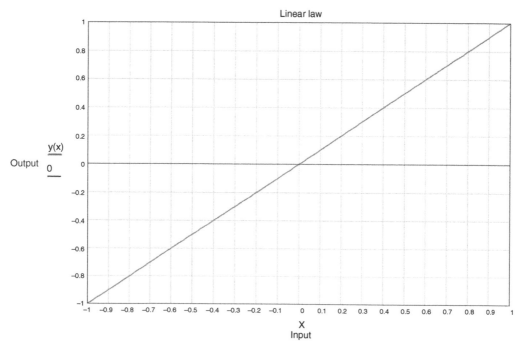

Figure 2.1. The input-output relationship for a linear amplifier with $V_{out} = V_{in}$, giving a gain of unity.

implement pretty much any sort of non-linearity you can dream up, as we shall see. In SPICE text, everything after a semi-colon is a comment and is ignored by the simulator.

The procedure is simply to run a transient simulation with a sinewave input and then do a Fourier analysis of the output. This gives you the level of each harmonic generated, and the Total Harmonic Distortion, which is the RMS sum of the harmonics.

Why is Rdummy1 there? SPICE flatly refuses to have anything to do with infinitely large quantities, as might be expected, but it is perhaps a little less obvious that it also stoutly objects to the infinitely small. In this sort of modelling, the voltage source output will most likely be connected to the control input of another VCVS, which draws no current. A voltage source with an open-circuit output has an infinitely large resistance across it, and so passes an infinitely small current. SPICE will have none of it, and it is necessary to put a resistor across the voltage source to allow a finite current to flow. I use 10 Gigohm resistors are they unlikely to be confused with really functional resistors. Names like Rdummy help with that.

Cubic Distortion

Distortion can be expressed as the effect of a power law on the signal. An example frequently used in textbooks is the 'cubic distorter', where the output is simply the cube of the input, for example, see.[1] This is an odd-order power so it gives a nice symmetrical result and there are no problems with the generation of DC components.

$$V_{out} = (V_{in})^3 \qquad \text{Equation 2.1}$$

The SPICE code for a cubic law like this is very simple:

```
E1 7 0 VALUE = {V(3)*V(3)*V(3)}
Rdummy1 7 0 10G
```

Here the input voltage at node 3 is cubed by multiplication rather than using the SPICE PWR function, to make it a bit clearer what is going on.

If a 1 V peak sine wave is applied to this model, then the output is the original frequency at three-quarters of the input amplitude (0.75 V), plus a new component, the third harmonic at three times the frequency, and one quarter the input amplitude (0.25 V). The third harmonic is always at one-third of the amplitude of the fundamental, no matter what input level you use. A THD calculation, which relates the level of the

harmonics to the level of the fundamental at the output (*not* the input), will always give 33.33%. The same result can be reached by mathematics; see.[1] It is a result of the way the cubic power law is self-similar — doubling the input amplitude always gives eight times the output, wherever you start. The constant-THD behaviour is quite unlike anything measured from real amplifiers, and that should give you a broad hint that the cubic law is not a realistic model, even if the rapid increase of output with input did not make it obvious; see Figure 2.2. Another problem with any simple power model is that the gain is always zero with 0 V input, which is hardly realistic; and in case you were wondering that does *not* make it a good model of crossover distortion.

A greater than proportional increase of output with input, as shown by the cubic law, is called an *expansive* distortion. In the real world, distortion is much more likely to be *compressive*; the output increases more slowly than proportionally with input. All real amplifiers are ultimately compressive, as the output cannot move outside the supply rails, and will clip if sufficient input is applied.

Cubic + Linear Distortion

A somewhat more sophisticated approach is to use a polynomial rather than a simple power law. A polynomial is a combination of integer, non-negative powers of a variable. Figure 2.3 shows the input/output law for the polynomial in Equation 2.2.

$$V_{out} = V_{in} - (V_{in})^3 \qquad \text{Equation 2.2}$$

The SPICE code for this law is:

```
E1 7 0 VALUE = {V(3)-(V(3)*V(3)*V(3))}
Rdummy1 7 0 10G
```

Once again, the voltage at node 3 is the input. Note that all the brackets inside the two curly brackets are ordinary round ones.

The subtraction of the cube from the linear term makes the input/output law level out nicely for input levels up to just over ± 0.5 V, giving us a compressive distortion, and also gives unity gain for zero input, which is much more plausible. However, you will note that the law does not stay levelled out, but heads in the opposite direction at ever-increasing speed; no real amplifier would do this. This is a good illustration of the typical problem when using polynomials to model anything. You can set things up to be well controlled

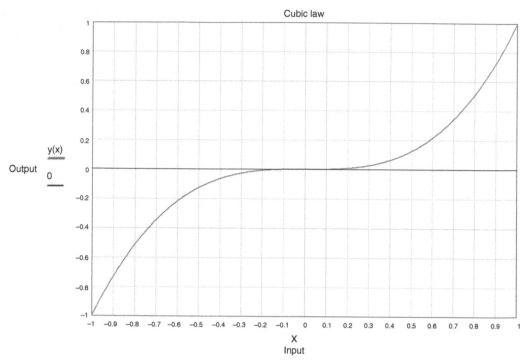

Figure 2.2. The input-output relationship for a cubic distorter with $V_{out} = (V_{in})^3$. This is an expansive distortion characteristic, and, as such, unrealistic.

within a specific region, but outside this the polynomial usually 'blows up' and heads rapidly off towards infinity.

Having written that, I recall an exception. I once encountered an amplifier design, built into an active speaker, that reversed the phase of the forward path when it clipped. Naturally it would then latch up solid, and as there was no DC-offset protection, the drive unit would catch fire at once. It was as fine an electronic trap as I have ever seen; it would work perfectly for months at low level but the first time ever it was clipped, it would destroy itself, and quite possibly set fire to the curtains. Regrettably, the 'designer' had already left the company, for otherwise we would have had an interesting discussion about this reckless design work. He was last heard of working in the medical technology field. (Really!)

To aid clarity, this model has been set up to create much more distortion than one would hope to find (except perhaps in a guitar amplifier) and is more characteristic of a fuzz-box. Figure 2.4 shows how the level of the third harmonic, the only one generated, rises increasingly rapidly with input level; in fact, it is roughly proportional to the square of input level. This is much more realistic than the constant level generated

by the pure cubic law. It is noticeable that there is not a sudden change in the third harmonic level when the input signal exceeds the limits of realistic modelling set by the levelling-out points; it just continues to increase at the same square law rate. This is not what happens when real amplifiers clip, softly or otherwise. In reality, there is more than one harmonic, and harmonic levels shoot up when clipping starts; almost vertically, in the case of hard clipping.

Square Law Distortion

We will now look at what happens with a square law. For the 'cubic distorter' we could just use $V_{out} = V_{in}^3$, but if we do that with the pure square law $V_{out} = V_{in}^2$, we get the parabola shown in Figure 2.5. This is not so much an amplifier as a full-wave rectifier. In passing, we note that the only harmonic generated is the second, as we might expect, and its amplitude is always half the amplitude of the input signal, because of the self-similar nature of the law. A DC component of half the amplitude of the input signal is also produced. Since the fundamental is completely suppressed by this circuit, it is not possible to calculate a meaningful THD figure.

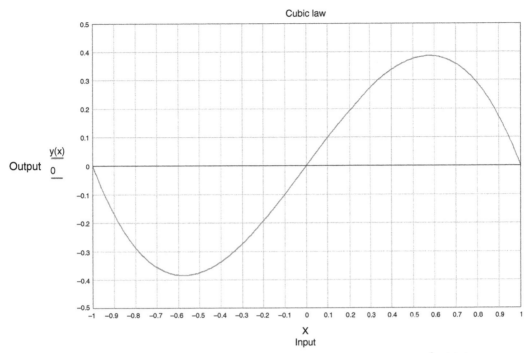

Figure 2.3. The input-output relationship for a linear + cubic distorter with $V_{out} = V_{in} - (V_{in})^3$. This is a compressive distortion so long as the input does not exceed $\pm$ 0.5V.

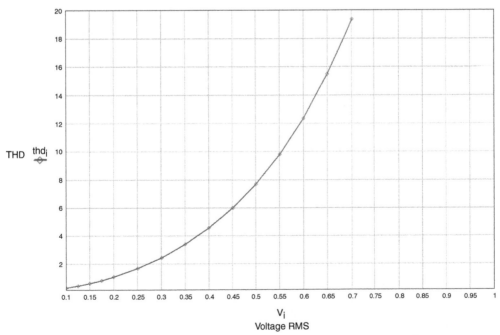

Figure 2.4. The level of the third harmonic (in %) generated by a model with $V_{out} = V_{in} - (V_{in})^3$. This is the only harmonic produced. The X-axis is linear.

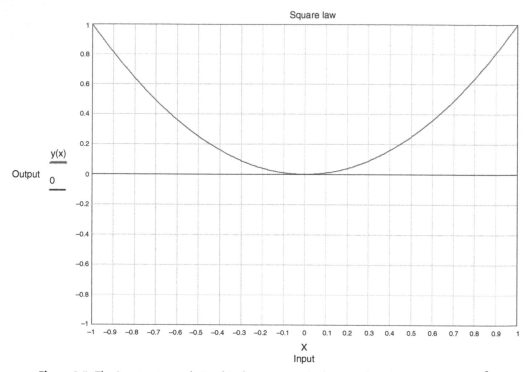

Figure 2.5. The input-output relationship for a symmetrical square law distorter $V_{out} = (V_{in})^2$.

The SPICE code for this law is simply:

```
E1 7 0 VALUE = {V(3)*V(3)}
Rdummy1 7 0 10G
```

but if you use the PWR function, it can also be written:

```
E1 7 0 VALUE = {PWR(V(3),2)}
Rdummy1 7 0 10G
```

which just means raise V(3) to the power of two.

We can attempt to make a more usable square law model by stitching together two halves of a square law, with the negative side inverted. This can be simply done in SPICE with the handy PWRS function. Like PWR, it raises a value to a power, but the difference is that it preserves the original sign of that value; our voltage-controlled voltage-source E1 therefore becomes:

```
E1 7 0 VALUE = {PWRS(V(3),2)}
Rdummy1 7 0 10G
```

and we get the law shown in Figure 2.6, which is rather different from the cubic law we examined first; note it is less flat around zero.

So, what harmonics does this law generate? The stitching together is a non-linear process, so we get all the odd harmonics. (No even harmonics, because the law is symmetrical.) The harmonics are constant with input level because the law is self-similar — doubling the input always gives four times the output, no matter what the input level. Table 2.1 shows the level of each harmonic relative to the fundamental in the output.

While the stitched square law model may be instructive, it is not otherwise useful. However, it is well known that some amplifier circuits generate large amounts of second-harmonic distortion that *is* level-dependent. This works because their characteristic is only a part of a square law and is not symmetrical about zero, as shown in Figure 2.7. This curve is produced by the law $V_{out} = (0.7V_{in})^2 + V_{in}$. Note that the gain is unity around zero.

The SPICE code is:

```
E1 7 0 VALUE = {0.7*(V(3)*V(3))+V(3)}  ;
                              y = 0.7x^2 + x
Rdummy1 7 0 10G
```

The result is that, as with the previous square law model, the second harmonic only is generated, and its amplitude is proportional to the input level (Figure 2.8). This is potentially a good model for amplifier stages

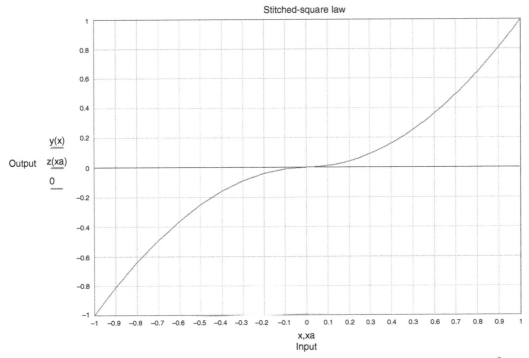

Figure 2.6. The input-output relationship for a stitched-together square law distorter $V_{out} = (V_{in})^2$.

such as the VAS, which tend to generate mostly second harmonic.

Square Root Distortion

We have so far not found a good model for soft clipping; the linear + cubic distorter we looked at earlier was unsatisfactory because of the way its input/output law turned over and the unrealistic way it generated only the third harmonic. How about a square root law? We know that has a decreasing slope as we move away from zero, as shown in Figure 2.9. Since you cannot (in this context, you certainly can in complex algebra) take the square root of a negative number, the input/output law is made up of two square root curves stitched together at zero, as we did for the square law earlier.

The SPICE code is:

```
E1 7 0 VALUE = {PWRS(V(3),0.5)}; stitched squares
Rdummy1 7 0 10G
```

Here we once more use the PWRS function to do the stitching. It is quite happy with an exponent of 0.5, to give the square root law.

The result is instructive, but no better at emulating a real amplifier. The model is no longer a polynomial, as it contains a fractional exponent (0.5), and so all odd-order harmonics are generated. Even-order harmonics do not appear because the law is symmetrical about zero. As for polynomials, the percentage level of harmonics is constant, whatever the input level. Table 2.2 shows the level of each harmonic relative to the fundamental in the output, and it can be seen that the level of the higher harmonics falls off much more slowly than for the stitched square law case.

Soft-clipping Distortion

We still need a good model for soft clipping, and to do that we need to use slightly more complicated

Table 2.1. Harmonic levels for the stitched square law characteristic

Harmonic	% ref output fundamental
3rd	20.0
5th	2.86
7th	0.955
9th	0.433

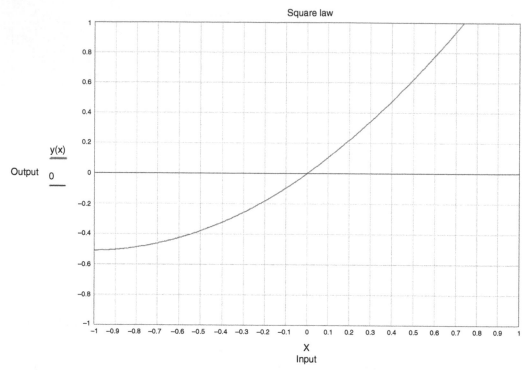

Figure 2.7. The input-output relationship using just part of a square law $V_{out} = (0.7V_{in})^2 + V_{in}$.

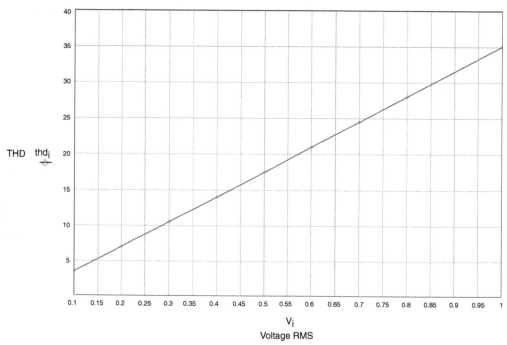

Figure 2.8. The level of the second harmonic (in %) generated by a model with the square law characteristic $V_{out} = (V_{in})^2$.

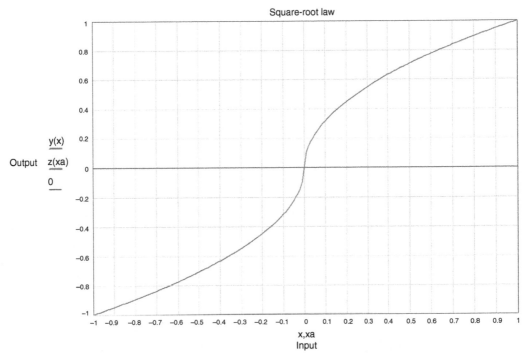

Figure 2.9. The input-output relationship for a symmetrical square root distorter with $V_{out} = \sqrt{V_{in}}$.

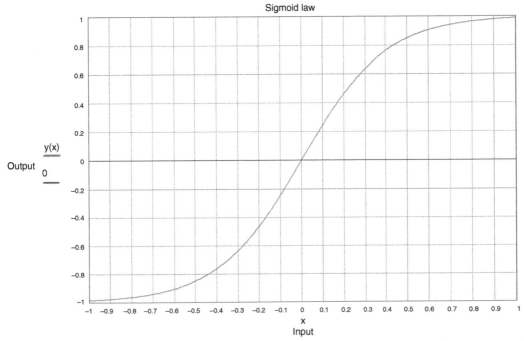

Figure 2.10. The input-output relationship for a sigmoid curve with $V_{out} = \sqrt{V_{in}}$.

Table 2.2. Harmonic levels for stitched square root law characteristic

Harmonic	% ref output fundamental
3rd	14.29
5th	6.49
7th	3.89
9th	2.66

$$V_{out} = \frac{1}{1 + e^{-V_{in}}}$$ Equation 2.3

where e is the base of natural logarithms, approximately equal to 2.71828.

I modified Equation 2.3 to give zero output with zero input (the -1 on the end), and scaled the input and output voltages (the factors of 5 and 2) to fit the input/output law into the same limits as the other distortion models; see Equation 2.4:

$$V_{out} = \frac{2}{1 + e^{-5V_{in}}} - 1$$ Equation 2.4

mathematics. The law shown in Figure 2.10 looks very promising; it is called a sigmoid curve. It is inherently symmetrical without any curve-stitching, and levels out to ±1. Actually it approaches ±1 asymptotically; in other words, it gets closer and closer without ever actually reaching it, though as you can see, an input voltage of ±1 gets it very close at ±0.987. The sigmoid curve is very similar to the transfer function of a degenerated differential pair. See Chapter 6 on input stages.

The sigmoid curve in Figure 2.10 is generated from a variation on what is called the Logistic Equation; it is shown in its basic form in Equation 2.3:

Note that the gain is about 2.4 times around zero. Figure 2.11 shows that all odd harmonics are generated, because the sigmoid law is symmetrical, and the level of each rises quickly as the input voltage increases and moves into the more curved part of the law. This is the first distortion model that has shown this behaviour, which is typical of a real amplifier. To the best of my knowledge, this is the first time the sigmoid curve has been used in this way.

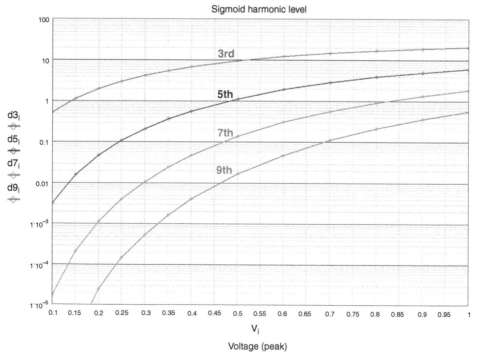

Figure 2.11. The levels of the first nine harmonics (in %) generated by a model with the sigmoid characteristic of Figure 2.10. Note that the vertical axis is logarithmic.

The SPICE code to implement Equation 2.4 is:

```
E1 7 0 VALUE = {2/(1+PWR(2.718,-5*V(3)))-1}      ;
                         modded logistic equation
Rdummy1 7 0 10G
```

Note that the very helpful PWR function is quite content to have a variable setting the exponent.

At low input voltages the higher harmonics are a long way below the level of the third, but the difference decreases as the input level goes up. However, even at 1V input, the contribution of the higher harmonics to a THD calculation is small; the third harmonic is at 20.67%, while the THD including harmonics up to the ninth is only slightly greater at 21.61%. This does not take into account the possible use of harmonic weighting to give a better estimate of the audible effect of distortion; see Chapter 1.

Hard-clipping Distortion: Symmetrical

While soft clipping is characteristic of some power amplifier stages, notably the input differential pair, when a negative feedback loop is closed around an amplifier with plenty of open-loop gain, it keeps the input/output law very linear until the output hits the rails, and is squared-off abruptly. This is called hard clipping or hard limiting, and a typical input/output law is shown in Figure 2.12, where the limiting values have been set at ± 0.5 V to give symmetrical clipping. The gain between these limits is unity.

As the amplitude of the input sinewave increases, this law is completely linear until it hits the limiting values. As Figure 2.13 shows, the third harmonic abruptly appears and increases very rapidly. The higher harmonics show more complicated behaviour, with the fifth harmonic level rising, dropping back to zero, then rising steadily after that. The seventh harmonic shows two nulls, and the ninth harmonic three. This is nothing like soft clipping, where the harmonic levels simply increase.

When the input is 1 V peak in amplitude, the top and bottom halves of the waveform are removed; this is often described as '50% clipping'. As Figure 2.13 shows, there is nothing special about this level in term of the harmonic levels.

SPICE has a function called LIMIT that implements this very neatly; the code is:

```
E1  7  0  VALUE  =  {LIMIT(V(3),-0.5,0.5)}     ;
                                         limiting
Rdummy1 7 0 10G
```

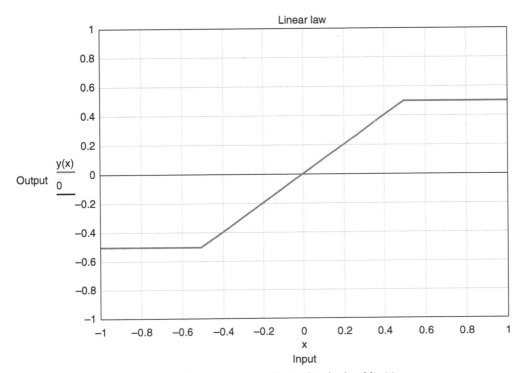

Figure 2.12. The input-output relationship for hard limiting.

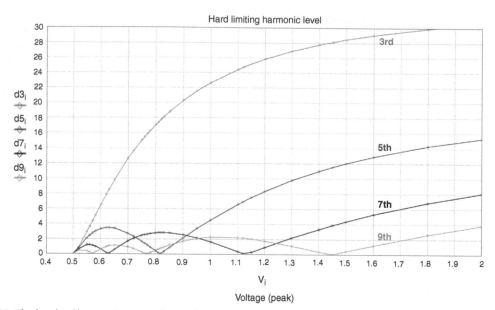

Hard limiting harmonic level

Figure 2.13. The levels of harmonics up to the ninth (in %) generated by hard limiting with the characteristic of Figure 2.12. Note horizontal scale starts at 0.4.

The lower limit is set to -0.5 V and the upper limit to 0.5 V, as in Figure 2.12.

As the input amplitude increases above 1 V peak, the output waveform is clipped harder and looks more and more like a square wave, which is known to contain only odd integer harmonics, in the ratio of 1/3, 1/5, 1/7, 1/9, etc. relative to the fundamental; see Table 2.3. It is clear from Figure 2.13 that each of the harmonic levels, once they have finished bouncing off the bottom axis, is heading in that direction. The higher the harmonic, the more slowly it approaches its asymptote.

To demonstrate this, Figure 2.14 shows how the harmonic levels for greater inputs slowly approach the theoretical values (the dashed lines). For an input of 4 V peak, the limiting levels are at one-eighth of the peak amplitude, and the output waveform looks very

much like a square wave, with only a slight deviation from vertical on the rising and falling edges.

Hard-clipping Distortion: Asymmetrical

In practice, the clipping of an amplifier is unlikely to be exactly symmetrical; asymmetrical clipping is the norm, though the amount of asymmetry is usually small. The distinctive sound of asymmetrical clipping is generally supposed to be the explanation for the famous Fuzzface guitar effect, though there heavy clipping would have been used.

The same LIMIT function is used but the parameter setting the negative and positive limits has changed:

```
E1 7 0 VALUE={LIMIT(V(3),-0.4,0.6)}   ;limiting
Rdummy1 7 0 10G
```

The lower limit is set to -0.4 V and the upper limit to 0.6 V to give asymmetrical clipping.

We now find that all harmonics are generated, because the distortion law is no longer symmetrical. In Figure 2.15, the second harmonic rises steadily once negative clipping starts at -0.4 V, but begins a slow decline when positive clipping at $+0.6$ V also begins. Note the definite kink in the third harmonic, as it increases in slope as positive clipping starts. The fourth harmonic drops to zero at an input voltage of 0.7 V peak, in the same way that the higher harmonics did in the symmetrical clipping case,

Table 2.3. The harmonic levels in a square wave

Harmonic	Fraction	Percentage
Fundamental	1	100
3	1/3	33.33
5	1/5	20.00
7	1/7	14.29
9	1/9	11.11

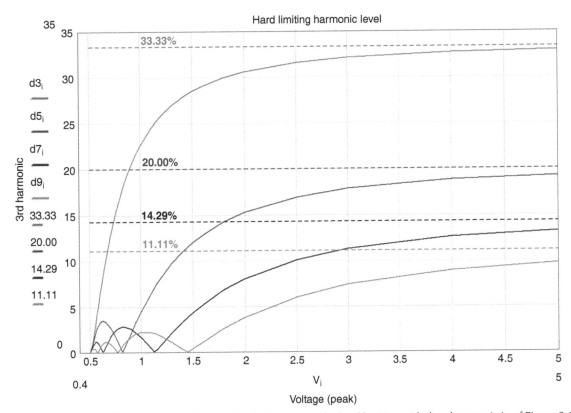

Figure 2.14. The levels of harmonics (in %) up to the ninth generated by hard limiting with the characteristic of Figure 2.12, for greater input levels.

and the higher harmonics show even more complex behaviour, with multiple nulls as input level increases.

Crossover Distortion Modelling

It is all very well having these models for the various distortion mechanisms in amplifiers, but the one that is missing is the toughest problem of the lot: crossover distortion. This appears as an innocent-looking wobble in the voltage gain of the output stage, but it creates high-order harmonics that are not well linearised by negative feedback that falls off with increasing frequency. The law is roughly symmetrical so odd harmonics are much higher in level than the even harmonics.

However, things did not go well when I tackled this problem. The distortion models described above are all made by plugging together mathematical black boxes with simple properties. To emulate the more complex crossover gain-wobble, the obvious method is to simulate an actual output stage, with appropriate transistor types and output loading. I duly did that, but made the

unwelcome discovery that as soon as a transistor model was introduced to the simulation, the numerical accuracy dropped by orders of magnitude, and the results were useless for Fourier analysis.

I therefore tried to approximate the crossover gain-wobble by a piece-wise-linear model, but this has sharp corners in the gain characteristic that give rise to unacceptable errors in the levels of high-order harmonics. And there it stands for the moment; possibly using a much larger number of line segments would reduce the generation of spurious high-order harmonics to acceptable levels.

Other Distortion Models

There is currently interest in controlled non-linearities for generating harmonics to give virtual bass; the idea is not exactly new, going back at least to 1951.[2] A recent *JAES* article on this by Woon-Seng Gan and Hawksford contains a large collection of distortion generators of various kinds.[3]

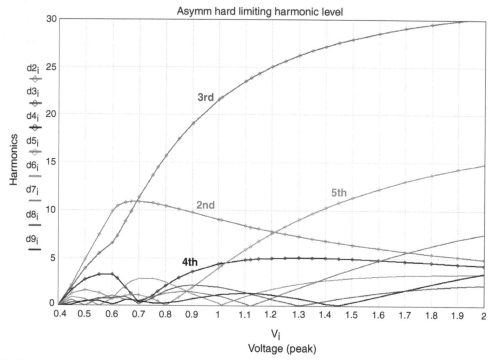

Figure 2.15. The levels of harmonics (in %) up to the ninth generated by asymmetrical hard limiting. Same input levels as Figure 2.13.

Choosing a Distortion Model

The mathematical input/output law of an input differential pair is well known, and determined by definite circuit parameters such as the tail current and the amount of emitter degeneration (see Chapter 6). The input/output law of a VAS is a much more uncertain matter, depending on Early effect, the non-linear loading of the output stage, and dynamically upon the variation of transistor collector-base capacitance with Vce. This is dealt with in detail in Chapter 7 on the VAS stage.

SPICE Models for Passive Components

So far we have looked at how various mathematical models induce the generation of distortion harmonics. While these are useful for application to complete amplifier stages, they do not fit well with the non-linearity sometimes shown by passive components such as resistors and capacitors.

SPICE provides excellent facilities for modelling the non-linearity of passive components, but the techniques are somewhat less than obvious. Most if not all SPICE simulators include facilities for Analog Behavioural

Modelling, which is the simplest way to do it. Running a transient simulation followed by Fourier analysis gives the level of the harmonics and allows the THD to be calculated. I am going to stick to SPICE text format here, as the details of schematic entry vary greatly between different brands of simulator.

In SPICE it is very easy to define a fixed resistor, for example:

```
R1 8 9 1K
```

gives you a 1K Ω resistor R1 connected between nodes 8 and 9. It is of course a perfect resistor in that its value is exact, it is perfectly linear, it can dissipate GigaWatts if necessary, and it doesn't cost anything. To make a non-linear resistance is a little more complicated. One method is to treat the resistor as a voltage source with a value proportional to the current going through it; this is just a statement of Ohm's Law.

We will start off by modelling a linear resistor in this way. Figure 2.16 shows how it is done. E1 uses the SPICE prefix E to denote a voltage-controlled voltage source (VCVS), while V1 is a plain voltage source, here set to zero volts and effectively used as an ammeter to measure the current through E1 and V1.

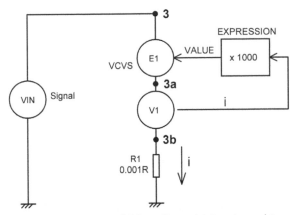

Figure 2.16. SPICE model for a linear 1 kΩ resistor, driven by signal source V$_{in}$.

This value is then used to control the voltage of E1; if E1 gives 1 Volt for a 1 Amp current, then we have built a 1 Ω resistor. If we multiply the ammeter reading by 1000 before applying it to E1, using the box marked 'expression' in Figure 2.16, we get 1 Volt for a 1 mA current, and so we have made a 1 kΩ resistor. In SPICE text format, the code looks like this:

```
Eres1 3 3a VALUE = {I(Vsense1)*1K}
Vsense1 3a 3b
R1 3b 0 0.001R
```

Note that the outside brackets around the VALUE expression must be curly brackets, and that it is quite alright to call nodes 3a, 3b, and so on. The synthesised resistance is shown with one end connected to ground, but it works equally well with both ends floating.

You are no doubt wondering what that 1/1000 of an Ohm resistor R1 is doing in there. The answer is that it prevents some SPICE simulators from throwing an error complaining about the two voltage sources. If the resistor is omitted, my version of PSPICE says:

'ERROR — Voltage source and/or inductor loop involving E1. You may break the loop by adding a series resistance.'

Which is a bit opaque as to what the exact problem is but does at least tell you a good fix. I added 1/1000 of an Ohm, which makes no detectable difference to circuit operation here but appears to be a completely robust solution. This may not be true with other models, and it is essential to check for correct operation by doing a DC sweep, as well as a transient simulation.

This sort of model is a powerful tool, and like all powerful tools it must be used with care. It is very easy to come up with a model resistor that generates power out of nowhere — in other words, current will flow when the voltage across it is zero. This is obviously wrong and must be carefully checked for every time you modify the model.

Obviously we have made our 1 kΩ resistor the hard way, but the great advantage of this approach is that we have the expression box in which we can manipulate the current signal before applying it to E1. This is part of the Analog Behavioural Modelling facility in SPICE. We can put in, using a BASIC-like syntax, whatever mathematical equation we like in the expression box, and if it is non-linear, then we synthesise a correspondingly non-linear resistance. Let's do it.

First-order Voltage-coefficient Distortion Model

The non-linearity of passive components is usually specified, if it is specified at all, in the form of a voltage coefficient that describes how the value of the component changes with the applied voltage. If a resistor with a value of 1000 Ω with no voltage across it has a voltage coefficient of +0.001 (which is much larger than you would really get with a decent quality component), then the value will increase to 1001 Ω with 1 Volt across it. Note that there is usually an assumption of component symmetry here; in other words −1 Volt across the resistor will also give 1001 Ω, and not 999 Ω. This assumption is usually applied to resistors and non-electrolytic capacitors. It should not be used for electrolytic capacitors, though it might be valid for non-polarised electrolytics; I have no information on this at present.

Here the change in component value is proportional to the voltage across it; this is first-order voltage-coefficient distortion, or it might be described as linear voltage-coefficient distortion, though the behaviour it leads to is certainly not linear.

At this point you might feel a bit suspicious of this highly convenient model for non-linearity. The world being the way it is, the real truth might be that the value is not really controlled by a constant voltage coefficient, but some more complex relationship. This is certainly true for some components, such as carbon-composition resistors, but in general a first-order voltage-coefficient distortion model works well for more common components. It is assumed that the voltage coefficient is very small compared with unity. If it is not, the waveforms become grossly distorted and the neat relationships between the harmonic levels described below will no longer apply.

If a resistance has a constant value, i.e., it has a zero voltage coefficient, then we have the familiar Ohm's Law as in Equation 2.5, where i is the current and V the voltage across a resistor of value R Ohms.

$$i = \frac{V}{R}$$
Equation 2.5

A non-zero voltage coefficient means that R is replaced by Equation 2.6, where R_0 is the resistance at zero voltage, and ρ is the voltage coefficient. A plus sign is shown here, indicating that the resistance increases in value with the voltage across it when the coefficient is positive, but the coefficient may also be negative. The two lines on either side of V mean the absolute value of V, which prevents a change of sign for negative voltages. This absolute value operation makes our component model symmetric.

$$R = R_0(1 + \rho|V|)$$
Equation 2.6

Plugging this new definition into Equation 2.5, we get Equation 2.7

$$i = \frac{V}{R_0(1 + \rho|V|)}$$
Equation 2.7

which can of course also be written as Equation 2.8, which perhaps makes it a little clearer that this is not a linear equation, and so harmonics will be generated.

$$i = \frac{V}{R_o}(1 + \rho|V|)^{-1}$$
Equation 2.8

The fact that this equation is not a simple power of V implies that there will be not just a single harmonic produced, such as the second or third, but multiple harmonics, and this is indeed the case. A voltage coefficient model that assumes component symmetry will give only odd harmonics, so our model can be expected to give multiple odd harmonics, probably going on up to an infinite frequency.

Equation 2.8 in particular could be dealt with mathematically, using the binomial expansion, which does indeed demonstrate that the harmonics form an infinite series. It is, however, very often less work to use a simulator, and the great advantage of this approach is that a simulator offers much less opportunity for mistakes,

while in mathematics any line of algebra is vulnerable to non-obvious errors.

In SPICE text format, this looks like:

```
E1  3  3a  VALUE  =  {I(Vsense3)*1K*(1-(coeff)
                              *ABS(V(3,3a)))}
V1 3a 3b
R1 3b 0 0.001R
```

Here the expression for the VALUE variable that controls E1 is in two parts: I(V1)*1K creates a linear 1 kΩ resistance, while the (coeff)*ABS(V(3,3a)) part reads the voltage across E1, which is V(3,3a), takes the absolute value of it with the ABS function, multiplies it by the voltage coefficient, and then subtracts it from the linear part of the resistance. Thus the value of the resistance synthesised varies with the voltage across it. See Figure 2.17 for the model.

Figure 2.18 shows the results for the first nine harmonics, (which is as far up as the version of SPICE I used would go with its FFT output) using a voltage coefficient of -0.001. All the harmonics increase proportionally to signal level. For twice the signal, you get twice the harmonic level, or +6 dB. You will note that the ratios between the harmonic levels are constant, as demonstrated by the constant spacing of the traces on the log-log plot.

As expected, the levels of even harmonics were at the level of simulator numerical noise, being at least 10,000 times lower than the odd harmonics. The THD is not plotted as it would lie on top of the third-harmonic line. This is because THD is calculated as the RMS sum of the harmonic levels, so harmonics at lower levels make very little contribution to the final figure.

At any level, the fifth harmonic is 17 dB below the third harmonic. The seventh harmonic is lower than the fifth by 9.5 dB, and the ninth harmonic lower than the seventh by 6.8 dB. The THD figure, using the harmonics up to and including the ninth, at an input of 15 Vrms is 0.3678%. Eliminating all but the third harmonic only reduces this to 0.3649%, which I suggest is negligible in terms of human perception.

You may, however, object that it is generally accepted that the higher the order of a harmonic, the more unpleasant its effect. This can be compensated for by multiplying the level of each harmonic by a weighting factor. The most commonly used is that suggested by Shorter in 1950,[4] which multiplies the level of the nth harmonic by $n^2/4$. This leaves the second harmonic unchanged, but raises the third by $9/4 = 2.25$ times, the fourth by $16/4 = 4$ times, and so on. (A more sophisticated version was proposed by Lidia Lee and Earl Geddes in 2003.[5]) If we use the Shorter weighting to generate the

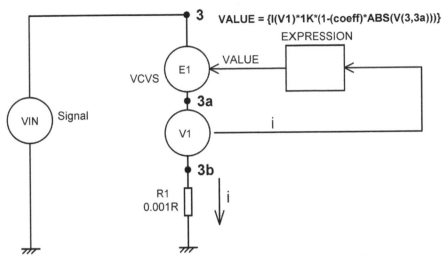

VALUE = {I(V1)*1K*(1-(coeff)*ABS(V(3,3a)))}

Figure 2.17. SPICE model for voltage coefficient non-linearity.

THD figure for the same conditions, we get 0.5648%, the increase largely being due to the scaling up of the third harmonic. If we once more eliminate all but the third harmonic, we get 0.5473%, a reduction that is larger than in the unweighted case but still very small and unlikely to have significant perceptual effects. It therefore seems that in assessing the effect of these distortion models we can actually neglect all but the third harmonic.

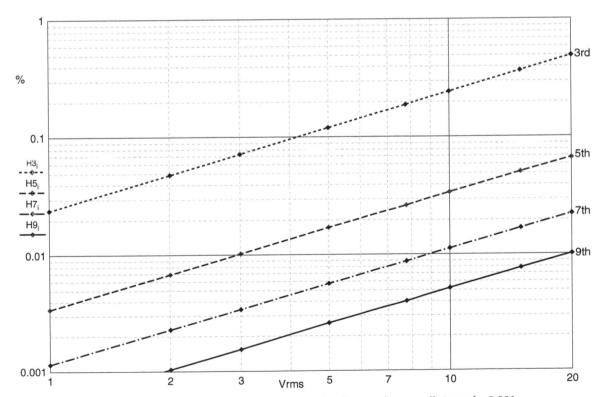

Figure 2.18. Harmonic percentages versus signal voltage; voltage coefficient of −0.001.

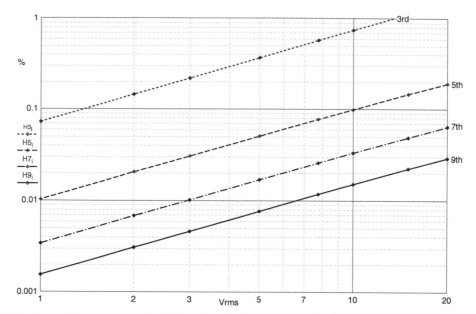

Figure 2.19. Harmonic percentages (3rd, 5th, 7th and 9th) versus signal voltage; voltage coefficient of −0.003.

Changing the voltage coefficient from −0.001 to −0.003 increases the amount of distortion generated, as would be expected, but the interesting thing is that the harmonics have the same relative level compared with each other, and the same rate of increase; 6 dB for each doubling of input signal level.

While Figures 2.18 and 2.19 give a good overview of the levels at which the harmonics are generated, it is well known that many laws look like a straight line on a log-log plot. Figure 2.20 shows the same information on a lin-lin plot for voltage coefficient of −0.001, and it is clear that the level of each harmonic is proportional to the signal level applied.

Figure 2.21 shows how the harmonic levels vary with the voltage coefficient.

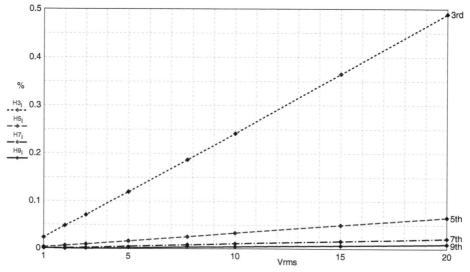

Figure 2.20. Harmonic percentages (3rd, 5th, 7th and 9th) versus signal voltage plotted on linear axes; voltage coefficient of −0.001.

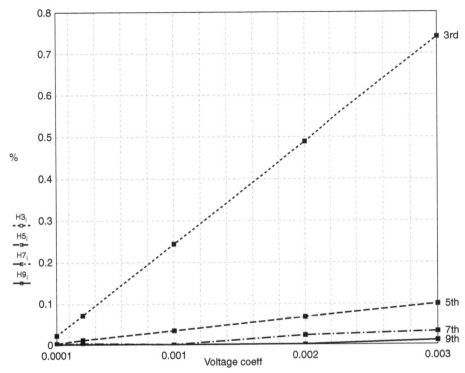

Figure 2.21. Harmonic percentages (3rd, 5th, 7th and 9th) versus voltage coefficient plotted on linear axes for signal level of 10 Vrms.

It is sometimes recommended that the effects of passive distortion be reduced by using multiple components. Thus, a required resistance may be made up of two resistors of half the value in series. If a first-order voltage coefficient model is appropriate for resistors, then putting two resistors in series only causes the THD/third-harmonic level to be halved. Three resistors drop it to a third, and so on, with N resistors in series reducing the THD/third harmonic in proportion to N. I have checked this in SPICE up to N = 5, using the model described above. This means that putting resistors in series gives a relatively poor return. While using, say, 10 resistors in series does not cost much money (unless you're using exotic parts, in which case you shouldn't be having this trouble in the first place), it is not exactly an elegant solution and takes up a lot of PCB area.

Lower resistance values tend to have a lower voltage coefficient so putting two in series gives an extra benefit apart from the halving of the signal voltage across the component. Putting the resistors in parallel gains you nothing as the voltage across each one is unchanged.

What sort of values for voltage coefficients should we expect? Table 2.4 shows the generally accepted ranges for different types of resistor in ppm/V.

Second-order Voltage-coefficient Distortion Model

Another possible model for non-linearity is a square law variation with voltage, with the characteristic of Equation 2.9.

$$i = \frac{V}{R_0(1 + \rho V^2)} \qquad \text{Equation 2.9}$$

Table 2.4. Voltage coefficients for different types of resistor

Type	ppm/V
Carbon composition	200 − 500
Carbon film	<100
Metal oxide	< 10
Metal film	Approx 1
Metal foil	<0.1
Wirewound	Less than 1[*]

[*]Note: Wirewound resistors are normally considered to be completely free of voltage effects.

Since we are squaring the voltage, the result is always positive and there is no need to use an absolute-value function to give symmetrical behaviour. In SPICE text format, this looks like:

```
E1  3  3a  VALUE  =  {I(Vsense3)*1K*(1-(coeff)*
                             V(3,3a)*V(3,3a))}
V1 3a 3b
R1 3b 0 0.001R
```

where now the square of the voltage between nodes 3 and 3a is used (Figure 2.22).

As before, the symmetry of the model means that only odd harmonics are produced. For each doubling of input level, the third harmonic now increases by a factor of four (12 dB). In contrast to the first-order model, the higher harmonics go up at increasing rates with their order; the 5th harmonic increases by a factor of sixteen for input level doubling, the 7th by 64 times, and so on, as summarised in Table 2.5.

The rapid rates of increase for the higher harmonics could have serious consequences were they not all well below the third harmonic. There is no longer a constant dB difference between them, as for the first-order model, but if we select the levels at 10 Vrms, the fifth

Table 2.5. Harmonic increase ratio for doubled input voltage: square-law voltage coefficient model.

Harmonic	Ratio	dB
Third	4	12
Fifth	16	24
Seventh	64	36
Ninth	256	48

harmonic is 36.7 dB below the third, the seventh is 36.7 dB below the fifth, and the ninth 36.7 dB below the seventh. The effect of such low levels of higher harmonics on the THD figures is negligible, and the THD is effectively just the level of the third harmonic.

Other Voltage Coefficient Distortion Models

A third-order voltage coefficient distortion would have a cubic component law. As for the first-order case, it will be necessary to use an absolute-value function to obtain models with symmetrical behaviour about zero.

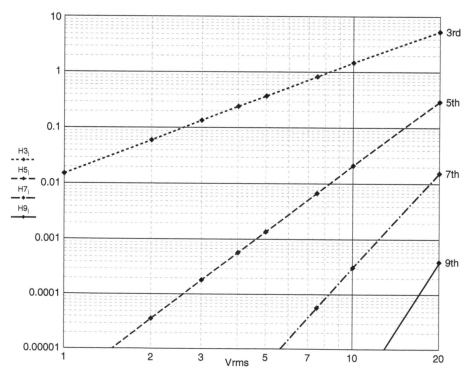

Figure 2.22. Harmonic levels versus signal voltage with a square law model; voltage coefficient is −0.0003.

There is no requirement that the exponent used in the distortion models be an integer. Equation 2.10 shows an exponent of 1.5. We have seen that an exponent of 1 causes the third harmonic to double in amplitude (+6 dB) for a doubling of signal level, while an exponent of 2 causes it to increase by four times (+12 dB) for a doubling of signal level.

$$i = \frac{V}{R_0(1 + \rho V^{1.5})} \qquad \text{Equation 2.10}$$

As you might expect, an exponent of 1.5 gives an intermediate slope, the third harmonic increasing at almost exactly 9 dB per level-doubling. Slopes for the third harmonic with varying exponents are given in Table 2.6. The levels of the higher harmonics remain negligible when it comes to calculating the THD figures.

Measuring Resistor Distortion

Having looked at the various techniques for simulating distortion, we move on to see how we can measure the reality.

Table 2.6. Slopes of the third harmonic level for various exponents in the voltage coefficient model

Exponent	Third-harmonic slope (dB/level doubling)
1	6
1.2	7.2
1.5	9.02
1.8	10.82
1.99	11.95
2	12

Metal Film, Metal Foil, and Wirewound Resistors

Metal film (MF), metal foil and wirewound resistors produce so little distortion that no meaningful measurements can be done by straightforward THD methods. Metal film resistors have a voltage coefficient of 1 ppm/V or less, metal foil resistors have a voltage coefficient of 0.1 ppm/V or less, while wirewounds certainly show a coefficient less than 1 ppm/V but are normally considered completely linear. A bridge method for measuring metal film resistor non-linearity using an Audio Precision THD analyser was suggested by Ed Simon in *Linear Audio*[6] and an improvement to it that removed some common-mode difficulties was introduced by Samuel Groner.[7]

Metal Oxide Resistors

This type of resistor is made with a film of metal oxides such as tin oxide. This results in a higher operating temperature and greater stability/reliability than metal film under adverse conditions. However, such conditions are unlikely to exist in audio equipment. The voltage coefficient should be less than 10 ppm/V.

Carbon Film Resistors

Carbon film (CF) resistors typically have a voltage coefficient of around 100 ppm/V, and this generates enough distortion for simple THD methods to work, given low audio frequencies and some pretty hefty signal levels. Figure 2.23 shows a simple test circuit consisting of a 2:1 attenuator. The upper resistor is a metal film type and the lower a carbon film resistor of the same value.

Figure 2.24 shows how the distortion of a CF resistor, mostly third harmonic, increases at an accelerating rate, though here it does not clear the noise floor until about 10 Vrms. In this case the divider values were 180 Ω,

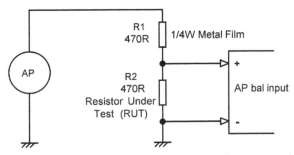

Figure 2.23. Simple potential divider for examining resistor distortion, with typical values.

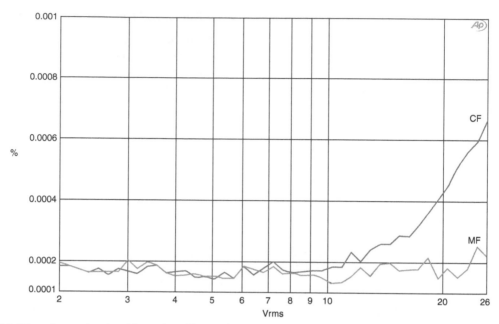

Figure 2.24. Distortion against level for carbon-film (CF) and metal-film (MF) resistors in the RUT position, at 400 Hz. The MF trace is the same as the testgear output.

giving a high dissipation of 939 mW in each resistor at the maximum level of 26 Vrms. This is well above the component ratings (250 mW) and should expose any non-linearity in the MF resistor – which it fails to do. Clearly MF resistors are highly linear. 26 Vrms is not going to be encountered except in a power amplifier or possibly a balanced line output.

The distortion from the CF resistor in Figure 2.24 rises from 0.00043% to 0.0065% as the input level goes up from 20 to 26 Vrms. However, there is actually more to it than that. Figure 2.25 shows that the distortion produced by a CF resistor is strongly frequency-dependent, increasing as frequency falls. (I know it was a CF resistor because I scraped the coating off and found the spiral film was a shiny black, rather than metallic.) Distortion increasing with falling frequency is much less common in electronics than the reverse, and usually results from excessive signal voltage across electrolytic capacitors, or saturating magnetics. Neither is possible here, so the likeliest remaining possibility is a thermal effect. CF resistors have much larger temperature coefficients than MF ones, and if the signal applied changes slowly enough to give cyclic changes in temperature, the resulting changes in resistance will cause distortion. Figure 2.25 shows that distortion increases by a factor of $\sqrt{2}$ when frequency halves, irrespective of the signal level. This test used 470 Ohm resistors, so

the dissipation in each resistor at 20 Vrms input is 213 mW; a 26 Vrms input gives 359 mW, somewhat outside the rating of a ¼-W resistor.

A complicating factor is that the distortion of CF resistors drops with time. This is not like the polyester capacitor self-improvement effect,[8] in which there is a mainly irreversible fall in distortion over time when a large signal is applied. With CF resistors the improvement is wholly reversible, and it looks like a thermal effect again. Figure 2.26 shows the effect; when the input signal is turned on, the THD drops quickly at first, then levels out, presumably as the resistor reaches thermal equilibrium. At 25 seconds, the resistor was blasted with canned air for a couple of seconds, and the THD rapidly went back up; it started falling again when the air blast ceased.

It should be possible to model this behaviour in SPICE, using electrical analogues to represent temperature, thermal capacity, etc. Finding figures for the effective thermal capacity of the resistor film and body may be the hardest part. In any sort of resistor, the thermal capacity of the resistive element is going to be much smaller than that of the substrate, and the effect of the latter cannot be neglected. It sounds plausible to assume that the resistive film is isothermal – in other words, all of it is at the same temperature because the power dissipation is uniform in it. The

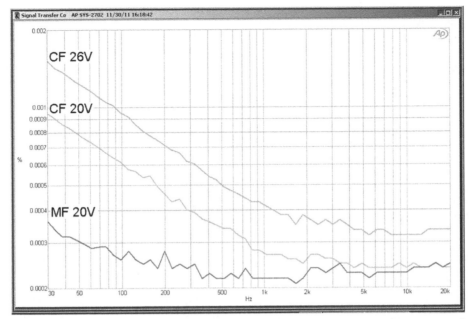

Figure 2.25. THD vs frequency for an MF-CF 470 Ω divider at 20 Vrms and 26 Vrms. Distortion goes up by a factor approximately √2 when frequency is halved. The bottom trace is for an MF-MF 470 Ω divider at 20 Vrms. It is indistinguishable from the test gear output.

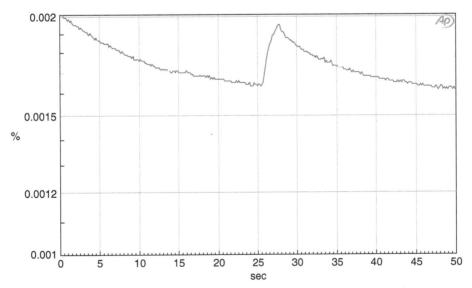

Figure 2.26. THD vs time for a 180 Ω − 180 Ω divider. at 26 Vrms and 400 Hz. The resistor-under-test was cooled with an air blast from 25 to 27 seconds.

substrate, however, will have significant temperature differences in its bulk, so might need to be modelled with an electrical analogue made up of interactive RC networks.

You might think carbon-film resistors are history, but when I was in San Francisco for Burning Amp in October 2011, I found a Radio Shack selling them, at 5% tolerance. Good grief!

For any sort of distortion modelling to be useful, the consistency of the components is important. Figure 2.27 shows six 470 Ω CF resistors of the same type at 20 Vrms input. The resistance value was surprisingly accurate, all being between 466 and 469 Ω; they are nominally 2% and have a cheapish look about them, but the plot shows they are pretty consistent. The bottom trace is an MF resistor for comparison and this is the same as a testgear output (gen-mon) plot.

On the other hand, Figure 2.28 shows the same test for a different type of 470 Ω resistor. They are probably CF but I am not sure — certainly they date from 1970 or so. They are marked as 1% and generally have a more up-market look to them than the first lot, but as you can see, the results are all over the place, even though I'm pretty sure they are all from the same batch. There would be no point in trying to model the distortion of these parts.

Carbon Film Resistor Usage

The decreasing usage of carbon-film resistors and the rise of metal-film parts in audio amplifiers can be tracked through the parts lists in manufacturers' service manuals. Table 2.7 gives a few data points.

Carbon Composition Resistors

Carbon composition resistors are obsolete for most purposes, but they do sometimes appear in valve amplifiers. They are usually regarded as having a voltage coefficient of around 200–500 ppm/V, but, as you will see, they do not actually fit well with a voltage coefficient distortion model.

Figure 2.29 shows THD against level for a 330 Ω nominal carbon composition resistor. (Its actual value was 344 Ω.) The provenance of this part is that I removed it from an old dual-standard television in something like 1965 — I knew it would come in handy one day. The distortion is serious, with third and fifth harmonics, and higher. There was, however, no trace of thermal distortion, the THD remaining constant at 0.038% across the audio band for a 26 Vrms input level.

The variation of THD with level, which as before is almost the same as the variation of the third harmonic, is

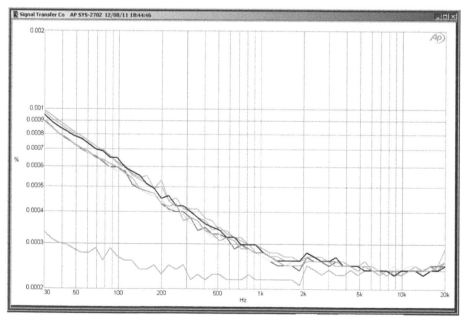

Figure 2.27. THD vs frequency for an MF-CF 470 Ω divider at 20 Vrms, six different specimens. The bottom trace is for an MF-MF 470 Ω divider at 20 Vrms. It is indistinguishable from the test gear output.

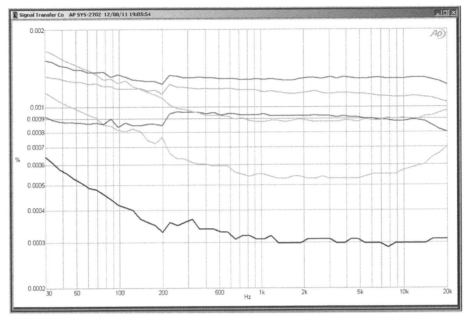

Figure 2.28. As Figure 2.27 but using a different type of CF resistor.

Table 2.7. Replacement of carbon-film resistors with metal-film resistors in amplifier designs

Pioneer SA-8500 Mk2 (1977)	Almost all CF, a few metal oxide
Pioneer M73 (1977)	Almost all CF, a few fusibles
Hitachi HA-3700 and HA-4700 (July 1980)	Almost all CF 5%, a few metal oxide, metal = fusible
Denon POA-2200 (1986)	Mostly CF
Yamaha AX-470 65Wx2 (1993)	50:50 CF and metal oxide
Pioneer A407 FET (May 1998)	Some MF, mostly CF
Pioneer A509 FET (Apr 2000)	Some MF, mostly CF
Rotel RA-01 (2003)	All MF

shown in Figure 2.29, with a linear THD% scale. Clearly the distortion is rising faster than proportionally, and we might think that the second-order or square law voltage coefficient model would be appropriate. The residual is visually pure third harmonic up to 6 Vrms at least, and above that its deviation from a sine wave is really quite subtle.

Table 2.8 shows how the THD varies with a doubling of level. It is almost the same as the third harmonic levels, confirmed by inspection of the residuals.

As you can see, the rate of increase is not constant, but slows, which rules out simple voltage coefficient models. First-order voltage-coeff gives 6 dB rise in 3rd harmonic for level doubling, and square law voltage-coeff gives 12 dB. You can get intermediate slopes by using non-integer exponents, as described in the section 'Other voltage-coefficient distortion models' above; an exponent of somewhere between 1.5 and 1.8 would be an approximate fit only. No simple coefficient model can give a varying slope, and so something more sophisticated would be needed to model the behaviour of carbon composition resistors.

It has been suggested[9] that tube aficionados prefer the 'warm' sound of composition resistors and dislike the 'sterile' neutral sound of the better quality types. Figure 2.29 certainly suggests that the distortion from a composition resistor in a critical position could be audible, given the high signal voltages present in valve equipment.

If for some reason you feel compelled to use carbon composition resistors but find their distortion excessive, putting two resistors in series will in this case give a great improvement; (unlike the case with a voltage-coefficient law). Figure 2.29 shows that going from one resistor with 20 Vrms across it to two with 10 Vrms across them will reduce the distortion from 0.030% to 0.013%.

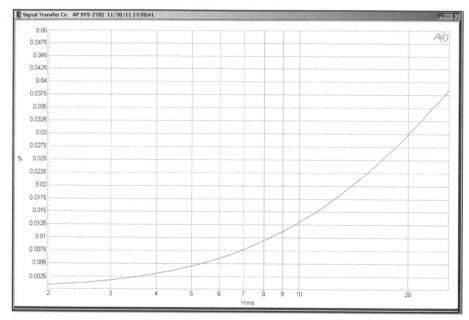

Figure 2.29. Distortion against level for a carbon-composition resistor in a 330 Ω − 330 Ω divider at 400 Hz.

Table 2.8. Changing slope of THD as input levels increase

Level range	Slope
2 Vrms−4 Vrms	10.92 dB
3 Vrms−6 Vrms	10.49 dB
4 Vrms−8 Vrms	9.78 dB
5 Vrms−10 Vrms	9.25 dB
6 Vrms−12 Vrms	8.85 dB
8 Vrms−16 Vrms	8.05 dB

Resistors in the Feedback Network

Most resistors in a conventional power amplifier have only small signal voltages across them. The significant exceptions are those in the feedback network. The upper feedback resistor is the critical component because it has almost the whole output voltage across it; a metal-film resistor, with its very low voltage coefficient, is essential for low distortion.

If you are afflicted with mysterious distortion that is mostly third harmonic, is not affected by adjustments to the amplifier forward path, and increases steadily with falling frequency, it is worth checking that a carbon-film resistor has not been used in the feedback path by mistake. The carbon-film resistors I have measured show much more thermal distortion than voltage coefficient distortion, giving odd-harmonic distortion that rises as frequency falls. The thermal nature of the effect causes the THD level to fall slowly after switch-on, as the resistor warms up. A short blast of canned air on the resistor will cause the THD to rise and then fall again, as demonstrated above, and this is a very good non-intrusive test for carbon-film resistor misbehaviour.

The upper feedback resistor value will be kept as low as is feasible to reduce Johnson and current noise in the feedback network, and so it may run quite hot at high output powers, and this must be taken into account when mounting it. It must not be too close to the input transistor pair as heating them will cause drift in the DC offset voltage at the output.

Using SMD resistors in the feedback network of a power amplifier has been shown to be a bad idea.[10]

Modelling Distortion from other Passive Components

The same techniques can be used to effectively model the distortion generated by some types of capacitor, such as those with polyester dielectrics. These capacitors show easily measurable distortion when operated with several volts of signal across them. This behaviour is of importance in active filters and equalisation circuits but not in power amplifier technology.

References

1. Hartmann, W M, *Signals*, *Sound & Sensation*, American Institute of Physics Press, 1997, p. 497. ISBN 1-56396-283-7.

2. Exley K, A Bass Without Big Baffles, *Wireless World*, April 1951, p. 132.

3. Gan, Nay Oo Woon-Seng and Hawksford M J, Perceptually-Motivated Objective Grading of Nonlinear Processing in Virtual-Bass Systems, *JAES*, 59(11), Nov. 2011, p. 806.

4. Shorter, D E L, The Influence of High-Order Products in Non-Linear Distortion, *Electronic Engineering*, 22(266), April 1950, pp. 152−153.

5. Geddes, E and Lee, L, Auditory Perception of Nonlinear Distortion, paper presented at 115th AES Convention, Oct. 2003.

6. Simon, E, Resistor Non-Linearity, *Linear Audio* 1, p. 138.

7. Groner S, http://www.linearaudio.net/userfiles/file/letters/Volume_1_LTE_ES.pdf (accessed 10 April 2012).

8. Self, D, Self-Improvement for Capacitors: Linearisation Over Time, *Linear Audio* 1, p. 156.

9. http://www.diyaudio.com/forums/parts/178941-resistor-distortion.html Posted Dec. 2010, (accessed 1 Dec. 2011).

10. http://www.diyaudio.com/forums/solid-state/160061-slones-11-4-blameless-13.html Posted Feb 2010 (accessed 1 Dec. 2011).

Chapter **3**

Negative Feedback

Where would rock and roll be without feedback?
David Gilmour, *Live at Pompeii* video, 1972

Negative Feedback in Amplifiers

It is not the role of this book to step through elementary theory which can be easily found in any number of textbooks. However, correspondence in audio and technical journals shows that considerable confusion exists on negative feedback as applied to power amplifiers; perhaps there is something inherently mysterious in a process that improves almost all performance parameters simply by feeding part of the output back to the input, but inflicts dire instability problems if used to excess. I therefore deal with a few of the less obvious points here; more information is provided in Chapter 13.

The main use of NFB in power amplifiers is the reduction of harmonic distortion, the reduction of output impedance, and the enhancement of supply-rail rejection. There are also analogous improvements in frequency response and gain stability, and reductions in DC drift.

The basic feedback equation is dealt with in a myriad of textbooks, but it is so fundamental to power amplifier design that it is worth a look here. In Figure 3.1, the open-loop amplifier is the big block with open-loop gain A. The negative feedback network is the block marked β; this could contain anything, but for our purposes it simply scales down its input, multiplying it by β and is usually in the form of a potential divider. The circle with the cross on is the conventional control theory symbol for a block that adds or subtracts with unity gain and does nothing else.

Firstly, it is pretty clear that one input to the subtractor is simply V_{in}, and the other is V_{out} times β,

so subtract these two, multiply by A, and you get the output signal V_{out}.

$$V_{out} = A(V_{in} - \beta.V_{out})$$

Collect the Vouts together, and you get:

$$V_{out}(1 + A\beta) = A.V_{in}$$

So the closed-loop gain is:

$$\frac{V_{out}}{V_{in}} = \frac{A}{1 + A\beta} \qquad \text{Equation 3.1}$$

This is the feedback equation, and it could not be more important. The first thing it shows is that negative feedback stabilises the gain. In real life, circuitry A is a high but uncertain and variable quantity, while β is firmly fixed by resistor values. The product Aβ appears a great deal in feedback analysis and is called the loop gain. The quantity (1 + Aβ) also appears a lot and is sometimes called the 'improvement factor', or the 'desensitivity factor'. Looking at Equation 3.1, you can see that the higher A is, the less significant the 1 on the bottom is; the As cancel out, and so with a high A, the equation can be regarded as simply:

$$\frac{V_{out}}{V_{in}} = \frac{1}{\beta} \qquad \text{Equation 3.2}$$

This is demonstrated in Table 3.1, where β is set at 0.04 with the intention of getting a closed loop gain of 25 times. With a low open-loop gain of 100, the closed-loop gain is only 20, a long way short of 25. But as the open-loop gain increases, the closed-loop gain gets

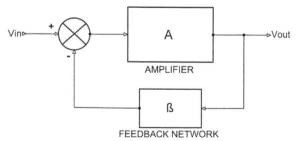

Figure 3.1. A simple negative feedback system with an amplifier with open-loop gain A and a feedback network with a 'gain', less than one, of β.

Table 3.1. How the closed-loop gain gets closer to the target as the open-loop gain increases

1	2	3	4	5	6	7
Desired C/L gain	β NFB fracn	A O/L gain	NFB factor	C/L gain	O/L Error	C/L Error
25	0.04	100	5	20.00	1	0.2
25	0.04	1000	41	24.39	1	0.0244
25	0.04	10000	401	24.94	1	0.0025
25	0.04	40000	1601	24.98	1	0.0006
25	0.04	100000	4001	24.99	1	0.0002

closer to the target. If you look at the bottom two rows, you will see that an increase in open-loop gain of more than a factor of two only alters the closed-loop gain by a trivial second decimal place.

Negative feedback is, however, capable of doing much more than stabilising gain. Anything untoward happening in the amplifier block A, be it distortion or DC drift, or any of the other ills that electronics is heir to, is also reduced by the negative feedback factor (NFB factor for short). This is equal to:

$$NFB factor = \frac{1}{1 + A\beta} \qquad \text{Equation 3.3}$$

and it is tabulated in the fourth column in Table 3.1. To show why this factor is vitally important, Figure 3.2 shows the same scenario as Figure 3.1, with the addition of a voltage Vd to the output of A; this represents noise, DC drift or anything that can cause a voltage error, but what is usually most interesting to the pursuivants of amplifier design is its use to represent distortion.

Repeating the simple algebra we did before, and adding in Vd, we get:

$$V_{out} = A(V_{in} - \beta.V_{out}) + Vd$$

$$V_{out}(1 + A\beta) = A.V_{in} + Vd$$

$$\frac{V_{out}}{V_{in}} = \frac{A}{1 + A\beta} + \frac{Vd}{1 + A\beta}$$

So the effect of Vd has been decreased by the NFB factor:

$$\frac{1}{1 + A\beta} \qquad \text{Equation 3.4}$$

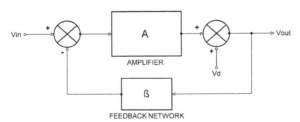

Figure 3.2. The negative feedback system with an error signal Vd added to the output of the amplifier.

In other words, the higher the open-loop gain A compared with the gain demanded by β, the lower the distortion. Since we are usually dealing with high values of A, the 1 on the bottom of the fraction has very little effect and it can be taken that doubling the open-loop gain halves the distortion. This effect is illustrated in the sixth and seventh columns of Table 3.1 above, which adds an error of magnitude 1 to the output of the amplifier; the closed loop error is then simply the reciprocal of the NFB factor for each value of open-loop gain.

In simple circuits with low open-loop gain, you just apply negative feedback and that is the end of the matter. In a typical power amplifier, which cannot be operated without NFB, if only because it would be saturated by its own DC offset voltages, there are several stages which may accumulate phase-shift, and simply closing the loop usually brings on severe Nyquist oscillation at HF. This is a serious matter, as it will not only burn out any tweeters that are unlucky enough to be connected, but can also destroy the output devices by overheating, as they may be unable to turn off fast enough at ultrasonic frequencies.

The standard cure for this instability is compensation. A capacitor is added, usually in Miller integrator format, to roll-off the open-loop gain at 6 dB per octave, so it reaches unity loop-gain before enough phase-shift can build up to allow oscillation. This means the NFB factor varies strongly with frequency, an inconvenient fact that many audio commentators seem to forget.

It is crucial to remember that a distortion harmonic, subjected to a frequency-dependent NFB factor as above, will be reduced by the NFB factor corresponding to its own frequency, not that of its fundamental. If you have a choice, generate low-order rather than high-order distortion harmonics, as the NFB deals with them much more effectively.

Negative feedback can be applied either locally (i.e., to each stage, or each active device) or globally, in other words, right around the whole amplifier. Global NFB is more efficient at distortion reduction than the same amount distributed as local NFB, but places much stricter limits on the amount of phase-shift that may be allowed to accumulate in the forward path. More on this later in this chapter.

Above the dominant pole frequency, the VAS acts as a Miller integrator, and introduces a constant 90° phase lag into the forward path. In other words, the output from the input stage must be in quadrature if the final amplifier output is to be in phase with the input, which to a close approximation it is. This raises the question of how the 90° phase shift is accommodated by the negative feedback loop; the answer is that the input and feedback signals applied to the input stage are there subtracted, and the small difference between two relatively large signals with a small phase shift between them has a much larger phase shift. This is the signal that drives the VAS input of the amplifier.

Solid-state power amplifiers, unlike many valve designs, are almost invariably designed to work at a fixed closed-loop gain. If the circuit is compensated by the usual dominant-pole method, the HF open-loop gain is also fixed, and therefore so is the important negative feedback factor. This is in contrast to valve amplifiers, where the amount of negative feedback applied was regarded as a variable, and often user-selectable parameter; it was presumably accepted that varying the negative feedback factor caused significant changes in input sensitivity. A further complication was serious peaking of the closed-loop frequency response at both LF and HF ends of the spectrum as negative feedback was increased, due to the inevitable bandwidth limitations in a transformer-coupled forward path. Solid-state amplifier designers go cold at the thought of the customer tampering with something as vital as the NFB factor, and such an approach is only acceptable in cases like valve amplification where global NFB plays a minor role.

Common Misconceptions about Negative Feedback

All of the comments quoted below have appeared many times in the hi-fi literature. All are wrong.

Negative feedback is a bad thing. Some audio commentators hold that, without qualification, negative feedback is a bad thing. This is of course completely untrue and based on no objective reality. Negative feedback is one of the fundamental concepts of electronics, and to avoid its use altogether is virtually impossible; apart from anything else, a small amount of local NFB exists in every common-emitter transistor because of the internal emitter resistance. I detect here distrust of good fortune; the uneasy feeling that if something apparently works brilliantly, then there must be something wrong with it.

A low negative-feedback factor is desirable. Untrue; global NFB makes just about everything better, and the sole effect of too much is HF oscillation, or poor transient behaviour on the brink of instability. These effects are painfully obvious on testing and not hard to

avoid unless there is something badly wrong with the basic design.

In any case, just what does *low* mean? One indicator of imperfect knowledge of negative feedback is that the amount enjoyed by an amplifier is almost always baldly specified as *so many dB* on the very few occasions it is specified at all — despite the fact that most amplifiers have a feedback factor that varies considerably with frequency. A dB figure quoted alone is meaningless, as it cannot be assumed that this is the figure at 1 kHz or any other standard frequency.

My practice is to quote the NFB factor at 20 kHz, as this can normally be assumed to be above the dominant pole frequency, and so in the region where open-loop gain is set by only two or three components. Normally the open-loop gain is falling at a constant 6 dB/octave at this frequency on its way down to intersect the unity-loop-gain line and so its magnitude allows some judgement as to Nyquist stability. Open-loop gain at LF depends on many more variables such as transistor beta, and consequently has wide tolerances and is a much less useful quantity to know. This is dealt with in more detail in Chapter 7 on Voltage-Amplifier Stages.

Negative feedback is a powerful technique, and therefore dangerous when misused. This bland truism usually implies an audio Rake's Progress that goes something like this: an amplifier has too much distortion, and so the open-loop gain is increased to augment the NFB factor. This causes HF instability, which has to be cured by increasing the compensation capacitance. This is turn reduces the slew-rate capability, and results in a sluggish, indolent, and generally bad amplifier.

The obvious flaw in this argument is that the amplifier so condemned no longer has a high NFB factor, because the increased compensation capacitor has reduced the open-loop gain at HF; therefore, feedback itself can hardly be blamed. The real problem in this situation is probably unduly low standing current in the input stage; this is the other parameter determining slew-rate.

NFB may reduce low-order harmonics but increases the energy in the discordant higher harmonics. A less common but recurring complaint is that the application of global NFB is a shady business because it transfers energy from low-order distortion harmonics — considered musically consonant — to higher-order ones that are anything but. This objection contains a grain of truth, but appears to be based on a misunderstanding of one article in an important series by Peter Baxandall[1] in which he showed that if you took an amplifier with only second-harmonic distortion, and then introduced

NFB around it, higher-order harmonics were indeed generated as the second harmonic was fed back round the loop. For example, the fundamental and the second-harmonic intermodulate to give a component at third-harmonic frequency. Likewise, the second and third intermodulate to give the fifth harmonic. If we accept that high-order harmonics should be numerically weighted to reflect their greater unpleasantness, there could conceivably be a rise rather than a fall in the weighted THD when negative feedback is applied.

This important issue is dealt with in detail at the end of this chapter. For the time being we will just note that the higher harmonics appear as soon as the tiniest amount of negative feedback is applied, and the answer to that is to apply a lot of negative feedback.

You could try to argue that the use of negative feedback has polluted the purity of square-law gain devices that would otherwise produce only a mellifluous second harmonic. This argument falls down because there are no purely square-law devices, and that includes FETs, which are sometimes erroneously thought to be so. They all generate small amounts of high-order harmonics. Feedback could and would generate these from nothing, but in practice they are already there.

The vital point is that if enough NFB is applied, all the harmonics can be reduced to a lower level than without it. The extra harmonics generated, effectively by the distortion of a distortion, are at an extremely low level providing a reasonable NFB factor is used. This is a very powerful argument against low feedback factors like 6 dB, which are most likely to increase the weighted THD. For a full understanding of this topic, a careful reading of the Baxandall series is absolutely indispensable.

A low open-loop bandwidth means a sluggish amplifier with a low slew-rate. Great confusion exists in some quarters between open-loop bandwidth and slew-rate. In truth, open-loop bandwidth and slew-rate are nothing to do with each other, and may be altered independently. Open-loop bandwidth is determined by compensation Cdom, VAS β, and the resistance at the VAS collector, while slew-rate is set by the input stage standing current and Cdom. Cdom affects both, but all the other parameters are independent. (See Chapter 5 for more details.)

In an amplifier, there is a maximum amount of NFB you can safely apply at 20 kHz; this does not mean that you are restricted to applying the same amount at 1 kHz, or indeed 10 Hz. The obvious thing to do is to allow the NFB to continue increasing at 6 dB/octave — or faster if possible — as frequency falls, so that the amount of NFB applied doubles with each octave as we move down in frequency, and we derive as much benefit as we can.

This obviously cannot continue indefinitely, for eventually open-loop gain runs out, being limited by transistor beta and other factors. Hence the NFB factor levels out at a relatively low and ill-defined frequency; this frequency is the open-loop bandwidth, and for an amplifier that can never be used open-loop, has very little importance.

It is difficult to convince people that this frequency is of no relevance whatever to the speed of amplifiers, and that it does not affect the slew-rate. Nonetheless, it is so, and any first-year electronics textbook will confirm this. High-gain opamps with sub-1 Hz bandwidths and blindingly fast slewing are as common as the grass (if somewhat less cheap) and if that does not demonstrate the point beyond doubt, then I really do not know what will.

Limited open-loop bandwidth prevents the feedback signal from immediately following the system input, so the utility of this delayed feedback is limited. No linear circuit can introduce a pure time-delay; the output must begin to respond at once, even if it takes a long time to complete its response. In the typical amplifier the dominant-pole capacitor introduces a 90° phase-shift between input-pair and output at all but the lowest audio frequencies, but this is not a true time-delay. The phrase *delayed feedback* is often used to describe this situation, and it is a wretchedly inaccurate term; if you really delay the feedback to a power amplifier (which can only be done by adding a time-constant to the feedback network rather than the forward path), it will quickly turn into the proverbial power oscillator as sure as night follows day.

Negative Feedback and Amplifier Stability

In controlling amplifier distortion, there are two main weapons. The first is to make the linearity of the circuitry as good as possible before closing the feedback loop. This is unquestionably important, but it could be argued it can only be taken so far before the complexity of the various amplifier stages involved becomes awkward. The second is to apply as much negative feedback as possible while maintaining amplifier stability. It is well known that an amplifier with a single time-constant is always stable, no matter how high the feedback factor. The linearisation of the VAS by local Miller feedback is a good example. However, more complex circuitry, such as the generic three-stage power amplifier, has more than one time-constant, and these extra poles will cause poor transient response or instability if a high feedback factor is maintained up to the higher frequencies where they start to take effect. It is therefore clear that if these higher poles can be eliminated or moved upward in frequency, more feedback can be applied and distortion will be less for the same stability margins. Before they can be altered – if indeed this is practical at all – they must be found and their impact assessed.

The dominant pole frequency of an amplifier is, in principle, easy to calculate; the mathematics is very simple (see Chapter 5). In practice, two of the most important factors, the effective beta of the VAS and the VAS collector impedance, are only known approximately, so the dominant pole frequency is a rather uncertain thing. Fortunately this parameter in itself has no effect on amplifier stability. What matters is the amount of feedback at high frequencies.

Things are different with the higher poles. To begin with, where are they? They are caused by internal transistor capacitances, and so on, so there is no physical component to show where the roll-off is. It is generally regarded as fact that the next poles occur in the output stage, which will use power devices that are slow compared with small-signal transistors. Taking the Class-B design in Chapter 10, the TO-92 MPSA06 devices have an Ft of 100 MHz, the MJE340 drivers about 15 MHz (for some reason this parameter is missing from the data sheet) and the MJ802 output devices an Ft of 2.0 MHz. Clearly the output stage is the prime suspect. The next question is at what frequencies these poles exist. There is no reason to suspect that each transistor can be modelled by one simple pole.

There is a huge body of knowledge devoted to the art of keeping feedback loops stable while optimising their accuracy; this is called Control Theory, and any technical bookshop will yield some intimidatingly fat volumes called things like 'Control System Design'.

Inside, system stability is tackled by Laplace-domain analysis, eigenmatrix methods, and joys like the Lyapunov stability criterion. I think that makes it clear that you need to be pretty good at mathematics to appreciate this kind of approach.

Even so, it is puzzling that there seems to have been so little application of Control Theory to audio amplifier design. The reason may be that so much Control Theory assumes that you know fairly accurately the characteristics of what you are trying to control, especially in terms of poles and zeros.

One approach to appreciating negative feedback and its stability problems is SPICE simulation. Some SPICE simulators have the ability to work in the Laplace or s-domain, but my own experiences with this have been deeply unhappy. Otherwise respectable simulator packages output complete rubbish in this mode. Quite what the issues are here I do not know, but it does seem that s-domain methods are best avoided. The approach suggested here instead models poles directly as poles, using RC networks to generate the time-constants. This requires minimal mathematics and is far more robust. Almost any SPICE simulator – evaluation versions included – should be able to handle the simple circuit used here.

Figure 3.3 shows the basic model, with SPICE node numbers. The scheme is to idealise the situation enough to highlight the basic issues and exclude distractions like non-linearities or clipping. The forward gain is simply the transconductance of the input stage multiplied by the transadmittance of the VAS integrator. An important point is that with correct parameter values, the current from the input stage is realistic, and so are all the voltages.

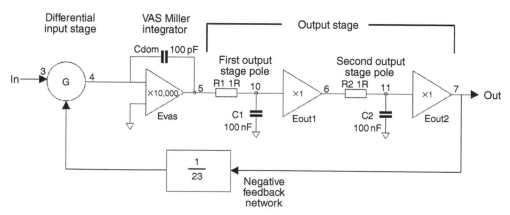

Figure 3.3. Block diagram of system for SPICE stability testing.

The input differential amplifier is represented by G. This is a standard SPICE element − the VCIS, or voltage-controlled current source. It is inherently differential, as the output current from Node 4 is the scaled difference between the voltages at Nodes 3 and 7. The scaling factor of 0.009 sets the input stage transconductance (gm) to 9 mA/V, a typical figure for a bipolar input with some local feedback. Stability in an amplifier depends on the amount of negative feedback available at 20 kHz. This is set at the design stage by choosing the input gm and Cdom, which are the only two factors affecting the open-loop gain. In simulation, it would be equally valid to change gm instead; however, in real life. it is easier to alter Cdom as the only other parameter this affects is slew-rate. Changing input stage transconductance is likely to mean altering the standing current and the amount of local feedback, which will in turn impact input stage linearity.

The VAS with its dominant pole is modelled by the integrator Evas, which is given a high but finite open-loop gain, so there really is a dominant pole P1 created when the gain demanded becomes equal to that available. With Cdom = 100pF, this is below 1 Hz. With infinite (or as near-infinite as SPICE allows) open-loop gain, the stage would be a perfect integrator. As explained elsewhere, the amount of open-loop gain available in real versions of this stage is not a well-controlled quantity, and P1 is liable to wander about in the 1−100 Hz region; fortunately this has no effect at all on HF stability. Cdom is the Miller capacitor that defines the transadmittance, and since the input stage has a realistic transconductance, Cdom can be set to 100 pF, its usual real-life value. Even with this simple model we have a nested feedback loop. This apparent complication here has little effect, so long as the open-loop gain of the VAS is kept high.

The output stage is modelled as a unity-gain buffer, to which we add extra poles modelled by R1, C1 and R2, C2. Eout1 is a unity-gain buffer internal to the output stage model, added so the second pole does not load the first. The second buffer Eout2 is not strictly necessary as no real loads are being driven, but it is convenient if extra complications are introduced later. Both are shown here as a part of the output stage but the first pole could equally well be due to input stage limitations instead; the order in which the poles are connected makes no difference to the final output. Strictly speaking, it would be more accurate to give the output stage a gain of 0.95, but this is so small a factor that it can be ignored.

The component values used to make the poles are of course completely unrealistic, and chosen purely to make the maths simple. It is easy to remember that 1 Ω and 1 µF make up a 1 µsec time-constant. This is a pole at 159 kHz. Remember that the voltages in the latter half of the circuit are realistic, but the currents most certainly are not.

The feedback network is represented simply by scaling the output as it is fed back to the input stage. The closed-loop gain is set to 23 times, which is representative of many power amplifiers.

Note that this is strictly a linear model, so the slew-rate limiting which is associated with Miller compensation is not modelled here. It would be done by placing limits on the amount of current that can flow in and out of the input stage.

Figure 3.4 shows the response to a 1 V step input, with the dominant pole the only time element in the circuit. (The other poles are disabled by making C1, C2 0.00001 pF, because this is quicker than changing the actual circuit.) The output is an exponential rise to an asymptote of 23 V, which is exactly what elementary theory predicts. The exponential shape comes from the way that the error signal which drives the integrator becomes less as the output approaches the desired level. The error, in the shape of the output current from G, is the smaller signal shown; it has been multiplied by 1000 to get mA onto the same scale as volts. The speed of response is inversely proportional to the size of Cdom, and is shown here for values of 50 pF and 220 pF as well as the standard 100 pF.

This simulation technique works well in the frequency domain, as well as the time domain. Simply tell SPICE to run an AC simulation instead of a TRANS (transient) simulation. The frequency response in Figure 3.5 exploits this to show how the closed-loop gain in a NFB amplifier depends on the open-loop gain available. Once more elementary feedback theory is brought to life. The value of Cdom controls the bandwidth, and it can be seen that the values used in the simulation do not give a very extended response compared with a 20 kHz audio bandwidth.

In Figure 3.6, one extra pole P2 at 1.59 MHz (a time-constant of only 100 ns) is added to the output stage, and Cdom stepped through 50, 100 and 200 pF as before. 100pF shows a slight overshoot that was not there before; with 50 pF there is a serious overshoot that does not bode well for the frequency response. Actually, it's not that bad; Figure 3.7 returns to the frequency-response domain to show that an apparently vicious

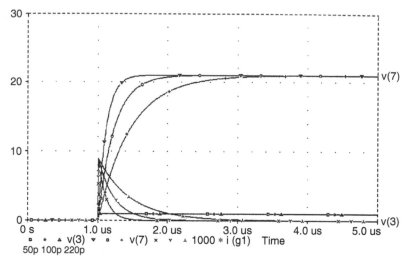

Figure 3.4. SPICE results in the time domain. As Cdom increases, the response V(7) becomes slower, and the error g(i) declines more slowly. The input is the step-function V(3) at the bottom.

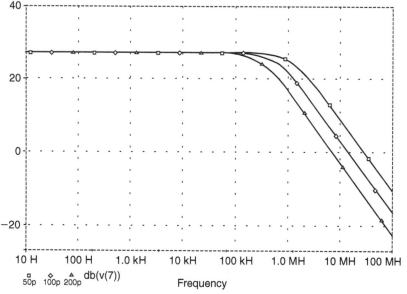

Figure 3.5. SPICE simulation in the frequency domain. As the compensation capacitor is increased, the closed-loop bandwidth decreases proportionally.

overshoot is actually associated with a very mild peaking in the frequency domain.

From here on, Cdom is left set to 100 pF, its real value in most cases. In Figure 3.6, P2 is stepped instead, increasing from 100 ns to 5 μs, and while the response gets slower and shows more overshoot, the system does not become unstable. The reason is

simple: sustained oscillation (as opposed to transient ringing) in a feedback loop requires positive feedback, which means that a total phase shift of 180° must have accumulated in the forward path, and reversed the phase of the feedback connection. With only two poles in a system, the phase shift cannot reach 180°. The VAS integrator gives a dependable 90° phase shift

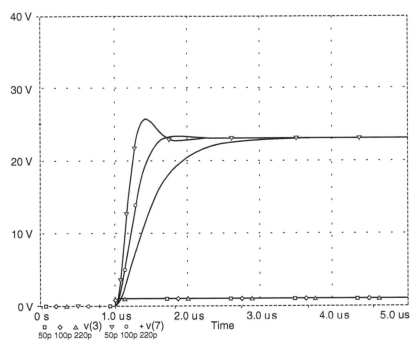

Figure 3.6. Adding a second pole P2 causes overshoot with smaller values Cdom, but cannot bring about sustained oscillation.

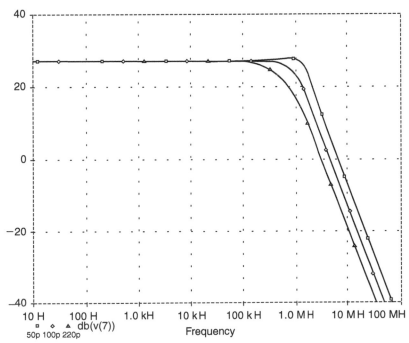

Figure 3.7. The frequency responses that go with the transient plots of Figure 3.6. The response peaking for Cdom = 50 pF is very small compared with the transient overshoot.

above P1, being an integrator, but P2 is instead a simple lag and can only give 90° phase lag at infinite frequency. So, even this very simple model gives some insight. Real amplifiers do oscillate if Cdom is too small, so we know that the frequency response of the output stage cannot be meaningfully modelled with one simple lag.

As President Nixon is alleged to have said: 'Two wrongs don't make a right − so let's see if three will do it!' Adding in a third pole P3 in the shape of another simple lag gives the possibility of sustained oscillation. This is case A in Table 3.2.

Stepping the value of P2 from 0.1 to 5 μsec with P3 = 500 nsec in Figure 3.8 shows that damped oscillation is present from the start. Figure 3.9 also shows over 50 μsec what happens when the amplifier is made very unstable (there are degrees of this) by setting P2 = 5 μsec and P3 = 500 nsec. It still takes time for the oscillation to develop, but exponentially diverging oscillation like this is a sure sign of disaster. Even in the short time examined here the amplitude has exceeded a rather theoretical half a kilovolt. In reality, oscillation cannot increase indefinitely, if only because the supply rail voltages would limit the amplitude. In practice, slew-rate limiting is probably the major controlling factor in the amplitude of high-frequency oscillation.

We have now modelled a system that will show instability. But does it do it right? Sadly, no. The oscillation is about 200 kHz, which is a rather lower frequency than is usually seen when an amplifier misbehaves. This low frequency stems from the low P2 frequency we have to use to provoke oscillation; apart from anything else this seems out of line with the known Ft of power transistors. Practical amplifiers are likely to take off at around 500 kHz to 1 MHz when Cdom is reduced, and this seems to suggest that phase shift is accumulating quickly at this sort of frequency. One possible explanation is that there are a large

number of poles close together at a relatively high frequency.

A fourth pole can simply be added to Figure 3.3 by inserting another RC−buffer combination into the system. With P2 = 0.5 μsec and P3 = P4 = 0.2 μsec, instability occurs at 345 kHz, which is a step towards a realistic frequency of oscillation. This is case B in Table 3.2.

When a fifth output stage pole is grafted on, so that P3 = P4 = P5 = 0.2 μsec, the system just oscillates at 500 kHz with P2 set to 0.01 μsec. This takes us close to a realistic frequency of oscillation. Rearranging the order of poles so P2 = P3 = P4 = 0.2 μsec, while P5 = 0.01 μsec, is tidier, and the stability results are of course the same; this is a linear system so the order does not matter. This is case C in Table 3.2.

Having P2, P3 and P4 all at the same frequency does not seem very plausible in physical terms, so case D shows what happens when the five poles are staggered in frequency. P2 needs to be increased to 0.3 μsec to start the oscillation, which is now at 400 kHz. Case E is another version with five poles, showing that if P5 is reduced, P2 needs to be doubled to 0.4 μsec for instability to begin.

In the final case F, a sixth pole is added to see if this permitted sustained oscillation is above 500 kHz. This seems not to be the case; the highest frequency that could be obtained after a lot of pole-twiddling was 475 kHz. This makes it clear that this model is of limited accuracy (as indeed are all models − it is a matter of degree) at high frequencies, and that further refinement is required to gain further insight.

The greatest inaccuracy in the model as it stands may be that it does not include transit times in semiconductors and stages. The poles used above give the beginning of an output instantly, though it may take a long time to reach full amplitude. In real amplifiers there will be a period in which there is no output at all. Modelling this with a pure time-delay in the forward path is likely to lead to better results.

You may be wondering at this point why I have devoted so much space to an experiment that basically ends in a failure to come up with a good model for high-frequency output stage behaviour. The answer is that no experiment which yields information is a failure. Many people have proposed alternative methods of amplifier compensation, relying wholly on simulations and without attempting to build proof-of-concept hardware or make any measurements. Very often the output stage is modelled as having a single time-constant or pole. As I hope I have demonstrated, this is hopelessly unrealistic, and simulation-based

Table 3.2. Instability onset: P2 is increased until sustained oscillation occurs

Case	Cdom	P2	P3	P4	P5		P6
A	100p	0.45	0.5	−	−		200kHz
B	100p	0.5	0.2	0.2	−		345kHz
C	100p	0.2	0.2	0.2	0.01		500kHz
D	100p	0.3	0.2	0.1	0.05		400kHz
E	100p	0.4	0.2	0.1	0.01		370kHz
F	100p	0.2	0.2	0.1	0.05	0.02	475kHz

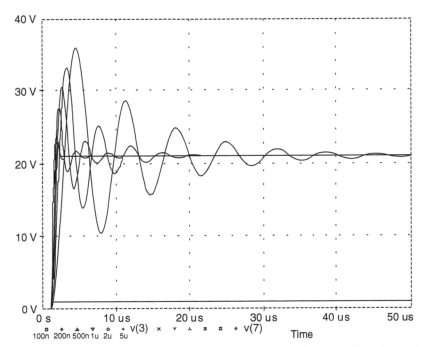

Figure 3.8. Manipulating the P2 frequency can make ringing more prolonged but it is still not possible to provoke sustained oscillation.

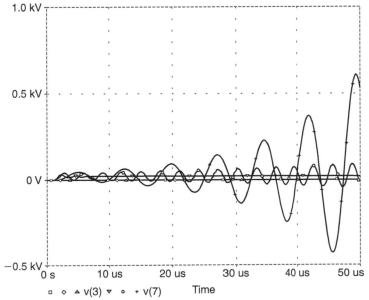

Figure 3.9. Adding a third pole makes possible true instability with exponentially increasing amplitude of oscillation. Note the unrealistic voltage scale on this plot.

proposals containing this assumption should be looked at with a very wary eye.

Feedback Intermodulation Distortion

It is an awkward but indisputable fact that applying negative feedback can create higher harmonics that did not previously exist. If we have an amplifier with a purely square-law characteristic in the forward path that generates only second harmonic distortion, and feed into it a sinewave (the fundamental), then when we close the feedback loop, that second harmonic is fed back into the amplifier where, due to the square-law non-linearity, it intermodulates with the fundamental to generate the third harmonic. This, when fed back, will generate the fourth harmonic, and so on, as high as we care to go. Note that all this happens simultaneously so far as audio frequencies are concerned. This phenomenon appears to lack a name, and I would suggest Feedback Intermodulation Distortion (FID) as being suitably descriptive.

Quite when this effect was recognised is uncertain, but it was certainly described mathematically by M. G. Scroggie in *Wireless World* in April 1961[2]; the article was reprinted in October 1978.[3] It was demonstrated by measurement by Peter Baxandall in December 1978,[1] using an FET to approximate a square-law amplifier. As Peter showed, that is indeed only an approximation, as FETs do create higher harmonics even without feedback.

I decided to confirm this by simulation, which allows a completely pure square law to be used. The model chosen uses a section of a parabola described by the law $V_{out} = (0.7V_{in})^2 + V_{in}$, as described earlier in Chapter 2. A distorter input level of 0.2 V peak was chosen as it gives a second harmonic level of 7.00%, which is close to what Peter measured from his FET with no feedback. Figure 3.10 shows the set-up. The VCVS E1 does nothing except perform the feedback subtraction. E2 is the square-law distorter, and E3 gives us a forward gain of 100 times. This is an unrealistically low figure for a typical solid-state power amplifier (and ignores the fact that the open-loop gain is usually frequency-dependent) but it works well for our demonstration. E4 simply scales the output to set the negative feedback factor β.

The SPICE code to implement the block diagram is:

```
E1  4  0  VALUE = {V(3)-V(8)}        ; feedback
                                       subtractor, 1x gain
Rdummy1 4 0 10G
E2  5  0  VALUE = {(0.7*(V(4)*V(4)))+V(4)}  ;
                                       y = 0.7x^2 + x
Rdummy2 5 0 10G
E3 7 0 VALUE = {100*V(5)}             ; stage with 100x
                                                  gain
Rdummy3 7 0 10G
E4 8 0 VALUE = {nfb*V(7)}             ; NFB attenuator
                                          scales output
Rdummy4 8 0 10G        ; by the factor 'nfb'
```

As before, the input node is 3 and the output node is 7.

The procedure is:

1. First, using a 0.2 V peak input level, make sure that the simulation gives the expected results with no negative feedback, i.e., β = 0. The second harmonic should be at 7.00% and the level of all others negligible. For the simulator I used here, the numerical noise floor was around 0.000001%.

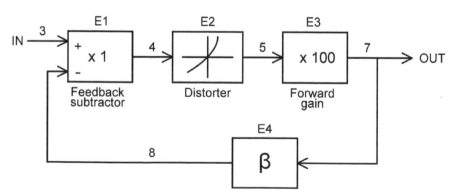

Figure 3.10. A conceptual amplifier with square law non-linearity and a variable negative feedback factor of β, with SPICE node numbers. Dummy resistors are not shown.

2. Apply the desired amount of negative feedback by setting β to a non-zero value.
3. With a 0.2 V peak input, find the level of fundamental output.
4. From this, calculate the input level required to give the fundamental at the output a level of 20 V peak. This keeps the signal level at the input to the distorter at 0.2 V peak, (because of the 100 times amplifier) and keeps the amount of distortion being generated inside the loop constant.
5. Enter the new input level into the simulator and run it.
6. Record the level of each harmonic and of the THD.
7. Set a new feedback factor β, then rinse and repeat by going back to Step 3.

It has to be said, this is a somewhat tedious business. Fortunately I have done it all for you and the result is shown in Figure 3.11. If you are familiar with Peter's famous graph, you may have noted with disquiet that this looks nothing like it. That is because his graph had the NFB improvement factor 1+Aβ as the X-axis, though it was rather confusingly labelled 'dB of feedback'. My Figure 3.11 has instead the feedback factor β as the X-axis, because this spreads the curves out much more for small amounts of feedback, and makes it clearer that higher harmonics are generated as soon as even a hint of negative feedback is applied. The data is plotted with 1+Aβ as the X-axis later in this section as Figure 3.14 to reassure you.

Figure 3.11 starts off at the left-hand side with a very low feedback factor β of 0.0001; in other words, only one ten-thousandth of the output is being fed back. This has very little effect on our overall gain, only reducing it from 100 to 99 times; it is not at all the sort of situation you get with real negative feedback amplifiers, and would be quite useless for reducing distortion. The second harmonic has only fallen from 7.00% to 6.933%. But, as if from nowhere, we now have a third harmonic at 0.01%. The fourth harmonic has also appeared, but only at 0.000016%, where it will be well below the noise in almost any system you can think of.

But things get worse, as they are wont to do. Let us increase the feedback factor β to 0.01, i.e., with one hundredth of the output fed back. Once again this is not a practical design for the real world. A β of 0.01 implies that we are looking for a gain of 100 times, but our forward gain is only 100 times, so we are going to be sorely disappointed; the actual closed-loop gain is only 49.9 times, and the second harmonic has

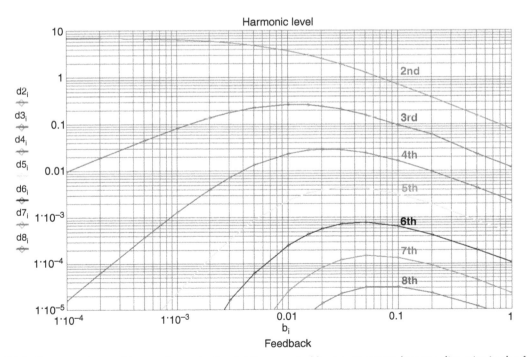

Figure 3.11. The percentages of harmonics up to the eighth generated by a pure square law non-linearity in the forward path as more negative feedback is applied. 200 mV peak at distorter input.

only fallen to 3.6%. However, what we have now collected is a whole bunch of higher harmonics. The third harmonic has risen alarmingly to 0.26%, and the fourth harmonic to 0.023%, where it will be well above the noise. The fifth harmonic has heaved itself up out of the swamp to reach 0.0023%; the sixth is up to 0.00025% but that is going to be hard to detect, and the seventh and eight are too low to cause concern. The ninth harmonic is off the bottom of the graph, and lost in the numerical noise of the simulator.

Using the maximum feedback factor of $\beta = 1$ with an open-loop gain of 100 times gets us into something more like conventional amplifier territory. The second harmonic has been substantially reduced from 7.00% to 0.077%. The higher harmonics, though now all reduced from their maxima by the extra feedback, are still very much with us, the third at 0.012%, the fourth at 0.0022%, and the fifth at 0.00046%. The sixth harmonic is at 0.0001%, and that and even higher-order harmonics can be quietly forgotten about.

The higher the order of harmonic, the greater the feedback factor at which its level reaches a maximum and then starts to come down again, exactly as reported by Peter.

Peter was able to argue that negative feedback was always a good thing, because no matter what the level of a harmonic coming from his FET with no feedback, it could always ultimately be reduced to an even lower level (after it had been initially increased), as more and more feedback was applied. That does not apply in this theoretical case, where without feedback there are no harmonics at all apart from the second. No finite amount of feedback will reduce the level of the new harmonics generated back to zero. This is, in a sense, a fundamental limitation of negative feedback. It is, however, very important to realise that in practice the higher harmonics are at very low levels, and will fall below the noise floor. Note that Feedback Intermodulation Distortion is not confined to global feedback around multi-stage amplifiers; it also applies to local feedback, such as emitter degeneration resistors.

Another point that Peter made was that the system is rather sensitive to the amount of open-loop non-linearity present. This can effectively be altered by changing the signal level at the input of the distorter, as signal level affects nothing else in the system. Figure 3.12 was produced by halving the distorter input level to 100 mV peak.

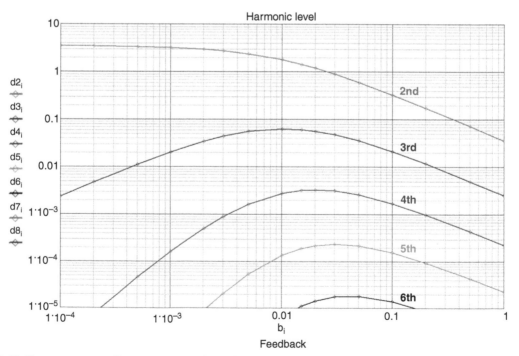

Figure 3.12. The percentages of harmonics up to the sixth generated by a pure square law non-linearity in the forward path as more negative feedback is applied. 100 mV peak at distorter input.

The level of the second harmonic roughly halves, as would be expected, dropping from 0.07724% to 0.03552%, but the higher the harmonics, the greater the reduction in amplitude; you can see that the seventh and eighth harmonics have completely disappeared off the bottom of the graph. In a practical system it is unlikely that even the sixth harmonic would be detectable in the noise by THD methods. In his article Peter stated that as the signal level was reduced, the third harmonic would fall with the square of the level, the fourth harmonic as the cube of the level, and so on. This means that on graphs like Figure 3.11 and 3.12, 'All the curves remain of the same shape, but each curve shifts downwards by a distance proportional to (n-1) where n is the order of the harmonic, so that the spacing between the curves becomes wider.'

While I hesitate to question any statement made by Peter Baxandall, plotting the harmonic level against the amount of distortion (as set by the input level to the distorter) shows that these relationships are only accurate for relatively small amounts of non-linearity. Figure 3.13 shows how the curves rise at an increasing rate as the input level reaches 300 mV, which corresponds to 10.5% second-harmonic distortion with no feedback applied.

Table 3.3. Reduction of harmonics by halving distorter input

Harmonic	100 mV Pk distn (%)	200 mV Pk distn (%)	100mV/ 200mV ratio	Expected ratio
2nd	0.03552	0.07724	2.174	2
3rd	0.002507	0.01162	4.646	4
4th	0.0002214	0.002190	9.891	8
5th	0.00002240	0.0004632	20.68	16

The important point, however, is that as the amount of non-linearity is reduced, the higher harmonics fall with increasing speed as their order increases. This is demonstrated by Table 3.3, which shows how the higher harmonics are in fact reduced by rather more than predicted by (n-1), when the non-linearity is halved by changing the distorter input level from 200 to 100 mV peak.

It is very clear that halving the amount of square-law non-linearity has a much greater effect on the higher harmonics; the third is reduced to a quarter, the fourth to about a tenth, and the fifth by more than twenty

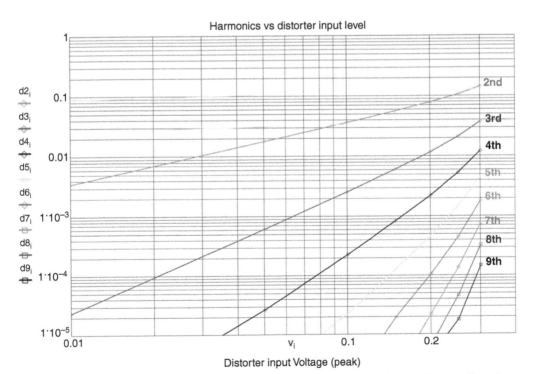

Figure 3.13. The higher the order of the harmonic, the faster its level falls as the amount of square law non-linearity is reduced.

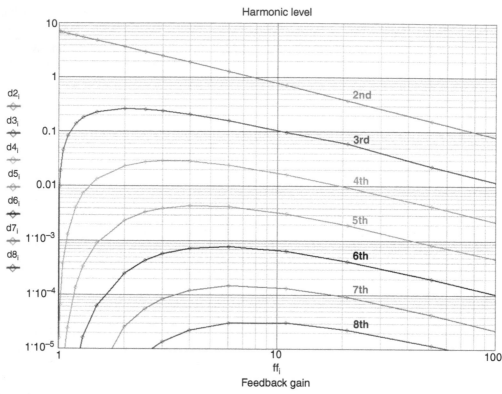

Figure 3.14. This is Figure 3.11 replotted with feedback improvement factor $(1 + A\beta)$ on the x-axis instead of feedback fraction β. This gives the familiar graph published by Peter Baxandall in 1978. 200 mV peak at distorter input.

times. It has often been said that the proper way to design a power amplifier is to make the open-loop response as linear as possible before applying any feedback, the rationale being no doubt that only a limited amount of feedback can be applied before HF instability sets in. We see here a more powerful and fundamental reason; the higher harmonics, which are generally held to be more unpleasant, are suppressed by an enormous amount if open-loop linearity is even mildly improved.

Finally, as promised, here is the data for the 200 mV case replotted with the NFB improvement factor $1+A\beta$ as the X-axis, and the result looks more familiar (Figure 3.14).

Maximising the Amount of Negative Feedback

Having hopefully freed ourselves from Fear of Feedback, and appreciating the dangers of using only a little of it, the next step is to see how much can be used. It is my view that the amount of negative feedback applied should be maximised at all audio frequencies to maximise linearity, and the only limit is the requirement for reliable HF stability. In fact, global or Nyquist oscillation is not normally a difficult design problem in power amplifiers; the HF feedback factor can be calculated simply and accurately, and set to whatever figure is considered safe. (Local oscillations and parasitics are beyond the reach of design calculations and simulations, and cause much more trouble in practice.)

In classical Control Theory, the stability of a servomechanism is specified by its Phase Margin, the amount of extra phase-shift that would be required to induce sustained oscillation, and its Gain Margin, the amount by which the open-loop gain would need to be increased for the same result. These concepts are not very useful in audio power amplifier work, where many of the significant time-constants are only vaguely known. However, it is worth remembering that the phase margin will never be better than 90°, because of the phase-lag caused by the VAS Miller capacitor; fortunately this is more than adequate.

In practice, the designer must use his judgement and experience to determine an NFB factor that will give reliable stability in production. My own experience leads me to believe that when the conventional three-stage architecture is used, 30 dB of global feedback at 20 kHz is safe, providing an output inductor is used to prevent capacitive loads from eroding the stability margins. I would say that 40 dB was distinctly risky, and I would not care to pin it down any more closely than that.

The 30 dB figure assumes simple dominant-pole compensation with a 6 dB/octave roll-off for the open-loop-gain. The phase and gain margins are determined by the angle at which this slope cuts the horizontal unity-loop-gain line. (I am being deliberately terse here; almost all textbooks give a very full treatment of this stability criterion.) An intersection of 12 dB/octave is definitely unstable. Working within this, there are two basic ways in which to maximise the NFB factor:

1. While a 12 dB/octave gain slope is unstable, inter-mediate slopes greater than 6 dB/octave can be made to work. The maximum usable is normally consid-ered to be 10 dB/octave, which gives a phase margin of 30°. This may be acceptable in some cases, but I think it cuts it a little fine. The steeper fall in gain means that more NFB is applied at lower frequen-cies, and so less distortion is produced. Electronic circuitry only provides slopes in multiples of 6 dB/octave, so 10 dB/octave requires multiple over-lapping time-constants to approximate a straight line at an intermediate slope. This gets complicated, and this method of maximising NFB is not popular.

2. Make the gain slope vary with frequency, so that maximum open-loop gain and hence NFB factor are sustained as long as possible as frequency increases; the gain then drops quickly, at 12 dB/octave or more, but flattens out to 6 dB/octave before it reaches the critical unity loop-gain intersection. In this case, the stability margins should be relatively unchanged compared with the conventional situation.

These approaches are dealt with in detail in Chapter 13 on compensation.

Overall Feedback Versus Local Feedback

It is one of the fundamental principles of negative feed-back that if you have more than one stage in an ampli-fier, each with a fixed amount of open-loop gain, it is more effective to close the feedback loop around all the stages, in what is called an overall or global feed-back configuration, rather than applying the feedback locally by giving each stage its own feedback loop. I hasten to add that this does not mean you cannot or should not use local feedback *as well* as overall feed-back — indeed one of the main themes of this book is that it is a very good idea, and indeed probably the only practical route to very low distortion levels. This is dealt with in more detail in Chapter 6 on input stages and Chapters 7 and 8 on voltage-amplifier stages.

It is worth underlining the effectiveness of overall feedback because some of the less informed audio commentators have been known to imply that overall feedback is in some way decadent or unhealthy, as opposed to the upright moral rigour of local feedback. The underlying thought, insofar as there is one, appears to be that overall feedback encloses more stages each with their own phase shift, and therefore requires compensation which will reduce the maximum slew-rate. The truth, as is usual with this sort of moan, is that this could happen if you get the compensation all wrong; so get it right, it isn't hard.

It has been proposed on many occasions that if there is an overall feedback loop, the output stage should be left outside it. I have tried this, and believe me, it is not a good idea. The distortion produced by an output stage so operated is jagged and nasty, and I think no one could convince themselves it was remotely accept-able if they had seen the distortion residuals.

Figure 3.15 shows a negative feedback system based on that in Figure 3.1 at the start of the chapter, but with two stages. Each has its own open loop gain A, its own

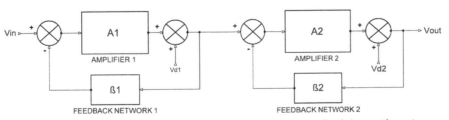

Figure 3.15. A negative feedback system with two stages, each with its own feedback loop. There is no overall negative feedback path.

NFB factor β, and its own open-loop error Vd added to the output of the amplifier. We want to achieve the same closed-loop gain of 25 as in Table 3.1, and we will make the wild assumption that the open-loop error of 1 in that table is now distributed equally between the two amplifiers A1 and A2. There are many ways the open- and closed-loop gains could be distributed between the two sections, but for simplicity we will give each section a closed-loop gain of 5; this means the conditions on the two sections are identical. The open-loop gains are also equally distributed between the two amplifiers so that their product is equal to column 3 in Table 3.1 above. The results are shown in Table 3.4; columns 1–7 show what's happening in each identical loop, and columns 8 and 9 give the results for the output of the two loops together, assuming for simplicity that the errors from each section can be simply added together; in other words, there is no partial cancellation due to differing phases, and so on.

This final result is compared with the overall feedback case of Table 3.1 in Table 3.5, where column 1 gives total open-loop gain, and column 2 is a copy of column 7 in Table 3.1 and gives the closed-loop error for the overall feedback case. Column 3 gives the closed-loop error for the two-stage feedback case.

It is brutally obvious that splitting the overall feedback situation into two local feedback stages has been a bad move. With a modest total open-loop gain of 100, the local feedback system is barely half as effective. Moving up to total open-loop gains that are more realistic for real power amplifiers, the factor of deterioration is between six and forty times – an amount that cannot be ignored. With higher open-loop gains the ratio gets even worse. Overall feedback is totally and unarguably superior at dealing with all kinds of amplifier errors, though in this book distortion is often the one at the front of our minds.

Table 3.5. Overall NFB gives a lower closed-loop error for the same total open-loop gain. The error ratio increases as the open-loop gain increases

1	2	3	4
A Total O/L gain	Overall NFB C/L Error	2-stage NFB C/L error	Error Ratio
100	0.2000	0.3333	1.67
1000	0.0244	0.1365	5.60
10000	0.0025	0.0476	19.10
40000	0.0006	0.0244	39.05
100000	0.0002	0.0156	62.28

While there is space here to give only one illustration in detail, you may be wondering what happens if the errors are not equally distributed between the two stages; the signal level at the output of the second stage will be greater than that at the output of the first stage, so it is plausible (but by no means automatically true in the real world) that the second stage will generate more distortion than the first. If this is so, and we stick with the assumption that open-loop gain is equally distributed between the two stages, then the best way to distribute the closed-loop gain is to put most of it in the first stage so we can get as high a feedback factor as possible in the second stage. As an example, take the case where the total open-loop gain is 40,000.

Assume that all the distortion is in the second stage, so its open-loop error is 1 while that of the first stage is zero. Now redistribute the total closed-loop gain of 25 so the first stage has a closed-loop gain of 10 and the second stage has a closed-loop gain of 2.5. This gives a closed-loop error of 0.0123, which is about half of 0.0244, the result we got with the closed-loop gain equally distributed. Clearly things have been improved

Table 3.4. Open-loop gain and closed-loop errors in the two loops

1	2	3	4	5	6	7	8	9
Desired C/L gain	β1 NFB fracn	A1 O/L gain	NFB factor	C/L gain	O/L Error	C/L Error	Total C/L gain	Total C/L error
5	0.2	10.00	3.00	3.333	0.5	0.1667	11.11	0.3333
5	0.2	31.62	7.32	4.317	0.5	0.0683	18.64	0.1365
5	0.2	100	21.00	4.762	0.5	0.0238	22.68	0.0476
5	0.2	200	41.00	4.878	0.5	0.0122	23.80	0.0244
5	0.2	316.23	64.25	4.922	0.5	0.0078	24.23	0.0156

by applying the greater part of the local negative feed-back where it is most needed. But … our improved figure is still about twenty times worse than if we had used overall feedback.

In a real power amplifier, the situation is of course much more complex than this. To start with, there are usually three rather than two stages, the distortion produced by each one is level-dependent, and in the case of the voltage-amplifier stage the amount of local feedback (and hence also the amount of overall feed-back) varies with frequency. Nonetheless, it will be found that overall feedback always gives better results.

Maximising Linearity before Feedback

Make your amplifier as linear as possible before applying NFB has long been a cliché. It blithely ignores the difficulty of running a typical solid-state amplifier without any feedback, to determine its basic linearity.

In the usual three-stage amplifier, the linearity of the input stage can be improved by making sure its collector currents are exactly balanced, and increasing the input tail current so local feedback can be applied by emitter-degeneration resistors while maintaining the same transconductance. This is fully covered in Chapter 6.

The VAS already has local feedback through the dominant-pole Miller capacitor; the most important additional linearisation is done by preventing non-linear local feedback via the VAS transistor collector-base capacitance, Cbc. This can be done either by adding an emitter-follower or using a cascode structure. This is dealt with in Chapter 7.

The output stage does not have equivalent techniques for simple and effective linearisation, and in Class-B amplifiers exhibits the stubborn problem of crossover distortion. It is often already an emitter-follower config-uration with 100% voltage feedback, so no more can be readily applied. Distortion can sometimes be reduced by using a CFP rather than an EF output stage. It can be reduced dependably by using the lowest practicable emitter resistors, which smooth the inherent crossover distortion of the output stage with optimal biasing, and also minimise the extra distortion introduced if the output stage strays into Class-AB operation due to over-biasing. Using more output devices in parallel also reduces crossover distortion; see Chapters 9 and 10.

The use of these linearising techniques gives a Blameless power amplifier.

Positive Feedback in Amplifiers

Just as negative feedback improves most parameters, positive feedback makes them worse. It therefore has few if any applications in modern amplifier design. It was, however, sometimes useful in valve amplifiers. Open-loop gain was typically in short supply, due to the high cost of valves, and it was sometimes worth-while to apply positive feedback to a relatively linear early stage, to permit more feedback to be used to straighten out a high-distortion output stage.[4] For a prac-tical design using this technique, see a 1950 5-Watt design by John Miller Jr.[5] This includes measurements that show that the method effectively reduces distortion.

There seems, however, to be little application for this technique in solid-state amplifiers, as usually all the open-loop gain that can be safely used is available.

References

1. Baxandall, P, Audio Power Amplifier Design: Part 5 *Wireless World*, Dec. 1978, pp. 53—56. This series of articles had 6 parts and ran on roughly alternate months, starting in Jan. 1978. It has been republished by Jan Didden under the *Linear Audio* imprint.

2. Scroggie, M G, Negative Feedback and Non-linearity, *Wireless World*, April 1961, p. 225.

3. Scroggie, M G, Negative Feedback and Non-linearity, *Wireless World*, October 1978, pp. 47—50 (under the name 'Cathode Ray').

4. Valley, G and Wallman, H (eds), *Vacuum Tube Amplifiers*, Radiation Laboratory Series, Volume 18, New York: McGraw-Hill 1948, pp. 477—479.

5. Miller, J Jr, Combining Positive and Negative Feedback, *Electronics*, Mar., 1950, pp. 106—109.

Chapter 4

Amplifier Architecture, Classes, and Variations

Amplifier Architectures

This grandiose title simply refers to the large-scale structure of the amplifier; that is, the block diagram of the circuit one level below that representing it as a single white block labelled *Power Amplifier*. Almost all solid-state amplifiers have a three-stage architecture as described below, though they vary in the detail of each stage. Two-stage architectures have occasionally been used, but their distortion performance is not very satisfactory. Four-stage architectures have been used in significant numbers, but they are still much rarer than three-stage designs, and usually involve relatively complex compensation schemes to deal with the fact that there is an extra stage to add phase-shift and potentially imperil high-frequency stability.

The Three-stage Amplifier Architecture

The vast majority of audio amplifiers use the conventional architecture, shown in Figure 4.1, and so it is dealt with first. There are three stages, the first being a transconductance stage (differential voltage in, current out), the second a transimpedance stage (current in, voltage out), and the third a unity-voltage-gain output stage. The second stage clearly has to provide all the voltage gain and I have therefore called it the voltage-amplifier stage or VAS. Other authors have called it the *pre-driver stage* but I prefer to reserve this term for the first transistors in output

triples. This three-stage architecture has several advantages, not least being that it is easy to arrange things so that interaction between stages is negligible. For example, there is very little signal voltage at the input to the second stage, due to its current-input (virtual-earth) nature, and therefore very little on the first stage output; this minimises Miller phase shift and possible Early effect in the input devices.

Similarly, the compensation capacitor reduces the second stage output impedance, so that the non-linear loading on it due to the input impedance of the third stage generates less distortion than might be expected. The conventional three-stage structure, familiar though it may be, holds several elegant mechanisms such as this. They will be fully revealed in later chapters.

Since the amount of linearising global NFB available depends upon amplifier open-loop gain, how the stages contribute to this is of great interest. The three-stage architecture always has a unity-gain output stage — unless you really want to make life difficult for yourself — and so the total forward gain is simply the product of the transconductance of the input stage and the transimpedance of the VAS, the latter being determined solely by the Miller capacitor Cdom, except at very low frequencies. Typically, the closed-loop gain will be between +20 and +30 dB. The NFB factor at 20 kHz will be 25 to 40 dB, increasing at 6 dB per octave with falling frequency until it reaches the dominant-pole frequency P1, when it flattens out. What matters for the control of distortion is the amount of negative feedback (NFB) available, rather than the open-loop bandwidth, to which it has no direct relationship. In my *Electronics World* Class-B design, the input stage gm is about 9 ma/V, and Cdom is 100pF, giving an NFB factor of 31 dB at 20 kHz. In other designs I have used as little as 26 dB (at 20 kHz) with good results.

Compensating a three-stage amplifier is relatively simple; since the pole at the VAS is already dominant, it can easily be increased to lower the HF negative-feedback factor to a safe level. The local NFB working on the VAS through Cdom has an extremely valuable linearising effect.

The conventional three-stage structure represents at least 99% of the solid-state amplifiers built, and I make no apology for devoting much of this book to its behaviour. You might think that a relatively simple configuration, essentially composed of 13 transistors, would have had every nuance of its behaviour completely determined many years ago. As the amount of new material in this book shows, that is not the case; even now, I am quite sure I have not exhausted its subtleties. It therefore appears certain that the

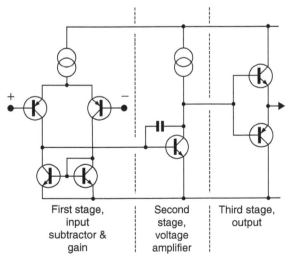

| First stage, input subtractor & gain | Second stage, voltage amplifier | Third stage, output |

Figure 4.1. The three-stage amplifier structure. There is a transconductance stage, a transimpedance stage (the VAS) and a unity-gain buffer output stage.

configuration was in a sense stumbled upon rather than the result of any carefully directed research program. As Chapter 30 on the history of amplifiers explains, the three-stage amplifier is not based on the original Lin circuit of 1956, because the Lin was a two-stage amplifier. The first three-stage amplifier in Chapter 30 is the Tobey & Dinsdale design of 1961, which used a single input transistor rather than a differential pair. The first three-stage design with a differential input is the Hardcastle & Lane 15 W Amplifier of 1969.

It will emerge that the three-stage amplifier, in its usual form with a constant-current load VAS, is the best configuration available for economy, predictability, the decoupling of one stage from the behaviour of another, ease of compensation, thermal stability, and low distortion. With a few inexpensive refinements the THD at 50 W/8 Ω can be kept below 0.001% up to 12 kHz and below 0.002% at 20 kHz.

The Two-stage Amplifier Architecture

In contrast with the three-stage approach, the architecture in Figure 4.2 is a two-stage amplifier, the first stage being once more a transconductance stage, though now without a guaranteed low impedance to accept its output current. The second stage combines the VAS and output stage in one block; it is inherent in this scheme that the VAS must double as a phase

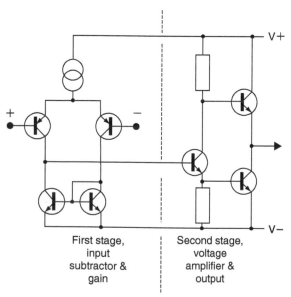

| First stage, input subtractor & gain | Second stage, voltage amplifier & output |

Figure 4.2. A two-stage amplifier structure. A voltage-amplifier output stage follows the same transconductance input stage.

splitter as well as a generator of raw gain. There are then two quite dissimilar signal paths to the output, and it is not at all clear that trying to break this block down further will assist a linearity analysis. The use of a phase-splitting stage harks back to valve amplifiers, where it was inescapable, as a complementary valve technology has so far eluded us.

Paradoxically, a two-stage amplifier is likely to be more complex in its gain structure than a three-stage. The forward gain depends on the input stage gm, the input stage collector load (because the input stage can no longer be assumed to be feeding a virtual earth) and the gain of the output stage, which will be found to vary in a most unsettling manner with bias and loading. Choosing the compensation is also more complex for a two-stage amplifier, as the VAS/phase-splitter has a significant signal voltage on its input and so the usual pole-splitting mechanism that enhances Nyquist stability by increasing the pole frequency associated with the input stage collector will no longer work so well. (I have used the term Nyquist stability, or Nyquist oscillation throughout this book to denote oscillation due to the accumulation of phase-shift in a global NFB loop, as opposed to local parasitics, etc.)

The LF feedback factor is likely to be about 6 dB less with a 4 Ω load, due to lower gain in the output stage. However, this variation is much reduced above the dominant-pole frequency, as there is then increasing local NFB acting in the output stage.

Here are two examples of two-stage amplifiers; Linsley-Hood[1]. and Olsson.[2] The two-stage amplifier offers little or no reduction in parts cost, is harder to design and in my experience invariably gives a poor distortion performance. A third configuration with a single-ended input/VAS stage is described in Chapter 30 on the history of amplifiers. The folded-cascode configuration could be considered as a two-stage amplifier; see Chapter 8.

The Four-stage Amplifier Architecture

The best-known example of a four-stage architecture is probably that published by Lohstroh and Otala in a paper which was confidently entitled 'An Audio Power Amplifier for Ultimate Quality Requirements' and appeared in December 1973.[3] A simplified circuit diagram of their design is shown in Figure 4.3. One of their design objectives was the use of a low value of overall feedback, made possible by heavy local feedback in the first three amplifier stages, in the form of emitter degeneration; the closed-loop gain was 32 dB (40 times) and the feedback factor 20 dB, allegedly flat

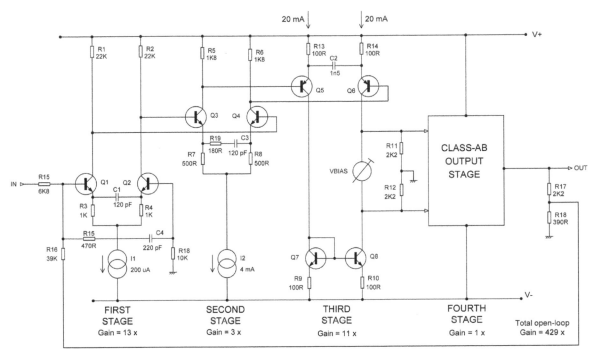

Figure 4.3. A simplified circuit diagram of the Lohstroh and Otala four-stage power amplifier. The gain figures for each stage are as quoted in the original paper.

across the audio band. Another objective was the elimination of so-called Transient Intermodulation Distortion, which after many years of argument and futile debate has at last been accepted to mean nothing more than old-fashioned slew-rate limiting. To this end, dominant-pole compensation was avoided in this design. The compensation scheme that was used was complex, but basically the lead capacitors C1, C2 and the lead-lag network R19, C3 were intended to cancel out the internal poles of the amplifier. According to Lohstroh and Otala, these lay between 200 kHz and 1 MHz, but after compensation the open-loop frequency response had its first pole at 1 MHz. A final lag compensation network R15, C4 was located outside the feedback loop. An important point is that the third stage was heavily loaded by the two resistors R11, R12. The EF-type output stage was biased far into Class-AB by a conventional Vbe-multiplier, drawing 600 mA of quiescent current. As explained later in Chapter 9, this gives poor linearity when you run out of the Class-A region.

You will note that the amplifier uses shunt feedback; this certainly prevents any possibility of common-mode distortion in the input stage, as there is no common-mode voltage, but it does have the frightening drawback of going berserk if the source equipment is disconnected, as there is then a greatly increased feedback factor, and

high-frequency instability is pretty much inevitable. Input common-mode non-linearity is dealt with in Chapter 6, where it is shown that in normal amplifier designs, it is of negligible proportions, and certainly not a good reason to adopt overall shunt feedback.

Many years ago I was asked to put a version of this amplifier circuit into production for one of the major hi-fi companies of the time. It was not a very happy experience. High-frequency stability was very doubtful and the distortion performance was distinctly unimpressive, being in line with that quoted in the original paper as 0.09% at 50 W, 1 kHz.[3] After a few weeks of struggle the four-stage architecture was abandoned and a more conventional (and much more tractable) three-stage architecture was adopted instead.

Another version of the four-stage architecture is shown in Figure 4.4; it is a simplified version of a circuit used for many years by another of the major hi-fi companies. There are two differential stages, the second one driving a push-pull VAS Q8,Q9. Once again the differential stages have been given a large amount of local negative feedback in the form of emitter degeneration. Compensation is by the lead-lag network R14, C1 between the two input stage collectors and the two lead-lag networks R15, C2 and R16, C3 that shunt the collectors of Q5, Q7 in the second differential

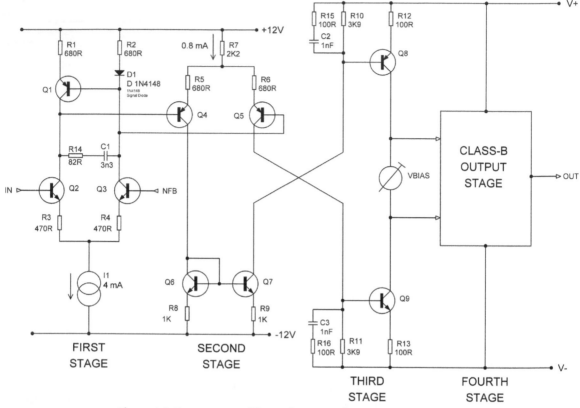

Figure 4.4. Four-stage amplifier architecture of a commercial amplifier.

stage. Unlike the Lohstroh and Otala design, series overall feedback was used, supplemented with an opamp DC servo to control the DC offset at the output.

Having had some experience with this design (no, it's not one of mine), I have to report that while in general the amplifier worked soundly and reliably, it was unduly fussy about transistor types and the distortion performance was not of the best.

The question now obtrudes itself: what is gained by using the greater complexity of a four-stage architecture? So far as I can see at the moment, little or nothing. The three-stage architecture appears to provide as much open-loop gain as can be safely used with a conventional output stage; if more is required, then the Miller compensation capacitor can be reduced, which will also improve the maximum slew-rates. The compensation of a three-stage amplifier is very simple; there is one, and only one, place to apply dominant compensation. In contrast, four-stage compensation is considerably more difficult, and likely to involve complicated compensation schemes. If low distortion is required, it is simpler to linearise the first two stages of a three-stage amplifier.

A four-stage architecture does, however, present some interesting possibilities for using nested Miller compensation, a concept which has been extensively used in opamps.

Though greatly out-numbered by three-stage designs, the four-stage architecture has been used in a number of Japanese commercial amplifiers. Almost all follow the Lohstroh and Otala configuration. A sample of these is briefly described in Table 4.1.

Table 4.1 gives an interesting perspective, though it does not pretend to be a complete or even representative selection. You can see that a balanced VAS (see Chapter 8) is common, as it can be driven from both outputs of the second differential stage. Alternatively phase-summing is performed by a current-mirror in the collectors of the second stage, as in the Technics SU-V2. The output stages are all in the EF format, with no CFP designs at all. Non-switching means preventing the output device from turning off at any point; see later in this chapter. The ZDR (Zero Distortion Rule) technology is dealt with in Chapter 13. Service manuals and schematics for all of these

Table 4.1. Commercial four-stage amplifier technology

Model	First stage	Second stage	VAS	Output stage	Tech features	Introduced
Denon PMA-1060	FET	BJT	Balanced	Double EF		1991
Denon POA-T2	BJT	BJT	Balanced	Double EF		1986?
Kenwood KA-501	BJT	BJT	Balanced	Double EF		1975?
Kenwood KA-56	BJT	BJT	Balanced	Double EF		1987
Nikko NA-990	FET	BJT	Balanced	Double EF		1983?
Nikko NA-2090	FET	BJT	Balanced	Double EF		1983
Pioneer M3	BJT	BJT	Balanced	Triple EF		1973
Pioneer M90	FET	BJT	Balanced	Triple EF	Non-switching	1986
Pioneer M91	FET	BJT	Balanced	Triple EF	Non-switching	1988
Pioneer A-616	FET	BJT	Balanced	Triple EF	Non-switching	1989
Sansui AU-D22	FET	BJT	Balanced	Triple EF	Feedforward	1982?
Sony TA-F70	FET	BJT	Balanced	Double EF		1979
Technics SU-V2	BJT	BJT	Simple	IC module	Synchronous bias	1981
Yamaha A-720	FET	BJT	Simple	Double EF	ZDR	1986

designs can be found on the Internet, though what their legal status is I do not know.

The Five-stage Amplifier Architecture

Amplifiers with five stages or more are very rare, no doubt because the extra gain available is of doubtful use, and the problems of compensation are definitely increased. The Pioneer A8 (1981) is a rare example of a five-stage amplifier, having an JFET differential input stage, then two BJT differential gain stages before the VAS. Whether this design uses Cherry's NDFL (see Chapter 13), as do the Pioneer A-5 and the A-6, is hard to say from looking at the complex and enigmatic schematic.

Power Amplifier Operating Classes

For a long time the only amplifier classes relevant to high-quality audio were Class-A and Class-AB. This is because valves were the only active devices, and Class-B valve amplifiers generated so much distortion that they were barely acceptable even for public address purposes. All amplifiers with pretensions to high fidelity operated in push-pull Class-A. Class-C was for radio work only.

The appearance of solid-state amplifiers allowed the use of a true Class-B, aided by the ability to apply more negative feedback than in a valve amplifier. Class-D

also only appeared when transistors were available, probably because of the difficulties of combining ultrasonic switching with output transformers.

If we put ultrasonic Class-D aside for the moment, as having little in common with the rest, we are still left with the need to describe an output device that is switched hard on and off at audio speeds; you will see why later. I have therefore used Class-D for this as well. We therefore have Classes A, AB, B, C, and D. You will see why I have included Class-C in a moment. These classes are defined by the portion of a cycle that the output devices conduct:

CLASS-A	Conducts for 100% of the cycle
CLASS-AB	Conducts less than 100% but more than 50% of the cycle
CLASS-B	Conducts very nearly 50% of the cycle
CLASS-C	Conducts less than 50% of the cycle
CLASS-D	Either on or off; conduction period not specified

You will note that Class-B is *not* the same as Class-AB or Class-C. Class-B is that unique amount of bias which gives the smoothest transfer of conduction between the two output devices, and so generates minimal crossover distortion. More on this later.

Despite the simple definitions above, there are apparently many more types of amplifier out there. There is Class-G, Class-H, Class-S, Class XD, Current-Dumping, Error-Correction, and so on. How do these fit in with the existence of only five classes of operation? What about the mysterious Classes E and F? It occurred to me that it might be better to classify all of them as combinations of the five basic classes. When I published the idea in 1999, I wrote: 'It may be optimistic to think that this proposal will be adopted overnight, or indeed ever. Nevertheless, it should at least stimulate thought on the many different kinds of power amplifier and the relationships between them.' Rather to my surprise, the idea has in fact gained some usage.

I am much against trying to popularise a new Class-letter every time a novel amplifier concept comes along, proliferating amplifier classes on through the alphabet. (The term 'Class XD' was not my idea, I hasten to add.) No matter how complex the amplifier, every active device in it will be working in Class-A, AB, B, C, or D, as I will describe below.

Combinations of Amplifier Classes

The basic five classes have been combined in many ways to produce the amplifier innovations that have appeared since 1970. Since the standard two-device output stage could hardly be simplified, all of these involve extra power devices that modify how the voltage or current is distributed. Assuming the output stage is symmetrical about the central output rail, then above and below it there will be at least two output devices connected together, in either a series or parallel format. Since these two devices may operate in different classes, two letters are required for a description, with a symbol between them to indicate parallel (•) or series (+) connection.

In parallel (shunt) connection, output currents are summed, the intention being either to increase power capability (which does not affect basic operation) or to improve linearity. A subordinate aim is often the elimination of the Class-B bias adjustment. The basic idea is usually a small high-quality amplifier correcting the output of a larger and less linear amplifier. For a parallel connection, the two Class letters are separated by a bullet (•).

In a series connection the voltage drop between supply rail and output is split up between two or more devices, or voltages are otherwise summed to produce the output signal. Since the collectors or drains of active devices are not very sensitive to voltage, such configurations are usually aimed at reducing overall power dissipation rather than enhancing linearity. Series connection is denoted by a plus sign (+) between the two Class letters.

The order of the two Class letters is significant. The first letter denotes the class of that section of the amplifier that actually controls the output voltage. Such a section must exist, if only because the global negative feedback must be taken from one specific point, and the voltage at this point is the controlled quantity.

First of all, a quick summary of how a good number of named types of amplifier can be represented is given in Table 4.2.

Some combinations cannot appear; for example, there is no Class-D in the parallel connection group because that signifies an on-off device connected directly to the output.

We will now take a closer look at the basic five classes, and go on to see how other named amplifier types can be built up from them.

Class-A

In a Class-A amplifier, current flows continuously in all the output devices, which enables the non-linearities of turning them on and off to be avoided. They come in two rather different kinds, although this is rarely explicitly stated, which work in very different ways. The first kind is simply a Class-B stage (i.e., two emitter-followers working back-to-back) with the bias voltage increased so that sufficient current flows so that neither device will cut off under normal loading. The great advantage of this approach is that it cannot abruptly run out of output current; if the load impedance becomes lower than specified, then the amplifier simply takes brief excursions into Class-AB, hopefully with a modest increase in distortion and no seriously audible distress.

The other kind could be called the controlled-current-source (VCIS) type, which is in essence a single emitter-follower with an active emitter load for adequate current-sinking. If this latter element runs out of current capability, it makes the output stage clip much as if it had run out of output voltage. This kind of output stage demands a very clear idea of how low an impedance it will be asked to drive before design begins.

Class-A is very inefficient; with musical signals its efficiency can be as low as 1%. Consequently, there have been many attempts to combine the linearity of Class-A with the efficiency of Class-B. One such scheme was the so-called 'Super-Class-A' introduced by Technics in 1978.[4] See Figure 4.5.

Table 4.2. Combinations of the five basic classes of device operation

	PARALLEL CONNECTION	
A•B	Sandman Class-S, parallel error-correction	Figure 4.11, 4.13
A•C	Quad current-dumping	
B•A	Class-XD (crossover displacement)	Figure 18.2
B•B	Self Load-Invariant amplifier, parallel output devices	Figure 10.12
B•C	Edwin and Crown amplifiers	
B•C	Class-G shunt. (Commutating) 2 rail voltages	Figure 4.6
B•C•C	Class-G shunt. (Commutating) 3 rail voltages	Figure 4.7
	SERIES CONNECTION	
A+B	Technics 'Super-Class-A'	Figure 4.5
A+B	Series error-correction, including Stochino	
A+D	A possible approach for cooler Class-A	
B+B	Series (totem-pole, cascade) output. No extra rails	Figure 9.20
B+C	Classical series Class-G, 2 rail voltages	Figure 4.6
B+C+C	Classical series Class-G, 3 rail voltages	
B+D	Class-G with outer devices in Class-D	Figure 4.9
B+D	Class-H	Figure 4.8

The Class-A controlling section A1 is powered by two floating supplies of relatively low voltage (± 15 V), but handles the full load current. The floating supplies are driven up and down by a Class-B amplifier A2, which must sustain much more dissipation as the same current is drawn from much higher rails, but which need not be very linear as in principle its distortion will have no effect on the output of A1. The circuit complexity and cost are more than twice that of a conventional amplifier, and the floating supplies are awkward; this seems to have limited its popularity. Because of the series amplifier connection, this system is described as Class-A+B.

Another A+B concept is the error-correction system of Stochino.[5] The voltage summation (the difficult bit) can be performed by a small transformer, as only the flux due to the correction signal exists in the core. This flux cancellation is enforced by the correcting amplifier feedback loop. Complexity and cost are at least twice that of a normal amplifier. More on this later in the chapter.

Class-A is examined in detail in Chapter 17.

Class-AB

This is perhaps not really a separate class of its own, but a combination of A and B. If an amplifier is biased into Class-B, and then the bias further increased, it becomes Class-AB. For outputs below a certain level, both output devices conduct, and operation is Class-A. At higher levels, one device will be turned completely off as the other provides more current, and the distortion jumps upward at this point as AB action begins. Each device will conduct between 50% and 100% of the time, depending on the degree of excess bias and the output level.

Class-AB is less linear than either A or B, and in my view its only legitimate use is as a fallback mode to allow Class-A amplifiers to continue working reasonably when faced with a low-load impedance.

Valve textbooks will be found to contain enigmatic references to classes of operation called AB1 and AB2; in the former grid, current did not flow for any part of the cycle, but in the latter it did. This distinction was important because the flow of output-valve grid

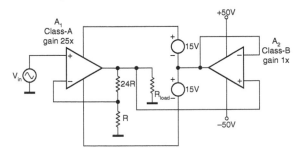

Figure 4.5. The 'Super-Class-A' concept. A1 runs in Class-A, while high-power Class-B amp A2 drives the two floating supplies up and down. Described as Class-A+B.

current in AB2 made the design of the previous stage much more difficult. AB1 or AB2 has no relevance to semiconductors, for in BJTs base current always flows when a device is conducting, while in power FETs gate current never does, apart from charging and discharging internal capacitances.

Class-AB is examined in detail in Chapters 9 and 10.

Class-B

Class-B is by far the most popular mode of operation, and probably more than 99% of the amplifiers currently made are of this type. Most of this book is devoted to it. My definition of Class-B is that unique amount of bias voltage which causes the conduction of the two output devices to overlap with the greatest smoothness and so generate the minimum possible amount of crossover distortion. For a single pair of output devices the quiescent current will be of the order of 10 ma for a CFP output stage, or 100 mA for the EF version. With bipolar transistors, collector current tails off exponentially as Vbe is reduced, and so the conduction period is rather arguable, depending on what current you define as 'conducting'. So-called 'non-switching' Class-B amplifiers, which maintain a small current in the output devices when they would otherwise be off, are treated as essentially Class-B.

Class-B is examined in detail in Chapters 9 and 10.

Class-C

Class-C implies device conduction for significantly less than 50% of the time, and is normally only usable in radio work, where an LC circuit can smooth out the current pulses and filter harmonics. Current-dumping amplifiers can be regarded as combining Class-A (the correcting amplifier) with Class-C (the current-dumping devices). The outer transistors in Class-G that switch in the higher supply rails also work in Class-C. It is, however, hard to visualise how an audio amplifier using devices in Class-C only could be built. A push-pull Class-B stage with the bias voltage removed works in Class-C. An EF output stage with no bias has a fixed dead-band of approx ±1.2 V, so clearly the exact conduction period varies with supply voltage for maximum output; ±40 V rails and a 1 mA criterion for conduction give 48.5% of the cycle. This looks like a trivial numerical deviation from 50%, but the gross crossover distortion prevents direct audio use.

The use of Class-C as part of Class-G is covered in Chapter 19.

Class-D

As usually defined, Class-D amplifiers continuously switch the output from one rail to the other at a supersonic frequency, controlling the mark/space ratio to give an average representing the instantaneous level of the audio signal; this is alternatively called Pulse Width Modulation (PWM). Great effort and ingenuity have been devoted to this approach, for the efficiency is in theory very high, but the practical difficulties are severe, especially so in a world of tightening EMC legislation, where it is not at all clear that a 200 kHz high-power square wave is a good place to start. Distortion is not inherently low,[6] and the amount of global negative feedback that can be applied is severely limited by the pole due to the effective sampling frequency in the forward path. A sharp cut-off low-pass filter is needed between amplifier and speaker, to remove most of the RF; this will require at least four inductors (for stereo) and will cost money, but its worst feature is that it will only give a flat frequency response into one specific load impedance.

Chapter 20 in this book is devoted to ultrasonic Class-D. Important references to consult for further information are Goldberg and Sandier[7] and Hancock.[8]

I have also used Class-D to describe an output device that is either on or off, but switching at signal frequencies. The conduction period is not specified.

Class-E

This is an extremely ingenious RF technique for operating a transistor so that it has either a small voltage across it or a small current through it almost all the time, so that the power dissipation is kept very low.[9] Regrettably it seems to have no sane application to audio.

Class-F

Class-F is another RF technique, related to Class-E. By using a more complex load network than Class-E to manipulate the harmonics, amplifier efficiency can be made to approach 100%.[10] It has no obvious audio application.

We will now look at amplifiers made by putting together the four basic classes, starting with those that have their own Class-letters.

Class-G

This concept was introduced by Hitachi in 1976 with the aim of reducing amplifier power dissipation. Musical signals have a high peak/mean ratio, spending most of the time at low levels, so internal dissipation is much reduced by running from low-voltage rails for small outputs, switching to higher rails current for larger excursions.[11,12] The relevant US patent appears to be 4,100,501.[13] Hitachi called it 'Dynaharmony' and first applied it to the HMA-8300 power amplifier in 1977. The Yamaha M-80 (1984) had an 'Auto-Class-A' feature which switched to Class-AB and increased the supply rails to deal with large signals. This could be considered as a form of Class-G.

The basic series Class-G with two rail voltages (i.e., four supply rails, as both voltages are ±) is shown in Figure 4.6. Current is drawn from the lower ± V1 supply rails whenever possible; should the signal exceed ± V1, TR6 conducts and D3 turns off, so the output current is now drawn entirely from the higher ± V2 rails, with power dissipation shared between TR3

and TR6. The inner stage TR3, 4 is usually operated in Class-B, although AB or A are equally feasible if the output stage bias is suitably increased. The outer devices are effectively in Class-C as they conduct for significantly less than 50% of the time. This Hitachi HMA-8300 was a powerful beast, giving 200 W into 8 Ω; its two supply voltages were ± 39 V and ± 96 V.

In principle, movements of the collector voltage on the inner device collectors should not significantly affect the output voltage, but in practice they do due to Early voltage effects. Class-G has been considered to have poorer linearity than Class-B because of glitching due to charge storage in commutation diodes D3, D4. However, if glitches occur, they do so at moderate power, well displaced from the crossover region, and so appear relatively infrequently with real signals. The use of Schottky power diodes eliminates this problem. Since the outer power devices conduct for less — usually much less — than 50% of the time, they are working in Class-C. As they are in series with the inner power devices, which maintain control of the output voltage, this is classified as Class-B+C.

An obvious extension of the basic Class-G principle is to increase the number of supply voltages, typically to three. Dissipation is reduced and efficiency increased, as the average voltage from which the output current is drawn is kept closer to the minimum. The inner devices will operate in Class-B or AB as before, the middle devices will be in Class-C, conducting for significantly less than 50% of the time. The outer devices are also in Class-C, conducting for even less of the time. Because there are three sets of output devices, three letters with intervening plus signs are required to describe this.

I am not aware of a commercial three-level series Class-G amplifier; perhaps because in series mode the cumulative voltage drops become too great, and compromise the efficiency gains. The extra complexity is significant, as there are now six supply rails and at least six power devices, all of which must carry the full output current. It seems most unlikely that this further reduction in power consumption could ever be worthwhile for domestic hi-fi. If it exists, such an amplifier would be described as operating in Class-B+C+C.

A closely related type of amplifier is Class-G-Shunt.[14] Figure 4.7 shows the principle; at low outputs only Q3, Q4 conduct, delivering power from the low-voltage rails. Above a threshold set by Vbias3 and Vbias4, D1 or D2 conduct and Q6, Q8 turn on, drawing current from the high-voltage rails, with D3, 4 protecting Q3, Q4 against reverse bias. The conduction periods of the Q6, Q8 Class-C devices are variable,

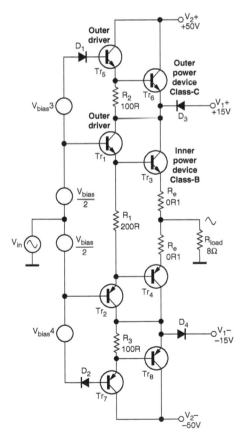

Figure 4.6. Class-G-Series output stage. When the output voltage exceeds the transition level, D3 or D4 turn off and power is drawn from the higher rails through the outer power devices. Class-B+C.

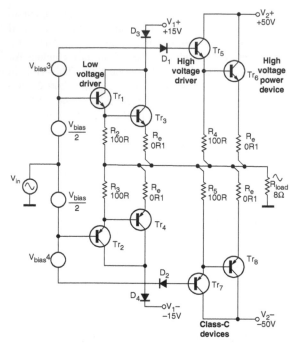

Figure 4.7. A Class-G-Shunt output stage, composed of two EF output stages with the usual drivers. Vbias3, 4 set the output level at which power is drawn from the higher rails. Described as Class-B•C.

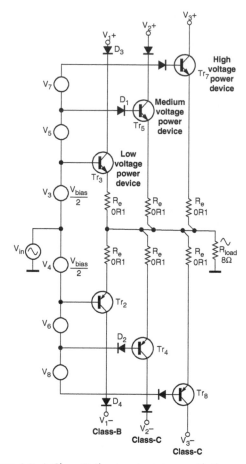

Figure 4.8. A Class-G-Shunt output stage with three pairs of supply rails, described as Class-B•C•C.

but inherently less than 50%. Normally the low-voltage section runs in Class-B to minimise dissipation. Such shunt Class-G arrangements are often called 'commutating amplifiers'. Since the two sets of output devices are in parallel, this is described as Class-B•C.

Some of the more powerful Class-G-Shunt PA amplifiers have three sets of supply rails, as shown in Figure 4.8, to further reduce the average voltage drop between rail and output. This is very useful in large PA amplifiers. This is described as Class-B•C•C.

Since the outer power devices in a Class-G-Series amplifier are not directly connected to the load, they need not be driven with waveforms that mimic the output signal. In fact, they can be simply driven by comparators so they are banged hard on and off, so long as they are always on when the output voltage is about to hit the lower supply rail. The abrupt voltage changes are likely to give worse glitching than classic Class-G. The inner power devices are in Class-B with the outer in Class-D (nothing to do with ultrasonic Class-D). Some of the more powerful amplifiers made by NAD (e.g., Model 340) use this approach, shown in Figure 4.9. This is Class-B+D.

When efficiency is important, Class-G can provide a viable alternative to the difficulties of ultrasonic Class-D. Chapter 19 in this book is devoted to Class-G.

Class-H

Class-H is once more basically Class-B, but with a method of dynamically boosting the supply rails (as opposed to switching to another one) in order to increase efficiency. The usual mechanism is a form of bootstrapping. Class-H is occasionally used to describe Class-G as above; this sort of confusion we can do without. Class-H is also used to describe a system where a single set of supply rails are continuously modulated to keep the voltage-drop across the output devices low.

One version of Class-H is shown in Figure 4.10; it uses a charge-pump for short-term boosting of the supply voltage. This approach was used by Philips.[15]in

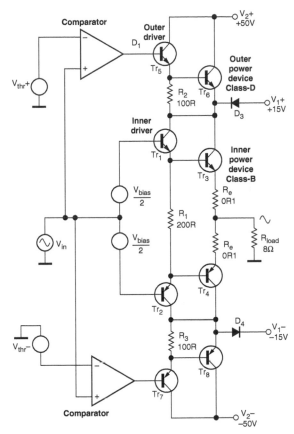

Figure 4.9. A Class-G output stage with the outer power devices in Class-D. Described as Class-B+D.

a single-rail car audio system. The low supply voltage (12 V) requires a bridged configuration, which in turn uses a clever floating-feedback system. At low outputs TR5 is on, keeping C charged from the rail via D. During large output excursions, TR5 is off and TR6 turns on, boosting the supply to TR3. The opposite half works in the same way.

Full circuitry has not been released, but it appears the charge-pump is an on/off subsystem, so could be regarded as Class-D. It is in series with the output device, so the amplifier is classified as another example of Class-B+D.

Class-S

Class-S, so named by Doctor Sandman,[16] uses a Class-A stage with very limited current capability, backed up by a Class-B stage connected so as to make the load appear as a higher resistance that is within the Class-A amplifier's capability. In Figure 4.11, A1 is the Class-A

controlling amplifier, while A2 is the Class-B heavy-weight stage. As far as the load is concerned, these two stages are delivering current in parallel. The aim was improved linearity, with the elimination of the bias preset of the Class-B stage as a secondary goal. Class-S can be reclassified as Class-A•B.

If A2 is unbiased and therefore working in Class-C, A1 has much greater errors to correct. This would put the amplifier into another category, Class-A•C.

The method used by the Technics SE-A100 and other amplifiers is apparently extremely similar.[17]

Class XD

Class XD, so named by the marketing department at Cambridge Audio, refers to my Crossover Displacement concept. It consists of a Class-B output stage combined with a Class-A output stage, the current from the latter displacing the crossover region away from the zero-voltage point so that the combination operates in pure Class-A at low levels, moving into Class-B at higher levels without the gain-steps associated with Class-AB. The two output stages are in parallel so this is described as Class-B•A.

Class XD is fully described in Chapter 18.

Edwin Amplifiers

An Edwin amplifier has zero bias for its output devices. The name is derived from the Edwin amplifier, published in *Elektor* in 1975.[18] It had a conventional EF output stage, but with low-value emitter resistors (33 Ω) for the drivers, so the output devices only turned on at significant output levels. It was claimed this had the advantage of zero quiescent current in the main output devices, though why this might be an advantage was not stated; in simulation, linearity appears worse than usual, as one might expect. This approach appears to have been introduced by Crown (Amcron) around 1970.[19] In my system it is described as Class-B•C.

The Limits of Classification

While I believe that my classification system gives about as much information as can be usefully stored in three characters, I have to say it is by no means fully comprehensive. It does not allow for amplifiers that are not symmetrical about the output rail, such as those with quasi-complementary output stages. Quad-style Current-Dumping can be handily specified as Class-A•C, which is accurate but that says nothing about the

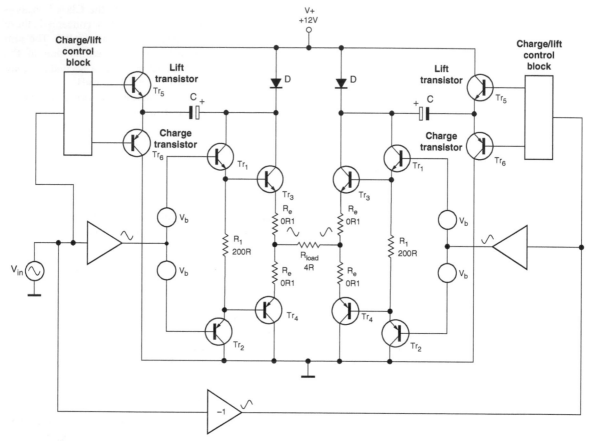

Figure 4.10. The Class-H principle applied to a bridged output stage for automotive use. Class-B+D.

error-correction principle of operation, which allows Class-C (in this case unbiased Class-B) to be used.

The test of any classification system is its gaps. When the Periodic Table of Elements was evolved, the obvious gaps spurred the discovery of new elements. This was convincing proof the Table was valid.

Table 4.2 is restricted to class combinations that are or have been in actual use, but a full matrix showing all the possible combinations has several intriguing gaps; some, such as C•C and C+C are of no obvious use to anybody, but others like A+C are more promising; this would be a form of Class-G with a Class-A inner stage. Glitches permitting, this approach might save a lot of heat.

Amplifier Variations

The solid-state Class-B amplifier has proved to be both successful and flexible. A good number of different

approaches to its basic operation have been described in the previous section. Nonetheless many attempts have been made to improve it further; one of the great challenges is to combine the efficiency of Class-B with the linearity of Class-A. It would be difficult to give a comprehensive list of the changes and improvements attempted, many of which got nowhere. Here I deal only those that have been either commercially successful or particularly thought-provoking to the amplifier-design community. While this section is of limited length, each topic could easily fill the entire book.

Error-correcting Amplifiers

This approach uses error-cancellation strategies rather than negative feedback. It can, in theory, cancel out distortion completely, while negative feedback can only reduce it because the amount that can be used is

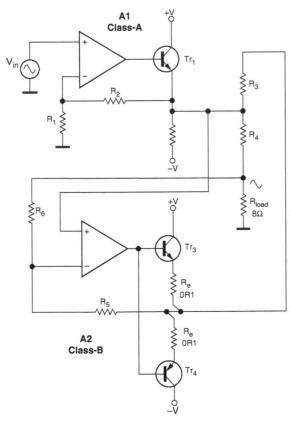

Figure 4.11. Sandman 'Class-S' scheme. R3,4,5,6 implement the feedback loop that controls amplifier A2 so as to raise the load impedance seen by A1.

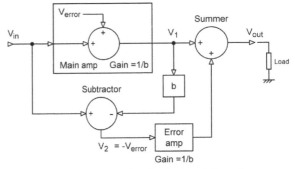

Figure 4.12. Error correction by 'output-injected error-feedforward'. The summing of the main output and the correction signal is done at the output power level and this is difficult.

limited by the need for stability. It is a complex field, and there is only space here for a brief overview.

Error-correction was brought to prominence in 1975 by the ground-breaking Quad 405 current-dumping amplifier, which was presented by Peter Walker and Michael Albinson at the 50th AES convention.[20]and described in *Wireless World*[21] The Quad USA patent is 3,970,953[22] Some years later there was considerable debate as to whether current-dumping was truly an error-feedforward system or not.[23,24,25] There were many letters to the editor; that by Peter Baxandall.[26] is particularly worth reading.

Feedforward error correction may be generally thought of as a modern idea, but in fact it was invented by Harold Black in 1928,[27] *before* he invented negative feedback. The simplest form of error correction is shown in Figure 4.12. This is called 'output-injected error-feedforward'.

The imperfect main amplifier is represented by a perfect non-inverting amplifier with Verror added to

give its output V1; this error represents every deviation from perfection; distortion, noise, and, very importantly, loss of gain due to output loading. Its voltage gain is defined as 1/b for reasons that will emerge. We will assume that Verror is added before the gain happens, and the output V1 is therefore:

$$V_1 = (V_{in} + V_{error}) \cdot (1 \backslash b) \qquad \text{Equation 4.1}$$

The input and output of the output stage are subtracted to isolate the error signal. Because the main amplifier has a gain of 1/b, we have to attenuate the output signal by b before subtracting it. The b's then cancel and the subtractor output V2 is:

$$V_2 = V_{in} - (V_{in} + V_{error}) \qquad \text{Equation 4.2}$$

and so:

$$V_2 = -V_{error} \qquad \text{Equation 4.3}$$

We have thus isolated and phase-inverted the error, and we can apply it to the non-inverting error amplifier. This also has a gain of 1/b so the error signals reaching the summer will in theory cancel completely, leaving just the scaled-up version of V_{in}.

The subtraction can be done accurately without any difficulty, probably by using a low-distortion opamp such as the LM4562. The inverted error signal is then 'output-injected', i.e., added to the output, cancelling the in-phase error from the output stage. This works brilliantly both in theory and in simulation, but in practice there are difficulties with the innocent phrase 'added to the output'. This summing has to be done at high power, with low losses, and maintaining a low output impedance at the load. Figure 4.13a shows the problem; R1 has to be very low, while R2 has to be reasonably high to prevent power from the main amplifier being dissipated in the error amplifier. This, however, means that the error amplifier has to have considerable extra gain as the correction signal is attenuated by the high value of R2 and the low value of R1.

The error amplifier in Figure 4.13a has to provide a significant amount of power to correct gain errors, without even considering distortion. Assume for the moment that we have an output stage with exactly unity gain and an output impedance of 0.1 Ω, plus 0.1 Ω for a summing resistance R1 as in Figure 4.13a. If the output is at 20 Vrms with no load, an 8 Ω load will pull it down to 19.5 Vrms. The error amplifier will seek to pull that back up to 20 Vrms, and to do so it will have to provide 61 mA and match the output voltage swing of the main output stage. With a 4 Ω load the error amplifier will have to give four times that (244 mA) as the output is pulled down twice as much but it is twice as hard to pull it back up again. At these current levels it will be necessary to use output-sized rather than driver-sized transistors in the error amplifier. The signal levels required to correct distortion rather than gain will be lower unless the main amplifier is seriously non-linear.

A fundamental assumption of error-feedforward is that the error amplifier, because it is only providing a fraction of the output power, can be made to have much less distortion than the main amplifier. There is often a tacit assumption that the error amplifier will work in Class-A, because if it was Class-B, its own crossover distortion would be intrusive despite the lower signal levels handled. Running the error amplifier

off lower supply rails will reduce the amount of heat-sinking required but the extra PSU components are likely to outweigh this. Amplifier clipping can cause trouble. If the main amplifier clips, the error amplifier will try to correct this error also, by attempting to drive large amounts of power into the load. Arrangements have to be made to handle this situation. In Figure 4.13a the two amplifier outputs are in parallel so this would be classified as Class-A•B.

One way of achieving 'perfect summing' at a high power level is to drive the bottom of the load from the error amplifier, with suitable adjustment of the error phase. The summation of the amplifier voltages across the load is simple addition with complete accuracy. The downside is that the error amplifier now has to equal the current capability of the main output stage; it does not have to have the same output voltage capability, but we still end up with two big power amplifiers instead of one big one and one small one. It is a clumsy solution. If we assume the main amplifier is in Class-B, and the correction amplifier is in Class-A, then because the amplifiers are in series so far as the load is concerned, the arrangement would be classified as Class-A+B.

Another possibility is to recognise that the distortion from a Blameless amplifier is only measurable at high frequencies, say, above 2 kHz. Above this frequency signal amplitudes are relatively low and so hopefully the feedforward correction signal is also small. Since the correction is only required at high frequencies, it can be coupled through a non-electrolytic capacitor to the output side of the existing main amplifier output inductor, as in Figure 4.13b. Thus a second DC-offset protection system is not required and money is saved. This L-C summing approach was used in the Sansui 'Super Feedforward' system described below; it was a feedforward system, but not an error-feedforward system.

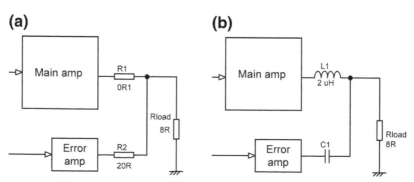

Figure 4.13. Two possible methods of output summing for error correction: (a) resistive; (b) L-C. Class-A•B.

For a highly ingenious treatment of the output-injected error-feedforward method, see a design by Giovanni Stochino.[5] Here the correction signal is added to the output of the main amplifier by a transformer. If this was done in the obvious way, the transformer would have to be huge and yet would still degrade the frequency response and LF distortion. However, in the Stochino system, the transformer can be very small because the flux in its core resulting from the main signal is cancelled by the error amplifier. Despite this ingenuity, the distortion results reported (at 20 kHz only) are no better than is attainable with a Blameless amplifier. This is another example of Class-A+B.

It has occurred to many people that the problem of summing at high power could be avoided by cancelling the error when the signals are still small. This thought leads to the 'input-injected error-feedforward' topology shown in Figure 4.14. It is sometimes called Hawksford Error Correction (HEC) as it was put forward by Malcolm Hawksford in 1980.[28]

The intended operation of this version is less obvious, but looking at the diagram we can write:

$$V_1 = V_{in} + V_2 \qquad \text{Equation 4.4}$$

$$V_{out} = (V_1 + V_{error}) \cdot (1/b) \qquad \text{Equation 4.5}$$

$$V_2 = V_1 - (V_1 + V_{error}) \qquad \text{Equation 4.6}$$

From that we easily work out:

$$V_1 = V_{in} + V_1 - (V_1 + V_{error}) = V_{in} - V_{error}$$
$$\text{Equation 4.7}$$

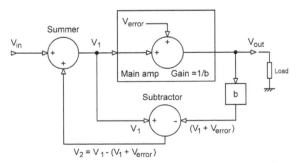

$$V_2 = V_1 - (V_1 + V_{error})$$

Figure 4.14. A fatally flawed attempt at 'input-injected error-feedforward'.

and then:

$$V_{out} = (V_{in} - V_{error} + V_{error}) \cdot (1/b) \qquad \text{Equation 4.8}$$

so the error terms cancel out and there is no distortion. This can be verified by DC simulation, and that has misled a lot of people. Before actually constructing an amplifier as in Figure 4.14, it is instructive to note that the circuitry can be simplified as in Figure 4.15, which is exactly equivalent apart from the order in which the additions and subtractions are done. It is revealed that 'input-injected error-feedforward' is just ordinary negative feedback with an added very doubtful-looking summer stage that implements 100% positive feedback, and therefore has infinite gain. This gives infinite negative feedback, and so a DC simulation, which takes no account of HF stability, gives apparently brilliant results.

The infinite-gain issue makes the HEC arrangement of Figure 4.15 difficult to apply. I tried it out once, using 5532s for the amplifier, summer, and subtractor, just to see what would happen. I coulda been a contender, but all I got was a one-way ticket to Oscillation City.

Another misleading error-correction concept is sometimes called 'active error feedback' in which normal negative feedback is applied via an active stage. Unlike error-feedforward, in which there is no return path and so no possibility of oscillation, stability is likely to be a serious issue. The advantages are not obvious. One example appears to be the 'distortion servo' used in the Hitachi HMA-7500 Mk II (1980). This design also uses non-switching technology — more on that below.

So far I have assumed that the amplifiers involved are complete amplifiers with their own internal negative feedback. It is, however, clear that it is — or should be — only the output stage that needs straightening out, and there have been various attempts at applying correction to this stage only.

See Cordell.[29] A most interesting recent design has recently been published by Jan Didden.[30]

While this section can only scratch the surface of the subject, I think I have shown that error-correction is in no way an easy option when it comes to reducing distortion. There are significant technical problems, and in many cases a plausible error-feedforward design is going to be twice as complex and twice as costly as a straightforward feedback amplifier like the Blameless design.

For further reading, two papers by Danyuk, Pilko, and Renardson[31,32] are well worth pursuing.

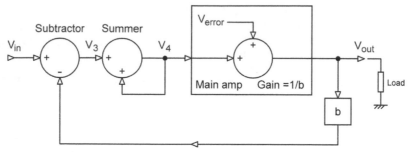

Figure 4.15. The attempt at 'input-injected error-feedforward' in Figure 4.14 rearranged to show that it conceals a 100% positive feedback loop.

Auxiliary Amplifiers

This section deals with amplifiers in which not all the output devices are driven by the same signal. Often there is a main amplifier and an auxiliary amplifier that handles less power.

The Yamaha version was called Linear Transfer Bias (LTB), and the central principle was that one of the two output device pairs was driven by a scaled-down version of the signal to the other pair. The intention was presumably to spread out the crossover region

so that it would show a smaller gain-wobble, and the harmonics generated by it would be both of lower order and better linearised by negative feedback that falls as frequency rises. This notion occurred to me independently, but not until the 1990s. The first known use of LTB is in 1979, in the Yamaha M-2 amplifier; the M-4, M-40 and M-60 did not have LTB. Figure 4.16a shows how it was applied to the M-50 (1982); the output stage is a triple, and the second pair of output devices was driven from tappings

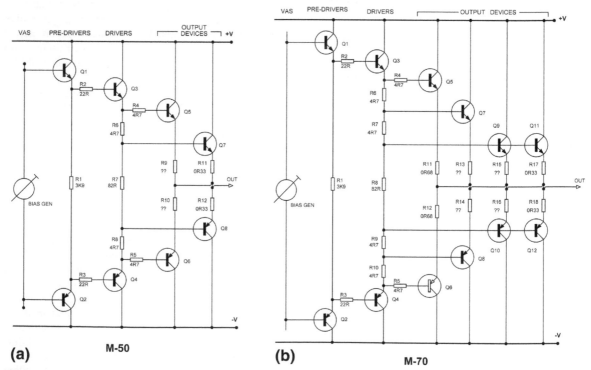

Figure 4.16. The principle of the Yamaha 'Linear Transfer Bias' system: (a) one level of driver tapping in the M-50; (b) two levels of driver tapping in the M-70.

on the driver emitter resistance chain R6-R8. Both output pairs Q5, Q6 and Q7, Q8 were of the same transistor type.

The idea was also applied to the Yamaha M-70 (1982) in a more complex form (Figure 4.16b) that more closely resembles the M-2 in that there are two levels of tapping; here the lowest level taps drive two pairs of output devices. The values of some of the output emitter resistors have proved impossible to determine which is unfortunate as it is likely they were carefully chosen to optimise the crossover region; note that in Figure 4.16b, R11, R12 are different in value from R17, R18. This seems like a good idea but it does not seem to have lived long or prospered. The M-50 and M-70 had good distortion specs (0.0005% at 1 kHz at rated output) but as they additionally used the Yamaha ZDR (Zero Distortion Rule) system (dealt with in Chapter 13 on compensation), it is difficult to assess how useful the Linear Transfer Bias concept was.

A rather different but equally intriguing concept along the same lines was introduced by Sansui in 1980. They called it 'Super Feedforward' and explicitly stated it was based on Black's 1928 invention; however, there is no subtractor to derive an error to be fed forward, so it is not a true error-feedforward correction system as described above. This is confirmed by an explanatory brochure published by Sansui,[33] and by a *JAES* paper by Takahashi and Tanaka in 1981.[34] The relevant US patent appears to be 4,367,442.[35]

The AES paper seems to say the feedforward system is based on the assumption that the signal just upstream of the output stage, i.e., at the VAS, is at a lower distortion level than the output of the imperfect main output stage. A small extra amplifier was driven from a scaled-down version of the VAS signal, and its contribution was summed with the output through an LC arrangement so the small-scale amplifier only has control at high frequencies. The arrangement seems wide open to the objection that the signal at the VAS is not purer than that at the main output; in fact it is much more distorted because the global negative feedback is trying to make the signal at the main amplifier output correct. Possibly the advantage, if any, lay in some advantageous modification of the gain changes in the crossover region. It is hard to be specific because the paper by T & T is not a model of clarity. They muddy the waters by using two-pole compensation, and for reasons best known to themselves only give measured results in the 20–100 kHz range; the improvement in THD at 20 kHz is unimpressive. 4 Ω loads are not considered.

Figure 4.17 shows a simplified version of its application in the AU-D22 amplifier. The output stage is basically a triple, with pre-drivers Q1, Q2, drivers Q3, Q4, and output devices Q5, Q6. The small feedforward amplifier is Q7, Q8. Note that the negative feedback is taken from the output of the main amplifier, not the combined output. This puts the feedforward amplifier outside the NFB loop, which does not strike me as a good idea.

The 'Super Feedforward' system was introduced by Sansui in their AU D707-F amplifier (1980) and later used in the AU-D11 amplifier (1981). It was also used in a slightly simpler form in the lower-power AU-D22, AU-D33 (both 1983), AU-D101, and AU-D55X amplifiers. The AU-D22 had a distortion spec of 0.004% at full power (no frequency stated), which is not impressive compared with a simple Blameless amplifier. The AU-D series was replaced by the AU-G series in 1984, with the 'Super Feedforward' feature being quietly dropped.

Both the Yamaha and Sansui concepts are essentially two Class-B amplifiers with their outputs in parallel. This is described in my classification scheme as Class-B●B. One wonders if a second bias adjustment would have been useful, if only to accommodate Vbe tolerances.

Non-switching Amplifiers

In the late 1970s and the early 1980s there was an astonishing flowering of creativity as Japanese amplifier manufacturers sought to combine Class-B efficiency with Class-A linearity.

Most of the distortion in Class-B is crossover distortion, and results from gain changes in the output stage as the power devices turn on and off. Several researchers have attempted to avoid this by ensuring that each device is clamped to pass a certain minimum current at all times. As we have seen in earlier chapters the current in an output device turning off is more of a glide down to zero rather than an abrupt stop, and the real problem is that the gain of the output stage undergoes a mild wobble while control of the output voltage is transferred from one output device to the other. It is not intuitively obvious (to me, anyway) that halting the diminishing device current in its tracks and clamping it to a fixed value will give better gain characteristics and so less crossover distortion.

Nevertheless, a good deal of ingenuity has been expended on the idea. I tried it out at Cambridge Audio in 1975, but the results were unpromising. The idea also surfaced in the 'Circuit Ideas' section of

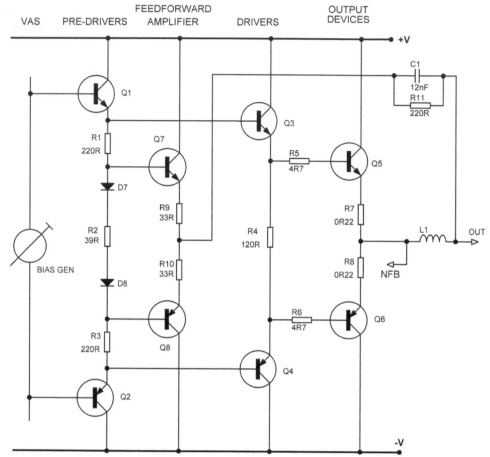

Figure 4.17. The principle of the Sansui 'Super Feedforward' system, based on the AU-D22 amplifier.

Wireless World from time to time. Non-switching schemes work by dynamically varying the bias voltage of the output stage. When one output device is turning on more, the other is usually turning off. To prevent this, the voltage bias is increased. This clearly needs to be done in a closely-controlled way to prevent damagingly large flows of quiescent current. Evaluating these schemes by simulation is complicated by the fact that they are intimately connected with the VAS and it is not possible to simulate the output stage alone.

It is sometimes said of non-switching schemes that they can eliminate the need for quiescent current adjustments. This appears not to be the case in practice. All the versions examined below have at least one preset, while the Hitachi 'super linear' non-switching circuit in Figure 4.18 actually has *two* adjustments. In commercial production a pre-set adjustment is really not a problem.

Non-switching gave scope for the generation of what I call Named Technical Features (NTFs) by the various Japanese companies involved. These terms, often abbreviated, refer to what are at least intended to be genuine technical improvements, but since they appear to be always dreamed up by the marketing department, they often have a very tenuous connection with the technical principles involved. A classic example was 'Spontaneous Twin Drive' applied by Sony to their TA-N77ES amplifier and to many other models. Now I like power amplifiers to be predictable. In my experience the only thing they do 'spontaneously' is explode and catch fire; it was not a well-chosen word. It appears that what it actually meant was that there were separate secondary windings on the mains transformer for the small-signal and the output stages. Assuming the small-signal rails had a slightly higher

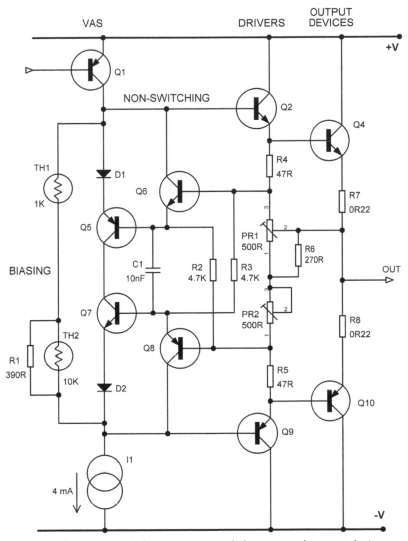

Figure 4.18. The Hitachi 'super linear' non-switching circuit, intended to prevent the output devices turning off at any point in the cycle.

voltage, this would have had a genuine benefit in terms of increasing amplifier efficiency.

The marketing departments had a field day thinking up NTFs for the various non-switching technologies.

Nelson Pass and Threshold

The Nelson Pass patent of 1976,[36] assigned to the Threshold Corporation, seems to have been the start of serious work on the subject. It may not be entirely a coincidence that only two years after this, many of the foremost Japanese amplifier manufacturers were taking a great interest in non-switching systems, and various

non-switching amplifiers reached the market in 1979. Non-switching was still in use by JVC in 1995, but in general it seems to have quietly faded away after 1990.

Hitachi 'super linear' non-switching

The Hitachi approach to non-switching was used in the HA-3700 and HA-4700 (1980) integrated amplifiers. These have BJT outputs and give 35 W/chan and 50 W/chan respectively. The service manual simply describes it as a 'super linear circuit' in lower case. The extra circuitry consists of four discrete transistors and is fully disclosed with a description of its operation.

The HA-3500 had conventional power amplifiers with no non-switching feature. The Hitachi HMA-7500 Mk2 (MOSFET outputs) had the 'super linear' feature, and also a 'distortion servo' which appears to be some kind of error-correction. It is claimed that the latter reduces output-stage distortion by a factor of three. The HMA-9500 Mk2 (1980) used a 'super linear' non-switching circuit in a different form.

A simplified version of the Hitachi is shown in Figure 4.18. Temperature-compensated output biasing is by the thermistor network to the left, and the four extra transistors Q5–Q8 only implement the non-switching.

When the output goes positive, there is a voltage drop across emitter resistor R7 due to the current flowing into the load; if the bias voltage is fixed, this turns off the lower output device. In the non-switching circuit the voltage on Q6 base is increased, and this tends to turn off Q5, and the bias voltage is increased so the lower output device does not cease conducting. The circuit is symmetrical and works similarly for negative output voltages. You will note that there are two presets; the bias adjustment procedure in the service manual is a rather awkward iterative process. I appreciate that the direct connection of Q5 and Q7 collectors looks rather suspect, but it does work. The question is, how well? The HA-3700 was spec'd at 0.05% THD at full output and 1 kHz, and the HA-4700 at 0.02% THD. These are not impressive figures and are easily surpassed by any sort of Blameless amplifier. The small number of models known to use the idea may indicate that Hitachi weren't too impressed themselves. The relevant US patent (4,345,215) was granted to Amada and Inoue in 1982.[37]

JVC 'Super-A' non-switching

The JVC non-switching system was referred to as 'Super-A' and was introduced in 1979, being applied to the A-X series of amplifiers; the A-X1 (1980, 30 W), A-X2 (1980, 40 W), A-X3 (1979, 55 W), A-X4 (1979, 60 W), A-X7 (1980, 90 W), and A-X9 (1979, 100 W). Due to an inexplicable oversight there was no A-X8. It was also applied to the M-7050 power amplifier (1979). By 1983 JVC were advertising 'Dynamic Super A-Amplification' which was clearly supposed to be an improvement to the concept; what that improvement was is at present obscure.

The additional Super-A circuitry was packaged is in a 9-pin SIL IC called the VC5022. The internal circuitry, which differs somewhat from that given in the Okabe US patent of 1981,[38] is disclosed in the service manual for the AX-70BK amplifier (1986). Two years later the AX-90VBK (1988) was using an upgraded IC labelled VC5022-2 IC. Although the basic number is the same, the internal circuitry is rather different. This chip was also used in the AX-F1GD amplifier as late as 1995.

A simplified version of the original Super-A circuit is given in Figure 4.19. When the output goes positive, there is an increased voltage drop across emitter resistor R14 due to the current flowing into the load, and therefore also at the base of Q2; this tends to turn on Q12, and increases the current through the mirror Q8, Q9. The collector current of Q8 tends to turn off Q6, increasing the total bias voltage, and so preventing the lower output device from turning off. The opposite effect occurs simultaneously in the lower half of the circuit, but its values are so arranged that this is insufficient to cancel out the action of the top half. The total bias voltage is therefore increased whenever the output voltage moves away from zero and thus the output devices do not turn off.

JVC appear to have persisted with the non-switching approach for longer than other manufacturers, still using it in 1995.

Pioneer non-switching

Pioneer non-switching technology appears to fall into two phases. In the first phase, the patent by Kawanabe in 1981[39] looks similar to the Hitachi and JVC approaches, dynamically altering the bias generator to prevent cutoff. The first amplifier to use it is believed to be the SA-7800 (1979) which was explicitly described as a 'Non-Switching Amplifier' and claimed an unimpressive 0.009% THD at 65 W. It was followed by the SA-9800 (1979) and the SA-8800 (1980). Other models were the A-9 amplifier (1980), described as having 'high-speed servo bias' which was a six-transistor version of the original four-transistor non-switching circuit. The A-5 and A-6 integrated amplifiers (1981) reverted to the four-transistor version, as did the A-8 (1981), and the A-60, A-70, and A-80 (all three 1983).

The A-5 and A-6 power amplifiers are actually most interesting designs. As well as the non-switching feature, they have four stages, with what they call 'Nested Feedback Loops' but which are better known as Cherry's 'Nested Differentiating Feedback Loops' (see Chapter 13). On top of all this, the global negative feedback loop implements the tone controls. The A-8 is a rare example of a five-stage amplifier, with three differential gain stages before the VAS; whether it uses NDFL is hard to say from looking at what can only be called an enigmatic schematic.

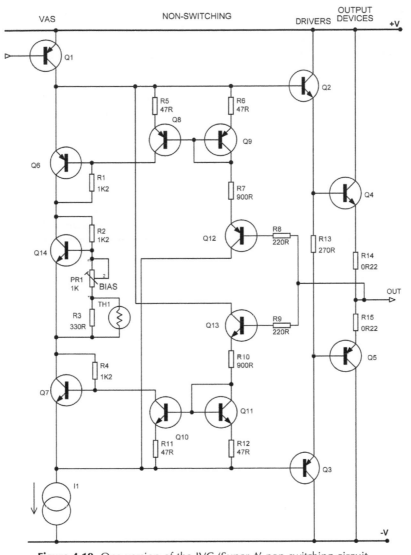

Figure 4.19. One version of the JVC 'Super-A' non-switching circuit.

The second Pioneer non-switching approach was based on a complex IC containing two 3-input opamps, (such things do exist: in this case, there is one inverting input and two non-inverting inputs per opamp), four controlled current-sources, and two voltage references. There was no discrete version. There is no space to explore its operation here, but the relevant US patents are 4,520,323[40] and 4,595,883[41] by Nakayama, both featuring two of those three-input opamps. The first products to which it was applied are believed to be the A-77X and A-88X integrated amplifiers (both 1985), both of which claimed 0.003% total harmonic distortion 20 Hz−20 kHz into 8 Ω. It would appear this system worked better than the first one, though 0.003% at LF is nothing special.

Models to which this technology was applied include the M90 amplifier (1986). It is simply described as 'Non-switching type II'. The M-91 (1988) also has it, though strangely the service manual does not mention non-switching at all, though the IC is plainly there on the schematic. It was also applied to the Pioneer A-91D (1987) and the A-616 (1989). The idea appears to have been dropped by 1991.

Sansui non-switching

Sansui brought to the party what they described as a 'super linear A-class circuit' based on a 1983 patent by Tanaka.[42] Once again the bias generator was dynamically modulated to avoid device cutoff. The method was described in a *JAES* paper,[43] but as for the Sansui 'Super Feedforward' plan described earlier, measurements were only given in the 20–100 kHz range, and the 20 kHz THD improvement is unexciting.

Unusually, the operation of the non-switching circuit is fully described in the service manual for the AU-D5 and AU-D7 (both 1981). The circuitry closely follows the patent. The basic circuit is shown in Figure 4.20. If the voltage drop across either output emitter resistor begins to fall too low, then either Q8 or Q9 will decrease conduction and increase the bias voltage.

Regrettably there appears to be a consensus that, in the AU-D5 at least, the non-switching circuit required extremely careful setting up or it would enter a latch-up state that would destroy the output devices.

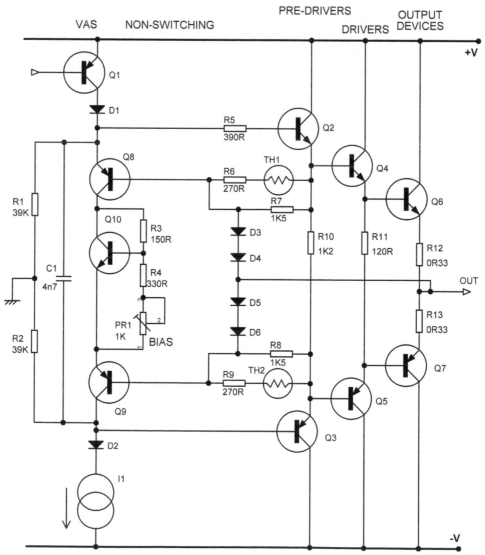

Figure 4.20. The Sansui 'super linear A-class' non-switching circuit. The value of the two thermistors is unknown.

Sony non-switching

Sony had their own version of non-switching, based on a patent by Manfred Schwarz,[44] who was R&D manager of Wega and also worked with Sony. It was used in the Sony ESPRIT amplifier range, specifically the TA-N900, TA-N901 and TA-N902 (all 1983). Details of the circuitry of these products have proved hard to come by; several implementations are shown in the patent and it is not currently known which were used in production. The general principle is once again that the bias voltage is dynamically modulated to stop the output devices turning off.

It is known that the TA-N900 used MOSFET output devices and was claimed to have 'non-switching type Class-A action' and also to use no global feedback. Another notable feature of the TA-N900 was the use of a heat pipe to connect the output devices and heatsinks.

Sony also had a technical feature called 'Legato Linear' or 'Super Legato Linear'. Since 'Legato' is a musical term meaning 'moving smoothly from note to note' you might think that referred to a non-switching system. However, it actually appears to mean the use of fast output transistors and 0.1 Ω output emitter resistors (lower than usual at the time) to reduce crossover distortion. I can warrant that the second feature will do some good. The feature seems to have been something of a reaction by Sony against the complicated non-switching biasing of other makers. Two amplifiers using it were the TA-F442E amplifier and the TA-N77ES (2002).

Technics non-switching

Between 1978 and 1987 Technics used several technical features. The most used was 'Synchronous bias' also called 'New Class-A' which was introduced with the V-series: V2, V4, V6 and the flagship V8 (1978). The V2 and V4 used STK output modules, but the V6 and V8 had discrete transistor outputs. In 1979/80 they were replaced with the V3, V5, V7, and flagship V9, all having discrete outputs with synchronous bias. (It occurs to me that asynchronous bias would probably not be a good idea.)

Later the range became V1X, V2X, V4X, V6X and V8X. The V1X introduced Technics' own hybrid output IC, the SVI2003. Later the V4X used the SVI2004A output IC, while the V6X and V8X had discrete outputs; all used synchronous bias. In 1986, the V7X and V10X were late additions to the synchro bias series. Other amplifiers in which it was used were the SU-V505, the SE-A3 K, and the SU-Z65.

The basic principle of synchronous bias is delightfully simple compared with some of the cross-coupled enigmas we saw above. Figure 4.21 shows a conventional VAS with fixed bias generator. The output stage is driven by the pre-driver stage Q3, Q4 via diodes D2 and D3. Diodes D6, D7 supply current to keep the relevant output device conducting when D2 or D3 turn off. D6, D7 are biased by diodes D4, D5; other versions had different methods of biasing D6, D7.

It is believed that D2, D3 and D6, D7 were germanium diodes to reduce the forward voltage drop, but this is so far unconfirmed. No relevant patent has been discovered so far.

Technics' 'New Class-A' should not be confused with their 'Class AA', which was introduced in 1986. It is an error-correction concept with separate voltage and current amplifiers, and is sometimes said to have been inspired by Sandman's Class S. Synchronous bias appears to have been dropped at this time.

Trio-Kenwood non-switching

Two US patents for non-switching technology were granted to Trio-Kenwood in 1982: 4,334,197[45] and 4,342,966[46]. It is not currently known if either version was used in production amplifiers; I have found no examples so far.

Yamaha non-switching

Yamaha did not want to be left out of the non-switching business. Their version was called Hyperbolic Conversion Amplification (HCA). The implication seems to be that two hyperbolic curves can be combined to give a straight line; this is not true mathematically as far as I am aware, though certainly two *parabolas* (square law curves) can be combined to result in a straight line. The relevant patent appears to be 4,803,441 by Noro,[47] which references an earlier patent by Yamaguchi,[48] and confirms that HCA is based on square law characteristics. More hyperbole than hyperbolic, I feel.

HCA was applied to the Yamaha MX-1000 amplifier in 1989; it was also applied to the MX-2000 (1988) though the details of the circuitry are rather different. A simplified version of the MX-1000 application is shown in Figure 4.22. A and B are sub-rails which are driven up and down with the output by C2 and C3. Biasing diodes D1, D2 set up a constant current in the current-mirror Q9, Q7, while D3, D4 do the same for current-mirror Q10, Q8. The signal is applied through the voltage amplifier, which has a low output impedance, and is converted from voltage to current by R15,

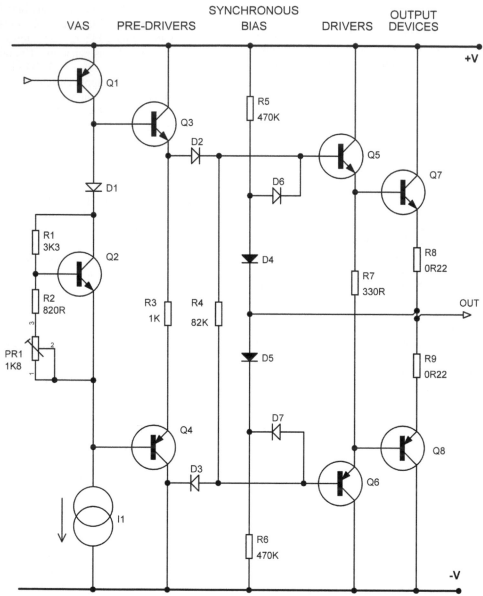

Figure 4.21. The principle of Technics 'Synchronous bias'. Diodes D, D prevent the output devices from turning off.

which sees a low impedance at the emitters of Q7 and Q8. The output current from mirror Q9, Q7 is bounced off the upper sub-rail A by mirror Q5. Q3, and likewise the output of mirror Q10, Q8 is bounced off the lower sub-rail B by mirror Q6, Q4. The output current from Q3 is fed to amplifier Q1, which is also fed with a version of the output signal, and Q1 controls driver Q11. Likewise for the lower half of the circuit. The output stage itself is a conventional Type II emitter-follower configuration.

The fundamental principle is that the product (not the sum) of the output currents from Q7 and Q8 is constant. Therefore, no matter how hard the circuit is driven, the smaller current never reaches zero, and therefore output devices never turn off.

Some of the Yamaha amplifiers with discrete HCA circuitry were the MX-630, MX-800, and MX-1000. The MX-1000 (260 W/8Ω) also had what Yamaha called Advanced Power Supply Circuitry (APS), which as far as I can tell from the schematic was

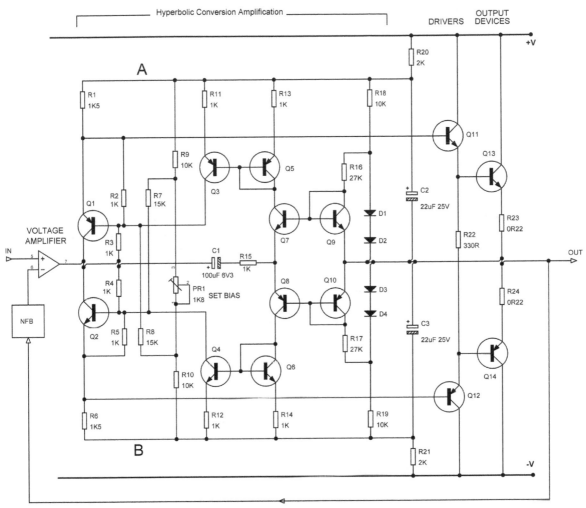

Figure 4.22. Simplified diagram of one version of the Yamaha Hyperbolic Conversion Amplification (HCA) circuit.

a form of Class-G. Later the HCA circuitry was incorporated in an IC called the BA3122 N, being applied to the MX-1 and MX-2 in 1993. The service manuals for these amplifiers give the internal circuit of the IC, revealing that it has four linked current-mirrors as shown in Figure 4.22, but offer no component values. The Yamaha AX-730, AX-930, AX-1050, and AX-1070 used the BA3122 N HCA IC and it is believed it was used in the AX-1090, but this is not so far confirmed.

Non-switching Conclusions

The fact that that non-switching technology quietly faded away after a few years, despite what was clearly a major effort by several manufacturers, seems to indicate that it was not satisfactory in practice. It is notable that in all the promotional literature, emphasis is laid on how it prevents *switching* distortion, i.e., that caused at HF by output devices turning off slowly, rather than on how crossover distortion in general is reduced. For the latter to be true, the non-switching action would have to make the two halves of Class-B conduction splice together in a better way than occurs with a fixed bias voltage, and I have yet to see any evidence that was achieved.

Geometric Mean Class-AB

This technique can be considered as a form of non-switching. In optimal Class-B operation there is a fairly sharp transfer of control of the output voltage between the two output devices, stemming from an equally abrupt switch in conduction from one to the other. This is not an inescapable result of the basic Class-B principle, but is caused by the presence of output emitter resistors, needed to improve quiescent-current stability and allow current-sensing for overload protection. It is these emitter resistances that to a large extent make classical Class-B what it is.

However, if the emitter resistors are omitted, and the stage biased with two matched diode junctions, then the diode and transistor junctions form a *translinear loop*[49] around which the junction voltages sum to zero. This links the two output transistor currents I_p, I_n in the relationship $l_n * l_p = $ constant, which in opamp practice is known as Geometric-Mean Class-AB operation. This gives smoother changes in device current at the cross-over point, but this does not necessarily mean lower THD. Such techniques are not very practical for discrete power amplifiers; first, in the absence of the very tight thermal coupling between the four junctions that exists in an IC, the quiescent-current stability will be atro-cious, with thermal runaway and spontaneous combus-tion a near-certainty. Second, the output device bulk emitter resistance will probably give enough voltage drop to turn the other device off anyway, when current flows. The need for drivers, with their extra junction-drops, also complicates things.

A new extension of this technique is to redesign the translinear loop so that $1/I_n + 1/I_p = $ constant, this being known as Harmonic-Mean AB operation.[50] It is too early to say whether this technique (assuming it can be made to work outside an IC) will be of use in reducing crossover distortion and thus improving amplifier performance.

The LT1166 Bias IC

The LT1166 is an integrated circuit made by Linear Technology. It is a bias generator intended to control the currents in a Class-B output stage using power MOSFET devices. The IC does not simply generate a fixed bias voltage, but senses the current in each output device by measuring the voltage drop across the source resistors (what in a bipolar transistor output would be the emitter resistors). A high-speed feedback control loop modifies the amount of drive applied to each power device to keep the product of the two currents in the two output devices constant; as one current increases, the other decreases. This relationship is the same as that described in the previous section, and is held over many decades of current, apparently also by the use of a translinear loop of transistor junc-tions. It is claimed that the LT1166 eliminates all

quiescent current adjustments and critical transistor matching. The Linear Tech data sheet quotes 0.3% THD at 10 kHz and 350 W/8 Ω.

I have no experience with this chip and cannot comment on how well it works, or if it can be usefully applied to bipolar output stages.

The Blomley Principle

A very different approach to preventing output transis-tors from turning off completely was introduced by Peter Blomley in 1971[51]; here the positive/negative splitting is done by circuitry ahead of the output stage, which can then be designed so that a minimum idling current can be separately set up in each output device. However, to the best of my knowledge, this approach has not achieved any commercial exploitation.

The circuit published in *Wireless World* consists of a single-transistor input stage (not ideal, if low distor-tion is the goal, see Chapter 6), followed by a VAS-like amplifying stage that drives two current sources in push-pull. Their combined output currents are applied to a complementary pair of the transistors connected as diodes; since these are current driven and they give a snappy and hopefully very precise turn-on and turn-off. The output stage also consists of two complementary voltage-controlled current sources (VCIS), each driven separately from the splitting section. A point that it often overlooked by those mentioning the Blomley principle in passing is the way that after the two halves of a cycle are split apart, they are then rejoined in the shape of *currents* delivered from the output stage into the load. The source impedance of the output stage without global feedback is therefore high, rather than low as it is with conventional stages that are basically emitter-followers. While applying feedback will certainly reduce the output impedance markedly, this unconven-tional kind of operation might cause problems under some loading conditions.

I have built Blomley amplifiers twice (way back in 1975) and on both occasions I found that there were still unwanted artefacts at the crossover point, and that transferring the crossover function from one part of the circuit to another did not seem to have achieved much. Possibly this was because the discontinuity was narrower than the usual crossover region and was there-fore linearised even less effectively by negative feed-back that reduces as frequency increases. I did not have the opportunity to investigate very deeply and this is not to be taken as a definitive judgement on the Blomley concept.

Ribbon Loudspeaker Amplifiers

Just as electrostatic loudspeakers present a much higher impedance than the standard 8 Ω, ribbon loudspeakers, which consist of a very thin metal sheet suspended in a magnetic field, present a much lower one. Both can be driven by transformers, but this has all the usual disadvantages. Ribbon loudspeakers typically have impedances of the order of 0.2−1 Ω, and so amplifiers to drive them directly must be specially designed to supply large amounts of current at low voltages. The large currents will cause more problems than usual with Large Signal Non-linearity (LSN) distortion in the output stage; see Chapter 10.

One of the very few designs published is that by Ton Giesberts in *Elektor* in 1992.[52] This uses a BJT output stage, ± 15 V supply rails, and many paralleled output emitter resistors to make up very low values. Only two output pairs are used but they are types with low beta-droop. The quoted distortion performance is good (0.004% at LF) even into a 0.4 Ω load.

Ribbon loudspeakers are held to give a good transient response because of the very low mass of the moving element, but this also means they have very little thermal capacity and so are very vulnerable to overload. A fairly sophisticated power output control system, probably based on modelling of the ribbon's thermal capacity, would seem to be a very good idea.

Power Amplifiers Combined with Tone-controls

An interesting variation in power amplifier design is to include tone-controls in the negative-feedback loop. This is, or was, relatively common in Japanese amplifiers. One example picked at random is the Yamaha AX-500 (1987) but there are many, many more. It would appear to make amplifier design, which is quite challenging enough already, a good deal more difficult because the power amplifier must both be stable and give an adequate distortion performance with a feedback actor that varies by 20 dB or so. The amount of money saved by eliminating a separate tone control stage is small, and I must admit I find the whole business puzzling. It is, however, a major theme in Japanese amplifier design.

Opamp Array Amplifiers

It occurred to me a long time ago that you could make an interesting power amplifier by connecting enough 5532s in parallel. The 5532 is not exactly a brand-new design, but it is a very capable device combining low noise and distortion with a good load-driving ability. An equally important point is that the 5532 is available at a remarkably low price − in fact it is usually the cheapest opamp you can buy, because of its wide use in audio applications.

It is therefore possible to build a very simple amplifier that consists of a large number of 5532 opamps with their outputs combined. Voltage-followers give maximum negative feedback to reduce distortion and use no passive feedback components which would also need to be multiplied. The excellent linearity, the power-supply rejection, and inbuilt overload protection of the 5532 are retained, reducing the external circuitry required to a minimum. The heat dissipation is spread out over many opamp packages, so no heatsink at all is required.

The obvious limitation with using opamps to drive loudspeakers is that the output voltage swing is limited, and using a single-ended array of 5532s gives only 15 Wrms into 8 Ω. This can be greatly extended by using two such amplifiers in bridge mode; one being driven with an inverted input signal so the voltage difference between the two amplifier outputs will be doubled, and power output quadrupled to about 60 Wrms into 8 Ω, enough for most domestic hi-fi situations. The block diagram of a bridged version is seen in Figure 4.23.

Each 15 W power amplifier consists of thirty-two 5532 dual opamps (i.e., 64 opamp sections) working as voltage-followers, with their outputs joined by 1 Ω current-sharing resistors. These combining resistors are outside the 5532 negative-feedback loops, but have little effect on the output impedance of the amplifier, as 64 times 1 Ω resistors in parallel gives an output impedance of only 0.0156 Ω. The wiring to the loudspeaker sockets will have more resistance than this. In bridged mode the number of opamps in parallel must be doubled to supply the extra load current; this presents no technical problems.

Since the power stage has unity gain, a preceding voltage amplifier with about 23 dB of gain is needed. I built this from 5532s to maintain the theme, and so I spread the gain over three stages to get more negative feedback and keep the distortion down; the first gain stage handles lower signal levels and so can carry a larger share of the gain. An LM4562 might be able to do the job in one stage. Overload protection is inherent in the opamps, but output relays are necessary for on/off muting and to protect loudspeakers against a DC fault.

The amplifier gives the low distortion expected from 5532s. There is no detectable crossover distortion. An

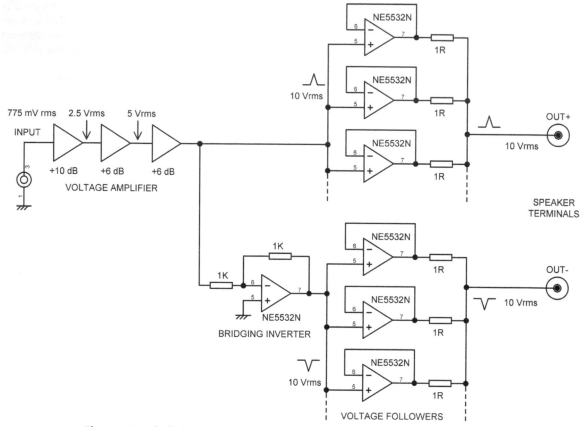

Figure 4.23. Block diagram of the bridged version of the 5532-array power amplifier.

LM4562-array amplifier could be built simply by swapping the opamps, and would give even better results, approaching the fabled Distortionless Power Amplifier. Unfortunately it would be an expensive item to construct.

My 5532-array design was published in *Elektor* in 2010,[53] and exhibited at the Burning Amp festival in San Francisco that year. It may not have been the first amplifier of its type, but I am pretty sure it is the first such design to be published.

Current-drive Amplifiers

Almost all power amplifiers aspire to be voltage sources of zero output impedance. This minimises frequency response variations caused by the peaks and dips of the impedance curve, and gives a universal amplifier that can drive any loudspeaker directly.

The opposite approach is an amplifier with a sufficiently high output impedance to act as a constant-current source. This eliminates some problems — such as rising

voice-coil resistance with heat dissipation — but introduces others, such as control of the cone resonance. Current amplifiers therefore appear to be only of use with active crossovers and velocity feedback from the cone.[54]

It is relatively simple to design an amplifier with any desired output impedance (even a negative one) and so any compromise between voltage and current drive is attainable. The inescapable snag is that loudspeakers are universally designed to be driven by voltage sources, and higher amplifier impedances demand tailoring to specific speaker types.[55]

Amplifier Bridging

When two power amplifiers are driven with anti-phase signals and the load connected between their outputs, with no connection to ground, this is called bridging. It is a convenient and inexpensive way to turn a stereo amplifier into a more powerful mono amplifier. It is called bridging because if you draw the four output transistors with the load connected between them, it looks something like the four arms of a Wheatstone bridge; see Figure 4.24. Doubling the voltage across a load of the same resistance naturally quadruples the output power — in theory. In harsh reality the available power will be considerably less, due to the power supply sagging and extra voltage losses in the two output stages. In most cases you will get something like three times the power rather than four, the ratio depending on how seriously the bridge mode was regarded when the initial design was done. It has to be said that in many designs the bridging mode looks like something of an afterthought.

In Figure 4.24 an 8 Ω load has been divided into two 4 Ω halves, to underline the point that the voltage at their centre is zero, and so both amplifiers are effectively driving 4 Ω loads to ground, with all that that implies for increased distortion and increased losses in the output stages. A unity-gain inverting stage is required to generate the anti-phase signal; nothing fancy is required and the simple shunt-feedback stage shown does the job nicely. I have used it in several products. The resistors in the inverter circuit need to be kept as low in values as possible to reduce their Johnson noise contribution, but not of course so low that the opamp distortion is increased by driving them; this is not too hard to arrange as the opamp will only be working over a small fraction of its voltage output capability, because the power amplifier it is driving will clip a long time before the opamp does. The capacitor assures stability — it causes a roll-off 3 dB down at 5 MHz, so it does not in any way imbalance the audio frequency response of the two amplifiers.

You sometimes see the statement that bridging reduces the distortion seen across the load because the push-pull action causes cancellation of the distortion products. In brief, it is not true. Push-pull systems can only cancel even-order distortion products, and in a well-found amplifier these are in short supply. In such an amplifier the input stage and the output stage will both be symmetrical, (it is hard to see why anyone would choose them to be anything else) and produce only odd-order harmonics, which will not be cancelled. The only asymmetrical stage is the VAS, and the distortion contribution from that is, or at any rate should be, very low. In reality, switching to bridging mode will almost certainly increase distortion, because as noted above, the output stages are now in effect driving 4 Ω loads to ground instead of 8 Ω.

Fractional Bridging

I will now tell you how I came to invent the strange practice of 'fractional bridging'. I was tasked with designing a two-channel amplifier module for a multi-channel unit. Five of these modules fitted into the chassis, and if each one was made independently bridge-able, you got a very flexible system that could be configured for anywhere between five and ten channels of amplification. The normal output of each amplifier was 85 W into 8 Ω, and the bridged output was about 270 W into 8 Ω as opposed to the theoretical 340 W. And now the problem. The next unit up in the product line had modules that gave 250 W into 8 Ω unbridged, and the marketing department felt that having the small modules giving more power than the large ones

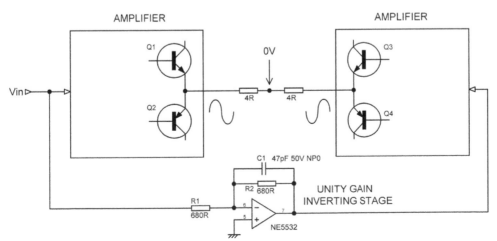

Figure 4.24. Bridging two power amplifiers to create a single more power amplifier.

was really not on; I'm not saying they were wrong. The problem was therefore to create an amplifier that only doubled its power when bridged. Hmm!

One way might have been to develop a power supply with deliberately poor regulation, but this implies a mains transformer with high-resistance windings that would probably have overheating problems. Another possibility was to make the bridged mode switch in a circuit that clipped the input signal before the power amplifiers clipped. The problem is that building a clipping circuit that does not exhibit poor distortion performance below the actual clipping level is actually surprisingly difficult — think about the non-linear capacitance of signal diodes. I worked out a way to do it, but it took up an amount of PCB area that simply wasn't available. So the ultimate solution was to let one of the power amplifiers do the clipping, which it does cleanly because of the high level of negative feedback, and the fractional bridging concept was born.

Figure 4.25 shows how it works. An inverter is still used to drive the anti-phase amplifier, but now it is configured with a gain G which is less than unity. This means that the in-phase amplifier will clip when the anti-phase amplifier is still well below maximum output, and the bridged output is therefore restricted. Double output power means an output voltage increased by root-two or 1.41 times, and so the anti-phase amplifier is driven with a signal attenuated by a factor of 0.41, which I call the bridging fraction, giving a total voltage swing across the load of 1.41 times. It worked very well, the product was a considerable success, and no salesmen were plagued with awkward questions about power output ratings.

There are two possible objections to this cunning plan, the first being that it is obviously inefficient compared with a normal Class-B amplifier. Figure 4.26 shows how the power is dissipated in the pair of amplifiers; this is derived from basic calculations and ignores output stage losses. PdissA is the power dissipated in the in-phase amplifier A, and varies in the usual way for a Class-B amplifier with a maximum at 63% of the maximum voltage output. PdissB is the dissipation in anti-phase amplifier B which receives a smaller drive signal and so never reaches its dissipation maximum; it dissipates more power because it is handling the same current but has more voltage left across the output devices, and this is what makes the overall efficiency low. Ptot is the sum of the two amplifier dissipations. The dotted lines show the output power contribution from each amplifier, and the total output power in the load.

The bridging fraction can of course be set to other values to get other maximum outputs. The lower it is, the lower the overall efficiency of the amplifier pair, reaching the limiting value when the bridging fraction is zero. In this (quite pointless) situation the anti-phase amplifier is simply being used as an expensive alternative to connecting one end of the load to ground, and so it dissipates a lot of heat. Figure 4.27 shows how the maximum efficiency (which always occurs at maximum output) varies with the bridging fraction. When it is unity, we get normal Class-B operation and the maximum efficiency is the familiar figure of 78.6%; when it is zero, the overall efficiency is halved to 39.3%, with a linear variation between these two extremes.

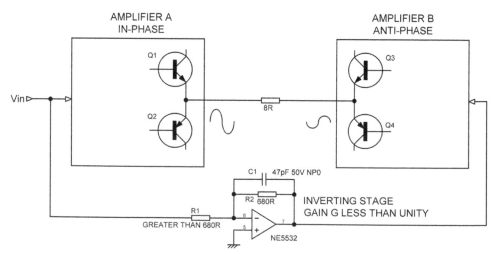

Figure 4.25. Fractional bridging of two power amplifiers to give doubled rather than quadrupled power output.

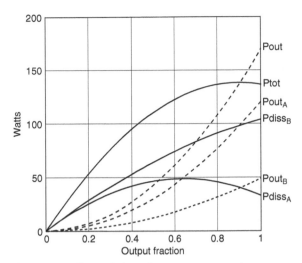

Figure 4.26. The variation of power output and power dissipation of two fractionally bridged power amplifiers, with a bridging fraction of 0.41 to give doubled rather than quadrupled power output.

The second possible objection is that you might think it is a grievous offence against engineering ethics to deliberately restrict the output of an amplifier for marketing reasons, and you might be right, but it kept people employed, including me. Nevertheless, given the current concerns about energy, perhaps this sort of thing should not be encouraged. Chapter 26 gives another example of devious engineering, where I describe how an input clipping circuit (the one I thought up in an attempt to solve this problem, in fact) can be used to emulate the performance of

a massive low-impedance power supply or a complicated regulated power supply. I have given semi-serious thought to writing a book called How to Cheat with Amplifiers.

Eliminating the Bridging Inverter

It is possible to use the second power amplifier to perform the bridging inversion itself, by using it in shunt feedback mode. This simplifies the circuitry and eliminates a possible cause of signal degradation. Figure 4.28 shows an effective way to do this; the system in shown in the unbridged (i.e., stereo) state. To convert to the bridged configuration SW1:A is closed, and the Right power amplifier now works as inverting stage with unity gain because R4 = R5. SW1:B grounds the non-inverting input of the Right power amplifier to keep out noise and hum from the unused (and probably unterminated) Right input. A vital point is that R3 remains connected to keep the noise gain and the feedback factor of the Right power amplifier almost the same on switching, and so HF stability is not affected. Fractional bridging could be implemented by raising the value of R5.

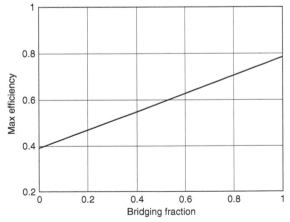

Figure 4.27. The variation of maximum efficiency of a fractionally bridged power amplifier with the bridging fraction.

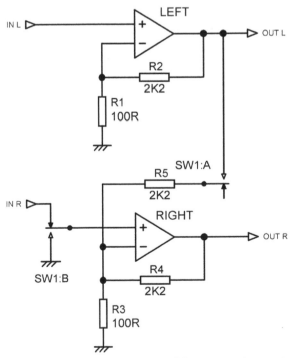

Figure 4.28. Using one power amplifier in inverting mode to eliminate the need for a bridging inverter.

Be aware of the possibility of left-to-right crosstalk through the inter-contact capacitance of switch SW1:A in the stereo configuration. A commercial example of this technique is seen in the Hitachi HMA-7500 Mk II (1980).

Increasing Bridging Reliability

Since a bridged amplifier consists of two separate amplifiers, it is possible to arrange matters so that if one fails, the other continues to supply signal to the load. The output is reduced by 6 dB (possibly more if you are using fractional bridging and you lose the amplifier with the greater gain) but very often this is a great deal more desirable than silence. Figure 4.29 shows two separate output relays, with their normally-closed contacts grounded. Should one amplifier fail, the offset-protection circuitry will open its relay, isolating it from the load, but the grounded back-contact gives a path to ground for the load current provided by the remaining amplifier.

It is of course essential to have separate protection and relay-control systems for the two power amplifiers.

AC- and DC-coupled Amplifiers

All power amplifiers are either AC-coupled or DC-coupled. The first kind have a single supply rail, with the output biased to be halfway between this rail and ground to give the maximum symmetrical voltage swing; a large DC-blocking capacitor is therefore used in series with the output. The second kind have positive and negative supply rails, and the output is biased to be

at 0 V, so no output DC-blocking is required in normal operation.

The Advantages of AC-coupling

1. The output DC offset is always zero (unless the output capacitor is leaky).
2. It is very simple to prevent turn-on thump by purely electronic means; there is no need for an expensive output relay. The amplifier output must rise up to half the supply voltage at turn-on, but providing this occurs slowly, there is no audible transient. Note that in many designs, this is not simply a matter of making the input bias voltage rise slowly, as it also takes time for the DC feedback to establish itself, and it tends to do this with a snap-action when a threshold is reached. The last AC-coupled power amplifier I designed (which was in 1980, I think) had a simple RC time-constant and diode arrangement that absolutely constrained the VAS collector voltage to rise slowly at turn-on, no matter what the rest of the circuitry was doing. Cheap but very effective.
3. No protection against DC faults is required, providing the output capacitor is voltage-rated to withstand the full supply rail. A DC-coupled amplifier requires an expensive and possibly unreliable output relay for dependable speaker protection.
4. The amplifier should be easier to make short-circuit proof, as the output capacitor limits the amount of electric charge that can be transferred each cycle, no matter how low the load impedance. This is speculative; I have no data as to how much it really helps in practice.

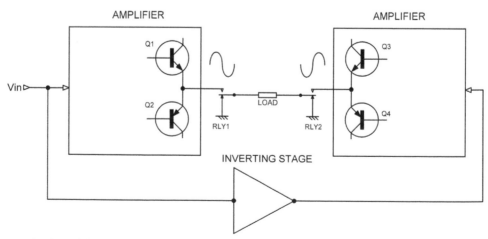

Figure 4.29. Bridged amplifiers with separate output relays to allow continued operation at reduced output if one amplifier fails.

5. AC-coupled amplifiers do not in general appear to require output inductors for stability. Large electrolytics have significant equivalent series resistance (ESR) and a little series inductance. For typical amplifier output sizes the ESR will be of the order of 100 mΩ; this resistance is probably the reason why AC-coupled amplifiers rarely had output inductors, as it is often enough resistance to provide isolation from capacitive loading and so gives stability. Capacitor series inductance is very low and probably irrelevant, being quoted by one manufacturer as 'A few tens of nanoHenrys'. The output capacitor was often condemned in the past for reducing the low-frequency damping factor (DF), for its ESR alone is usually enough to limit the DF to 80 or so. As explained above, this is not a technical problem because 'damping factor' means virtually nothing.

The Advantages of DC-coupling

1. No large and expensive DC-blocking capacitor is required. On the other hand the dual supply will need at least one more equally expensive reservoir capacitor, and a few extra components such as fuses.
2. In principle, there should be no turn-on thump, as the symmetrical supply rails mean the output voltage does not have to move through half the supply voltage to reach its bias point — it can just stay where it is. In practice, the various filtering time-constants

used to keep the bias voltages free from ripple are likely to make various sections of the amplifier turn on at different times, and the resulting thump can be substantial. This can be dealt with almost for free, when a protection relay is fitted, by delaying the relay pull-in until any transients are over. The delay required is usually less than a second.
3. Audio is a field where almost any technical eccentricity is permissible, so it is remarkable that AC-coupling appears to be the one technique that is widely regarded as unfashionable and unacceptable. DC-coupling avoids any marketing difficulties.
4. Some potential customers will be convinced that DC-coupled amplifiers give better speaker damping due to the absence of the output capacitor impedance. They will be wrong, as explained in Chapter 1, but this misconception has lasted at least 40 years and shows no sign of fading away.
5. Distortion generated by an output capacitor is avoided. This is a serious problem, as it is not confined to low frequencies, as is the case in small-signal circuitry; see Chapter 4. For a 6800 μF output capacitor driving 40 W into an 8 Ω load, there is significant mid-band third harmonic distortion at 0.0025%, as shown in Figure 4.30. This is at least five times more than the amplifier generates in this part of the frequency range. In addition, the THD rise at the LF end is much steeper than in the small-signal case, for reasons that are not yet clear. There

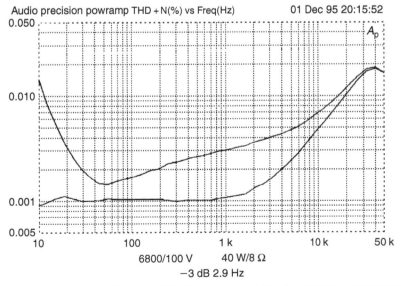

Figure 4.30. The extra distortion generated by an 6800 μF electrolytic delivering 40 W into 8 Ω. Distortion rises as frequency falls, as for the small-signal case, but at this current level there is also added distortion in the mid-band.

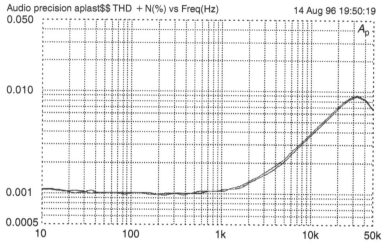

Figure 4.31. Distortion with and without a very large output capacitor, the BHC Aerovox 100,000 μF/40 V (40 watts/8 Ω). Capacitor distortion is eliminated.

are two cures for output capacitor distortion. The straightforward approach uses a huge output capacitor, far larger in value than required for a good low-frequency response. A 100,000μF/40 V Aerovox from BHC eliminated all distortion, as shown in Figure 4.31. An allegedly 'audiophile' capacitor gives some interesting results; a Cerafine Supercap of only moderate size (4700μF/63 V) gave Figure 4.32, where the mid-band distortion is gone, but the LF distortion rise remains. What special

audio properties this component is supposed to have are unknown; as far as I know, electrolytics are never advertised as 'low mid-band THD', but that seems to be the case here. The volume of the capacitor case is about twice as great as conventional electrolytics of the same value, so it is possible the crucial difference may be a thicker dielectric film than is usual for this voltage rating.

Either of these special capacitors costs more than the rest of the amplifier electronics put

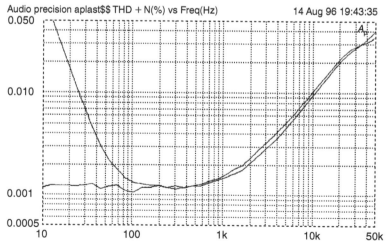

Figure 4.32. Distortion with and without an 'audiophile' Cerafine 4700 μF/63V capacitor. Mid-band distortion is eliminated but LF rise is much the same as the standard electrolytic.

together. Their physical size is large. A DC-coupled amplifier with protective output relay will be a more economical option.

A little-known complication with output capacitors is that their series reactance increases the power dissipation in the output stage at low frequencies. This is counter-intuitive as it would seem that any impedance added in series must reduce the current

drawn and hence the power dissipation. Actually the dominant factor is the load phase-shift and this increases the amplifier dissipation.

6. The supply currents can be kept out of the ground system. A single-rail AC amplifier has half-wave Class-B currents flowing in the 0 V rail, and these can have a serious effect on distortion and crosstalk performance.

References

1. Linsley-Hood, J, Simple Class-A Amplifier, *Wireless World*, April 1969, p. 148.

2. Olsson, B, Better Audio from Non-Complements? *Electronics World*, Dec. 1994, p. 988.

3. Lohstroh, J and Otala, M, An Audio Power Amplifier for Ultimate Quality Requirements, *IEEE Transactions on Audio & Electroacoustics*, AU-21, No. 6, Dec. 1973, pp. 545–551.

4. Sano, N et al., A High-Efficiency Class-A Audio Amplifier.' Preprint #1382 for 61st AES convention, Nov 1978.

5. Stochino, G, Audio Design Leaps Forward? *Electronics World*, Oct. 1994, p. 818.

6. Attwood, B, Design Parameters Important for the Optimisation of PWM (Class-D) Amplifiers *JAES*, 31, Nov. 1983, p. 842.

7. Goldberg, J and Sandler, M, Noise Shaping and Pulse-Width Modulation for All-Digital Audio Power Amplifier, *JAES*, 39, Feb. 1991, p. 449.

8. Hancock, J A, Class-D Amplifier Using MOSFETS with Reduced Minority Carrier Lifetime, *JAES*, 39, Sept. 1991, p. 650.

9. Peters, A, Class E RF Amplifiers, *IEEE J. Solid-State Circuits*, June 1975, p. 168.

10 Kee, S, http://www.rfic.eecs.berkeley.edu/~niknejad/ee242/pdf/eecs242_class_EF_PAs.pdf, Excerpt from PhD dissertation (accessed Oct. 2012).

11. Feldman, L, Class-G High-Efficiency Hi-Fi Amplifier, *Radio-Electronics,* Aug. 1976, p. 47.

12. Raab, F, Average Efficiency of Class-G Power Amplifiers, *IEEE Transactions on Consumer Electronics*, Vol CE-22, May 1986, p. 145.

13. Nakagaki, H, et al., *Audio Frequency Power Amplifier*, US patent 4,100, 501, July 1978.

14. Sampei, T et al., Highest Efficiency & Super Quality Audio Amplifier Using MOS-Power FETs in Class-G, *IEEE Transactions on Consumer Electronics*, Vol CE-24 Aug.1978, p. 300.

15. Buitendijk, P A, 40 W Integrated Car Radio Audio Amplifier, paper presented at IEEE Conference on Consumer Electronics, 1991 Session THAM 12.4, p. 174 (Class-H).

16. Sandman, A, Class S: A Novel Approach to Amplifier Distortion, *Wireless World*, Sept. 1982, p. 38.

17. Sinclair, R (ed.), *Audio and Hi-Fi Handbook*, Oxford: Newnes, 1993, p. 541.

18. Walker, P and Albinson, M, Current Dumping Audio Amplifier, paper presented at 50th AES Convention, London, March 1975.

19. Walker, P J, Current Dumping Audio Amplifier, *Wireless World*, Dec. 1975, p. 560.

20. Anon, Edwin Amplifier *Elektor*, Sept. 1975, p. 910.

21. Linsley-Hood, J, The Straight Wire with Gain? *Studio Sound*, April 1975, p. 22.

22. Walker, P and Albinson, M, *Distortion-free Amplifiers*, US patent 3,970,953, July 1976.

23. Vanderkooy, J and Lipshitz, S P, Current-Dumping: Does It Really Work? *Wireless World*, June 1978, pp. 38—40.

24. McLoughlin, M, Current Dumping Review- 1, *Wireless World*, Sept. 1983, pp. 39—43.

25. McLoughlin, M, Current Dumping Review- 2, *Wireless World*, Oct. 1983, pp. 35—41.

26. Baxandall, P, Current Dumping Audio Amplifier, *Wireless World*, July 1976, pp. 60—61.

27. Black, H, *Translating System*, US patent 2,245,598, June 1941.

28. Hawksford, M, Distortion Correction in Audio Power Amplifiers, *JAES*, 29(1/2), Jan./Feb. 1980, pp. 503—510.

29. Cordell, R A, MOSFET power amplifier with error-correction, *JAES* 32 No 1/2, Jan/Feb. 1984, pp. 5—17.

30. Didden, J, paX- a power amplifier with error correction *Elektor*, April, p. 24—32 May 2008

31. Danyuk and Pilko Error Correction in Audio Amplifiers, *JAES* 44 No 9 Sept. 1996, pp. 721—728.

32. Danyuk, D G and Renardson, M J, Error Correction in Class AB Power Amplifiers, Paper presented at 114th AES Convention, Amsterdam, Mar. 2003.

33. Sansui, *The End of Distortion Debates: Sansui Super Feedforward System*, Brochure. Date unknown.

34. Takahashi, S and Tanaka, S, Design and Construction of a Feedforward Error-correction Amplifier *JAES* 29(1/2), Jan./Feb. 1981, pp. 31—47.

35. Tanaka, S, *Distortion Correction Circuit for a Power Amplifier*. US patent, 4,367,442, Jan. 1983. Assigned to Sansui.

36. Pass, N, *Active Bias Circuit for Operating Push-pull Amplifiers in Class-A Mode*, US patent 3,995,228, Nov. 1976. Assigned to Threshold Corporation.

37. Amada, N and Inoue, S, *Audio Frequency Power Amplifier Circuit*, US patent 4,345,215, Aug. 1982. Assigned to Hitachi.

38. Okabe, Y, *Single Ended Push Pull Amplifier*, US patent 4,274,059. June 1981. Assigned to Victor (JVC).

39. Kawanabe, Y, *Push-Pull Amplifier Circuit*, US patent 4,254,379, Mar. 1981 Assigned to Pioneer.

40. Nakayama, K, *Emitter-Follower Type SEPP Circuit*, May 1985, US patent 4,520,323, Assigned to Pioneer.

41. Nakayama, K, *Emitter-Follower Type Single-Ended Push-Pull Circuit*, US patent 4,595,883, June 1986. Assigned to Pioneer.

42. Tanaka, S, *Bias Circuit for Use in a Single Ended Push-Pull Circuit*, US patent 4,401,951, Aug. 1983. Assigned to Sansui.

43. Tanaka, S, New Biasing Circuit for Class-B Operation, *JAES*, 29(3), Mar. 1981, pp. 148—152.

44. Schwarz, M, *Biasing Circuit for Power Amplifier*, US patent 4,439,743, Mar. 1984, (Filed Apr 1981).

45. Otao, K, *Power Amplifier Circuitry,* US patent 4,334,197, June 1982. Assigned to Trio-Kenwood.

46. Tamura, E, *Power Amplifier Circuitry*, US patent 4,342,966, Aug. 1982. Assigned to Trio-Kenwood.

47. Noro, M, *Amplifying Circuit*, US patent 4,803,441, Feb. 1989. Assigned to Yamaha.

48. Yamaguchi, H, *Power Amplifier* , US patent 4,404,528, Feb. 1983.

49. Gilbert, B, Current Mode Circuits from a Translinear Viewpoint, Ch 2, in *Analogue IC Design: The Current-Mode Approach*, ed. Toumazou, C, Lidgey, J and Haigh, D, London: IEE, 1990.

50. Thus Compact Bipolar Class AB Output Stage, *IEEE Journal of Solid-State Circuits*, Dec. 1992, p. 1718.

51. Blomley, P A, New Approach to Class-B, *Wireless World*, Feb. 1971, p. 57.

52. Giesberts, T, Output Amplifier for Ribbon Loudspeakers, *Elektor*, Nov. 1992, pp. 22–25.

53. Self, D, The 5532 OpAmplifier, *Elektor*, Oct. and Nov. 2010.

54. Mills, P G and Hawksford, M, Transconductance Power Amplifier Systems for Current-Driven Loudspeakers, *JAES* 37, March 1989, p. 809.

55. Evenson, R, *Audio Amplifiers with Tailored Output Impedances*, Preprint for Nov. 1988 AES Convention, Los Angeles.

Further reading

Baxandall, P, Audio Power Amplifier Design: Part 5, *Wireless World*, Dec 1978, p. 53.

This superb series of articles had six parts and ran on roughly alternate months, starting in Jan. 1978. A full reproduction can be found in *Baxandall & Self on Audio Power*, published by Linear Audio in 2011.

Chapter 5

General Principles and Distortion Mechanisms

Gain and Feedback in the Three-stage Amplifier

Figure 5.1 shows a very conventional power amplifier circuit; it is as standard as possible. A great deal has been written about this configuration, though the subtlety and quiet effectiveness of the topology are usually overlooked, and the explanation below therefore touches on several aspects that seem to be almost unknown. The circuit has the merit of being docile enough to be made into a functioning amplifier by someone who has only the sketchiest of notions as to how it works.

The input differential pair implements one of the few forms of distortion cancellation that can be relied upon

to work reliably without adjustment — this is because the transconductance of the input pair is determined by the physics of transistor action rather than matching of ill-defined parameters such as beta; the logarithmic relation between Ic and Vbe is proverbially accurate over some eight or nine decades of current variation.

The voltage signal at the Voltage-Amplifier Stage (hereafter VAS) transistor base is typically a couple of milliVolts, looking rather like a distorted triangle wave. Fortunately the voltage here is of little more than academic interest, as the circuit topology essentially consists of a transconductance amp (voltage-difference input to current output) driving into a transresistance (current-to-voltage converter) stage. In the first case the exponential

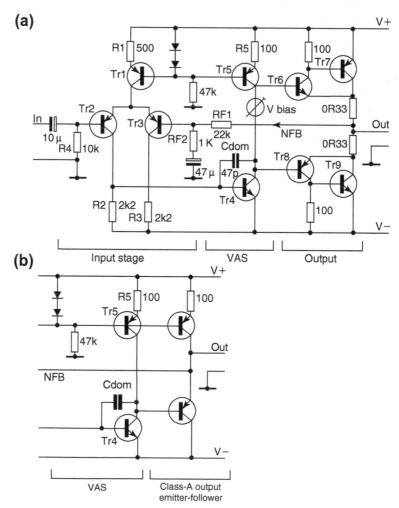

Figure 5.1. (a) A conventional Class-B power amp circuit; (b) with small-signal Class-A output emitter-follower replacing Class-B output to make a model amplifier.

Vbe/lc law is straightened out by the differential-pair action, and in the second, the global (overall) feedback factor at LF is sufficient to linearise the VAS, while at HF shunt Negative Feedback (hereafter NFB) through Cdom conveniently takes over VAS-linearisation while the overall feedback factor is falling.

The behaviour of Miller dominant-pole compensation in this stage is actually exceedingly elegant, and not at all a case of finding the most vulnerable transistor and slugging it. As frequency rises and Cdom begins to take effect, negative feedback is no longer applied globally around the whole amplifier, which would include the higher poles, but instead is seamlessly transferred to a purely local role in linearising the VAS. Since this stage effectively contains a single gain transistor, any amount of NFB can be applied to it without stability problems.

An amplifier with dominant-pole compensation operates in two regions; the LF, where open-loop (o/l) gain is substantially constant, and HF, above the dominant-pole breakpoint, where the gain is decreasing steadily at 6 dB/octave. This is shown in Figure 5.2, with realistic data for the gains and frequencies. In this section, all the LF gains given are illustrative and subject to some variation.

It may not be obvious, but the predictable part of the plot is the 6 dB/octave slope as gain falls with frequency. This is very convenient as it is this section that determines the closed-loop stability. The gain at any frequency in this region is, with beautiful simplicity:

$$HFgain = \frac{g_m}{\omega \cdot C_{dom}} \qquad \text{Equation 5.1}$$

Where $\omega = 2 \cdot \pi \cdot frequency$

The open-loop gain depends only on the input stage transconductance g_m, and the Miller dominant-pole capacitance C_{dom}. This is the basis of the dependable stability of the classic three-stage amplifier. As the frequency falls, there comes a point at which there is no longer enough open-loop gain to allow the value of C_{dom} to control it, and so it levels out. The LF gain is:

$$LFgain = g_m \cdot \beta \cdot R_c \qquad \text{Equation 5.2}$$

The open-loop gain now no longer depends on Cdom, but on the beta of the VAS transistor and its collector impedance Rc. Neither of these quantities are well controlled and so the LF open-loop gain of the amplifier is to a certain extent a matter of pot-luck; fortunately this does not matter, so long as it is high enough to give a suitable level of NFB to effectively eliminate LF distortion. The use of the word *eliminate* is deliberate, as will be seen later. The LF gain, or HF

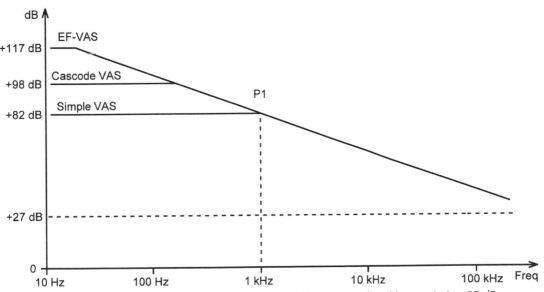

Figure 5.2. How amplifier open-loop gain varies with frequency. Closed-loop gain is +27 dB.

local feedback-factor is improved over that given by a simple resistive collector load by increasing the effective value of R_c, either by the use of a current-source collector-load, or bootstrapping, so a simple one-transistor VAS gives the trace that starts at +82 dB in Figure 5.2. This is a typical value of LF gain, though, as noted, it is subject to variation. The intersection of the LF and HF gain regions is the dominant-pole frequency P1, which can be calculated thus:

$$P1 = \frac{1}{C_{dom} \cdot \beta \cdot R_c}$$ Equation 5.3

With a simple VAS, P1 will be in the area of 1 to 2 kHz.

Figure 5.2 shows the open-loop gain for the simple VAS levelling out at +82 dB from 1 kHz downwards. The horizontal dotted line represents the closed-loop gain, which is +27 dB as for all the amplifier examples in this book. The vertical distance between that line and the open-loop line is the feedback factor in dB; this determines the amount of distortion reduction that occurs. The simple VAS can only provide $82 - 27 = 55$ dB of feedback in the LF region, and this is not enough to make the LF distortion invisible below the noise.

The amount of LF gain can be increased, and the P1 frequency lowered, by increasing either beta or R_c in Equation 5.2. Adding an emitter-follower inside the Miller compensation loop effectively multiplies the VAS beta by the EF beta, as in a Darlington connection, and gives the upper gain trace, which now flattens out at the much higher value of +117 dB, and at the much lower frequency of 20 Hz or so. The extra feedback factor gained by using an EF-VAS rather than a simple VAS is 4 dB at 1 kHz, 21 dB at 100 Hz, and 34 dB at 10 Hz. It thus gives some useful distortion reduction at 1 kHz, and much more at lower frequencies. It is, however, hard to quantify this as the added EF in the EF-VAS has two roles; to increase the LF gain, and also eliminate distortion caused by local feedback though the non-linear Vbc of the VAS transistor. It is not easy to separate these effects when examining midband distortion.

Alternatively, R_c can be increased by cascoding the VAS transistor, eliminating the Early effect that limits its voltage gain. The improvement in LF open-loop gain is less, due to the loading effect of the output stage on the VAS collector, as can be seen in Figure 5.2, and it is more complicated than the EF-VAS approach.

This is a brief summary of the fairly subtle operation of a three-stage amplifier. Much more detail on input

stages, the VAS, and compensation, is given in Chapters 6, 7, and 13 respectively.

In the HF region, things are more difficult as regards distortion, for while the VAS is locally linearised, the global feedback-factor available to linearise the input and output stages is falling steadily at 6 dB/octave. For the three-stage amplifier that is the main focus of this book, much experience has shown that a feedback factor of 30 dB at 20 kHz will assure stability with practical loads and component variations, and give an excellent distortion performance. The HF gain, and therefore both HF distortion and stability margin, are set by the simple combination of the input stage transconductance and one capacitor, and most components have no effect on it at all.

It is often said that the use of a high VAS collector impedance provides a current drive to the output devices, often with the implication that this somehow allows the stage to skip quickly and lightly over the dreaded crossover region. This is a misconception − the collector impedance falls to a few kilohms at HF, due to increasing local feedback through Cdom, and in any case it is very doubtful if true current drive would be a good thing − calculation shows that a low-impedance voltage drive minimises distortion due to beta-unmatched output halves,[1] and it certainly eliminates the effect of Distortion 4, described below.

The Advantages of the Conventional

It is probably not an accident that the generic three-stage configuration is by a long way the most popular, though in the uncertain world of audio technology it is unwise to be too dogmatic about this sort of thing. The generic configuration has several advantages over other approaches:

- The input pair not only provides the simplest way of making a DC-coupled amplifier with a dependably small output offset voltage, but can also (given half a chance) completely cancel the second-harmonic distortion which would be generated by a single-transistor input stage. One vital condition for this must be met; the pair must be accurately balanced by choosing the associated components so that the two collector currents are equal. (The *typical* component values shown in Figure 5.1 do *not* bring about this most desirable state of affairs.)
- The input devices work at a constant and near-equal Vce, giving good thermal balance.
- The input pair has virtually no voltage gain so no low-frequency pole can be generated by the Miller effect in

the TR2 collector-base capacitance. All the voltage gain is provided by the VAS stage, which makes for easy compensation. Feedback through Cdom lowers VAS input and output impedances, minimising the effect of input-stage capacitance, and the output-stage capacitance. This is often known as pole-splitting;[2] the pole of the VAS is moved downwards in frequency to become the dominant pole, while the input-stage pole is pushed up in frequency.

- The VAS Miller compensation capacitance smoothly transfers NFB from a global loop that may be unstable, to the VAS local loop that cannot be. It is quite wrong to state that *all* the benefits of feedback are lost as the frequency increases above the dominant pole, as the VAS is still being linearised.

The Distortion Mechanisms

My original series of articles on amplifier distortion listed seven important distortion mechanisms, all of which are applicable to any Class-B amplifier, and do not depend on particular circuit arrangements. As a result of further experimentation and further thought, I have now increased this to ten.

In the typical amplifier THD is often thought to be simply due to the Class-B nature of the output stage, which is linearised less effectively as the feedback factor falls with increasing frequency. This is, however, only true when all the removable sources of distortion have been eliminated. In the vast majority of amplifiers in production, the true situation is more complex, as the small-signal stages can generate significant distortion of their own, in at least two different ways; this distortion can easily exceed output stage distortion at high frequencies. It is particularly inelegant to allow this to occur given the freedom of design possible in the small-signal section.

If all the ills that a Class-B stage is heir to are included, then there are eleven major distortion mechanisms. Note that this assumes that the amplifier is not suffering from:

- Overloading not affecting protection circuitry (for example, insufficient current to drive the output stage due to a VAS current source running set to too low a value).
- Slew-rate limiting. (Unlikely.)
- Defective or out-of-tolerance components.

It also assumes the amplifier has proper global or Nyquist stability and does not suffer from any parasitic oscillations; the latter, if of high enough frequency, cannot be seen on the average oscilloscope and tend to manifest themselves only as unexpected increases in distortion, sometimes at very specific power outputs and frequencies.

In Figure 5.3 an attempt has been made to show the distortion situation diagrammatically, indicating the location of each mechanism within the amplifier. Distortion 8 is not shown as there is no output capacitor.

The first four distortion mechanisms are inherent to any three-stage amplifier.

Distortion One: Input Stage Distortion

Non-linearity in the input stage. If this is a carefully balanced differential pair, then the distortion is typically only measurable at HF, rises at 18 dB/octave, and is almost pure third harmonic. If the input pair is unbalanced (which from published circuitry it usually is), then the HF distortion emerges from the noise floor earlier, as frequency increases, and rises at 12 dB/octave as it is mostly second harmonic.

This mechanism is dealt with in Chapter 6.

Distortion Two: VAS Distortion

Non-linearity in the simple Voltage-Amplifier Stage (VAS) composed of one transistor is due to Early effect at LF and non-linear local feedback through the transistor Cbc at HF. Both effects can be much reduced by either adding an emitter-follower inside the VAS compensation loop, or cascoding the VAS transistor. VAS distortion is reduced at HF by local feedback through the Miller dominant pole capacitor. Hence if you crank up the local VAS open-loop gain, for example, by cascoding or putting more current-gain in the local VAS-Cdom loop, and attend to Distortion Four below, you can usually ignore VAS distortion.

This mechanism is dealt with in Chapters 7 and 8.

Distortion Three: Output Stage Distortion

Non-linearity in the output stage, which is naturally the obvious source. This in a Class-B amplifier will be a complex mix of large-signal distortion and crossover effects, the latter generating a spray of high-order harmonics, and in general rising at 6 dB/octave as the amount of negative feedback decreases. Large-signal THD worsens with 4 Ω loads and worsens again at 2 Ω. The picture is complicated by dilatory switch-off in the relatively slow output devices, ominously signalled by supply current increasing in the top audio octaves.

These mechanisms are dealt with in Chapters 9 and 10.

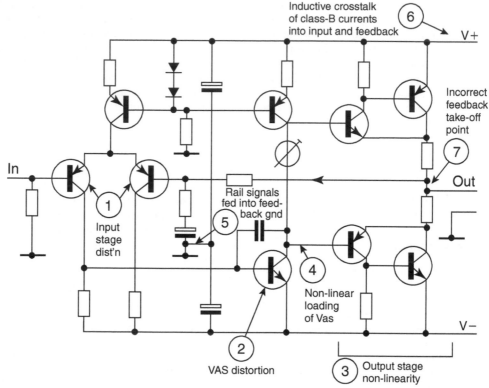

Figure 5.3. The location of the first seven major distortion mechanisms. The eighth (capacitor distortion) is omitted for clarity.

Distortion Four: VAS Loading Distortion

Loading of the VAS by the non-linear input impedance of the output stage. When all other distortion sources have been attended to, this can be the limiting distortion factor at LF (say, below 2 kHz); it is simply cured by buffering the VAS from the output stage. Magnitude is essentially constant with frequency, though overall effect in a complete amplifier becomes less as frequency rises and feedback through Cdom starts to linearise the VAS.

This mechanism is dealt with in Chapter 7.

The next three distortion mechanisms are in no way inherent; they may be reduced to unmeasurable levels by simple precautions. They are what might be called topological distortions, in that they depend wholly on the arrangement of wiring and connections, and on the physical layout of the amplifier:

Distortion Five: Rail Decoupling Distortion

Non-linearity caused by large rail decoupling capacitors feeding the distorted signals on the supply lines into the signal ground. This seems to be the reason that many

amplifiers have rising THD at low frequencies. Examining one commercial amplifier kit, I found that rerouting the decoupler ground-return reduced the THD at 20 Hz by a factor of three.

This mechanism is dealt with in Chapter 11.

Distortion Six: Induction Distortion

This is non-linearity caused by induction of Class-B supply currents into the output, ground, or negative-feedback lines. This was highlighted by Cherry[3] but seems to remain largely unknown; it is an insidious distortion that is hard to remove, though when you know what to look for on the THD residual, it is fairly easy to identify. I suspect that a large number of commercial amplifiers suffer from this to some extent.

This mechanism is dealt with in Chapter 11.

Distortion Seven: NFB Takeoff Distortion

This is non-linearity resulting from taking the NFB feed from slightly the wrong place near where the power-transistor Class-B currents sum to form the output. This may well be another very prevalent defect.

This mechanism is dealt with in Chapter 11.

The next two distortion mechanisms relate to circuit components that are non-ideal or poorly chosen.

Distortion Eight: Capacitor Distortion

In its most common manifestation it is caused by the non-ideal nature of electrolytic capacitors. It rises as frequency falls, being strongly dependent on the signal voltage across the capacitor. The most common sources of non-linearity are the input DC-blocking capacitor or the feedback network capacitor; the latter is more likely as it is much easier to make an input capacitor large enough to avoid the problem. It causes serious difficulties if a power amplifier is AC-coupled, i.e., has a series capacitor at the output, but this is rare these days.

It can also occur in ceramic capacitors that are nominally of the NPO/COG type but actually have a significant voltage coefficient, when they are used to implement Miller dominant pole compensation.

This mechanism is dealt with in detail in Chapter 11.

Distortion Nine: Magnetic Distortion

This arises when a signal at amplifier output level is passed through a ferromagnetic conductor. Ferromagnetic materials have a non-linear relationship between the current passing through them and the magnetic flux it creates, and this induces voltages that add distortion to the signal. The effect has been found in output relays, and also speaker terminals. The terminals appeared to be made of brass but were actually plated steel.

This mechanism is also dealt with in detail in Chapter 11.

Distortion Ten: Input Current Distortion

This distortion is caused when an amplifier input is driven from a significant source impedance. The input current taken by the amplifier is non-linear, even if the output of the amplifier is distortion-free, and the resulting voltage-drop in the source impedance introduces distortion.

This mechanism is purely a product of circuit design, rather than layout or component integrity, but it has been put in a category of its own because, unlike the inherent Distortions One to Four, it is a product of the interfacing between the amplifier and the circuitry upstream of it.

This mechanism is dealt with in Chapter 6.

Distortion Eleven: Premature Overload Protection Distortion

The overload protection of a power amplifier can be implemented in many ways, but without doubt the most popular method is the use of VI limiters that shunt signal current away from the inputs to the output stage. In their simplest and most common form, these come into operation relatively gradually as their set threshold is exceeded, and introduce distortion into the signal long before they close it down entirely. It is therefore essential to plan a sufficient safety margin into the output stage so that the VI limiters are never near activation in normal use. Other methods of overload protection that trigger and then latch the amplifier into a standby state cannot generate this distortion, but if this leads to repeated unnecessary shutdowns, it will be a good deal more annoying than occasional distortion.

This issue is examined more closely in Chapter 24.

Non-existent or Negligible Distortions

Having set down what might be called The Eleven Great Distortions, we must pause to put to flight a few paper tigers …

The first is common-mode distortion in the input stage, a spectre that haunts the correspondence columns. Since it is fairly easy to make an amplifier with less than 0.00065% THD (1 kHz) without paying any attention at all to this issue, it cannot be too serious a problem. It is perhaps a slight exaggeration to call it non-existent, as under special circumstances it can be seen, but it is certainly unmeasurable under normal circumstances.

If the common-mode voltage on the input pair is greatly increased, then a previously negligible distortion mechanism is indeed provoked. This increase is achieved by reducing the C/L gain to between 1 and 2 times; the input signal is now much larger for the same output, and the feedback signal must match it, so the input stage experiences a proportional increase in common-mode voltage.

The distortion produced by this mechanism increases as the square of the common-mode voltage, and therefore falls rapidly as the closed-loop gain is increased back to normal values. It therefore appears that the only precautions required against common-mode distortion are to ensure that the closed-loop gain is at least five times (which is no hardship, as it almost certainly is anyway) and to use a tail current-source for the input pair, which again is standard practice. This issue is dealt with in more detail in Chapter 6 on power amplifier input stages.

The second distortion conspicuous by its absence in the list is the injection of distorted supply-rail signals directly into the amplifier circuitry. Although this putative mechanism has received a lot of attention,[4] dealing with Distortion Five above by proper grounding seems to be all that is required; once more, if triple-zero THD can be attained using simple unregulated supplies and without paying any attention to the Power Supply Rejection Ratio beyond keeping the amplifier free from hum (which it reliably can be), then there seems to be no problem. There is certainly no need for regulated supply rails to get a good performance. PSRR does need careful attention if the hum/noise performance is to be of the first order, but a little RC filtering is usually all that is needed. This topic is dealt with in Chapter 26.

A third mechanism of very doubtful validity is thermal distortion, allegedly induced by parameter changes in semiconductor devices whose instantaneous power dissipation varies over a cycle. This would surely manifest itself as a distortion rise at very low frequencies, but it simply does not happen. There are several distortion mechanisms that can give a THD rise at LF, but when these are eliminated, the typical distortion trace remains flat down to at least 10 Hz. The worst thermal effects would be expected in Class-B output stages where dissipation varies wildly over a cycle; however, drivers and output devices have relatively large junctions with high thermal inertia. Low frequencies are of course also where the NFB factor is at its

maximum. This contentious issue is dealt with at greater length in Chapter 9.

To return to our list of distortion mechanisms, note that only Distortion Three is directly due to O/P stage non-linearity, though Distortions Four to Seven all result from the Class-B nature of the typical output stage. Distortions Eight, Nine and Ten can happen in any amplifier, whatever its operating class.

The Performance of a Standard Amplifier

The THD curve for the standard amplifier is shown in Figure 5.4. As usual, distortion increases with frequency, and as we shall see later, would give grounds for suspicion if it did not. The flat part of the curve below 500 Hz represents non-frequency-sensitive distortion rather than the noise floor, which for this case is at the 0.0005% level. Above 500 Hz the distortion rises at an increasing rate, rather than a constant number of dB/octave, due to the combination of Distortions One, Two, Three and Four. (In this case, Distortions Five, Six and Seven have been carefully eliminated to keep things simple; this is why the distortion performance looks good already, and the significance of this should not be overlooked.) It is often written that having distortion constant across the audio band is a Good Thing; a most unhappy conclusion, as the only practical way to achieve this with a normal Class-B amplifier is to increase the distortion at LF, for example, by allowing the VAS to distort significantly.

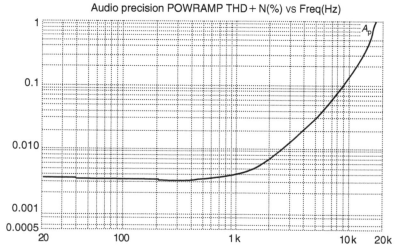

Figure 5.4. The distortion performance of the Class-B amplifier in Figure 5.1.

It should now be clear why it is hard to wring linearity out of such a snake-pit of contending distortions. A circuit-value change is likely to alter at least two of the distortion mechanisms, and probably change the o/l gain as well; in the coming chapters I shall demonstrate how each distortion mechanism can be measured and manipulated in isolation.

Open-loop Linearity and How to Determine It

Improving something demands measuring it, and thus it is essential to examine the open-loop linearity of power-amp circuitry. This cannot be done directly, so it is necessary to measure the NFB factor and calculate open-loop distortion from closed-loop measurements. The closed-loop gain is normally set by input sensitivity requirements.

Measuring the feedback-factor is at first sight difficult, as it means determining the open-loop gain. Standard methods for measuring opamp open-loop gain involve breaking feedback loops and manipulating closed-loop (c/l) gains, procedures that are likely to send the average power-amplifier into fits. Nonetheless the need to measure this parameter is inescapable, as a typical circuit modification, e.g., changing the value of R2 changes the open-loop gain as well as the linearity, and to prevent total confusion it is essential to keep a very clear idea of whether an observed change is due to an improvement in o/l linearity or merely because the o/l gain has risen. It is wise to keep a running check on this as work proceeds, so the direct method of open-loop gain measurement shown in Figure 5.5 was evolved.

Direct Open-loop Gain Measurement

The amplifier shown in Figure 5.1 is a differential amplifier, so its open-loop gain is simply the output divided by the voltage difference between the inputs. If output voltage is kept constant by providing a constant swept-frequency voltage at the +ve input, then a plot of open-loop gain versus frequency is obtained by measuring the error voltage between the inputs, and referring this to the output level. This gives an upside-down plot that rises at HF rather than falling, as the differential amplifier requires more input for the same output as frequency increases, but the method is so quick and convenient that this can be lived with. Gain is plotted in dB with respect to the chosen output level (+16 dBu in this case) and the actual gain at any frequency can be read off simply by dropping the

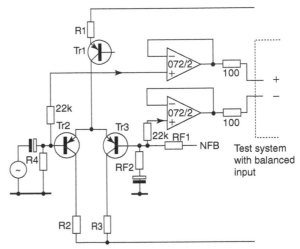

Figure 5.5. Test circuit for measuring open-loop gain directly. The accuracy with which high o/l gains can be measured depends on the testgear CMRR.

minus sign. Figure 5.6 shows the plot for the amplifier in Figure 5.1.

The HF-region gain slope is always 6 dB/octave unless you are using something special in the way of compensation, and by the Nyquist rules must continue at this slope until it intersects the horizontal line representing the feedback factor, if the amplifier is stable. In other words, the slope is not being accelerated by other poles until the loop gain has fallen to unity, and this provides a simple way of putting a lower bound on the next pole P2; the important P2 frequency (which is usually somewhat mysterious) must be above the intersection frequency if the amplifier is seen to be stable.

Given testgear with a sufficiently high Common Mode Rejection Ratio balanced input, the method of Figure 5.5 is simple; just buffer the differential inputs from the cable capacitance with TL072 buffers, which place negligible loading on the circuit if normal component values are used. In particular, be wary of adding stray capacitance to ground to the -ve input, as this directly imperils amplifier stability by adding an extra feedback pole. Short wires from power amplifier to buffer IC can usually be unscreened as they are driven from low impedances.

The testgear input CMRR defines the maximum open-loop gain measurable; I used an Audio Precision System-1 without any special alignment of CMRR. A calibration plot can be produced by feeding the two buffer inputs from the same signal; this will probably be found to rise at 6 dB/octave, being set by the

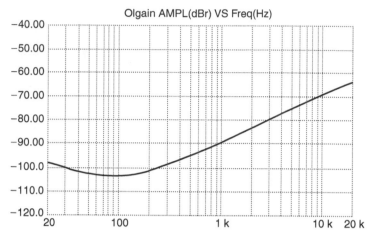

Figure 5.6. Open-loop gain versus freq plot for Figure 5.1. Note that the curve rises as gain falls, because the amplifier error is the actual quantity measured.

inevitable input asymmetries. This must be low enough for amplifier error signals to be above it by at least 10 dB for reasonable accuracy. The calibration plot will flatten out at low frequencies, and may even show an LF rise due to imbalance of the test-gear input-blocking capacitors; this can make determination of the lowest pole P1 difficult, but this is not usually a vital parameter in itself.

Using Model Amplifiers

Distortions 1 and 2 can dominate amplifier performance and need to be studied without the manifold complications introduced by a Class-B output stage. This can be done by reducing the circuit to a *model* amplifier that consists of the small-signal stages alone, with a very linear Class-A emitter-follower attached to the output to allow driving the feedback network; here *small-signal* refers to current rather than voltage, as the model amplifier should be capable of giving a full power-amp voltage swing, given sufficiently high rail voltages. From Figure 5.3 it is clear that this will allow study of Distortions One and Two in isolation, and using this approach it will prove relatively easy to design a small-signal amplifier with negligible distortion across the audio band, and this is the only sure foundation on which to build a good power amplifier.

A typical plot combining Distortions One and Two from a model amp is shown in Figure 5.7, where it can be seen that the distortion rises with an accelerating slope, as the initial rise at 6 dB/octave from the VAS is contributed to and then dominated by the 12 dB/octave rise in distortion from an unbalanced input stage.

The model can be powered from a regulated current-limited PSU to cut down the number of variables, and a standard output level chosen for comparison of different amplifier configurations; the rails and output level used for the results in this work were ±15 V and +16 dBu. The rail voltages can be made comfortably lower than the average amplifier HT rail, so that radical bits of circuitry can be tried out without the creation of a silicon cemetery around your feet. It must be remembered that some phenomena such as input-pair distortion depend on absolute output level, rather than the proportion of the rail voltage used in the output swing, and will be increased by a mathematically predictable amount when the real voltage swings are used.

The use of such model amplifiers requires some caution, and gives no insight into BJT output stages, whose behaviour is heavily influenced by the sloth and low current gain of the power devices. As a general rule, it should be possible to replace the small-signal output with a real output stage and get a stable and workable power amplifier; if not, then the model is probably dangerously unrealistic.

The Concept of the Blameless Amplifier

Here I introduce the concept of what I have chosen to call a *Blameless* audio power amplifier. This is an amplifier designed so that all the easily-defeated distortion mechanisms have been rendered negligible. (Note that the word *Blameless* has been carefully chosen to *not* imply Perfection, but merely the avoidance of known

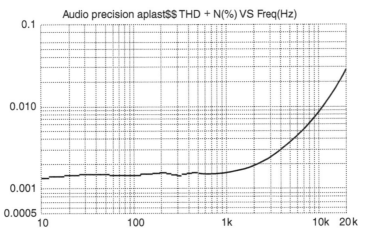

Figure 5.7. The distortion from a model amplifier, produced by the input pair and the Voltage-Amplifier Stage — note increasing slope as input pair distortion begins to add to VAS distortion

errors.) Such an amplifier gives about 0.0005% THD at 1 kHz and approximately 0.003% at 10 kHz when driving 8 Ω. This is much less THD than a Class-B amplifier is normally expected to produce, but the performance is repeatable, predictable, and definitely does not require large global feedback factors.

Distortion One cannot be totally eradicated, but its onset can be pushed well above 20 kHz by the use of local feedback. Distortion Two (VAS distortion) can be similarly suppressed by cascoding or beta-enhancement, and Distortions Four to Seven can be made negligible by simple topological methods. All these measures will be detailed later. This leaves Distortion Three, which includes the intractable Class-B problems, i.e., crossover distortion (Distortion 3b) and HF switch-off difficulties (Distortion 3c). Minimising Distortion 3b requires a Blameless amplifier to use a BJT output rather than FETs.

A Blameless Class-B amplifier essentially shows crossover distortion only, so long as the load is no heavier than 8 Ω; this distortion increases with frequency as the amount of global NFB falls. At 4 Ω loading an extra distortion mechanism (Distortion Three a) generates significant third harmonic.

The importance of the Blameless concept is that it represents the best distortion performance obtainable from straightforward Class-B. This performance is stable and repeatable, and varies little with transistor type as it is not sensitive to variable quantities such as beta.

Blamelessness is a condition that can be defined with precision, and is therefore a standard other amplifiers can be judged against. A Blameless design represents a stable point of departure for more radical designs, such as the Trimodal concept in Chapter 16. This may be the most important use of the idea.

References

1. Olive, B M, Distortion in Complementary-Pair Class-B Amplifiers, *Hewlett-Packard Journal*, Feb 1971, p. 11.

2. Feucht, D L, *Handbook of Analog Circuit Design*, San Diego: Academic Press, 1990, p. 256 (Pole-splitting).

3. Cherry, E A, New Distortion Mechanism in Class-B Amplifiers, *JAES*, May 1981, p. 327.

4. Ball, G, Distorting Power Supplies, *Electronics World+WW*, Dec. 1990, p. 1084

The Input Stage

A beginning is the time for taking the most delicate care that the balances are correct.

                    Frank Herbert, *Dune*.

The Role of the Input Stage

The input stage of an amplifier performs the critical duty of subtracting the feedback signal from the input, to generate the error signal that drives the output. It is almost invariably a differential transconductance stage; a voltage-difference input results in a current output that is essentially insensitive to the voltage at the output port. Its design is also frequently neglected, as it is assumed that the signals involved must be small, and that its linearity can therefore be taken lightly compared with that of the VAS or the output stage. This is quite wrong, for a misconceived or even mildly wayward input stage can easily dominate the HF distortion performance.

The input transconductance is one of the two parameters setting HF open-loop (o/l) gain, and therefore has a powerful influence on stability and transient behaviour as well as distortion. Ideally the designer should set out with some notion of how much o/l gain at 20 kHz will be safe when driving worst-case reactive loads (this information should be easier to gather now there is a way to measure o/l gain directly) and from this a suitable combination of input transconductance and dominant-pole Miller capacitance can be chosen.

Many of the performance graphs shown here are taken from a *model* (small-signal stages only) amplifier with a Class-A emitter-follower output, at +16 dBu on

±15 V rails; however, since the output from the input pair is in current form, the rail voltage in itself has no significant effect on the linearity of the input stage; it is the current swing at its output that is the crucial factor.

Distortion from the Input Stage

The motivation for using a differential pair as the input stage of an amplifier is usually its low DC offset. Apart from its inherently lower offset due to the cancellation of the Vbe voltages, it has the important added advantage that its standing current does not have to flow through the feedback network. However, a second powerful reason, which seems less well known, is that linearity is far superior to single-transistor input stages. Figure 6.1 shows three versions, in increasing order of sophistication. The resistor-tail version at 1a has poor CMRR and PSRR and is generally a false economy of the shabbiest kind; it will not be further considered here. The mirrored version at 1c has the best balance, as well as twice the transconductance of 1b.

At first sight, the input stage should generate a minimal proportion of the overall distortion because the voltage signals it handles are very small, appearing as they do upstream of the VAS that provides almost all the voltage gain. However, above the first pole frequency P1, the current required to drive Cdom dominates the proceedings, and this remorselessly doubles with each octave, thus:

$$i_{pk} = \omega \cdot C_{dom} \cdot V_{pk} \qquad \text{Equation 6.1}$$

where $\omega = 2 \cdot \pi \cdot \text{frequency}$

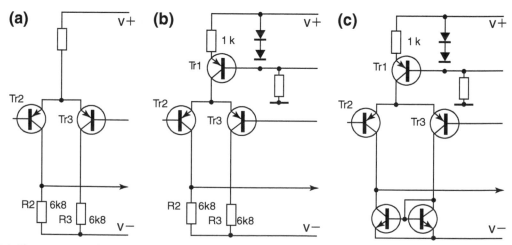

Figure 6.1. Three versions of an input pair: (a) simple tail resistor; (b) tail current-source; (c) with collector current-mirror to give inherently good Ic balance.

For example, the current required at 100 W (8 Ω) and 20 kHz, with a 100 pF Cdom is 0.5 mA peak, which may be a large proportion of the input standing current, and so the linearity of transconductance for large current excursions will be of the first importance if we want low distortion at high frequencies.

Curve A in Figure 6.2 shows the distortion plot for a model amplifier (at +16 dBu output) designed so the distortion from all other sources is negligible compared with that from the carefully balanced input stage; with a small-signal Class-A stage, this reduces to making sure that the VAS is properly linearised. Plots are shown for both 80 kHz and 500 kHz measurement bandwidths, in an attempt to show both HF behaviour and the vanishingly low LF distortion. It can be seen that the distortion is below the noise floor until 10 kHz, when it emerges and heaves upwards at a precipitous 18 dB/octave. This rapid increase is due to the input stage signal current doubling with every octave, to feed Cdom; this means that the associated third harmonic distortion will quadruple with every octave increase. Simultaneously the overall NFB available to linearise this distortion is falling at 6 dB/octave since we are almost certainly above the dominant-pole frequency P1, and so the combined effect is an octuple or 18 dB/octave rise. If the VAS or the output stage were generating distortion, this would be rising at only 6 dB/octave, and so would look quite different on the plot.

This non-linearity, which depends on the rate-of-change of the output voltage, is the nearest thing that exists to the late unlamented TID (Transient Intermodulation Distortion), an acronym that has now fallen out of fashion. It was sometimes known by the alias TIM (Transient InterModulation). SID (Slew-Induced Distortion) is a better description of the effect, but implies that slew-limiting is responsible, which is not the case.

If the input pair is *not* accurately balanced, then the situation is more complex. Second as well as third harmonic distortion is now generated, and by the same reasoning this has a slope nearer to 12 dB/octave; this vital point is examined more closely below.

All the input stages in this book are of the PNP format shown in Figure 6.1. One reason for this is that PNP bipolar transistors are claimed to have lower recombination noise than their NPN complements, though how much difference this makes in practice is doubtful. Another reason is that this puts the VAS transistor at the bottom of the circuit diagram and its current source at the top, which somehow seems the visually more accessible arrangement.

BJTs vs FETs for the Input Stage

At every stage in the design of an amplifier, it is perhaps wise to consider whether BJTs or FETs are the best devices for the job. I may as well say at once that the

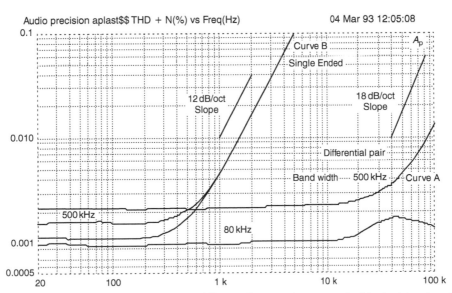

Figure 6.2. Distortion performance of model amplifier-differential pair at A compared with singleton input at B. The singleton generates copious second-harmonic distortion.

predictable Vbe/lc relationship and much higher trans-conductance of the bipolar transistor make it, in my opinion, the best choice for all three stages of a generic power amplifier. To quickly summarise the position:

Advantages of the FET Input Stage

There is no base current with FETs, so this is eliminated as a source of DC offset errors. However, it is wise to bear in mind that FET gate leakage currents increase very rapidly with temperature, and under some circumstances may need to be allowed for.

Disadvantages of the FET Input Stage

1. The undegenerated transconductance is low compared with BJTs. There is much less scope for linearising the input stage by adding degeneration in the form of source resistors, and so a FET input stage will be very non-linear compared with a BJT version degenerated to give the same low transconductance.
2. The Vgs offset spreads will be high. Having examined many different amplifier designs, it seems that in practice it is essential to use dual FETs, which are relatively very expensive and not always easy to obtain. Even then, the Vgs mismatch will probably be greater than Vbe mismatch in a pair of cheap discrete BJTs; for example, the 2N5912 N-channel

dual FET has a specified maximum Vgs mismatch of 15 mV. In contrast, the Vbe mismatches of BJTs, especially those taken from the same batch (which is the norm in production) will be much lower, at about 2–3 mV, and usually negligible compared with DC offset caused by unbalanced base currents.
3. The noise performance will be inferior if the amplifier is being driven from a low-impedance source, say, 5 kΩ or less. This is almost always the case.

Singleton Input Stage Versus Differential Pair

Using a single input transistor (Figure 6.3a) may seem attractive, where the amplifier is capacitor-coupled or has a separate DC servo; it at least promises strict economy. However, any cost saving would be trivial, and the snag is that this singleton configuration has no way to cancel the second-harmonics generated in copious quantities by its strongly-curved exponential V_{in}/I_{out} characteristic.[1] The result is shown in Figure 6.2 curve-B, where the distortion is much higher, though rising at the slower rate of 12 dB/octave.

The Input Stage Distortion in Isolation

Examining the slope of the distortion plot for the whole amplifier is instructive, but for serious research we need

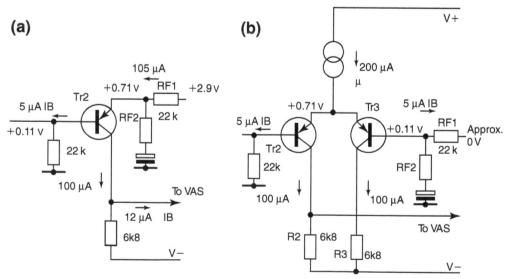

Figure 6.3. Singleton and differential pair input stages, showing typical DC conditions. The large DC offset of the singleton is mainly due to all the stage current flowing through the feedback resistor RF1.

to measure input-stage non-linearity in isolation. This can be done with the test circuit of Figure 6.4. The opamp uses shunt feedback to generate an appropriate AC virtual-earth at the input-pair output. Note that this current-to-voltage conversion opamp requires a third −30 V rail to allow the i/p pair collectors to work at a realistic DC voltage, i.e., about one diode-worth above the −15 V rail. The opamp feedback resistor can be scaled as convenient, to stop opamp clipping, without the input stage knowing anything has changed. The DC balance of the pair can be manipulated by the potentiometer, and it is instructive to see the THD residual diminish as balance is approached, until at its minimum amplitude it is almost pure third harmonic.

The differential pair has the great advantage that its transfer characteristic is mathematically highly predictable.[2] The output current is related to the differential input voltage V_{in} by:

$$I_{out} = I_e \bullet \tanh(- V_{in}/2V_t) \qquad \text{Equation 6.2}$$

where:

Vt is the usual thermal voltage of about 26 mV at 25°C

Ie is the tail current

Two vital facts derived from this equation are that the transconductance (gm) is maximal at $V_{in} = 0$, when the two collector currents are equal, and that the value of this maximum is proportional to the tail current *l*e. Device beta does not figure in the equation, and so the performance of the input pair is not significantly affected by transistor type. The 'tanh' in the equation is shorthand for the hyperbolic tangent function. There is no need to worry about its derivation or significance; you just look up the value as with an ordinary tangent function.

Figure 6.5a shows the linearising effect of local feedback or degeneration on the voltage-in/current-out law; Figure 6.5b plots transconductance against input voltage and shows clearly how the peak transconductance value is reduced, but the curve made flatter and linear over a wider operating range. Simply adding emitter degeneration markedly improves the linearity of the input stage, but the noise performance is slightly worsened, and of course the overall amplifier feedback factor has been reduced, for as previously shown, the vitally-important HF closed-loop gain is determined solely by the input transconductance and the value of the dominant pole.

Input Stage Balance

Exact DC balance of the input differential pair is absolutely essential in power amplifiers. It still seems almost unknown that minor deviations from equal I_c in the pair seriously upset the second-harmonic cancellation, by

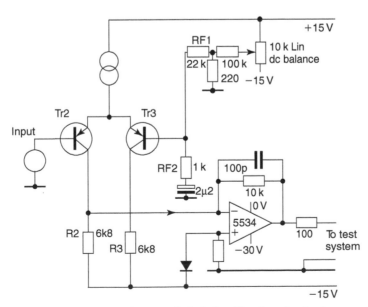

Figure 6.4. Test circuit for examining input stage distortion in isolation. The shunt-feedback opamp is biased to provide the right DC conditions for TR2.

(a)

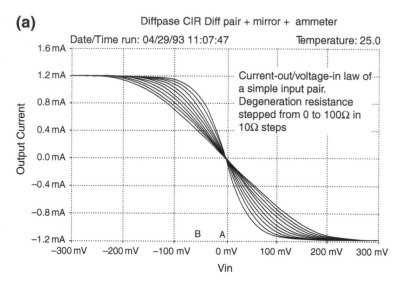

(b)

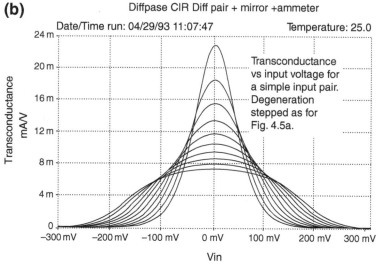

Figure 6.5. Effect of degeneration on input pair V/I law, showing how transconductance is sacrificed in favour of linearity (SPICE simulation).

moving the operating point from A to B in Figure 6.5a. The average slope of the characteristic is greatest at A, so imbalance also reduces the open-loop gain if serious enough. The effect of small amounts of imbalance is shown in Figure 6.6 and Table 6.1; for an input of −45 dBu a collector-current imbalance of only 2% gives a startling worsening of linearity, with THD increasing from 0.10% to 0.16%; for 10% imbalance this deteriorates badly to 0.55%. Unsurprisingly, imbalance in the other direction ($Ic1 > Ic2$) gives similar results.

Imbalance is defined as deviation of Ic (per device) from that value which gives equal currents in the pair.

This explains the complex distortion changes that accompany the apparently simple experiment of altering the value of R2.[3] We might design an input stage like Figure 6.7a, where R1 has been selected as 1 kΩ by uninspired guesswork and R2 made highish at 10 kΩ in a plausible but wholly misguided attempt to maximise o/l gain by minimising loading on Q1 collector. R3 is also 10 k to give the stage a notional 'balance', though unhappily this is a visual rather than an electrical

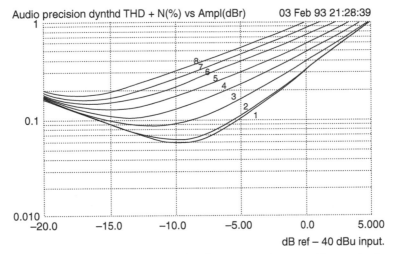

Audio precision dynthd THD + N(%) vs Ampl(dBr) 03 Feb 93 21:28:39

dB ref – 40 dBu input.

Figure 6.6. Effect of collector-current imbalance on an isolated input pair; the second harmonic rises well above the level of the third if the pair moves away from balance by as little as 2%.

balance. The asymmetry is shown in the resulting collector currents; the design generates a lot of avoidable second harmonic distortion, displayed in the 10 kΩ curve of Figure 6.8.

Recognising the crucial importance of DC balance, the circuit can be rethought as Figure 6.7b. If the collector currents are to be roughly equal, then R2 must be about $2 \times R1$, as both have about 0.6 V across them. The dramatic effect of this simple change is

Table 6.1. Key to Figure 6.6

Curve No.	Ic Imbalance (%)	Curve No.	Ic Imbalance (%)
1	0	5	5.4
2	0.5	6	6.9
3	2.2	7	8.5
4	3.6	8	10

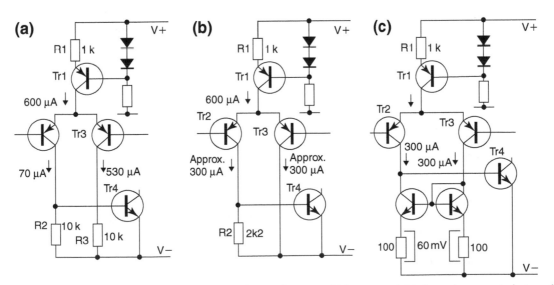

Figure 6.7. Improvements to the input pair: (a) poorly designed version; (b) better; partial balance by correct choice of R2; (c) best; near-perfect Ic balance enforced by mirror.

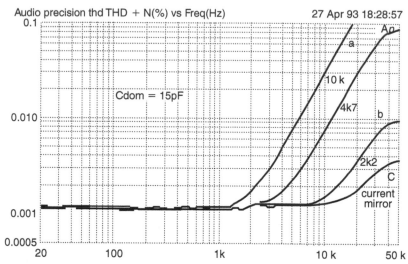

Figure 6.8. Distortion of model amplifier: (a) unbalanced with R2 = 10k; (b) partially balanced with R = 2k2; (c) accurately balanced by current-mirror.

shown in the 2k2 curve of Figure 6.8; the improvement is accentuated as the o/l gain has also increased by some 7 dB, though this has only a minor effect on the closed-loop linearity compared with the improved balance of the input pair. R3 has been excised as it contributes very little to input stage balance.

There are very few references in the literature to the importance of collector-current balance in differential pairs; one worth looking up is an article in *Wireless World* by Eric Taylor, that appeared in August 1977.[4]

The Joy of Current-mirrors

Although the input pair can be approximately balanced by the correct values for R1 and R2, we remain at the mercy of several circuit tolerances. Figure 6.6 shows that balance is critical, needing an accuracy of 1% or better for optimal linearity and hence low distortion at HF, where the input pair works hardest. The standard current-mirror configuration in Figure 6.7c forces the two collector currents very close to equality, giving correct cancellation of the second harmonic; the great improvement that results is seen in the current-mirror curve in Figure 6.8. There is also less DC offset due to unequal base currents flowing through input and feedback resistances; I often find that a power-amplifier improvement gives at least two separate benefits.

It will be noticed that both the current-mirror transistors have a very low collector-emitter voltages; the diode-connect one has just its own Vbe, while the other sustains the Vbe of the VAS transistor, or two Vbes if the VAS has been enhanced with an emitter-follower. This means that they can be low-voltage types with a high beta, which improves the mirror action.

The hyperbolic-tangent law also holds for the mirrored pair,[5] though the output current swing is twice as great for the same input voltage as the resistor-loaded version. This doubled output is given at the same distortion as for the unmirrored version, as input-pair linearity depends on the input voltage, which has not changed. Alternatively, we can halve the input and get the same output, which with a properly balanced pair generating third harmonic only will give one-quarter the distortion. A most pleasing result.

The input mirror is made from discrete transistors, regretfully foregoing the Vbe-matching available to IC designers, and so it needs its own emitter-degeneration resistors to ensure good current-matching. A voltage-drop across the current-mirror emitter-resistors in the range 30–60 mV will be enough to make the effect of Vbe tolerances on distortion negligible; if degeneration is omitted, then there is significant variation in HF distortion performance with different specimens of the same transistor type. Current-mirrors generate significant current noise, and this can be minimised by using the maximum practicable degeneration resistance. There is more on this later in the chapter.

Current-mirrors can be made using a signal diode such as the 1N4148 instead of the diode-connected transistor, but this gives poor matching, saves little if any money, and is generally to be deprecated.

Putting a current-mirror in a well-balanced input stage increases the total o/l gain by 6 dB; the increase might be as much as 15 dB if the stage was previously poorly balanced, and this needs to be taken into account in setting the compensation. Another happy consequence is that the slew-rate is roughly doubled, as the input stage can now source and sink current into Cdom without wasting it in a collector load. If Cdom is 100 pF, the slew-rate of Figure 6.7b is about 2.8 V/μsec up and down, while 4.7c gives 5.6 V/μsec. The unbalanced pair at 4.7a displays further vices by giving 0.7 V/μsec positive-going and 5 V/μsec negative-going.

In the world of opamp design, the utilisation of both outputs from the input differential stage is called 'phase summing'. Herpy[6] gives some interesting information on alternative ways to couple the input stage to the VAS, though some of them look unpromising for power amplifier use.

Better Current-mirrors

The simple mirror has well-known residual base-current errors, as demonstrated in Figure 6.9 (emitter degeneration resistors are omitted for clarity, and all transistors are assumed to be identical to keep things simple). In Figure 6.9a, Q1 turns on as much as necessary to absorb the current Ic1 into its collector, and Q2, which perforce has the same Vbe, turns on exactly the same current. But ... Ic1 is not the same as Iin, because two helpings of base current Ib1 and Ib2 have been siphoned off it. (It is helpful at this point to keep a firm grip on the idea

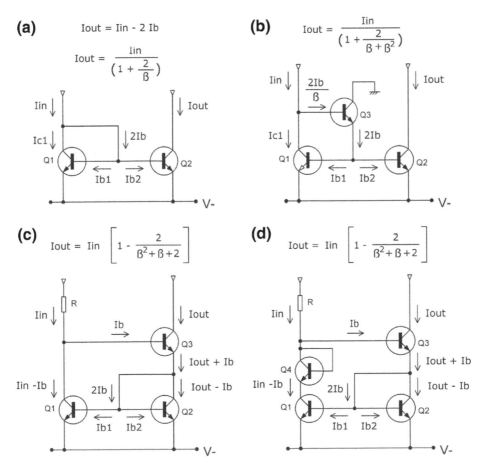

Figure 6.9. Current-mirrors and their discontents: (a) the basic mirror has base-current errors; (b) the EFA circuit reduces these; (c) the Wilson mirror greatly reduces these; (d) a further improvement to Wilson by equalising the Vce's of Q1 and Q2.

that a bipolar transistor is a voltage-operated device NOT a current-operated device, and the base currents are not 'what turns the transistors on' but the unwanted effect of finite beta. It does not help that beta is sometimes called 'current gain' when it is nothing of the sort.) Therefore, Iout is going to be less than Iin by twice Ib, and Ib will not be a linear function of Iin, because beta varies with collector current. Note that this problem occurs even though our transistors have been assumed to be perfectly matched for both beta and Vbe.

A great deal of effort has been put into making good current-mirrors by opamp designers, and we can cheerfully exploit the results. A good source is.[7] One way to reduce the base current problem is to add a third transistor, as in Figure 6.9b. This reduces the base current bled away from the input current by a factor of beta; the beta of Q3. If this configuration has an official name I don't know it, and I have always called it the EFA (Emitter-follower Added) circuit. Another way is shown in Figure 6.9c. This is the famous Wilson current-mirror, which unlike the previous versions, uses negative-feedback. Q3 is a voltage-follower, and Q1, Q2 a basic current-mirror. If the current through Q3 should tend to increase, the current-mirror pulls current away from Q3 base and turns it off a bit. We are assuming that R1 exists in some form; (for it would make little sense to have a low impedance feeding a current-mirror), in our case, it is the collector impedance of one of the input pair transistors. An important feature of the Wilson is the way the base current errors cancel, as shown in Figure 6.9c. It really is a beautiful sight. The input/output equations are given for each version, and it is clear that the EFA and the Wilson have beta-squared terms that make the fractions much closer to unity. The calculated results are shown in Table 6.2, and it is clear that both the EFA and the Wilson are far superior to the simple mirror, but this superiority lessens as beta increases. The Wilson comes out slightly better than the EFA at very low betas, but at betas of 25 or more (and hopefully the beta won't be lower than that in small-signal transistors, even if they are high-voltage types), there is really very little between the two of them.

So far we have not looked at the influence of Early effect on mirror accuracy; for our purposes it is probably very small, but it is worth noting that in Figure 6.9c Q1 has a Vce of two Vbe drops while Q2 has a Vce of only one Vbe. If you are feeling perfectionist, the mirror in Figure 6.9d has an added diode-connected transistor Q4 that reduces the Vce of Q1 to a single Vbe drop.

So how much benefit can be gained by using more sophisticated current-mirrors? In some studies I have

Table 6.2. Current-mirror accuracy

Beta	Simple mirror Iout/Iin	EFA Iout/Iin	Wilson Iout/Iin
1	0.33333	0.50000	0.60000
2	0.50000	0.75000	0.80000
5	0.71429	0.93750	0.94595
10	0.83333	0.98214	0.98361
25	0.92593	0.99693	0.99705
50	0.96154	0.99922	0.99923
100	0.98039	0.99980	0.99980
150	0.98684	0.99991	0.99991
200	0.99010	0.99995	0.99995
250	0.99206	0.99997	0.99997
500	0.99602	0.99999	0.99999

made of advanced input stages with very good linearity (not ready for publication yet, I'm afraid), I found that a simple mirror could introduce more non-linearity than the input stage itself, that the three-transistor Wilson improved things greatly, and the four-transistor version even more.

In practical measurements, when I tried replacing the standard mirror with a Wilson in a Blameless amplifier the improvement in the distortion performance was marginal at best, for as usual most of the distortion was coming from the output stage. That does not mean we should never look at ways of improving the small-signal stages; when the Bulletproof Distortionless Output Stage finally appears, we want to be ready.

Improving Input Stage Linearity

Even if the input pair has a current-mirror, we may still feel that the HF distortion needs further reduction; after all, once it emerges from the noise floor, it octuples with each doubling of frequency, and so it is well worth postponing the evil day until as far as possible up the frequency range. The input pair shown has a conventional value of tail-current. We have seen that the stage transconductance increases with Ic, and so it is possible to increase the gm by increasing the tail-current, and then return it to its previous value (otherwise Cdom would have to be increased proportionately to maintain stability margins) by applying local NFB in the form of emitter-degeneration resistors. This ruse powerfully improves input linearity, despite its rather unsettling flavour of something-for-nothing. The transistor non-linearity can here be regarded as an internal non-linear emitter resistance *re*, and what we have done is to reduce the value of this (by increasing Ic) and replaced the missing part of it with a linear external resistor Re.

For a single device, the value of re can be approximated by:

$$r_e = 25/I_c \ \Omega \ \text{(for Ic in mA)} \qquad \text{Equation 6.3}$$

Our original stage at Figure 6.10a has a per-device Ic of 600 μA, giving a differential (i.e., mirrored) gm of 23 mA/V and *re* = 41.6 Ω. The improved version at Figure 6.10b has *Ic* = 1.35 mA and so *re* = 18.6 Ω; therefore emitter degeneration resistors of 22 Ω are required to reduce the gm back to its original value, as 18.6 + 22 = 40.6 Ω, which is near enough. The distortion measured by the circuit of Figure 6.4 for a −40 dBu input voltage is reduced from 0.32% to 0.032%, which is an extremely valuable linearisation, and will translate into a distortion reduction at HF of about five times for a complete amplifier; for reasons that will emerge later, the full advantage is rarely gained. The distortion remains a visually pure third-harmonic, so long as the input pair remains balanced. Clearly this sort of thing can only be pushed so far, as the reciprocal-law reduction of re is limited by practical values of tail current. A name for this technique seems to be lacking; *constant-gm degeneration* is descriptive but rather a mouthful.

The standing current is roughly doubled so we have also gained a higher slew-rate; it has theoretically increased from 10 V/μsec to 20 V/μsec, and once again we get two benefits for the price of one inexpensive modification.

It is, however, not all benefit when we add emitter-degeneration resistors. The extra resistances will generate Johnson noise, increasing the total noise from the input stage. Differing values for the two resistors due to the usual tolerances will increase the input offset voltage. If the resistor matching is α%, the tail current is Itail, and the degeneration resistors have the value Re, the extra offset voltage Voff is given by:

$$V_{off} = \left(\frac{\alpha}{100}\right) \times \left(\frac{I_{tail} \times R_e}{2}\right) \qquad \text{Equation 6.4}$$

Thus for 100 Ω 1% resistors and a tail current of 6 mA, the extra offset voltage is 3 mV, which is small compared with the offsets due to the base currents flowing in the input and feedback resistances. This looks like one issue you need not worry about.

When a mirrored input stage is degenerated in this way, it is important to realise that its transconductance can only be very roughly estimated from the value of the emitter resistors. An input pair with a tail current of 4 mA and 22 Ω emitter resistors has a gm of 25.6 mA/V, which represents an effective V/I conversion resistance of 39.0 Ω, the extra resistance being the internal *res* of the transistors. (Remember that the input voltage is shared between two emitter resistors, apparently halving the current swing, but it is doubled again by the presence of the current-mirror.) In this case the value of the emitter resistors gives a very poor estimate of the gm. When 100 Ω emitter resistors are used with a tail current of 4 mA, the gm is 8.18 mA/V, representing an effective V/I conversion resistance of 122 Ω, which makes the estimate somewhat better but still more than 20% out. Increasing the tail current to 6 mA, which is the value used in the designs in this book, changes those values to 34.2 Ω and 118 Ω, because the internal *re*'s are reduced, but the estimates are still some way off. If more accurate figures are wanted at the design stage, then SPICE simulation will usually be faster and better than manual calculation.

Further Improving Input-Stage Linearity

If we are seeking still better linearity, various techniques exist, but before deploying them, we need to get an idea of the signal levels the input stage will be handling; the

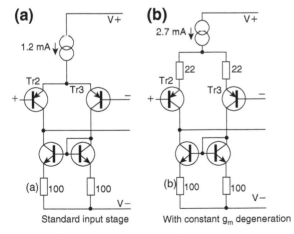

Figure 6.10. Input pairs before and after constant-gm degeneration, showing how to double stage current while keeping transconductance constant; distortion is reduced by about ten times.

critical factor is the differential input voltage across the input stage — the error voltage created by the global feedback. Using Equation 5.2 (in Chapter 5) it is straightforward to work out the input voltage to the input stage for a given input stage gm, Cdom value, and frequency; these parameters give us the open-loop gain, and we can then work back from the output voltage to the input voltage. The closed-loop gain and resulting global feedback factor are not involved except insofar as the feedback factor determines how much the input stage distortion is reduced when the loop is closed. However much the reduction by global feedback, an input stage that is twice as linear remains twice as linear.

Let us take an example typical of the designs in this book, with 100 Ω input degeneration resistors and a resulting gm of 8.18 mA/V, and a Cdom of 100 pF. The worst-case frequency is 20 kHz, and we will assume a 50 W/8 Ω output level. This gives an input voltage of −28.3 dBu, which for the discussion below is rounded to −30 dBu. For a 100 W/8 Ω output level, the input voltage would be −25.3 dBu.

Whenever it is needful to increase the linearity of a circuit, it is often a good approach to increase the *local* feedback factor, because if this operates in a tight local NFB loop, there is often little effect on the overall global-loop stability. A reliable method is to replace the input transistors with complementary-feedback (CFP or Sziklai) pairs, as shown in the stage of Figure 6.11a. If an isolated input stage is measured using the test circuit of Figure 6.4, the constant-gm degenerated version shown in Figure 6.10b yields 0.35% third-harmonic distortion for a −30 dBu input voltage, while the CFP version of Figure 6.11a gives 0.045%, a very valuable improvement of almost eight times. (Note that the input level here is 10 dB up on the -40 dBu input level used for the example in the previous section, which is both more realistic and gets the distortion well clear of the noise floor.) When this stage is put to work in a model amplifier, the third-harmonic distortion at a given frequency is roughly halved, assuming all other distortion sources have been appropriately minimised; the reason for the discrepancy is not currently known. However, given the high-slope of input-stage distortion, this only extends the low-distortion regime up in frequency by less than an octave. See Figure 6.12.

A compromise is required in the CFP circuit on the value of Rc, which sets the proportion of the standing current that goes through the NPN and PNP devices on each side of the stage. A higher value of Rc gives better linearity (see Table 6.3 below for more details

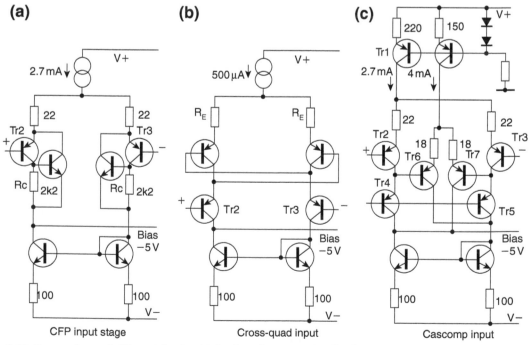

Figure 6.11. Some enhanced differential pairs: (a) the Complementary-Feedback Pair; (b) the Cross-quad; (c) the Cascomp.

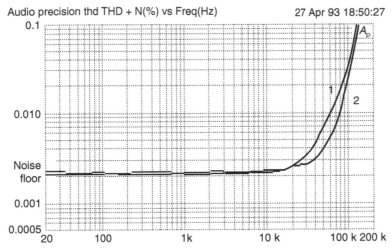

Figure 6.12. Whole-amplifier THD with normal and CFP input stages; input stage distortion only shows above noise floor at 20 kHz, so improvement occurs above this frequency. The noise floor appears high as the measurement bandwidth is 500kHz.

on this), but potentially more noise, due to the lower collector current in the PNP devices that are the inputs of the input stage, as it were, causing them to perform less well with the relatively low source resistances. 2k2 seems to be a good compromise value for Rc; it gives a collector current of 320 uA.

Several other elaborations of the basic input pair are possible, although almost unknown in the audio community. We are lucky in power-amp design as we can tolerate a restricted input common-mode range that would be unusable in an opamp, giving the designer great scope. Complexity in itself is not a serious disadvantage as the small-signal stages of the typical amplifier are of almost negligible cost compared with mains transformers, heatsinks, etc.

Two established methods to produce a linear input transconductance stage (often referred to in opamp literature simply as a transconductor) are the cross-quad[8] and the cascomp[9] configurations. The cross-quad input stage (Figure 6.11b) works by imposing the input voltage to each half across two base-emitter junctions in series, one in each arm of the circuit. In theory, the errors due to non-linear r_e of the transistors are divided by beta, but in practice the reduction in distortion is modest. The cross-quad nonetheless gives a useful reduction in input distortion when operated in isolation, but is hard to incorporate in a practical amplifier because it relies on very low source-resistances to tame the negative conductances inherent in its operation. If you just drop it into a normal power amplifier circuit with the usual source-resistances in the input

and feedback arms, it will promptly latch up, with one side or the other turning hard on. This does not seem like a good start to an amplifier design, despite the seductive simplicity of the circuit, and with some lingering regret it will not be considered further here.

The cascomp (Figure 6.11c) does not have problems with negative impedances, but it is significantly more complicated, and significantly more complex to design. Tr2, Tr3 are the main input pair as before, delivering current through cascode transistors Tr4, Tr5 (this cascoding does not in itself affect linearity), which, since they carry almost the same current as Tr2, Tr3 duplicate the input Vbe errors at their emitters. These error voltages are sensed by error diff-amp Tr6, Tr7, whose output currents are summed with the main output in the correct phase for error-correction. By careful optimisation of the (many) circuit variables, distortion at −30dBu input can be reduced to about 0.016% with the circuit values shown, which handily beats the intractable cross-quad. Sadly, this effort provides very little further improvement in whole-amplifier HF distortion over the simpler CFP input, as other distortion mechanisms are coming into play, one of which is the finite ability of the VAS to source current into the other end of Cdom.

Much more information on the Cascomp concept can be found in a remarkable book by Staric and Margan.[10] This also covers the general topic of feedforward error-correction in differential input stages.

Table 6.3 summarises the performance of the various types of input stage examined so far.

Table 6.3. Summary of measured input stage linearity

Type	Input level dBu	Rdegen Ω	THD (%)	Notes	Figure
Simple	−40	0	0.32		Fig 6.10a
Simple	−40	22	0.032		Fig 6.10b
Simple	−30	22	0.35		Fig 6.10b
CFP	−30	22	0.045	Rc = 2K2	Fig 6.11a
CFP	−30	39	0.058	Rc = 1K5	Fig 6.11a
CFP	−30	39	0.039	Rc = 2K2	Fig 6.11a
CFP	−30	39	0.026	Rc = 4K7	Fig 6.11a
CFP	−30	39	0.022	Rc = 10K	Fig 6.11a
Cascomp	−30	50	0.016		Fig 6.11c

Another approach to making a very linear transconductance stage is the 'Multi-tanh principle'. Looking back at Equation 6.2, the non-linear part of it is the hyperbolic tangent, written 'tanh'. If multiple input transistor pairs are used, with a different voltage offset applied to each pair, then the input voltage range with linear transconductance can be very much extended.[11] The original concept goes back to 1968, but it does not appear to have ever been used in a power amplifier. The reference covers both double and triple input pairs, and demonstrates how the rather awkward offset voltages required can be replaced by using different emitter areas for the transistors; this is obviously directed at IC fabrication, but it is an interesting point if this could be emulated in discrete form by using transistors in parallel to increase the effective emitter areas. There seems at least the possibility that the greater number of transistors used would improve the noise performance due to the partial cancellation of non-correlated noise components.

Increasing the Output Capability

The standing current in the input pair tail is one of the parameters that defines the maximum slew rate, the other being the size of the dominant-pole Miller capacitor. The value of this capacitor is usually fixed by the requirements of stability, but increasing the tail current can increase slew-rate without directly affecting stability so long as the degeneration resistors are adjusted to keep the input stage transconductance at the desired value.

Unfortunately there are limits to how much this current can be increased; the input bias currents increase, as do the voltage drops across the degeneration resistors, and both these factors increase the spread of DC offset voltage. The ultimate limit is of course the power dissipation in the input stage; if you take a 6 mA tail current, which is the value I commonly use, and ± 50 V supply rails, the dissipation in each input transistor is 150 mW to a close approximation, and there is clearly not a vast amount of scope for increasing this. There is also the point that hot input devices are more susceptible to stray air currents and therefore we can expect more drift.

Opamp designers face the same problems, exacerbated by the need to keep currents and dissipations to a much lower level than those permissible in a power amplifier. Much ingenuity has therefore been expended in devising input stages that do not work in Class-A, like the standard differential pair, but operate in what might be called Class-AB; they have a linear region for normal input levels, but can turn on much more than the standing current when faced with large inputs. Typically there is an abrupt change in transconductance, and linearity is much degraded as the input stage enters the high-current mode. The first input stage of this type was designed by W. E. Hearn in 1970, and it appeared in the Signetics NE531 opamp.[12] Another such stage was put forward by Van de Plassche.[13] Both types have been used successfully in the standard 3-stage architecture by Giovanni Stochino.[14]

The rest of this chapter deals only with the standard input differential amplifier.

Input Stage Cascode Configurations

Cascoding is the addition of a common-base stage to the collector of a common-emitter amplifier, to prevent the stage output from affecting the common-emitter stage, or to define its operating collector voltage. The word is

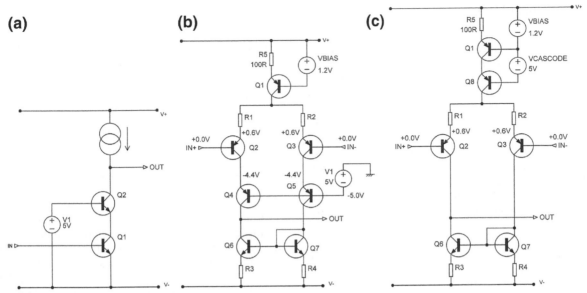

Figure 6.13. Cascode configurations: (a) the basic cascode concept; (b) cascoding applied to the input devices of a differential input stage, with DC conditions shown; (c) cascoding applied to the tail current source.

a contraction of 'cascade to cathode', which tells you at once that, like so many circuit techniques, it dates back to the valve era.[15] It can often be usefully applied to the standard input differential amplifier. The basic principle of a cascode amplifier stage is shown in Figure 6.13a. There is a common-emitter amplifier Q1, directly coupled to a common-base stage Q2. The common-base-stage gives no increase in the transconductance of the overall stage, as it simply passes the collector current of Q1 onto the current-source collector load I1, less a small amount that is the base current of Q2. The important job that it does do is to hold the collector of Q1 at a substantially constant voltage; the voltage of biasing voltage source V1, minus the Vbe voltage of Q2. This constant collector voltage for Q1 gives two benefits; the frequency response of the stage is improved because there is no longer local negative feedback through the collector-base capacitance of Q1, and the stage gain is potentially both greater and more linear because the Early effect (the modulation of collector current Ic by Vce) can no longer occur in Q1. The Vce of Q1 is now both lower and constant — a Vce of 5 V is usually quite enough — and the consequent reduction of heating in Q1 can have indirect benefits in reducing thermal drift. This configuration will be met with again in Chapter 7 on Voltage-Amplifier stages.

When the cascoding principle is applied to the input stage of a power amplifier, we get the configuration shown in Figure 6.13b, with the DC conditions indicated. The circuit is inverted compared with the single-transistor example so it corresponds with the other input stages in this book. If the bases of the input devices Q1, Q2 are at 0 V, which is usually the case, their collectors need to be held at something like −5 V for correct operation.

Cascoding an input stage does nothing to improve the linearity of the stage itself, as there is no appreciable voltage swing on the input device collectors due to the low-impedance current input of a typical VAS stage; it can, however, in some circumstances reduce input current distortion as it allows high-beta low-Vce input devices to be used. See later in this chapter, in the section on input current distortion, where it is shown that sometimes there are real benefits in hum rejection to be obtained by cascoding the input pair tail current source, as shown in Figure 6.13c. In specialised circumstances, for example, where the closed-loop gain of the amplifier is lower than usual, cascoding the input stage can actually make linearity worse; see the section on Input Common-mode Distortion below.

Isolating the input device collector capacitance from the VAS input sometimes allows Cdom to be slightly reduced for the same stability margins, but the improvement is marginal. A more significant advantage is the reduction of the high Vce that the input devices work at. This allows them to run cooler, and so be less susceptible to drift caused by air currents. This is dealt with in more detail in Chapter 23 on DC servos.

Double Input Stages

Two input stages, one the complement of the other, are quite often used to drive both the top and bottom of a push-pull VAS. See Chapter 8. Their operation is just the same as for a single input stage, and both emitter degeneration and the use of current-mirrors are recommended as before. If the input bases are connected directly together, as is usual, there should be first-order cancellation of the input currents (see the section on input current distortion later in this chapter on why this can be important), but complete cancellation will not occur because of the poor beta-matching of discrete transistors, and those of differing polarity at that. Chapter 8 includes an interesting example of a series input stage using one NPN and one PNP transistor.

The use of double input stages could in theory give 3 dB less noise due to arithmetical summing of the signals but RMS-summing of the input-stage noise, but I haven't had the opportunity to test this myself.

Input Stage Common-mode Distortion

This is distortion generated by the common-mode voltage on the input stage, if there is any, as opposed to that produced by the differential input voltage in the normal course of business, as it were. In a typical power amplifier with series feedback, the common-mode voltage is equal to the input signal voltage. If shunt feedback is used, the common-mode voltage is negligible, but this is not a good reason to use it; if the input becomes disconnected, the likely result is horrifying instability. There has been much speculation about the importance or otherwise of common-mode distortion, but very little actual measurement. You will find some here.

Common-mode distortion does not appear to exist at detectable levels in normal amplifier circuitry, and by this I mean I am assuming that the input stage has emitter degeneration resistors and a current-mirror, as previously described. A much higher common-mode (CM) voltage on the input stage than normally exists is required to produce a measurable amount of distortion. If an amplifier is operated at a low closed-loop gain such as 1 or 2 times, so that both input and feedback signals are much larger than usual, this puts a large CM voltage on the input stage, and distortion at HF is unexpectedly high, despite the much increased negative-feedback factor. This distortion is mainly second-harmonic. The immediate cause is clearly the increased CM voltage on the input devices, but the exact mechanism is at present unclear.

Table 6.4. How distortion varies with common-mode voltage

Closed-loop gain Times	CM voltage V rms	15 kHz THD meas (%)	15 kHz THD calc (%)
1.00	10.00	.0112	.00871
1.22	8.20	.00602	.00585
1.47	**6.81**	**.00404**	**.00404**
2.00	5.00	.00220	.00218
23	0.43	-------	0.000017

Table 6.4 shows distortion increasing as closed-loop gain is reduced, with input increased to keep the output level constant at 10 V rms. A model amplifier (i.e., one with the output stage replaced by a small-signal Class-A stage, as described in Chapter 5) was used because the extra phase shift of a normal output stage would have made stability impossible to obtain at such low closed-loop gains; the basic circuit without any input stage modifications is shown in Figure 6.14. This is an excellent illustration of the use of a model amplifier to investigate input stage distortions without the extra complications of a Class-B output stage driving a load. For some reason long forgotten, NPN input devices were used, so the diagrams appear upside-down relative to most of the amplifiers in this book.

This version of a model amplifier has a couple of points of interest; you will note that the input degeneration resistors R2, R3 have been increased from the usual value of 100R to 220R to help achieve stability by reducing the open-loop gain. The output stage is a push-pull Class A configuration which has twice the drive capability than the usual constant-current version, chosen so it could drive the relatively low value feedback resistance R5, (needed to keep the noise down as the lower feedback arm is higher in value than usual because of the low gain) without an increase in distortion; there was no other load on the output stage apart from the distortion analyser. I have used this configuration extensively in the past in discrete-component preamplifiers, for example, in.[16] It is very linear, because the push-pull action halves the current swing in the emitter-follower, and it is thoroughly stable and dependable but does require regulated supply rails to work properly.

Tests were done at 10 Vrms output and data taken at 15 kHz, so the falling global negative-feedback factor with frequency allowed the distortion to be far enough above the noise floor for accurate measurement. The

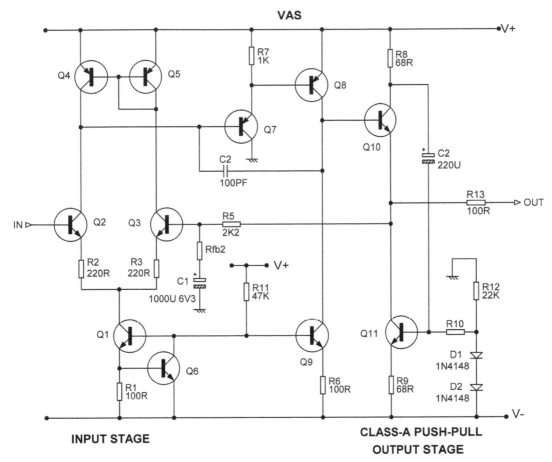

Figure 6.14. A model amplifier (output stage replaced by a small-signal Class-A stage) with low closed-loop gain.

closed-loop gain was altered by changing the lower feedback arm Rfb2. The THD plots can be seen in Figure 6.15, which provided the data for Table 6.4. It can be seen that the distortion goes up at 6 dB per octave.

It appears THD is proportional to CM voltage squared. Taking the measured THD at a closed-loop gain of 1.47x as a reference, scaling it by the square of the gain gives the figures in the rightmost column, which correspond quite nicely with the measured THD figures. Some extra higher-order distortion was coming in at a closed-loop gain of 1.00, so the square law is less accurate there.

Thus, assuming the square law, the THD at 1.47 times gain (.00404%) when scaled down for a more realistic closed-loop gain of 23, is reduced by a factor of $(23/1.47)^2 = 245$, giving a negligible 0.000017% at 15 kHz. In terms of practical amplifier design, there are other things to worry about.

And yet ... I was curious as to the actual distortion mechanism, and decided to probe deeper, without any expectation that the answer would be directly useful in amplifier design until we had made a lot more progress in other areas of non-linearity; however, there was always the possibility that the knowledge gained would be applicable to other problems, and it would certainly come in handy if it was necessary to design a low-gain power amplifier for some reason. Giovanni Stochino and I therefore investigated this issue back in 1996, and at the end of lot of thought, experimentation, and international faxing, I felt I could put down the following statements:

- Reducing the Vce of the input devices by inserting capacitively decoupled resistors into the collector circuits, as shown in Figure 6.16a, makes the CM distortion worse. Altering the V+ supply rail (assuming NPN input devices) has a similar effect;

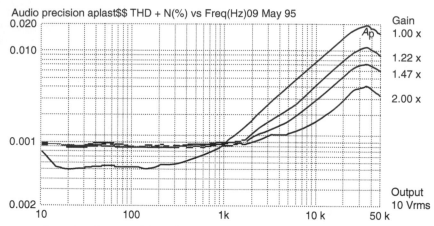

Figure 6.15. THD plots from model amplifier with various low closed-loop gains. Output 10 Vrms.

less Vce means more distortion. Vce has a powerful effect on the HF THD. This seems to indicate that the non-linearity is due to either the Early effect (an increase in the effective beta of a BJT as the Vce increases, due to narrowing of the effective base width) or the modulation of Vbc, the base-collector capacitance, or very possibly a mixture of the two. Very little seems to have been published on these sort of non-linearities; but Taylor[17] is well worth reading.

- The effect cannot be found in SPICE simulation, which is a bit disconcerting. It must therefore originate either in imbalances in transistor parameters, which don't exist in SPICE unless you put them in, *or* in second-order effects that are not modelled by SPICE. A particular suspect here is the fact that SPICE models the Early effect as linear with Vce. I was told by Edward Cherry[18] that that SPICE would need to include the second-order term in the Early voltage model, rather than use a linear law, before the effect could be simulated.

- The HF distortion does not alter AT ALL when the input devices are changed, so the THD mechanism cannot depend on beta or Early voltage, as these would vary between device samples. I know this finding makes little sense, but I checked it several times, always with the same result.

Giovanni and I therefore concluded that if the problem is due to the Early effect, it should be possible to eliminate it by cascoding the input device collectors and driving the cascode bases with a suitable CM voltage so that the input device Vce remains constant. I tried this, and found that if the CM voltage was derived from the amplifier output, via a variable attenuator, it allowed only partial nulling of the distortion. With a CM voltage

of 6.81 V rms, the drive to the cascode for best second-harmonic nulling was only 131 mV rms, which made little sense.

A more effective means of reducing the common-mode non-linearity was suggested to me by Giovanni Stochino.[19] Driving the input cascode bases directly from the input tail, rather than an output derived signal, completely eliminates the HF distortion effect. It is not completely established as to why this works so much better than driving a bootstrap signal from the output, but Giovanni feels that it is because the output signal is phase-shifted compared with the input, and I suspect he is right. At any rate it is now possible to make a low-gain amplifier with very, very low HF distortion, which is rather pleasing. The bootstrapped cascode input configuration is shown in Figure 6.16b, and the impressive THD results are plotted in Figure 6.17.

And there, for the moment, the matter rests. If the closed-loop gain of your amplifier is low, you need to worry about common-mode distortion, but there is a fix. However, in most cases it is too low to worry about.

Input Current Distortion

When power amplifiers are measured, the input is normally driven from a low impedance signal generator. Some testgear, such as the much-loved but now obsolete Audio Precision System-1, has selectable output impedance options of 50, 150, and 600 Ω. The lowest value available is almost invariably used because:

1. It minimises the Johnson noise from the source resistance.

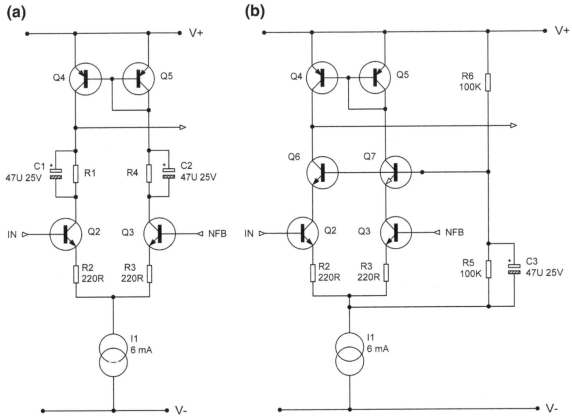

Figure 6.16. (a) Method of reducing input device Vce; (b) a method of driving the bases of an input cascode structure directly from the input tail.

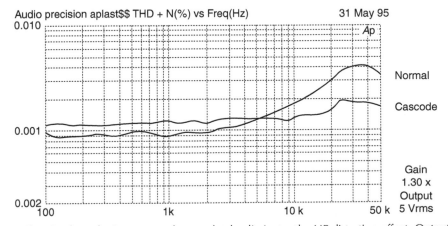

Figure 6.17. Showing how the input cascode completely eliminates the HF distortion effect. Output 5 Vrms.

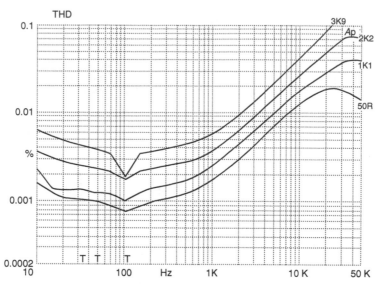

Figure 6.18. Second-harmonic distortion and 100 Hz ripple get worse as the source impedance rises from 50 Ω to 3.9 K. 50 Watts into 8 Ω.

2. It minimises level changes due to loading by the amplifier input impedance.
3. It minimises the possibility of hum, etc., being picked up by the input.

This is all very sensible, and exactly the way I do it myself — 99% of the time. There are, however, two subtle effects that can be missed if the amplifier is always tested this way. These are:

1. Distortion caused by the non-linear input currents drawn by the typical amplifier.
2. Hum caused by ripple modulation of the input bias currents.

Note that (1) is not the same effect as the excess distortion produced by FET-input opamps when driven from significant source impedances; this is due to their non-linear input capacitances to the IC substrate, and has no equivalent in power amplifiers made of discrete bipolar transistors.

Figure 6.18 shows both the effects. The amplifier under test was a conventional Blameless design with an EF output stage comprising a single pair of sustained-beta bipolar power transistors; the circuit can be seen in Figure 6.19. The output power was 50 Watts into 8 Ω. The bottom trace is the distortion+noise with the usual source impedance of 50 Ω, and the top one shows how much worse the THD is with a source impedance of 3.9 k. The intermediate traces are for 2.2 k and 1.1 k source resistances. The THD residual

shows both second-harmonic distortion and 100 Hz ripple components, the ripple dominating at low frequencies, while at higher frequencies the distortion dominates. The presence of ripple is signalled by the dip in the top trace at 100 Hz, where distortion products and ripple have partially cancelled, and the distortion analyser has settled on the minimum reading. The amount of degradation from both ripple and distortion is proportional to the source impedance.

The input currents are not a problem in many cases, where the preamplifier is driven by an active preamplifier, or by a buffer internal to the power amplifier. Competent active preamplifiers have a low output impedance, often around 50–100 Ω, and sometimes less — there are no great technical difficulties involved in reducing it to a few ohms. This is to minimise high-frequency losses in cable capacitance. (I have just been hearing of a system with 10 metres of cable between preamp and power amp.) However, some active designs seem to take this issue less seriously than they should and active preamp output impedances of up to 1 k are not unknown. To the best of my knowledge, preamp output impedances have never been made deliberately low to minimise power amplifier input current distortion, but it would certainly be no bad thing.

There are two scenarios where the input source resistance is considerably higher than the desirable 50–100 Ω. If a so-called 'passive preamp' is used, then

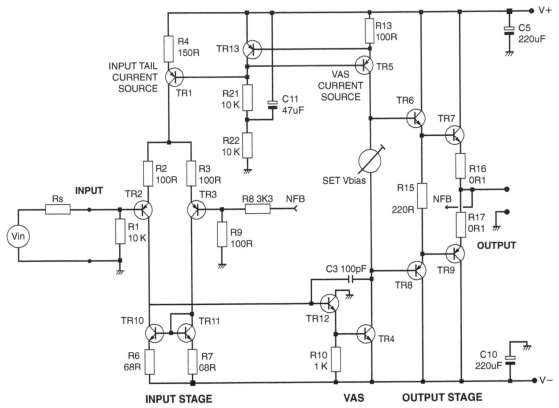

Figure 6.19. Simplified circuit of a typical Blameless power amplifier, with negative-feedback control of VAS current source TR5 by TR13. The bias voltage generated is also used by the input tail source TR1.

the output impedance is both much higher and volume-setting dependent. A 10 k volume potentiometer, which is the lowest value likely to be practical if the loading on source equipment is to be kept low, has a maximum output impedance of one-quarter the track resistance, i.e., 2.5 k, at its mid-point setting.

It is also possible for significant source resistance to exist inside the power amplifier unit, for example, there might be a balanced input amplifier, which, while it has a very low output impedance itself, may have a resistive gain control network between it and the power amp. The value of this potentiometer is not likely to be less than 5 k, and is more likely to be 10 k, so once again we are faced with a maximum 2.5 k source resistance at the mid-point setting. (Assuming the input amplifier is a 5532 or equally capable opamp, there would be no difficulty in driving a 2 k or even a 1 k pot without its loading introducing measurable extra distortion; this would reduce the source resistance and also the Johnson noise generated. However, I digress, more on amplifier input circuitry in Chapter 27.)

So, Houston, we have a problem, or rather two of them, in the form of extra ripple and extra distortion, and the first step to curing it is to understand the mechanisms involved. Since the problems get worse in proportion to the source impedance, it seems very likely that the input transistor base currents are directly to blame for both, so an obvious option is to minimise these currents by using transistors with the highest available beta in the input pair. In this amplifier the input pair were originally ZTX753, with a beta range of 70–200. Replacing these with BC556B input devices (beta range 180–460) gives Figure 6.20 which shows a useful improvement in THD above 1 kHz; distortion at 10 kHz drops from 0.04% to 0.01%. Our theory that the base currents are to blame is clearly correct. The bottom trace is the reference 50 Ω source plot with the original ZTX753s, and the gap between this and our new result demonstrates that the problem has been reduced but certainly not eliminated.

The power amplifier used for the experiments here is very linear when fed from a low source impedance, and it

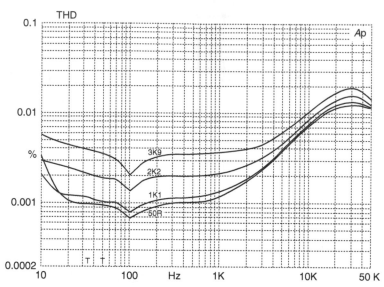

Figure 6.20. There is less introduction of ripple and distortion with high-beta input transistors and the same set of source resistances as Figure 6.17.

might well be asked why the input currents drawn are distorted if the output is beautifully distortion-free. The reason is of course that global negative feedback constrains the output to be linear – because this is where the NFB is taken from – but the internal signals of the amplifier are not necessarily linear, but whatever is required to keep the output linear. The output stage is known to be significantly non-linear, so if the amplifier output is sinusoidal, the signals internal to the amplifier will not be. The collector currents of the input pair clearly are not perfectly linear. Even if they were, the beta of the input transistors is not constant so the base currents drawn by them would still be non-linear.

It is also possible to get a reduction in hum and distortion by reducing the input pair tail current, but this very important parameter also affects input stage linearity and the slew-rate of the whole amplifier. Figure 6.21 shows the result. The problem is reduced – though far from eliminated – but the high-frequency THD has actually got worse because of poorer linearity in the input stage. This is not a promising route to follow: no matter how much the tail current is reduced, the problem will not be eliminated.

Both the ripple and THD effects consequent on the base currents drawn could be eliminated by using FETs instead of bipolars in the input stage. The drawbacks are:

1. Poor Vgs matching, which means that a DC servo becomes essential to control the amplifier output DC

offset. Dual FETs do exist but they are discouragingly expensive.
2. Low transconductance, which means the stage cannot be linearised by local feedback as the raw gain is just not available.
3. Although there will be negligible DC gate currents, there might well be problems with non-linear input capacitance, as there are with FET-input opamps.

Once again, this is not a promising route; the use of FETs will create more problems than it solves.

The distortion problem looks rather intractable; one possible total cure is to put a unity-gain buffer between input and amplifier. The snag (for those seeking the highest possible performance) is that any opamp will compromise the noise, and almost any will compromise the distortion, of a Blameless amplifier. It is quite correct to argue that this doesn't matter, as any preamp hooked up to the power amp will have opamps in it anyway, but the preamp is a different box, a different project, and possibly has a different designer, so philosophically this does not appeal to everyone. If a balanced input is required, then an opamp stage is mandatory (unless you prefer transformers, which of course have their own problems).

The best choice for the opamp is either the commonplace but extremely capable 5532 (which is pretty much distortion-free, but not, alas, noise-free, though it is very quiet) or the very expensive but very quiet AD797. A relatively new alternative is the LM4562, which has

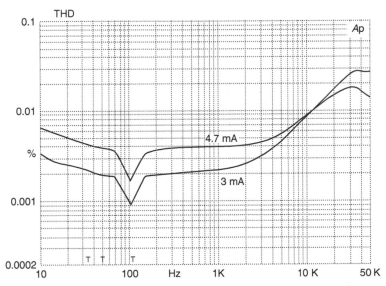

Figure 6.21. Reducing the tail current improves things at low frequencies but increases HF distortion above 10 kHz. The notches at 100 Hz indicate that the ripple content is still substantial.

lower noise than a 5532, but at present they are a good deal more expensive.

The ripple problem, however, has a more elegant solution. If there is ripple in the input base current, then clearly there is some ripple in the tail current. This is not normally detectable because the balanced nature of the input stage cancels it out. A significant input source impedance upsets this balance, and the ripple appears.

The tail is fed from a simple constant-current source TR1, and this is clearly not a mathematically perfect circuit element. Investigation showed that the cause of the tail-current ripple contamination is Early effect in this transistor, which is effectively fed with a constant bias voltage tapped off from the VAS negative-feedback current source (TR13 in Figure 6.19), the problem is *not* due to ripple in the bias voltage. (Early effect is the modulation of transistor collector current caused by changing the Vce; as it is in most cases a relatively minor aspect of bipolar transistor behaviour, it is modelled by SPICE simulators in a rather simplistic way, by assuming a linear Vce/Ic relationship.) Note that this kind of negative-feedback current-source could control the tail current instead of the VAS current, which might well reduce the ripple problem, but the biasing system is arranged this way as it gives faster positive slewing. Another option is two separate negative feedback current-sources.

The root cause of our hum problem is therefore the modulation of the Vce of TR1 by ripple on the positive

rail, and this variation is easily eliminated by cascoding, as shown in Figure 6.22. This causes the TR1 emitter and collector to move up and down together, preventing Vce variations. It completely eradicates the ripple components, but leaves the input-current distortion unaltered, giving the results in Figure 6.23, where the upper trace is now degraded only by the extra distortion introduced by a 2 k source impedance; note that the 100 Hz cancellation notch has disappeared. The reference 50 Ω source plot is below it.

The voltage at A that determines the Vce of TR1 is not critical. It must be sufficiently below the positive supply rail for TR1 to have enough Vce to conduct properly, and it must be sufficiently above ground to give the input pair enough common-mode range. I usually split the biasing chain R21, R22 in half, as shown, so C11 is working with the maximum resistance to filter out rail noise and ripple, and biasing the cascode transistor from the mid-point works very well. Note that this is preferable to biasing the cascode transistor with a fixed voltage (e.g., from a Zener diode) for a non-obvious reason. It means that an untried amplifier will start up earlier when you are cautiously increasing the supply rail voltages by nervous manipulation of a variable transformer, and the earlier it starts, the less damage will be done if there is something wrong.

An alternative, though rather less elegant, approach to preventing ripple injection is simply to smooth the positive rail with an RC filter before applying it to the tail-current source. The resulting voltage drop in the R

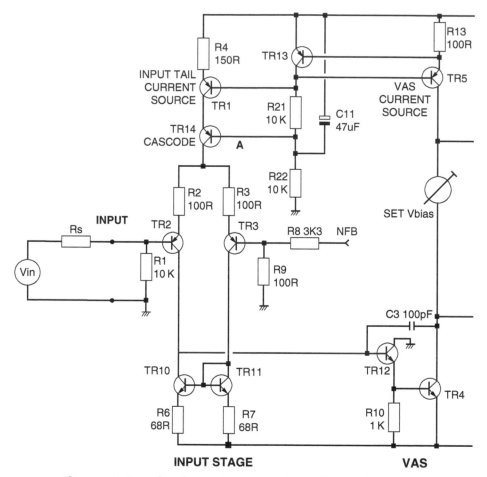

Figure 6.22. Cascoding the input tail; one method of biasing the cascode.

part means that a separate tail-current source biasing voltage must now be generated, and the C will have to be a high-voltage component as it has to withstand almost all the positive rail voltage.

At the end of the day the cascode approach will probably be cheaper as well as more elegant. And you can always put 'cascoded input stage!' in your publicity material.

It may have occurred to the reader that simply balancing the impedances seen by the two input devices will cancel out the unwanted noise and distortion. This is not very practical, as with discrete transistors there is no guarantee that the two input devices will have the same beta. (I know there are such things as dual bipolars, but once more the cost is depressing.) This also implies that the feedback network will have to have its impedance raised to equal that at the input,

which would give unnecessarily high levels of Johnson noise. It is of course impractical where the source resistance is variable, as when the amplifier is being fed from a volume-control potentiometer.

Another line of enquiry is cancelling out the input current by applying an equal and opposite current, generated elsewhere in the input stage, to the input. This kind of stratagem is used in some BJT-input opamps, where it is called 'input bias-current cancellation'; it is hard to see how to apply it to an input stage made with discrete transistors because creating the cancellation currents relies on having closely matched betas in all the devices. Even if it were possible, there would almost certainly be a penalty in the shape of increased noise. Opamps such as the OP27, which has input bias cancellation, have gained a certain notoriety for giving disappointing noise results. At first sight it

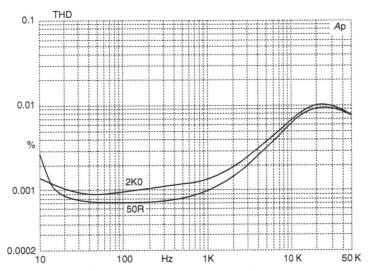

Figure 6.23. Cascoding the input tail removes the ripple problem, but not the extra distortion.

appears that the OP27 is quieter than the 5534/5532; its e_n is 3.2 nV/rtHz compared with 4 nV/rtHz for the 5534. However, on practical measurement, the OP27 is often slightly noisier, and this is believed to be because the OP27 input bias-current circuitry generates common-mode noise. When the impedances on the two inputs are equal, all is well, but when they are different, the common-mode noise does not cancel, and this effect seems to be enough to degrade the noise performance significantly. If you want to pursue the matter of input bias cancellation further (and it has to be said that some of the circuitry is most ingenious and well worth studying) a good reference is Dostal.[20]

Since neither of these approaches looks very promising, what else can be done? It seems likely that the CFP input stage described earlier in this chapter would give lower values of input current distortion, as the base currents of the NPN transistors that can potentially flow in external source resistances (or indeed, the feedback network source impedance) are much lower. A simple differential pair draws an input current from 0 to 49 uA over the input voltage range (from SPICE using MPSA42/MPSA92, tail current 6 mA) while the CFP draws 0 to 5.3 uA. I have not yet assessed the comparative linearity of the two currents but it looks as though there might be an order of magnitude improvement here.

The discussion above has focused on the effects of a significant source impedance at the input to the power amplifier. But a power amplifier, like an opamp, has two inputs, and that not used for the signal input is used for the feedback connection. The current that this input draws from the feedback network will also lead to extra distortion, by exactly the same mechanism. If the feedback network consisted of, say, a 47 k upper arm and a 2k2 lower arm, giving a closed-loop gain of 22.4 times, the source impedance seen by the input will be 2k1, and we can expect to see some serious extra distortion, as shown in Figure 6.11a above. This is an important point; if this problem exists in an amplifier design, then no amount of work that attempts to improve the linearity of input stage transconductance or the VAS will improve matters in the slightest, and I suspect that in many cases this has been a source of intractable grief for amplifier designers.

In the next part of this chapter, I emphasise that the impedance of the feedback network should be kept as low as practicable to minimise the Johnson noise it generates and to minimise offset voltages. If this philosophy is followed, the feedback network source impedance as seen by the amplifier input will be too low (around 100 Ω) for the input current distortion from this part of the circuit to be measurable above the noise floor.

To summarise, if the system design requires or permits an opamp at the input, then both the hum and distortion problems that the input currents create are removed with no further effort. If a significant source resistance is inescapable, for whatever reason, then cascoding the input pair tail cures the ripple problem but not the extra distortion. Using high-beta input transistors reduces both problems but does not eliminate them. When considering input current distortion, don't forget the feedback network has its own source impedance.

Noise

The noise performance of a power amplifier is defined by its input stage, and so the issue is examined here. Power-amp noise is not an irrelevance; a powerful amplifier will have a high voltage gain, and this can easily result in a faint but irritating hiss from efficient loudspeakers even when all volume controls in the system are fully down. In the Blameless designs considered in this book, the input stages all consist of a pair of discrete PNP transistors with 100 Ω of emitter degeneration each, and a current-mirror collector load with 68 Ω degeneration resistors. This technology gives an Equivalent Input Noise (EIN) of around -122 dBu, which is only 5 or 6 dB worse than a first-class microphone preamplifier connected to a 200 Ω source impedance. This means that a power amplifier alone is likely to be quieter than almost anything you can put in front of it. This issue is examined in detail in Chapter 27 on amplifier input systems.

As a specific example, we will use a model amplifier, (a power amplifier with the full-scale output stage replaced by a small signal Class-A output stage) and a regulated PSU, to eliminate the possibility of hum and ripple getting in from unregulated supply rails. Model amplifiers usually have a high-impedance feedback network of 22 kΩ and 1 kΩ, to keep its impedance within the drive capabilities of the Class-A output stage, and in that form I measured -91.8 dBu of noise at the output. Subtracting the +27.2 dB of gain, we find the Equivalent Input Noise (EIN) is −119.0 dBu. That is worse than the figure quoted just above because of the high-impedance feedback network. If we replace it with the usual low-impedance power amplifier feedback network of 2k2 and 100 Ω, noting that is valid for noise but not for THD measurements, the EIN drops by 3.6 dB to −122.6 dBu.

This is the basic amplifier noise. The external source resistance noise is negligible as at 40 Ω its Johnson noise is only −136.2 dB. If we eliminate it, by shorting the input, then because of the way rms-addition works, the EIN only drops to −122.8 dBu.

Internal amplifier noise is generated by the active devices at the input and the surrounding resistances, so we can list the obvious noise sources:

1. The input transistor pair. A differential pair of transistors has two transistors in series from the

point of noise, and so is inevitably 3 dB noisier than a single transistor input stage. The advantages of the differential pair in DC balance and distortion cancellation are, however, so overwhelming that this compromise is gladly accepted by almost everybody.

2. The input stage degeneration resistors. Johnson noise from the input degeneration resistors R2, R3 is the price we pay for linearising the input stage by running it at a high current, and then bringing its transconductance down to a useable value by adding linearising local negative feedback.

3. The impedance of the negative-feedback network. We just saw that a low-impedance feedback network is essential for good noise performance; 2k2 and 100 Ω have a combined source impedance of 96 Ω. An inconvenient consequence is that input resistor R1 must also be reduced to 2k2 to maintain DC balance, and this is too low an input impedance for direct connection to the outside world. Some sort of buffering or input bootstrapping is required.

Another consequence is the need for feedback capacitor C2 to be proportionally increased to maintain LF response, and prevent capacitor distortion from causing a rise in THD at low frequencies; it is the latter requirement that determines the value, and sets a value of 1000 μF, necessitating a low voltage rating such as 6V3 if the component is to be of reasonable size. This in turn means that C2 needs protective shunt diodes in both directions, because if the amplifier fails, it may saturate in either direction. Examination of the distortion residual shows that the onset of conduction of back-to-back diodes will cause a minor increase in THD at 10 Hz, from less than 0.001% to 0.002%, even at the low power of 20 W/8 Ω. It is not my practice to tolerate such gross non-linearity, and therefore four series-parallel diodes are used in the final circuit, and this eliminates the distortion effect. It could be argued that a possible reverse-bias of 1.2 V does not protect C2 very well, but at least there will be no explosion.

A low-impedance feedback network also significantly improves the output DC offset performance, as described later in this chapter.

Noise Sources in Power Amplifiers

It is instructive to go a little deeper into the sources of noise inside a power amplifier, to see what determines it and how (and if) it can be improved. You might well object that since the power amplifier noise is already so low that almost any active element put in front of it will be noisier, there is very little practical point in this. My reasons are firstly, that I simply want to know what's going on in there, and to find out if my notions of what determines the noise performance are correct. Another excellent reason is that lowering the noise floor will allow small amounts of distortion to be better seen, and hopefully better understood. A very-low-noise power amplifier is a useful research tool.

We have our model amplifier with a measured Equivalent Input Noise of -122.6 dBu, and the three obvious noise sources listed in the previous section. We will attempt to calculate the contribution of each source, and then examine the options for reducing it.

It is easy to calculate what proportion of this comes from Johnson noise in the circuit resistances that are clearly relevant. These are the feedback network resistance and the emitter degeneration resistors.

1. The input transistor pair. The operating conditions of the input transistors are set by the demands of linearity and slew-rate, so there is little freedom of design here; however, the collector currents are already high enough to give near-optimal noise figures with the low source impedances (a few hundred Ohms) that we have here. The noise characteristics of bipolar transistors are dealt with in detail in the next section of this chapter, and we will just note here that the calculated noise from a single transistor is −135.7 dBu. That includes both the transistor voltage noise and the result of the transistor current noise flowing in a 100 Ω resistor; that resistance is an approximation but it will do for now. Since there are two transistors effectively in series, the total transistor noise is 3 dB higher, at −132.7 dBu.

   Input transistor noise (both) = − 132.7 dBu.

2. The input stage degeneration resistors. The emitter degeneration resistors are 100 Ω, each of which generates −132.2 dBu of Johnson noise for the usual bandwidth of 22 kHz at 25 °C. They are effectively in series so their noise output sums in the usual RMS manner, and gives a total noise for the degeneration resistors of −129.2 dBu. The only way to reduce the noise from these resistors is to reduce their value; if the HF NFB factor is to remain constant, then Cdom will have to be proportionally increased, reducing the maximum slew-rate.

   Input stage degeneration resistor noise (both)

   = − 129.2 dBu.

3. The impedance of the negative-feedback network. The effective resistance of the feedback network is that of both its resistors in parallel, which comes to 96 Ω. This gives a Johnson noise voltage of −132.4 dBu. Once again, the only way to reduce this is to reduce the resistor values. The capacitor at the bottom of the feedback network is already 1000 μF (see Figure 6.27 on p. 158), so any radical decrease in resistance is clearly going to lead to some cumbersome components. 2200 μF at a low voltage is quite do-able, so let's see what happens. If we divide the value of both resistors by 2.2 times, giving 1 kΩ and 45 Ω as a feedback network. The effective resistance of that is 43 Ω, with a Johnson noise of −135.9 dBu, a healthy improvement of 3.5 dB. Unfortunately the feedback network contribution was already some 3 dB quieter than the degeneration resistor noise, so the actual improvement is small. An unwanted consequence of the alteration is that a good deal more power is being wasted in the feedback network. A 100 W/8 Ω amplifier at full throttle will dissipate 332 mW in the original 2k2 feedback resistor (which means a 1/2 W part) and 729 mW in the new 1 kΩ resistor. That means a rather bulky 1 W resistor, or more conveniently, four 1/4 W resistors in parallel dissipating 182 mW each. The accuracy of the 1 kΩ value is improved by a factor of two by the use of four resistors.[21]

If we add these three noise levels together, rms-fashion, we get a calculated amplifier EIN of − 126.4 dBu. We have to admit that does not fit well with our measured EIN of −122.6 dBu; the calculations are 3.8 dB too optimistic. Noise calculations for transistors are never super-accurate, because of the variation in device characteristics, but they should be better than that. The Johnson noise calculations are much more accurate, affected only by the accuracy of the resistor values.

The likeliest explanation is that we have neglected a significant noise source in the amplifier. A little calculation shows that there is a missing contribution of −125.0 dBu, which is larger than any of the known contributions. What the ...?

Examining other possible noise sources, we might consider how much noise the VAS generates. In some low-noise opamps the noise from the second stage is significant. However, it seems very unlikely that the VAS would produce more noise than the input stage. And that only leaves ...

The answer is given in Samuel Groner's tour-de-force commentary on my previous power amplifier books.[22] You cannot assume that the signal is safe once it has left the collectors of the input pair. Lurking just below it is the current-mirror, and this creates more noise than you might think. I am very grateful to Samuel Groner for drawing this to my attention.

The noise generated by a current-mirror decreases as the value of the degeneration resistors is increased. This may appear counterintuitive, because when we want to reduce noise we normally reduce resistor values, but here it works the other way around because the noise from the mirror is in the form of a current. Samuel makes the point that the reason that the amplifier is sensitive to mirror noise is because the input pair is heavily degenerated.[23] One of the few sources of information on mirror noise is Bilotti.[24] I have so far had little success in making his equations line up with my measurements.

Figure 6.24 shows the measured effect on the amplifier EIN. Starting from my usual value of 68 Ω, the mirror degeneration resistors are increased in steps up to 470 Ω. The EIN falls from −122.6 to −124.6 dBu, an improvement of 2.0 dB. If we assume that there are no other significant noise sources apart from those already listed, it can be calculated that the mirror noise contribution falls from −125.0 dBu to −129.4 dBu. That is quite an improvement, but only gives 2.0 dB lower noise overall because the other noise sources are of comparable level. These results agree with Samuel Groner's measurements.

Raising the mirror degeneration resistances increases the voltage drop across them, and so reduces the voltage elbow-room in which the current-mirror has to operate. When the resistance reached 100 Ω, it was necessary to insert a diode in the VAS emitter to increase the voltage on the current-mirror output. This worked up to 470 Ω but not for 620 Ω, the maximum value used by Samuel Groner. I tried putting a suitable resistance, decoupled with a big capacitor, in the VAS emitter. It gave an unchanged distortion performance, but made the amplifier prone to a really horrible latch-up process with excessive current flowing through the VAS. Switching on the amplifier with a reasonably large input signal present will trigger it. This line of thought was then abandoned as it was clear that even if the latch-up effect could be defeated, (which it no

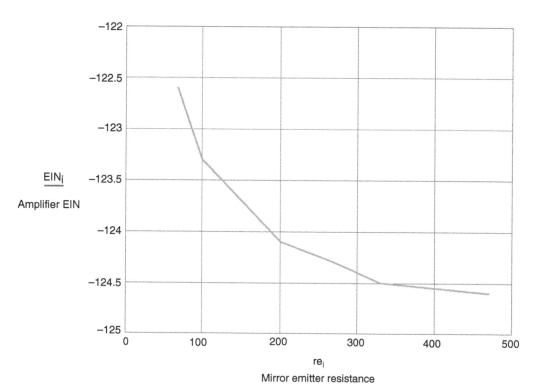

Figure 6.24. The measured improvement in amplifier noise performance (EIN) as the current-mirror degeneration resistors are increased from 68 Ω to 470 Ω. The improvement is 2.0 dB.

doubt could be with a bit of work), the likely noise improvement would be very small.

Given that almost any conceivable piece of electronics put in front of the present Blameless amplifier will be noisier, there is little incentive to put in a lot of work to make the amplifier quieter still. It would unquestionably be an interesting intellectual exercise, but that does not make it a priority.

Noise in Bipolar Transistors

To understand the noise behaviour of discrete bipolar transistors, it is necessary to delve a little deeper into their internal operation than is usually required, and take account of imperfections that do not appear in the simplest transistor models. I give here a quick summary rather than a thorough analysis; the latter can be found in many textbooks. Two important transistor parameters for understanding noise are r_{bb}, the base spreading resistance, and r_e, the intrinsic emitter resistance. r_{bb} is a real physical resistance — what is called an *extrinsic* resistance. The second parameter r_e is an expression of the V_{be}/I_c slope and not a physical resistance at all, and it is therefore called an *intrinsic* resistance.

Noise in bipolar transistors is best dealt with by assuming we have a noiseless transistor with a theoretical noise voltage source in series with the base and a theoretical noise current source connected from base to ground. These sources are usually just described as the 'voltage noise' and the 'current noise' of a transistor.

The voltage noise v_n has two components, one of which is the Johnson noise generated in the base spreading resistance r_{bb}, the other is the collector current (I_c) shot noise creating a noise voltage across r_e, the intrinsic emitter resistance. Shot noise occurs simply because an electric current is a stream of discrete electric charges, and not a continuous fluid, and it increases as the square root of the current. The two components can be represented thus:

$$\text{Voltage noise density } v_n = \sqrt{4kTR_{bb} + 2(kT)^2/\left(qIc\right)}$$

$$\text{in V/rtHz (usually nV/rtHz)} \qquad \text{Equation 6.5}$$

The first part of this equation is the usual expression for Johnson noise, and is fixed for a given transistor type by the physical value of r_{bb}; the lower this is, the better. The absolute temperature is obviously a factor but there is not usually much you can do about this.

The second (shot noise) part of the equation decreases as collector current I_c increases; this is because as I_c increases, r_e decreases proportionally while the shot noise only increases as the square root of I_c. These factors are all built into the second part of the equation. The overall result is that the total v_n falls — though relatively slowly — as the collector current increases, approaching asymptotically the level of noise set by first part of the equation. There is no way you can reduce that except by changing to another type of transistor with a lower R_{bb}.

There is an extra voltage noise source resulting from flicker noise produced by the base current flowing through r_{bb}; this is only significant at high collector currents and low frequencies due to its $1/f$ nature, and is usually not included in design calculations unless low frequency quietness is a special requirement.

The current noise i_n, which is mainly produced by the shot noise of the steady current I_b flowing through the transistor base. This means it increases as the square root of I_b increases. Naturally I_b increases with I_c. Current noise is given by

$$\text{Current noise density } i_n = \sqrt{2qI_b} \text{ in A/rtHz}$$

$$\text{(usual values are in pA)} \qquad \text{Equation 6.6}$$

So, for a fixed collector current, you get less current noise with high-beta transistors because there is less base current. Such transistors usually have a V_{ce} (max) that is too low for use in most power amplifiers; one solution to this would be a cascode input stage, as described earlier, which would take most of the voltage strain off the input devices. However, as we shall see, at the kind of source resistances we are dealing with, the current noise makes only a minor contribution to the total, and cascoding is probably not worthwhile for this reason alone.

The existence of current noise as well as voltage noise means that in general it is not possible to minimise transistor noise just by increasing the collector current to the maximum value the device can take. Increasing I_c certainly reduces voltage noise, but it increases current noise. Hence there is an optimum collector current for each value of source resistance, where the contributions are equal, and the total thus at a minimum. Because both voltage and current mechanisms are proportional to the square root of I_c, they change relatively slowly as it is altered, and the noise curve is rather flat at the bottom. See Figure 6.25. There is no need to control collector

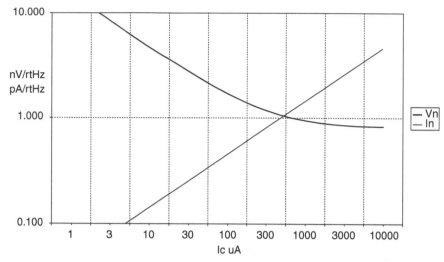

Figure 6.25. Showing how voltage noise density Vn and current noise density In vary with collector current Ic in a generic transistor. As Ic increases, the voltage noise falls to a lower limit while the current noise continuously increases.

current with great accuracy to obtain the optimum noise performance.

I want to emphasise here that this is a simplified noise model, not least because in practice both voltage and current noise densities vary with frequency. I have also ignored 1/f noise. However, it gives the essential insight into what is happening and leads to the right design decisions so we will put our heads down and press on.

A quick example shows how this works. In an audio power amplifier we want the source impedances seen by the input transistors to be as low as possible, to minimise Johnson noise, and to minimise the effects of input current distortion, as described elsewhere in this chapter. The output impedance of the source equipment will, if we are lucky, simply be the value of the output

resistor required to give stability when driving cable capacitance, i.e., about 100 Ω. It is also usually possible to design the negative feedback network so it has a similar source impedance; see Figure 6.23 above, for an example. So let us look at optimising the noise from a single transistor faced with a 100 Ω source resistance.

A few assumptions need to be made. The temperature is 25°C, the bandwidth is 22 kHz, and the Rbb of our transistor is 40 Ω, which seems like an average value. (Why don't they put this on spec sheets any more?) The beta (hfe) is 150. Set the Ic to 1 mA, which is plausible for an amplifier input stage, step the source resistance from 1 to 100,000Ω and the calculations come out like this, in Table 6.5.

Table 6.5. The summation of Johnson noise from the source resistance with transistor noise

1	2	3	4	5	6	7	8	9
Rsource Ohms	Rsource Johnson nV/rtHz	Rsource Johnson BW nV	Rsource Johnson BW dBu	Transistor noise incl In in Rs nV/rtHz	Transistor noise plus Rs Johnson nV/rtHz	Noise in BW nV	Noise in BW dBu	Noise Fig dB
1	0.128	19.0	−152.2	0.93	0.94	139.7	−134.9	17.3
10	0.406	60.2	−142.2	0.93	1.02	150.9	−134.2	8.0
100	1.283	190.3	−132.2	0.94	1.59	236.3	−130.3	1.9
1000	4.057	601.8	−122.2	1.73	4.41	654.4	−121.5	0.7
10000	12.830	1903.0	−112.2	14.64	19.46	2886.9	−108.6	3.6
100000	40.573	6017.9	−102.2	146.06	151.59	22484.8	−90.7	11.4

The first column shows the source resistance, and the second column the Johnson noise density it generates by itself. Factor in the bandwidth, and you get the third and fourth columns which show the actual noise voltage in two different ways. The fifth column is the aggregate noise density from the transistor, obtained by taking the RMS sum of the voltage noise and the voltage generated by the current noise flowing in the source resistance. The sixth column gives total noise density when we sum the source resistance noise density with the transistor noise density. Factor in the bandwidth again, and the resultant noise voltage is given in columns seven and eight. The last column gives the Noise Figure (NF), which is the amount by which the combination of transistor and source resistance is noisier than the source resistance alone. In other words, it tells how close we have got to theoretical perfection, which would be a Noise Figure of 0 dB.

The results are, I hope, instructive. The results for 100 Ω show that the transistor noise is less than the source resistance noise, and we know at once that the amount by which we can improve things by twiddling the transistor operating conditions is pretty limited. The results for the other source resistances are worth looking at. The lowest noise output (−134.9 dBu) is achieved by the lowest source resistance of 1 Ω, as you would expect, but the NF is very poor at 17.3 dB; this gives you some idea why it is hard to design quiet moving-coil head amplifiers. The best noise figure, and the closest approach to theoretical perfection is with 1000 Ω, but this is attained with a *greater* noise output than 100 Ω. As the source resistance increases further, the NF begins to get worse again; a transistor with an Ic of 1 mA has relatively high current noise and does not perform well with high source resistances.

You will note that we started off with what in most areas of electronics would be a high collector current:

1 mA. In fact, this is too low for amplifier input stages designed to my philosophy, and most of the examples in this book have a 6 mA tail current, which splits into 3 mA in each device; this value is chosen to allow linearisation of the input pair and give a good slew-rate, rather than from noise considerations. So we dial an Ic of 3 mA into our spreadsheet, and we find there is a slight improvement for our 100 Ω source resistance case; but only a marginal 0.2 dB. See Table 6.6, which this time skips the intermediate calculations and just gives the results.

For 1 Ω things are 0.7 dB better, due to slightly lower voltage noise, and for 100,000 Ω they are worse by no less than 9.8 dB as the current noise is much increased. So ... let's get radical and increase Ic to 10 mA. Unfortunately this makes the 100 Ω noise worse, and we have lost our slender 0.2 dB improvement. This theoretical result is backed up by practical experience, where it is found that increasing the tail current from 6 mA (3 mA per device) to 20 mA (10 mA per device) gives no significant reduction in the noise output.

For 1 Ω, the noise is 0.3 dB better − hardly a triumph − and for the higher source resistances things get rapidly worse, the 100,000 Ω noise increasing by another 5.2 dB. It therefore appears that a collector current of 3 mA is actually pretty much optimal for noise with our 100 Ω source resistance, even though it was originally chosen for other reasons.

Let us now pluck out the 'ordinary' transistor and replace it with a specialised low-Rbb part like the much-lamented 2SB737 (now regrettably obsolete), which has a superbly low Rbb of 2 Ω. The noise output at 1 Ω plummets by 10 dB, showing just how important low Rbb is under these conditions; for a more practical 100 Ω source resistance noise drops by a useful 1.0 dB. As you might expect.

Table 6.6. How input device collector current affects noise figure

Rsource Ohms	Ic = 3 mA		Ic = 10 mA		Ic = 10 mA, 2SB737		Ic = 100 uA	
	Noise dBu	Noise Fig dB	Noise dBu	Noise Fig dB	Noise dBu	Noise Fig dB	Noise dBu	Noise Fig dB
1	−135.6	16.6	−135.9	16.3	−145.9	6.3	−129.9	22.3
10	−134.8	7.4	−135.1	7.1	−140.9	1.3	−129.7	12.5
100	−130.5	1.7	−130.3	1.9	−131.5	0.7	−127.9	4.3
1000	−120.6	1.6	−118.5	3.7	−118.6	3.6	−121.5	0.7
10000	−105.3	6.9	−100.7	11.4	−100.7	11.4	−111.6	0.6
100000	−86.2	16.0	−81.0	21.2	−81.0	21.2	−98.6	3.6

As an aside, let's go back to the ordinary transistor and cut its Ic right down to 100 uA, giving the last two columns in Table 6.6. Compared with Ic = 3 mA, noise with the 1 Ω source degrades by 5.7 dB, and with the 100 Ω source by 2.6 dB, but with the 100,000 Ω source, there is a hefty 12.4 dB improvement, showing why BJT inputs for high impedances use low collector currents.

If you're stuck with a high source impedance, JFETs can give better noise performance than BJTs; JFETs are not dealt with here for reasons already explained; their low transconductance and poor Vgs matching.

We therefore conclude that our theoretical noise output with Ic = 3 mA and Rs = 100 Ω will be −130.5 dBu, with a Noise Figure of 1.7 dB. However, these calculations are dealing with a single transistor and a single source resistance; in a differential input stage, there are two transistors, and if we assume equal source resistances of 100 Ω for each one, as explained above, the noise output has to be increased by 3 dB as we are adding two non-correlated noise voltages. This gives us a theoretical noise output of −127.5 dBu, which, it has to be said, does not match up particularly well with the practical figure of −124.2 dBu that we deduced in the previous section. There are several reasons for this; to make the explanation manageable in the space available we have had to ignore some minor sources of extra noise, the frequency dependence of the voltage and current noise sources, and we have used a generic transistor. There is not much choice about the latter as manufacturers tend not to publish Rbb or noise data for the high-voltage transistors that are used in audio power amplifiers.

It seems pretty clear that we are not going to get any significant improvement in power amplifier noise by altering the input device conditions. It could of course be argued that there is no point in making it any quieter, because a pair of discrete transistors with a low source impedance are about as quiet as it gets, and pretty much anything you put in front of it is going to dominate the noise situation. This issue is developed further in Chapter 27 on power amplifier input systems, which deals with balanced input amplifiers and so on. On the other hand …

Reducing Input Transistor Noise

Let's assume that we wish to reduce the transistor contribution to the amplifier noise. A reliable method of doing this, often used in moving-coil preamplifiers, is the use of multiple transistors in parallel. The gain will sum arithmetically but the noise from each transistor will be uncorrelated and therefore subject to rms-summing. Two transistors will be 3 dB quieter than one, three transistors 4.8 dB quieter, and four transistors 6 dB quieter. There are obvious practical limits to this, and you soon start thinking about grains of corn on chessboards, but putting four transistors into each side of an input stage is quite feasible. The cost of the small-signal part of the amplifier will still be a very small fraction of the cost of power devices, heatsinks, mains transformer, and so on. The main thing that needs to be taken into account is current-sharing between the devices.

Figure 6.26 shows two different ways that this could be implemented, assuming that it is desired to keep the emitter degeneration resistors at their usual value of 100 Ω. In Figure 6.26b, three 27 Ω resistors effectively in parallel give 9 Ω, which with a series 91 Ω resistor very

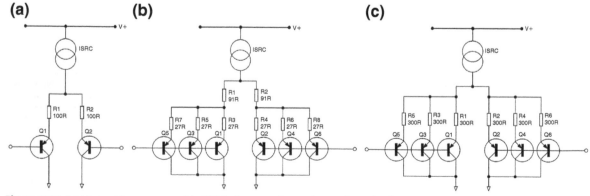

Figure 6.26. (a) Normal input stage with 100 Ω degeneration resistors; (b) multiple input devices with small current-sharing resistors; (c) multiple input devices with split emitter degeneration resistors.

handily keeps the total degeneration resistance at exactly 100 Ω. In Figure 6.26c, the values work out equally neatly, and two resistors are saved. The tail current source may have to be increased in value, but not necessarily trebled, for, as we have seen, the noise performance varies quite slowly as collector current changes.

Offset and Match: The DC Precision Issue

The same components that dominate amplifier noise performance also determine the output DC offset. Looking at Figure 6.27, if R9 is reduced to minimise the source resistance seen by TR3, then the value of R8 must be scaled to preserve the same closed-loop gain, and this reduces the voltage drops caused by input transistor base currents.

Most of my amplifier designs have assumed that a ±50 mV output DC offset is acceptable. This allows DC trim-pots, offset servos, etc. to be gratefully dispensed with. However, it is not in my nature to leave well enough alone, and it could be argued that ±50 mV is on the high side for a top-flight amplifier. I have therefore reduced this range as much as possible without resorting to a servo; the required changes have already been made when the NFB network was reduced in impedance to minimise Johnson noise (see earlier in this chapter).

With the usual range of component values, the DC offset is determined not so much by input transistor Vbe mismatch, which tends to be only 5 mV or so, but more by a second mechanism − imbalance in beta. This causes imbalance of the base currents (lb) drawn thorough input bias resistor R1 and feedback resistor R8, and the cancellation of the voltage-drops across these components is therefore compromised.

A third source of DC offset is non-ideal matching of input degeneration resistors R2, R3. Here they are 100 Ω, with 300 mV dropped across each, so two 1% components at opposite ends of their tolerance bands could give a maximum offset of 6 mV. In practice, this is most unlikely, and the error from this source will probably not exceed 2 mV.

There are several ways to reduce DC offset. First, low-power amplifiers with a single output pair must be run from modest HT rails and so the requirement for high-Vce input transistors can be relaxed. This allows higher beta devices to be used, directly reducing lb. The 2SA970 devices used in this design have a beta range of 350−700, compared with 100 or less for MPSA06/56. Note the pinout is not the same.

Earlier, we reduced the impedance of the feedback network by a factor of 4.5, and the offset component due to lb imbalance is reduced by the same ratio. We might therefore hope to keep the DC output offset for the improved amplifier to within ±15 mV without trimming or servos. Using high-beta input devices, the lb errors did not exceed ±15 mV for 10 sample pairs (*not* all from the same batch) and only three pairs exceeded ±10 mV. The lb errors are now reduced to the same order of magnitude as Vbe mismatches, and so no great improvement can be expected from further reduction of circuit resistances. Drift over time was measured at less than 1 mV, and this seems to be entirely a function of temperature equality in the input pair.

Figure 6.27 shows the ideal DC conditions in a perfectly balanced input stage, assuming $\beta = 400$, compared with a set of real voltages and currents from the prototype amplifier. In the latter case, there is a typical partial cancellation of offsets from the three different mechanisms, resulting in a creditable output offset of -2.6 mV.

The Input Stage and the Slew-rate

This is another parameter which is usually assumed to be set by the input stage, and has a close association with HF distortion. A brief summary is therefore given here, but the subject is dealt with in much greater depth in Chapter 15.

An amplifier's slew-rate is proportional to the input stage's maximum-current capability, most circuit configurations being limited to switching the whole of the tail current to one side or the other. The usual differential pair can only manage half of this, as with the output slewing negatively half the tail current is wasted in the input collector load R2. The addition of an input current-mirror, as advocated above, will double the slew rate in both directions as this ineffi-ciency is abolished. With a tail current of 1.2 mA, a mirror improves the slew-rate from about 5 V/μsec to 10 V/μsec (for Cdom = 100pF). The constant-gm degeneration method of linearity enhancement in Figure 6.9 further increases it to 20 V/μsec.

In practice, slew-rates are not the same for positive- and negative-going directions, especially in the conven-tional amplifier architecture which is the main focus of this book; this issue is examined in Chapters 7 and 15.

Input Stage Conclusions

Hopefully this chapter has shown that input stage design is not something to be taken lightly if low noise, low distortion, and low offset are desired. A good design

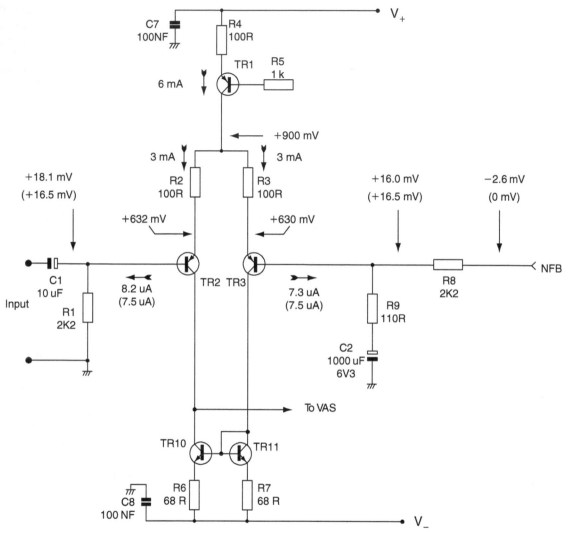

Figure 6.27. The measured DC conditions in a real input stage. Ideal voltages and currents for perfectly matched components are shown in brackets.

choice even for very high quality requirements is a constant-gm degenerated input pair with a degenerated

current-mirror; the extra cost of the mirror will be trivial.

References

1. Gray, P R and Meyer, R G, *Analysis and Design of Analog Integrated Circuits* Wiley 1984, p. 172 (exponential law of singleton).

2. Gray, P R and Meyer, R G, *Analysis and Design of Analog Integrated Circuits* Wiley 1984, p. 194 (tanh law of simple pair).

3. Self, D, Sound Mosfet Design, *Electronics and Wireless World*, Sept. 1990, p. 760 (varying input balance with R2).

4. Taylor, E, Distortion in Low Noise Amplifiers, *Wireless World*, Aug. 1977, p. 32.

5. Gray, P R and Meyer, R G, *Analysis and Design of Analog Integrated Circuits*, Chichester: John Wiley 1984, p. 256 (tanh law of current-mirror pair).

6. Herpy, M, *Analog Integrated Circuits*, New York: Wiley-Interscience, 1987, p.118.

7. Geiger, N, *Allen Strader VLSI Design Techniques for Analog and Digital Circuits* New York: McGraw-Hill 1990.

8. Feucht, D L, *Handbook of Analog Circuit Design*, San Diego: Academic Press, 1990, p. 432 (Cross-quad).

9. Quinn, P, *IEEE International Solid-State Circuits Conference*, THPM 14.5, p. 188. (Cascomp).

10. Staric, P and Margan, E, *Wideband Amplifiers*, New York: Springer, 2006, pp. 5.014-5.116.

11. Gilbert, B, The Multi-tanh Principle: A Tutorial Overview, *IEEE Journal Of Solid-State Circuits*, 33, No 1, Jan. 1998, pp. 2–17.

12. Hearn, W E, Fast Slewing Monolithic Operational Amplifier, *IEEE Journal of Solid State Circuits* Vol SC6, Feb. 1971, pp. 20–24 (AB input stage).

13. Van de Plassche, R J, A Wide-band Monolithic Instrumentation Amplifier, *IEEE Journal of Solid State Circuits*, Vol SC,10 Dec. 1975, pp. 424–431.

14. Stochino, G, Ultra-Fast Amplifier, *Electronics and Wireless World*, Oct. 1996, p. 835.

15. Hickman, F V and Hunt, R W, On Electronic Voltage Stabilizers, *Review of Scientific Instruments* 10, Jan. 1939, pp. 6–21.

16. Self, D, A High Performance Preamplifier, *Wireless World*, Feb. 1979, p. 40.

17. Taylor, E, Distortion in Low Noise Amplifiers, *Wireless World*, Aug. 1977, p. 29.

18. Edward Cherry, Private communication, June 1996.

19. Giovanni Stochino, Private communication, May 1996.

20. Dostal, J, *Operational Amplifiers*, Oxford: Butterworth-Heinemann, 1993, p. 65.

21. Self, D, *The Design of Active Crossovers*, Burlington, MA: Focal Press, 2011, pp. 341–349.

22. Groner, S, *Comments on Audio Power Amplifier Design Handbook*, 2011, pp. 2–5, SG Acoustics, http://www.sg-acoustics.ch/analogue_audio/power_amplifiers/index.html (accessed Aug. 2012).

23. Ibid., p5.

24. Bilotti, A, Noise Characteristics of Current Mirror Sinks/Sources *IEEE J. Solid-State Circuits*, vol. SC-10, no. 6, pp. 516–524, December 1975.

The Voltage-Amplifier Stage

You have to have extra voltage, some extra temperament to reach certain heights.
                              Laurence Olivier, interview, 1979

The Voltage-Amplifier Stage (VAS)

The Voltage-Amplifier Stage (VAS) has often been regarded as the most critical part of a power amplifier, since it not only provides all the voltage gain but also must give the full output voltage swing. The input stage may give substantial transconductance gain, but the output is in the form of a current, and the signal voltage on the VAS input is only a few milliVolts. However, as is not uncommon in audio, all is not quite as it appears. A VAS designed with a few simple precautions will contribute relatively little to the overall distortion total of an amplifier, and if even the simplest steps are taken to linearise it further, its contribution sinks out of sight.

VASs can be divided into two types: single-ended (where the collector load is a current-source or equivalent); and push-pull, with actively-driven transistors at top and bottom. This chapter deals extensively with the single-ended type; the push-pull VAS is examined in Chapter 8.

In February 2011, Samuel Groner sent me an advance copy of a commentary he had written on my previous power amplifier design books. It is an impressive document, covering 55 pages and going beyond commentary to describe some very penetrating experiments that went a good deal further into the details of amplifier operation than I had previously done. As a result Samuel and I spent a few months working together to extend the scope of his tour-de-force document, and a good deal of that work is used in this chapter. The original commentary can be found at.[1] I strongly recommend it to anyone deeply interested in power amplifiers. The topics in it that I have worked on intensively myself are mainly current-mirror noise and VAS distortion. I have yet to find that Samuel has anywhere made a mistake.

The measurements illustrating Samuel's points in this chapter are all mine, as are any mistakes. However, since everything has been tested independently by two people, making the same real measurements, I hope you will find that this chapter is an exceptionally solid source of audio information.

The Naming of Parts

I have been roundly criticised in some quarters for referring to the second stage of a three-stage amplifier as the

Voltage-Amplifier Stage, it being pointed out that this implies a voltage-in voltage-out amplifier. The stage is actually a current-in voltage-out amplifier, which makes it a transimpedance amplifier. (Not a transresistance amplifier, as that implies the shunt feedback is a resistance, whereas it is actually the compensation capacitor Cdom.) I did not call the stage a transimpedance amplifier for the simple reason that this term is not widely used and would mystify a lot of people. Describing the input stage as a transconductance amplifier is more common but most people still refer to it as 'the input stage'.

Therefore, when I first started writing about power amplifiers, I decided to go with Voltage-Amplifier Stage or VAS. My justifications are:

1. While it may be a transimpedance amplifier, the VAS does provide all the voltage gain in the amplifier. (The signal voltage at the input stage output is less than the differential error voltage above 6 kHz in a typical amplifier design with a simple VAS.)
2. It is the first stage in the amplifier that has the full voltage swing at its output.

From these considerations it is clear that the VAS has a demanding role, and that is mainly because it gives a high-amplitude voltage output. I think that Voltage-Amplifier Stage is as good a name as any. Please fight down the temptation to talk about a 'VAS stage'.

The Basic Single-ended VAS

There are three basic ways to use a transistor: common-emitter, common-base, and common-collector (better known as emitter-follower). Since voltage gain is required, the common-collector configuration is off the menu. A common-base amplifier can give voltage gain, but the need for the input to be at a lower voltage than the base makes it awkward to use; nonetheless some amplifiers with a common-base VAS have been designed, though they are usually referred to as 'folded-cascode' types, and these are dealt with at the end of this chapter. That leaves the common-emitter stage as overwhelmingly the most popular form of VAS.

The simplest sort of common-emitter amplifier has just a resistor for a collector load, as in Figure 7.1a, which shows a model amplifier with a small-signal constant-current Class-A output stage. This output stage gives very little distortion of its own. The mirrored input pair that drives the VAS is heavily degenerated with emitter resistors and also contributes very little.

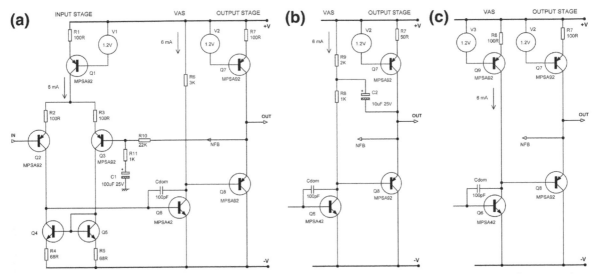

Figure 7.1. Common-emitter VAS in a model amplifier with a small-signal Class-A output stage; (a) simple VAS with resistive load; (b) simple VAS with bootstrapped resistive load; (c) simple VAS with current-source load.

The VAS is not elaborated by cascoding or adding an emitter-follower, so it is called a 'simple VAS'. It will generate significant distortion, as we will see.

The collector load in Figure 7.1a is chosen as 3 kΩ so the VAS collector current is 6 mA with ±20 V supply rails. (The normal design range is 6 to 10 mA.) The distortion performance is shown by the "no bootstrap" trace in Figure 7.2. The LF distortion (below 5 kHz in this case) is third-harmonic; the HF distortion (above 5 kHz) is a mixture of second and third harmonics.

Despite the very simple resistive load the VAS generates a fairly low level of distortion. This is because at LF, global feedback linearises the whole amplifier, while at HF the VAS is linearised by local NFB through Cdom, with a smooth transition between the two. It is therefore important that the local open-loop gain of the VAS (that existing inside the local feedback loop closed by Cdom) be high, so that the VAS can be well linearised. In Figure 7.1a, increasing the value of Rc in an attempt to increase the gain will decrease the collector current of the VAS transistor, reducing its transconductance and getting you back where you started; the gain will be low. An active load is very desirable to increase the effective collector impedance of the VAS and thus increase the raw voltage gain; either bootstrapping or a current-source works, but the current source is more dependable, and almost the universal choice for hi-fi or professional amplifiers.

Another serious disadvantage of the simple resistive load is its very limited ability to source current into the output stage on positive half-cycles; this is a particular problem with sub-8 Ω loads which draw disproportionally more current. This is why a simple resistor load is hardly ever used in power amplifiers, the only example coming to mind being a design by Locanthi,[2] which used a rather high 9 kΩ resistor; the published distortion performance was unimpressive, to put it mildly. An active load ensures the VAS stage can source enough current to drive the upper half of the output stage right up to the supply rail.

It may not be immediately obvious how to check that impedance-enhancing measures are working properly, but it is actually fairly simple. The VAS collector impedance can be determined by the simple expedient of shunting the VAS collector to ground with decreasing resistance until the open-loop gain reading falls by 6 dB, indicating that the collector impedance is equal to the current value of the test resistor.

Bootstrapping the VAS

The collector impedance of the VAS, and its current-sourcing abilities can both be improved by dividing the resistive load and bootstrapping the central point, as in Figure 7.1b. It works in most respects as well as a current source load, for all its old-fashioned look. In a model amplifier. It is necessary to double the standing current in the Class-A output stage by reducing R7 to 50 Ω, because apart from any external load it is now driving a 2 kΩ resistor effectively connected to ground

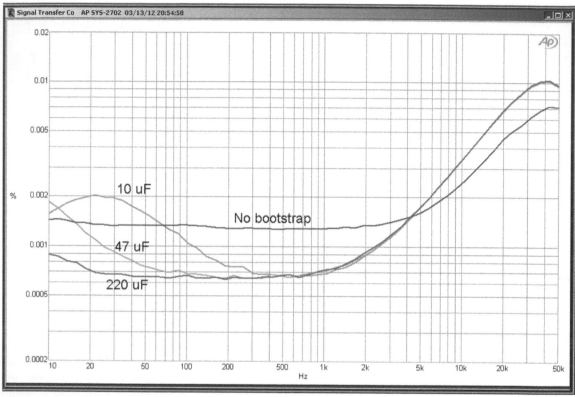

Figure 7.2. Distortion resulting from no bootstrapping, and bootstrapping with capacitors of 10 uF; 47 uF; and 220 uF; +20 dBu output, ±20V rails

via the +V supply rail. In a full-scale power amplifier the output stage would naturally have ample drive capability.

The bootstrap appears to promise more output swing, as the VAS collector can in theory soar like a lark above the V+ rail; under some circumstances this can be the overriding concern, and bootstrapping is alive and well in applications such as automotive power-amps that must make the best possible use of a restricted supply voltage.[3] However, in general, this is of little help as clipping tends to occur first on the negative half-cycle, and it is also necessary to make sure that the output stage cannot be overdriven to the point where it is damaged.

Correct bootstrapping is more subtle than it looks. In the circuit of Figure 7.1b, a 2 kΩ load needs to be driven and the capacitor driving it must be of adequate size. If we blithely assume that a bootstrap signal −3 dB down at 10 Hz will be satisfactory, we can use a 10 uF capacitor; that actually gives −3 dB at 8 Hz. But ... the distortion performance is shown by the "10 uF" trace in Figure 7.2. There is a significant improvement from

5 kHz down, but the distortion starts to increase at 300 Hz and by 60 Hz is actually worse. This is because effective bootstrapping depends on the bootstrapping signal being the same as the signal on the VAS collector, and even a very small LF roll-off messes that up. Abandoning calculation and turning to good old trial-and-error, we find that even 47 uF is not adequate, giving a trace which starts to rise from 60 Hz down. We need to use 220 uF (Trace 4) to keep the distortion flat down to 20 Hz, and to keep it flat to 10 Hz would require 470 uF, almost fifty times larger than the first value we arrived at simply by considering the frequency response. The bootstrap capacitor has become a sizeable component.

Bootstrapping like this has been criticised in the past for prolonging recovery from clipping because of the charge stored on the bootstrapping capacitor; I have no evidence to offer on this myself, but it is a point to keep an eye on.

Bootstrapping from the output stage in the usual way holds another subtle drawback. The LF open-loop gain is dependent on amplifier output loading. As we have just seen, the effectiveness of bootstrapping depends crucially on the bootstrapping signal being at the same level as the signal on the VAS collector, or at any rate very close to it; however, the finite transconductance of the output-transistors, and the presence of their emitter resistors mean that there will be a load-dependent gain loss in the output stage. This in turn significantly alters the amount by which the VAS collector impedance is increased; hence the LF feedback factor is dynamically altered by the impedance characteristics of the loudspeaker load and the spectral distribution of the source material. This has a special significance if the load is an 'audiophile' speaker that may have impedance dips down to 2 Ω, in which case, the loss of gain is serious. If anyone needs a new audio-impairment mechanism to fret about, then I humbly offer this one in the confident belief that its effects, while measurable, are not of audible significance. Possibly this is a more convincing reason for avoiding bootstrapping than any alleged difficulties with recovery from clipping.

Another potential problem with bootstrapping is that the standing DC current through the VAS, and hence the bias generator, varies with rail voltage. Setting and maintaining the quiescent conditions is quite difficult enough already, so this source of variation is apparently wholly unwelcome. However, for reasons fully explained in Chapter 22, it cannot be said that the ideal VAS would have an absolutely constant standing current. If it varies in a controlled way with rail voltage, this can be used to compensate for output bias variations due to changing rail voltages acting directly on the drivers and output devices so the overall effect is constant Vq.

The conclusion must be that classical bootstrapping may look simple but it has a lot of subtle drawbacks. It is now rarely used. However, bootstrapping using a DC shift rather than a capacitor can be effective; see Figure 7.27b later in this chapter.

The Current-source VAS

A better collector load for the VAS is a current source, and these are almost universally used. VAS gain is not dependent on the output stage gain, as it is with bootstrapping, and the collector current is stabilised against supply rail variations. Figure 7.1c shows the arrangement. The current source requires a bias voltage, and this can be provided either by a pair of diodes or by a negative-feedback system. (See Chapter 15.) There is no evidence that anything more sophisticated than the simple source of Figure 7.1c (such as a configuration

with a very high output impedance) will give any benefits.

The current source VAS has its collector impedance limited by the effective output resistance Ro of the VAS and the current source transistors,[4] which is another way of saying that the improvement is limited by Early effect. It has been stated that this topology provides current-drive to the output stage; this is only very partly true. Once the local NFB loop has been closed by adding Cdom, the impedance at the VAS output falls at 6 dB/octave for frequencies above P1. With typical values the impedance is only a few kΩ at 10 kHz, and I think this hardly qualifies as current-drive.

VAS Operation and Open-loop Gain

The typical VAS topology as shown in Figure 7.1 is a classic common-emitter voltage-amplifier stage, with current-drive input into the base from the input stage. The small-signal characteristics, which set open-loop gain and so on, can be usefully simulated by the SPICE model shown in Figure 7.3, of a VAS reduced to its conceptual essentials. G is a current-source whose value is controlled by the voltage-difference between Rin and RF2, and represents the differential transconductance input stage. F represents the VAS transistor, and is a current-source yielding a current of beta times that sensed flowing through 'ammeter' VA which by SPICE convention is a voltage-source set to 0 V; the value of beta, representing current-gain as usual, models the relationship between VAS collector current and base current. Rc represents the total VAS collector impedance, a typical real value being 22 kΩ. With suitable parameter values, this simple model provides a good demonstration of the relationships between gain, dominant-pole frequency, and input stage current that were introduced in Chapter 5. Injecting a small signal current into the output node

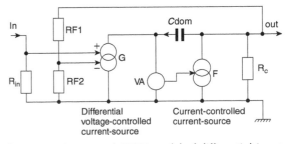

Figure 7.3. Conceptual SPICE model of differential input stage (G) and VAS (F). The current in F is Beta times the current in VA.

from an extra current-source also allows the fall of impedance with frequency to be examined.

Before Cdom is connected, the overall voltage-gain clearly depends linearly on beta, which in real transistors varies widely. Working on the trusty engineering principle that what cannot be controlled must be made irrelevant, adding Cdom gives local shunt NFB that provides a predictable 6 dB/octave slope as gain falls with frequency, thus setting the crucial HF gain that controls Nyquist stability. As we saw in Chapter 5, the HF open-loop gain depends only on the input stage transconductance g_m, and the Miller dominant pole capacitance C_{dom}, and this is at the root of the dependable stability of the classic three-stage amplifier. The gain at any frequency in this region is simply:

$$HFgain = \frac{g_m}{\omega \cdot C_{dom}}$$ Equation 7.1

where $\omega = 2 \cdot \pi \cdot$ frequency

The LF open-loop gain below the dominant-pole frequency P1 remains variable, and therefore so does frequency P1. This is of lesser importance, as it does not affect HF stability, but it is still very desirable to make sure there is enough LF open-loop gain to push the distortion below the noise floor. As we saw in Chapter 5, the LF gain is proportional to both the VAS transistor beta and the VAS collector impedance Rc:

$$LF\ gain = g_m \cdot \beta \cdot R_c$$ Equation 7.2

Cdom no longer comes into it; the open-loop gain now depends on the beta of the VAS transistor and its collector impedance Rc.

With a simple VAS and typical components, we get a typical LF open-loop gain of +82 dB, with P1 just above 1 kHz. Both these answers will vary somewhat as they depend on transistor parameters. Increasing the open-loop gain without changing the LF gain means P1 must be lower; see Figure 5.2.

Further SPICE simulations were run with a complete amplifier, including an EF output stage driving an 8 Ω load. These were to demonstrate the factors affecting the LF open-loop gain. The first suspect is the Early voltage of the VAS transistor, as this influences the effective value of Rc. The default Early voltage for the MPSA42 model is 45.1 Volts, set by the parameter VAF. This is rather lower than the 100 Volts that the majority of SPICE models use, and is perhaps not ideal for a VAS; I am not sure if it derives from the high-voltage capability of the part. Remember that the higher the Early voltage, the less Early effect there is.

Increasing VAF to 1000 Volts increased the LF open-loop gain from +82.8 dB to +93.3 dB, a substantial improvement of 10.5 dB, more than three times. Increasing VAF again to 10,000 Volts further increased the LF gain to +98.9 dB, while VAF = 100,000 Volts gave +99.8 dB. Clearly we have pretty much disabled the Early effect completely, but as a final check, VAF was set to 1,000,000 Volts, giving +99.9 dB; we have done all we can with the Early voltage. In each case the frequency of P1 moved down as expected.

Having reached the limits of what we can do with the Early effect, the only thing left to tamper with is the beta in the LF gain equation. We cannot alter the input stage gm because that will affect HF stability. The basic parameter that sets beta in SPICE transistor models is BF, though there are other parameters that affect how it varies with collector current, etc. The MPSA42 model has a BF of 70.4 which reflects its generally low beta. If we keep VAF at 1,000,000, then we find that increasing BF from 70.4 to 200 gives a further 7.9 dB increase in LF open-loop gain, with P1 once more reduced as expected. That is a ratio of 2.48 times, while we increased BF by a ratio of 2.84 times. The correspondence is only approximate, but it confirms the role of the VAS transistor beta in setting the LF open-loop gain.

If we put VAF back to its default of 45.1, but keep BF at 200, the LF gain is reduced to +91.9 dB. This is still 9.1 dB higher than with the default value of BF = 70.4, and demonstrates again the part VAS beta plays in determining the LF open-loop gain.

In real life the only way to change the Early effect or the beta of the VAS transistor is to use a different type. However, one quantity that we can change easily is the VAS collector current. We will see later in this chapter that the Early effect can be approximately modelled by adding to the basic transistor model a collector-emitter resistance r_o which is proportional to the Early voltage of the transistor and inversely proportional to its collector current Ic. This suggests that we could reduce the Early effect, and so increase the LF gain, simply by reducing Ic. Table 7.1 shows that this is true; dropping Ic from 7.9 mA to 2.6 mA increases the LF gain by 8.9 dB. The final column shows the result of multiplying Ic with the LF open-loop gain; the result is almost constant, demonstrating that to a reasonable approximation LF gain is indeed inversely proportional to collector current Ic.

However, the VAS standing current has to be kept reasonably high so it can drive the output stage easily, and also give an adequate slew-rate, so as a means of increasing LF open-loop gain, this is of limited use.

Table 7.1. LF gain against simple VAS collector current Ic. SPICE simulation

Re Ohms	Ic mA	LF gain dB	LF gain x	Ic x LF gain
47	11.176	79.9	9886	110
51	10.352	80.6	10715	111
56	9.480	81.3	11614	110
62	8.613	82.1	12735	110
68	7.895	82.8	13804	109
82	6.615	84.3	16406	109
100	5.462	85.8	19498	107
120	4.612	87.2	22909	106
150	3.733	88.9	27861	104
180	3.135	90.2	32359	101
200	2.836	91.0	35481	101
220	2.590	91.7	38459	100

You will have noted that in this section all the attention has been focused on the VAS transistor. It is, however, directly connected to its constant-current load. What effect do the characteristics of the current-source transistor have? Fortunately, almost none. In the unmodified circuit the LF open-loop gain is +82.8 dB. Changing the VAF of the MPSA92 current-source transistor from the default value of 260 Volts to 1000 Volts only increases the LF open-loop gain to +83.0 dB. Further VAF increases to 10,000, to 100,000, and even 1,000,000 Volts only give us +83.1 dB. The effect of the current-source transistor characteristics is negligible, and this is yet another reason why the three-stage amplifier with a single-ended VAS is so predictable and dependable.

The Simple VAS in a Model Amplifier

To isolate the VAS distortion, we can use the model amplifier in Figure 7.4. The small-scale Class-A output stage is not required to drive a low-impedance

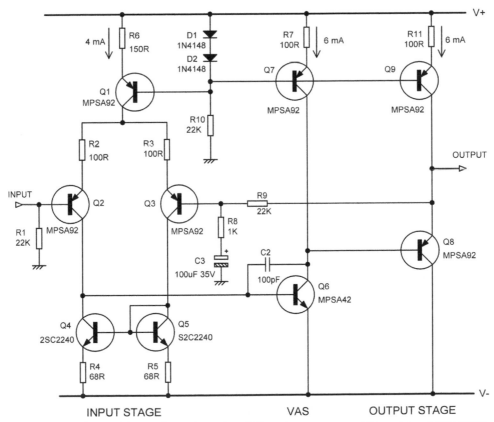

Figure 7.4. Model amplifier with simple VAS and Class-A output stage. All transistors MPSA42/92.

load; it only has to be able to drive the feedback network and the measuring equipment, and so it is easy to make it very linear. A simple emitter-follower with a constant-current emitter load has negligible distortion compared with a simple VAS. If this is ever in doubt the constant-current Class-A stage can be replaced by a push-pull Class-A stage[5] which has reliably lower distortion again. If there is no change in the THD reading, then the distortion contribution of the output stage can be neglected.

The supply rails and internal bias currents remain unchanged. Note that while the signal levels on the VAS terminals are modelled correctly, the circuit cannot allow for the effect of loading, and, in particular, non-linear loading on the VAS by the output stage. Note too that the components defining the open-loop gain and thus the feedback-factor, which are the input pair emitter-resistors and the Miller compensation capacitor, have the same values as for a full-scale power amplifier.

The power supply rails are in general ±20 V, which is a little low for power amplifier use but minimises embarrassing explosions and domino-theory disasters if you get a connection wrong. The rail voltages can always be turned up later in the development process.

The output signal level generally used here was +20 dBu (7.75 Vrms, 11 V peak) which exercises the linearity without getting near clipping, which might bring in extra effects irrelevant to the main line of enquiry.

The output stage needs to have much lower distortion than the VAS, and so does the input stage. In fact, applying the same amount of input emitter degeneration (100 Ω resistors), as in the full-size power amplifiers does the job very nicely, and there is the great advantage that parameters like the input stage transconductance and noise are the same.

When this amplifier is built mostly with MPSA42/92 transistors (these are high-voltage TO-92 devices very commonly used in the small-signal stages of power amplifiers), the distortion is as the lower trace in Figure 7.5. The THD at 10 kHz is 0.0050%, composed almost entirely of the second harmonic.

You are no doubt wondering why the current-mirrors are not MPSA42. The reason is not, as you might think,

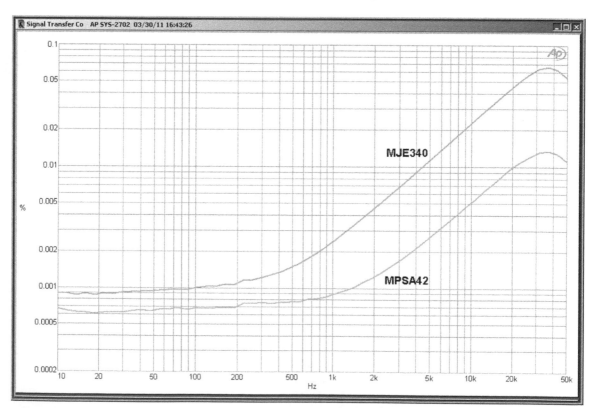

Figure 7.5. Distortion with MJE340 VAS (upper) compared with MPSA42 VAS (lower) +20 dBu output, ±20V rails.

that the 2SC2240 has a much higher beta and this helps mirror accuracy (though it has, and that would be a valid reason). It is because if the mirror is made up of MPSA42, there is rapidly rising distortion above 10 kHz, rising at 15.6 dB/octave (i.e., by a ratio of 6 times in an octave) which at first was rather puzzling. All other distortion mechanisms in a model amplifier rise at multiples of 6 dB/octave; 6 dB/octave for simple VAS second-harmonic distortion, 12 dB/octave for second-harmonic from a poorly balanced input stage, and 18 dB/octave for third-harmonic from a well-balanced input stage.

I will cut to the chase. The problem is that the MPSA42 requires an unusually high Vce to work properly, and in this amplifier Q4 in the mirror has only the Vbe of the VAS Q6 as Vce, less the 135 mV dropped across the mirror degeneration resistor R4. For every other transistor I have tried that is enough, but the MPSA42 is an exception, and I imagine this has something to do with its high-voltage capabilities. There is more on this issue in the section on the cascode VAS. When it first appeared, this issue took some time to sort out, and I called it The Current-Mirror Conundrum; at least that was the printable version.

It is common to find that TO-92 devices are reaching their dissipation limits in the more powerful amplifier designs, and the obvious solution is to replace both the VAS transistor Q6 and the current-source Q7 with driver-type transistors such as MJE340/350. Unfortunately this apparently straightforward modification makes the distortion dramatically worse, as shown by the upper trace in Figure 7.5. What appeared to be a straightforward replacement has increased the distortion at 10 kHz to 0.023%, a ratio of 4.6 times. Experiment soon shows that it is the replacement of the VAS transistor Q6, and not the current source, that has done the damage. The current source transistor type has very little if any effect and will not be looked at further.

It can be seen that there are two separate regimes of operation. At low frequencies (LF), the THD traces are flat, but at high frequencies (HF), the distortion increases with frequency, the percentage doubling for each octave of frequency increase. Both regimes have been affected by replacing the TO-92 VAS transistor with a medium-power type.

I am sure many amplifier designers have been aware of this behaviour, but may (like me) have previously dismissed it as a result of the generally lower beta of medium-power transistors. However, this hypothesis does not hold water, as the beta of high voltage TO-92 transistors is also low.

Figure 7.6 shows the results of replacing the VAS transistor not with a driver type, but a power transistor. You might expect this to end badly, and it does. The distortion is very much higher, 0.54% at 10 kHz. The HF distortion regime now covers the whole of the audio band, and is 108 times greater at 10 kHz than for the MPSA42. It is clear that the transistor types tried show very different performance when used in a VAS.

The Mechanisms of VAS Distortion

We have just seen convincing evidence that small-signal devices, such as those in TO-92 packages, in the VAS transistor position give a better distortion performance than medium-power types in TO-220 packages, and are far superior to power transistors. The question is why.

There are at least four plausible mechanisms for distortion from the VAS transistor:

1. The basic non-linearity of the VAS transistor Vbe−Ic relationship.
2. Variation of the VAS transistor beta with Ic.
3. Variation of the VAS transistor collector-base junction capacitance Cbc with Vce.
4. Variation of the VAS transistor Ic with Vce (Early effect).

In judging which of these are being demonstrated by the VAS in our model amplifier, we note that:

Mechanism (1) is implausible, because the non-linear VAS transistor Vbe−Ic relationship (often called the transistor equation) is noted for its predictability over a wide range of different device geometries and collector currents. It cannot account for the large variation in distortion seen between different types of VAS transistor. I speculated in the past that the Vbe−Ic non-linearity was the main cause of distortion in the simple VAS, with the transfer characteristic being a portion of an exponential.[6] It is my duty to tell you that I was quite wrong. Before, however, you start to gather the stockpile of First Stones, consider that the Vbe−Ic non-linearity surely must come into it at some deep level, for the non-linearity clearly exists; I suspect the effects of it may be below the threshold of actual THD measurements.

Mechanism (2) is also unconvincing, because VAS distortion in no way correlates with the transistor beta characteristics. Table 7.2 shows some relevant parameters for transistor types commonly used as a VAS, roughly in order of power capability, and the measured distortion at 10 kHz. The HF distortion of an MJE340 was 4.6 times greater than for the MPSA42, and the HF distortion of an MJL3281A was 108 times greater. The beta of the latter device is certainly not 108 times

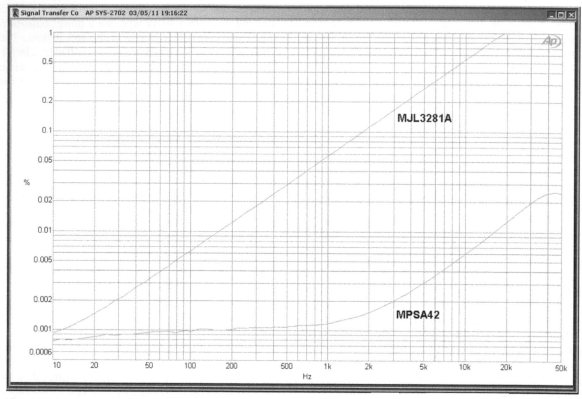

Figure 7.6. Distortion with MJL3281A VAS (upper) compared with MPSA42 VAS (lower) +20 dBu output, ±20V rails.

worse than that of the MPSA42; in fact, its minimum beta is nearly twice as high.

That leaves mechanisms (3) and (4). It was suggested to me by Samuel Groner[7] that the variation of the collector-base junction capacitance (Cbc) with Vcb was the main cause of the distortion in the HF region, and that Early effect was the main cause of the distortion in the LF region.

Table 7.2 shows that the only transistor parameter that varies by anything like 108 times between the MPSA42 and the MJL3281A is Cbc, the actual ratio being 300/2.6 = 115 times. Likewise the ratio for the MPSA42 and the MJE340 is 15/2.6 = 5.8 times. This is a less persuasive match to the HF distortion ratio of 4.6 times, and is probably because the Cbc data for the MJE340 is not very dependable.

Table 7.2. Data for different types of VAS transistor (all NPN)

Type	Vce max	Cbc or Cob at Vce	beta	Package	THD % 10 kHz
MPSA42	300V	2.6 pF at 20V	40 min	TO-92	0.0060
MPSA06	80V	4pF typ at 20V	100 min	TO-92	0.0055
2SA2240BL	120V	3pF typ at 10V	200–700	TO-92	0.0040
BD139	80V	Not known	25 min	TO-126	0.018
MJE340	300V	15 pF?	30–240	TO-225AA	0.023
BC441	60V	25pF max at 10V	40 min	TO-39	0.020
BFY50	35V	7pF typ, 12pF max at 10V	30 min	TO-39	0.011
BSS15	75V	Not known	30 min	TO-39	0.0085
MJL3281A	260V	600pF max at 10V	75–150	TO-264	0.55

We shall now see what evidence there is to support the hypothesis that Cbc causes the distortion in the HF region, and then go on to examine the role of the Early effect in LF region distortion.

HF Distortion from the VAS Transistor Cbc

All bipolar transistors have some collector-base junction capacitance appearing across the reverse-biased junction. There is also an emitter-base capacitance Ceb, but that usually has little effect on anything and is rarely mentioned.

Unfortunately, Cbc varies with applied voltage, as the junction width is affected. Figure 7.7 shows the variation of Cbc with Vcb for three small-signal (TO-92) transistors. This variation is great at the low-voltage end but flattens out markedly as Vcb increases. The curves are a little wobbly, as the numbers are derived from quite small graphs on the manufacturer's datasheets. Cbc is not usually specified in much detail and is sometimes omitted altogether.

Figure 7.7 emphasises that the change in Cbc with Vcb is highly asymmetrical about the operating point of our model amplifier, which has ±20 V rails and so corresponds to Vcb = 20V with no signal. Cbc changes much more rapidly at the low-voltage end.

The curves are essentially power laws. For example, the 2SC2240 curve shown can be closely approximated by the equation:

$$Cbc = 5.5(Vcb)^{-0.22} \qquad \text{Equation 7.3}$$

SPICE simulation uses this sort of power law to approximate the variation of junction capacitance for both diodes and BJTs.[8] In the diode case:

$$Cbc = \frac{CJO}{\left(1 - \dfrac{V}{\phi}\right)^{M}} \qquad \text{Equation 7.4}$$

Where:

CJO = SPICE parameter for zero-bias capacitance
Φ = the junction barrier potential (approx 0.7V)
M = the junction grading coefficient (in the range 0.5−0.33)

Similar models are used for the base-emitter and base-collector junction capacitances of BJTs, with MJE and MJC corresponding to M in each case.[9]

Equation 3 is clearly only an approximation; for example, it predicts infinite capacitance if V = Φ, but I think it does give some insight into how Cbc varies in real life. More detail on junction capacitances can be found in Tietze and Schenk.[10]

Changing the Amplifier Operating Point

Figure 7.7 suggests that if the Cbc hypothesis is correct, when the operating point of the model amplifier is

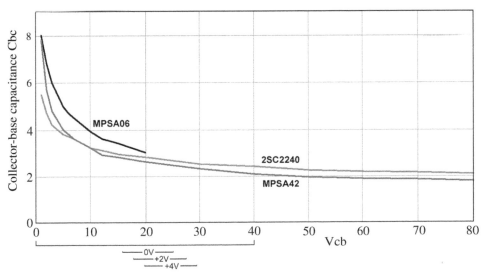

Figure 7.7. Cbc against Vcb for the three TO-92 transistors used. The wide bar represents the ±20 V rails of the model amplifier, while the short bars show the voltage range of a 3 Vrms signal, centred on operating points at 0 V, +2 V, and +4V.

shifted away from 0 V, a positive shift will reduce the amount of Cbc variation and so reduce the HF distortion, while a negative shift will increase the Cbc variation and increase the HF distortion. For positive shifts the Cbc variations become less as the shift increases, while for a negative shift, Cbc changes more rapidly as the shift increases, until it increases very quickly when Vcb is in the range 0–10 V.

The results are as predicted. Figure 7.8 shows how the distortion is reduced at a decreasing rate with increasing positive shifts of the amplifier operating point. Similarly, Figure 7.9 shows how distortion rises with increasing negative shifts of the amplifier operating point.

Changing the Supply Rails

The action of Cbc can also be explored by changing the supply rails of the model amplifier. Figure 7.10 shows how distortion decreases as the supply rails increase. This is because the voltage swing at the VAS collector now does not extend so far towards the negative supply rail, so the strongly curved portion of Figure 7.7 is not traversed.

The Dual VAS

The power-handling capacities of a TO-92 package VAS transistor are marginal for high-power amplifiers. As a compromise between TO-92 distortion performance and the markedly worse distortion of TO-5 or TO-220 devices, the use of two TO-92 transistors in parallel is a possibility, doubling the power dissipation capability.

If the total VAS Ic remains the same, Ic for each transistor will be halved, and so will its transconductance, since that is directly proportional to Ic. Since the collectors of the two transistors are connected together, the total transconductance remains unchanged, and so should have no effect on the VAS distortion. The collector voltage of the transistors is unchanged by doubling the transistors, and so there should be no change in whatever part of the VAS distortion is due to the Early effect.

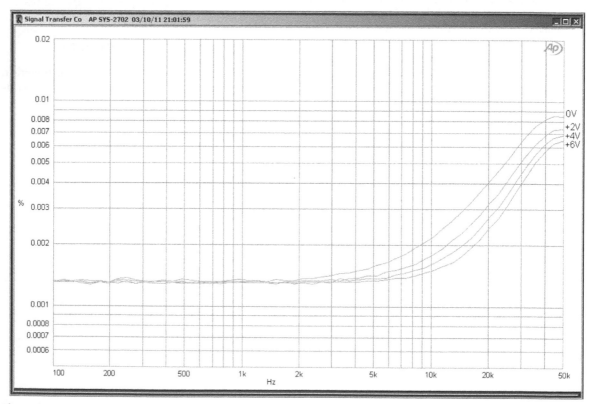

Figure 7.8. VAS distortion decreases as the amplifier operating point is shifted positive of 0 V. Simple MPSA42 VAS. 3 Vrms output, ±20 V rails.

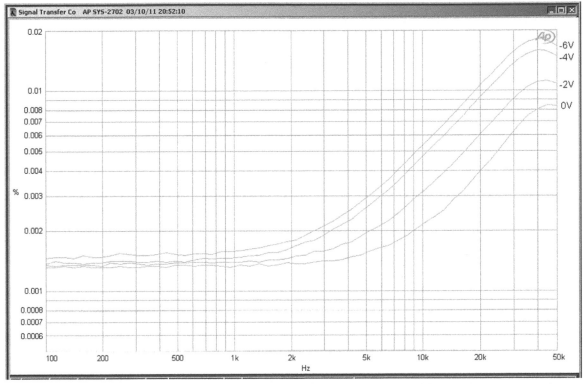

Figure 7.9. VAS distortion increases as the amplifier operating point is shifted negative of 0 V. MPSA42 simple VAS. 3 Vrms output, ±20V rails.

However, Cbc is doubled (approximately, due to tolerances in its value), and so we expect the HF distortion to also double. Figure 7.11 shows with beautiful clarity the HF distortion exactly doubling when the second VAS transistor is added. This is further evidence that Cbc is the cause of the HF VAS distortion.

Although a dual TO-92 VAS generates twice as much HF distortion as the single-transistor version, this is preferable to using a TO-220 device such as the MJE340, which gives 4.6 times more distortion.

VAS Distortion from Clamp Diodes

Some amplifier designs put a reverse-biased clamp diode across the Miller capacitor Cdom, with the intention of making negative clipping occur more cleanly. This is not the place to debate whether that is a useful technique, but what is certain is that it has unexpected and unhelpful consequences.

Diodes also have non-linear junction capacitance, and this acts in the same way as the non-linear Cbc of the VAS transistor to cause distortion; see Figure 7.12.

This is one more piece of evidence that Cbc causes VAS HF distortion. The junction capacitance Cj of the 1N4148 diode is specified as 4.0 pF max at zero bias. The value at 20V reverse-bias (the average bias here) does not seem to be quoted but by extrapolation from data sheet graphs is about 3.2 pF. This compares with Cbc = 2.6 pF at Vce = 20V for the MPSA42. You would therefore expect the distortion to increase considerably— in fact, more than double — when a diode is placed across Cdom. This did not happen, and at present the only explanation seems to be that the diode junction capacitance is specified rather pessimistically.

The History of Non-linear Cbc Distortion

The first reference to this problem I have found so far is in a famous amplifier design project by Arthur Bailey, entitled '30W High Fidelity Amplifier' which appeared in *Wireless World* in 1968.[11] He states that the distortion was initially worse than expected, and attributes this to Early effect in the VAS, and went on to say '... the

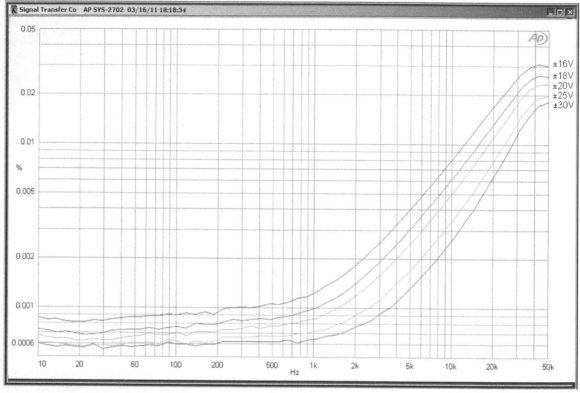

Figure 7.10. MPSA42 simple VAS. +20 dBu output, ±16 V, ±18 V, ±20 V, ±25 V, ±30 V rails.

high-frequency distortion was found to increase more rapidly than expected and this was traced to the modulation of the collector-base capacitance of this transistor. The high collector voltage swing was causing non-linear capacitive feedback, and this in turn was increasing the high-frequency distortion.' That's pretty specific. Unfortunately Arthur gave no details as to how the Cbc problem was traced, and the only solution he offered was choosing the best available transistor type. He used an RCA 40362; I have not so far been able to find any Cbc specs for this device.

I once did some consultancy work in the early 1980s (on a preamplifier design) for a well-respected power amplifier company where a lot of trouble was taken to select good specimens of the VAS transistor, which was a 2N5415 (PNP, TO-39). They were chosen for linearity by exercising them in a test-rig amplifier rather than by measuring transistor parameters, and I recall plastic drawers in the stores labelled 'Low distortion' and 'Medium distortion'. At the time I assumed that they were effectively being selected for high beta, but with the clarity of hindsight Cbc was almost

certainly the real variable that mattered. From his comments, I am absolutely certain the designer had no idea what he was selecting for. The 2N5415 data sheet shows a Cbc of 15 pF max, which would have made them comparable in HF distortion performance with the MJE340 in Figure 7.5 (i.e., much worse than a TO-92 device), and this is consistent with the known performance of the amplifier.

LF Distortion Due to VAS Transistor Early Effect

We have just seen that the HF distortion from a simple VAS is caused by the non-linear Cbc of the VAS transistor. We now look at the LF distortion regime. We have already seen that it is likely that Early effect causes this. Early effect is the increase in collector current Ic as the collector-emitter voltage Vce increases, with all other conditions constant.

Early effect can be modelled by adding to the basic transistor model two notional resistances: a collector-emitter resistance r_o and a collector-base resistance r_μ.[12] The values of these resistances are roughly

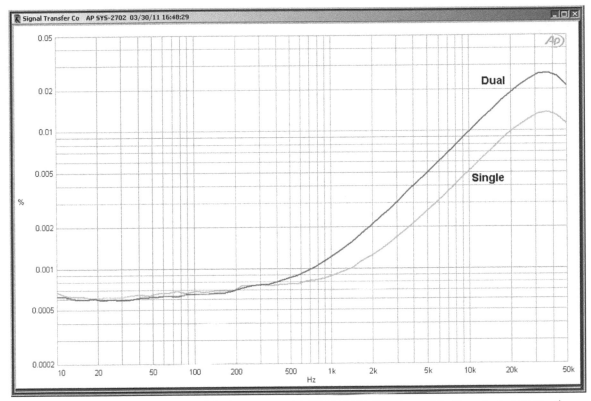

Figure 7.11. Single (lower trace) and dual (upper trace) MPSA42 VAS distortion. +20 dBu output, ±20 V rails.

proportional to Early voltage V_A and inversely proportional to collector current Ic, thus:

$$r_0 = \frac{V_A + V_{ce}}{I_c} \qquad \text{Equation 7.5}$$

$$r_\mu = \beta \cdot r_0 \qquad \text{Equation 7.6}$$

These equations show that both r_0 and r_μ vary with the collector voltage Vce. At low frequencies Ic varies very little with output voltage as the combined impedance of the compensation capacitor Cdom, the constant current collector load, and that looking into the output stage is high. However, Vce clearly varies with output voltage so the notional resistors will vary, modulating the gain because r_μ appears as a local feedback path between the collector and base of the VAS transistor.

The distortion mechanism of Vce modulating r_μ is very similar to that of Vce modulating the collector-base junction capacitance Cbc, which we have just

examined. However, the result here is distortion that is independent of frequency, rather than rising at 6 dB/octave. This is because as the signal frequency increases, the impedance of Cdom falls; it is in parallel with r_μ so the effect of the latter is reduced. On the other hand, the global feedback factor is also falling, so the effects cancel and the distortion is independent of frequency.

From this we can predict that if Early effect is the cause of LF distortion, the amount will be proportional to the VAS Ic, and will also vary with the Early voltage of the VAS transistor type.

Early Effect in the Simple VAS

Early effect is almost always modelled as linear across the Vce range, and that includes in SPICE. As we have just seen, that does not stop it creating distortion. If Early effect was linear, the amount of distortion resulting from it would not show changes with the amplifier operating point shift. All the textbooks say 'modelled as' or something equivalent; actual data on how Early effect changes with Vce is very hard to

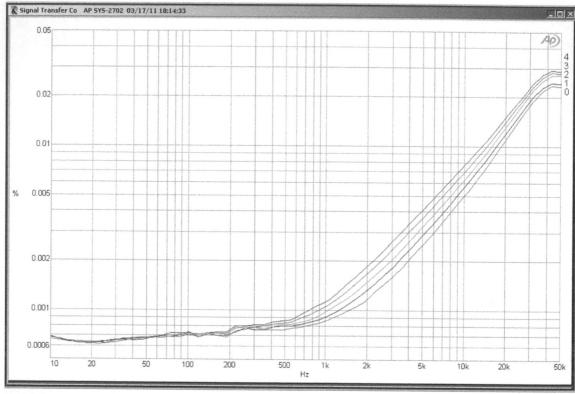

Figure 7.12. Increased HF distortion from a simple VAS with 0, (lower) 1, 2, 3, and 4 (upper) 1N4148 diodes across Cdom. +20 dBu out, ±20 V rails.

find. The only graph I have discovered so far is shown in Figure 7.13.

The Ic vs Vce curves in Figure 7.13 show greater curvature as Vce increases, suggesting that a positive shift in the operating point of the model amplifier will *increase* the LF distortion, while a negative shift will *decrease* it. This is the opposite of what you get for Cbc and HF distortion. Note, however, that our experiments use only a small part of the curves in Figure 7.13, about 11 V peak either side of Vce = 20V.

I did some measurements to see how LF distortion altered with varying VAS collector currents from 3 mA to 12 mA. There was no VAS emitter resistor. The results are shown in Figure 7.14, where LF distortion clearly increases with Ic. The distortion appears to be a mixture of second and third harmonics; the third is present even for Ic = 3 mA. The distortion level is roughly proportional to Ic, for example, on going from 6 mA to 8 mA, but distortion increases faster than this for the higher collector currents, and it is likely that another cause of non-linearity is contributing. The

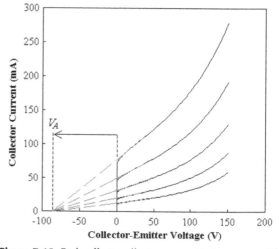

Figure 7.13. Early effect: collector current Ic increases with collector voltage Vce. The curves above 0 V are usually drawn as straight lines converging on a point VA along on the negative part of the Vce axis. VA is defined as the Early Voltage. (after Van Zeghbroeck, 2007).

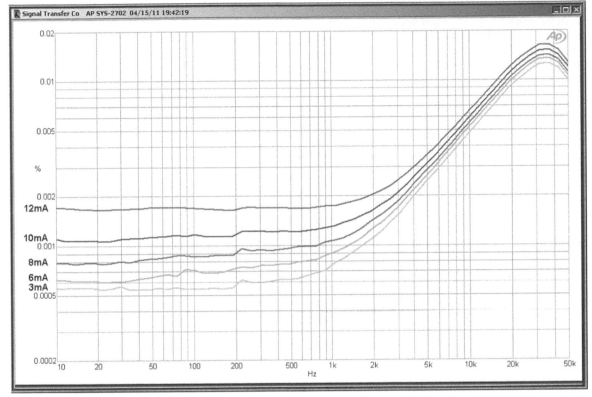

Figure 7.14. MPSA42 VAS. Ic is 3 mA, 6 mA, 8 mA, 10 mA, 12 mA +20 dBu output ±20 V rails.

proportionality at lower currents is obscured by noise at this measurement bandwidth (80 kHz).

There is also a smaller variation in HF distortion with Ic. Above 5 kHz the traces are a constant space apart, which shows that the variation is not due to the addition of frequency-independent distortion; if that were the case, the traces would tend to close up as frequency increased, due to the logarithmic vertical scale used. The spacing may be due to secondary effects that alter the effective non-linear Cbc as Ic is changed.

The LF distortion was remeasured at 100 Hz, with a measurement bandwidth of 22 kHz to reduce noise; see Table 7.3. Ic and THD are normalised to be 1.0 at Ic = 6 mA.

The assumption that Early effect LF distortion is proportional to Ic is reasonable from 3 mA to 8 mA, if we allow for the fact that the noise contribution to the 3 mA result is still significant. At Ic = 10 mA and 12 mA, distortion has increased faster than proportionally, and some other non-linearity is beginning to intrude.

While VAS LF distortion can be reduced by reducing the VAS standing current, this is not a practical

way to make a low-distortion amplifier. The standing current needs to be around 8 mA as a minimum if it is to give adequate slew-rates and provide enough current to the output stage when the latter is driving low-impedance loads.

The Simulation of Simple VAS Early Effect Distortion

While the measurement evidence is persuasive, SPICE simulation provides more information, despite the fact

Table 7.3. Proportionality of LF distortion to Ic

Ic	Ic normalised	THD (%)	THD (%) normalised
3	0.50	0.00032	0.67
6	1.00	0.00048	1.00
8	1.33	0.00067	1.39
10	1.67	0.00095	1.98
12	2.00	0.00125	2.60

that its modelling of Early effect is simplistic. SPICE allows the separation of the distortion from non-linear Cbc and from Early effect by altering the transistor model parameters. Simulation conditions are as for the real measurements; with a +20 dBu output and ±20 V supply rails.

The circuit simulated is shown in Figure 7.15. VCIS is a current source controlled by the difference between its two inputs, and represents the transconductance input stage; gm was set to 9 mA/V, approximating that of a differential pair with 100 Ω emitter degeneration and a current-mirror. The simple VAS is composed of a transistor and a collector current-source. VCVS is a voltage-controlled voltage source with a gain of one, representing the output stage. Here it is only required to prevent the feedback resistors Rnfb1, Rnfb2 from loading the VAS. The resistor Rdummy is required to prevent PSPICE complaining bitterly that there is no connection to the V1 input voltage source. MJC is a SPICE parameter controlling the base-collector junction capacitance, and the default value for the MPSA42 model is 0.5489. VAF is the Early voltage, and as usual is set to the default 100 V.

The first set of simulations in Table 7.4 demonstrates how distortion rises with frequency for a simple VAS using the standard MPSA42 model.

Table 7.4. Distortion from Cbc and Early effect: MPSA42 as VAS

Freq	Mjc	Vaf	2nd harmonic (%)	3rd harmonic (%)	THD (%)
1 kHz	Default	Default	0.00452	0.000483	0.00463
5 kHz	Default	Default	0.00545	0.000679	0.00551
10 kHz	Default	Default	0.00790	0.00147	0.00804
20 kHz	Default	Default	0.0139	0.00303	0.0142

That looks quite reasonable, with the results at 10 kHz and 20 kHz in line with the MPSA42 measurements in Figures 7.5 and 7.6 above. However, distortion is too high at 1 kHz and 5 kHz, measurement giving 0.001% and 0.003% respectively.

As we saw earlier in this chapter, the non-linearity of Cbc is controlled by the SPICE parameter MJC. Setting MJC = 0 gives a constant Cbc with no non-linearity. This eliminates the HF distortion, confirming that is due to Cbc variation, and leaves distortion of about 0.004% that is flat with frequency, just as predicted. See Table 7.5.

This confirms that Early distortion is flat with frequency. However, the general level is too high — the

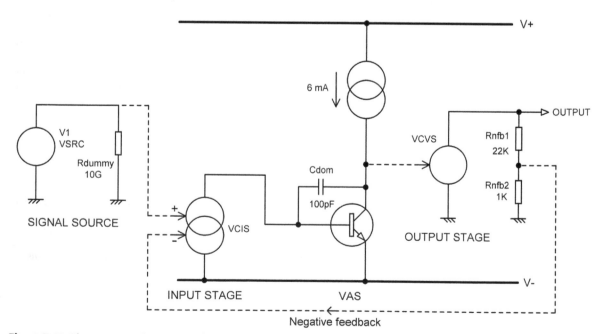

Figure 7.15. The conceptual circuit used to examine sources of distortion in a simple VAS. Only the VAS transistor itself is modelled as a real active component. +20 dBu output ±20V rails.

Table 7.5. Distortion from Early effect only: MPSA42 as VAS

Freq	Mjc Vaf	2nd harmonic (%)	3rd harmonic (%)	THD (%)
1 kHz	0 Default	0.00402	0.000245	0.00405
5 kHz	0 Default	0.00437	0.000274	0.00442
10 kHz	0 Default	0.00422	0.000274	0.00424
20 kHz	0 Default	0.00432	0.000234	0.00433

measurements give about 0.001%. I suspect that the Early voltage in the SPICE model may be wrong — it is set to exactly 100, along with very many transistor types in the official libraries. Early voltage is rarely if ever included on transistor data sheets, and I have a dark suspicion that in many cases no one has ever measured it.

Having switched off the effect of Cbc, we can alter the Early effect in steps by altering VAF, getting Table 7.6.

Increasing the Early voltage, so that Vce has less effect on Ic, reduces the distortion as expected. The lowest levels are close to the numerical noise floor, showing that in SPICE at least there is no need to go looking for further distortion mechanisms. There is very little non-linearity left to attribute to either varying beta or VAS reference problems.

I also ran simulations using the MJE340 as a VAS, and as expected I got a lot more distortion. Using the standard SPICE model the distortion is flat with frequency and much higher than measured in Figure 7.5 above. It is not frequency-sensitive and so seems to come from Early effect — the Early voltage in the MJE340 SPICE model is only 27 V, and that would certainly increase the distortion. See Table 7.7.

Once more, we set MJC to 0 to remove Cbc non-linearity; but this time there is very little change. Early effect is predominating. See Table 7.8.

Table 7.6. Distortion from variable Early effect: MPSA42 as VAS

Freq	Mjc	Vaf	2nd harmonic (%)	3rd harmonic (%)	THD (%)
20 kHz	0	Default	0.00432	0.000234	0.00433
20 kHz	0	1000	0.000154	0.0000629	0.000228
20 kHz	0	10000	0.0000428	0.0000764	0.000195
20 kHz	0	100000	0.0000494	0.0000765	0.000197

Table 7.7. Distortion from Cbc and Early effect: MJE340 as VAS

Freq	Mjc	Vaf	2nd harmonic (%)	3rd harmonic (%)	THD (%)
1 kHz	Default	Default	0.0527	0.00664	0.0531
10 kHz	Default	Default	0.0538	0.00709	0.0543
20 kHz	Default	Default	0.0551	0.000597	0.0555

Table 7.8. Distortion from variable Early effect: MJE340 as VAS

Freq	Mjc	Vaf	2nd harmonic (%)	3rd harmonic (%)	THD (%)
1 kHz	0	Default	0.0529	0.00699	0.0534
1 kHz	0	100	0.0118	0.000914	0.0119
1 kHz	0	1000	0.0000203	0.000229	0.000454

Again we reduce the Early effect in steps by increasing VAF, and this causes a dramatic reduction in distortion, with what remains being very low indeed.

In the light of all this evidence, I think it can be concluded that Early effect is indeed the source of the measurable LF distortion in the simple VAS. It can be minimised by selecting the right transistor type, i.e., that with the highest Early voltage and therefore showing the least Early effect. However, as we have seen, Early voltage rarely shows its face on data sheets. So how to pick the type? I can provide some help in the shape of Table 7.9.

Table 7.9. Measured LF distortion due to Early effect for different types of VAS transistor (all NPN)

Type	Vce (max)	Package	THD (%) 200 Hz
2SA2240BL	120V	TO-92	0.00083
MPSA06	80V	TO-92	0.00095
MPSA42	300V	TO-92	0.0010
MJE340	300V	TO-225AA	0.0016
BSS15	75V	TO-39	0.0018
BD139	80V	TO-126	0.0019
BC441	60V	TO-39	0.00275
BFY50	35V	TO-39	0.0025−0.0069

Samuel Groner has shown[13] that there is a fairly good correlation between Early voltage and the collector-emitter breakdown voltage Vceo, which always appears on datasheets. Table 7.9 shows in descending order of linearity my THD measurements at 200 Hz (under the standard conditions of +20 dBu output, ±20 V rails). I think it's clear there is some correlation between LF distortion and Vce (max) but it is not enough to use as a design guide by itself.

Table 7.9 also shows why the MPSA42 is a popular VAS transistor. The MPSA06 is rather short of voltage capability, and, as we saw earlier, the 2SC2240 has only half the power capability. Most of these measurements are for one sample only; the exceptions being the MPSA42, where five samples gave almost identical results, and the BFY50, where the results were all over the place, with almost a 3:1 range.

Methods for The Reduction of VAS Distortion

We have seen that while the distortion from the simple VAS can be kept under control by informed transistor choice, it is still too high. You may wonder why so much space was devoted to it, when better versions are well known. Firstly, it is essential to understand the shortcomings of the simple VAS, or it will be impossible to understand why the better versions *are* better. Secondly, sometimes you don't need Blameless performance but you do need a discrete power amplifier with a minimum PCB footprint. I am thinking here of multi-channel AV amplifiers that typically have to cram seven or more amplifier channels around a single heatsink. Space is in very short supply, and if a simple VAS can be made to do the job by informed transistor selection, that is a definite advantage. These multi-channel AV amplifiers are of considerable economic importance.

While many VAS configurations of greater complexity than the simple VAS have been both proposed and put into production, the most common two methods of improving linearity are the emitter-follower VAS (EF-VAS) and the cascode VAS. Both are dealt with in detail below, but more space is given to the EF-VAS as it is usually more economical, more effective, and easier to apply in practice.

The Emitter-follower VAS

The simple VAS has too much distortion, even when the best TO-92 devices are used. The two methods of linearisation I have described in earlier editions of this book are the addition of an emitter-follower inside the Cdom loop, and cascoding the VAS. All my public designs

have used the emitter-follower method. Cambridge Audio used the cascode VAS (buffered) in the 640A amplifier, but this was not implemented by me − just copied out of an earlier edition by someone else.

While I did not explicitly say so, the implication in earlier editions was that both methods were equally effective at linearisation. I wanted to find out if this is really true, and see if the operation of the EF-VAS throws more light on the distortion mechanisms of the simple VAS. The EF-VAS model amplifier used is shown in Figure 7.16.

Figure 7.17 compares a simple MPSA42 VAS with the EF-VAS shown in Figure 7.16. It is very satisfying to see that both LF and HF distortion are much reduced − in fact, the improvement is startling, with distortion at 20 kHz reduced from 0.009% to less than 0.0005%, an improvement of more than 18 times. That extra transistor and extra resistor must represent one of the best bargains in power amplifier design. The EF-VAS distortion plot is indistinguishable from that of the testgear alone. The LF region is now simply the noise floor. All the transistors are MPSA42/92 except for those in the current-mirror, which are high-gain 2SC2240BL, for reasons explained earlier.

I examined whether it mattered if the EF collector was connected to 0V or V+; I could find no difference at all in the distortion results. It is always connected to 0 V from now on. For the effect of the EF collector connection on PSRR see Chapter 26.

When the MJE340 was used as a simple VAS, it gave more than twice as much LF distortion as an MPSA42, and more than four times as much HF distortion. See Figure 7.5. Using it in the EF-VAS configuration, as in Figure 7.18 (note scale change from Figure 17.7), gives indistinguishable LF performance from an EF-VAS made of two MPSA42s (which is essentially the testgear floor), and the HF distortion is only very slightly worse. The THD improvement at 20 kHz is no less than 56 times; this is a higher ratio than for a simple MPSA42 VAS because the simple VAS distortion is much higher at 0.034% but the EF-VAS figure is mostly noise, and has barely changed. The EF-VAS allows driver-type transistors, with their much greater power capability than TO-92s, to be used with very good results. It is an extremely valuable configuration.

The addition of the emitter-follower gives us only one component to choose the value of: the emitter resistor of the emitter-follower (R32 in Figure 7.16). This doesn't give another degree of freedom for design, because altering it over a wide range, with 1 kΩ as a minimum, has very little effect on the distortion. There is no measurable change when the VAS

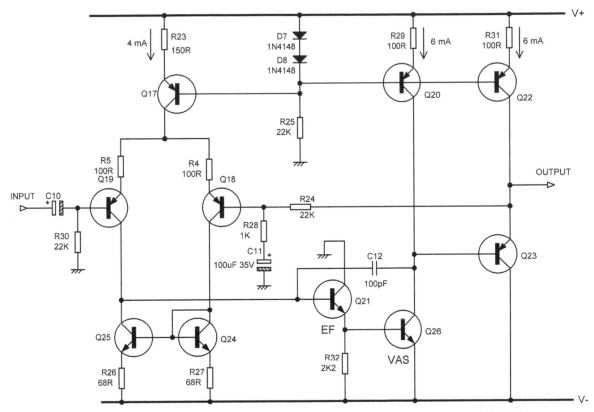

Figure 7.16. Model amplifier with EF-VAS. All transistors except current-mirror are MPSA42/92.

transistor is a MPSA42, but there is a slight effect when a MJE340 with its greater Cbc is used. This is shown in Figure 7.19, where reducing the EF emitter resistor from 2k2 to 1 kΩ gives a small improvement in distortion at 20 kHz. The upper trace shows what happens if the emitter resistor is omitted altogether. The increase in distortion at 20 kHz is not dreadful, but even if you're penny-pinching, I don't recommend this, because you might find it depends on transistor betas. Also, we are using relatively small signals here and I strongly suspect that, in a full-scale power amplifier, omitting the resistor altogether would cause problems with slewing. The effect of the EF emitter resistance value is even smaller when the VAS Ic is increased to 12 mA; this is perhaps a more likely VAS current.

Does the use of an EF-VAS allow us to use the MJL3281 power transistor in an EF-VAS? You will recall that in a simple VAS its distortion was grotesque (see Figure 7.6). I can't think of any reason why you would want to do this, but hang on in there because the experiment is instructive. Adding the emitter-follower reduces the 20 kHz

distortion from 1.0% to 0.03%, as in Figure 7.20. While this is a big improvement of 33 times, the end-result is still uninspiring, reaching 0.006% at 10 kHz. Furthermore, just after 20 kHz the THD shoots upwards as slew-rate-limiting occurs on the positive-going sides of the sinewave. Clearly the emitter-follower is having trouble pulling current out of the very large MJL3281 Cbc (approx 600 pF). Reducing the emitter-follower emitter resistor (R32 in Figure 7.16) from 2k2 to 1 kΩ helpfully postpones the onset of slew limiting from 20 kHz to 37 kHz, but does not affect the HF distortion below 10 kHz. This is probably due to non-linearity in the emitter-follower when sourcing and sinking significant currents.

How the EF-VAS Works

How does the addition of a simple emitter-follower (EF) inside the Miller loop improve the VAS so dramatically? It would be hard to argue that the inherent non-linearity of the VAS transistor Vbe-Ic curve has been

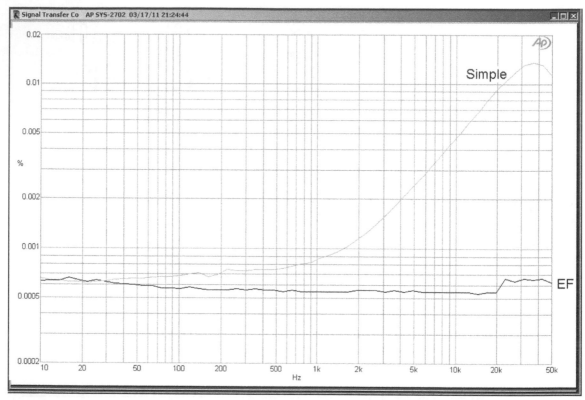

Figure 7.17. Simple VAS with MPSA42: (upper) EF-VAS with EF = MPSA42, VAS = MPSA42 (lower) +20 dBu output, ±20 V rails.

improved as we now have two base-emitter junctions instead of one. While it enhances the overall beta of the VAS stage, it has been shown that this plays little if any role in distortion. It has been said that the EF increases the transconductance of the stage, but this seems unlikely as the EF gain is slightly less than one, so slightly smaller voltage changes are passed on to the VAS transistor, and the overall transconductance would actually be slightly reduced.

I believe that the EF is simply working as a buffer to prevent the non-linear Cbc of the VAS from causing distortion. The same non-linear current still flows through the VAS Cbc, but since it is sourced by the EF, its effect on the circuit node where the input stage output current meets Cdom, i.e., the EF base, will be reduced by the beta of the EF; see Figure 7.21. The emitter-follower Cbc has a very small signal on its base end and no signal at all on its collector end, and so has negligible effect.

I also believe that the reason why the cascoded-VAS (see later section) gives low distortion is related. It is

because the absence of signal voltage on the VAS collector means that there is no signal current through the non-linear Cbc of the VAS. The two methods deal with the non-linear Cbc problem at the opposite ends of it, so to speak.

Note that the collector current of the VAS emitter-follower (approx 275 uA) is quite low. It seems to be enough, though, as the emitter-follower is not required to charge/discharge Cdom, just the much smaller Cbc of the VAS transistor. Another point to note is that the collector of the VAS transistor is still exposed to the full signal voltage on its collector, unlike the situation in a cascoded-VAS, and therefore we might wonder if Early effect would still be an issue.

The statement is sometimes seen that the emitter-follower acts as a buffer between the input stage and the VAS. This is of course nonsense because the input stage output is a current, so it prefers to drive into a low impedance like the VAS input, and in any case the emitter-follower is *inside* the VAS stage and its local feedback loop.

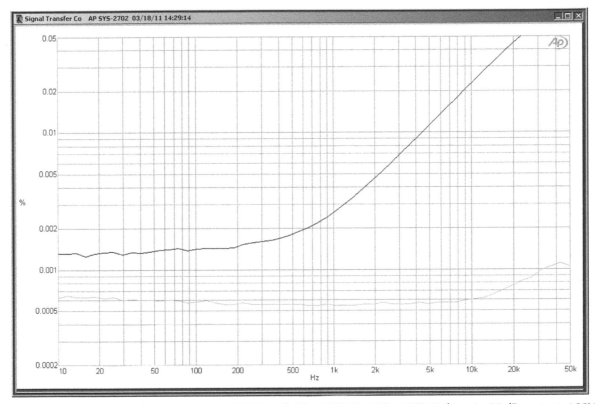

Figure 7.18. Simple VAS with MJE340: (upper) EF-VAS with EF = MPSA42, VAS = MJE340 (lower) +20 dBu output, ±20V rails.

Adding the emitter-follower has a second beneficial effect. It considerably increases the LF gain of the VAS and therefore the open-loop gain of the complete amplifier. When a simple VAS is combined with the usual input stage having 100 Ω emitter degeneration and a current-mirror, the LF open-loop gain is +82 dB and the dominant pole frequency P1 is just above 1 kHz. Adding the EF increases the gain to +117 dB, with a correspondingly lower pole around 20 Hz. The extra feedback factor is 4 dB at 1 kHz, 21 dB at 100 Hz, and 34 dB at 10 Hz. It thus gives a useful distortion reduction at 1 kHz, and a good deal more at lower frequencies. The importance of this is hard to assess because of the major role of the added EF in eliminating distortion caused by local feedback though the non-linear Cbc of the VAS transistor.

The EF-VAS works equally effectively when a dual VAS is used to allow greater dissipation, as in Figure 7.22. HF distortion is very slightly higher above 20 kHz, presumably because the emitter-follower has twice the VAS Cbc to deal with. On the other hand, the EF collector current is increased because of the presence of the 68 Ω current-sharing resistors means a greater voltage drop across R1, and this is likely to make it more linear.

It is often stated that adding a resistance in the VAS emitter connection introduces local feedback. This is just not true, because the VAS accepts a current input rather than a voltage input, so the voltage developed across the emitter resistor does not cause negative feedback. Putting 68 Ω in the VAS emitter of an EF-VAS built with two MPSA4 makes absolutely no difference to the residual; it is still the same as the testgear output.

In the section above on Early effect distortion we saw that doubling the VAS Ic from 6 mA to 12 mA caused a considerable increase in LF distortion. This is not the case with the EF-VAS; no effect on the very low distortion is measurable; see Figure 7.23.

A Brief History of the EF-VAS

The EF-VAS configuration reduces VAS distortion remarkably, especially considering its great simplicity. I started wondering who invented it. My first encounter

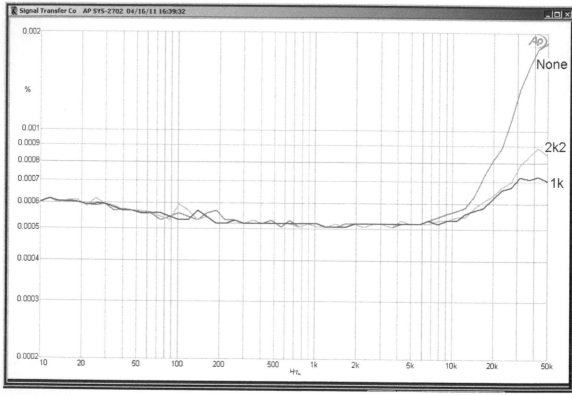

Figure 7.19. EF-VAS with EF = MPSA42, VAS = MJE340 (lower) Emitter-follower emitter resistor is 1 kΩ, 2k2, or absent (upper) VAS Ic = 6 mA +20 dBu output, ±20V rails.

with it was in 1975 in the Cambridge Audio P60, where the EF emitter resistor was 470 Ω.

Before that, the Cambridge Audio P50 (in Versions 2 and 3; Version 1 had a quite different power amplifier configuration) had an emitter-follower but with no emitter resistor, so it would not have been fully effective at reducing HF distortion. My P50 drawing is dated 16th May 1972.

An EF-VAS was also used in the Armstrong 621 power amplifier in the 1970s – I cannot give a closer date than that at the moment. The EF emitter resistor was the popular value of 2k2. The emitter-follower was also used to bootstrap the collector load of the single input transistor, but I am doubtful whether that can have done any good.

Clamp Diodes and the EF-VAS

Figure 7.12 above (in the simple VAS section) showed that adding 1N4148 diodes, with their non-linear junction capacitance, across Cdom caused only a minor increase in HF distortion, because of the large amount already being generated by the Cbc of the simple VAS. The EF-VAS produces negligible HF distortion, so we will try the experiment again.

Adding even one 1N4148 diode across Cdom now wrecks the HF distortion performance, with the addition of mostly second harmonic; the greatest disturbance of the residual occurs at the negative peaks, as would be expected. As more diodes are added, the THD gets worse, but not by a constant increment or ratio. This may be due to variations in the diode samples. The important point is that clamp diodes across Cdom are no more acceptable with the EF-VAS than they are with the simple VAS. The EF cannot help because the non-linear diode capacitance is added to the Miller capacitance, and is not inside the Miller loop. If you feel you must clamp the VAS voltage excursions, you will need to find another method.

The Benefits of the EF-VAS

We have, I think, thoroughly proved that the EF-VAS can give a truly excellent distortion performance, the

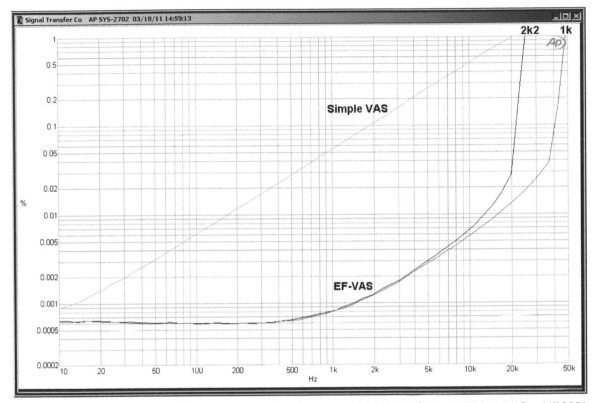

Figure 7.20. Simple VAS using a MJL3281 power transistor: (upper trace) EF-VAS with EF = MPSA42, VAS = MJL3281 (lower traces) +20 dBu output, ±20V rails.

THD plot being indistinguishable from that of the test-gear alone; see Figures 7.17 and 7.23. There are no snags at all and the extra cost is tiny. When wrestling with these kinds of financial decisions, it is as well to remember that the cost of a small-signal transistor is often less than a fiftieth of that of an output device, and the entire small-signal section of an amplifier usually represents less than 1% of the total amplifier cost, when heavy metal such as the mains transformer and heatsinks are included.

I think this justifies my contention that VAS distortion, like input stage distortion, can be made negligible; we have all but eliminated Distortions One and Two from the list given in Chapter 5.

The Cascode VAS

The cascoded-VAS is a well-known alternative to the EF-VAS for reducing VAS distortion. A model amplifier with a cascoded-VAS is shown in Figure 7.24.

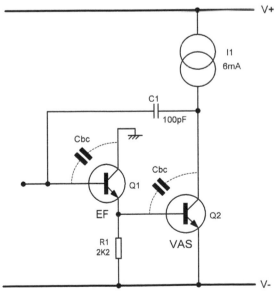

Figure 7.21. EF-VAS showing the two Cbc's involved.

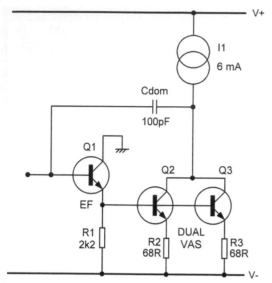

Figure 7.22. An EF-dual-VAS; the VAS emitter resistors are essential for current-sharing.

Figure 7.25 compares a simple VAS, an EF-VAS, and a cascoded-VAS as in Figure 7.24. The cascoded-VAS is the middle trace, is an improvement on the simple VAS, but gives very poor results at HF compared with the EF-VAS. The basic problem is the same as we saw at the start of the chapter — there is not enough Vce for an MPSA42 transistor as Q30 in the current-mirror, because we no longer have an emitter-follower which causes there to be an extra 0.6V of Vce. The MPSA42 appears to need considerably more Vce to work properly than the data sheet implies.

Figure 7.26 shows how fixing this reduces the distortion from the cascode VAS. The upper trace is reduced to the lower by the simple expedient of putting a 68 Ω resistor in the VAS emitter circuit. The voltage drop across this increases the Vce of Q30, and the HF distortion drops remarkably. Another way to achieve the same end is simply to make Q30 a 2SC2240 or MPSA06. These more 'normal' transistors work well with a much lower Vce than the MPSA42 will tolerate. Note that Figure 7.26 is also one of the few plots in this chapter on which the testgear noise floor

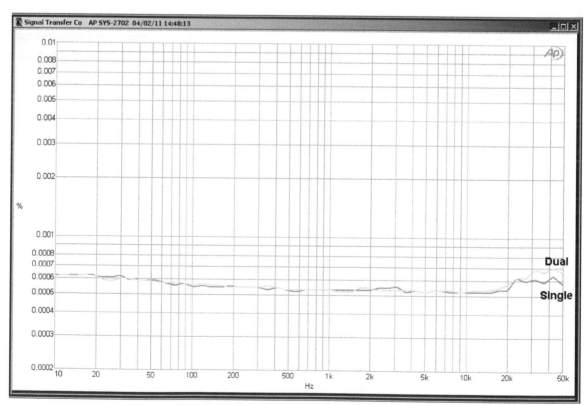

Figure 7.23. EF-VAS using MPSA42: single-VAS (lower) dual-VAS (upper).

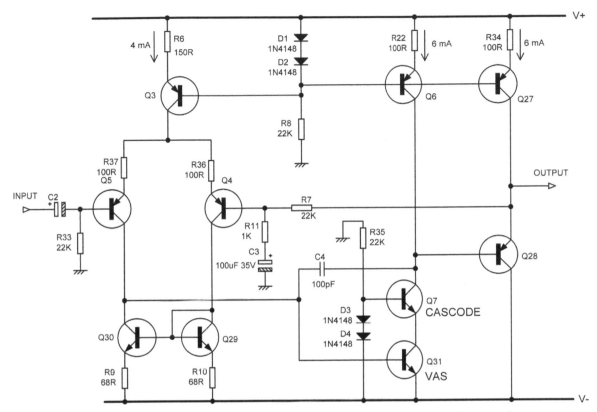

Figure 7.24. Model amplifier with cascode VAS. All transistors MPSA42/92 except current-mirror transistors are 2SC2240BL.

(GenMon) is shown; very often it would obscure more important detail.

An immediate word of caution. The extra voltage-drop across the 68 Ω resistor also reduces the Vce of the cascode transistor Q7 in Figure 7.24. That device will almost certainly be an MPSA42 because it sees the whole rail-to-rail voltage, and it too will need more Vce than expected; in fact, here it will stop working altogether. Increasing the cascode bias to 3 series diodes, giving Q7 a Vce of about 1.20 V, gets things working again but there is still a rapid rise in HF distortion above 10 kHz. Rather disconcertingly, it takes 5 diodes, giving Q7 Vce of 2.63 V, to get the gratifyingly low distortion shown by the '5d' trace in Figure 7.26. Increasing the cascode bias to 6 diodes-worth does not reduce the HF distortion further — in fact, it increases it very slightly. The obvious snag is that such a high cascode bias will cause earlier clipping for negative signal peaks. This is, however, not a problem if you are using a lower negative sub-rail for the small-signal stages of the amplifier.

How the Cascode VAS Works

We saw in the section on the EF-VAS that one reason why the added emitter-follower gave a dramatic reduction in distortion was that it buffered the base end of the non-linear Cbc from the rest of the circuit. It also considerably increases the LF open-loop gain by multiplying the effective beta of the VAS transistor. The cascode also renders Cbc harmless, but in a quite different way. The cascode transistor Q7 has a constant voltage set up on its emitter, and so on the VAS collector and the collector end of Cbc. Therefore negligible signal flows through Cbc and its non-linearity has no effect.

The cascode also increases the LF open-loop gain, but to a lesser extent. It does this by increasing the VAS collector impedance Rc rather than increasing the beta. (These are the two quantities that, together with the input stage transconductance, determine the LF gain, see Equation 5.2.) The problem is that the increase in collector impedance is limited by the loading of the output stage. Thus in a simulation of

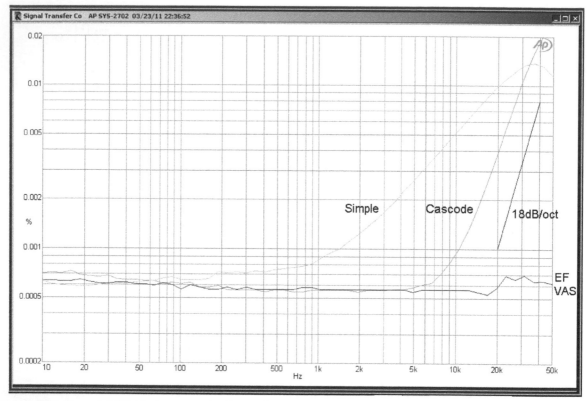

Figure 7.25. Cascode VAS (blue) compared with simple VAS (green) and with EF-VAS (2x MPSA42) (yellow). The straight line demonstrates an 18dB/octave slope +20 dBu output, ±20V rails.

a complete cascode VAS amplifier, with an optimally-biased EF output stage driving an 8 Ω load, the LF gain only increased from +82 dB to +98 dB, with P1 at 200 Hz, an improvement of 16 dB. The EF-VAS under the same conditions gave an LF gain of +117 dB, with P1 at about 20 Hz, and that is 19 dB more than the cascode. See Figure 5.2 for a summary of this.

To prove the point about loading the cascode VAS collector, the simulation was repeated with the full-scale output stage replaced by a Class-A constant-current emitter-follower, with a 6 mA quiescent current, and no load except the negative feedback network. The LF gain increased to -103 dB, a 4 dB improvement, and P1 was consequently lowered to 120 Hz. This indicates that it might be useful to use a unity-gain buffer stage between the VAS collector and the output stage. Adding the Class-A constant-current emitter-follower described above as a buffer in front of the EF output stage increased the simulated LF open-loop gain to +107 dB, which is greater than that of the model amplifier. This suggests that the NFB network loading on the model amplifier is significant, emphasising the great sensitivity of a cascoded

VAS to even light loading. The topic of the VAS buffer is dealt with in more detail below.

A detailed mathematical analysis of cascode operation can be found in Tietze and Schenk.[14]

A Brief History of the Cascode VAS

The cascode configuration itself dates back to 1939 (see Chapter 6) and is rooted in valve technology. It was quickly adapted to transistors when they appeared. Research has not yet revealed when a cascode VAS (or, come to that, a cascode input stage) was first used in an audio power amplifier, but I can testify that the technique was well known in the mid-1970s. To pick examples almost at random, a cascode VAS was used in the Marantz SM-80 (1990), the Onkyo A-8190, the Sony TA-F555ES (1983), and the Technics SU-V5.

The Benefits of the Cascode VAS

While the cascode VAS has considerably reduced distortion, it is not as low as the EF-VAS; compare

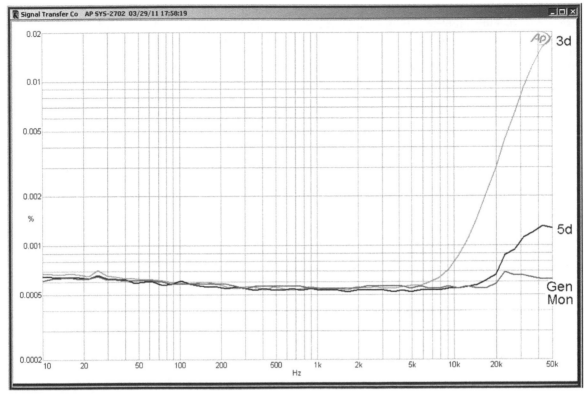

Figure 7.26. Cascode VAS with VAS = MPSA42 Cascode = MPSA42. Cascode bias is 3 diodes, (3d) or 5 diodes (5d) above the V- rail. The lowest trace is the testgear floor (GenMon) 20 dBu output, ±20V rails.

the '5d' trace in Figure 7.26 with the EF-VAS trace in Figure 7.17. The extra component cost is small, but not as low as the EF-VAS because a cascode bias voltage must be provided.

A cascode VAS allows the use of a high-beta transistor for the lower device; it will typically have a lower Vce (max) that cannot withstand the supply rail voltages of a high-power amplifier. When cascoded, there are only a few volts across it.

A definite disadvantage is the loss of available voltage swing. Since output power depends on voltage squared, losing even a Volt of output swing can make a significant difference to the figures. As we have seen, using the MPSA42 increases this considerably. Unfortunately the headroom loss is in the negative direction, which usually clips first whatever VAS configuration is used. This is not an objection if an extra negative sub-rail is used.

Another snag is that a cascode VAS will draw more base current from the input stage, because of the absence of the EF transistor, possibly unbalancing it. This is

made worse by the low Vce on the VAS transistor (Early effect).

A further disadvantage is that the increase in LF open-loop gain over that of a simple VAS is less than with the EF-VAS, and is variable according to the loading on the VAS collector.

The conclusion must be that the cascode approach is not as good as the EF-VAS method.

My experiments show that there is nothing to be gained by cascoding the current-source collector load. In theory, this should further increase the open-loop gain by isolating the VAS output from the r_0 of the current source as well as the r_0 of the VAS transistor; this works in opamps but requires a very high impedance buffer as a next stage to get the benefit.

The VAS Buffer

It is possible to insert a unity-gain Class-A buffer stage between the VAS and the output stage, to eliminate any possibility of loading effects on the VAS, without causing any stability problems. In previous sections we saw that output stage loading did not appear to be a problem, but this may not be the case if a cascode VAS with a high output impedance is used. Adding a stage with gain would require a complete redesign of the compensation scheme, but a simple unity-gain stage such as an emitter-follower is usually stable

without further modification. Figure 7.27a shows a simple VAS buffered by an emitter-follower with a 10 mA constant-current emitter load.

A PNP emitter-follower can be used instead, as in Figure 7.27b, and this allows a less well-known but very usable form of bootstrapping to be used instead of a current-source load for the VAS. R1 is the collector load, and the collector current of VAS Q1 is determined

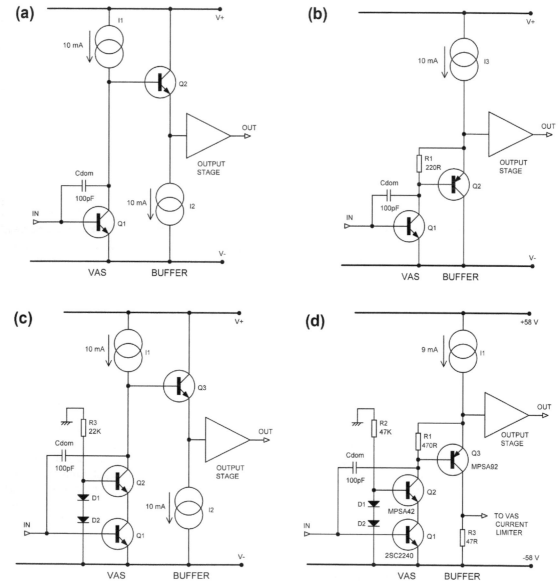

Figure 7.27. The VAS buffer: (a) simple VAS with buffer; (b) simple VAS bootstrapped by buffer; (c) cascode VAS with buffer; (d) cascode VAS bootstrapped by buffer.

by the Vbe of the buffer transistor across it; in this case it is 3 mA. Since both ends of R1 go up and down together due to the buffer action, R1 is bootstrapped and appears to the VAS collector as a constant-current source. In this topology a VAS current of 3 mA is quite sufficient, compared with the 6 mA standing current in the buffer stage. The VAS in fact works fairly well with collector currents down to 1 mA, but this tends to compromise linearity at the high-frequency, high-voltage corner of the operating envelope, as the VAS collector current is the only source for driving current into Cdom. This method is more dependable than conventional bootstrapping from the output because the signal level at the buffer output is unaffected by loading on the output stage, and so the amount of bootstrapping does not vary.

There is more incentive to buffer the VAS when it is cascoded, as in Figure 7.27c, as the collector impedance is potentially higher, giving more open-loop gain.

A cascoded VAS can also be bootstrapped by a buffer, as in Figure 7.27d. This variation may look a little unlikely, but I used it for economy of parts in the Soundcraft Powerstation amplifier series which was made in its tens of thousands, so I can assure the doubtful that it very much works as advertised. The component values were those used in production. Note the relatively low VAS current (1.4 mA) and the low-voltage high-beta transistor in the Q1 position.

It is assumed that the VAS buffer will work in Class-A because it would not be sensible to introduce a second source of crossover distortion by using a separate Class-B buffer. On the other hand, a triple output stage will also reduce the load on the VAS and could be regarded as having built-in buffering.

There are other potential benefits to VAS buffering apart from the reduction of distortion. It has been claimed that the effect of beta mismatches in the output stage halves is minimised,[15] though I cannot say that I have myself ever found such mismatches to be a problem. Voltage drive also promises the highest fT from the output devices, and therefore potentially greater stability, though I have no data of my own to offer on this point. A VAS buffer for a cascode VAS in the form of a constant-current emitter-follower was used on the Cambridge Audio 540A and 640A amplifiers (2003), and the Cambridge Audio 740A (2007).

VAS Distortion Due to Output Stage Loading

As explained earlier, it is important to linear VAS operation that the collector impedance (before Cdom is connected) is high, permitting a large amount of local negative feedback. The obviously non-linear Class-B output stage, with large input impedance variations around the crossover point, would appear to be about the worst thing you could connect to it, and it is a tribute to the general robustness of the conventional three-stage amplifier configuration that it handles this internal unpleasantness gracefully. In one experiment on a Blameless 100 W/8 Ω amplifier, the distortion degraded only from 0.0008% to 0.0017% at 1 kHz, so the loading effect exists but is not massive. The degradation increases as the global feedback-factor is reduced, as expected. There is little deterioration at HF, where other distortions dominate.

Clearly the vulnerability of the VAS to loading depends on the impedance at its collector. A major section of Samuel Groner's commentary examines the issue of VAS distortion caused by presumed non-linear loading by the output stage,[16] and he gives Equation 7.7 as an expression for VAS output impedance Z_0.

$$Z_0 = \frac{1}{R_L \cdot 2\pi f \cdot C_{dom} \cdot g_m} + \frac{1}{g_m} \qquad \text{Equation 7.7}$$

where: R_L is the lumped resistance across the VAS input (i.e., with all resistances lumped into one)

f is the frequency
C_{dom} the Miller compensation capacitor
g_m is the transconductance of the VAS transistor or transistors.

Equation 7.7 shows that if we want to reduce Z_0, we can alter either R_L or g_m, as the frequency range is fixed for audio and C_{dom} is set by stability considerations. That basically means our only accessible parameter is R_L unless we want to concoct a compound VAS with higher transconductance than a single transistor. R_L is determined by the input resistance of the VAS in parallel with the output resistance of the input stage. The input resistance of the VAS can be increased by adding the emitter-follower that makes it an EF-VAS; given its ability to also eliminate the effects of non-linear Cbc, that transistor and resistor must be one of the better bargains in amplifier design. The output resistance of the input stage is again composed of two parallel elements— the output resistance of the input stage itself and the output resistance of its associated current-mirror. The output resistance of the input stage can be increased by cascoding its collectors, as in Figure 6.13b, and the output resistance of the current-mirror can be increased by using a more sophisticated configuration such as the

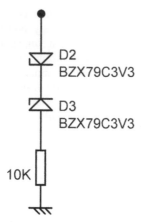

Figure 7.28. Samuel Groner's non-linear Zener network, intended to approximate the loading on a VAS by an output stage.

Wilson mirror (Figure 6.9). Samuel Groner deals with this in more detail in.[16] I have also worked with him on this topic, but our investigations are by no means complete.

To emulate the current demands of an output stage, Samuel used the non-linear VAS loading network shown in Figure 7.28. When this is connected between the VAS collector and ground of the model amplifier with simple VAS in Figure 7.4, the effect on the distortion is remarkably small, as seen in Figure 7.29.

Another warning. This very desirable result is *only* obtained if there is no VAS emitter resistor. Samuel Groner points this out in.[17] Adding even a few Ohms there makes the VAS much more susceptible to non-linear loading. This is demonstrated in Figure 7.30 where the setup is identical except for the insertion of 10, 20, 33, 47, and 68 Ω in the VAS emitter; the level of both LF and HF distortion increases with disturbing rapidity as the resistance increases. The LF distortion at 200 Hz is approximately proportional to the value of the added emitter resistor but the agreement is not very good. What is going on?

Equation 7.3 gives the answer. Adding a VAS emitter resistor reduces the VAS transconductance, and so increases the VAS output impedance Z_0, and the non-linear loading therefore has more effect.

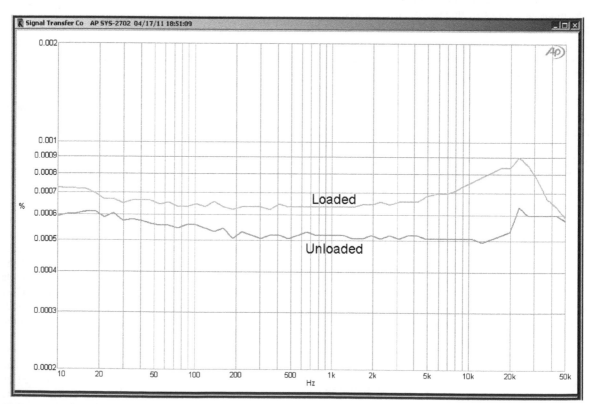

Figure 7.29. The relatively small effect of the Zener network on an EF-VAS made from two MPSA42 (loaded is upper). No VAS emitter resistor. +20 dBu output, ±20V rails.

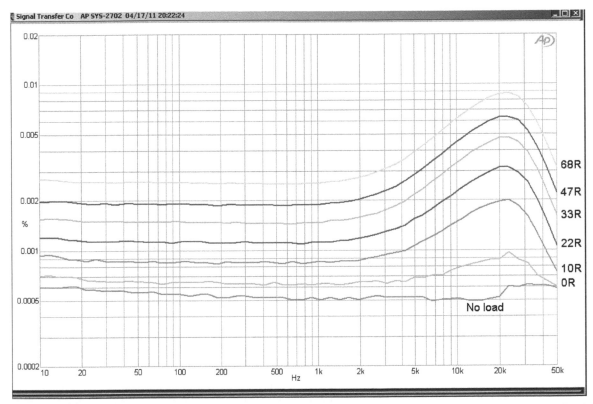

Figure 7.30. Effect of Zener loading network on EF-VAS with 2x MPSA42. VAS emitter resistor is 0 10 20 33 47 68R +20 dBu output, ±20V rails.

Since a VAS emitter resistor, typically in the range 47 Ω–68 Ω is very often necessary for VAS current-limiting, Figure 7.30 is a worrying diagram. However, we know it is straightforward to produce amplifiers with much less distortion than this. The reason is that the non-linear Zener load may be a good theoretical tool, as it gives enough distortion to measure easily, so instructive plots like Figure 7.30 can be produced, but it was never intended to be representative of real output stage loading. Firstly, it has a dead-band, where there is no conduction at all, of ±3.9 V (7.8 V total) which is much greater than a wholly unbiased Class-B output stage, which would have a dead-band of ±2.4 V for the EF type, (four Vbe's) or ±1.2 V for the CFP type (two Vbe's). But amplifiers are not operated in zero-bias conditions, and if we assume optimal Class-B biasing, with a smooth handover of conduction from one half of the output stage to the other, we would expect no dead-band in the loading at all.

The simulation results in Figure 7.31 show this to be the case. The variation in current drawn from the input

(in this simulation an ideal voltage source rather than a simulated VAS) is completely smooth over the central region, (+30 V) being quite free from kinks or wobbles around the crossover point, and inside the limits of ±1 mA. Only outside this region does the current drawn begin to increase rapidly, taking on an S-shaped curve. The curve is not symmetrical, reaching −3.5 mA for negative signals and 5 mA for positive signals, due to differing betas in the two halves of the output stage, and for the same reason it does not go through zero at zero input voltage.

The rapid rise in current for large signals is due to the reduced Vce on the output transistors. Increasing the supply rails from ±50 V to ±60 V considerably reduces the currents drawn to -0.72 mA and +2.0 mA. Going to ±70 V reduces them still further to −0.28 mA and +1.2 mA. This appears to be due to Early effect.

It must be remembered that, as described in Chapter 9, in a typical EF output stage the driver transistors never turn off; therefore base currents always flow

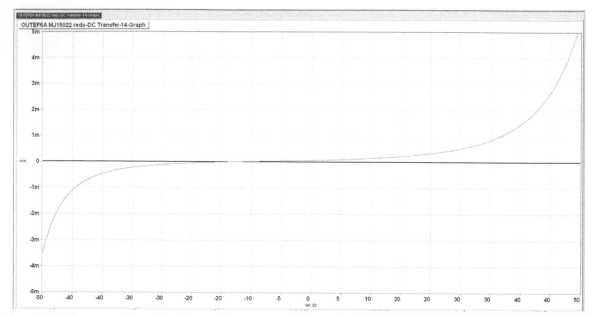

Figure 7.31. The current drawn from the VAS by an optimally-biased EF output stage driving a ±45 V signal into 8 Ω. ±50 V supply rails.

into the upper driver and out of the lower driver and it is their difference that determines the current drawn from the VAS.

The result is very similar for a 4 Ω load, though the maximum currents become −20 mA and +30 mA; much more than twice the currents for the 8 Ω case; presumably due to beta decreasing with increasing collector current (beta-droop).

The trace in Figure 7.31 starts to curve noticeably outside the range ±20 V. This is not reflected in the distortion behaviour of Blameless amplifiers. There is not a steady rise in third-harmonic distortion above medium powers; all that can be seen is crossover artefacts that remain at a relatively constant level until clipping occurs. This suggests that VAS loading is not a significant effect on linearity.

Some More VAS Variations

The different kinds of VAS we have looked at so far by no means exhaust the possibilities. Here are two more.

Figure 7.32a shows an inverted version of the EF-VAS, where the emitter-follower Q1 is now a PNP device. Its operating Vce is limited to the Vbe voltage of Q2, about 0.6V; that should be enough for most transistor types passing a low Ic; the MPSA42 is a notable exception. The collector current of Q1 is set by the

value of R1, and is here 33 kΩ, giving approximately the same 275 uA Ic for the emitter-follower as in the conventional EF-VAS, in an attempt at a fair comparison. R1 might alternatively have been connected to the V+ rail, as has been done by Yamaha in some recent designs; an interesting variation was used on the Yamaha M-73 power amplifier (1977) where the emitter-follower load was a cascoded current-source connected to the opposite supply rail.

An important feature of this somewhat strange-looking configuration as shown in Figure 7.32a is that the voltage-shift due to the Vbe of the inverted emitter-follower is now in the opposite direction, so an 120 Ω emitter resistance R2 has to be added to give the current-mirror enough Vce to work properly (here 920 mV). To help with this, the current-mirror transistors were 2SC2240BL; all other devices in the model amplifier were MPSA42/92.

The same excellent measured THD performance is obtained as for the conventional EF-VAS (see Figure 7.17). However, this configuration seems to have no unique advantages, and the need to use a large emitter resistor to cancel the reversed voltage drop in the emitter-follower will lead to earlier clipping on negative peaks.

Figure 7.32b shows a variation on the cascoded VAS suggested by Hawksford in.[18] The intention is

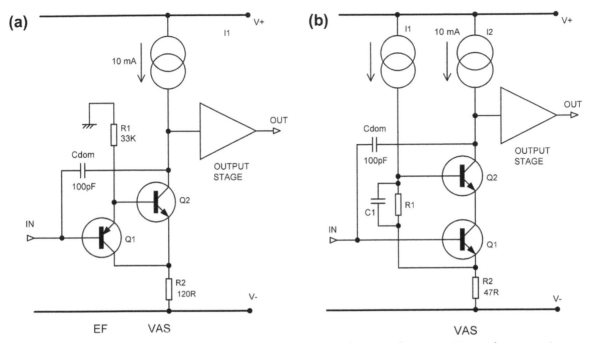

Figure 7.32. VAS variations: (a) EF-VAS with inverted EF; (b) cascode VAS with constant Vce on lower transistor.

apparently to reduce the Vce variation on the VAS transistor Q1 by bootstrapping the cascode transistor Q2 from the emitter of Q1. Note that the emitter resistor R2 is not present to introduce local negative feedback; it is normally put there to allow current-sensing and so over-current protection of the VAS transistor when output stage overload circuitry is operating. I remember trying this scheme out many years ago, but found no improvement in the overall distortion performance of the amplifier. This was almost certainly because, as previously stated, the distortion produced by a linearised VAS *without* using this enhancement is already well below the much more intractable distortion produced by a Class-B output stage. A current-source I1 and resistor R1 are used to voltage-shift the signal on the emitter resistor.

VAS Operating Conditions

It is important to operate the VAS at a sufficiently high quiescent current. If a non-balanced VAS configuration is used, then this current is fixed by the current-source load; it must be high enough to allow enough current drive for the top half of the output stage when the lowest load impedance contemplated is being driven to full output. The value of the current required obviously depends on the design of the output stage;

a triple output will draw much less current than a dual EF output. A high VAS quiescent current also has the potential to improve the maximum slew-rate, but as described in Chapter 15, there are several important provisos to this. Typical VAS quiescent current values are 5 to 20 mA. Note must be taken of the base current the VAS transistor will draw from the input stage; it must not be allowed to unbalance the input transistor collector currents significantly. A high VAS beta helps, but will not be found in company with a high Vce(max). The EF-VAS is in my experience wholly free of this problem because of the extra stage of current amplification.

The primary limitation on the VAS quiescent current is the dissipation of the transistors that make up this part of the circuit. There is a strong motivation to use TO-92 package transistors for, as we have seen, they show much less distortion than larger devices. The need to withstand high collector voltages in powerful amplifiers requires high-Vce devices which have significantly lower beta than lower-voltage transistors. As an example, the MPSA42 is often used in the VAS position; it can sustain 300V but the minimum beta is only a humble 25. Its maximum dissipation is 625 mW.

Table 7.10 compares the important parameters for an NPN VAS transistor such as the BC337, the MPSA42, and the medium-power MJE340 which comes in

Table 7.10. Parameters of possible VAS transistors

Type	Beta	Vce max	Pdiss max	Package
BC337-25	160–400	50 V	625 mW	TO-92
2SC2240BL	350–700	120 V	300 mW	TO-92
MPSA42	25 min	300 V	625 mW	TO-92
MJE340	30–240	300 V	20 W	TO225

a TO-225 format. Note the minimum beta spec for MPSA42 is 25, which is actually less than the minimum of 30 for the much more power-capable MJE340.

Let us examine these options, assuming the VAS quiescent current is chosen to be 10 mA.

If you choose the BC337-25, then the supply rails are limited to +/−25 V by Vce max, and this restricts the theoretical maximum output power to 39 W into 8 Ω; allow for output-stage voltage-drops and you'd be lucky to get 30 W. A BC337-25 could of course be cascoded with a higher-power device, to shield it from the high voltage; the extra complication is not great, and the cascoding itself may improve linearity, but the need to give the lower high-beta transistor a couple of volts of Vce to work in will lead to an asymmetrical voltage swing capability. There seems little point in this.

The 2SC2240BL transistor is noted for its high beta (BL is the highest beta grade) and this can sometimes be useful. The 120 V figure for Vce max allows a theoretical maximum output of 225 W into 8 Ω, but the maximum permitted dissipation is less than half that of its competitors, so it is not generally suitable for high-power designs.

The MPSA42 has a Vce max of 300V, so supply rail voltages as such are not going to be a problem; the theoretical maximum output is 1406W. into 8 Ω. However, if the VAS quiescent current is 10 mA, the amplifier supply rails will be limited to about ±50 V if the maximum package dissipation is taken as 500 mW, to provide some margin of safety. As it is, a TO-92 package dissipating 500 mW is disconcertingly (and painfully) hot, but this sort of operation does not in my experience lead to reliability problems; I have used it many times in commercial designs and it works and keeps working. Small bent-metal heatsinks that solder into the PCB are available for the TO-92, and these are well worth using if you are pushing the dissipation envelope; I have a packet of brass ones in front of me that are labelled with a thermal resistance of 36 °C/Watt. It is also good practice to use substantial PCB pads with thick

tracks attached, so as much heat as possible can be lost down the legs of the transistor.

If that doesn't do it, another possibility is the use of a dual VAS, as described earlier (Figure 7.22). Using two TO-92 VAS transistors in parallel, using the small resistors (circa 56 Ω) that are usually placed in the VAS emitter circuit for current-limiting (*not* for local negative feedback) to ensure proper current-sharing, will allow twice the dissipation and this will be a good option in some circumstances. As we have seen earlier in this chapter, this doubles the total Cbc, and the EF-VAS version is advisable to keep distortion down. The same approach can be used for the VAS current-source; two transistors sharing a bias generator. With separate emitter resistors, good current-sharing is inherent.

Should a dual-VAS be inadequate, the next step is a driver-sized device like the MJE340. A heatsink will probably not be required though the transistor will not of course dissipate anything like its Pd (max) of 20 Watts without one. In a simple VAS the distortion from the VAS stage will be considerably higher because of the larger non-linear Cbc of the MJE340. Adding an emitter-follower to make it an EF-VAS is strongly recommended.

The VAS current-source load naturally dissipates as much power as the VAS as it carries the same current, and must be upgraded accordingly, but this will not significantly affect distortion.

VAS Current Limiting

It is not always appreciated that a VAS transistor is in a vulnerable position. Firstly, most overload protection works by diverting away the current from the VAS to the output stage. The collector current-source is inherently safe, but the VAS may turn on large currents as it attempts to pull the output to where the negative feedback loop thinks it should be. More subtly, the VAS can expire even under normal loading. When the amplifier hits negative clipping, the VAS will again pass high currents as it attempts to pull the output down further than it will go. Either scenario can destroy a TO-92 VAS transistor fairly quickly − though not usually instantaneously.

For this reason every amplifier I design has VAS current-limiting. The simple circuit of Figure 7.33 is that used in the Signal Transfer power amplifiers[19] and it is highly effective. The current that has to be sunk by Q3 is limited by the current output of the input stage. For commercial production, the resistor R2 must be an approved fusible type that is designed to

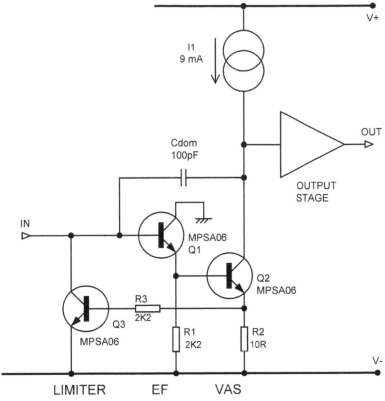

Figure 7.33. Over-current protection for the VAS.

fail without producing flame; this is because safety testing will involve shorting the current source I1 and disabling the limiter Q3, producing a heavy current through Q2 and R2. This requirement exists for all VAS emitter resistors.

The Class-AB VAS and Further Developments

A push-pull circuit is generally regarded as the most efficient available in terms of current delivery, unless you give up the linear Class-A operation of the circuit devices and opt for some form of Class-AB operation. (The push-pull Class-A VAS is thoroughly explored in Chapter 8.) The use of Class-AB implies inferior distortion performance unless the high-current mode is strictly reserved for slew-rate testing and not used during the normal operation of the amplifier.

An excellent starting point for the study of this sort of stage is Giovanni Stochino's fine article in *Electronics World*[20] in which he described input stages and a VAS that gave very high slew-rates by operating in Class-AB. Samuel Groner gives a detailed description of a new push-pull stage in Jan Didden's *Linear*

Audio,[21] that uses push-pull Class-A rather than Class-AB.

Manipulating Open-loop Bandwidth

In the past there has been much discussion of the need for an amplifier to have a wide open-loop bandwidth before the global negative-feedback loop is closed, and the seemingly logical statement has been made that open-loop bandwidth should match or exceed the audio bandwidth, so that the amount of NFB applied is constant with frequency. Even if this is not the aim, 'Open-loop gain held constant up to 20 kHz' reads better than 'Open-loop bandwidth restricted to 20 Hz' although these two statements could describe near-identical amplifiers, both having the same feedback factor at 20 kHz. The difference is that if the second amplifier has a 20 Hz open-loop bandwidth, its open-loop gain continues to increase as frequency falls, levelling off 60 dB up at 20 Hz; meanwhile the first amplifier, with its flat response, has the same feedback factor at 20 Hz as 20 kHz. See Figure 7.34. The second amplifier, with much more feedback across the audio band, is

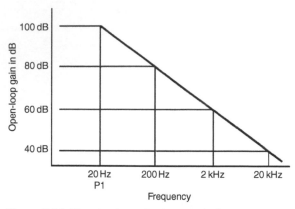

Figure 7.34. Showing how dominant-pole frequency P1 can be altered by changing the LF open-loop gain; the gain at HF, which determines Nyquist stability and HF distortion, is unaffected.

clearly superior, but it could easily read as sluggish and indolent to the uninformed.

If, however, you feel that for marketing reasons the open-loop bandwidth of your amplifier design must be increased, here is how to do it. Assuming a reliably safe feedback factor at 20 kHz, you cannot increase the HF open-loop gain without losing that stability, so you must reduce the LF open-loop gain. That means reducing the LF feedback factor, which will reveal more of the output stage distortion in the mid-band. Since, in general, NFB is the only weapon we have to deal with this, blunting its edge seems ill-advised.

It is of course simple to reduce o/l gain by degenerating the input pair, but this diminishes it at HF as well as LF. To alter it at LF only it is necessary to

tackle the VAS instead, and Figure 7.35 shows two ways to reduce its gain. Figure 7.35a reduces gain by reducing the value of the collector impedance, having previously raised it with the use of a current-source collector load. This is no way to treat a gain stage; loading resistors low enough to have a significant effect cause unwanted current variations in the VAS as well as shunting its high collector impedance, and serious LF distortion appears. While this sort of practice has been advocated in the past,[22] it seems to have nothing to recommend it as it degrades VAS linearity at the same time as siphoning off the feedback that would try to minimise the harmonic.

Figure 7.35b also reduces overall o/l gain, but by adding a frequency-insensitive component to the local shunt feedback around the VAS. The value of Rnfb is too high to load the collector significantly and therefore the full gain is available for local feedback at LF, even before Cdom comes into action.

Figure 7.36 shows the effect on the open-loop gain of a model amplifier for several values of Rnfb; this plot is in the format described in Chapter 5, where error-voltage is plotted rather than gained directly, and so the curve once more appears upside down compared with the usual presentation. Note that the dominant pole frequency is increased from 800 Hz to above 20 kHz by using 220 kΩ for Rnfb; however, the gain at higher frequencies is unaffected and so is the stability. Although the amount of feedback available at 1 kHz has been decreased by nearly 20 dB, the distortion at +16 dBu output is only increased from less than 0.001 to 0.0013%; most of this reading is due to noise.

In contrast, reducing the open-loop gain even by 10 dB by loading the VAS collector to ground requires

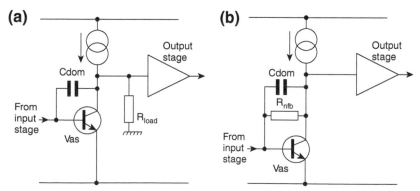

Figure 7.35. Two ways to reduce o/l gain: (a) by simply loading down the collector. This is a cruel way to treat a VAS, current variations cause extra distortion. (b) Local NFB with a resistor in parallel with Cdom. This looks crude, but actually works very well.

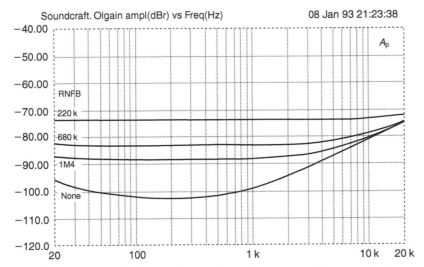

Figure 7.36. The result of VAS gain-reduction by local feedback, the dominant pole frequency is increased from about 800 Hz to about 20 kHz, with high-frequency gain hardly affected.

a load of 4k7, which under the same conditions yields distortion of more than 0.01%.

If the value of Rnfb required falls below about 100 kΩ, then the standing current flowing through it can become large enough to upset the amplifier operating conditions (Figure 7.35b). This is revealed by a rise in distortion above that expected from reducing the feedback factor, as the input stage becomes unbalanced as a result of the global feedback straightening things up. This effect can be simply prevented by putting a suitably large capacitor in series with Rnfb. A 2μ2 non-electrolytic works well, and does not cause any strange response effects at low frequencies.

Samuel Groner's commentary includes a section on bandwidth manipulation[23] in which he suggests a method of maintaining the input stage balance, and the connection of the bandwidth-manipulating resistor to the amplifier output so that a degree of output-inclusive compensation (see Chapter 13) will offset the reduction in feedback factor. I haven't tried it but it sounds like a jolly good idea. He further shows how adding a second resistor between the current-mirror and ground can restore the PSRR almost back to the value it had before bandwidth-manipulation.

An unwelcome consequence of reducing the global negative feedback is that power-supply rejection is impaired (see Chapter 26). To prevent negative supply-rail ripple reaching the output, it is necessary to increase the filtering of the V-rail that powers the input stage and the VAS. Since the voltage drop in an

RC filter so used detracts directly from the output voltage swing, there are severe restrictions on the highest resistor value that can be tolerated. The only direction left to go is increasing C, but this is also subject to limitations as it must withstand the full supply voltage and rapidly becomes a bulky and expensive item. However, this method has the advantage of being totally dependable.

That describes the 'brawn' approach to improving PSRR. The 'brains' method is to use the input cascode compensation scheme described in Chapter 13. This solves the problem by eliminating the change of reference at the VAS, and works extremely well with no compromise on HF stability. No filtering at all is now required for the negative supply rail − it can feed the input stage and VAS directly.

Conclusions

This chapter showed how the strenuous efforts of the input circuitry can be best exploited by the voltage-amplifier stage following it. At first it appears axiomatic that the stage providing all the voltage gain of an amplifier, at the full voltage swing, is the prime suspect for generating a major part of its non-linearity. In actual fact, though the distortion behaviour of the simple VAS may be complicated, it can be made too linear to measure by adding a transistor and a resistor to make it an EF-VAS. This is my preferred solution. The cascode VAS solution is also viable, but gives less LF open-loop gain, requires more parts and will impair negative clipping unless a sub-rail is used. The effect on linearity of output-stage loading on the VAS appears to be very small. The second of our distortion mechanisms can be cheaply and easily made negligible.

References

1. Groner, S, *Comments on Audio Power Amplifier Design Handbook*, 2011. SG Acoustics http://www.sg-acoustics.ch/analogue_audio/power_amplifiers/index.html (accessed Aug 2012).

2. Locanthi, B, Operational Amplifier Circuit for Hifi, *Electronics World* (USA), Jan. 1967, p. 39.

3. Antognetti, P (ed), *Power Integrated Circuits*, New York: McGraw-Hill, 1986, p. 9.31.

4. Gray, P R and Meyer, R G, *Analysis and Design of Analog Integrated Circuits*, 2nd edition, New York: John Wiley & Sons, Ltd, 1984, p. 252. (Ro limit on VAS gain).

5. Self, D, *Small Signal Audio Design,* Burlington, MA: Focal Press, 2010, pp. 70–72.

6. Gray, P R and Meyer, R G, *Analysis and Design of Analog Integrated Circuits*, 2nd edition, New York: John Wiley & Sons, Ltd, 1984, p. 251. (VAS transfer characteristic).

7. Groner, S, *Comments on Audio Power Amplifier Design Handbook*, 2011 p.8 onwards, SG Acoustics, http://www.sg-acoustics.ch/analogue_audio/power_amplifiers/index.html (accessed Aug, 2012).

8. Tuinenga, P, *Spice*, 2nd edition, New Jersey: Prentice Hall, 1992, p. 158. ISBN 0-13-747270-6.

9. Ibid., p. 166.

10. Tietze, U and Schenk, Ch, *Electronic Circuits*, Berlin: Springer, 2nd edition, 2002, pp. 69–72. ISBN: 978-3-540-00429-5.

11. Bailey, A, 30W High Fidelity Amplifier, *Wireless World*, May 1968, pp. 94–98.

12. Gray, P R and Meyer, R G, *Analysis and Design of Analog Integrated Circuits*. 3rd edition, New York: John Wiley & Sons, 1993.

13. Groner, S, *Comments on Audio Power Amplifier Design Handbook*, 2011, pp. 27–28. SG Acoustics, http://www.sg-acoustics.ch/analogue_audio/power_amplifiers/index.html (accessed Aug. 2012).

14. Tietze, U and Schenk, Ch, *Electronic Circuits*, pp. 312–327.

15. Oliver, B, Distortion in Complementary-Pair Class-B Amplifiers, *Hewlett-Packard Journal*, Feb. 1971, p. 11.

16. Groner, S, *Comments on Audio Power Amplifier Design Handbook*, 2011, pp. 34–44. SG Acoustics, http://www.sg-acoustics.ch/analogue_audio/power_amplifiers/index.html (accessed Aug. 2012).

17. Ibid., p36.

18. Hawksford, M, Reduction of Transistor Slope Impedance Dependent Distortion in Large-Signal Amplifiers *J. Audio Eng. Soc*, 36, No 4, April 1988, pp. 213–222. (enhanced cascode VAS).

19. The Signal Transfer Company, http://www.signaltransfer.freeuk.com/ (accessed Aug. 2012).

20. Stochino, G, Ultra-Fast Amplifier, Electronics and Wireless *World*, Oct. 1996, p. 835.

21. Groner, S, A New Amplifier Topology with Push-Pull Transimpedance Stage, *Linear Audio* 2, 2011, pp. 89–114.

22. Hefley, J, High Fidelity, Low Feedback, 200W, *Electronics and Wireless World*, June 1992, p. 454.

23. Groner, S, *Comments on Audio Power Amplifier Design Handbook*, 2011, pp. 44–47 SG Acoustic, http://www.sg-acoustics.ch/analogue_audio/power_amplifiers/index.html (accessed Aug. 2012).

The Push-pull Voltage-Amplifier Stage

Push answers with pull and pull with push.
Hans Hofmann (1880–1966), abstract
Expressionist painter

The Push-pull VAS

A criticism of previous editions of this book was that it focused mainly on one configuration, that with a mirrored input pair driving a constant-current VAS, which is enhanced by adding either an emitter follower inside the Cdom loop, or a cascode transistor to the VAS collector. I have also been accused of glutinous adherence to the three-stage architecture that forms the basis of the Blameless amplifier. To address this regrettable situation I have set out to analyse several types of push-pull VAS, and to the best of my knowledge, this is the first time it has been done. This is a somewhat more challenging proposition than the single-ended VAS in Chapter 7 because there are more variables to grapple with. As John Causebrook said, 'An art is a science with more than seven variables' but we cannot afford to let VAS design become an art.

In earlier editions, VAS configurations that had a signal-varying operating current (as opposed to a fixed operating current set by a constant-current source) were referred to as a 'balanced VAS' but on mature consideration I have decided that the phrase 'push-pull VAS' is more accurate and more descriptive, and I have changed the nomenclature in this section accordingly.

When we are exhorted to 'make the amplifier linear before adding negative feedback', one of the few specific recommendations made — if there are any at all — is usually the employment of some sort of push-pull VAS. A statement is usually made that a push-pull VAS 'cancels out the even-order harmonics' but put like that, the statement is completely untrue. A differential-pair input stage is inherently push-pull, and will cancel the second harmonic distortion effectively if the collector currents are designed to be very closely equal, but it does not follow that any old push-pull circuit will cancel even-order harmonics.

A push-pull VAS usually requires two drive signals; one for the active element at the top of the VAS, and one for that at the bottom. There are broadly two popular methods of doing this. In the first case there is one input stage differential pair, and the push-pull signals are taken from the two collectors. One of the drive signals is therefore at the wrong end of the VAS, and must be transferred to the other supply rail in some way. In the second case, there are *two* complementary input stages, and a single output from each one drives

the top or bottom of the VAS. The circuitry involved in both methods is described below.

A possible difficulty is that there are now two signal paths from the input stage to the VAS output, and it is difficult to ensure that these have exactly the same bandwidth; if they do not, then a pole-zero doublet is generated in the open-loop gain characteristic that will markedly increase settling-time after a transient. This seems likely to apply to all balanced VAS configurations, as they must have two signal paths in one way or another. Whether this is in any way audible is another matter; it seems most unlikely. If you want to dig deeper into the matter of frequency doublets — which have nothing to do with Tudor clothing — then Dostal[1] is an excellent reference.

Single Input Stages with a Push-pull VAS

We will begin by looking at push-pull VAS stages that only require a single input stage. It would no doubt be possible to devise many configurations that would meet this description, but undoubtedly the most popular in both commercial manufacture and designs by audio enthusiasts is what I call the Hitachi circuit. This is analysed in this chapter.

Another push-pull VAS of this type that has received some attention is that originated by R Lender in 1974. That too is analysed here.

The Hitachi Push-pull VAS

Figure 8.1 shows what is probably the most common type of push-pull VAS. While its exact origin is currently obscure, it was used in the Hitachi HMA-7500, one of the first MOSFET power amplifiers, released in 1978. Whether the use of the push-pull VAS was a deliberate attempt to ease the pumping of large currents in and out of the substantial gate capacitances of the MOSFETs is not known; it seems likely. The configuration was widely used, with minor component value changes, in MOSFET power amplifier kits from about 1980 onwards, and for many people was their first exposure to a push-pull VAS. This configuration needs a name so I will call it the Hitachi circuit.

The component values shown are derived from a kit design that was widely sold at the time and make at least a starting point for study. The original circuit had a resistor in series with the collector of Q4, but this has been omitted for reasons that will be later made plain. A similar push-pull VAS was used in the notorious Lohstroh and Otala amplifier in 1973 (see

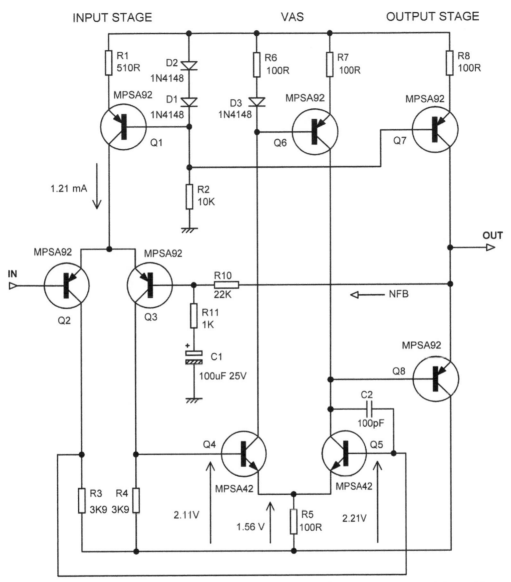

Figure 8.1. Push-pull VAS as used in power amplifier to demonstrate Hitachi power FETs. Here it is used in a model amplifier with a small-signal Class-A output stage.

Chapter 4, where there is also another example of a push-pull VAS). A VAS using a superficially similar configuration to that of Figure 8.1 was put forward by Bart Locanthi in 1967[2] but the output from Q4 collector was simply sent to ground and there was no current-mirror, so this does not qualify as a push-pull VAS. Very strangely, the VAS collector load was a simple resistor; there was no bootstrapping or current source. The main point of the article was the so-called T-circuit output stage.

Figure 8.1 has an input pair with no emitter degeneration resistors. The tail current source is set at about 1.2 mA and is biased by the usual two diodes; the VAS standing current is 8 mA. There is no phase-summing current-mirror in the input stage because two drive signals are required for the VAS. This is a model amplifier with a constant-current Class-A output stage that is highly linear because it only has to drive the feedback network and the testgear; the current source is biased by the same diodes as the input tail source.

Figure 8.1 shows voltages around the VAS when the circuit was built using unselected MPSA42 and MPSA92 transistors. It is immediately obvious that the voltages across the input-pair collector resistors R3, R4 are unequal, and that the collector currents of Q2 and Q3 are therefore not equal. As we saw in Chapter 6 on input stages, quite accurate equality of collector currents is required if the second-harmonic distortion from the input devices is to cancel out properly. This is illustrated in Figure 8.2, where Trace 1 shows the distortion as found in the prototype.

The first tests were done at with ±30 V supply rails to make the VAS distortion negligible and show the input stage distortion clearly; more on the VAS distortion later. The THD residual was a mixture of second and third harmonics, as would be expected from an unbalanced input stage, and its slope in the 25 kHz–50 kHz region was about a factor of 6 in an octave. The input stage was then manually balanced by trimming the value of

R4; when it was shunted with 56 kΩ the second harmonic disappeared, giving Trace 2 in Figure 8.2, which has a slope of a factor of 8 times per octave, as expected from a input stage with balanced Ic's. It was noticeable that the point of input stage Ic balance drifted after switch-on as the transistors heated. There is more on this below.

Be aware that as the distortion is only measurable at relatively high frequencies, the measurement bandwidth was 500 kHz. This raises the noise floor and results in the relatively high readings below 10 kHz. This region shows only noise, and remains flat down to 10 Hz with no distortion visible.

The important question here is why the input pair balance is poor, and the quick answer is that it depends on many more factors than it does in the case of a mirrored input-pair driving a constant-current VAS. There the input stage current-mirror forces equality of the input transistor collector currents no matter what else is going on in the circuit.

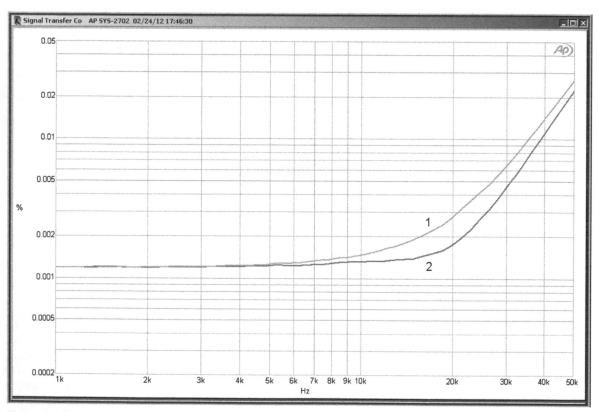

Figure 8.2. Distortion from push-pull VAS in Figure 8.1 (Trace 1) is reduced at HF by adding a resistor to balance the input pair collector currents (Trace 2). Second harmonic distortion from the input pair is eliminated, so the slope is steeper at 8 times per octave. +20 dBu, ±30 V rails, 500 kHz bandwidth.

Here the input transistor collector currents depend on:

1. The exact values of the input-pair collector resistors R3, R4, which are likely to be of 1% tolerance.
2. The Vbe's of VAS transistors Q4 and Q5.
3. The betas of VAS transistors Q4 and Q5.
4. The accuracy of the current-mirror at the top of the VAS.

The factors affecting balance are best studied by simulation, where you can be sure that the betas, Vbe's and Early voltages are identical for each transistor of the same type, and there are no resistor tolerances. This eliminates (1), (2) and (3) of the issues listed, but not (4).

When the circuit of Figure 8.1, which we will call Version 1, is simulated we get the results in Table 8.1. The input pair collector currents are out of balance by 2.38%, which is quite enough to mess up the cancellation of the second harmonic. With no component mismatches, we need to look at the operating conditions for the cause of the imbalance. It is at once apparent that the Vce of Q4 is nearly twice that at Q5. In real life this has a heating effect, described below, but in SPICE everything stays at the default temperature unless you tell it otherwise, so that is not the cause.

Early effect is a more likely suspect. Q4 and Q5 collector currents differ by 12%, and there is a corresponding imbalance in the base currents they draw from the input stage. If we approximately equalise the Vce's of Q4 and Q5 by putting a 3.9K resistor in the Q4 collector circuit (Version 2), the base currents become more equal, and the input pair collector current imbalance falls to 1.05%.

Table 8.1 shows that despite the equalisation of the Vce's of Q4 and Q5, their collector currents are still far from equal. The reason is the rather crude current-mirror Q6. If we replace the 1N4148 diode with a diode-connected MPSA92 transistor, giving Version 3, then we get a much more accurate current-mirror. Q4 and Q5 collector currents now differ by only 1%, and the base currents they draw are correspondingly better matched, so the input pair currents now show an imbalance of only 0.19%, which is good enough to give an excellent cancellation of the second harmonic.

Note that in these simulations there is negligible variation in the sum of Q2 and Q3 collector currents because this is closely fixed by current-source Q1. Also, the presence or absence of 100 Ω input degeneration resistors makes no detectable difference to the DC conditions we are looking at here.

Informal tests plugging in different transistor samples showed very little change, suggesting that variations in Vbe and Early voltage are not of great importance.

The Hitachi Push-pull VAS: Heating and Drift

The distortion measurements were complicated by thermal drift. Figure 8.1 is shown, as noted, with no resistor in the collector of Q4. This means that the collector of Q4 is almost at the full +V supply voltage, while Q5 sees 0 V at its collector when the amplifier is quiescent. Thus Q4 runs substantially hotter than Q5, and so Q4 Vbe falls as it heats up. This is the reason for the drift in the input pair balance after turn-on, mentioned above.

Inserting the 3.9 K resistor into the Q4 collector circuit brings the voltage on it down to 2.7 V, roughly matching the power dissipated in Q4 and Q5, and eliminating this source of error, as shown in Figure 8.3.

The Hitachi Circuit: AC Gain

The calculation of the HF AC gain of this circuit looks complex but it is the same as for the constant-current VAS.

The AC voltage on the emitters of Q4 and Q5 is very low until the frequency is reached at which Cdom begins to take charge; the signal begins to rise at the dominant pole frequency, the same frequency at which the error voltage between the bases of the input pair starts to rise. The version with no input pair degeneration has 10 dB more gain at HF (20 kHz) and would almost certainly not be stable if combined with a full-scale output stage.

Table 8.1. Transistor collector currents

	Q2 Ic (uA)	Q3 Ic (uA)	Q2 + Q3 Ic (uA)	Q2 Q3 bal (%)	Q4 Ic (mA)	Q5 Ic (mA)	Q4 + Q5 Ic (mA)
Version 1	582.8	555.6	1138.4	2.38	6.399	7.205	13.604
Version 2	575.2	563.1	1138.3	1.05	6.280	7.072	13.352
Version 3	570.3	568.0	1138.3	0.19	6.702	6.631	13.333

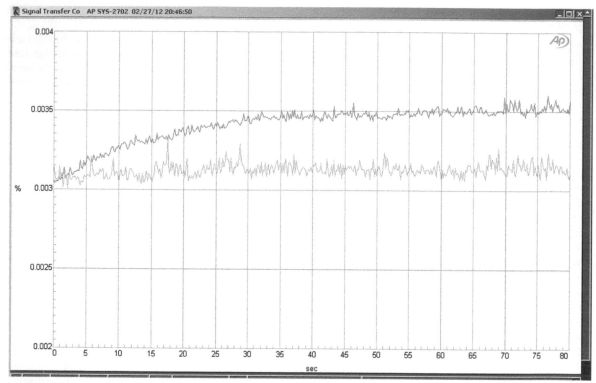

Figure 8.3. Drift of distortion from input pair over 80 seconds after turn-on, without (Trace 1) and with (Trace 2) R4 = 3K9 in series with Q4 collector. At 25 kHz +20 dBu out, ±30V rails.

The Hitachi push-pull VAS: distortion

We saw that adding 100 Ω input degeneration resistors reduces both LF and HF gain by exactly 10 dB, with the dominant pole frequency unchanged. Increasing these resistors to 430 Ω reduces both LF and HF gain by another 10 dB and exposes the VAS distortion, because (1) the extra degeneration reduces the input stage distortion, and (2) the amount of global feedback around the VAS is reduced. Figure 8.4 shows the distortion under these conditions, with three values of supply rail. It is not very impressive; distortion is about five times greater than a simple VAS at 10 kHz with ±20 V rails. The distortion is all second harmonic, and this demolishes the claim that a push-pull VAS (or at any rate this version of it) automatically cancels even-order harmonics.

As we saw earlier with the constant-current VAS, the distortion in both the flat LF and the upward-sloping HF sections of the plot are increased by reducing the supply rails. The slopes in the HF region are all 6 dB/octave, confirming that the distortion seen is from the VAS and not the input stage. Furthermore, the distortion in

both is visually pure second harmonic, so the same distortion mechanisms are operating – Early effect in the LF region and non-linear Cbc in the HF region.

For the constant-current VAS, the two sovereign remedies against Cbc distortion were adding an emitter-follower inside the Miller dominant pole loop, and cascoding. Figure 8.5 shows the configuration, with cascode transistor Q9 added to one side of the VAS; Figure 8.6 shows the results, which are similar to those obtained with a constant-current VAS. Both LF and HF distortion are much reduced, but there is a rapid rise above 20 kHz. Note that Q5 was been replaced by an MPSA06 as this works better with a low Vce than the MPSA42.

One problem with this configuration is that adding the single cascode transistor Q9 has unbalanced the input stage collector currents. If we go back to the simulation used to create Table 8.1 above, we find that the Q2 and Q3 collector currents are now 595.4 uA and 542.9 uA respectively. This is because the Vce of Q5 is much reduced, so as a result of Early effect, the base current drawn by Q5 is 88.5 uA, more than twice that of Q4, which is 41.4 uA.

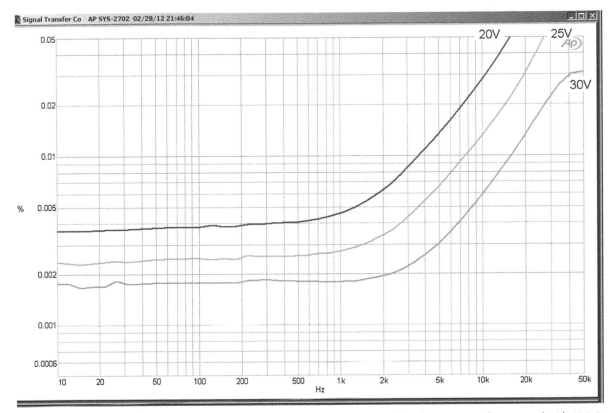

Figure 8.4. Hitachi push-pull VAS distortion with ±20 V, ±25 V and ±30 V supply rails. Input pair degenerated with 430 Ω, +20 dBu out.

The obvious cure for this is to cascode the other half of the VAS as well, to make the Vces of Q4 and Q5 equal. This works well in simulation, with the Q2 and Q3 collector currents much better balanced at 568.9 uA and 569.4 uA respectively. However, at this point I decided that this configuration really was not worth further investigation.

The Hitachi Push-pull VAS: Asymmetrical Clipping

The circuit clips significantly earlier on negative peaks, because of the significant voltage drop of about 1.5 V across R5. The Hitachi HMA-7500 MOSFET power amplifier mentioned earlier had higher supply rails (±58 V) for the small-signal circuitry than for the output devices, (±52 V) partly no doubt to accommodate the large Vgs voltage required to turn the output devices on. This would have avoided any problems with asymmetrical clipping.

Asymmetrical clipping is undesirable not only because it reduces efficiency (as one polarity is clipping

early) but also because it never looks good when revealed in product reviews.

The Lender Push-pull VAS

This configuration was originated by a Mr R. Lender at Motorola in Switzerland. It appears to have been documented solely in a 1974 internal report that has never been published.[3] It became widely known through a notable article by Erno Borbely in *Audio Amateur* in 1982,[4] where two Lender circuits are used back-to-back to drive MOSFETs. Unfortunately the design equation for the VAS standing current given there is not correct; it says that it is determined solely by the ratio of the input pair collector resistors R3, R4. This is, however, only true if the global negative feedback forces the voltages across these resistors to be equal, which is generally not the case. The main determinant of the VAS current is actually the value of R5, which sets the current through Q4 and the input of the current-mirror. Note that the top end of this resistor is effectively connected to the +V supply rail,

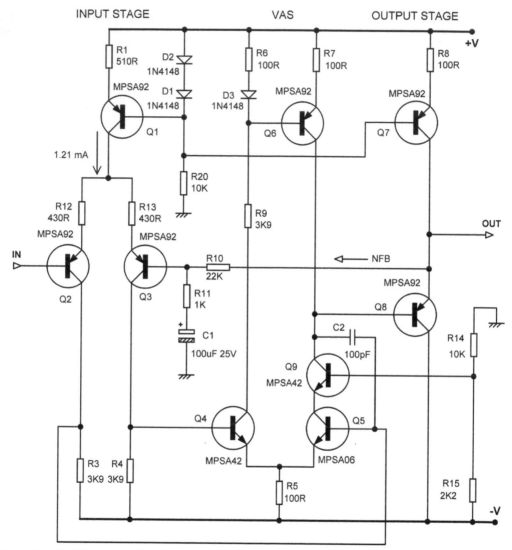

Figure 8.5. Push-pull VAS Hitachi VAS with input stage heavily linearised by R12, R13, and transistor Q5 cascoded by Q9, working in a model amplifier.

so unless precautions are taken, the VAS standing current will be modulated by ripple.

Figure 8.7 shows a Lender configuration with R5 set to 7K5, which gives a VAS standing current of 8 mA with ±30 V rails, to allow comparison with the other VAS types in this chapter. The input pair collector resistors R3, R4 are set to the 11:1 ratio suggested by Borbely and his values of 2K2 and 200 Ω are used. As with the previous circuit, the input pair is initially left undegenerated, to expose any problems with the input stage.

The first step was to run a simulation, which showed that the input pair collector currents were

not balanced: Q2 Ic was 762 uA while Q3 Ic was 1131 uA. This suggests that the input pair will generate significant second-harmonic distortion, and so it did when I built and measured it. Figure 8.8 shows no measurable distortion, just noise, up to 2 kHz but then Trace 1, with R4 = 200 Ω, increases rapidly at 4 times per octave; the distortion is second harmonic. Adjusting R4 to 240 Ω, which simulation suggested should give approximate input pair balance, delays the appearance of second harmonic distortion until 4 kHz in Trace 2 but the input balance is clearly still far from perfect. Setting

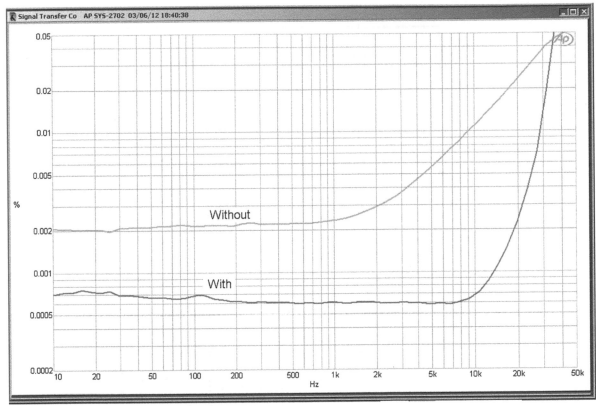

Figure 8.6. Hitachi push-pull VAS distortion with and without cascode. Input pair degenerated with 430 Ω. +20 dBu out, ±30 V rails.

R4 to 270 Ω gives Trace 3, which is almost all third harmonic, with a steeper slope, and very close to input balance.

Figure 8.9 is Figure 8.8 rerun with a measurement bandwidth of 500 kHz rather than 80 kHz, and this gives a true picture of the distortion slopes at HF. The measured slope of Trace 3 is 7.5 times per octave, very close to the 8 times that we expect from a balanced input.

Balancing the input pair by adjusting R4 still leaves us with distortion that is rapidly rising above 10 kHz. We will now increase the input tail current to 4 mA and degenerate the input pair with the usual value of 100 Ω, to reduce the feedback factor to a practical value for use with a full-size output stage, and we get Figure 8.10; the distortion from the input stage has been pushed up in frequency and out of the audio band, but unhappily the reduced global feedback has revealed a constant 0.002% of second-harmonic distortion at LF, and this must be from the VAS.

This seemed like a good point to check just how much pushing and pulling was really going on. The

answer from simulation is: very little. At a +20 dBu output the current-mirror output is only modulated by a negligible 0.01 mA rms. This is not really a push-pull stage at all.

There is another interesting point about this circuit; its HF distortion is much better than that of the Hitachi circuit, without any cascoding. This is because it acts like an EF-VAS, as described in Chapter 7. Q4 provides a low-impedance drive to the non-linear base-collector capacitance of Q5. A well-designed full-size output stage would allow this difference to be clearly visible, and I speculate that this might be why Borbely describes the Lender configuration as being 'very linear'. It is certainly better than a simple VAS, but *not* because of any push-pull action.

The Lender Push-pull VAS: Heating and Drift

Having earlier enjoyed the drift problems of the Hitachi circuit, I was expecting something similar

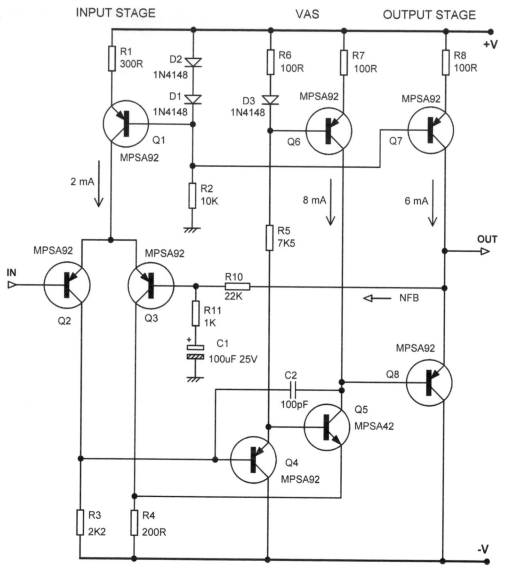

Figure 8.7. The Lender Push-pull VAS. Here it is used in a model amplifier with a small-signal Class-A output stage.

here, and so it was. The circuit was adjusted for input balance after it had been running for a long time (hours), and was then switched off and allowed to cool. Figure 8.11 shows how the input pair starts off unbalanced at cold, with very visible second harmonic generated. After 30 seconds it is back in balance and only third harmonic is produced. After that there is a very slow long-term drift.

The reason for this drift in input pair balance is the differing power dissipation of Q4 and Q5. They have approximately the same collector currents, but Q4 has

only 3 Volts across it, while Q5 has the full voltage between supply rail and ground, which here is 30 V.

Single Input Stages with a One-input Push-pull VAS

Both the Hitachi and Lender push-pull VAS circuits are open to the objection that since they use both outputs from a single input pair to create a push-pull action, they do not have the precision to enforce an accurate

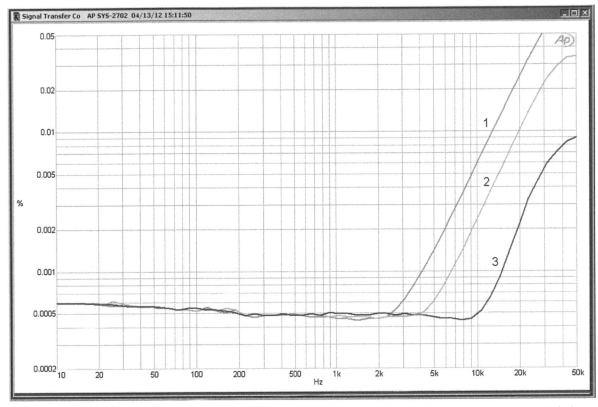

Figure 8.8. Distortion performance of the Lender Push-pull VAS in Figure 8.7 showing improvement in input pair balance as R4 is adjusted. R4 = 200 Ω (1) 240 Ω (2) and 270 Ω (3). Bandwidth 80 kHz. No input pair degeneration. +20 dBu out, ±30 V rails.

input collector current balance in the same way that a current-mirror does.

The only answer to this seems to be that if push-pull action is required in the VAS, then it must provide this itself. It ought to be possible to come up with a configuration that generates an anti-phase drive signal by sensing its own internal current flow. An example of this idea is the push-pull Class-A output stage used in the model amplifiers in Chapters 6 and 7. This is an extremely useful and trouble-free form of push-pull output; I have used it many times in discrete preamplifiers, mixers, etc., and it has served me very well. I derived the notion from the valve-technology White cathode-follower, described by Nelson-Jones in a long-ago *Wireless World*.[5] The original reference is British patent 564,250 taken out by Eric White in 1940.[6]

Figure 8.12a shows such a push-pull emitter-follower. When the output is sourcing current, there is a voltage drop through the upper sensing resistor R1, so its lower end goes downwards in voltage. This is

coupled to the current-source Q2 through C1, and tends to turn it off. Likewise, when the current through Q1 falls, Q2 is turned on more. This is essentially a negative-feedback loop with an open-loop gain of unity, and so by simple arithmetic the current variations in Q1, Q2 are halved, and this stage can sink twice the current of an emitter-follower where Q2 is a constant-current source, while running at the same quiescent current. The effect of output loading on the linearity is much reduced, and only one resistor and one capacitor have been added. This configuration needs fairly clean supply rails to work, as any upper-rail ripple or disturbance is passed directly through C3 to the current-source, modulating the quiescent current and disrupting the operation of the circuit.

It seems plausible that the same principle could be applied to a VAS, and indeed it has been done. A bit of research unearthed the use of the configuration shown in Figure 8.12b in the Pioneer SA-610 (1979) integrated amplifier.

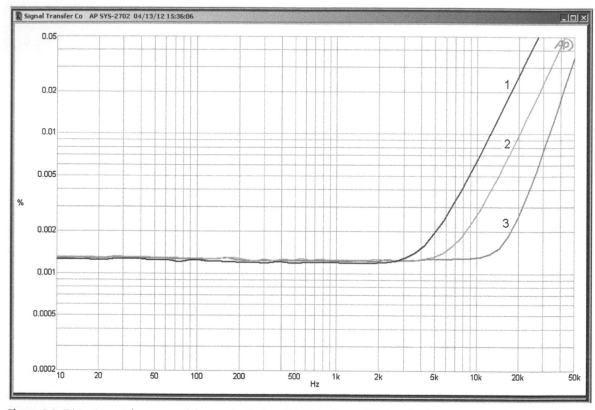

Signal Transfer Co   AP SYS-2702 04/13/12 15:36:06

Figure 8.9. Distortion performance of the Lender Push-pull VAS in Figure 8.7 showing improvement in input pair balance as R4 is adjusted. R4 = 200 Ω (1) 240 Ω (2) and 270 Ω (3) Bandwidth now 500 kHz. No input pair degeneration. +20 dBu out, ±30 V rails.

According to the SA-610 service manual, the push-pull VAS stage was used to increase open-loop gain and prevent capacitive loading on the VAS from degrading the HF response. The second reason sounds a little strange to me as I would have thought that the global negative feedback would have dealt with that issue, and anyway the capacitive loading appears to be only about 300 pF, which would cause no problems at all. The first reason makes more sense as the SA-610 is one of those amplifiers that follow the strange Japanese practice of putting the bass and treble controls in the power amplifier global feedback loop. More open-loop gain might well come in handy when applying bass or treble boost.

There are one or two unexplained features about the SA-610 circuit. Firstly, the current mirror has its AC operation completely suppressed by C1, so while the mirror will presumably ensure the DC balance of the input stage collector currents, it could not double the input stage transconductance and the VAS slew-rate in the usual way. Secondly, the arrangement of

the current-control feedback loop is rather different from that of the emitter-follower in Figure 8.12a. The current-sense resistor R7 in the VAS emitter is less than a third of the value of the resistor R5 in the current-source emitter. There is a diode D2 in the bias network for the current source. Both of these differences would appear to disrupt the correct push-pull operation of the VAS but it is impossible to be definite on this point without a complete analysis, which has at present not been done. Another anomaly appears to be the small value of C2, which when taken with the value of R4 at 2.2 kΩ, implies that push-pull action will only occur above 7 kHz.

Slight variations of this push-pull VAS were used in other Pioneer integrated amplifiers, such as the SA-620 (1981) which omits C2 and D2. The Pioneer SA-608 and SA-6800 (which differed only in external appearance) replaced R6 with a Zener diode; here an explicit claim was made in the service manual that even harmonics were cancelled and gain increased.

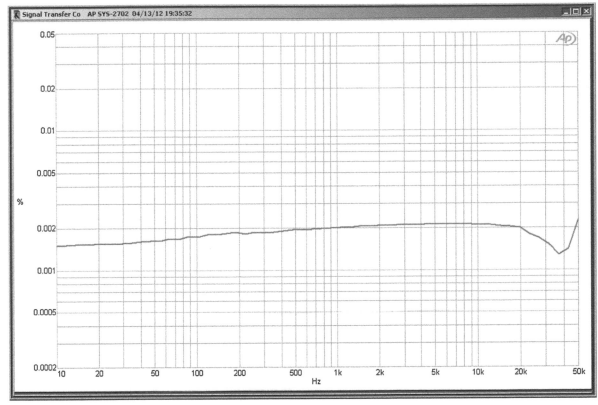

Figure 8.10. Distortion performance of the Lender Push-pull VAS with input pair degeneration of 100 Ω and tail current increased to 4 mA. Bandwidth 500 kHz +20 dBu out, ±30 V rails.

There are other ways of making a push-pull VAS that generate an anti-phase drive signal for a variable current source by means of a current-sensing resistor.

The Series Input Stage Push-pull VAS

Another method of generating push-pull drive signals from a single input stage is possible, and has been used occasionally in commercial equipment. Here there are still two input transistors, but they are complementary rather than the same type, and are connected in series rather than in the parallel format of the conventional long-tailed pair. The two output signals are conveniently referenced to the top and bottom supply rails. See Figure 8.13, which has representative component values and operating currents. Note that two complementary level-shifting emitter-followers Q1, Q3 are required at the input, so that Q2 and Q4 have enough Vce to operate. The version shown here has the collector currents from the input emitter-followers fed into the emitters of Q2, Q4. A very similar arrangement was used by Sony in their massive TA-N1 power amplifier (2000).

A serious objection to this configuration is that it tries to cancel non-linearity and temperature effects in two transistors that are not of the same type. Even so-called complementary pairs are not exact mirror-images of each other. It is significant that in every example of this configuration that I have seen, a DC servo has been fitted to give an acceptable output offset voltage. The complementary emitter-followers in front of the gain devices are also expected to cancel the Vbe's of the latter, which introduces more questions about accuracy.

This approach presents some interesting problems with the definition of the operating conditions. Note the degeneration resistors R8, R9. These are essential to define the collector current passing through Q2, Q4; the current through the series input stage depends on a relatively small voltage established across these two low-value resistors. In contrast, in a conventional differential pair, the value of the emitter-degeneration resistors has no effect at all on the operating current, which is set by the tail current-source.

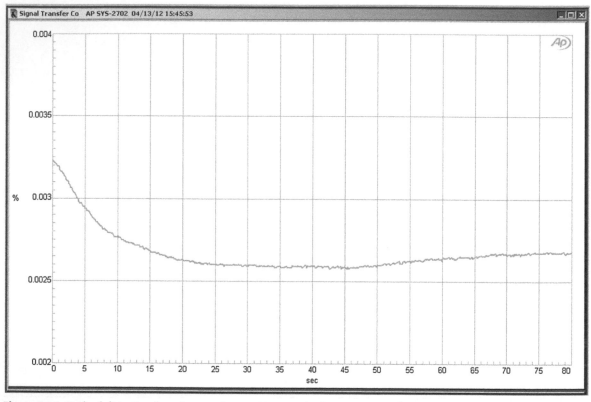

Figure 8.11. Drift of distortion from input pair over 80 seconds after turn-on, without (Trace 1) and with (Trace 2) R4 (= 3K9) in series with Q4 collector. At 25 kHz +20 dBu out, ±30V rails.

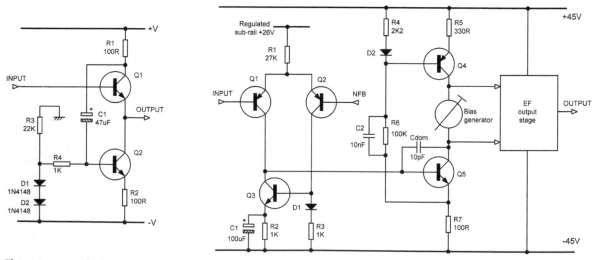

Figure 8.12. (a) The basic push-pull emitter-follower; (b) simplified version of Pioneer SA-610 push-pull VAS that generates its own anti-phase signal to modulate current-source Q4 from sensing resistor R7.

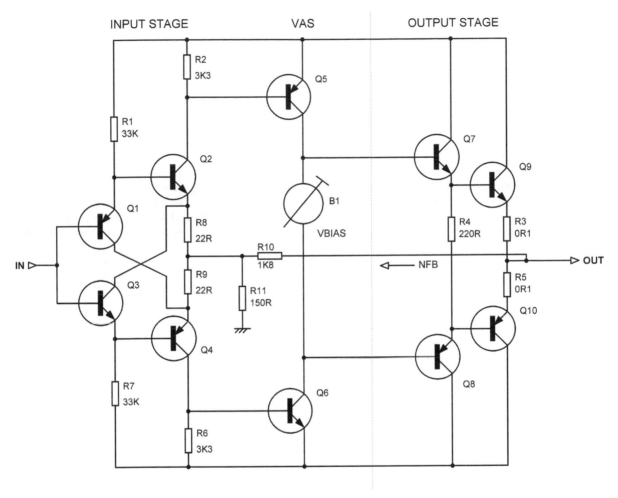

Figure 8.13. Series differential input configuration.

It is not easy to assess the linearity of this input stage configuration in isolation as it does not give a single current output, but two that are combined at the output of the push-pull VAS stage. It is therefore more difficult to separate the non-linearities of the input and VAS stages in practical measurement. It should be easier in SPICE simulation as the two collector currents can be subtracted mathematically.

Given the unpromising prospects for this arrangement, I have not at present analysed it further.

Single-input Push-pull VAS Circuits: Conclusions

In both the Hitachi and Lender configurations we have seen that the vital input pair balance is intimately bound up with the internal conditions of the VAS, and is therefore at the mercy of many more factors than is the simpler arrangement of a mirrored input pair and a constant-current VAS. In both cases undesirable drift in the distortion performance occurs because transistors in the VAS are working with very different power dissipations.

For this reason no further analysis or development of these configurations has been attempted.

Since a differential pair (when not combined with a current-mirror) inherently offers two anti-phase outputs, it seems like an obvious course to use these to drive a push-pull VAS directly. However, this means we lose the guaranteed balance of an input pair with a current-mirror, and a more satisfactory arrangement may be to retain the mirror in the input stage, and interface this with the VAS by a single current output, as for the Blameless amplifier.

The Double Input Stage Push-pull Simple VAS

If two differential input stages are used, with one the complement or mirror-image of the other, then the two VAS drive signals are conveniently referenced to the top and bottom supply rails. I call this a Double Input Stage Push-Pull VAS, which I think we'd better shorten to DIS-PP-VAS. The basic circuit is shown in Figure 8.14 with representative component values and operating currents. I have tried to make these correspond as closely as possible with the circuit conditions in the single-ended VAS amplifiers examined in Chapter 7,

to allow meaningful comparisons. The output stage is a Class-A constant-current emitter-follower Q9, Q10 which gives very low distortion as it only has to drive the light loading of the negative-feedback network R10, R11.

This configuration works rather differently from a single-ended VAS. There is now an amplifying transistor Q7 at the top, and another Q8 at the bottom of the VAS, and no fixed current-source to define the VAS standing current. This is set by the voltages across R9 and R12, and those voltages are in turn set by the voltage drops across the input stage collector resistors R1 and R3. These voltage drops are determined by the values of R1 and R3, which have half the input stage tail current flowing through them. Each tail

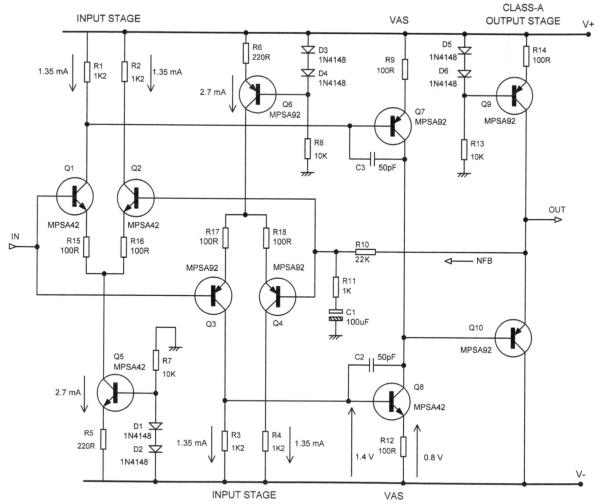

Figure 8.14. Model amplifier with double differential input configuration giving drive to top and bottom of push-pull VAS.

current depends on the resistor value in the current-source and the value of the bias voltage. You will note that this is a rather long chain of reasoning compared with the single-ended VAS, where the VAS standing current depends only on one bias voltage and one resistor value. Here there are many more places where component tolerances can have an effect.

Note also that the differential pairs have their collector currents approximately balanced by correct resistor values, but not held exactly correct by current-mirrors.

Because the loading on the VAS is (or should be) relatively light, the currents flowing through Q7 and Q8 must be almost exactly the same. The only way the global negative feedback loop can bring this about is to adjust the DC offset at the bases of Q2 and Q4, so the accumulated inaccuracies in the two paths setting the VAS current will result in a greater DC offset at the output. There is no comparable problem with the single-ended VAS.

To put some figures to this, I measured the result of reducing R12 in Figure 8.14 by 10%. The existing offset at Q2, Q4 bases went further negative by 11 mV. You can see why; to keep the current in Q7 and Q8 the same, the voltage across R12 must be reduced, while that across R9 increases. The increased negative offset tends to turn on Q4 which tends to turn off Q3, and so reduce the voltage across R12. At the same time Q2 tends to turn off, so Q1 tends to turn on, and the voltage across R9 is increased.

In Chapter 6 we noted that we got a dramatic reduction in input-stage distortion on going from a single transistor input to a differential pair, as the latter cancels out the second-order non-linearities of the input stage and eliminated the second harmonic. This obviously cannot be cancelled out twice over, so using two differential inputs instead of one is not likely to give radical improvements in linearity. However, there may be something to be gained in terms of input stage distortion; if the drive signals to the VAS are correctly proportioned, then it should be possible to have each input differential pair working only half as hard as a single one. This would halve the input voltage seen by each pair, reducing its distortion (which is effectively all third-order) by a factor of four.

The model amplifier in Figure 8.14 is completely symmetrical, apart from the Class-A output stage which in a full-scale amplifier would be replaced by a symmetrical Class-B output stage. This symmetry is, however, more apparent than real, for even transistors that are sold as complementary pairs are by no means mirror-images. More on that later.

The Double Input Stage Push-pull Simple VAS: Open-loop Gain

In Chapters 5 and 7 we saw that the open-loop gain of a three-stage amplifier with a single-ended VAS had two regions in its open-loop gain characteristic: a portion flat with frequency at LF and a region falling at 6 dB/octave under the control of the dominant pole capacitor Cdom. The gain in the HF region could be calculated very simply from the input stage transconductance and the value of Cdom. The LF gain was determined by the input stage transconductance, the VAS transistor beta, and the VAS collector impedance, the last two being rather shifty and uncertain quantities.

The DIS-PP-VAS has the same two gain regions. Simulation and measurement show that the HF gain appears to follow the same equation as in Chapter 5. If simulation and measurement agree, we can only hope that theory will fall into line.

$$HF gain = \frac{g_m}{\omega \cdot C_{dom}} \qquad \text{Equation 8.1}$$

Where $\omega = 2 \cdot \pi \cdot$ frequency

This naturally requires a bit of interpretation if we are going to apply it to a configuration that looks radically different. Firstly, what about the transconductance g_m? There are now not one but two input stages with 100 Ω emitter degeneration resistors. Their current outputs are effectively summed, as they feed into the same VAS. This gives us twice the transconductance of a single input stage. However, the double inputs in Figure 8.14 do not have current-mirrors in their collectors (as do the single-ended VAS configurations examined in Chapter 7), so their effective transconductance is halved and we conclude that the overall input stage transconductance is the same as for a single input stage that does have a current-mirror.

Secondly, how do we treat the value of Cdom? Figure 8.14 shows equal values of Cdom on the top and bottom VAS transistors. Given the total symmetry of the amplifier circuit, we might guess that both capacitors are equally important for setting the HF gain. SPICE simulation quickly shows that almost the same HF gain within a tenth of a dB is obtained with one 100 pF capacitor at the top, *or* one 100 pF capacitor at the bottom, *or* 50 pF in each position, as shown in Figure 8.14. Furthermore, this gives the same HF gain at a given frequency, to within a dB, as the single-ended VAS configuration with 100 Ω input emitter degeneration resistors and a 100 pF Cdom on the VAS.

This is gratifyingly simple; it might have been much more complicated to deal with.

The LF gain of a single-ended VAS is:

$$LF gain = g_m \cdot \beta \cdot R_c \qquad \text{Equation 8.2}$$

We have dealt with the gm already, so we have only to grapple with the VAS beta and the VAS collector impedance Rc.

In Chapter 7 we saw that in a simulation of a complete amplifier, with an optimally-biased EF output stage driving an 8 Ω load, the LF gain with a simple VAS was +82 dB, with the transition to the HF gain regime P1 at about 1 kHz. The EF-VAS in the same circuit gave a much larger LF gain of +117 dB, with P1 at about 20 Hz, while the cascode VAS gave a less impressive +98 dB, with P1 at 200 Hz. Figure 8.15 includes a summary of this; note that the LF gains depend on transistor parameters and so will be somewhat variable.

Since the push-pull VAS has two amplifying transistors rather than one, we might hope that we would get a healthy amount of LF open-loop gain. In fact, we get less. The circuit of Figure 8.14 has an open-loop gain of only +72.4 dB at LF, as shown in Figure 8.15. We might expect it to be less than that of the EF-VAS and cascode VAS, but it is also 10 dB less than the single-ended simple VAS. We saw in Chapter 7 how the Early voltage of the VAS transistor had a major effect on the LF open-loop gain, so I ran some SPICE simulations to see how that works when there are two in push-pull.

Table 8.2 shows very clearly that the LF open-loop gain depends on the Early voltage of both transistors, which, in view of the symmetry of the circuit, is exactly what we would expect. However, this symmetry does not extend to the characteristics of the MPSA42 and MPSA92 transistors. They may be officially 'complementary' in voltage ratings, etc., but they are certainly not mirror-images. The Early voltage of the MPSA42 is only 45.1 Volts, while that of the MPSA92 is 260 Volts, nearly six times greater. If we alter the SPICE model of the lower VAS transistor MPSA42 to make its Early voltage 1000 Volts, the LF open-loop gain jumps up from +72 dB to +81 dB. If we do that to the upper VAS transistor MPSA92, leaving the MPSA42 model at the default 45.1 Volts, the gain only increases to +74 dB. The transistor with the lower Early voltage has the greatest effect in limiting the LF open-loop gain.

In Chapter 7 we also saw that the beta term in Equation 8.2 also has a powerful effect on the LF open-loop gain. That is not so with the push-pull simple VAS. If we set the Early voltage of both VAS transistors to 100,000 Volts, to get rid of Early effect entirely, we have an LF open-loop gain of +103.7 dB, still much inferior to the +117 dB that the EF-VAS gives with real transistors that have not had their models tweaked. If we now modify the beta of both VAS transistors by altering the BF parameter in the SPICE models, the result is disappointing. The MPSA42 has a default BF of 70.4, while the MPSA92 has a default BF of 98. Setting both BFs to 400 only increased the LF gain to +105.9 dB,

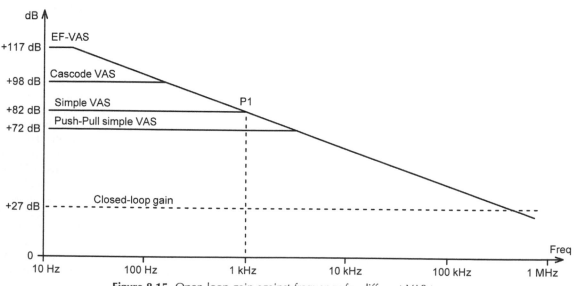

Figure 8.15. Open-loop gain against frequency for different VAS types.

Table 8.2. Push-pull VAS open-loop gain and VAS transistor Early voltages. SPICE simulation

Lower VAS Early voltage MPSA42	Upper VAS Early voltage MPSA92	LF gain dB
45.1	260	+72.4
1000	260	+81.5
10,000	260	+83.2
100,000	260	+83.4
100,000	1000	+92.1
100,000	10,000	+101.5
100,000	100,000	+103.7
45.1	1000	+74.1
45.1	10,000	+74.9
45.1	100,000	+75.0

an increase of only 2.2 dB. It looks as if the use of different transistor types with high beta for the VAS is not going to do much to improve the LF open-loop gain.

The Double Input Stage Push-pull Simple VAS: Distortion

In Figure 8.16 the DIS-PP-VAS model amplifier of Figure 8.14 is measured under the same conditions as the single-ended VAS configurations examined in Chapters 7; +20 dBu out and with ±20V supply rails. The results for ±25V and ±30V supply rails are also shown.

The ±20V trace in Figure 8.16 should be compared with the THD results for the single-ended simple VAS with MPSA42 in Figure 7.5. The distortion in the HF region is very similar, and in fact exactly the same at 20 kHz, but the LF distortion is significantly worse for the push-pull VAS. The push-pull LF distortion is 0.002%, and appears to be mainly a mixture of second and third harmonics. The LF distortion for the single-ended VAS is approximately 0.0007%, but this

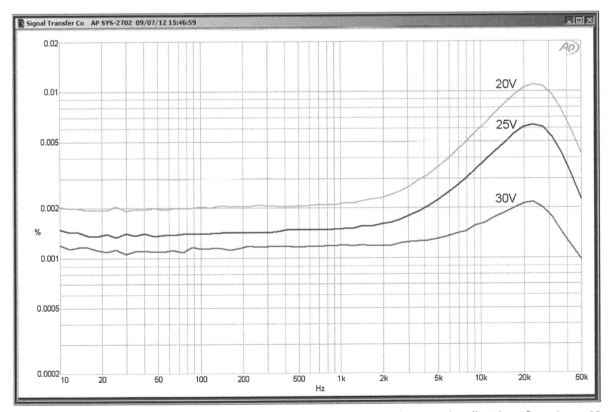

Figure 8.16. Distortion performance of model amplifier with double differential input push-pull VAS configuration. +20 dBu output. ±20V, ±25V, and ±30V rails

figure includes some noise. The ratio between these distortion figures is 2.86 times or 9.1 dB, while the ratio between the LF open-loop gain for push-pull and single-ended is 10.4 dB. If we allow for some noise contamination of the single-ended figure, it seems that the extra LF distortion of the push-pull VAS is accounted for wholly by its lower open-loop gain.

As usual, increasing the supply rails reduces the distortion. In Figure 8.16 both the LF and HF distortion are reduced, the HF distortion by a greater amount. Compare this with a similar plot for the single-ended VAS in Figure 7.10. The push-pull VAS has much higher LF distortion, and this is presumably because there are two transistors in the VAS experiencing Early effect as opposed to one in the single-ended VAS.

The tail currents of the two input stages are set by the need for good linearity (a high tail current allows emitter-degeneration resistors to be used) and the need for an adequate slew-rate. The VAS standing current can be set independently by changing the VAS emitter resistor R9 and R12. Figure 8.17 shows the results of an experiment on this.

In Chapter 7 it was established that in the single-ended simple VAS, the LF distortion was due to Early effect, and this was confirmed by noting that LF distortion decreased as the VAS standing current was reduced; see Figure 7.14. The mechanism is fully explained in Chapter 7. There seems to be no reason why this should not also be the case in the push-pull simple VAS, and Figure 8.17 confirms that it is so. The VAS standing current was reduced from 7.8 mA to 5.5 mA by changing R9 and R12 from 100 Ω to 150 Ω, and as in Chapter 7, it was found that the LF distortion was (within the limits of measurement) proportional to the standing current. As with the single-ended VAS, the HF distortion is very little affected by the VAS standing current.

It has been said many times that a virtue of the DIS-PP-VAS is that it cancels out its own second-harmonic distortion, but I do not recall that anyone ever offered any solid evidence that this was the case. (No doubt it might also cancel second harmonic distortion from the input stages, but if they are correctly balanced the level of this should be negligible.) I don't

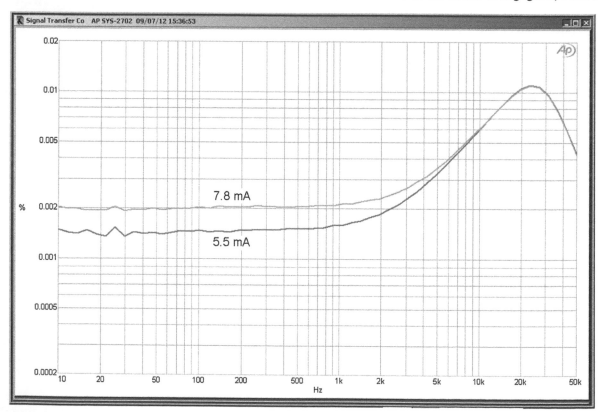

Figure 8.17. Distortion performance of model amplifier for two values of push-pull VAS standing current. +20 dBu output. ±20V rails.

pretend that this analysis offers a definitive answer to the question of whether the DIS-PP-VAS cancels second harmonics, but I think it is clear that it is worse in its general distortion performance than the single-ended VAS, so perhaps the question is not that important after all.

The Double Input Stage Push-pull Simple VAS: Noise

It occurred to me that there might be a possible noise advantage in the use of double input stages. If we can assume that the gain of the two signal paths is simply summed, with equal contributions from each, then noise from the input stage should be reduced by 3 dB as two uncorrelated noise sources add by RMS summation rather than simple addition. We also saw in Chapter 6 that a major contribution to the noise of an amplifier with a single-ended VAS is the current-mirror that keeps the input stage balanced. Since the model amplifier of Figure 8.14 has no current-mirrors, we might expect a substantial reduction in the noise output. (See Chapter 6 for more on the disconcerting amounts of noise produced by current-mirrors.)

Unfortunately that does not seem to occur. The EIN of Figure 8.14 was measured at -122.7 dBu, only 0.1 dB lower than the single-ended VAS amplifier. There seems to be more here than meets the eye, and it will be investigated.

The Double Input Stage Push-pull Simple VAS: PSRR

In Chapter 26, the issue of power supply rejection, the measure of how well an amplifier ignores the ripple and signal voltages on its supply rails, is examined in detail. The power supply is assumed to be the simple unregulated type as this is almost always the best solution. The attenuation that rail disturbances receive is known as the Power Supply Rejection Ratio, or PSRR. One of the most important points that emerge there is that in a standard three-stage amplifier with an NPN VAS sitting on the negative rail, the PSRR is much worse for negative rail disturbances than for those on the positive rail. This is because the VAS transistor has its emitter connected more or less directly to the negative rail, and so this is the 'ground' reference for the stage. The only reason that the hum is not appalling is that the effect of the entry of ripple at this point is counteracted by the global negative feedback loop, which is trying to enforce a hum-free output signal. This means that the greater the LF open-loop gain, the

better the PSRR, and this is one reason why the factors affecting LF open-loop gain have been examined quite closely in Chapter 7 and in this chapter. The amount of LF gain available at the minimum ripple frequency of 100 Hz is what matters.

However, even if the LF open-loop gain is increased by using an EF-VAS, the negative PSRR is nothing like good enough for a high-quality amplifier. The most direct solution to this problem, with guaranteed effectiveness, is to provide RC filtering of the negative supply to the input stage and VAS. Unfortunately even if large capacitors and small resistors are used, there is a voltage drop which makes negative clipping occur earlier, which is unfortunate as it already occurs earlier than positive clipping, due to the need for VAS current-limiting sense resistors, and so on. It is therefore sometimes worthwhile to provide an extra negative sub-rail a few volts below the main one, and this requires another transformer tap, a rectifier, fuse, reservoir capacitor and clamp diode so the extra cost and PCB area used is significant. See Chapter 26 for the details of providing a sub-rail.

There is no such reference issue with the current-source at the top of the VAS stage, and so the PSRR of the positive rail is much better, and can be made effectively perfect by very simple capacitor filtering of the reference voltage for the current-source.

The DIS-PP-VAS has a second amplifying transistor with the positive supply rail as its emitter reference, and so for a good PSRR a positive sub-rail will have to be provided as well, and that will need its own rectifier, fuse, reservoir capacitor, and clamp diode. This seems a definite point against the DIS-PP-VAS configuration when it is compared with a single-ended VAS.

In Chapter 7 we saw the need for VAS current limiting, to deal with not only the operation of overload protection circuitry, but also the effect of clipping, when the VAS will turn on more and more current in an attempt to make the output stage reach an unattainable voltage. The DIS-PP-VAS clearly needs a current-limiter on each VAS, to handle both positive and negative clipping, and so circuit complexity and the PCB area used are increased again.

A Brief History of the Double Input Stage Push-pull VAS

The first use of double input stages to drive a push-pull VAS that I am aware of was by Dan Meyer of SWTPC, who published an amplifier design called the Tiger-saurus in 1973.[7] The input and VAS configuration used was much the same as that of Figure 8.14. The output stage, however, was a very complicated affair combining a CFP stage with gain and stacked output device pairs to halve the Vces on the output devices. There seems to be a consensus that this design was not as stable at HF as it might have been, though that was probably nothing to do with the push-pull VAS configuration and a lot more to do with the output stage.

The DIS-PP-VAS was used in Jim Bongiorno's Ampzilla design in 1974[8] though he has said the idea was conceived some time earlier. In 1977 Marshall Leach used a complementary simple VAS driven by two non-mirrored complementary input pairs.[9] As previously mentioned, Erno Borbely used double complementary input stages in 1982,[4] but he employed two Lender circuits back-to-back for the VAS.

The Double Input Stage Push-pull EF-VAS

In Chapter 7 it became very clear that adding an emitter-follower inside the Miller loop, making it an EF-VAS was a very straightforward and effective way of reducing distortion by making the non-linear VAS Cbc harmless, and also increasing LF open-loop gain. In exactly the same way, two emitter-followers can be added to the push-pull VAS, also with great advantage. I suppose the appropriate acronym would be DIS-PP-EF-VAS though I can't say I feel that much enthusiasm for it.

Figure 8.18 shows the circuit of a model amplifier with a push-pull EF-VAS. Apart from adding the two emitter-followers Q11 and Q12, the only other change required is a change in the value of R9 and R12 to keep the VAS standing current the same as it was.

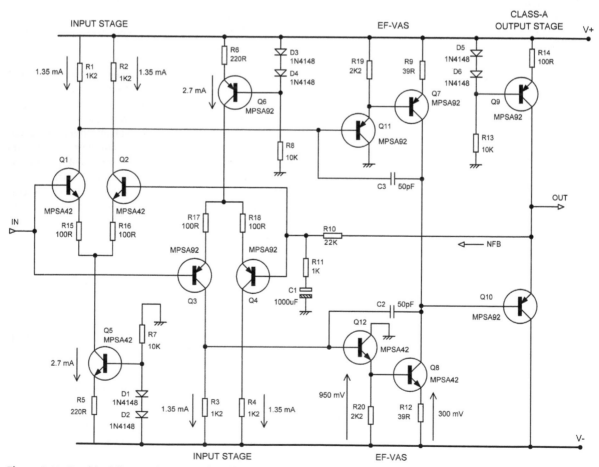

Figure 8.18. Double differential input push-pull EF-VAS in a model amplifier. Emitter-followers have been added inside the upper and lower VAS Miller loops. R9 and R12 have been altered to keep the VAS standing current at 7.8 mA.

Since there is an extra 0.6 V voltage drop due to the emitter-follower Vbes, R9 and R12 are reduced from 100 Ω to 39 Ω.

The Double Input Stage Push-pull EF-VAS: Open-loop Gain

When we added the emitter-follower to the single-ended VAS, to create an EF-VAS, we found that the LF open-loop gain increased prodigiously, from +82 dB to +117 dB. There was naturally a corresponding reduction in the frequency P1 at which the open-loop gain changed from the HF to LF regime; this happens automatically as the HF gain/frequency relationship is fixed by the input stage transconductance and the size of the dominant pole Miller capacitor Cdom.

I regret to say that we get no such impressive increase in LF open-loop gain. We are starting from a worse position of having only 72 dB of LF gain, and adding the two emitter-followers only increases it to 83.5 dB, which is only 1.5 dB better than a single-ended simple VAS. The

reason for this disappointing result is not yet fully investigated, but is probably something to do with Early effect acting on two amplifying VAS transistors rather than one and a fixed current-source.

Here the VAS standing current has only a very minor effect on the LF gain, but as we have seen before, in the direction of gain decreasing as the standing current increases.

The Double Input Stage Push-pull EF-VAS: Distortion

As for the single-ended VAS, adding the emitter-follower transforms the distortion performance. Figure 8.19 shows the THD for ±20V, ±25V, and ±30V supply rails. The measurement floor of the Audio Precision SYS-2702 test system was at 0.00055%, and the THD trace for ±30V is only just above that.

These distortion results are very gratifying, and compare well with the equivalent results for a single-ended EF-VAS in Figures 7.17, 7.18, and 7.19.

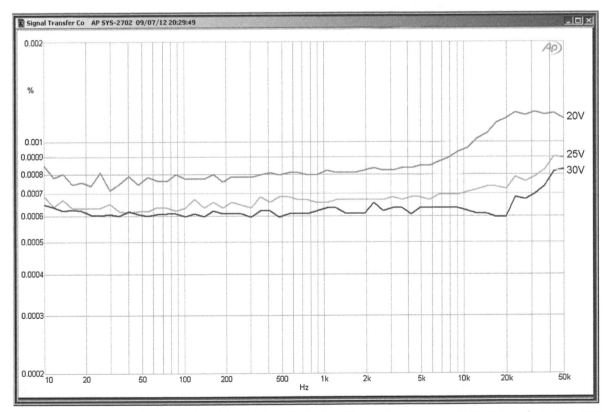

Figure 8.19. Distortion performance of model amplifier with double differential input push-pull EF-VAS configuration. +20 dBu output. ±20V, ±25V, and ±30V rails.

The Double Input Stage Push-pull EF-VAS: Slew-rate

One of the main advantages claimed for the general idea of a push-pull VAS is that it allows a much improved maximum slew-rate, as there is no fixed current-source that puts an absolute limit on the slew-rate in one direction (positive slewing, for the usual arrangement with an NPN VAS sitting on the negative supply rail). The assumption is that since the DIS-PP-VAS has amplifying transistors at top and bottom, either will be able to turn on as much current as is required to charge and discharge the Miller capacitor Cdom.

However, when Figure 8.14 is simulated, we get only 21.1 V/usec for positive slewing and 19.7 V/usec for negative slewing. This is actually slower than a single-ended version which gives 34.2 V/usec positive and 33.5 V/usec negative; you will note that this is reasonably symmetrical because care has been taken to prevent the VAS current-source from putting a premature limit on the speed. (The slewing behaviour of the single-ended VAS is examined in detail in Chapter 15.) The simulation tested slewing from +20 V to -20 V, with ±35 V supply rails; the load was 8 Ω.

There is actually some difficulty in coming up with truly comparable circuits of each type for slew-rate testing. The push-pull amplifier in Figure 8.14 has two input stages, each with a tail current of 2.6 mA, but no current-mirrors. In contrast the single-ended version has a single input with a tail current of 6 mA and a current-mirror, which doubles the maximum current that can flow in and out of the input stage. The VAS standing currents were the same at 8 mA. The reason why the push-pull circuit of Figure 8.14 is slower is simply that its tail currents are lower, and are placing the limit on how fast current can be moved in and out of the two Miller capacitors.

We can make the comparison more meaningful by increasing both tail currents of the push-pull amplifier to 6 mA, to match the single-ended version. If we do that, then the input stage collector resistors have to be reduced to 520 Ω to keep the VAS standing current at 8 mA. The slew-rates improve to 38.9 V/usec positive and 33.4 V/usec negative, somewhat better than the single-ended version. You will notice that they are not notably symmetrical; the schematic may be absolutely symmetrical but the transistors that make it up are not.

Another obvious difference is that the single-ended version has a current-mirror, doubling its current capability. If we replace the collector resistors in our push-pull circuit (keeping the tail currents at 6 mA each) with current-mirrors as used in the single-ended version, with 68 Ω mirror degeneration resistors, then as expected the slew-rates increase dramatically to 100 V/usec positive and 104 V/usec negative. Speed and symmetry are much improved. However, we have also doubled the transconductance of the input stage, so changes must be made elsewhere to maintain the HF gain at the same level and so maintain the stability margins. Doubling the Miller capacitors will get us pretty much back where we started, so it would probably be better to double the input stage degeneration resistors to 200 Ω.

This is necessarily only a brief description of the slewing behaviour of the push-pull VAS; its full investigation would fill another chapter. It is of course very necessary to back up simulator results with real measurements, as slew-rate can depend on factors like beta that are only approximately modelled by SPICE, and by variations in device parameters. In the absence of clear performance advantages for the push-pull VAS over the single-ended VAS in other areas, and the fact that the slew rates obtained from a single-ended VAS are more than adequate, that has not so far reached the top of the to-do list.

The Double Input Stage with Mirrors and Push-pull Simple VAS

The addition of a current-mirror to the input stage of an amplifier using a single-ended VAS gives many advantages; input stage balance is enforced so that second-order distortion is suppressed, the input transconductance is doubled, and so is the slew-rate. The downside is that the noise performance is worsened unless you can work in some big mirror degeneration resistors (see Chapter 6). It is, however, not wholly clear that the same advantages can be gained when mirrors are added to a double input push-pull VAS circuit. In the first place, the exact balance of the input stage depends on the global negative feedback loop adjusting the input stage currents. With only *one* feedback signal, it is not possible to enforce balance in *two* separate input stages, each with their own tolerances and device variations. There is also a widely held belief that adding current-mirrors makes it impossible to determine the VAS standing current, as there is no explicit mechanism to set it. However, rather than giving up at once, let us examine the possibilities. If we replace the 1.2 kΩ collector resistors with current mirrors we get Figure 8.20 if we use a simple VAS, and Figure 8.21 if we use an EF-VAS. Note that the Miller capacitors have been increased from 50 pF to

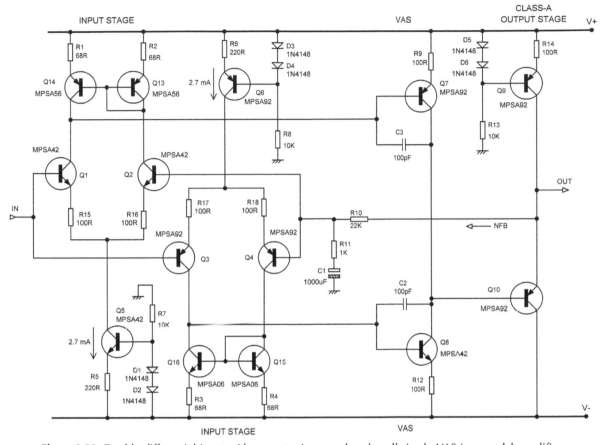

Figure 8.20. Double differential input with current-mirrors and push-pull simple VAS in a model amplifier.

100 pF to allow for the doubling of the input stage transconductance by the mirrors.

As we saw earlier, in Figure 8.14 (simple VAS) and Figure 8.18 (EF-VAS) the tail sources set the voltage drop across the input stage collector resistors, and they in turn set the voltage drop across the VAS emitter resistor. But now we have installed current-mirrors, and their outputs are high resistance current nodes that are effectively floating in that they have no power to impose bias conditions on the VAS. In these conditions, the standing current of the VAS will not be well defined, as it depends on the device characteristics of various transistors. The betas of the upper and lower VAS transistors would appear to be the strongest influence.

Desiring to know what this indeterminacy would mean in practice, I first tried simulation. I tried two versions of Figure 8.20, with very different circuit values, and in both cases the DC bias point was found at once with no problems at all. Undoubtedly the answers depend on device characteristics, and so

ultimately on the device models, but an answer can be found. In both cases the VAS standing currents came out at reasonable values, which was rather unexpected, and led me to think there might be a bit more to this business than met the eye.

Simulation is a wonderful tool, but measurement is reality. I therefore built a model version with a push-pull EF-VAS and it is working very nicely about a foot from my elbow at the moment. THD is 0.00064% for 11 Vrms out, 1 kHz (testgear genmon = 0.00041%). The transistors were all MPSA42/92 except for the mirrors which were MPSA06/56 to avoid problems with the need of the MPSA42 for more Vce than you might think (see Chapter 7). There was no tweaking or adjustment or selecting of devices. The VAS standing current appears to have adjusted itself to a very low value (42 uA) but despite this, the linearity is extraordinarily good. The fox in the ointment is that there are some nasty HF stability problems when it is driven with higher frequencies, e.g., 20 kHz, and

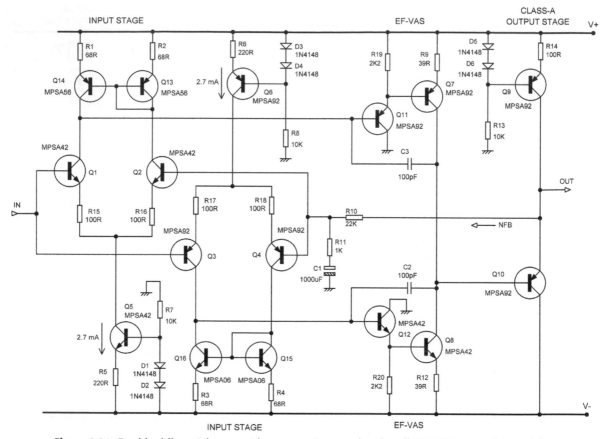

Figure 8.21. Double differential input with current-mirrors and push-pull EF-VAS in a model amplifier.

this may well be rooted in problems of current indeterminacy. Clearly more research is needed …

The use of a servo to control the VAS standing current has been suggested on more than one occasion, but the extra complication is most unwelcome, and the need for such assistance confirms that the basic configuration is not sound.

If the arguments above are correct, there is nothing to stop us making a double input push-pull stage that has only *one* current-mirror. The non-mirror input stage will define the VAS standing current and the single mirror will accommodate itself to that, though not necessarily being able at the same time to enforce the balance of the input pair to which it is attached. A model amplifier was built along the same lines as Figure 8.21, with the mirror applied to the NPN input pair, and the 1.2 kΩ collector resistors retained in the PNP pair. The Miller capacitors were 50 pF. It worked very nicely indeed, with the EF-VAS current properly defined, and the excellent distortion performance can

be seen in Figure 8.22. This is rather better than that of the non-mirror EF-VAS shown in Figure 8.19.

Another version of the one-mirror push-pull VAS amplifier was tried, using a simple VAS. The result was as seen earlier in this chapter; the simple-VAS had higher LF distortion than the EF-VAS (see Figures 8.16 and 8.19) and in this case was about 0.0092%.

Further enquiry will be needed to determine if the new one-mirror version has real advantages over other circuitry; it certainly seems to have good linearity.

The Double Input Stage Push-pull VAS: Conclusions

Looking at both the DIS-PP-VAS and the DIS-PP-EF-VAS, it is clear that the first of the two has relatively poor linearity and is not suitable for a high-quality amplifier. The second, however, with its added emitter-followers, is definitely a possible rival to the single-ended EF-VAS. The basic configuration has the

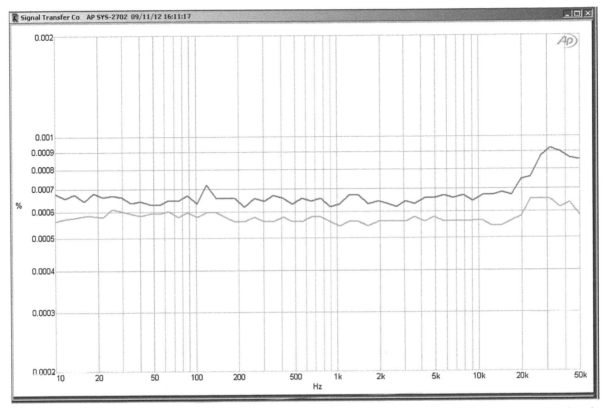

Figure 8.22. Distortion performance of double differential input with *one* current-mirror (on the NPN input pair) and push-pull EF-VAS in a model amplifier. (Upper trace) Lower trace is testgear output (genmon) +20 dBu output. ±20V rails.

merit that it can be modified and enhanced in exactly the same way as the more familiar single-ended VAS; you will note that adding the two emitter-followers to create the EF-VAS version was a wholly straightforward operation. With other VAS configurations this is not the case. Faster slewing than a single-ended VAS is certainly possible but the extra speed is of no practical benefit, though it may make for good-looking specs.

Its disadvantages compared with a single-ended EF-VAS are that imbalances in the VAS will feed through to increased output offset voltages, the LF open-loop gain is much lower, and the general complexity is greater, which may make fault-finding a bit more difficult. Twice the amount of circuitry is required to provide sub-rails for a good PSRR, and twice the circuitry is required for VAS current limiting.

A More Advanced Push-pull VAS

While I strongly suspect that this chapter is the most detailed description of the push-pull VAS ever put to paper, it absolutely does not exhaust the subject. Samuel Groner gives a detailed description of a new push-pull VAS stage in Jan Didden's *Linear Audio*, Volume 2.[10] This highly ingenious design not only addresses the ability of a push-pull stage to provide more current, but also solves the PSRR issue of having the VAS emitter referenced to a supply rail, not by adding inconvenient extra sub-rails, but by some very cunning use of cascoding. I have no space to describe it further here, but consider it required reading if you are concerned with audio power amplifiers. It can be obtained via Jan Didden's website.[11]

The Folded-cascode VAS

The folded-cascode configuration shown in Figure 8.23 is not just another variation on the push-pull VAS driven by a single input stage. It is usually called the folded-cascode configuration, because Q4, Q5 are effectively cascoding the collectors of input stage Q2, Q3. The distinguishing characteristic of this configuration is

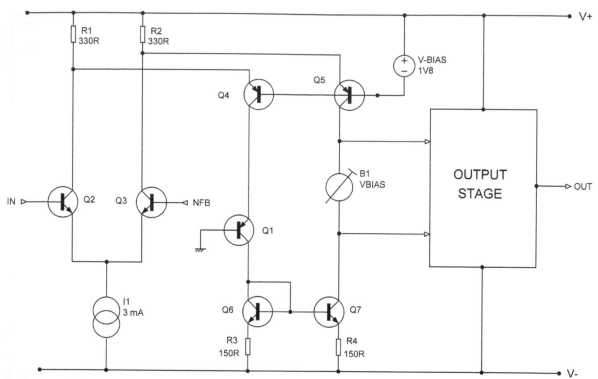

Figure 8.23. Folded-cascode configuration giving drive to top and bottom of the push-pull VAS stage Q5, Q7 via the current-mirror Q6, Q7.

that the two transistors Q4, Q5 are common base stages and their emitters are driven from the collectors of input stage Q2, Q3. Q1 is a cascode transistor for the collector of Q4 (sort of a cascode of a cascode); it is not an essential part of the folded-cascode concept. The current output of Q1 is bounced off the V-rail by the current-mirror Q6, Q7 and provides the lower part of the push-pull drive to the VAS stage Q5, Q7.

The really important point about this approach is that it is not a three-stage architecture; it is a two-stage architecture with input stage and VAS combined in one stage. The current variations in the input pair are passed unamplified to the VAS output by the folded-cascode transistors Q4, Q5, which give no current gain. These relatively small current variations have to do the best they can to generate a high voltage gain although subject to non-linear loading from the output stage. This is bound to lead to a lack of overall open-loop gain. The only obvious way to increase it is to raise the impedance seen at this point by putting a unity-gain buffer between the input/VAS and the output stage. This approach can work well in opamps,[11] but there seems to be little advantage in a power amplifier.

We are already using more transistors than in a three-stage Blameless amplifier. Another objection is that there is no obvious way to apply the Miller dominant pole compensation that is so very useful in linearising a conventional VAS.

While the folded-cascode configuration has been used extensively in opamps (in opamp usage the two resistors R1, R2 are normally replaced by constant-current sources), it has only rarely been applied to audio power amplifiers, and I have no practical experience with it.

I am not sure if any commercial amplifiers have been built using the folded-cascode structure, but at least one such design has been published for amateur construction by Michael Bittner,[12] and the circuit values here are based on that.

The Push-pull VAS: Final Conclusions

In the first part of this chapter we saw that the two versions of the push-pull VAS that used a single input stage had various unsettling problems that made them unattractive as alternatives to the single-ended EF-VAS. The double input push-pull VAS is more promising, but still has some distinct difficulties that do not afflict the single-ended alternative. If I was starting the design of an amplifier tomorrow (which at the time of writing is a distinct possibility), I would almost certainly use a single-ended EF-VAS. This remains my preferred topology.

References

1. Dostal, J, *Operational Amplifiers*, Oxford: Butterworth-Heinemann, 1993, p. 195.

2. Locanthi, B, Operational Amplifier Circuit for Hifi, *Electronics World*, (USA), Jan. 1967, p. 39.

3. Lender, R, Power Amplifier with Darlington Output Stage, Motorola internal report 7 Sept. 1974. This report was not published. It is *not* Motorola Application Note AN-483B, as seems to be widely believed.

4. Borbely, E, A 60W MOSFET Power Amplifier, *The Audio Amateur*, Issue 2, 1982, p. 9.

5. Nelson-Jones, L, Wideband Oscilloscope Probe, *Wireless World*, Aug. 1968, p. 276.

6. White, E, Improvements in or Relating to Thermionic Valve Amplifier Circuit Arrangements, British patent No. 564,250 (1940).

7. Meyer, D, Tigersaurus: build this 250-watt hifi amplifier, *Radio-Electronics*, Dec. 1973, pp. 43—47.

8. Bongiorno, J, http://www.ampzilla2000.com/Amp_History.html (accessed Mar. 2012).

9. Leach, M, Build a Low TIM Amplifier, *Audio*, Feb 1976, p. 30.

10. Groner, S, A New Amplifier Topology with Push-Pull Transimpedance Stage, *Linear Audio*, 2, 2011, pp. 89—114.

11. Didden, J, Linear Audio, website http://www.linearaudio.nl/ (accessed Sept. 2012).

12. Bittner, M, http://www.diyaudio.com/forums/solid-state/60918-explendid-amplifier-designed-michael-bittner-our-mikeb.html (accessed Mar. 2012)

The Output Stage

Classes and Devices

The almost universal choice in semiconductor power amplifiers is for a unity-gain output stage, and specifically a voltage-follower. Output stages with gain are not unknown − see Mann[1] for a design with ten times gain in the output section − but they have significantly failed to win popularity. Most people feel that controlling distortion while handling large currents is quite hard enough without trying to generate gain at the same time. Nonetheless, I have now added a section on output stages with gain to this chapter.

In examining the small-signal stages of a power amplifier, we have so far only needed to deal with one kind of distortion at a time, due to the monotonic transfer characteristics of such stages, which usually (but not invariably[2]) work in Class-A. Economic and thermal realities mean that most output stages are Class-B, and so we must now also consider crossover distortion (which remains the thorniest problem in power amplifier design) and HF switch-off effects.

We must also decide what *kind* of active device is to be used; JFETs offer few if any advantages in the small-current stages, but power FETS in the output appear to be a real possibility, providing that the extra cost proves to bring with it some tangible benefits.

The most fundamental factor in determining output-stage distortion is the Class of operation. Apart from its inherent inefficiency, Class-A is the ideal operating mode, because there can be no crossover or switch-off distortion. However, of those designs which have been published or reviewed, it is notable that the large-signal distortion produced is still significant. This looks like an opportunity lost, as of the distortions enumerated in Chapter 5, we now only have to deal with Distortion One (input-stage), Distortion Two (VAS), and Distortion Three (output-stage large-signal non-linearity). Distortions Four, Five, Six and Seven, as mentioned earlier, are direct results of Class-B operation and therefore can be thankfully disregarded in a Class-A design. However, Class-B is overwhelmingly of the greater importance, and is therefore dealt with in detail below.

Class-B is subject to much misunderstanding. It is often said that a pair of output transistors operated without any bias are 'working in Class-B', and therefore 'generate severe crossover distortion'. In fact, with no bias, each output device is operating for slightly less than half the time, and the question arises as to whether it would not be more accurate to call this Class-C and reserve Class-B for that condition of quiescent current which eliminates, or rather minimises, the crossover artefacts.

There is a further complication; it is not generally appreciated that moving into what is usually called Class-AB, by increasing the quiescent current, does *not* make things better. In fact, if the output power is above the level at which Class-A operation can be sustained, the THD reading will certainly increase as the bias control is advanced. This is due to what is usually called *gm-doubling* (i.e., the voltage-gain increase caused by both devices conducting simultaneously in the centre of the output-voltage range, that is, in the Class-A region) putting edges into the distortion residual that generate high-order harmonics much as under-biasing does. This vital fact seems almost unknown, presumably because the gm-doubling distortion is at a relatively low level and is completely obscured in most amplifiers by other distortions.

This phenomenon is demonstrated in Figure 9.1a, b and c, which shows spectrum analysis of the distortion residuals for under-biasing, optimal, and over-biasing of a 150 W/8 Ω amplifier at 1 kHz. As before, all non-linearities except the unavoidable Distortion Three (output stage) have been effectively eliminated. The over-biased case had the quiescent current increased until the gm-doubling edges in the residual had an approximately 50:50 mark/space ratio, and so it was in Class-A about half the time, which represents a rather generous amount of quiescent current for Class-AB. Nonetheless, the higher-order odd harmonics in Figure 9.1c are at least 10 dB greater in amplitude than those for the optimal Class-B case, and the third harmonic is actually higher than for the under-biased case as well. However, the under-biased amplifier, generating the familiar sharp spikes on the residual, has a generally greater level of high-order odd harmonics above the fifth; about 8 dB higher than the Class-AB case.

Since high-order odd harmonics are generally considered to be the most unpleasant, there seems to be a clear case for avoiding Class-AB altogether, as it will always be less efficient and generate more high-order distortion than the equivalent Class-B circuit as soon as it leaves Class-A. Class distinction seems to resolve itself into a binary choice between A or B.

It must be emphasised that these effects are only visible in an amplifier where the other forms of distortion have been properly minimised. The RMS THD reading for Figure 9.1a was 0.00151%, for Figure 9.1b 0.00103%, and for Figure 9.1c 0.00153%. The tests were repeated at the 40W power level with very similar results. The spike just below 16 kHz is interference from the testgear VDU.

This is complex enough, but there are other and deeper subtleties in Class-B, which are dealt with below.

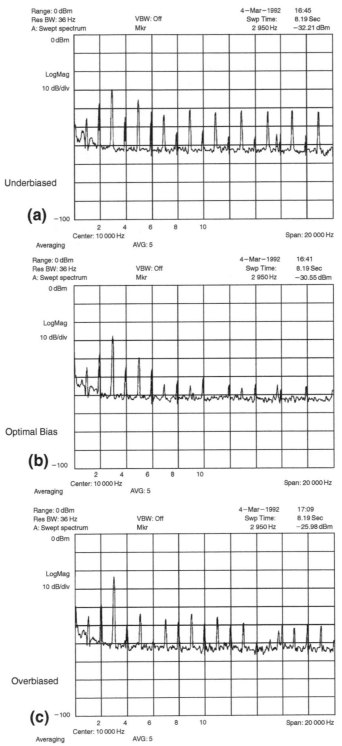

Figure 9.1. Spectrum analysis of Class-B and AB distortion residual.

The Distortions of the Output

I have called the distortion produced directly by the output stage Distortion Three (see p. 117) and this can now be subdivided into three categories. Distortion 3a describes the large-signal distortion that is produced by both Class-A and Class-B, ultimately because of the large current swings in the active devices; in bipolars, but not FETs, large collector currents reduce the beta, leading to drooping gain at large output excursions. I shall use the term 'LSN' for Large-Signal Non-linearity, as opposed to crossover and switch-off phenomena that cause trouble at all output levels.

These other two contributions to Distortion Three are associated with Class-B and Class-AB only; Distortion 3b is classic crossover distortion, resulting from the non-conjugate nature of the output characteristics, and is essentially non-frequency dependent. In contrast, Distortion 3c is switch-off distortion, generated by the output devices failing to turn off quickly and cleanly at high frequencies, and is very strongly frequency-dependent. It is sometimes called 'switching distortion', but this allows room for confusion, as some writers use the term 'switching distortion' to cover crossover distortion as well; hence I have used the term 'switch-off distortion' to refer specifically to charge-storage turn-off troubles. Since Class-B is almost universal, and

regrettably introduces all three kinds of non-linearity, in this chapter we will concentrate on this kind of output stage.

Harmonic Generation by Crossover Distortion

The usual non-linear distortions generate most of their unwanted energy in low-order harmonics that NFB can deal with effectively. However, crossover and switching distortions that warp only a small part of the output swing tend to push energy into high-order harmonics, and this important process is demonstrated here, by Fourier analysis of a SPICE waveform.

Taking a sinewave fundamental, and treating the distortion as an added error signal E, let the ratio WR describe the proportion of the cycle where E is non-zero. If this error is a triangle-wave extending over the whole cycle (WR = 1), this would represent large-signal non-linearity, and Figure 9.2 shows that most of the harmonic energy goes into the third and fifth harmonics; the even harmonics are all zero due to the symmetry of the waveform.

Figure 9.3 shows how the situation is made more like crossover or switching distortion by squeezing the triangular error into the centre of the cycle so that its value is zero elsewhere; now E is non-zero for only half the

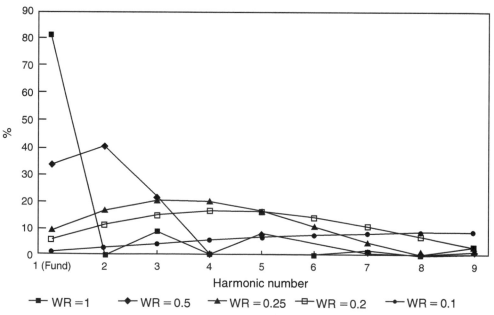

Figure 9.2. The amplitude of each harmonic changes with WR; as the error waveform gets narrower, energy is transferred to the higher harmonics.

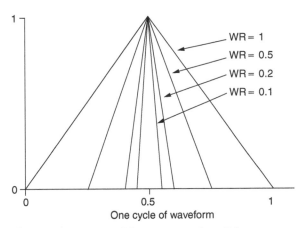

Figure 9.3. Diagram of the error waveform E for some values of WR.

cycle (denoted by WR = 0.5) and Figure 9.2 shows that the even harmonics are no longer absent. As WR is further decreased, the energy is pushed into higher-order harmonics, the amplitude of the lower falling.

The high harmonics have roughly equal amplitude, spectrum analysis (see Figure 9.1), confirming that even in a Blameless amplifier driven at 1 kHz, harmonics are freely generated from the seventh to the 19th at an equal level to a dB or so. The 19th harmonic is only 10 dB below the third.

Thus, in an amplifier with crossover distortion, the order of the harmonics will decrease as signal amplitude reduces, and WR increases; their lower frequencies allow them to be better corrected by the frequency-dependent NFB. This effect seems to work *against* the commonly assumed rise of percentage crossover distortion as level is reduced.

This is of course only a very crude approximation to crossover distortion. The difficulties involved in making a more accurate model are described in Chapter 2.

Comparing Output Stages

One of my aims in this book is to show how to isolate each source of distortion so that it can be studied (and hopefully reduced) with a minimum of confusion and perplexity. When investigating output behaviour, it is perfectly practical to drive output stages open-loop, providing the driving source-impedance is properly specified; this is difficult with a conventional amplifier, as it means the output must be driven from a frequency-dependent impedance simulating that at the VAS collector, with some sort of feedback mechanism incorporated to keep the drive voltage constant.

However, if the VAS is buffered from the output stage by some form of emitter-follower, as suggested in Chapter 7, it makes things much simpler, a straightforward low-impedance source (e.g., 50 Ω) providing a good approximation of conditions in a VAS-buffered closed-loop amplifier. The VAS-buffer makes the system more designable by eliminating two variables — the VAS collector impedance at LF, and the frequency at which it starts to decrease due to local feedback through Cdom. This markedly simplifies the study of output stage behaviour.

The large-signal linearity of various kinds of open-loop output stage with typical values are shown in Figures 9.6–9.13. These diagrams were all generated by SPICE simulation, and are plotted as incremental output gain against output voltage, with the load resistance stepped from 16 to 2 Ω, which I hope is the lowest impedance that feckless loudspeaker designers will throw at us. They have come to be known as *wing-spread* diagrams, from their vaguely bird-like appearance. The power devices are MJ802 and MJ4502, which are more complementary than many so-called pairs, and minimise distracting large-signal asymmetry. The quiescent conditions are in each case set to minimise the peak deviations of gain around the crossover point for 8Ω loading; for the moment it is assumed that you can set this accurately and keep it where you want it. The difficulties in actually doing this will be examined later.

If we confine ourselves to the most straightforward output stages, there are at least 16 distinct configurations, without including error-correcting,[3] current-dumping,[4] or Blomley[5] types. These are shown in Table 9.1.

The Emitter-follower Output

Three versions of the most common type of output stage are shown in Figure 9.4; this is the double-emitter-follower, where the first follower acts as driver to the second (output) device. I have deliberately called this an Emitter-Follower (EF) rather than

Table 9.1.

Emitter-Follower	3 types	Figure 9.4
Complementary-Feedback Pair	1 type	Figure 9.5
Quasi-Complementary	2 types	Figure 9.5
Output Triples	At least 7 types	Figure 9.6
Power FET	3 types	Chapter 21

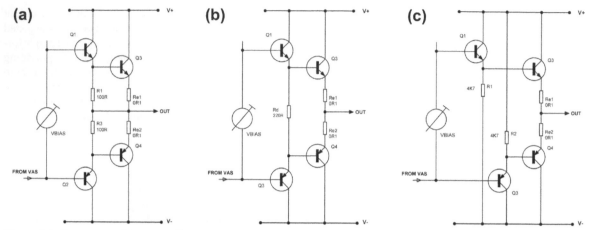

Figure 9.4. Three types of Emitter-Follower (EF) output stage: (a) with driver emitter-resistors connected to output rail (Type I); (b) with single driver emitter resistor (Type II); (c) with driver emitter-resistors connected to opposite supply rail (Type III).

a Darlington configuration, as this latter implies an integrated device that includes driver, output, and assorted emitter resistors in one ill-conceived package. (Ill-conceived for this application because the output devices heat the drivers, making thermal stability worse.) As for all the circuitry here, the component values are representative of real practice. Important attributes of this topology are:

1. The input is transferred to the output via two base-emitter junctions in series, with no local feedback around the stage (apart from the very local 100% voltage feedback that makes an EF what it is).
2. There are two dissimilar base-emitter junctions between the bias voltage and the emitter resistor Re, carrying different currents and at different temperatures. The bias generator must attempt to compensate for both at once, though it can only be thermally coupled to one. The output devices have substantial thermal inertia, and so any thermal compensation can only be a time-average of the preceding conditions. Figure 9.4a shows the most prevalent version (Type I) which has its driver emitter resistors connected to the output rail.

The Type II EF configuration in Figure 9.4b is at first sight merely a pointless variation on Type I, but in fact it has a valuable extra property. The shared driver emitter-resistor Rd, with no output-rail connection, allows the drivers to reverse-bias the base-emitter junction of the output device being turned off. Assume that the output voltage is heading downwards through the crossover region; the current through Re1 has dropped to zero, but that through Re2 is increasing, giving a voltage drop

across it, so Q4 base is caused to go more negative to get the output to the right voltage. This negative excursion is coupled to Q3 base through Rd, and with the values shown can reverse bias it by up to -0.5 V, increasing to -1.6V with a 4Ω load. A speed-up capacitor Cs connected across Rd markedly improves this action, preventing the charge-suckout rate being limited by the resistance of Rd. While the Type I circuit has a similar voltage drop across Re2, the connection of the mid-point of R1, R2 to the output rail prevents this from reaching Q3 base; instead Q1 base is reverse-biased as the output moves negative, and since charge-storage in the drivers is usually not a problem, this does little good. In Type II, the drivers are never reverse-biased, though they do turn off. The Type II EF configuration has of course the additional advantage that it eliminates a resistor. Just think of the money it saves …

The important issue of output turn-off and switching distortion is further examined in Chapter 10.

The Type III topology shown in Figure 9.4c maintains the drivers in Class-A by connecting the driver Res to the opposite supply rail, rather than the output rail. It is a common misconception[6] that Class-A drivers somehow maintain better low-frequency control over the output devices, but I have yet to locate any advantage myself. The driver dissipation is of course substantially increased, and nothing seems to be gained at LF as far as the output transistors are concerned, for in both Type I and Type II the drivers are still conducting at the moment the outputs turn off, and are back in conduction before the outputs turn on, which would seem to be all that matters. Type III is equally as good as Type II at reverse-biasing the output bases, and may give even cleaner HF turn-off as the carriers are being swept from the bases by a higher resistance terminated in a higher voltage, approximating constant-current drive; this remains to be determined by experiment. The Type III topology is used in the Lohstroh and Otala amplifier described in Chapter 4. A rare commercial example is the Pioneer Exclusive-M3.

The large-signal linearity of these three versions is virtually identical: all have the same feature of two base-emitter junctions in series between input and load. The gain/output voltage plot is shown in Figure 9.5; with BJTs the gain reduction with increasing loading is largely due to the Res. Note that the crossover region appears as a relatively smooth wobble rather than a jagged shape. Another major feature is the gain-droop at high output voltages and low loads, and this gives us a clue that high collector currents are the fundamental cause of this. A close-up of the crossover region gain for 8 Ω loading only is shown in Figure 9.6; note that no Vbias setting can be found to give a constant or even monotonic gain; the double-dip and central gain peak are characteristic of optimal adjustment. The region extends over an output range of about ±5V.

Multiple Output Devices: EF Output

As the power output required from an amplifier increases, a point is reached when a single pair of output devices is no longer adequate for reliable operation. Multiple output devices also reduce Large Signal Nonlinearity (Distortion 3a), as described below. Adding parallel output devices to an EF stage is straightforward, as shown in Figure 9.7a,

which is configured as a Type II EF stage. The only precaution required is to ensure there is proper sharing of current between the output devices. If they were simply connected in parallel at all three terminals, the Vbe tolerances could lead to unequal current-sharing and consequent over-dissipation of one or more devices. This is a potentially unstable situation as the Vbe of the hottest device will fall and it will take an even bigger share of the current until something bad happens.

It is therefore essential to give each transistor its own emitter resistor for local DC feedback; I have found 0.1 Ω to be large enough in all circumstances; as described in Chapter 10 in the section on crossover distortion, the value needs to be kept as low as possible to minimise crossover non-linearities. However, if you have an eccentric heatsink design that does not keep all the output devices at roughly the same temperature, it might be necessary to increase the value to give good current-sharing.

Another very important point made in Chapter 10 is that the use of multiple devices in the EF type of output stage reduces crossover distortion; this is quite a separate effect from the reduction of crossover distortion obtained by using the lowest practicable emitter resistors.

A triple-based EF output stage with three output pairs is shown in Figure 9.18.

The CFP Output

The other major type of bipolar complementary output is that using two Complementary-Feedback Pairs (hereinafter CFPs). These are sometimes called

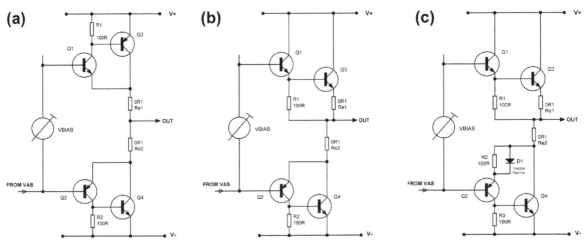

Figure 9.5. Output stages: (a) Complementary-Feedback-Pair (CFP); (b) quasi-complementary output; (c) quasi-complementary output with Baxandall diode.

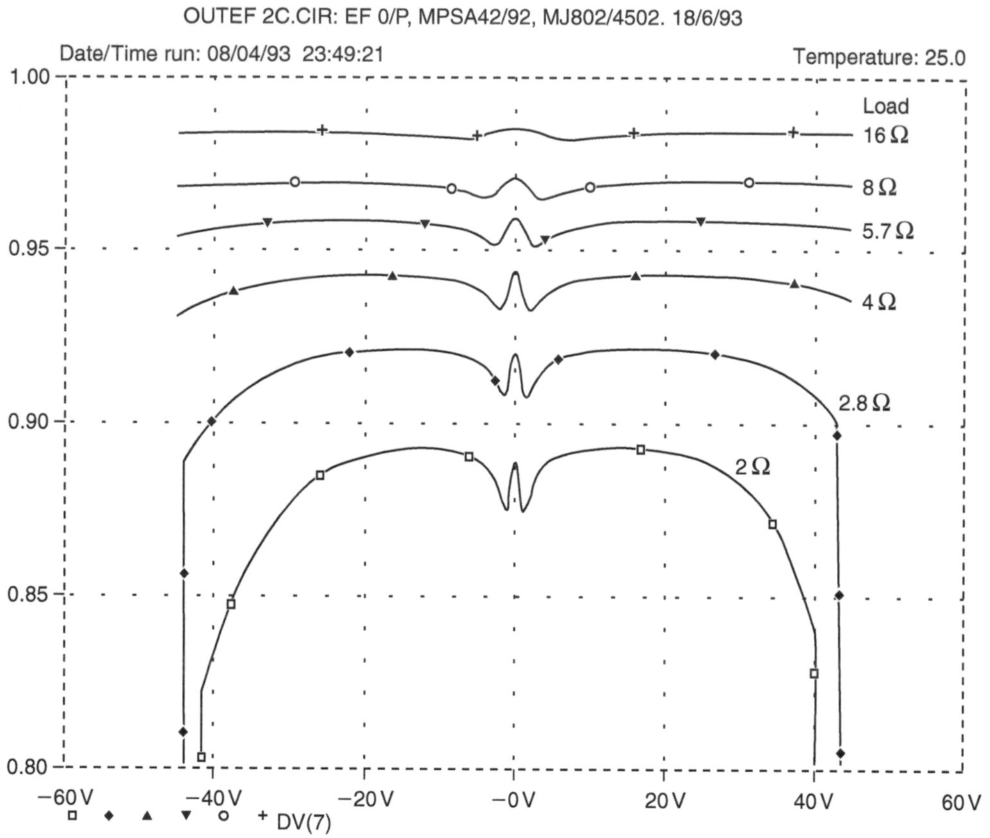

OUTEF 2C.CIR: EF 0/P, MPSA42/92, MJ802/4502. 18/6/93

Figure 9.6. Emitter-Follower large-signal gain versus output.

Sziklai-Pairs or conjugate pairs. The output stage can be seen in Figure 9.8a. There seems to be only one popular configuration, though versions with gain are possible, and have been used occasionally. The driver transistors are now placed so that they compare the output voltage with that at the input. Thus wrapping the outputs in a local NFB loop promises better linearity than emitter-follower versions with 100% feedback applied separately to driver and output transistors.

The CFP topology is generally considered to show better thermal stability than the EF, because the Vbe of the output devices is inside the local NFB loop, and only the driver Vbe has a major effect on the quiescent conditions. The true situation is rather more complex, and is explored in Chapter 22.

In the CFP output, like the EF, the drivers are conducting whenever the outputs are, so special arrangements to keep them in Class-A seem pointless.

The CFP stage, like EF Type I, can only reverse-bias the driver bases, and not the output bases, unless extra voltage rails outside the main ones are provided.

The output gain plot is shown in Figure 9.9; Fourier analysis of this shows that the CFP generates less than half the LSN of an emitter-follower stage (see Table 9.2). Given also the greater quiescent stability, it is hard to see why this topology is not more popular. One possible reason is that it can be more prone to parasitic oscillation.

Table 9.2 summarises the SPICE curves for 4 Ω and 8 Ω loadings; FET results from Chapter 21 are included for comparison; note the low gain for these. Each gain plot was subjected to Fourier analysis to calculate THD percentage results for a ±40 V input.

The crossover region is much narrower, at about ±0.3 V (Figure 9.10). When under-biased, this shows up on the distortion residual as narrower spikes than an emitter-follower output gives. The bad effects of gm-doubling as Vbias increases above optimal (here

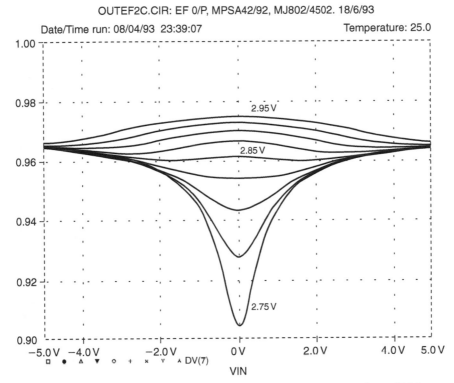

Figure 9.7. EF crossover region gain deviations, ±5 V range. Nine values of Vbias.

1.296 V) can be seen in the slopes moving outwards from the centre.

Multiple Output Devices: CFP Output

Adding parallel output devices for increased output to a CFP stage is straightforward, as shown in Figure 9.7b, but extra current-sharing resistors R5, R7 and R6, R8 must be inserted in the output device emitter circuits. The value of these resistors does not have any significant influence on the linearity of the CFP output

stage, as they are not emitter resistors in the same sense as Re1, Re2. I have always found that 0.1 Ω is large enough for current-sharing resistors. The actual 'emitter resistors' as used in EF stages, Re1, Re2, are still required and will need to be chosen to cope with the increased output permitted by the increased number of output devices. In the EF stage this occurs automatically as emitter resistors are added with each extra device pair.

Small emitter degeneration resistors R1, R2 are shown; these will not inevitably be required to ensure stability with this configuration, but in a CFP stage parasitic

Table 9.2. Summary of output distortion

	Emitter Follower	CFP	Quasi Simple	Quasi Bax	Triple Type1	Simple MOSFET	Quasi MOSFET	Hybrid MOSFET
8Ω THD	0.031%	0.014%	0.069%	0.44%	0.13%	0.47%	0.44%	0.052%
Gain:	0.97	0.97	0.97	0.96	0.97	0.83	0.84	0.97
4Ω THD	0.042%	0.030%	0.079%	0.84%	0.60%	0.84%	0.072%	0.072%
Gain:	0.94	0.94	0.94	0.94	0.92	0.72	0.73	0.94

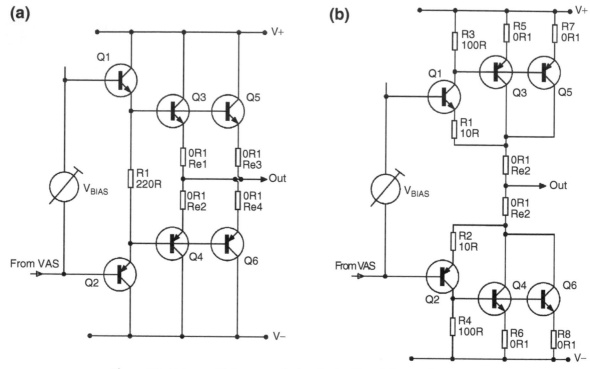

Figure 9.8. Using multiple output devices in the EF and CFP configurations.

oscillation is more likely with multiple output devices, so I strongly advise you to make provision for them.

As for the EF output stage, multiple output devices not only increase output capability but also reduce Large Signal Nonlinearity (Distortion 3a) as described in its own section below. However, multiple output devices in the CFP configuration do not necessarily decrease crossover distortion and can in some circumstances increase it. I discovered this when doing a design study for a new power amplifier.

The design aim was a more powerful version of an existing (and very satisfactory) design giving 120 W/8 Ω from a CFP output stage using two paralleled output devices, as in Figure 9.7b. The bigger version was required to produce 200 W/8 Ω with a particularly good ability to drive lower impedances, so four parallel output devices were used CFP in the output stage. The first measurements on the bench prototype showed unexpectedly high crossover distortion, and after everything conceivable had been checked, I had to conclude that it was inherent in the particular output stage structure; adding two more parallel output devices had made the linearity *worse*, which is the exact opposite of what happens in the EF output.

This did not seem exactly intuitively obvious, so a series of simulations were run with from 1 to 6 parallel output devices, to measure their incremental gain against output voltage. This type of plot is as displayed in Figure 9.6, though only the optimal bias case was examined. Optimal bias gives a central gain peak and a gain dip on either side, as in the fifth curve up in Figure 9.7, and the difference between the peak gain and the dip gain is a measure of how much crossover distortion the output stage will generate. For a single complementary pair of output devices, the peak gain was 0.9746 and the dip gain 0.9717. (The dip gains are not usually exactly equal, but the differences are small.) The peak − dip gain difference was therefore 0.9746 − 0.9717 = 0.0029. This gain difference was then calculated for each number N of output devices, and the results can be seen in Table 9.3. The gain difference increases with N up to N = 6, where it decreases again; that is definitely the figure I recorded, so perhaps another effect is coming into play.

And there the matter rests for the time being; though obviously it would be desirable to go into the matter more deeply and establish exactly how it works. It

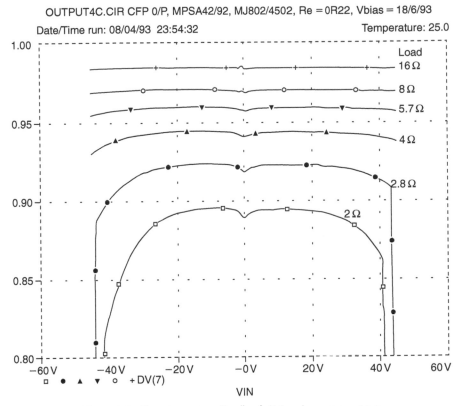

OUTPUT4C.CIR CFP 0/P, MPSA42/92, MJ802/4502, Re = 0R22, Vbias = 18/6/93

Date/Time run: 08/04/93 23:54:32 — Temperature: 25.0

Figure 9.9. Complementary-Feedback-Pair gain versus output.

appears to be inadvisable to use more than two parallel output devices in the CFP output stage.

Output Stages with Gain

It was explained at the start of this chapter that almost all output stages have a gain of unity, or to be precise, slightly less than unity. This is because, firstly, there are voltage losses in the emitter resistors Re, which

Table 9.3. Peak - dip gain differences for varying numbers of CFP output devices

Number of output devices N	Peak - dip gain difference
1	0.0029
2	0.0053
3	0.0059
4	0.0064
5	0.0093
6	0.0076

form the upper arm of a potential divider with the external load as the lower arm. Thus if you assume an amplifier stage with Res of 0.1 Ω, and an instantaneous operating point well away from the crossover region you get a gain of 0.988 times with an 8 Ω load, reducing to 0.976 for 4 Ω loads. Secondly, in the case of EF type output stages, the gain of an emitter-follower is always slightly less than one.

Output stages with significant gain (typically two times) have been advocated on the grounds that the lower voltage swing required to drive the stage would reduce VAS distortion. This is much misguided, for as we have seen in earlier chapters, the distortion produced by the small-signal stages can be made very low by simple methods, and there is no pressing need to seek radical ways of reducing it further. On the other hand, distortion in the output stage is a much more difficult problem, so making things worse by seeking voltage gain is not the way forward.

A slightly better justification for seeking voltage gain in the output stage is that it would allow more output voltage swing from the same supply rails, improving

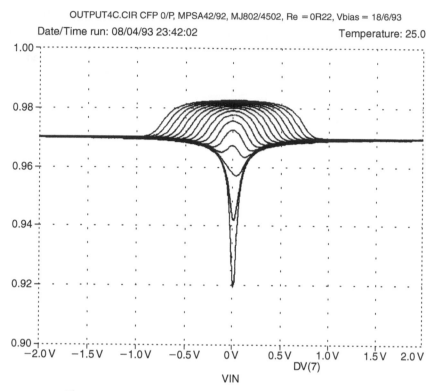

OUTPUT4C.CIR CFP 0/P, MPSA42/92, MJ802/4502, Re = 0R22, Vbias = 18/6/93
Date/Time run: 08/04/93 23:42:02   Temperature: 25.0

Figure 9.10. CFP crossover region ±2V, Vbias as a parameter.

efficiency. The VAS stage will have some saturation voltages, so it cannot swing fully between the rails (this point is looked at in detail in Chapter 17 on Class-A amplifiers, where you need to squeeze out every Watt you can) and so a little gain afterwards, say, 1.1 times, will allow the maximum swing the output stage can provide. However, even allowing for the fact that output power in Watts, which is the figure everyone looks at, goes up with the square of voltage, the advantage to be gained is small compared with the extra difficulties you are likely to get into in the output stage. Figure 9.12 shows a CFP output stage with a gain of two. For obvious reasons you cannot make an EF stage with gain — it is composed of emitter-followers which all have sub-unity gain.

The circuit in Figure 9.11 gives gain because two potential dividers R5, R1 and R6, R2 have been inserted in the local feedback path to the driver emitters; as you might expect, equal resistor values top and bottom give a gain of two. The value of these resistors is problematic. If they are too large, the source impedance seen by the driver emitter is unduly

increased, and this local degeneration reduces the loop gain in the CFP output structure, and distortion will increase. If the divider resistors are kept low to avoid this, they are going to dissipate a lot of power, as they are effectively connected between the amplifier output and ground. The value of 47 Ω shown here in Figure 9.11a is a reasonable compromise, giving the driver stage an open-loop voltage gain of ten times, while keeping the divider values up. However, a 100 W/8 Ω amplifier at full throttle is still going to dissipate 4.2 W in each of the 47 Ω divider resistors, requiring some hefty resistors that take up a lot of PCB space, and drawing in total a discouragingly large 16.8 W from the amplifier output.

If you are seeking just a small amount of gain, such as 1.1 times, to maximise the output swing, things are slightly easier. The example in Figure 9.11a has a source impedance of 47/2 = 23.5 Ω seen by the driver emitter; if we stick roughly to this figure, the divider values for a gain of 1.1 times become R5 = 22 Ω and R1 = 220 Ω in Figure 9.11b, the divider R6, R2 in the lower half of the output stage having corresponding values. This gives an impedance at the driver

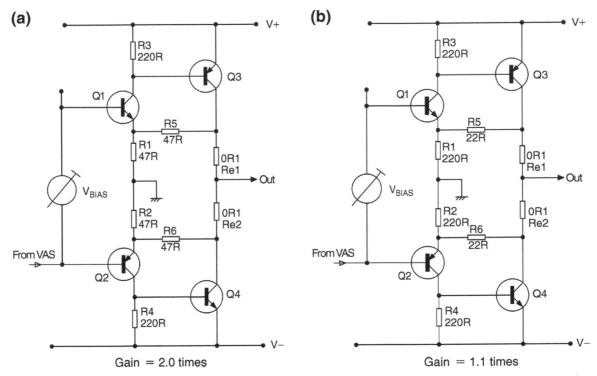

Figure 9.11. Examples of CFP output stages configured to give (a) a voltage gain of two times, and (b) a voltage gain of 1.1 times, by the addition of potential dividers R5, R1 and R6, R2 in the local feedback to the driver emitters.

emitter of 20 Ω. For the same 100 W/8 Ω amplifier, this reduces the dissipation in R5 to 298 mW, and in R1 to 2.98 W; the total extra power drawn from the amplifier output is reduced to 6.56 W, which is a bit more manageable.

Output stages with gain can be made to work, but ultimately my advice would be that you probably don't want to go this way.

Quasi-complementary Outputs

Originally, the quasi-complementary configuration [7] was virtually mandatory, as it was a long time before PNP silicon power transistors were available in anything approaching complements of the NPN versions. The standard quasi-complementary circuit shown in Figure 9.5b is well known for poor symmetry around the crossover region, as shown in Figure 9.13. Figure 9.14 zooms in to show that the crossover region is a kind of unhappy hybrid of the EF and CFP, as might be expected, and that there is no setting of Vbias that can remove the sharp edge in the gain plot.

A major improvement to symmetry can be made by using a Baxandall diode,[8] as shown in Figure 9.5c. Placing a diode in the driver circuit of the CFP (lower) part of the output stage gives it a more gradual turn-on, approximating to the EF section in the upper half of the output stage. This stratagem yields gain plots very similar to those for the true complementary EF in Figures 6.7 and 6.8, though in practice the crossover distortion seems rather higher. When this quasi-Baxandall stage is used closed-loop in an amplifier in which Distortions One and Two, and Four to Seven have been properly eliminated, it is capable of much better performance than is commonly believed; for example, 0.0015% (1 kHz) and 0.015% (10 kHz) at 100 W is straightforward to obtain from an amplifier with a moderate NFB factor of about 34 dB at 20 kHz.

Peter Baxandall introduced the concept of a diode in the CFP part of the output stage in response to an earlier proposal by Shaw,[9] which put a power diode in series with the output of the CFP stage, as shown in Figure 9.14, with the same intention of making it turn on more slowly. A serious disadvantage of the Shaw

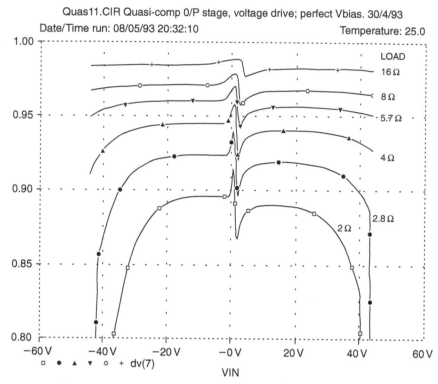

Figure 9.12. Quasi-complementary large-signal gain versus output.

scheme is that the added diode passes the full output stage current and therefore needs to be a hefty component. The Baxandall diode only passes the driver current and can therefore be a small part.

I received this communication[10] from Peter Baxandall, written not long before his untimely death:

> *It is slightly preferable to use a transdiode (a transistor with collector connected to base) rather than an ordinary diode such as 1N4148, since the transdiode follows the transistor equation much more accurately, matching better the Vbe characteristic of the top driver transistor. (As you probably know, most diodes follow, over a moderate current range, the transistor equation but with mkT in place of kT, where m is a constant in the region of 1.8, though varying somewhat with the type of diode. Consequently whereas the voltage across a transdiode at fairly small currents varies, at 20 degC, by a remarkably accurate 58 mV per decade of current change, that across an ordinary diode is nearer 100 mV per decade or just over.)*

The transistor equation is the well-known fundamental relationship that describes how transistors work. It is:

$$I_c = I_o \times e^{a \times V_{be}/kT - 1} \qquad \text{Equation 9.1}$$

where Ic is the collector current, and Io is the saturation current, q the charge on an electron, k is Boltzmann's constant, and T is absolute temperature in degrees Kelvin. kT/q is often called V_t, the 'thermal voltage'. It is 25.3 mV at 20 degrees C.

A Baxandall output stage with transdiode is shown in Figure 9.15. Peter did not discuss with me the type of transistor to be used as a transdiode, but I would guess that it should be the same as the lower driver Q2. He recommended the use of a transdiode rather than an ordinary diode in this reference.[11]

Fully complementary output devices have been available for many years now, and you may be wondering why it is worth examining configurations that are obsolete. The answer is that in the world of hi-fi, no circuit concept seems to ever quite die, and at the time of writing (mid-2012) at least one well-known company is still producing quasi-complementary power amplifiers.

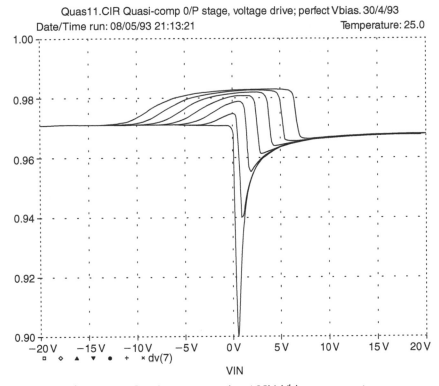

Quas11.CIR Quasi-comp 0/P stage, voltage drive; perfect Vbias. 30/4/93
Date/Time run: 08/05/93 21:13:21   Temperature: 25.0

Figure 9.13. Quasi crossover region ±20V, Vbias as parameter.

The best reason to use the quasi-Baxandall approach today is to save a little money on output devices, as PNP power BJTs remain somewhat pricier than NPNs.

Given the tiny cost of a Baxandall diode or trans-diode, and the absolutely dependable improvement it gives, there seems no reason why anyone should ever use the standard quasi-circuit. My experiments show that the value of R1 in Figure 9.15 is not critical; making it about the same as R3 seems to work well.

Triple-based Output Configurations

If we allow the use of three rather than two bipolar transistors in each half of an output stage, the number of circuit permutations leaps upwards, and I cannot provide even a rapid overview of every possible configuration in the space available. Here are some of the possible advantages if output triples are used correctly:

1. Better linearity at high output voltages and currents, due to increased local feedback in the triple loop.
2. More stable quiescent setting as the pre-drivers can be arranged to handle very little power indeed, and

to remain almost cold in use. This means they can be low-power TO92-type devices with superior beta, which enhances the local loop gain.
3. The extra current gain allows greater output power without undesirable increases in the operating currents of the VAS.

However, triples do not abolish crossover distortion, and they are, as usually configured, incapable of reverse-biasing the output bases to improve switch-off. Figure 9.16 shows two of the more useful ways to make a triple output stage − all of those shown have been used in commercial designs so they must be considered as practical in use. This is an important proviso as it is not hard to make a triple output stage which cannot be made to give reliable freedom from oscillation in the triple loop.

The most straightforward triple-based output stage is the triple-EF configuration of Figure 9.16a, which adds to the output stage of Figure 9.4b a pair of pre-driver emitter followers. This is dealt with in detail in its own section just below.

Figure 9.16b is the famous Quad-303 quasi-complementary triple. The Quad 303 amplifier was

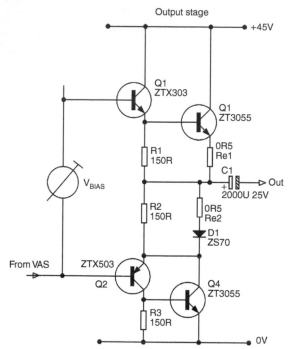

Figure 9.14. Quasi-complementary output stage with Shaw diode added to improve symmetry. The component values and transistor types are as used in Shaw's original circuit.

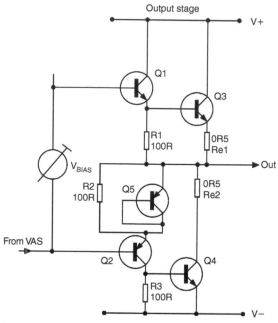

Figure 9.15. Quasi-complementary output stage with added Baxandall transdiode Q5 to give better symmetry than a simple diode. R1, R3 and R4 have the values used by Peter with the diode version.

introduced in 1967, when complementary silicon output transistors were not yet a practical proposition. This configuration uses the extra local negative feedback of the triple stage to give much better linearity than a conventional quasi-complementary output stage. Note that the Re resistors here are shown as 0.3 Ω, which was the value used in the original Quad 303 circuit.

The output stage in Figure 9.17a is, in contrast, a fully complementary output stage. Its top half consists of Q1 and Q3 configured as common-emitter voltage amplifiers, while output device Q5 is a common-collector emitter follower. The local negative feedback loop is closed by connecting the emitter of Q1 to the top of Re1, making the top triple effectively a 'super emitter-follower' with high loop-gain and a high degree of negative feedback, which gives it a very high input-impedance, a low output-impedance, and a gain of very nearly unity. The resistors R1, R2 limit the internal loop-gain of the triple, by applying what might be called 'very local feedback' or emitter-degeneration to the emitter of Q1; in my experience this is absolutely essential if anything like reliable stability is to be obtained with this configuration. In

some versions the driver stages Q3, Q4 also have small resistors in their emitter circuits (typically 10 Ω) to give more control of output stage loop gain. The bottom half of the output stage works in exactly the same way as the top. This configuration was used in the Lecson AP-1 (1975).

The output stage in Figure 9.17b is another variation on the triple output. In this case only the pre-driver transistor Q1 is configured as a common-emitter voltage amplifier, with the driver and output transistors being connected as cascaded emitter-followers. This gives less voltage gain inside the triple loop, less local feedback, and hence less chance of local oscillation. Note that Q1 and Q2 still have emitter resistors R1, R2 to control the transconductance of the pre-driver stages. The design and testing of triple-based output stages demand care, as the possibility of local HF instability in each output half is very real.

Given the number of possibilities for triple-based output stages, it might be useful to have a concise notation to describe them. The output stage in Figure 9.17a is composed of two common-emitter voltage amplifiers followed by a common-collector emitter-follower, making up a single local negative feedback loop. It

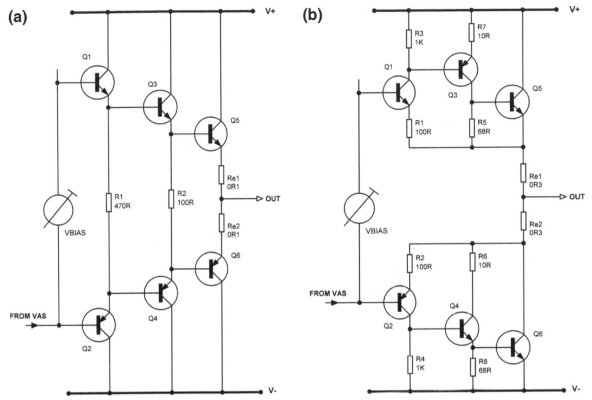

Figure 9.16. Two of the possible triple output configurations: (a) Triple-EF; (b) Quad 303 quasi-complementary triple.

could be written as a CE-CE-EF triple output stage. Likewise, the output stage in Figure 9.17b has common-emitter pre-driver, with both the driver and output transistors connected as emitter-followers, so it could be described as a CE-EF-EF configuration.

Quasi-complementary stages like that of Figure 9.6a are a bit less straightforward, but if we adopt the convention that the top half the output stage is always described first, it could be written as CE-CE-EF/CE-EF-EF.

Some more triple output stages are shown in Figure 9.18. These have the feature that the local negative feedback is only closed around two of the three devices in the triple. The first one is an emitter-follower feeding a CFP stage (which is in turn composed of two CE voltage-amplifier stages), which could be written EF-CE-CE, but EF-CFP is rather more indicative of its structure and operation. It can be regarded as a simple emitter-follower feeding a compound output device. This configuration has the potential disadvantage that the pre-driver emitter-follower is outside the local NFB loop; the same naturally applies

to the EF-EF-EF triple-emitter-follower output stage described in the next section.

Figure 9.18b shows another variation on the triple theme. This time we have a CFP stage feeding an emitter-follower. Once again it could be written CE-CE-EF, but CFP-EF is more instructive. The importance of this configuration is that it looks promising for reducing the effects of Large-Signal-Nonlinearity when driving low impedances.

An unconventional triple output staged used by Bryston is shown in Figure 9.18c. As with other triple outputs, the pre-driver stage Q1 is run at low power so it stays cool and gives good bias stability. The driver stage Q3 is now a phase-splitter; the output from its emitter drives a CE stage Q5 that feeds current directly into the output rail, while the output from its collector drives an EF stage Q7 that feeds current into the output rail via two emitter resistors Re1, Re2. An important feature of this stage is that it has a voltage gain of three, set up by the potential dividers R5, R1 and R6, R2.

The Bryston configuration has the further interesting property that it completely defeats the output stage

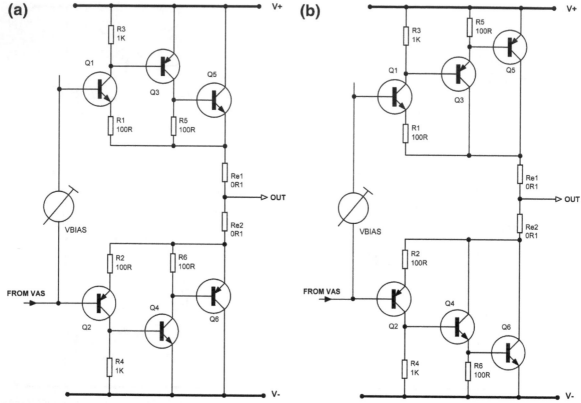

Figure 9.17. Two more output-triple configurations: (a) emitter-follower output devices; (b) common-emitter output devices.

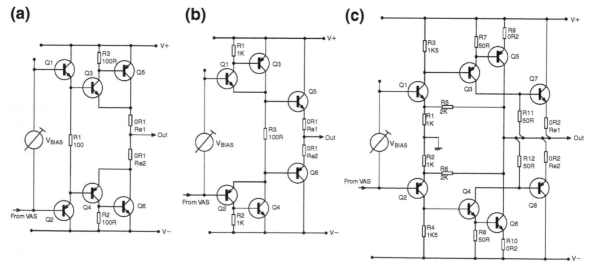

Figure 9.18. (a) and (b) Two more possible types of output triple; (c) the Bryston output stage.

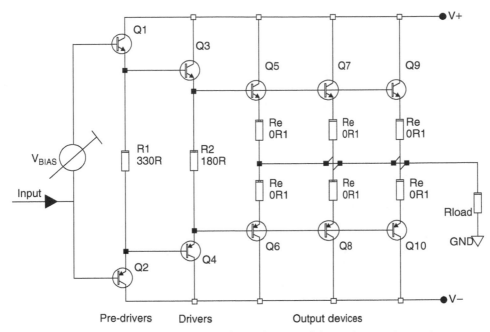

Figure 9.19. Triple-EF output stage. Both pre-drivers and drivers have emitter-resistors.

notation suggested only a few paragraphs ago. According to Bryston's own publicity material, their output stage configuration requires close matching of output transistor betas, not only between similar types, but between complementary devices. They say that since Bryston products are hand-built from selected components anyway, this is not a serious disadvantage in production.

Triple EF Output Stages

Sometimes it is necessary to use a triple output stage not because you are hoping to gain a distortion advantage (the triple-EF output does not provide this, as described below), but simply because the currents flowing in the output stage are too big to be handled by two transistors in cascade. If you are driving 2 Ω or 1 Ω loads, then typically there will be multiple output devices in parallel. Special amplifiers for driving ribbon loudspeakers have to drive into a 0.4 Ω load (see Chapter 4). Providing the base current for five or more output transistors, with their relatively low beta, will usually be beyond the power capability of normal driver types, and it is common to use another output device as the driver. This will certainly have the power-handling capability, but with this comes low beta once again. This means that the driver base currents in turn

become too large for a normal VAS stage to source. There are two solutions: (1) make the VAS capable of sourcing hundreds of mA, or (2) insert another stage of current-gain between VAS and drivers. The latter is much easier, and the usual choice. These extra transistors are usually called the pre-drivers (see Figure 9.19).

In this circuit the pre-drivers dissipate relatively little power, and TO-92 devices may be usable in some applications. For higher outputs, the pre-drivers can be medium-power devices in a TO-220 package; it is unlikely that they will need heatsinking to cope with the demands made on them. There is, however, another reason to fit pre-driver heatsinks — or at least make room at the layout stage so you have the option.

In Figure 9.19 there is about 1.2 V across R2, so Q3, 4 have to supply a standing current of about 7 mA. This has no effect on the drivers as they are likely to be well cooled to deal with normal load demands. However, the voltage across R1 is two Vbe's higher, at 2.4 V, so the standing current through it is actually higher at 7.3 mA. (The exact figures naturally depend on the values for R1, R2 that are chosen, but it is difficult to make them much higher than shown here without compromising the speed of high-frequency turn-off.) The pre-drivers are usually small devices, and so they are likely to get warm, and this leads to drift in the bias conditions after switch-on. Adding heatsinks cannot eliminate this effect, but does usefully reduce it.

In a triple-EF output stage like this, the Vbias generator must produce enough voltage to turn on six base-emitter junctions, plus the small standing voltage Vq across the emitter resistors, totalling about 3.9 V in practice. The Vbe of the bias transistor is therefore being multiplied by a larger factor, and Vbias will drop more for the same temperature rise. This should be taken into account, as it is easy with this kind of output stage to come up with a bias generator that is overcompensated for temperature.

Looking at the triple EF output stage in Figure 9.19, it is natural to wonder if the six rather than four base-emitter junctions between the VAS and the output might make the crossover region wider, and so cause it to generate lower-order harmonics that would be better dealt with by negative feedback that falls with frequency. SPICE simulation shows that regrettably this is not the case. The gain-wobble in the crossover region can be defined by the wobble-height (the difference in gain between the peak and the average of the two dips) and the wobble-width (the difference between the two input voltages at which the dips occur). For both a double-EF and triple-EF output stage the wobble-height is about 0.002 and the wobble-width about 9.0 V. What minor differences were seen appeared to be due solely to the difficulty of getting the bias conditions identical in the two configurations.

Another possibility that suggests itself is that careful adjustment of the pre-driver and driver emitter resistor values (R1 and R2 in Figure 9.19) might reduce the height or width of the gain-wobble. Unfortunately, simulation reveals that this is not so, and the values have very little effect on the static linearity. Changes of value do, however, require the bias to be re-adjusted if it is to remain optimal, due to changes in the Vbe drops.

An early and influential design using a fully-complementary triple-EF output stage was put forward by Bart Locanthi of JBL in 1967.[12] He called it the 'T-circuit' (apparently because of its shape when it was drawn sideways) and the output stage was essentially that of Figure 9.16a. An interesting feature was that the pre-drivers ran from ±45 V rails, the drivers from ±40V rails, and the output devices from ±35 V rails. I would have thought that might have caused problems when clipping, but no clamp diodes are visible on the schematic. This technique has not found favour. The VAS load was a simple resistor with no bootstrapping, which would not have been very linear (see Chapter 7).

Distortion figures appear to have been generally lower than the stated measurement limit of 0.015%.

A very large number of power amplifiers have been manufactured with triple-EF output stages; naturally they tend to be the more powerful designs where the extra current gain in the output stage is needed. To take a few examples pretty much at random, the Onkyo M-501 power amplifier (1991) used the configuration of Figure 9.19, with two pairs of paralleled output devices.

The Pioneer A-450R and 550R integrated amplifiers (1991) use the same configuration but with only one output pair; every device has a base-stopper resistor in series with the base. (This design, by the way, has some interesting asymmetrical crossfeed in the power amplifier that is clearly intended for crosstalk cancellation.)

The Pioneer A-604R has a very strange-looking push-pull VAS, enigmatically called a 'Current Miller' but the output stage is a standard triple, once again with base-stoppers everywhere. This might indicate some difficulties with parasitic oscillation.

In contrast, the Rotel RA935 (1993) integrated amplifier uses triples with two pairs of parallel output devices with not a base-stopper in sight. The rated output power is only 40 W/8 Ω so triples were not adopted to lighten the load on the VAS, but for some other reason. The Rotel RB-1050/70/90 power amplifier series shows a neat hierarchy of output device multiplication. The RB-1050 (2001) has an output of 70 W/8 Ω and uses triples with two pairs of paralleled output devices The RB-1070 (2001) has a greater output at 130 W/8 Ω and uses triples with three pairs of paralleled output devices. The RB-1090 (1999) provides 380 W/8 Ω and uses triples with four pairs of paralleled output devices. There was also an RB-1080 (2001) with an output of 200 W/8 Ω but its output arrangements are currently unknown.

The Pioneer Exclusive-M3 is a rare example of the use of Type III pre-drivers and drivers; their emitter resistors are connected not to the output rail but to the opposite supply rails, presumably to keep them conducting at all times. I have not found any benefit in this myself. Apart from this, the output stage is a conventional triple with three pairs of paralleled output devices. The quiescent current for the pre-drivers is 8 mA and for the drivers 27 mA.

Quadruple Output Stages

If three transistors in a triple output stage can be useful, it is only human (if you're a designer or engineer, anyway) to ponder if four transistors would be even better. As I have stressed above, triple stages where all three transistors are configured in a single local negative loop can be difficult to stabilise. Four must be worse, and while I have not tried the experiment, it seems highly unlikely that a quadruple output stage with single loops could be made reliably stable.

A quadruple output stage could be made from four emitter-followers in series. This is never likely to be required to deal with the output currents demanded by low-impedance loads, but might usefully spread the crossover region out more than the triple output stage does. Temperature compensation should be no more difficult as the first two emitter-follower sections should remain cold.

Another interesting possibility is a combination of the EF-CFP and CFP-EF output structures described above, that would give CFP-CFP; in other words, two cascaded local feedback loops, with each loop only encompassing two transistors. A possible arrangement of this is shown in Figure 9.20.

This configuration could be regarded as an enhancement of an EF output stage, in that instead of two cascaded emitter-followers, there are two cascaded 'super-emitter-followers' in the form of CFP stages, which will be a good deal more linear than simple emitter-followers because each has their own local feedback loop. Temperature compensation may present

some interesting challenges, as there are four layers of transistors, and on which one do you put the bias sensor?

I was going to call this a 'quad output stage', but you can see the opportunity for confusion there. It seems best to stick with 'quadruple output stage'.

Series Output Stages

It is commonplace to connect transistors in parallel to increase the current capability of the output stage, and in EF outputs this also gives a useful reduction in crossover distortion (see Chapter 10). It is also possible to put transistors in series to increase the voltage capability. This is largely a historical technique that was used for high powers beyond the voltage ratings of the transistors of the day. A typical and very popular complementary pair in the 1970s was the Motorola MJ802 (NPN) and MJ4502 (PNP). These were and are quite capable devices, capable of passing 30 Amps, but the Vceo is limited to 90 V. (Another snag is that beta falls off badly with increasing collector current by today's standards.) They were widely used, for example, in the Cambridge Audio P100 (1971), but the maximum supply rail voltage was clearly rather limited. Pushing the limits and using ±45 V rails would give a theoretical maximum output of 126 W/8 Ω; obviously in reality a safety margin would be required. More power could be obtained by connecting two transistors in series and sharing the voltage between them. Bridging was another option but that required two complete amplifiers rather than just more parts in the output stage.

A series output stage is sometimes known as a totem-pole (because of the devices stacked one upon another) or as a cascade output. In my classification system, it is Class-B+B.

Figure 9.21 shows the most common way in which this was done. The configuration is based on that used in the Sound Dragon SD400II and SD400IV amplifiers. The inner section of the amplifier (Q1 to Q6) is a conventional EF output stage with two output devices in parallel. The power for the upper half (Q1, Q3, Q5) is provided by a second EF output stage (Q7, Q9, Q11) which is driven from the output via the resistor network R1, R2, R3. This network shifts the signal to Q7 base up by about half of the rail voltage, and halves the amplitude of the AC signal. Allowance has to be made both for the Vbe drops in Q7, Q9, Q11, and their gain of slightly less than unity, and this is presumably the function of R3, shunted at AC by C2. The capacitor C1 no doubt carries out a stabilising role at HF, but it is clearly vital that the upper half of the output stage is not slowed down so

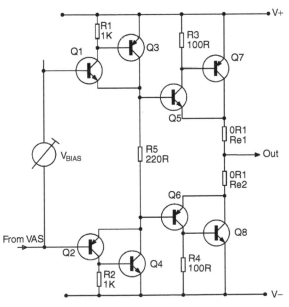

Figure 9.20. Quadruple CFP-CFP output stage.

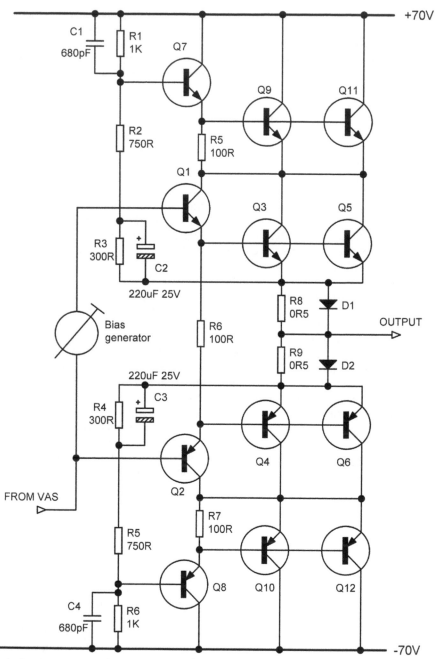

Figure 9.21. A typical series output stage. The networks R1, R2, R3 and R4, R5, R6 drive the outer part of the amplifier with a signal at half the output amplitude and with a half-rail-voltage DC shift.

much it is unable to provide current for the lower half in a timely fashion; the values shown give a pole at around 700 kHz. Judging by the ±70 V supply rails, the output power would have been around 270 W/8 Ω.

Power diodes are placed across the 0.5 Ω emitter resistors. These are not in any way essential to a series output stage, but were presumably added to reduce voltage losses when driving low-impedance loads.

They may also have gone some way to counteract LSN (see the use of feedforward diodes in Chapter 10).

We now have much improved output transistors, with considerably higher voltage ratings. The well-known MJ15024/15025 pair (still somewhat historical – they are packaged in TO-3) has a Vceo of 250 V. If that is translated into ±125 V rails, the theoretical maximum output is 977 W/8 Ω, which is quite a lot of power. A typical modern complementary pair like the 2SA1295/2SC3264 in the MT-200 package has a Vceo of 230 V. Voltage ratings like these make a series output configuration unnecessary. If you really need a lot of output voltage, it is of course also possible to double the voltage capability of an amplifier by using a bridged configuration, and this would normally be preferred to the uncertainties of a series output stage.

That does not mean the idea is completely dead. If you need a really high voltage output to drive an electrostatic loudspeaker directly, in other words, without a transformer, it is probably the only way to go unless you resort to valves. There is currently at least one commercial valve-based amplifier for the direct drive of electrostatics, the TA-3000 made by Innox.[13] However, if we stick with solid-state, we find the Vceos of bipolar transistors only go so high; two examples are the 2SC3892A with a Vceo of 600 V (TO-3P package) and the 2SC2979 with a Vceo of 800 V (TO-220 package). These are both NPN devices and there appear to be no comparable PNPs, so you will have to use a quasi-complementary output stage. Power FETs are available with higher voltage ratings than this, and may be useful in this application.

Even 800 V devices leave us rather short of the thousands of volts required by the typical electrostatic speaker. The Innox TA-3000 mentioned above has a maximum output of 3000 V peak-to-peak. For a projected solid-state amplifier that means ±1500 V supply rails, and 800 V transistors aren't going to be much good. But if we put four of them in series, we can handle ±1600 V rails, though the safety margin is not large, and five in series would be better. The voltage-sharing arrangements would be just an extension of those shown in Figure 9.20. I hasten to add that I have never constructed such an amplifier, and it may well bristle with technical problems, not to mention serious safety hazards. But the series configuration does make it possible. The potential is there ...

Selecting an Output Stage

Even if we stick to the most conventional of output stages, there are still an embarrassingly large number to choose from. The cost of a complementary pair of power FETs is currently at least twice that of roughly equivalent BJTs, and taken with the poor linearity and low efficiency of these devices, the use of them will require a marketing rather than a technical motivation.

Turning to BJTs, I conclude that there are the following candidates for Best Output Stage:

1. The Emitter-Follower Type II output stage is the best at coping with switch-off distortion but the quiescent-current stability needs careful consideration.
2. The CFP topology has good quiescent stability and low Large Signal Nonlinearity; it has the drawback that reverse-biasing the output device bases for fast switch-off is impossible without additional HT rails. The linearity appears to worsen if a number of output devices are used in parallel. CFP output stages are more prone to issues with parasitic oscillation.
3. The quasi-complementary-with-Baxandall-diode stage comes close to mimicking the EF-type stages in linearity, with a potential for some cost-saving on output devices. Quiescent stability is not as good as the CFP configuration.

In the last ten years most of my power amplifier designs, both commercial and those published in journals, have used the EF output stage. The last quasi-complementary output stage I designed was in 1980. The application did not require the highest possible performance, and at that date there was still a significant cost saving in avoiding the use of PNP output devices. A Baxandall diode was used.

Output Stage Conclusions

1. Class-AB is best avoided, unless you feel that the presence of a region of Class-A for small signals outweighs the greater distortion of medium and large signals. Class-AB will always have more distortion than either Class-A or Class-B. The practical efficiency of Class-A, i.e., when it is driven by music rather than sine waves, is very poor indeed (see Chapter 16). Class-A is impractical for large power outputs. The conclusion is – use Class-B, always. Its linearity can be made very good indeed, as described in Chapter 12.
2. FET outputs offer freedom from some BJT problems, but in general have poorer linearity and cost more.

3. The distortion generated by a Blameless amplifier driving an 8 Ω load is almost wholly due to the effects of crossover and switching distortion. This does not hold for 4 Ω or lower loads, where third harmonic on the residual shows the presence of large-signal non-linearity, caused by beta-loss at high output currents.

References

1. Mann, R, The Texan 20 + 20 Watt Stereo Amplifier, *Practical Wireless*, May 1972, p. 48 (Output stage with gain).

2. Takahashi, S et al., *Design and Construction of High Slew Rate Amplifiers* Preprint No. 1348 (A-4) for 60th AES Convention 1978 (Class-B small-signal stages).

3. Hawksford, M, Distortion Correction in Audio Power Amplifiers, *JAES*, Jan./Feb. 1981, p. 27 (Error-correction).

4. Walker, P, Current-Dumping Audio Amplifier, *Wireless World*, 1975, pp. 560–562.

5. Blomley, P, New Approach to Class-B, *Wireless World*, Feb. 1971, p. 57 and March 1971, pp. 127–131.

6. Lohstroh, J and Otala, M. An Audio Power Amplifier for Ultimate Quality Requirements, *IEEE Trans on Audio and Electroacoustics*, Dec. 1973, p. 548.

7. Lin, H, Quasi Complementary Transistor Amplifier, *Electronics*, Sept. 1956, pp. 173–175 (Quasi-comp).

8. Baxandall, P, Symmetry in Class B, Letters, *Wireless World*, Sept. 1969, p. 416 (Baxandall diode).

9. Shaw, I M, Quasi-Complementary Output Stage Modification, *Wireless World*, June 1969, p. 265.

10. P Baxandall, Private communication, 1995.

11. Baxandall, P, *Radio, TV & Audio Technical Reference Book*, ed. S W Amos, Oxford: Newnes-Butterworths, 1977.

12. Locanthi, B, Operational Amplifier Circuit for Hi-Fi, *Electronics World*, (USA magazine), Jan. 1967, pp. 39–41.

13. Innox Audio http://www.innoxx.com/english/main.html (accessed Oct. 2012).

Chapter 10

Output Stage Distortions

Output Stage Distortions and their Mechanisms

Subdividing Distortion Three into Large-Signal Non-linearity, crossover, and switch-off distortion provides a basis for judging which output stage is best. The LSN is determined by both circuit topology and device characteristics, crossover distortion is critically related to quiescent-conditions' stability, and switch-off distortion depends strongly on the output stage's ability to remove carriers from power BJT bases. I now look at how these shortcomings can be improved, and the effect they have when an output stage is used closed-loop.

In Chapters 6, 7, and 8 it was demonstrated that the distortion from the small-signal stages can be kept to very low levels that will prove to be negligible compared with closed-loop output-stage distortion, by the adroit use of relatively conventional circuitry. Likewise, Chapters 10 and 11 will reveal that Distortions Four to Eleven can be effectively eliminated by lesser-known but straightforward methods. This leaves Distortion Three, in its three components, as the only distortion that is in any sense unavoidable, as Class-B stages completely free from crossover artefacts are so far beyond us, despite much effort.

This is therefore a good place to review the concept of a 'Blameless' amplifier, introduced in Chapter 5: one designed so that all the easily defeated distortion mechanisms have been rendered negligible. (Note that the word 'Blameless' has been carefully chosen to not imply perfection.)

Distortion One cannot be totally eradicated, but its onset can be pushed well above 20 kHz. Distortion Two can be effectively eliminated by the use of the EF-VAS or cascoding, and Distortion Four to Distortion Seven can be made negligible by simple measures to be described later. This leaves Distortion Three, which includes the knottiest Class-B problems, i.e., crossover distortion (Distortion 3b) and HF switchoff difficulties (Distortion 3c).

The design rules presented here will allow the routine design of Blameless amplifiers. However, this still leaves the most difficult problem of Class-B unsolved, so it is too early to conclude that as far as amplifier linearity is concerned, the story is over …

Large-signal Distortion (Distortion 3a)

Amplifiers always distort more with heavier loading. This is true without exception so far as I am aware. Why? Is there anything we can do about it?

A Blameless Class-B amplifier typically gives an 8 Ω distortion performance that depends very little on variable transistor characteristics such as beta. At this load impedance output stage, non-linearity is almost entirely crossover distortion, which is a voltage-domain effect.

As the load impedance of the amplifier is decreased from infinity to 4 Ω, distortion increases in an intriguing manner. The unloaded THD is not much greater than that from the AP System-1 test oscillator, but as loading increases, so crossover distortion rises steadily: see Figure 10.1. When the load impedance falls below about 8 Ω, a new distortion begins to appear, overlaying the existing crossover non-linearities. It is essentially third harmonic. In Figure 10.1 the upper trace shows

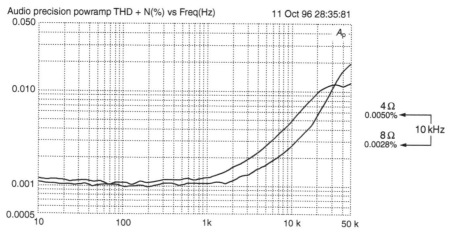

Figure 10.1. Upper trace shows distortion increase due to LSN as load goes from 8 Ω to 4 Ω. Blameless amplifier at 25 W/8 Ω.

the 4 Ω THD is consistently twice that for 8 Ω, once it appears above the noise floor.

I label this Distortion 3a, or Large Signal Non-linearity (LSN), where 'Large' refers to currents rather than voltages. Unlike crossover Distortion 3b, the amount of LSN generated is highly dependent on device characteristics. The distortion residual is basically third order because of the symmetric and compressive nature of the output stage gain characteristic, with some second harmonic because the beta loss is component-dependent and not perfectly symmetrical in the upper and lower output stage halves. Figure 10.2 shows a typical THD residual for Large Signal Non-linearity, driving 50 W into 4 Ω. The residual is averaged 64 times to reduce noise.

LSN occurs in both emitter-follower (EF) and Complementary-Feedback Pair (CFP) output configurations; this section concentrates on the CFP version. Figure 10.3 shows the incremental gain of a simulated CFP output stage for 8 Ω and 4 Ω; the lower 4 Ω trace has greater downward curvature, i.e., a greater fall-off of gain with increasing current. Note that this fall-off is steeper in the negative half, so the THD generated will contain even as well as odd harmonics. The simulated EF behaviour is very similar.

As it happens, an 8 Ω nominal impedance is a reasonably good match for standard power BJTs, though 16 Ω might be better for minimising LSN if loudspeaker technology permits. It is coincidental that an 8 Ω nominal impedance corresponds approximately to the heaviest load that can be driven without LSN appearing, as this value is a legacy from valve technology. LSN is an extra distortion component laid on top of others, and usually dominating them in amplitude, so it is obviously simplest to minimise the 8 Ω distortion first. 4 Ω effects can then be seen more or less in isolation when load impedance is reduced.

The typical result of 4 Ω loading was shown in Figure 10.1, for the modern MJ15024/25 complementary pair from Motorola. Figure 10.4 shows the same diagram for one of the oldest silicon complementary pairs, the 2N3055/2955. The 8 Ω distortion is similar for the different devices, but the 4 Ω THD is 3.0 times worse for the venerable 2N3055/2955. Such is progress.

Such experiments with different output devices throw useful light on the Blameless concept — from the various types tried so far, it can be said that Blameless performance, whatever the output device type, should not exceed 0.001% at 1 kHz and 0.006% at 10 kHz, when driving 8 Ω. The components existed to build sub-0.001% THD amplifiers in mid-1969, but not the knowledge.

Low-impedance loads have other implications beyond worse THD. The requirements for sustained long-term 4 Ω operation are severe, demanding more heatsinking and greater power supply capacity. For economic reasons the peak/average ratio of music is usually fully exploited, though this can cause real problems on extended sinewave tests, such as the FTC 40%-power-for-an-hour preconditioning procedure.

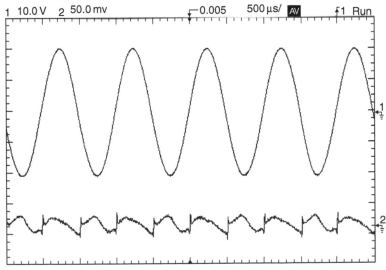

Figure 10.2. Distortion residual showing Large Signal Non-linearity, driving 50W into 4 Ω and averaged 64 times.

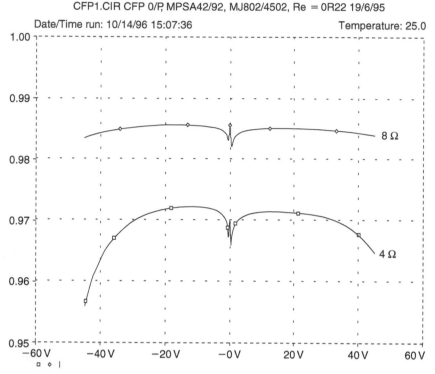

Figure 10.3. The incremental gain of a standard CFP output stage. The 4 Ω trace droops much more as the gain falls off at higher currents. PSpice simulation.

The focus of this section is the extra distortion generated in the output stage itself by increased loading, but there are other ways in which linearity may be degraded by the higher currents flowing. Of the amplifier distortion mechanisms (see p. 117), Distortions One, Two, and Eight are unaffected by output stage current magnitudes. Distortion Four might be expected to increase, as increased loading on the output stage is

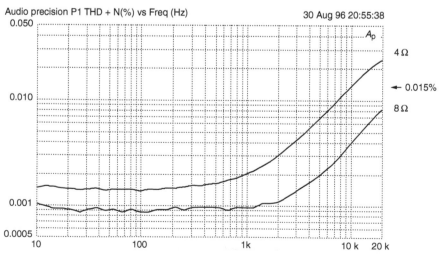

Figure 10.4. 4 Ω distortion is 3 times greater than 8 Ω for 2N3055/2955 output devices. Compare Figure 10.1.

reflected in increased loading on the VAS. However, both the beta-enhanced EF and buffered-cascode methods of VAS linearisation deal effectively with sub-8 Ω loads, and this does not seem to be a problem.

When a 4 Ω load is driven, the current taken from the power supply is greater, potentially increasing the rail ripple, which could worsen Distortion Five. However, if the supply reservoir capacitances have been sized to permit greater power delivery, their increased capacitance reduces ripple again, so this effect tends to cancel out. Even if rail ripple doubles, the usual RC filtering of bias supplies should keep it out of the amplifier, preventing intrusion via the input pair tail, and so on.

Distortion Six could worsen as the half-wave currents flowing in the output circuitry are twice as large, with no counteracting mechanism. Distortion Seven, if present, will be worse due to the increased load currents flowing in the output stage wiring resistances.

Of those mechanisms above, Distortion Four is inherent in the circuit configuration (though easily reducible below the threshold of measurement) while Distortions Five, Six, and Seven are topological, in that they depend on the spatial and geometrical relationships of components and wiring. The latter three distortions can therefore be completely eliminated in both theory and practice. This leaves only the LSN component, otherwise known as Distortion 3a, to deal with.

The Load-Invariant Concept

In an ideal amplifier the extra LSN distortion component would not exist. Such an amplifier would give no more distortion into 4 Ω than 8 Ω, and could be called 'Load-Invariant to 4 Ω'. The minimum load qualification is required because it will be seen that the lower the impedance, the greater the difficulties in aspiring to Load-Invariance. I assume that we start out with an amplifier that is Blameless at 8 Ω; it would be logical but quite pointless to apply the term 'Load-Invariant' to an ill-conceived amplifier delivering 1% THD into both 8 Ω and 4 Ω.

The LSN Mechanism

When the load impedance is reduced, the voltage conditions are essentially unchanged. LSN is therefore clearly a current-domain effect, a function of the magnitude of the signal currents flowing in drivers and output devices.

A 4 Ω load doubles the output device currents, but this does not in itself generate significant extra distortion. The crucial factor appears to be that the current drawn from the drivers by the output device bases *more* than doubles, due to beta fall-off in the output devices as collector current increases.

It is this *extra* increase of current that causes almost all the additional distortion. The exact details of this have not been completely clarified, but it seems that this 'extra current' due to beta fall-off varies very non-linearly with output voltage, and combines with driver non-linearity to reinforce it rather than cancel. Beta-droop is ultimately due to high-level injection effects, which are in the province of semi-conductor physics rather than amplifier design. Such effects vary greatly with device type, so when output transistors are selected, the likely performance with loads below 8 Ω must be considered.

There is good simulator evidence that LSN is entirely due to beta-droop causing extra current to be drawn from the drivers. To summarise:

- Simulated output stages with output devices modified to have no beta-droop (by increasing SPICE model parameter IKF) do not show LSN. It appears to be specifically that extra current taken due to beta-droop causes the extra non-linearity.

- Simulated output devices driven with zero-impedance voltage sources instead of the usual transistor drivers exhibit no LSN. This shows that LSN does not occur in the outputs themselves, and so it must be happening in the driver transistors.

- Output stage distortion can be treated as an error voltage between input and output. The double emitter-follower (EF) stage error is therefore: driver Vbe + output Vbe + Re drop. A simulated EF output stage with the usual drivers shows that non-linearity increases in the driver Vbe rather than in the output Vbe, as load resistance is reduced. The voltage drop across the emitter resistors Re is essentially linear.

The knowledge that beta-droop caused by increased output device Ic is at the root of the problem leads to some solutions. First, the per-device Ic can be reduced by using parallel output devices. Alternatively Ic can be left unchanged and output device types selected for those with the least beta-droop.

There is the possibility that increasing the current drawn from the drivers will in turn increase the current that they draw from the VAS, compromising its linearity. The investigations recorded here show that to be a very minor effect, if it exists at all. However, it is a possibility worth bearing in mind.

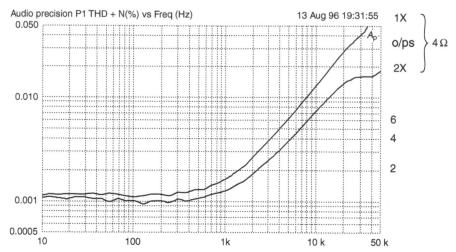

Figure 10.5. 4 Ω distortion is reduced by 1.9 × upon doubling standard (MJ15024/15025) output transistor: output 30 W/8 Ω.

LSN with Doubled Output Devices

LSN can be effectively reduced by doubling the output devices, when this is quite unnecessary for handling the rated power output. The fall-off of beta depends on collector current, and if two output devices are connected in parallel, the collector current divides in two between them. Beta-droop is much reduced.

From the above evidence, I predicted that this doubling ought to reduce LSN − and when measured, indeed it does. Such reality checks must never be omitted when using circuit simulators. Figure 10.5 compares the 4 Ω THD at 60 W for single and double output devices, showing that doubling reduces distortion by about 1.9 times, which is a worthwhile improvement.

The output transistors used for this test were modern devices, the Motorola MJ15024/15025. The much older 2N3055/2955 complementary pair gives a similar halving of LSN when their number is doubled, though the initial distortion is three times higher into 4 Ω. 2N3055 specimens with an H suffix show markedly worse linearity than those without. In my classification scheme this use of paralleled output devices could be considered as an example of Class-B•B.

No explicit current-sharing components were added when doubling the devices, and this lack seemed to have no effect on LSN reduction. There was no evidence of current hogging, and it appears that the circuit cabling resistances alone were sufficient to prevent this.

Doubling the number of power devices naturally increases the power output capability, though if this is exploited LSN will tend to rise again, and you are back where you started. Opting for increased power output will also make it necessary to uprate the power supply, heatsinks, and so on. The essence of this technique is to use parallel devices to reduce distortion long before power handling alone compels you to do so.

LSN with Better Output Devices

The 2SC3281 2SA1302 complementary pair are plastic TO3P devices with a reputation in the hi-fi industry for being 'more linear' than the general run of transistors. Vague claims of this sort arouse the deepest of suspicions; compare the many assertions of superior linearity for power FETs, which is the exact opposite of reality. However, in this case the core of truth is that 2SC3281 and 2SA1302 show much less beta-droop than average power transistors. These devices were introduced by Toshiba; the Motorola versions are MJL3281A, MJL1302A, also in TO3P package. Figure 10.6 shows the beta-droop for the various devices discussed here, and it is clear that more droop means more LSN.

The 3281/1302 pair is clearly in a different class from conventional transistors, as they maintain beta much more effectively when collector current increases. There seems to be no special name for this class of BJTs, so I have called them 'sustained-beta' devices here.

The THD into 4 Ω and 8 Ω for single 3281/1302 devices is shown in Figure 10.7. Distortion is reduced by about 1.4 times compared with the standard devices of Figure 10.1, over the range 2 to 8 kHz. Several

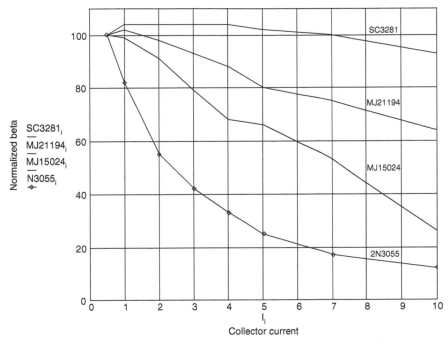

Figure 10.6. Power transistor beta falls as collector current increases. Beta is normalised to 100 at 0.5 A (from manufacturers' data sheets).

pairs of 3281/1302 were tested and the 4 Ω improvement is consistent and repeatable.

The obvious next step is to combine these two techniques by using doubled sustained-beta devices. The doubled-device results are shown in Figure 10.8 where the distortion at 80 W/4 Ω (15 kHz) is reduced from 0.009% in Figure 10.7 to 0.0045%; in other words, halved. The 8 Ω and 4Ω traces are now very close together, the 4 Ω THD being only 1.2 times higher than the 8 Ω case.

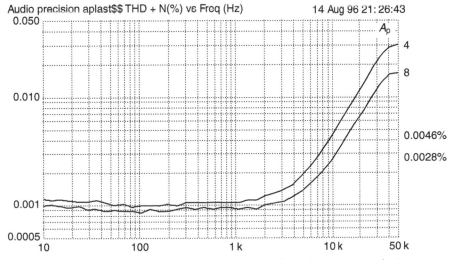

Figure 10.7. THD at 40 W/8 Ω and 80 W/4 Ω with single 3281/1302 devices.

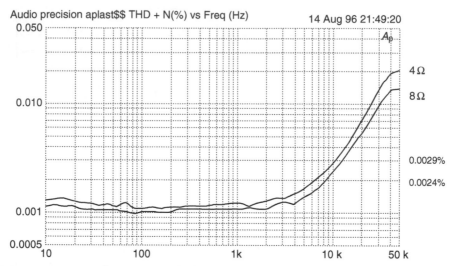

Figure 10.8. THD at 40 W/8 Ω and 80 W/4 Ω with doubled 3281/1302 output transistors. 4 Ω THD has been halved compared with Figure 10.1.

There are other devices showing less beta-droop than standard. In a very quick survey I unearthed the MJ21193, MJ21194 pair (TO3 package) and the MJL21193, MJL21194 pair (TO3P package), both from Motorola. These devices show beta-maintenance intermediate between the 'super' 3281/1302 and 'ordinary' MJ15024/25, so it seemed likely that they would give less LSN than ordinary power devices, but more than the 3281/1302. This prediction was tested and duly fulfilled.

It could be argued that multiplying output transistors is an expensive way to solve a linearity problem. To give this perspective, in a typical stereo power amplifier the total cost including heatsink, metal work and mains transformer will only increase by about 5% when the output devices are doubled.

LSN with Feedforward Diodes

The first technique I tried to reduce LSN was the addition of power diodes across OR22 output emitter resistors. The improvement was only significant for high power into sub-3 Ω loading, and was of rather doubtful utility for hi-fi. Feedforward diodes treat the symptoms (by attempting distortion cancellation) rather than the root cause, so it is not surprising this method is of limited effectiveness; see Figure 10.9.

It has been my practice for many years now to set the output emitter resistors Re at 0.1 Ω, rather than the more common 0.22 Ω. This change both improves voltage-

swing efficiency and reduces the extra distortion generated if the amplifier is erroneously biased into Class-AB. As a result, even low-impedance loads give a relatively small voltage drop across Re, which is insufficient to turn on a silicon power diode at realistic output levels.

Schottky diodes have a lower forward voltage drop and might be useful here. Tests with 50 A diodes have been made but have so far not been encouraging in the amount of distortion reduction achieved. Suitable Schottky diodes cost at least as much as an output transistor, and two will be needed.

LSN with Triple Output Stages

In electronics, as in many fields, there is often a choice between applying brawn (in this case multiple power devices) or brains to solve a given problem. The 'brains' option here would be a clever circuit configuration that reduced LSN without replication of expensive power silicon, and the obvious place to look is the output-triple approach. Note 'output triples' here refers to pre-driver, driver, and output device all in one local NFB loop, rather than three identical output devices in parallel, which I would call 'tripled outputs'. Getting the nomenclature right is a bit of a problem.

In simulation, output-triple configurations do reduce the gain-droop that causes LSN. There are many different ways to configure output triples, and they

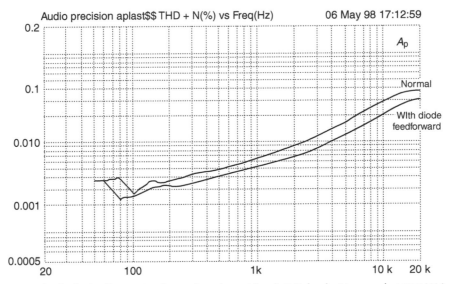

Figure 10.9. Simple diode feedforward reduces distortion with sub-8 Ω loads. Measured at 210 W into 2.7 Ω.

vary in their linearity and immunity to LSN. The true difficulty with this approach is that three transistors in a tight local loop are very prone to parasitic and local oscillations. This tendency is exacerbated by reducing the load impedances, presumably because the higher collector currents lead to increased device transconductance. This sort of instability can be very hard to deal with, and in some configurations appears almost insoluble. At present this approach has not been studied further.

Loads below 4 Ω

So far I have concentrated on 4 Ω loads; loudspeaker impedances often sink lower than this, so further tests were done at 3 Ω. One pair of 3281/1302 devices will give 50 W into 3 Ω for THD of 0.006% (10 kHz), see Figure 10.10. Two pairs of 3281/1302 reduce the distortion to 0.003% (10 kHz) as in Figure 10.11. This is an excellent result for such simple circuitry, and may well be a record for 3 Ω linearity.

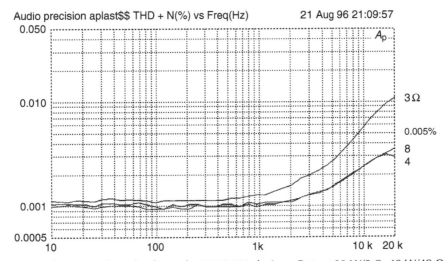

Figure 10.10. Distortion for 3, 4 and 8 Ω loads, single 3281/1302 devices. Output 20 W/8 Ω, 40 W/40 Ω and 60 W/3 Ω.

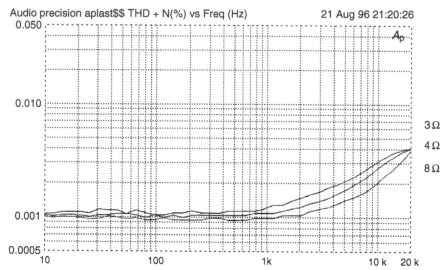

Figure 10.11. Distortion for 3, 4 and 8 Ω load, double 3281/1302 devices. Output 20 W/8 Ω, 40 W/40 Ω and 60 W/3 Ω.

It appears that whatever the device type, doubling the outputs halves the THD percentage for 4 Ω loading. This principle can be extended to 2 Ω operation, but tripled devices are required for sustained operation at significant powers. The resistive losses will be serious, so 2 Ω power output may be little greater than that into 4 Ω.

Better 8 Ω Performance

It was not expected that the sustained-beta devices would also show lower crossover distortion at 8 Ω, but they do, and the effect is once more repeatable. It may be that whatever improves the beta characteristic also somewhat alters the turn-on law so that crossover distortion is reduced; alternatively traces of LSN, not visible in the THD residual, may have been eliminated. The latter is probably the more likely explanation.

The plot in Figure 10.11 shows the improvement over the MJ15024/25 pair; compare the 8 Ω line in Figure 10.1. The 8 Ω THD at 10 kHz is reduced from 0.003% to 0.002%, and with correct bias adjustment, the crossover artefacts are invisible on the 1 kHz THD residual. Crossover artefacts are only just visible in the 4 Ω case, and to get a feel for the distortion being produced, and to set the bias optimally, it is necessary to test at 5 kHz into 4 Ω.

A Practical Load-Invariant Design

Figure 10.12 is the circuit of a practical Load-Invariant amplifier designed for 8 Ω nominal loads with 4 Ω

impedance dips; not for speakers that start out at 4 Ω nominal and plummet from there. The distortion performance is shown in Figures 10.26, 10.27, 10.29, and 10.30 for various fitments of output device. The supply voltage can be from ±20 to ±40 V; checking power capability for a given output device fit must be left to the constructor.

Apart from Load-Invariance, the design also incorporates two new techniques from the Thermal Dynamics section of this book in Chapter 22.

The first technique greatly reduces time lag in the thermal compensation. With a CFP output stage, the bias generator aims to shadow the driver junction temperature rather than the output junctions. A much faster response to power dissipation changes is obtained by mounting a bias generator transistor TR8 on top of the driver TR14, rather than on the other side of the heatsink. The driver heatsink mass is largely decoupled from the thermal compensation system, and the response is speeded up by at least two orders of magnitude.

The second innovation is a bias generator with an increased temperature coefficient (tempco), to reduce the static errors introduced by thermal losses between driver and sensor. The bias generator tempco is increased to −4.0 mV/°C. Distortion Five also compensates for the effect of ambient temperature changes.

This design is not described in detail because it closely resembles the Blameless Class-B amp described elsewhere. The low-noise feedback network is taken from the Trimodal amplifier in Chapter 17; note the requirement for input bootstrapping if a 10 k input impedance

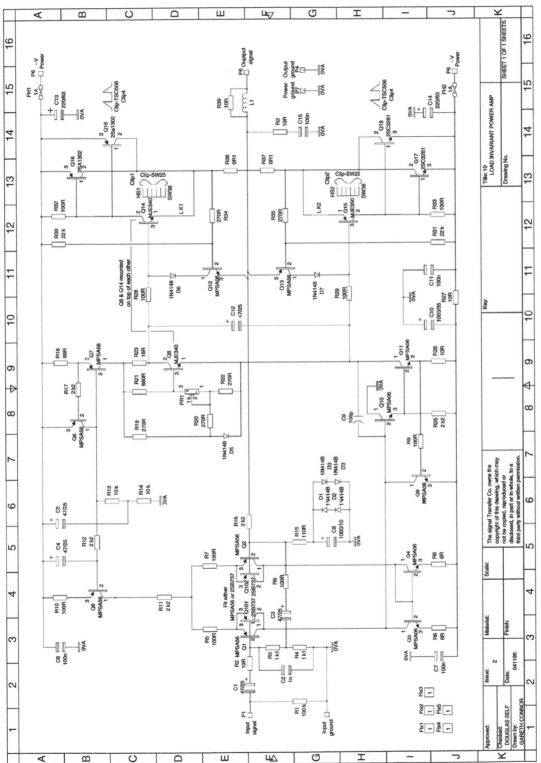

Figure 10.12. Circuit diagram of the Load-Invariant power amplifier.

is required. Single-slope VI limiting is incorporated for overload protection, implemented by TR12, 13. The global NFB factor is once more a modest 30 dB at 20 kHz.

More on Multiple Output Devices

I have done some further experiments with multiple devices, using three, four, five and six in parallel. The 2SC2922/2SA1612 complementary pair was used. In this case the circuit used was somewhat different; see Figure 10.13. With a greater number of devices I was now more concerned about proper current sharing, and so each device has its own emitter resistor. This makes it look much more like a conventional paralleled output stage, which essentially it is. This time I tried both double and the triple-EF output configurations, as I wished to prove:

1. that LSN theory worked for both of the common configurations EF and CFP — it does.
2. that LSN theory worked for both double and triple versions of the EF output stage — it does.

For reasons of space, only the triple-EF results are discussed here.

Figure 10.14 shows the measured THD results for one complementary pair of output devices in the triple-EF circuit of Figure 10.13. Distortion is slightly higher, and the noise floor relatively lower, than in previous graphs because of the higher output power of 50 W/8 Ω. Figure 10.15 shows the same measurement but there are now two pairs of output devices. Note that THD has halved at both 8 Ω and 4 Ω loads; this is probably due to the larger currents taken by 8 Ω loads at this higher power. Figure 10.16 shows the result for six devices; 8 Ω distortion has almost been abolished, and the 4 Ω result is almost as good. It is necessary to go down to a 2 Ω load to get the THD clear of the noise so it can be measured accurately. With six output devices, driving a substantial amount of power into a 2 Ω load is not a problem.

On a practical note, the more output devices you have, the harder the amplifier may be to purge of parasitic oscillations in the output stage. This is presumably due to the extra raw transconductance available, and can be a problem even with the triple-EF circuit, which has no local NFB loops. I do not pretend to be able to give a detailed explanation of this effect at the moment.

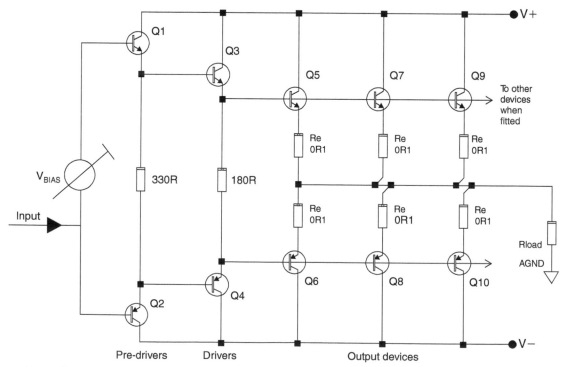

Figure 10.13. The triple-EF output stage used for the measurements described below. 'Triple' refers to the fact that there are three transistors from input to output, rather than the fact that there happen to be three output devices in parallel.

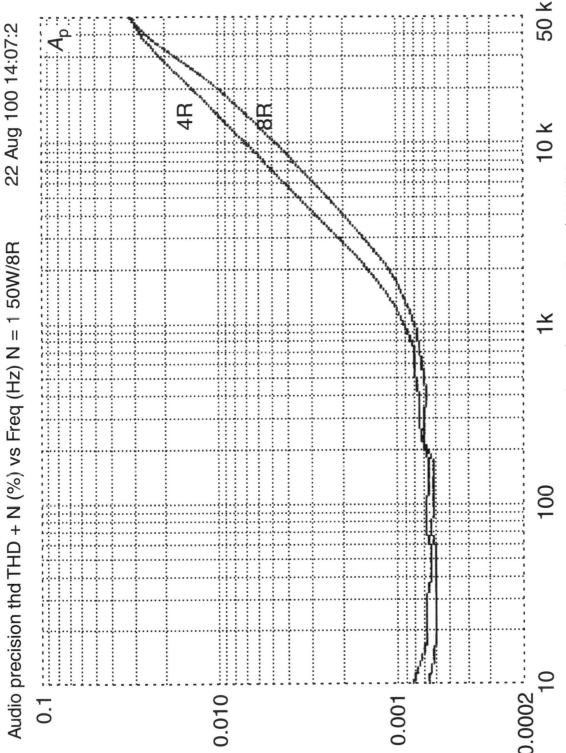

Figure 10.14. THD for one pair (*N* = 1) of output devices, at 50 W/8 Ω and 100 W/4 Ω.

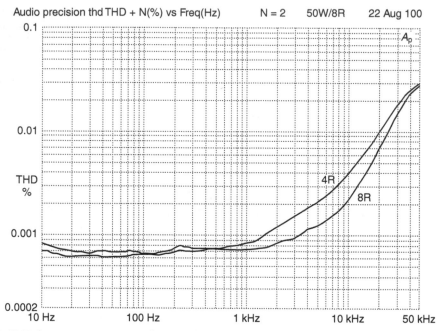

Figure 10.15. THD for two pairs (*N* = 2) of output devices, at 50 W/8 Ω and 100 W/4 Ω. A definite improvement.

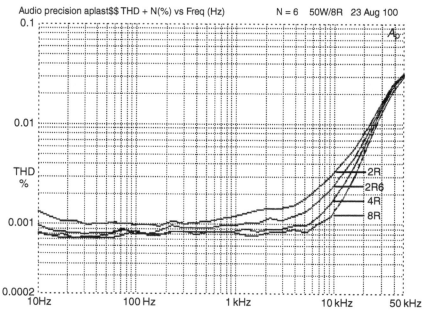

Figure 10.16. THD for six pairs (*N* = 6) of output devices, at 50 W/8 Ω, 100 W/4 Ω, 200 W/2 Ω. Note very low distortion at 8 Ω.

Having demonstrated that sustained-beta output devices not only reduce LSN but also unexpectedly reduce crossover distortion, it seemed worth checking if using multiple output devices would give a similar reduction at light loading. I was rather surprised to find they did.

Adding more output devices in parallel, while driving an 8 Ω load, results in a steady reduction in distortion. Figures 10.14 to 10.16 show how this works in reality. The SPICE simulations in Figure 10.17 reveal that increasing the number N of output devices not only flattens the crossover gain wobble, but spreads it out over a greater width. This spreading effect is an extra bonus because it means that lower-order harmonics are generated, and at lower frequencies there will be more negative feedback to linearise them. (Bear in mind also that a triple-EF output has an inherently wider gain wobble than the double-EF.) Taking the gain wobble width as the voltage between the bottoms of the two dips, this appears to be proportional to N. The amount of gain wobble, as measured from top of the peak to bottom of the dips, appears to be proportional to 1/N.

This makes sense. We know that crossover distortion increases with heavier loading, i.e., with greater currents flowing in the output devices, but under the same voltage conditions. It is therefore not surprising that reducing the device currents by using multiple devices has the same effect as reducing loading. If there are two output devices in parallel, each sees half the current variations, and crossover non-linearity is reduced. The voltage conditions are the same in each half and so are unchanged. This offers us the interesting possibility that crossover distortion – which has hitherto appeared inescapable – can be reduced to an arbitrary level simply by paralleling enough output transistors. To the best of my knowledge, this is a new insight.

Load Invariance: Summary

In conventional amplifiers, reducing the 8 Ω load to 4 Ω increases the THD by 2 to 3 times. The figure attained by the Load-Invariant amplifier presented here is 1.2 times, and the ratio could be made even closer to unity by tripling or further multiplying the output devices.

Crossover Distortion (Distortion 3b)

In a field like Audio where consensus of any sort is rare, it is a truth universally acknowledged that crossover distortion is the worst problem that can afflict Class-B

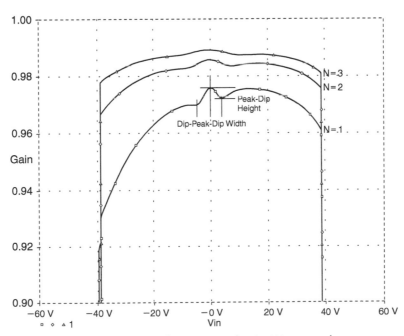

Figure 10.17. SPICE simulation of triple-EF output with $N = 1$, 2 and 3. As N increases the crossover gain wobble becomes flatter and more spread out laterally.

power amplifiers. The difficulty stems from the slight wobble in gain in the crossover region, where control of the output voltage is handed over from one device to another. Crossover distortion is rightly feared as it generates unpleasant high-order harmonics, and there is at least the possibility that it will increase in percentage as signal level is reduced.

The pernicious nature of crossover distortion is partly because it occurs over a small part of the signal swing, and so generates high-order harmonics. Worse still, this small range over which it does occur is at the zero-crossing point, so not only is it present at all levels and all but the lightest loads, but it is generally believed to increase as output level falls, threatening very poor linearity at the modest listening powers that most people use.

Unusually, there is something of a consensus that audible crossover distortion was responsible for the so-called 'transistor sound' of the 1960s. This is very likely true if we are talking about transistor radios, which in those days were prone to stop reproducing low-level signals completely as the battery voltage fell. However, the situation in transistor hi fi amplifiers was very different. To the best of my knowledge, the 'transistor sound' has never actually been investigated,

nor even its existence confirmed, by double-blind testing vintage equipment.

The amount of crossover distortion produced depends strongly on optimal quiescent adjustment, so the thermal compensation used to stabilise this against changes in temperature and power dissipation must be accurate. The provision of fast and accurate thermal compensation is to a large extent one of the great unsolved problems of amplifier design. Chapter 22 on thermal dynamics deals with the difficulties involved.

The Vbe-Ic characteristic of a bipolar transistor is initially exponential, blending into linear as the internal emitter resistance Re comes to dominate the transconductance. The usual Class-B stage puts two of these curves back-to-back, and Peter Blomley has shown[1] that these curves are non-conjugate, i.e., there is no way they can be shuffled sideways so they will sum to a completely linear transfer characteristic, whatever the offset between them imposed by the bias voltage. This can be demonstrated quickly and easily by SPICE simulation; see Figure 10.18. There is at first sight not much you can do except maintain the bias voltage, and hence quiescent current, at the optimal level for minimum gain deviation in the crossover region.

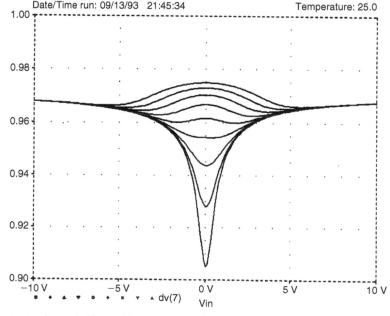

Figure 10.18. Gain/output voltage plot for an EF output shows how non-conjugate transistor characteristics at the crossover region cannot be blended into a flat line at any bias voltage setting. Bias varies 2.75 to 2.95 V in 25 mV steps, from too little to too much quiescent.

It should be said that the crossover distortion levels generated in a Blameless amplifier can be very low up to around 1 kHz, being barely visible in residual noise and only measurable with a spectrum-analyser. As an instructive example, if a Blameless closed-loop Class-B amplifier is driven through a TL072 unity-gain buffer, the added noise from this opamp will usually submerge the 1 kHz crossover artefacts into the noise floor, at least as judged by the eye on the oscilloscope. (It is most important to note that Distortions Four, Five, Six and Seven create disturbances of the THD residual at the zero-crossing point that can be easily mistaken for crossover distortion, but the actual mechanisms are quite different.) However, the crossover distortion becomes obvious as the frequency increases, and the high-order harmonics benefit less from NFB.

It will be seen later that in a Blameless amplifier driving 8 Ω, the overall linearity is dominated by crossover distortion, even with a well-designed and optimally biased output stage. There is an obvious incentive to minimise this distortion mechanism, but there seems no obvious way to reduce crossover gain deviations by tinkering with any of the relatively conventional stages considered so far.

Figure 10.19 shows the signal waveform and THD residual from a Blameless power amplifier with optimal Class-B bias. Output power was 25 W into 8 Ω, or 50 W into 4 Ω (i.e., the same output voltage) as appropriate,

for all the residuals shown here. The figure is a record of a single sweep so the residual appears to be almost totally random noise; without the visual averaging that occurs when we look at an oscilloscope, the crossover artefacts are much less visible than in real time.

In Figure 10.20, 64 times averaging is applied, and the disturbances around crossover become very clear. There is also revealed a low-order component at roughly 0.0004%, which is probably due to very small amounts of Distortion Six that were not visible when the amplifier layout was optimised.

Figure 10.21 shows Class-B slightly underbiased to generate crossover distortion. The crossover spikes are very sharp, so their height in the residual depends strongly on measurement bandwidth. Their presence warns immediately of underbiasing and avoidable crossover distortion.

In Figure 10.22 an optimally biased amplifier is tested at 10 kHz. The THD increases to approximately 0.004%, as the amount of global negative feedback is 20 dB less than at 1 kHz. The timebase is faster so crossover events appear wider than in Figure 10.20. The THD level is now well above the noise so the residual is averaged 8 times only. The measurement bandwidth is still 80 kHz, so harmonics above the eighth are now lost. This is illustrated in Figure 10.23, which is Figure 10.22 rerun with a 500 kHz measurement bandwidth. The distortion products now look much more jagged.

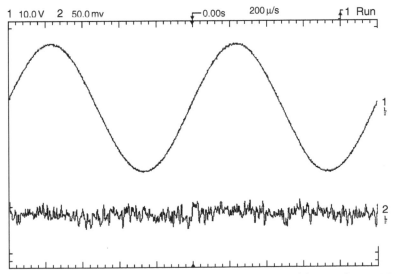

Figure 10.19. The THD residual from an optimally biased Blameless power amplifier at 1 kHz, 25 W/8 Ω is essentially white noise. There is some evidence of artefacts at the crossover point, but they are not measurable. THD 0.00097%, 80 kHz bandwidth.

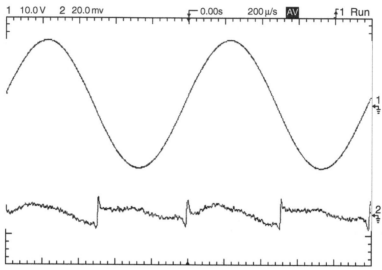

Figure 10.20. Averaging Figure 10.19 residual 64 times reduces the noise by 18 dB, and crossover discontinuities are now obvious. The residual has been scaled up by 2.5 times from Figure 10.19 for greater clarity.

Figure 10.24 shows the gain-step distortion introduced by Class-AB. The undesirable edges in the residual are no longer in close pairs that partially cancel, but are spread apart on either side of the zero crossing. No averaging is used here as the THD is higher.

It is commonplace in Audio to discover that a problem like crossover distortion has been written about and agonised over for decades, but the amount of technical investigation that has been done (or at any rate published) is disappointingly small. I decided to do some basic investigations myself.

I first looked to see if crossover distortion really *did* increase with decreasing output level in a Blameless amplifier; to attempt its study with an amplifier contaminated with any of the avoidable distortion mechanisms

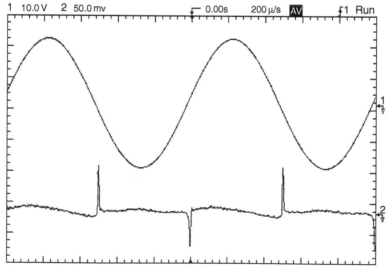

Figure 10.21. The results of mild underbias in Class-B. Note residual scale is changed back to 50mV.

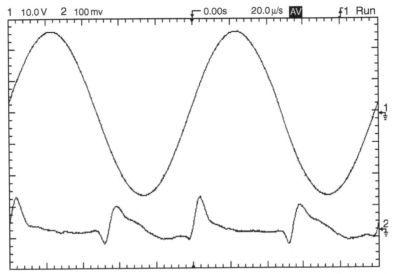

Figure 10.22. An optimally biased Blameless power amplifier at 10 kHz. THD approximately 0.004%, measurement bandwidth 80 kHz. Averaged 8 times.

is completely pointless. One problem is that a Blameless amplifier has such a low level of distortion at 1 kHz (0.001% or less) that the crossover artefacts are barely visible in circuit noise, even if low-noise techniques are used in the design of the amplifier. The measured percentage level of the noise-plus-distortion residual is bound to rise with falling output, because the noise voltage remains constant; this is the lowest line in Figure 10.25. To circumvent this, the amplifier was

deliberately underbiased by varying amounts to generate ample crossover spikes, on the assumption that any correctly adjusted amplifier should be less barbarous than this.

The answer from Figure 10.25 is that the THD percentage *does* increase as the level falls, but relatively slowly. Both EF and CFP output stages give similar diagrams to Figure 10.25, and whatever the degree of underbias, THD increases by about 1.6

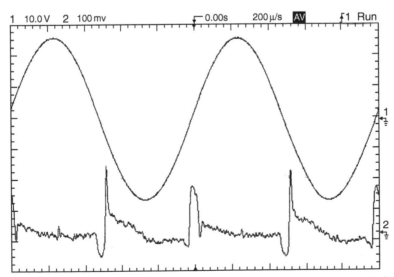

Figure 10.23. As Figure 10.22, but 500 kHz measurement bandwidth. The distortion products look quite different.

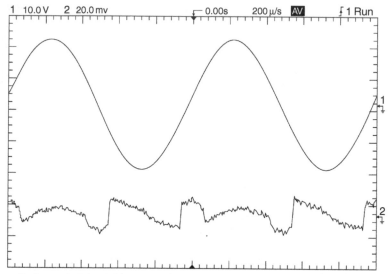

Figure 10.24. The so-called 'gm-doubling' distortion introduced by Class-AB. The edges in the residual are larger and no longer at the zero crossing, but displaced either side of it. Compare Figure 10.20.

times as the output voltage is halved. In other words, reducing the output power from 25 W to 250 mW, which is pretty drastic, only increases THD percentage by six times, and so it is clear that the *absolute* (as opposed to percentage) THD level in fact falls slowly with amplitude, and therefore probably remains imperceptible. This is something of a relief; but crossover distortion at any level remains a bad thing to have.

Distortion versus level was also investigated at high frequencies, i.e., above 1 kHz where there is more THD to measure, and optimal biasing can be used.

Figure 10.26 shows the variation of THD with level for the EF stage at a selection of frequencies; Figure 10.27 shows the same for the CFP. Neither shows a significant rise in percentage THD with falling level, though it is noticeable that the EF gives a good deal less distortion at lower power levels around 1 W. This is an unexpected observation, and possibly a new one.

It is almost certainly due to the fact that the crossover region for an EF output stage is much wider than that for the CFP version, and so creates lower-order distortion products. Attempts to verify this by SPICE simulation

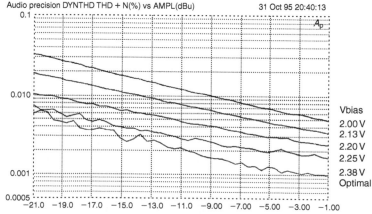

Figure 10.25. Showing how crossover distortion rises slowly as output power is reduced from 25 W to 250 mW (8 Ω) for optimal bias and increasingly severe underbias (upper lines). This is an EF type output stage. Measurement bandwidth 22 kHz.

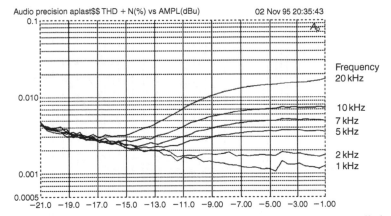

Figure 10.26. Variation of crossover distortion with output level for higher frequencies. Optimally biased EF output stage. Bandwidth 80 kHz.

have so far foundered on the insufficient accuracy of the Fourier output data (Feb. 2012).

To further get the measure of the problem, Figure 10.28 shows how HF distortion is greatly reduced by increasing the load resistance, providing further confirmation that almost all the 8 Ω distortion originates as crossover in the output stage.

Output Stage Quiescent Conditions

This section deals with the crossover region and its quiescent conditions; the specific issue of the effectiveness of the thermal compensation for temperature effects is dealt with in detail in Chapter 22.

Figure 10.29 shows the two most common types of output stage: the Emitter-Follower (EF) and the Complementary Feedback Pair (CFP) configurations. The manifold types of output stage based on triples will have to be set aside for the moment. The two circuits shown have few components, and there are equally few variables to explore in attempting to reduce crossover distortion.

To get the terminology straight: here, as in my previous writings, Vbias refers to the voltage set up across the driver bases by the Vbe-multiplier bias generator, and is in the range 1 to 3 V for Class-B operation. Vq is the quiescent voltage across the two emitter resistors (hereafter Re) alone, and is between 5 and 50 mV, depending on the configuration chosen. Quiescent current lq refers only to that flowing in the output devices, and does not include driver standing currents.

I have already shown that the two most common output configurations are quite different in behaviour,

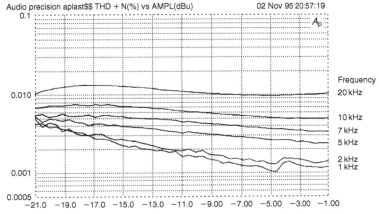

Figure 10.27. Variation of distortion with level for higher frequencies. Optimally biased CFP output stage. Bandwidth 80 kHz.

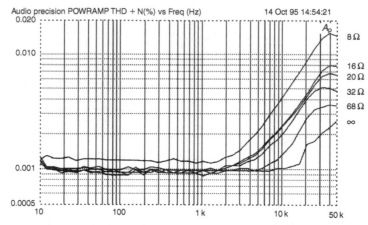

Figure 10.28. How crossover distortion is reduced with increasing load resistance. 20W into 8 Ω, 80 kHz bandwidth.

with the CFP being superior on most criteria. Table 10.1 shows that crossover gain variation for the EF stage is smoother (being some 20 times wider) but of four times higher amplitude than for the CFP version. It is not immediately obvious from this which stage will generate the least HF THD, bearing in mind that the NFB factor falls with frequency.

Table 10.1 also emphasises that a little-known drawback of the EF version is that its quiescent dissipation may be far from negligible.

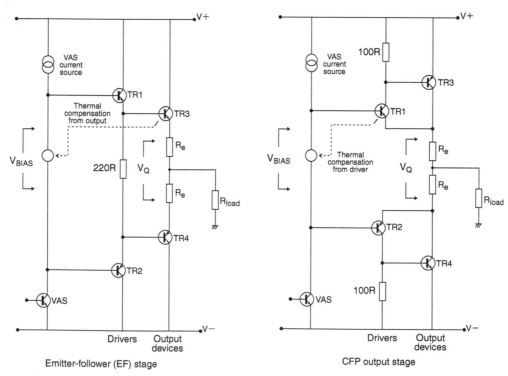

Figure 10.29. The two most popular kinds of output stage: the Emitter-Follower (EF) and Complementary Feedback Pair (CFP) Vbias and Vq are identified.

Table 10.1. Quiescent conditions compared (For Re = 0.22 Ω, 8 Ω load, and ±40V supply rails)

	Emitter-Follower	CFP
Vbias	2.930V	1.297V
Vq	50mV	5mV
Iq	114mA	11mA
Pq (per o/p device)	4.6W	0.44W
Average gain	0.968	0.971
Peak gain deviation from average	0.48%	0.13%
Crossover width*	±12V	±0.6V

*Note: Crossover-width is the central region of the output voltage range over which crossover effects are significant; I have rather arbitrarily defined it as the ± output range over which the incremental gain curves diverge by more than 0.0005 when Vbias is altered around the optimum value. This is evaluated here for an 8 Ω load only.

An Experiment on Crossover Distortion

Looking hard at the two output stage circuit diagrams, intuition suggests that the value of emitter resistor Re is worth experimenting with. Since these two resistors are placed between the output devices, and alternately pass the full load current, it seems possible that their value could be critical in mediating the handover of output control from one device to the other. Re was therefore stepped from 0.1 Ω to 0.47 Ω, which covers the practical range. Vbias was re-optimised at each step, though the changes were very small, especially for the CFP version.

Figure 10.30 shows the resulting gain variations in the crossover region for the EF stage, while Figure 10.31 shows the same for the CFP configuration. Table 10.2 summarises some numerical results for the EF stage, and Table 10.3 for the CFP.

There are some obvious features; first, Re is clearly not critical in value as the gain changes in the crossover region are relatively minor. Reducing the Re value allows the average gain to approach unity more closely, with a consequent advantage in output power capability. Similarly, reducing Re widens the crossover region for a constant load resistance, because more current must pass through one Re to generate enough voltage drop to turn off the other output device. This implies that as Re is reduced, the crossover products become lower-order and so of lower frequency. They should be better linearised by the frequency-dependent global NFB, and so overall closed-loop HF THD should be lower.

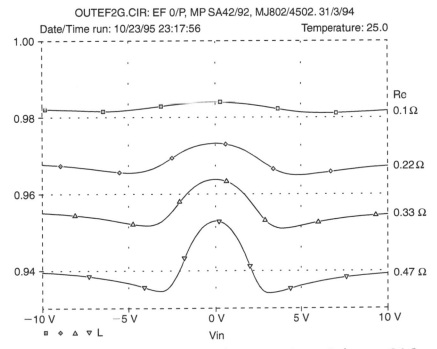

Figure 10.30. Output linearity of the EF output stage for emitter-resistance Re between 0.1 Ω and 0.47 Ω.

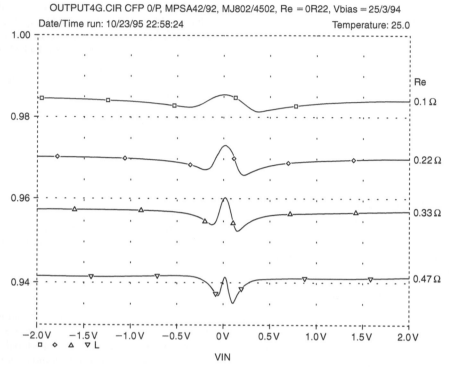

OUTPUT4G.CIR CFP 0/P, MPSA42/92, MJ802/4502, Re = 0R22, Vbias = 25/3/94
Date/Time run: 10/23/95 22:58:24   Temperature: 25.0

Figure 10.31. Output linearity of the CFP output stage for emitter-resistance Re between 0.1 Ω and 0.47 Ω.

Table 10.2. Emitter-Follower output (Type-1); data for 8 Ω load and EF o/p stage

Re Ω	Optimal Vbias Volts	Optimal Vq mV	Iq mA	X-Width Volts	Average Gain ratio
0.1	2.86	42.6	215	18	0.982
0.22	2.87	46.2	107	12	0.968
0.33	2.89	47.6	74	9	0.955
0.47	2.93	54.8	59	7	0.939

Table 10.3. CFP output: data for 8 Ω, load and CFP o/p stage

Re Ω	Optimal Vbias Volts	Optimal Vq mV	Iq mA	X-Width Volts	Average Gain ratio
0.1	1.297	3.06	15.3	1.0	0.983
0.22	1.297	4.62	11.5	0.62	0.971
0.33	1.297	5.64	8.54	0.40	0.956
0.47	1.298	7.18	7.64	0.29	0.941

The simulated crossover distortion experiment described earlier in this chapter showed that as the crossover region was made narrower, the distortion energy became more evenly spread over higher harmonics. A wider crossover region implies energy more concentrated in the lower harmonics, which will receive the benefit of more negative feedback. However, if the region is made wider, but retains the same amount of gain deviation, it seems likely that the total harmonic energy is greater, and so there are two opposing effects to be considered.

I conclude that selecting Re = 0.1 Ω will give adequate bias stability, maximum efficiency, and minimum distortion. This has the additional benefit that if the stage is erroneously over-biased into Class-AB, the resulting gm-doubling distortion will only be half as bad as if the more usual 0.22 Ω values had been used for Re. I have not attempted to use lower values for Re than 0.1 Ω as I have doubts about the resulting bias stability, but it might be an area worth looking at.

As Re is varied, Vq varies by only 29%, while Iq varies by 365%.

It would be easy to assume that higher values of Re must be more linear, because of a vague feeling that there is 'more local feedback', but this cannot be true as an emitter-follower already has 100% voltage feedback to its emitter, by definition. Changing the value of Re slightly alters the total resistive load seen by the emitter itself, and this does seem to have a small but measurable effect on linearity.

As Re is varied, Vq varies by 230% while Iq varies by 85%. However, the absolute Vq change is only 4mV, while the sum of Vbe's varies by only 0.23%. This makes it pretty plain that the voltage domain is what counts, rather than the absolute value of Iq.

The first surprise from this experiment is that in the typical Class-B output stage, quiescent current as such does not matter a great deal. This may be hard to believe, particularly after my repeated statements that quiescent conditions are critical in Class-B, but both assertions are true. The data for both the EF and CFP output stages show that changing Re alters the Iq considerably, but the optimal value of Vbias and Vq barely change.

The voltage across the transistor base-emitter junctions and Re's seems to be what counts, and the actual value of current flowing as a result is not in itself of much interest. However, the Vbias setting remains critical for minimum distortion; once the Re value is settled at the design stage, the adjustment procedure for optimal crossover is just as before.

The irrelevance of quiescent current was confirmed by the Trimodal amplifier, which was designed after the work described here was done, and where I found that changing the output emitter resistor value Re over a 5:1 range required no alteration in Vbias to maintain optimal crossover conditions.

The critical factor is therefore the voltages across the various components in the output stage. Output stages get hot, and when the junction temperatures change, both experiment and simulation show that if Vbias is altered to maintain optimal crossover, Vq remains virtually constant. This confirms that the task of thermal compensation is solely to cancel out the Vbe changes in the transistors; this may appear to be a blinding glimpse of the obvious, but it was worth checking as there is no inherent reason why the optimal Vq should not be a function of device temperature. Fortunately it is not, for thermal compensation that also dealt with a need for Vq to change with temperature might be a good deal more complex.

Vq as the Critical Quiescent Parameter

The recognition that Vq is the critical parameter has some interesting implications. Can we immediately start setting up amplifiers for optimal crossover with a cheap DVM rather than an expensive THD analyser? Setting up quiescent current with a milliammeter has often been advocated, but the direct measurement of this current is not easy. It requires breaking the output circuit so a meter can be inserted, and not all amplifiers react favourably to so rude an intrusion. (The amplifier must also have near-zero DC offset voltage to get any accuracy.) Measuring the total amplifier consumption is not acceptable because the standing-current taken by the small-signal and driver sections will, in the CFP case at least, swamp the quiescent current. It is possible to determine quiescent current indirectly from the Vq drop across the Re's (still assuming zero DC offset) but this can never give a very accurate current reading as the tolerance of low-value Re's is unlikely to be better than ±10%.

However, if Vq is the real quantity we need to get at, then Re tolerances can be blissfully ignored. This does not make THD analysers obsolete overnight. It would be first necessary to show that Vq was always a reliable indicator of crossover setting, no matter what variations occurred in driver or output transistor parameters. This would be a sizeable undertaking.

There is also the difficulty that real-life DC offsets are not zero, though this could possibly be side-stepped by measuring Vq with the load disconnected.

A final objection is that without THD analysis and visual examination of the residual, you can never be sure an amplifier is free from parasitic oscillations and working properly.

I have previously demonstrated that the distortion behaviour of a typical amplifier is quite different when driving 4 Ω rather than 8 Ω loads. This is because with the heavier load, the output stage gain-behaviour tends to be dominated by beta-loss in the output devices at higher currents, and consequent extra loading on the drivers, giving third-harmonic distortion. If this is to be reduced, which may be well worthwhile as many loudspeaker loads have serious impedance dips, then it will need to be tackled in a completely different way from crossover distortion.

It is disappointing to find that no manipulation of output-stage component values appears to significantly improve crossover distortion, but apart from this one small piece of (negative) information gained, we have in addition determined that:

1. quiescent current as such does not matter; Vq is the vital quantity;
2. a perfect thermal compensation scheme, that was able to maintain Vq at exactly the correct value, requires no more information than the junction temperatures of the driver and output devices. Regrettably none of these temperatures are actually accessible, but at least we know what to aim for. The introduction of the Sanken and ONsemi ThermalTrak transistors with integral temperature-sense diodes (see Chapter 22)

opens possibilities in this direction but it remains to be seen how best to exploit this new technology.

As an aside, there is anecdotal evidence that back when transistors were made of germanium, crossover distortion was less of a problem because germanium transistors turn on more gradually. I have no idea if this is true or not, and making a germanium-device power amplifier nowadays is hardly practical, but it is an interesting point.

Switching Distortion (Distortion 3c)

This depends on several variables, notably the speed characteristics of the output devices and the output topology. Leaving aside the semi-conductor physics and concentrating on the topology, the critical factor is whether or not the output stage can reverse-bias the output device base-emitter junctions to maximise the speed at which carriers are sucked out, so the device is turned off quickly. The only conventional configuration that can reverse-bias the output base-emitter junctions is the EF Type II, described earlier.

A second influence is the value of the driver emitter or collector resistors; the lower they are, the faster the stored charge can be removed. Applying these criteria can reduce HF distortion markedly, but of equal importance is that it minimises overlap of output conduction at high frequencies, which, if unchecked, results in an inefficient and potentially destructive increase in supply current. To illustrate this, Figure 10.32 shows a graph

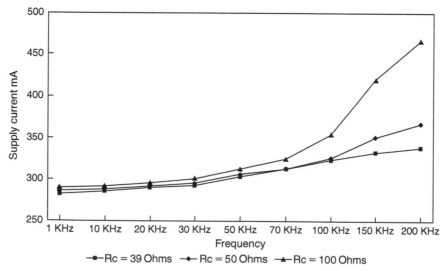

Figure 10.32. Power supply current versus frequency, for a CFP output with the driver collector resistors varied. There is little to be gained from reducing Rc below 50 Ω.

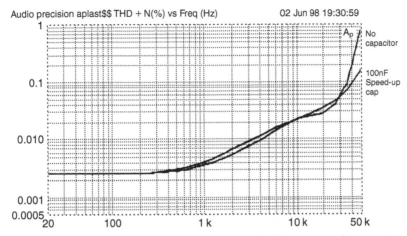

Figure 10.33. HF THD reduction by adding speed-up capacitance across the common driver resistance of a Type II EF output stage.

of current consumption versus frequency for varying driver collector resistance, for a CFP type output.

Figure 10.33 shows the reduction of HF THD by adding a speed-up capacitor across the common driver resistor of an EF Type II. At LF the difference is small, but at 40 kHz THD is halved, indicating much cleaner switch-off. There is also a small benefit over the range 300 Hz–8 kHz.

Thermal Distortion

Thermal distortion is that caused by cyclic temperature changes at signal frequency, causing corresponding modulation of device parameters. While it is certainly a real problem in IC opamps, which have input and output devices in very close thermal proximity, the situation in a normal discrete-component power amplifier is quite different, and thermal distortion cannot be detected. Having studied in detail distortion mechanisms that are all too real, it comes as some relief to find that one prospective distortion is illusory. Some writers appear to take it as given that such a distortion mechanism exists in power amplifiers, but having studied the subject in some depth, I have yet to see the effect, and quite frankly I do not think it exists.

While now and again there have been odd mentions of thermal distortion in power amps in some of the hi-fi press, you will never find:

1. any explanation of how it might work;
2. any estimate of the magnitude of the effect;
3. a circuit that will demonstrate its production.

In the usual absence of specific theories, one can only assume that the alleged mechanism induces parameter changes in semi-conductors whose power dissipation varies over a cycle. If this were to happen, it would presumably manifest itself as a rise in second or third harmonic distortion at very low frequencies, but this simply does not happen. The largest effects would be expected in Class-B output stages where dissipation varies wildly over a cycle; the effect is still wholly absent.

One reason for this may be that drivers and output devices have relatively large junctions with high thermal inertia — a few seconds with a hammer and chisel revealed that an MJE340 driver has a chip with four times the total area of a TL072. Given this thermal mass, parameters presumably cannot change much even at 10 Hz. Low frequencies are also where the global NFB factor is at its maximum; it is perfectly possible to design an amplifier with 100 dB of feedback at 10 Hz, though much more modest figures are sufficient to make distortion unmeasurably low up to 1 kHz or so. Using my design methodology, a Blameless amplifier can be straightforwardly designed to produce less than 0.0006% THD at 10 Hz (150 W/8 Ω) without even considering thermal distortion; this suggests that we have here a non-problem.

I accept that it is not uncommon to see amplifier THD plots that rise at low frequencies; but whenever I have been able to investigate this, the LF rise could be eliminated by attending to either defective decoupling or feedback-capacitor distortion. Any thermal distortion must be at a very low level as it is invisible at 0.0006%;

remember that this is the level of a THD reading that is visually pure noise, though there are real amplifier distortion products buried in it.

I have therefore done some deeper investigation by spectrum analysis of the residual, which enables the harmonics to be extracted from the noise. The test amplifier was an optimally biased Class-B design with a CFP output. The Audio Precision oscillator is very, very clean but this amplifier tests it to its limits, and so Table 10.4 shows harmonics in a before-and-after-amplifier comparison. The spectrum analyser bandwidth was 1 Hz for 10 Hz tests, and 4.5 Hz for 1 kHz, to discriminate against wideband noise.

This further peeling of the distortion onion shows several things; that the AP is a brilliant piece of machinery, and that the amplifier is really quite linear too. However, there is nothing resembling evidence for thermal distortion effects.

As a final argument, consider the distortion residual of a slightly under-biased power-amp, using a CFP output configuration so that output device junction temperatures do not affect the quiescent current; it therefore depends only on the driver temperatures. When the amplifier is switched on and begins to apply sinewave power to a load, the crossover spikes (generated by the deliberate underbiasing) will be seen to slowly shrink in height over a couple of minutes as the drivers warm up. This occurs even with the usual temperature compensation system, because of the delays and losses in heating up the Vbe-multiplier transistor.

The size of these crossover spikes gives in effect a continuous readout of driver temperature, and the slow variations that are seen imply time-constants measured in tens of seconds or more; this must mean a negligible response at 10 Hz.

There is no doubt that long-term thermal effects can alter Class-B amplifier distortion, because as I have written elsewhere, the quiescent current setting is critical for the lowest possible high-frequency THD. However, this is strictly a slow (several minutes) phenomenon, whereas enthusiasts for thermal distortion are thinking of the usual sort of per-cycle distortion.

The above arguments lead me to conclude that thermal distortion as usually described does not exist at a detectable level.

Thermal Distortion in a Power Amp IC

As explained above, thermal non-linearities would presumably appear as second or third harmonic distortion rising at low frequencies, and the largest effects should be in Class-B output stages where dissipation varies greatly over a cycle. There is absolutely no such effect to be seen in discrete-component power amplifiers.

But ... thermal distortion certainly does exist in IC power amplifiers. Figure 10.34 is a distortion plot for the Philips TDA 1522Q power amp IC, which I believe shows the effect. The power level was 4.4 W into 8 Ω, 8 W into 4 Ω. As is usual for such amplifiers, the distortion is generally high, but drops into a notch at 40 Hz; the only feasible explanation for this is

Table 10.4. Relative amplitude of distortion harmonic

	10 Hz APout (%)	Amp out (%)	1 kHz AP out (%)	Amp out (%)
Fundamental	0.00013	0.00031	0.00012	0.00035
Second	0.00033	0.00092	0.00008	0.00060
Third	0.00035	0.000050	0.000013	0.00024
Fourth	<0.000002	0.00035	<0.000008	0.00048
Fifth	<0.00025	<0.00045	0.000014	0.00024
Sixth	<0.000006	0.00030	0.000008	0.00021
Seventh	<0.000006	<0.00008	0.000009	0.00009
Eighth	<0.000003	0.000003	0.000008	0.00016
Ninth	<0.000004	0.00011	0.000007	<0.00008
AP THD reading (80 kHz bandwidth)	0.00046	0.00095	0.00060	0.00117

NoteB: The rejection of the fundamental is not perfect, and this is shown as it contributes to the THD figure.

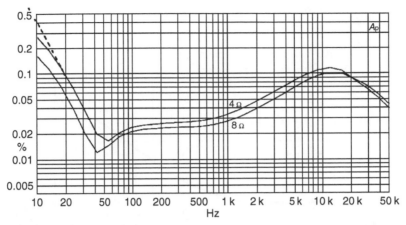

Figure 10.34. Distortion plot for the Philips TDA1522Q IC. Power out was 4.4W rms into 8 Ω, 8W rms into 4 Ω. The dotted line shows a 12 dB/octave slope.

cancellation of distortion products from two separate distortion sources. At frequencies below this notch, there is second-harmonic distortion rising at 12 dB/octave as frequency falls. The LF residual looks quite different from the midband distortion, which was a mixture of second and third harmonic plus crossover spikes.

The THD figure falls above 10 kHz because of the 80 kHz bandwidth limitation on the residual, and the high-order nature of the harmonics that make up crossover distortion.

All other possible sources of an LF distortion rise, such as inadequate decoupling, were excluded. There was no output capacitor to introduce non-linearity.

It seems pretty clear that the steep rise here is due to thermal distortion, in the form of feedback from the power output stage to earlier parts of the amplifier — probably the input stage. As would be expected, the effect is greater with a heavier load which causes more heating; in fact, halving the load doubles the THD reading for frequencies below the 40 Hz notch.

Closing the Loop: Distortion in Complete Amplifiers

In Chapters 6 and 7 it was shown how relatively simple design rules could ensure that the THD of the small-signal stages alone could be reduced to less than 0.001% across the audio band, in a thoroughly repeatable fashion, and without using frightening amounts of negative feedback. Combining this sub-system with one of the more linear output stages described in

Chapter 9, such as the CFP version which gives 0.014% THD open-loop, and bearing in mind that ample NFB is available, it seems we have all the ingredients for a virtually distortionless power amplifier. However, life is rarely so simple …

Figure 10.35 shows the distortion performance of such a closed-loop amplifier with an EF output stage, Figure 10.36 showing the same with a CFP output stage. Figure 10.37 shows the THD of a quasi-complementary stage with Baxandall diode. In each case Distortion One, Distortion Two, and Distortion Four to Distortion Seven have been eliminated, by methods described in past and future chapters, to make the amplifier Blameless.

(Note: the AP plots in Figures 10.35 to 10.37 were taken at 100 Wrms into 8 Ω, from an amplifier with an input error of −70 dB at 10 kHz and a C/L gain of 27 dB, giving a feedback factor of 43 dB at this frequency. This is well above the dominant pole frequency and so the NFB factor is dropping at 6 dB/octave and will be down to 37 dB at 20 kHz. My experience suggests that this is about as much feedback as is safe for general hi-fi usage, assuming an output inductor to improve stability with capacitive loads. Sadly, published data on this touchy topic seems to be non-existent.

It will be seen at once that these amplifiers are not distortionless, though the performance is markedly superior to the usual run of hardware. THD in the LF region is very low, well below a noise floor of 0.0007%, and the usual rise below 100 Hz is very small indeed. However, above 2 kHz, THD rises with frequency at between 6 and 12 dB/octave, and the

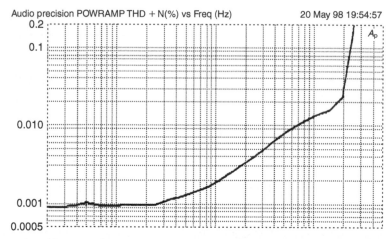

Figure 10.35. Closed-loop amplifier performance with Emitter-Follower output stage. 100 W into 8 Ω.

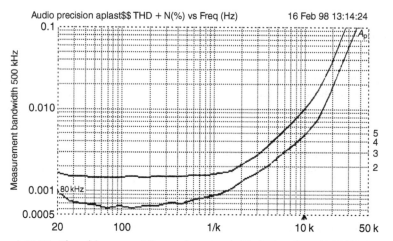

Figure 10.36. Closed-loop amplifier performance with CFP output. 100 W into 8.Ω.

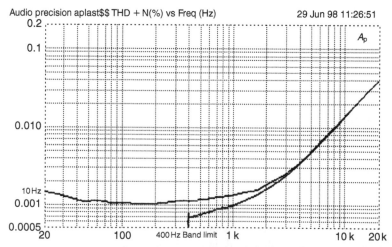

Figure 10.37. Closed-loop amplifier performance; quasi-complementary output stage with Baxandall diode. 100 W into 8 Ω.

Table 10.5. Summary of closed-loop amp performance

	1kHz (%)	10kHz (%)
EF	0.0019	0.013
CFP	0.0008	0.005
Quasi Bax	0.0015	0.015

distortion residual in this region is clearly time-aligned with the crossover region, and consists of high-order harmonics rather than second or third. It is intriguing to note that the quasi-Baxandall output gives about the same HF THD as the EF topology, which confirms my earlier statement that the addition of a Baxandall diode essentially turns a conventional quasi-complementary stage with serious crossover asymmetry into a reasonable emulation of a complementary EF stage. There is less HF THD with a CFP output; this cannot be due to large-signal non-linearity as this is negligible with an 8 Ω load for all three stages, and so it must be due to high-order crossover products. (See Table 10.5.)

The distortion figures given in this book are rather lower than usual. I would like to emphasise that these are not freakish or unrepeatable figures; they are the result of attending to all of the major sources of distortion, rather than just one or two. I have at the time of writing personally built 12 models of the CFP version, and performance showed little variation.

Here the closed-loop distortion is much greater than that produced by the small-signal stages alone; however, if the input pair is badly designed, its HF distortion can easily exceed that caused by the output stage.

Our feedback-factor here is a minimum of 70× across the band (being much higher at LF) and the output stages examined above are mostly capable of less than 0.1% THD open-loop. It seems a combination of these should yield a closed-loop distortion at least 70 times better, i.e., below 0.001% from 10 Hz to 20 kHz. This happy outcome fails to materialise, and we had better find out why …

First, when an amplifier with a frequency-dependent NFB factor generates distortion, the reduction is not that due to the NFB factor at the fundamental frequency, but the amount available at the frequency of the harmonic in question. A typical amplifier with open-loop gain rolling-off at 6 dB/octave will be half as effective at reducing fourth-harmonic distortion as it is at reducing the second harmonic. LSN is largely third (and possibly second) harmonic, and so NFB will deal with this effectively. However, both crossover and switchoff distortions generate high-order harmonics significant up to at least the nineteenth and these receive much less linearisation. As the fundamental moves up in frequency, the harmonics do too, and benefit from even less feedback. This is the reason for the differentiated look to many distortion residuals; higher harmonics are emphasised at the rate of 6 db/octave.

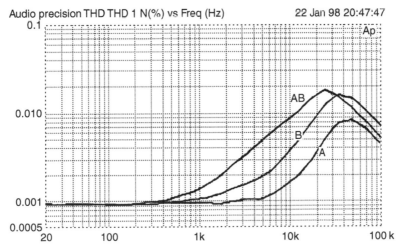

Figure 10.38. Closed-loop amplifier performance CFP output stage. Setting quiescent for Class-AB gives more HF THD than either Class- A or B.

Here is a real example of the inability of NFB to cure all possible amplifier ills. To reduce this HF distortion, we must reduce the crossover gain-deviations of the output stage before closing the loop. There seems no obvious way to do this by minor modifications to any of the conventional output stages; we can only optimise the quiescent current.

As I have stated many times, Class-AB is generally not a Good Thing, as it gives more distortion than Class-B, rather than less, and so will not help us. Figure 10.38 makes this very clear for the closed-loop case; Class-AB clearly gives the worst performance. (As before, the AB quiescent was set for 50:50 m/s ratio of the gm-doubling artefacts on the residual.)

Reference

1. Blomley P, New Approach to Class-B, *Wireless World*, Feb. 1971, p. 57 and March 1971, pp. 127–131.

Chapter 11

More Distortion Mechanisms

New insight begins when satisfaction comes to an end, when all that has been seen, said, or done looks like a distortion.

Abraham Heschel

Distortion Four: VAS Loading Distortion

Distortion Four is that which results from the loading of the Voltage-Amplifier Stage (VAS) by the non-linear input impedance of a Class-B output stage. This was looked at in Chapter 7 from the point of view of the VAS, where it was shown that since the VAS provides all the voltage gain, its collector impedance tends to be high. This renders it vulnerable to non-linear loading unless it is buffered or otherwise protected.

The VAS is routinely (though usually unknowingly) linearised by applying local negative feedback via the dominant pole Miller capacitor Cdom, and this is a powerful argument against any other form of compensation. If VAS distortion still adds significantly to the amplifier total, then the local open-loop gain of the VAS stage can be raised to increase the local feedback factor. The obvious method is to raise the impedance at the VAS collector, and so the gain, by cascoding. However, if this is done without buffering, the output stage loading will render the cascoding almost completely ineffective. Using a VAS buffer eliminates this problem.

As explained in Chapter 7, the VAS collector impedance, while high at LF compared with other circuit nodes, falls with frequency as soon as Cdom takes effect, and so Distortion Four is usually only visible at LF. It is also often masked by the increase in output stage distortion above dominant pole frequency P1 as the amount of global NFB reduces.

The fall in VAS impedance with frequency is demonstrated in Figure 11.1, obtained from the SPICE conceptual model in Chapter 7, but with values appropriate to real-life components; the input stage transconductance is set at 3 mA/V, and the VAS beta is assumed to be constant at 350. The LF impedance is basically that of the VAS collector resistance, but halves with each octave once P1 is reached. By 3 kHz the impedance is down to 1 kΩ, and still falling. Nevertheless, it usually remains high enough for the input impedance of a Class-B output stage to significantly degrade linearity, the actual effect being shown in Figure 11.2.

In Chapter 7, it was shown that as an alternative to cascoding, an effective means of linearising the VAS is to add an emitter-follower within the VAS local feedback loop. As well as good VAS linearity, this establishes a much lower VAS collector impedance across

the audio band, and is much more resistant to Distortion Four than the cascode version. VAS buffering is not essential, so this method has a lower component count.

The question remains as to whether, even so, buffering an EF-VAS may give some benefit in terms of lower distortion. Adding a unity-gain buffer between the VAS and the output stage gives the interesting possibility of taking the feed to the Miller compensation capacitor from three possible circuit nodes:

1. The VAS collector, as usual.
2. The output of the unity-gain buffer.
3. Partly from the output stage, to obtain output-inclusive compensation (see Chapter 13).

This experiment has not yet been performed; I expect the improvement, if any, will be small.

Figure 11.3 confirms that the input impedance of a conventional EF Type I output stage is highly non-linear; the data is derived from a SPICE output stage simulation with optimal *I*q. Even with an undemanding 8 Ω load, the impedance varies by 10:1 over the output voltage swing. The Type II EF output (using a shared drive emitter resistance) has a 50% higher impedance around crossover, but the variation ratio is rather greater. CFP output stages have a more complex variation that includes a precipitous drop to less than 20 kΩ around the crossover point. With all types, under-biasing produces additional sharp impedance changes at crossover.

This is only a brief summary of the VAS loading issue, which is actually quite complex; more details can be found in Chapter 7 on the VAS. At present it appears that if an EF-VAS is used, the effect of loading on the distortion performance is very small. This approach is recommended rather than cascoding.

Distortion Five: Rail Decoupling Distortion

Almost all amplifiers have some form of rail decoupling apart from the main reservoir capacitors; this is usually required to guarantee HF stability. Standard decoupling arrangements include small to medium-sized electrolytics (say, 10–470 μF) connected between each rail and ground, and an inevitable consequence is that rail-voltage variations cause the current to flow into the ground connection chosen. This is just one mechanism that defines the Power-Supply Rejection Ratio (PSRR) of an amplifier, but it is one that can seriously damage linearity.

If we use an unregulated power supply (and there are almost overwhelming reasons for using such a supply, detailed in Chapter 26), comprising transformer, bridge rectifier, and reservoir capacitors, then these

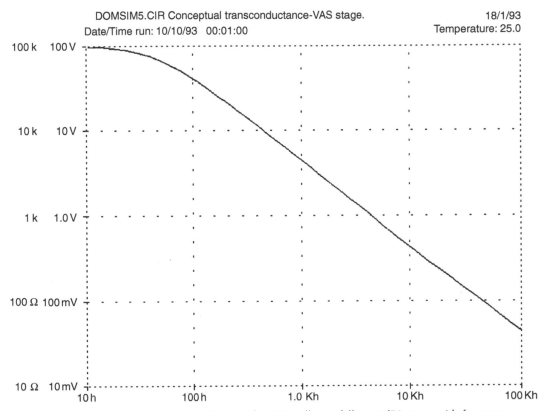

Figure 11.1. Distortion 4; the impedance at the VAS collector falls at 6 dB/octave with frequency.

rails have a non-zero AC impedance and their voltage variations will be due to amplifier load currents as well as 100 Hz ripple. In Class-B, the supply-rail currents are half-wave-rectified sine pulses with strong harmonic content, and if they contaminate the signal, then distortion is badly degraded; a common route for interaction is via decoupling grounds shared with input or feedback networks, and a separate decoupler ground is usually a complete cure. This point is easy to overlook, and attempts to improve amplifier linearity by labouring on the input pair, VAS, etc., are doomed to failure unless this distortion mechanism is eliminated first. As a rule it is simply necessary to take the decoupling ground separately back to the ground star-point, as shown in Figure 11.4. (Note that the star-point A is defined on a short spur from the heavy connection joining the reservoirs; trying to use B as the star-point will introduce ripple due to the large reservoir-charging current pulses passing through it.)

Figure 11.5 shows the effect on an otherwise Blameless amplifier handling 60 W/8 Ω, with 220 µF rail decoupling capacitors; at 1 kHz, distortion has increased by more than ten times, which is quite bad enough.

However, at 20 Hz, the THD has increased at least 100-fold, turning a very good amplifier into a profoundly mediocre one with one misconceived connection.

When the waveform on the supply rails is examined, the 100 Hz ripple amplitude will usually be found to exceed the pulses due to Class-B signal current, and so some of the *distortion* on the upper curve of the plot is actually due to ripple injection. This is hinted at by the phase-crevasse at 100 Hz, where the ripple happened to partly cancel the signal at the instant of measurement. Below 100 Hz, the curve rises as greater demands are made on the reservoirs, the signal voltage on the rails increases, and more distorted current is forced into the ground system.

Figure 11.6 shows a typical Distortion Five residual, produced by deliberately connecting the negative supply-rail decoupling capacitor to the input ground instead of properly giving it its own return to the far side of the star-point. THD increased from 0.00097% to 0.008%, appearing mostly as second harmonic. Distortion Five is usually easy to identify as it is accompanied by 100 Hz power-supply ripple; Distortions Six and Seven introduce no extra ripple. The ripple

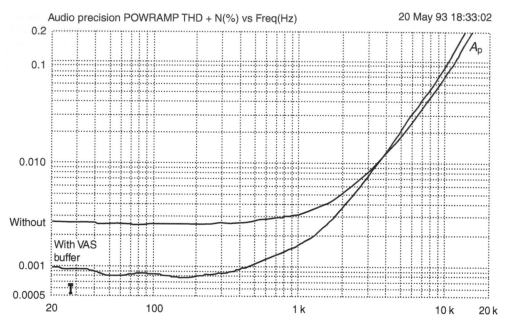

Figure 11.2. Distortion 4 afflicting a simple VAS; the lower trace shows the result of its elimination by the use of a VAS-buffer.

contamination here — the two humps at the bottom — is significant and contributes to the THD reading.

As a general rule, if an amplifier is made free from ripple injection under drive conditions, demonstrated by a THD residual without ripple components, there will be no distortion from the power-supply rails, and the complications and inefficiencies of high-current rail regulators are quite unnecessary.

There has been much discussion of PSRR-induced distortion in the literature recently, e.g., Greg Ball.[1] I part company with some writers at the point where they assume a power amplifier is likely to have 25 dB PSRR, making an expensive set of HT regulators the only answer. Greg Ball also initially assumes that a power amp has the same PSRR characteristics as an opamp, i.e., falling steadily at 6 dB/octave. There is absolutely no need for this to be so, given a little RC decoupling, and Ball states at the end of his article that 'a more elegant solution ... is to depend on a high PSRR in the amplifier proper'. Quite so. This issue is dealt with in detail in Chapter 26.

Distortion Six: Induction Distortion

The existence of this distortion mechanism, like Distortion Five, stems directly from the Class-B nature of the output stage. With a sine input, the output hopefully carries a good sinewave, but the supply-rail currents are half-wave-rectified sine pulses, which will readily crosstalk into sensitive parts of the circuit by induction. This is very damaging to the distortion performance, as Figure 11.7 shows.

The distortion signal may intrude into the input circuitry, the feedback path, the output inductor, or even the cables to the output terminals, in order of decreasing sensitivity. The result is a kind of sawtooth on the distortion residual that is very distinctive, and causes the THD to rise at 6 dB/octave with frequency. The induced distortion voltage in any part of the circuit is proportional to the rate of change of magnetic flux, and that is why it increases proportionally with frequency, and is inversely proportional to the amplifier load impedance.

Because of its multi-turn structure the output coil is quite sensitive to magnetic fields and it is a good idea to keep it as far away from the output devices as possible.

A Distortion Six residual is displayed in Figure 11.8. The V-supply rail was routed parallel to the negative feedback line to produce this diagram. THD is more than doubled, but is still relatively low at 0.0021%. 64-times averaging is used. Distortion Six is easily identified if

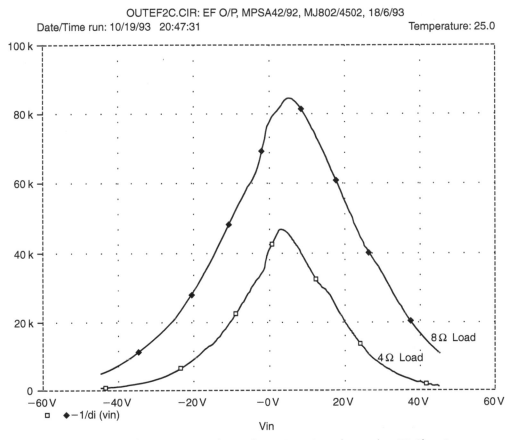

OUTEF2C.CIR: EF O/P, MPSA42/92, MJ802/4502, 18/6/93

Figure 11.3. Distortion 4 and its root cause; the nonlinear input impedance of an EF Class-B output stage.

the DC supply cables are movable, for altering their run will strongly affect the quantity generated.

This inductive effect appears to have been first formally publicised by Cherry[2] in a paper that deserves more attention. The effect has, however, been recognised and avoided by some practitioners for many years.[3] Nonetheless, having examined many power amplifiers with varying degrees of virtue, I feel that this effect could be better known, and is probably the most widespread cause of unnecessary distortion.

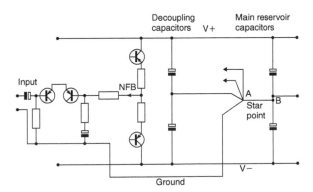

Figure 11.4. Distortion 5; the correct way to route decouple grounding to the star-point.

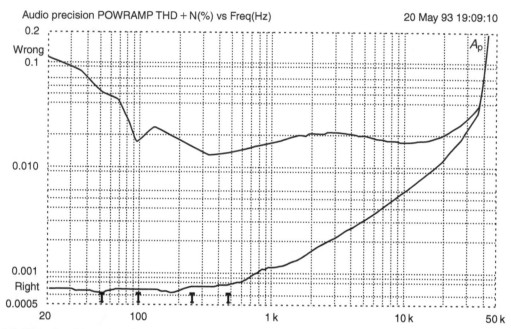

Figure 11.5. Distortion 5 in action; the upper trace was produced simply by taking the decoupler ground from the star-point and connecting it via the input ground line instead.

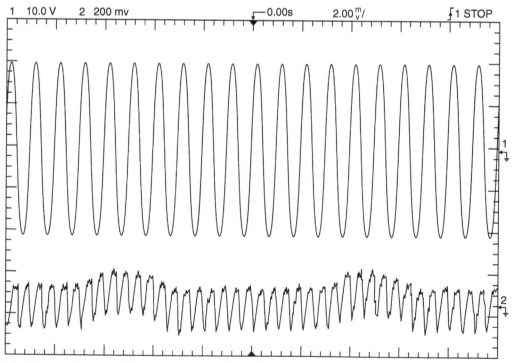

Figure 11.6. Distortion 5 revealed. Connecting the rail decoupler to input ground increases THD eight-fold from 0.00097% to 0.008%, mostly as second harmonic. 100 Hz ripple is also visible. No averaging.

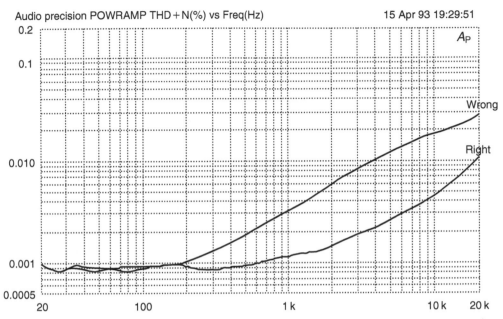

Figure 11.7. Distortion 6 exposed. The upper trace shows the effects of Class-B rail induction into signal circuitry.

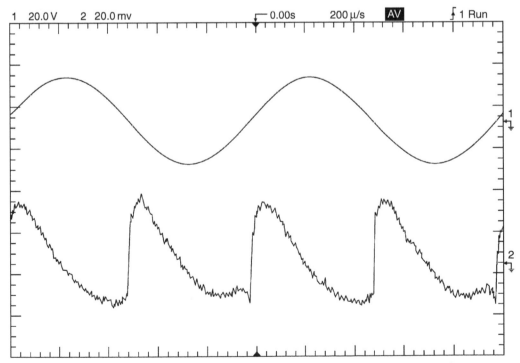

Figure 11.8. Distortion 6. Induction of half-wave signal from the negative supply rail into the NFB line increases THD to 0.0021%. Averaged 64 times. This should not be confused with crossover distortion.

The contribution of Distortion Six can be reduced below the measurement threshold by taking sufficient care over the layout of supply-rail cabling relative to signal leads, and avoiding loops that will induce or pick up magnetic fields. I wish I could give precise rules for layout that would guarantee freedom from the problem, but each amplifier has its own physical layout, and the cabling topology has to take this into account. However, here are some guidelines:

First, implement rigorous minimisation of loop area in the input and feedback circuitry; keeping each signal line as close to its ground return as possible. Keep the output coil away from the output stage if you can.

Second, minimise the ability of the supply wiring to establish magnetic fields in the first place.

Third, put as much distance between these two areas as you can. Fresh air beats metal shielding on price every time.

Figure 11.9a shows one straightforward approach to solving the problem; the supply and ground wires are tightly twisted together to reduce radiation. In practice this does not seem to be very effective, for reasons that are not wholly clear, but seem to involve the difficulty of ensuring exactly equal coupling between three twisted conductors. In Figure 11.9b, the supply rails are

twisted together but kept well away from the ground return; this will allow field generation, but if the currents in the two rails butt together to make a nice sinewave at the output, then they should do the same when the magnetic fields from each rail sum. There is an obvious risk of interchannel crosstalk if this approach is used in a stereo amplifier, but it does deal effectively with the induced-distortion problem in some layouts.

It is difficult to over-emphasise the importance of keeping a good look-out for this form of distortion when evaluating prototype amplifiers; trying to remove it by any method other than correcting the physical layout is quite futile, and I cannot help wondering how many unhappy man-hours have been spent trying to do just that. The sawtooth-like distortion residual is a dead give-away; another simple test is to move around the power supply cables if it is possible to do so, and see if the distortion residual varies. The output inductor is inherently fairly sensitive to unwanted magnetic fields, and you may have to change its orientation to avoid picking them up. There is no need for the inductor to be snuggled up to the output devices, and in fact it is a good idea to position it more towards the small-signal circuitry; because of the low closed-loop gain there is no possibility of feedback from output to

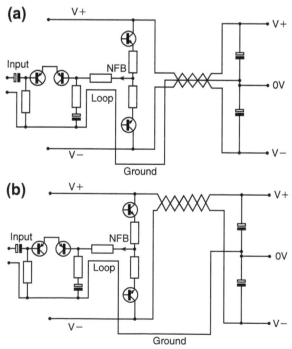

Figure 11.9. Distortion 6; countermeasures against the induction of distortion from the supply rails. 11.9b is usually the more effective.

input. In a recent case, an experimental amplifier was giving an excessive 0.0075% THD at 10 kHz (25 W/8 Ω) and squashing the output coil flat with an authoritative thumb reduced this at once to a fairly Blameless 0.0026%.

In cases of difficulty with induction distortion, a powerful tool is a small search coil connected to an audio analyser input (a spare output inductor works very well for this); it can be moved around to look for unsuspected current paths carrying half-wave-rectified sine pulses. This distortion mechanism does not, of course, trouble Class-A amplifiers.

Distortion Seven: NFB Takeoff Point Distortion

It has become a tired old truism that negative feedback is a powerful technique, and like all such, must be used with care if you are to avoid tweeter-frying HF instability.

However, there is another and much more subtle trap in applying global NFB. Class-B output stages are a maelstrom of high-amplitude half-wave-rectified currents, and if the feedback takeoff point is in slightly the wrong place, these currents contaminate the feedback signal, making it an inaccurate representation of the output voltage, and hence introducing distortion; Figure 11.10 shows the problem. At the current levels in question, all wires and PCB tracks must be treated

as resistances, and it follows that point C is not at the same potential as point D whenever TR1 conducts. If feedback is taken from D, then a clean signal will be established here, but the signal at output point C will have a half-wave rectified sinewave added to it, due to the resistance C–D. The actual output will be distorted but the feedback loop will do nothing about it as it does not know about the error.

Figure 11.11 shows the practical result for an amplifier driving 100 W into 8 Ω, with the extra distortion interestingly shadowing the original curve as it rises with frequency. The resistive path C–D that did the damage was a mere 6 mm length of heavy-gauge wire-wound resistor lead.

Figure 11.12 shows a THD residual for Distortion Seven, introduced by deliberately taking the NFB from the wrong point. The THD rose from 0.00097% to 0.0027%, simply because the NFB feed was taken from the wrong end of the leg of one of the output emitter resistors Re. Note this is not the wrong side of the resistor, or the distortion would have been gross, but a mere 10 mm along a very thick resistor leg from the actual output junction point.

Of the linearity problems that afflict generic Class-B power amplifiers, Distortions Five, Six and Seven all look rather similar when they appear in the THD residual, which is perhaps not surprising since all result from adding half-wave disturbances to the

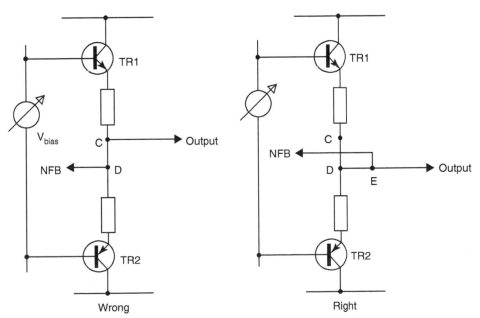

Figure 11.10. Distortion 7; wrong and right ways of arranging the critical negative-feedback takeoff point.

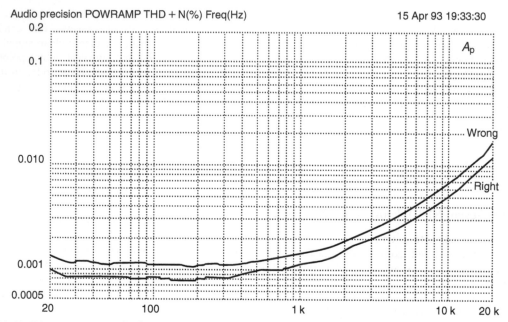

Figure 11.11. Distortion 7 at work; the upper (WRONG) trace shows the result of a mere 6 mm of heavy-gauge wire between the about output and the feedback point.

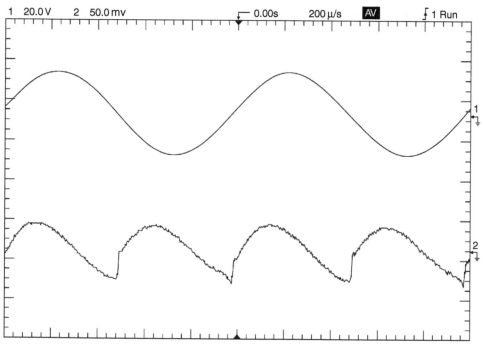

Figure 11.12. Distortion 7, caused by choosing an NFB takeoff point inside the Class-B output stage rather than on the output line itself. THD is increased from 0.00097% to 0.0027%, by taking the NFB from the wrong end of 10 mm of very thick resistor leg. Averaged 64 times.

signal. All show sharp edges on the residual at the cross-over point, like those in Figure 11.8, but these should not be confused with crossover distortion, which does not have the same sawtooth look.

To eliminate this distortion is easy, once you are alert to the danger. Taking the NFB feed from D is not advisable as D is not a mathematical point, but has a physical extent, inside which the current distribution is unknown. Point E on the output line is the right place, as the half-wave currents do not flow through this arm of the circuit. The correct output point E should also be used for the output-coil damping resistor, to prevent the introduction of distortion at high frequencies when the coil reactance is significant. The same precaution should be taken with the Zobel network which might introduce distortion into the ground at high frequencies.

Distortion Eight: Capacitor Distortion

When I wrote the original series on amplifier distortion,[4] I listed seven types of distortion that defined an amplifier's linearity. The number has since grown, and Distortion Eight refers to capacitor distortion. This has nothing to do with Subjectivist hypotheses about mysterious non-measurable effects; this phenomenon is all too real, though for some reason it seems to be almost unknown — or at any rate not talked about — among audio designers. Clearly this is the distortion that dare not speak its name.

It is, however, a sad fact that both electrolytic and non-electrolytic capacitors generate distortion whenever they are used in such a fashion that a significant AC voltage develops across them.

Standard aluminium electrolytics create distortion when they are used for coupling and DC blocking while driving a significant resistive load. Figure 11.13 is the test circuit; Figure 11.14 shows the resulting distortion for a 47 µF 25 V capacitor driving +20 dBm (7.75V rms) into a 680 Ω load, while Figure 11.15 shows how the associated LF roll-off has barely begun. The distortion is a mixture of second and third harmonics, and rises rapidly as frequency falls, at something between 12 and 18 dB/octave.

The great danger of this mechanism is that serious distortion begins while the response roll-off is barely detectable; here the THD reaches 0.01% when the response has only fallen by 0.2 dB. The voltage across the capacitor is 2.6 V peak, and this voltage is a better warning of danger than the degree of roll-off.

Further tests showed that the distortion roughly triples as the applied voltage doubles; this factor seems to vary somewhat between different capacitor rated voltages.

The mechanism by which capacitors generate this distortion is unclear. Dielectric absorption appears to be ruled out as this is invariably (and therefore presumably successfully) modelled by adding linear components, in the shape of resistors and capacitors, to the basic capacitor model. Reverse-biasing is not the problem, for capacitors DC biased by up to +15 V show slightly increased, not reduced distortion. Non-polarised electrolytics show the same effect but at a much greater AC voltage, typically giving the same distortion at one-tenth the frequency of a conventional capacitor with the same time-constant; the cost and size of these components generally rule out their use to combat this effect. Usually the best solution is simply to keep increasing the capacitor value until the LF distortion rise disappears off the left of the THD graph. Negligible roll-off in the audio band is not a sufficient criterion.

Electrolytics are therefore best reserved for DC filtering, and for signal coupling where the AC voltage across them will be negligible. If a coupling capacitor

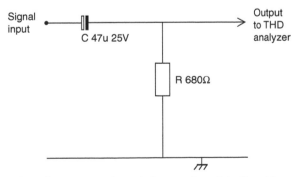

Figure 11.13. A very simple circuit to demonstrate electrolytic capacitor distortion. Measurable distortion begins at 100 Hz.

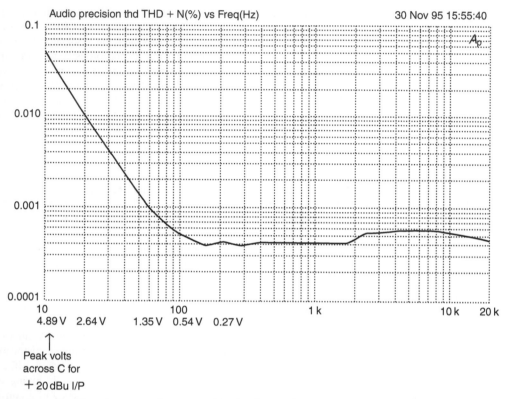

Figure 11.14. Capacitor distortion versus frequency, showing the rapid rise in THD once the distortion threshold is reached.

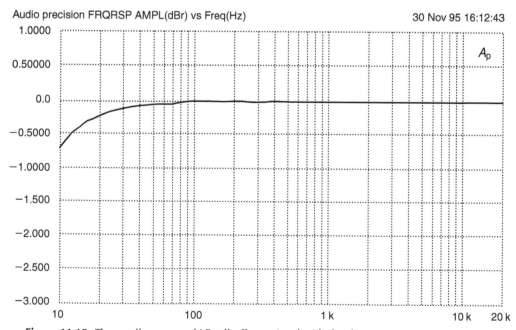

Figure 11.15. The small amount of LF roll-off associated with the distortion rise in Figure 11.14.

does have AC voltage across it, and drives the usual resistive load, then it must be acting as a high-pass filter. This is never good design practice, because electrolytics have large tolerances and make inaccurate filters; it is now clear they generate distortion as well.

It is therefore most undesirable to define the lower bandwidth limit simply by relying on the high-pass action of electrolytics and circuit resistances; it should be done with a non-electrolytic capacitor, made as large as possible economically in order to reduce the value of the associated resistance and so keep down circuit impedances, thus minimising the danger of noise and crosstalk.

Capacitor distortion in power amplifiers is most likely to occur in the feedback network blocking capacitor, assuming it is a DC-coupled amplifier; if it is AC-coupled, the output capacitor may generate serious distortion, as described in Chapter 4. The input blocking capacitor usually feeds a high impedance, but the feedback arm must have the lowest possible resistances to minimise both noise and DC offset. The feedback capacitor therefore tends to be relatively large, and if it is not quite large enough, the THD plot of the amplifier will show the characteristic kick up at the LF end. An example of this is dealt with in detail in Chapter 6.

It is common for amplifiers to show a rise in distortion at the LF end, but there is no reason why this should ever occur. Capacitor distortion is usually the reason, but Distortion Five (Rail Decoupling Distortion) can also contribute. These two mechanisms can be distinguished because Distortion Five typically rises by only 6 dB/octave as frequency decreases, rather than the 12−18 dB/octave of capacitor distortion.

Amplifiers with AC-coupled outputs are now fairly rare, and one reason may be that distortion in the output capacitor is a major problem, occurring in the mid-band as well as at LF. The reason for this mid-band problem is not obvious; probably it is due to the much higher levels of current passing through the output capacitor activating distortion mechanisms that are not otherwise visible. If an amplifier is driving 50 W into an 8 Ω load, and has a feedback resistor of 2K2 (which is probably about as low as is likely), then the peak current through the output capacitor will be 3.5 A, while the peak current through the feedback capacitor at the bottom of the feedback network is only 12.7 mA. See the section on AC-coupled amplifiers in Chapter 4 for more details of output capacitor distortion.

Non-electrolytic capacitors of middling value (say, 10 nF to 470 nF) with a polyester dielectric (the cheapest and most common type) also generate distortion when operated with significant signal voltages across them. This typically occurs when they are used to realise time-constants in filter and equalisation circuits, and this is not relevant to power amplifier design. Such capacitors may appear in DC servo circuitry, but there the signal voltage across them is very small because of the integrator action. The use of polypropylene capacitors instead eliminates the effect, but cost and component bulk are greater.

However, the linearity of non-electrolytic capacitors of small value (say, 10 pF to 220 pF) is very much of interest to the designer, as they are used for compensation and RF filtering purposes. This is of particular relevance to the capacitor Cdom used for dominant pole Miller compensation, which stabilises the overall feedback loop by converting global feedback into local feedback around the VAS transistor. It has the full output voltage of the amplifier impressed across it, and it is therefore vital that it is a completely linear component.

Its size, usually around 100 pF, means that it will almost certainly be a ceramic type. It is essential that a type with C0G or NP0 dielectric is used. These have the lowest capacitance/temperature dependence (NP0 stands for Negative-Positive zero), and the lowest losses, but for our purposes the important point is that they have the lowest capacitance/voltage coefficients. It is generally known that ceramics with X7R dielectrics have large capacitance/voltage coefficients and are quite unsuitable for any application where linearity matters; their value is that they pack a lot of capacitance into a small space, which makes them valuable for decoupling jobs. In general, all that is required is to specify a C0G or NP0 type; but this can go wrong, as I will now relate.

Figure 11.16 shows the THD plot for a Blameless amplifier delivering 180 W into 8 Ω. It had the usual mirrored input pair, an emitter-follower-enhanced VAS, and the EF output configuration with three pairs of output devices in parallel. This multiple-output approach can give excellent distortion performance, as shown in the lower trace. What, however, we actually got was the upper trace; between 1 kHz and 20 kHz there is about three times more distortion than there should be. Given that the product was on the very threshold of mass production, there was alarm, consternation, and worse. The culprit was quickly shown to be the 100 pF dominant pole Miller compensation capacitor, a Chinese-sourced component that in theory, but not in practice, was an NP0 part. Replacing it with an identically specified part from a more reputable Chinese manufacturer cured the problem at once, yielding the expected lower trace in Figure 11.16.

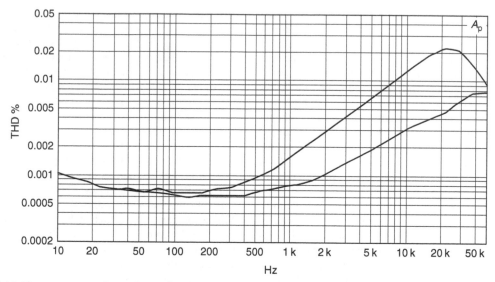

Figure 11.16. The upper trace shows the excess distortion generated by a substandard NP0 Miller compensation capacitor. The lower trace is the result with a good component (180 W into 8 Ω).

There are several interesting points here; the extra distortion was fairly pure second harmonic, and as the plot shows, the amount is rising at a steady 6dB/octave. 'Bad NP0 distortion' is here shown in action for the first time, I believe. Note that the capacitor was not an X7R type by mistake — if that had been the case, the distortion would have been gross. What we had was an attempt at making an NP0 capacitor that failed.

Distortion Nine: Magnetic Distortion

This arises when a signal at amplifier output level is passed through a ferromagnetic conductor. Ferromagnetic materials have a non-linear relationship between the current passing through them and the magnetic flux it creates, and this induces voltages that add distortion to the signal. The effect has been found in some types of output relays where the signal being switched passes through the soft-iron frame that makes up part of the magnetic circuit. That particular manifestation is dealt with in detail in Chapter 24, where output relays are examined.

The problem has also been experienced with loudspeaker terminals. The terminal pair in question was a classy-looking Chinese item with all its metal parts gold-plated, and had proved wholly satisfactory at the prototype stage. Once again, the product involved was trembling on the brink of mass production, and, once again, the pre-production batch showed more distortion

than expected. The THD residual showed third harmonic distortion that had certainly not been there before. Some rapid investigation revealed the hitherto unknown concept of non-linear loudspeaker terminals. The metal parts of the terminals appeared to be made of gold-plated brass (as they were in all the prototype samples) but were actually gold-plated steel, which is of course a cheaper material — brass has copper in it, and copper is expensive. Although the amplifier output currents were only passing through about 10 mm of steel (the current went through that length twice, on go and return), the non-linear magnetic effects were sufficient to increase the output distortion from 0.00120% to 0.00227% at 100 W into 8 Ω at 1 kHz. In other words, distortion nearly doubled. It is, however, highly likely that if the offending terminals had been used with a non-Blameless amplifier having rather more distortion of its own, the extra non-linearity would have gone completely unnoticed, and I can only presume that this was what the manufacturer hoped and expected. Parts incorrectly made from steel can of course be readily detected by the application of a small magnet.

It might be thought that ferromagnetic distortion might be most likely to affect the only part of the signal path that is deliberately inductive — the output coils. Amplifier chassis are very often made of steel, and the output coil is usually close to a large ferromagnetic component in the shape of the output relay.

While it is certainly good practice to keep the output coil away from ferrous metals as far as practical, in fact, there is very little to worry about; the effect of adjacent steel or iron parts on the coil is not as large as you might think. To put it into perspective, a little experiment was performed. A Blameless power amplifier driving 115 W into an 8 Ω load was yielding 0.00080% THD at 1 kHz. Inserting a steel screwdriver shaft 6 mm in diameter into the output coil, which consisted of 10 turns of heavy copper wire 24 mm in diameter, only degraded the THD to 0.0094%, which, while clearly undesirable, is not exactly a dramatic change. When the screwdriver shaft was replaced with a complete small signal relay tucked wholly inside the coil, the THD only worsened slightly to 0.0011%. The effect across the audio frequency band is seen in Figure 11.17; the worst effect is at about 7 kHz, where 0.006% is degraded to 0.010%. These tests put gross amounts of ferromagnetic material right inside the coil, so it is safe to assume (and, I hasten to add, further experiments prove) that metal chassis sections some centimetres away from the coil are going to have no detectable effect.

Some further tests showed that mounting an output relay (which contains substantially more ferrous metal than the small-signal relay alluded to above) so that the end of the output coil was in contact with the plastic relay casing caused no detectable degradation from 0.00080% THD under the same conditions. However, not everybody seems prepared to believe this, and there is a wide consensus that it 'looks wrong', so it's best avoided if humanly possible.

Other output coil issues – such as the crosstalk between two coils in a stereo amplifier – are dealt with in Chapter 14.

I don't want you to think that I am prejudiced against Chinese electronic components. I have used them extensively, and providing you take due care with suppliers, there are few difficulties. The worst problems I have had with components – none of them Chinese – were thus:

1. Defective electrolytic capacitors that generated their own DC voltage. Short them out, and it would disappear; remove the short and it would slowly return, like dielectric absorption only much, much worse. Result – big mixing consoles where every switch clicked when operated. Not good.
2. Batches of IC power amplifiers that died after a few weeks of normal domestic use. This caused mayhem. Every possible design-based reason was investigated, without result, and it took the manufacturer (the very well-known, apparently thoroughly reputable manufacturer) something like nine months to admit that they had made a large batch of thoroughly defective ICs.
3. IC voltage regulators with non-functional overload protection. The application was a power supply that could quite easily be short-circuited by the user, so it

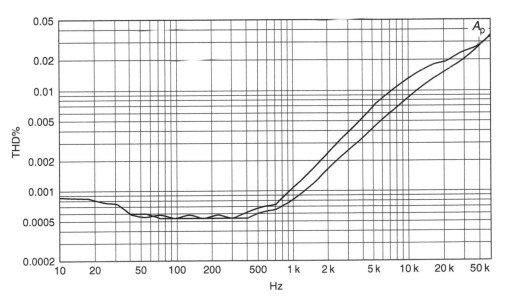

Figure 11.17. The not very dramatic effect of placing a complete relay inside the output coil (115 W into 8 Ω).

did matter. The manufacturer's response was not to offer to replace the parts, but to fly in a team of four people from Another European Country to convince us that we didn't *really* need overload protection after all. I need hardly say we remained unconvinced, and years after the event I'm still wondering about the mental state of whoever decided that was an appropriate reaction to the problem.

I won't tell you the manufacturers involved, as this might turn historical technical problems into contemporary legal ones, but the capacitors came from Japan and the ICs both came from very big Western semiconductor manufacturers.

Distortion Ten: Input Current Distortion

This distortion is caused when an amplifier input is driven from a significant source impedance. The input current taken by the amplifier is non-linear, even if the output of the amplifier is distortion-free, and the resulting voltage-drop in the source impedance introduces distortion.

This mechanism is dealt with in detail in Chapter 6, as it relates closely to the design of the amplifier input stage.

Distortion Eleven: Premature Overload Protection

The most common method of overload protection of a power amplifier is the use of VI limiters that shunt signal current away from the inputs to the output stage. In their most common form, these come into operation relatively gradually as their threshold is exceeded, and start introducing distortion into the signal long before they close it down entirely. This problem is made more serious because the simplest and most used VI limiter circuits show significant temperature sensitivity, coming into action sooner as they warm up in the internal environment of the amplifier. It is therefore vital to design an adequate safety margin into the output stage so that the VI limiters need never be near activation during normal use. This issue is examined more closely in Chapter 24.

Design Example: a 50 W Class-B Amplifier

Figure 11.18 shows a design example of a Class-B amplifier, intended for domestic hi-fi applications. Despite its relatively conventional appearance, the circuit parameters selected give much better than

a conventional distortion performance; this is potentially a Blameless design, but only if due care is given to wiring topology and physical layout will this be achieved.

With the supply voltages and values shown, it gives 50 W into 8 Ω, for 1 V rms input. In earlier chapters, I have used the word *Blameless* to describe amplifiers in which all distortion mechanisms, except the apparently unavoidable ones due to Class-B, have been rendered negligible. This circuit has the potential to be Blameless (as do we all), but achieving this depends on care in cabling and layout. It does not aim to be a cookbook project; for example, overcurrent and DC-offset protection are omitted.

In Chapter 21, power FETs amplifiers are examined, and the conclusion drawn that they are disappointingly expensive, inefficient, and non-linear. Therefore, bipolars it is. The best BJT configurations were the emitter-follower Type II, with least output switch-off distortion, and the complementary feedback pair (CFP), giving the best basic linearity.

The output configuration chosen is the emitter-follower Type II, which has the advantage of reducing switch-off non-linearities (Distortion 3c) due to the action of R15 in reverse-biasing the output base-emitter junctions as they turn off. A possible disadvantage is that quiescent stability might be worse than for the CFP output topology, as there is no local feedback loop to servo out Vbe variations in the hot output devices. Domestic ambient temperature changes will be small, so that adequate quiescent stability can be attained by suitable heatsinking and thermal compensation.

A global NFB factor of 30 dB at 20 kHz was chosen, which should give generous HF stability margins. The input stage (current-source TR1 and differential pair TR2, TR3) is heavily degenerated by R2, R3 to delay the onset of third harmonic Distortion One, and to assist this, the contribution of transistor internal Re variation is minimised by using the unusually high tail current of 4 mA. TR11, TR12 form a degenerated current-mirror that enforces accurate balance of the TR2, TR3 collector currents, preventing the generation of second harmonic distortion. Tail source TR1, TR14 has a basic PSRR 10 dB better than the usual two-diode version, though this is academic when C11 is fitted.

Input resistor R1 and feedback arm R8 are made equal and kept as low as possible, consistent with a reasonably high input impedance, so that base current mismatch caused by beta variations will give a minimal DC offset; this does not affect TR2−TR3

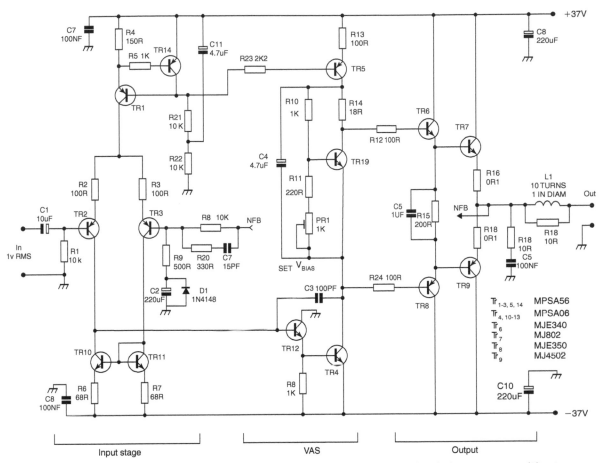

Figure 11.18. 50W Class-B amplifier circuit diagram. Transistor numbers correspond with the generic amplifier in Chapter 12.

Vbe mismatches, which appear directly at the output, but these are much smaller than the effects of Ib. Even if TR2, TR3 are high voltage types with low beta, the output offset should be within ±50 mV, which should be quite adequate, and eliminates balance presets and DC servos. A low value for R8 also gives a low value for R9, which improves the noise performance.

The value of C2 shown (220 μF) gives an LF roll-off with R9 that is -3 dB at 1.4 Hz. The aim is not an unreasonably extended sub-bass response, but to prevent an LF rise in distortion due to capacitor non-linearity; 100μF degraded the THD at 10 Hz from less than 0.0006% to 0.0011%, and I judge this unacceptable aesthetically if not audibly. Band-limiting should be done earlier, with non-electrolytic capacitors. Protection diode D1 prevents damage to C2 if the amplifier suffers a fault that makes it saturate negatively; it looks unlikely

but causes no measurable distortion.[5] C7 provides some stabilising phase-advance and limits the closed-loop bandwidth; R20 prevents it upsetting TR3.

The VAS stage is enhanced by an emitter-follower inside the Miller compensation loop, so that the local NFB that linearises the VAS is increased by augmenting total VAS beta, rather than by increasing the collector impedance by cascoding. This extra local NFB effectively eliminates Distortion Two (VAS non-linearity). Further study has shown that thus increasing VAS beta gives a much lower collector impedance than a cascode stage, due to the greater local feedback, and so a VAS-buffer to eliminate Distortion Four (loading of VAS collector by the non-linear input impedance of the output stage) appears unnecessary. *C*dom is relatively high at 100 pF, to swamp transistor internal capacitances and circuit strays, and make the design

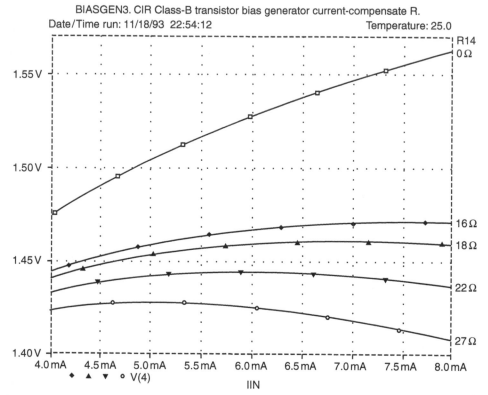

Figure 11.19. SPICE plot of the voltage-peaking behaviour of a current-compensated bias generator.

predictable. The slew-rate calculates as 40 V/μsec. The VAS collector load is a standard current source, to avoid the uncertainties of bootstrapping.

Since almost all the THD from a Blameless amplifier is crossover, keeping the quiescent conditions optimal is essential. Quiescent stability requires the bias generator to cancel out the Vbe variations of four junctions in series; those of two drivers and of two output devices. Bias generator TR8 is the standard Vbe-multiplier, modified to make its voltage more stable against variations in the current through it. These occur because the biasing of TR5 does not completely reject rail variations; its output current also drifts initially due to heating and changes in TR5 Vbe. Keeping Class-B quiescent stable is hard enough at the best of times, and so it makes sense to keep these extra factors out of the equation. The basic Vbe-multiplier has an incremental resistance of about 20 Ω; in other words its voltage changes by 1 mV for a 50 μA drift in standing current. Adding R14 converts this to a gently peaking characteristic that can be made perfectly flat at one chosen current; see Figure 11.19. Setting R14 to 22 Ω

makes the voltage peak at 6 mA, and the standing current now must deviate from this value by more than 500 μA for a 1 mV bias change. The R14 value needs to be altered if TR15 is run at a different current; for example, 16 Ω makes the voltage peak at 8 mA instead. If TO3 outputs are used, the bias generator should be in contact with the top or can of one of the output devices, rather than the heatsink, as this is the fastest and least attenuated source for thermal feedback.

The output stage is a standard double emitter-follower apart from the connection of R15 between the driver emitters without connection to the output rail. This gives quicker and cleaner switch-off of the outputs at high frequencies; switch-off distortion may significantly degrade THD from 10 kHz upwards, dependent on transistor type. Speed-up capacitor C4 noticeably improves the switch-off action, though I should say at this point that its use has been questioned because of the possibility of unhelpful charges building up on it during asymmetrical clipping. C6, R18 form the Zobel network (sometimes confusingly called a Boucherot cell) while L1, damped by R19, isolates the amplifier from load capacitance.

Figure 11.20 shows the 50 W/8 Ω distortion performance; about 0.001% at 1 kHz, and 0.006% at 10 kHz (see Table 11.1). The measurement bandwidth makes a big difference to the appearance, because what little distortion is present is crossover-derived, and so high-order. It rises at 6 dB/octave, at the rate the feedback factor falls, and it is instructive to watch the crossover glitches emerging from the noise, like Grendel from the marsh, as the test frequency increases above 1 kHz. There is no precipitous THD rise in the ultrasonic region.

(Most of the AP plots in this book were obtained from an amplifier similar to Figure 11.18, though with higher supply rails and so greater power capability. The main differences were the use of a cascode VAS with a buffer, and a CFP output to minimise distracting quiescent variations. Measurements at powers above 100 W/8 Ω, used a version with two paralleled output devices.)

The zigzags on the LF end of the plot are measurement artefacts, apparently caused by the Audio Precision system trying to winkle out distortion from visually pure white noise. Below 700 Hz the residual was pure noise with a level equivalent to approximately 0.0006% (yes, three zeros) at 30 kHz bandwidth; the

Table 11.1. Class-B amplifier performance

Power output	50 W rms into 8 Ω
Distortion	Below 0.0006% at 1 kHz and 50 W/8 Ω Below 0.006% at 10 kHz
Slew-rate	Approximately 35 V/μsec
Noise	−91 dBu at the output
EIN	−117 dBu (referred to input)
Freq Response	+0, −0.5 dB over 20 Hz−20 kHz

actual THD here must be microscopic. This performance can only be obtained if all seven of the distortion mechanisms are properly addressed; Distortions One, Two, Three and Four are determined by the circuit design, but the remaining three depend critically on physical layout and grounding topology.

It is hard to beat a well-gilded lily, and so Figure 11.21 shows the startling results of applying 2-pole compensation to the basic amplifier; C3 remains 100 pF, while CP2 was 220 pF and Rp 1 k (see Figure 13.13 on p. 339). The extra global NFB does its work extremely well, the 10 kHz THD dropping to 0.0015%, while the 1 kHz figure can only be guessed at. There were no unusual signs of instability, but, as

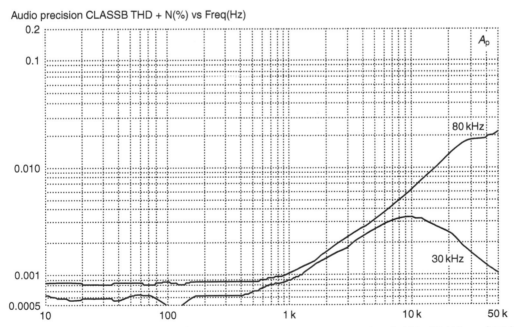

Audio precision CLASSB THD + N(%) vs Freq(Hz)

Figure 11.20. Class-B amplifier: THD performance at 50 W/8 Ω; measurement bandwidths 30 kHz and 80 kHz.

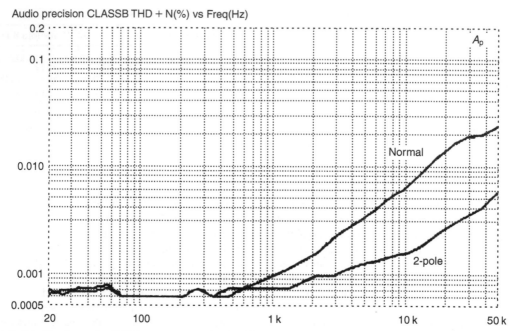

Figure 11.21. The dramatic THD improvement obtained by converting the Class-B amplifier to 2-pole compensation.

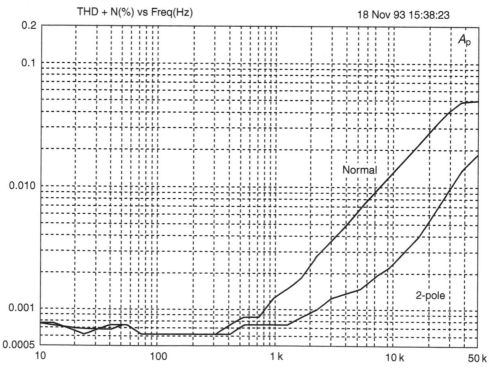

Figure 11.22. Class-B amplifier with simple quasi-complementary output. Lower trace is for two-pole compensation.

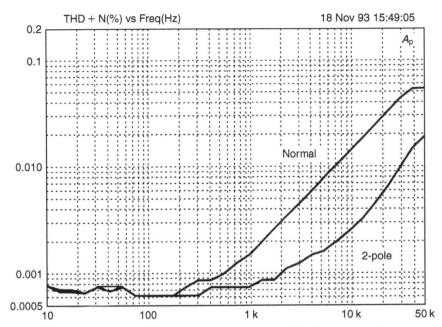

Figure 11.23. Figure 11.23 Class-B amplifier with quasi-comp plus Baxandall diode output. Lower trace is the two-pole case.

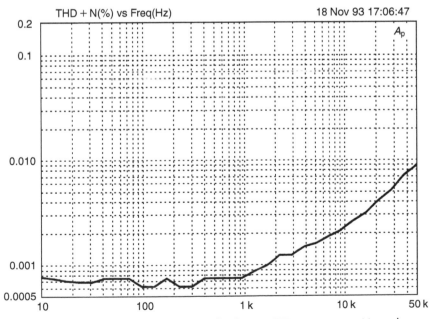

Figure 11.24. Class-B amplifier with Complementary-Feedback Pair (CFP) output stage. Normal compensation only.

always, unusual compensation schemes require careful testing. It does appear that a Blameless amplifier with 2-pole compensation takes us close to the long-sought goal of the Distortionless Amplifier.

The basic Blameless EF amplifier was experimentally rebuilt with three alternative output stages: the simple quasi-complementary, the quasi-Baxandall, and the CFP. The results for both single- and two-pole compensation are shown in Figures 11.22, 11.23, and 11.24. The simple quasi-complementary generates more crossover distortion, as expected, and the quasi-Baxandall version is not a lot better, probably due to remaining asymmetries around the crossover region. The CFP gives even lower distortion than the original EF-II output, with Figure 11.19 showing only the result for single-pole compensation; in this case the improvement with two-pole was marginal and the trace is omitted for clarity.

References

1. Ball, G, Distorting Power Supplies, *Electronics & Wireless World*, Dec. 1990, p. 1084.

2. Cherry, E, A New Distortion Mechanism in Class-B Amplifiers, *JAES*, May 1981, p. 327.

3. P Baxandall, Private communication, 1995.

4. Self, D, Distortion in Power Amplifiers, Series in *Electronics & Wireless World*, Aug. 1993 to March 1994.

5. Self, D, An Advanced Preamplifier, *Wireless World*, Nov. 1976, p. 43.

Chapter 12

Closely Observed Amplifiers: Design Examples

Amplifier Design Examples

In this chapter, five amplifier designs are closely observed, not only to evaluate their performance in detail, but also to see how that performance relates to the details of the design.

The first amplifier is the most recent version of the Blameless amplifier manufactured and sold by The Signal Transfer Company; it is known as the Compact Blameless Power Amplifier.[1] This has the usual heavily-degenerated input pair with current-mirror, an EF-VAS, and a single pair of output devices in a CFP configuration output stage. The modern power transistors used (2SA1295 and 2SC3264) have little beta-droop and so reduce Large-Signal Distortion, and are in big MT-200 packages that allow outputs of up to 100 W/8 Ω. This is Amplifier 1.

The second amplifier is simply the first with the emitter-follower removed from the EF-VAS, converting it to a simple-VAS design. The reduction in cost is trivial, and the difference in performance is startling; there is more distortion at every frequency. This design is purely to demonstrate the working of a simple VAS in a complete amplifier, and is most certainly not a recommended design. This is Amplifier 2.

The third amplifier is Amplifier 1 modified from standard Miller dominant pole compensation to output-inclusive compensation, which closes a semi-local feedback loop around the output stage, to great effect; there is a dramatic improvement in distortion performance. No other modifications are made. This is Amplifier 3, and I call it an Inclusive Amplifier.

The fourth amplifier is the Load-Invariant design pictured in Figure 10.12. The only significant difference between it and Amplifier 1 is that it has a CFP output stage with two output devices in parallel. It has standard Miller dominant pole compensation. This is Amplifier 4.

The fifth amplifier is Amplifier 4 modified from standard Miller dominant pole compensation to output-inclusive compensation. No other modifications are made. This is Amplifier 5.

Amplifier 1: EF-VAS, CFP Output Stage, Miller Compensation

The schematic of the Compact Blameless Power Amplifier is as shown in Figure 12.1. The input pair Q1, Q2 run at a collector current of 3 mA each, and are degenerated by R5, R7. Note: only one pair of transistors is used; two are shown on the schematic as they represent different pinouts (EBC and ECB) to allow alternative transistors to be used. The main VAS transistor is

Q11, and its associated emitter-follower is Q10. Note the low value of the output emitter resistors at 0.1 Ω.

Figure 12.2 shows the distortion performance of the Compact Blameless Power Amplifier at measurement bandwidths of 80 kHz and 30 kHz. In most cases, 80 kHz is the standard bandwidth for THD measurements as it allows through a reasonable number of harmonics (up to the fourth) of a maximum test frequency of 20 kHz. The 30 kHz bandwidth naturally excludes noise more effectively, but the THD results are not meaningful above a test frequency of 10 kHz. The THD traces turn over and head downward when the bandwidth definition filter is starting to work on the lower, larger amplitude harmonics, and this turnover alerts you that the THD figure is no longer valid.

At 1 kHz the distortion residual is almost all noise, though the visual averaging process that goes on when we view a noisy oscilloscope trace allows small disturbances to be seen at the crossover points even when the bias is optimally adjusted. This is illustrated in Chapter 9 and will not be repeated here. The THD at 10 kHz is 0.0029%, and at 20 kHz is 0.0057% (both 80 kHz bandwidth). Now I hope you will agree that is pretty good performance from a relatively simple and straightforward circuit. The specimen tested was pulled straight from Signal Transfer stock, and there was certainly no transistor selection or anything like that.

The performance is noticeably better than that of some other Blameless designs in this book, and you may wonder why that is. One reason is that all the new measurements in this edition were done with the wonderful Audio Precision SYS-2702, rather than the veteran AP System 1. In both cases the testgear distortion is well below that of the amplifiers; the advantage of the 2702 is that it has a significantly lower noise floor, allowing the limits of measurement in the LF region to be much reduced. The other main reason is that most of the earlier tests were done using output devices that are now outdated. The latest power transistors show lower crossover distortion and lower LSN.

To get the distortion clear of the noise and properly visible, we can increase the test frequency to 5 kHz, where the THD reading is 0.0012% with an 80 kHz measurement bandwidth. The distortion residual is shown in Figure 12.3. The large spikes are the disturbances at the crossover points; the small wobbles in between are just noise. The sine wave is simply to show the timing of the crossover points and does not relate to the vertical amplitude scale. None of the residual waveforms in this chapter have been subjected to averaging for noise reduction.

In case you're wondering, the little diamond in the middle of the picture indicates the scope trigger level.

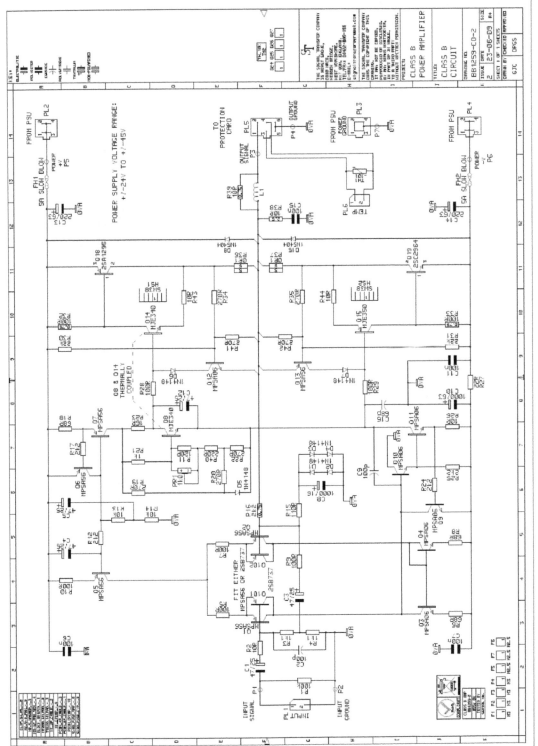

Figure 12.1. Schematic of EF-VAS and EF-output Blameless power amplifier.

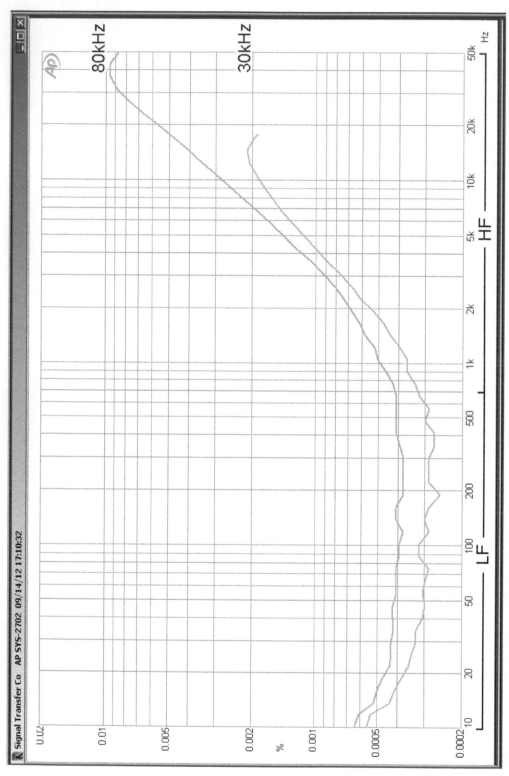

Figure 12.2. Distortion performance of EF-VAS, EF-output, Miller compensation Blameless power amplifier at 20 W/8 Ω. 30 kHz and 80 kHz measurement bandwidth. ±26V supply rails LF and HF distortion regimes labelled.

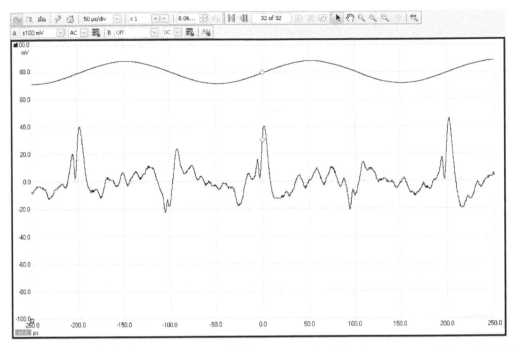

Figure 12.3. Distortion residual of EF-VAS, EF-output, Miller compensation Blameless power amplifier at 20 W/8 Ω. Test frequency 5 kHz, measurement bandwidth 80 kHz.

Figure 12.4 shows the distortion residual when the measurement bandwidth is increased to 500 kHz, and it looks very different. We inevitably get a much higher noise level, but what is significant is that we see the true nature of the crossover disturbances; they are much spikier and of considerably higher amplitude. The vertical scale has changed from ±100 mV to ±200 mV to accommodate the spikes. Since THD measurements are almost always done with an 80 kHz measurement bandwidth to reduce noise, it is easy to forget that such a bandwidth makes the distortion at higher test frequencies look a lot more benign than it actually is. On the other hand, you could of course argue that harmonics above 20 kHz can be ignored as they are inaudible, but that ignores the possibility of intermodulation distortion at high frequencies creating distortion products at lower frequencies that are very audible.

A very significant feature of this residual is that there is no distortion visible between the crossover spikes. All the distortion comes from the crossover problems of the output stage. It is noticeable that the modest wobble in gain that is visible on the incremental-gain plots (see Chapter 9) manifests itself as sharp spikes. This is because the open-loop gain, and so the amount of negative feedback, falls with frequency and so higher harmonics are suppressed less.

Increasing the test frequency to 20 kHz gives the distortion residual in Figure 12.5. The crossover spikes now look wider as the test frequency has increased with respect to the fixed measurement bandwidth, and have increased considerably in amplitude. The vertical scale has been increased again to ±1 V. (The internal automatic range-switching of the Audio Precision was switched to 'fixed gain' for this series of measurements.)

In Figure 12.6 we take a different look at the amplifier performance, this time comparing the THD at 20 W/8 Ω with that at 40 W/4 Ω. While I do not classify this design as a Load-Invariant amplifier, as it does not have a larger number of output devices than necessary simply to reduce the distortion into sub-8 Ω loads, it does use modern power transistors (Sanken 2SC3264 and 2SA1295 in MT-200 packages) that have little beta-droop. This is why the 4 Ω distortion shows only a modest increase on that for 8 Ω. The distortion at 5 kHz only increases from 0.0012% to 0.0028%, roughly doubling.

Figure 12.7 shows the residual (with reference sine wave) for the 40 W/4 Ω case, which should be compared with Figure 12.5 You will note the large extra wobble on the residual that coincides with the positive peak of the

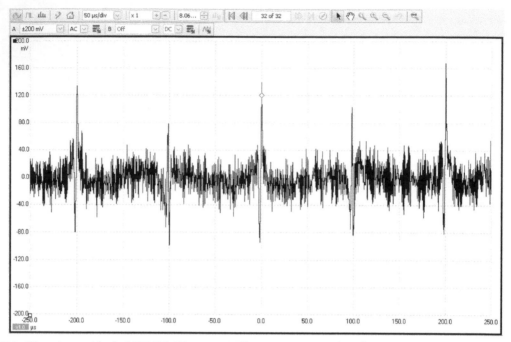

Figure 12.4. Distortion residual of EF-VAS, EF-output, Miller compensation Blameless power amplifier at 20 W/8 Ω. Test frequency 5 kHz, measurement bandwidth 500 kHz.

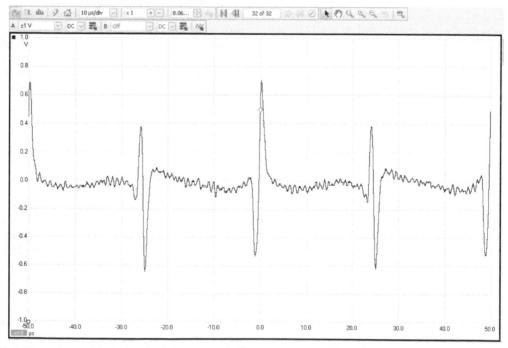

Figure 12.5. Distortion residual of EF-VAS, EF-output, Miller compensation Blameless power amplifier at 20 W/8 Ω. Test frequency 20 kHz, measurement bandwidth 500 kHz.

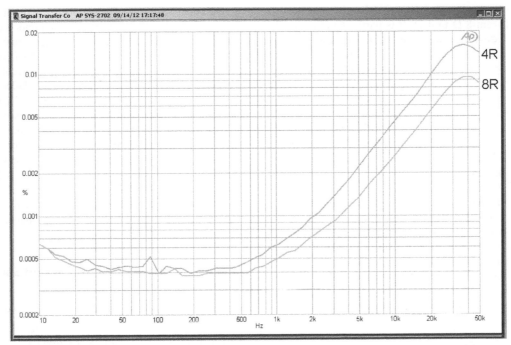

Figure 12.6. Distortion performance of EF-VAS, EF-output, Miller compensation Blameless power amplifier at 20 W/8 Ω and 40 W/4 Ω. 80 kHz measurement bandwidth ±26 V supply rails.

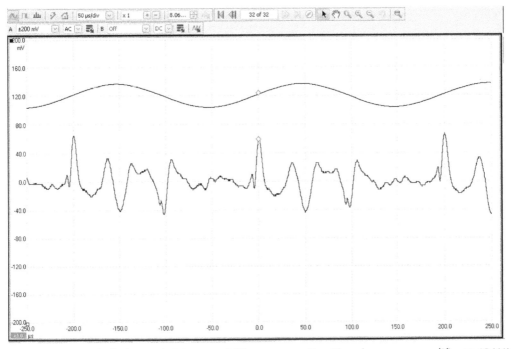

Figure 12.7. Distortion residual of EF-VAS, EF-output, Miller compensation Blameless power amplifier at 40 W/4 Ω. Test frequency 5 kHz, measurement bandwidth 80 kHz.

output waveform. This appears only when the load is switched from 8 Ω to 4 Ω. It is partly the cause of the increased distortion in Figure 12.6, but the crossover disturbances also increase in amplitude, and make a greater contribution. A similar wobble is introduced that coincides with the negative peak of the output waveform but it is much smaller in amplitude and not really visible in Figure 12.7.

Let's now take a look at how the distortion varies with the output level. Figure 12.8 shows the distortion at 50 W, 40 W, 30 W and 20 W into 8 Ω. The equivalent output voltages are 28.3 Vrms, 25.3 Vrms, 15.5 Vrms, and 12.7 Vrms. It is clear that there is not much difference – in fact, the traces are so close together it is impossible to label all of them.

The upper trace below 1 kHz is the 50 W result; it is higher than the other traces because a very small amount of third harmonic, 0.00032% at 50 Hz, is visible in the residual when the measurement bandwidth is reduced to 22 kHz to reduce noise. (The AP output residual is pure noise at 0.00022% under these conditions.) It is not visible when the output power is 40 W/8 Ω.

The origin of this tiny amount of third harmonic is so far unknown, but I am confident it will not spoil your listening pleasure.

The upper trace above 1 kHz is the 20 W result, and it is slightly worse than the others because of minor changes in bias level as the test proceeded. Adjusting the bias brought the trace into line with the other three. As I have said elsewhere (notably in Chapter 22), truly accurate bias compensation is something that has so far eluded us.

The most important lesson from Figure 12.8 is that the only significant distortion mechanism is relatively insensitive to level. This fact, plus the obvious presence of crossover artefacts in the THD residual, confirms that crossover distortion in the output stage is the origin of almost all the distortion when driving an 8 Ω load. This is not the case when driving loads of 4 Ω or lower, unless special measures have been made to make the amplifier Load-Invariant (see Chapter 10).

Amplifier 2: Simple VAS, CFP Output Stage, Miller Compensation

The second amplifier we will measure is a simple-VAS design, and was produced simply by taking the amplifier shown in Figure 12.1, removing emitter-follower Q10 and R25, and connecting the base of Q11 directly to Q3 collector. Thus Q11 becomes a simple VAS.

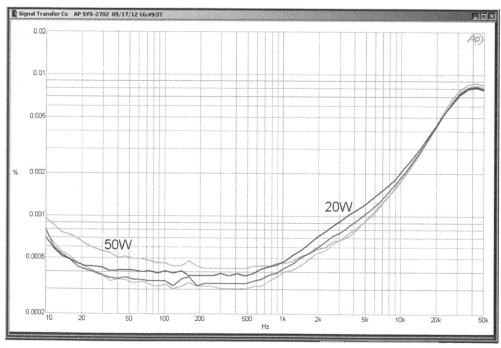

Figure 12.8. Distortion performance of EF-VAS, EF-output, Miller compensation Blameless power amplifier at 50 W, 40 W, 30 W and 20 W into 8 Ω. 80 kHz measurement bandwidth ±35V supply rails.

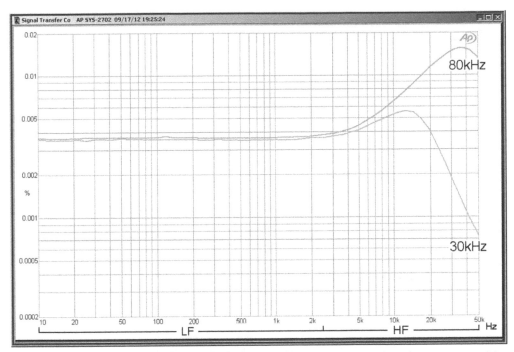

Figure 12.9. Distortion performance of simple-VAS, EF-output, Miller compensation Blameless power amplifier at 20 W/8 Ω. 30 kHz and 80 kHz measurement bandwidth. ±26 V supply rails. LF and HF distortion regimes labelled.

Figure 12.9 shows the distortion performance of the simple VAS amplifier at measurement bandwidths of 80 kHz and 30 kHz, and it is very much inferior to Figure 12.2. The flat low-frequency THD is at least ten times higher at 0.0036%, and very much above the noise floor. The 10 kHz THD is 0.0056% rather than 0.0029%, and the 20 kHz THD is 0.012% rather than 0.0057% (both 80 kHz bandwidth).

Unlike the first amplifier, the distortion residual at 1 kHz shows clear distortion at 0.0036% rather than just noise. Figure 12.10 shows why; superimposed on the crossover artefacts (which are still the same size, though now they are much less obvious) is a strong second harmonic component. This is exactly what we would expect after reading Chapter 7, where we saw that a simple VAS, placed in a model amplifier where other distortions were negligible, generated copious amounts of second harmonic in both the LF and HF regions. It is generated by VAS transistor Early effect in the LF region, and VAS transistor non-linear collector-base capacitance in the HF region. Chapter 7 describes how the emitter-follower in the EF-VAS eliminates this.

The distortion residual for a 5 kHz test signal is shown in Figure 12.11, for comparison with Figure 12.3 above for Amplifier 1. The THD has risen somewhat to 0.0048%, and the second harmonic

appears to still be much larger than the crossover artefacts, but this is an illusion caused by the 80 kHz measurement bandwidth. Switching to 500 kHz shows sharp spikes like those in Figure 12.4, of greater amplitude than the second harmonic content.

Let's stop and check how this compares with the results in Chapter 7. In Figure 7.5 the second-harmonic distortion of a simple VAS at 5 kHz is shown as 0.0027% for an output voltage of 7.75 Vrms (+20 dBu). From Amplifier 2 we have 0.0048% at 5 kHz for an output voltage of 12.7 Vrms (20 W/8 Ω). Since pure second harmonic increases in direct proportion to the signal level, if all is well, the THD numbers and output voltages should be in the same ratio. We find 48/27 = 1.77 and 12.7/7.75 = 1.64, which is about as accurate as we could hope for given that the Amplifier 2 results include some crossover distortion.

Figure 12.12 compares the THD of Amplifier 2 at 20 W/8 Ω with that at 40 W/4 Ω. As for Amplifier 1, the greater loading increases the HF distortion, but here it increases the LF distortion as well. In the LF region switching from 8 Ω to 4 Ω increases the amplitude of both the crossover artefacts and the second harmonic. This appears to be because the VAS collector impedance is now higher in the absence of the emitter-follower, and so is more vulnerable to non-linear

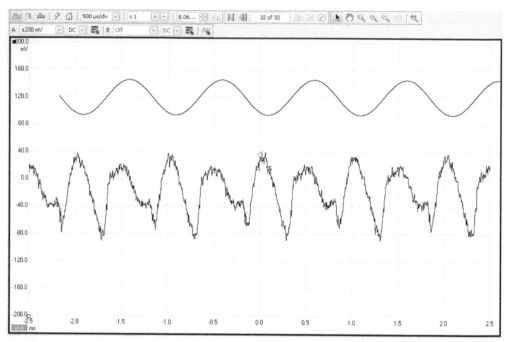

Figure 12.10. Distortion residual of simple-VAS, EF-output, Miller compensation Blameless power amplifier at 20 W/8 Ω. Test frequency 1 kHz, measurement bandwidth 80 kHz.

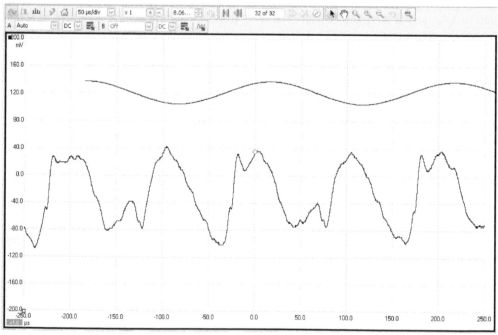

Figure 12.11. Distortion residual of simple-VAS, EF-output, Miller compensation Blameless power amplifier at 20 W/8 Ω. Test frequency 5 kHz, measurement bandwidth 80 kHz.

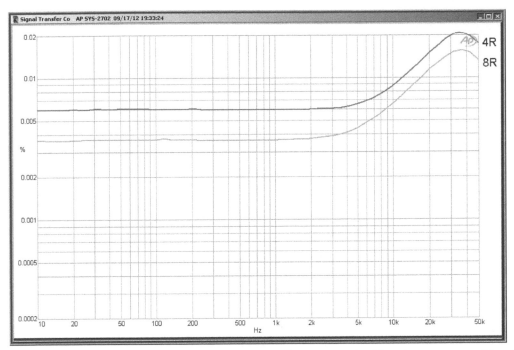

Figure 12.12. Distortion performance of simple-VAS, EF-output, Miller compensation Blameless power amplifier at 20 W/8 Ω and 40 W/4 Ω. 80 kHz measurement bandwidth ±26 V supply rails.

loading; see Samuel Groner's commentary on this issue.[2] That emitter-follower is doing a lot of work …

In the HF region the amplitude of the crossover spikes on the residual doubles as expected (this can only be seen properly with a 500 kHz measurement bandwidth), but there is only a minor effect on the second harmonic content.

Amplifier 1 showed only small variations in distortion as the output power varied from 20 W/8 Ω to 50 W/8 Ω. Amplifier 2 is different; Figure 12.13 shows significant increases in THD with output level, in both the LF and HF regions; note that the vertical scale has been changed. In the HF region increasing the output level increases both the crossover artefact amplitude and the level of second harmonic, as expected.

It is clear that the use of a simple VAS instead of an EF-VAS degrades the distortion performance badly. That little emitter-follower must be the best bargain in audio.

Amplifier 3: EF-VAS, CFP Output Stage, Inclusive Compensation

The normal Miller compensation scheme elegantly moves from global negative feedback to local feedback around the VAS as frequency increases. The output-inclusive compensation scheme, fully described in Chapter 13, is even more subtle. At low frequencies all the open-loop gain is used for global feedback, moving to semi-local feedback around the VAS and output stage as frequency rises, and then moving to purely local feedback around the VAS as in Miller compensation, as frequency rises further. This means that extra negative feedback is applied around the output stage, which is the one that needs it most because of the intractable problems of crossover distortion. The inclusive-compensation component values used here work well and give stability but it is not claimed that they are necessarily fully optimised.

Amplifier 1 was converted to output-inclusive compensation simply by replacing the Miller capacitor C9 in Figure 12.1 with two 220 pF capacitors in series, their junction being fed from the amplifier output via a 1 kΩ resistor. These values are the same as those used in Chapter 13; it is not yet clear if the resistor value is optimal. See Figure 13.13 for the circuit.

Figure 12.14 compares normal and inclusive compensation to show how extremely effective this technology is. (Note the high power output of 50 W/8 Ω compared with the 20 W/8 Ω used in other

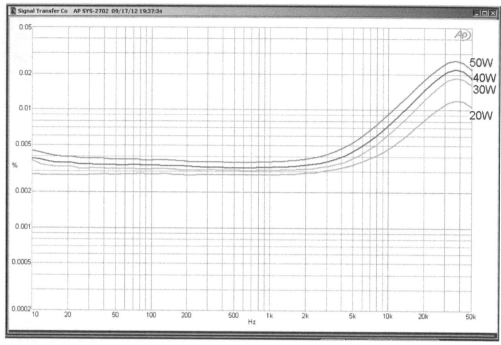

Figure 12.13. Distortion performance of EF-VAS, EF-output, Miller compensation Blameless power amplifier at 50 W, 40 W, 30 W and 20 W into 8 Ω. 80 kHz measurement bandwidth ±35 V supply rails. Note vertical scale change.

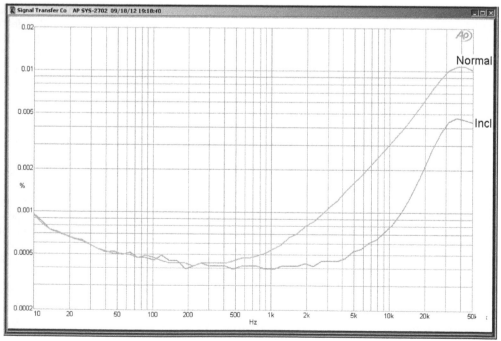

Figure 12.14. Distortion performance of EF-VAS, EF-output Blameless power amplifier with normal and with inclusive compensation, at 50 W/8 Ω. ±35 V supply rails.

parts of this chapter: I wanted to give inclusive compensation a good workout.) When the amplifier is switched from normal to inclusive compensation, which is done simply by connecting the 1 kΩ resistor, all traces of crossover artefacts disappear at 1 kHz. This makes it impossible to optimise the output stage bias by examining the THD residual, so providing means of connecting and disconnecting the resistor, perhaps by a push-on link, is essential. Crossover artefacts become just discernible on the residual at approximately 5 kHz, at a level of 0.0005%. With levels of distortion below 0.001% it can be more convenient to quote parts-per-million (ppm) rather than percent; 0.00050% is equivalent to 5.0 ppm. We note that the THD at 10 kHz drops from 0.0030% (30 ppm) with normal compensation to 0.00077% (7.7 ppm) with inclusive compensation, an improvement of almost four times. The THD reduction at 20 kHz is not so dramatic but it still falls from 0.0060% (60 ppm) to 0.0020% (20ppm), a three times improvement, and jolly well worth having. I don't think I'm going too far out on a limb if I say that it is going to be very difficult to argue that the distortion from Amplifier 3 is going to be audible. Since this goes well beyond the performance of a Blameless Amplifier, I think it might be time for

a new name, and as mentioned I plan to call it an Inclusive Amplifier.

Looking at Figure 12.14, it is clear that the only measurable distortion left is in the region 5—20 kHz. That's just a couple of octaves. Be assured I am working on it.

Figure 12.15 examines inclusive compensation only at 80 kHz and 30 kHz measurement bandwidths, to get a closer look at the distortion behaviour below 10 kHz. The measured THD at 1 kHz can now be said to be below 0.00028% (2.8 ppm) at 30 kHz bandwidth. The AP SYS-2702 alone gives a reading of 0.00022% (2.2 ppm) which is all noise in these conditions.

Figure 12.16 demonstrates that inclusive compensation also works very well with sub-8 Ω loads. With a 4 Ω load, the 1 kHz THD residual remains free from visible crossover artefacts for measurement bandwidths of both 80 kHz and 30 kHz. Comparing it with Figure 12.6 (for Amplifier 1), the 10 kHz THD with a 4 Ω load is reduced from 0.0048% to 0.0014%, and the 20 kHz THD is reduced from 0.0098% to 0.0046%. Once again these are very useful improvements in linearity.

Figure 12.17 shows how the distortion performance varies with output level. Compare it with Figure 12.8 for Amplifier 1; as before, there is little variation in

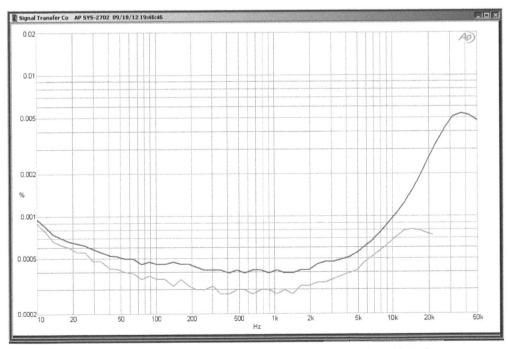

Figure 12.15. Distortion performance of EF-VAS, EF-output Blameless power amplifier with inclusive compensation, at 50 W/8 Ω. 30 kHz (lower trace) and 80 kHz (upper trace) measurement bandwidths, ±35 V supply rails.

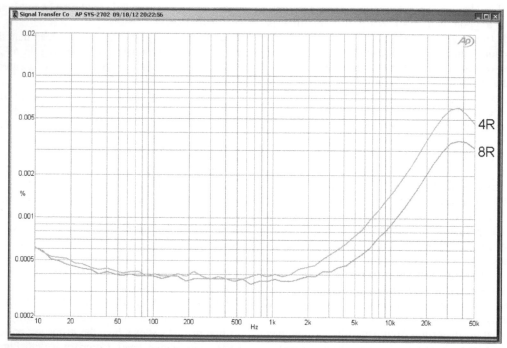

Figure 12.16. Distortion performance of EF-VAS, EF-output Blameless power amplifier with inclusive compensation, at 20 W/8 Ω and 40 W/4 Ω. ±35 V supply rails.

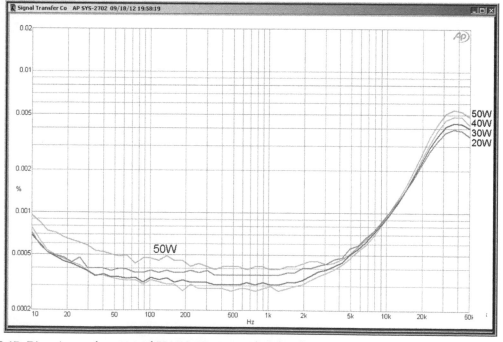

Figure 12.17. Distortion performance of EF-VAS, EF-output Blameless power amplifier with inclusive compensation, at 50 W, 40 W, 30 W and 20 W into 8 Ω. 80 kHz measurement bandwidth. ±35 V supply rails.

THD in the HF region, demonstrating once more that the distortion here originates in crossover artefacts. Likewise, as before, the 50 W/8 Ω trace is raised above the others in the LF region, due to the presence of a very small amount of third harmonic. The origin of this remains obscure for the time being.

The results of the output-inclusive compensation scheme given in Chapter 13 were the first obtained. These new results extend the validation of the concept to output levels of 50 W/8 Ω and 40 W/4 Ω. Behaviour into reactive loads has been crudely tested by shunting the 8 Ω load with capacitors between 100 nF and 2.1 uF, with no hint of any stability problems. This works.

Amplifier 4: EF-VAS, CFP Output Stage, Miller Compensation

The next example, Amplifier 4, differs only from Amplifier 1 in having a CFP output stage with two output devices in parallel, rather than a single output pair. It is essentially the Load-Invariant design shown in Figure 10.12. As Amplifier 4 it has standard Miller dominant pole compensation.

Figure 12.18 shows the distortion behaviour with 80 kHz and 30 kHz measurement bandwidths. The 80 kHz-bandwidth THD at 10 kHz is 0.0047% compared with 0.0030% for Amplifier 1. Likewise the THD at 10 kHz is 0.009% compared with 0.0056% for Amplifier 1. It has to be said at this point that Amplifier 1 is based on a much more recent design, though the basic Blameless amplifier principles are used unchanged.

Figure 12.19 shows the distortion performance of Amplifier 4 with 8 Ω and 4 Ω loads, with an 80 kHz measurement bandwidth.

Figure 12.20 shows the distortion performance of Amplifier 4 varies with output levels from 20 W to 50 W into an 8 Ω load. As for Amplifiers 1 and 3, the difference in the HF region is very small, indicating that again the overwhelming majority of the distortion here is due to crossover artefacts. However, things are a little different in the LF region, where the 50 W trace is no longer noticeably higher than those for lower output levels.

Amplifier 5: EF-VAS, CFP Output Stage, Inclusive Compensation

The final example is Amplifier 5, which is Amplifier 4 with the Miller compensation replaced by the same output-inclusive compensation scheme used in Amplifier 3. Figure 12.21 shows how distortion is lowered

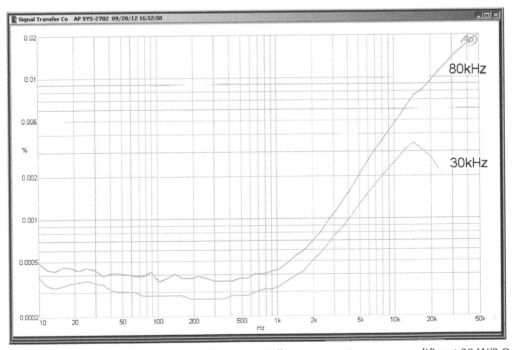

Figure 12.18. Distortion performance of EF-VAS, CFP output, Miller compensation power amplifier at 20 W/8 Ω. 30 kHz and 80 kHz measurement bandwidth, ±26 V supply rails.

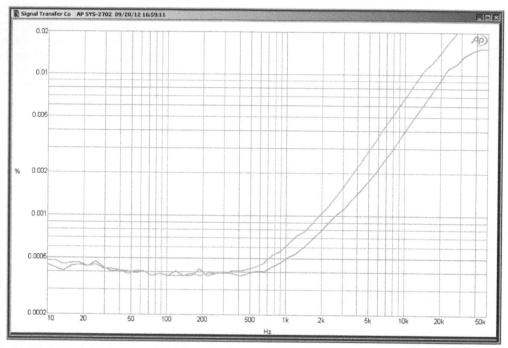

Figure 12.19. Distortion performance of EF-VAS and CFP output, Miller compensation Blameless power amplifier at 20 W/8 Ω and 40 W/4 Ω. 80 kHz measurement bandwidth. ±26 V supply rails.

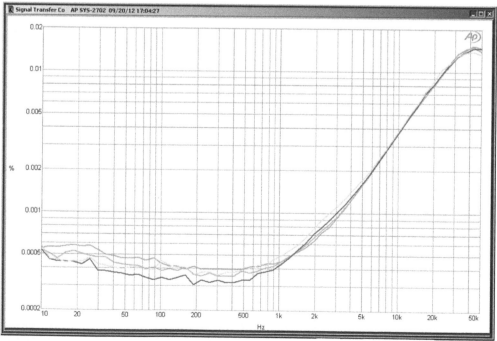

Figure 12.20. Distortion performance of EF-VAS, CFP output, Miller compensation Blameless power amplifier at 50 W, 40 W, 30 W and 20 W into 8 Ω. 80 kHz measurement bandwidth. ±35 V supply rails.

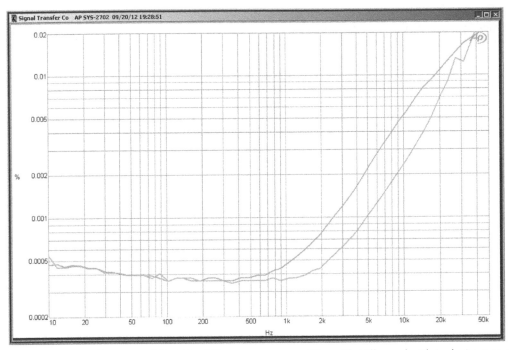

Figure 12.21. Distortion performance of EF-VAS, CFP ouput power amplifier with Miller and with inclusive compensation at 20 W/8 Ω. 80 kHz measurement bandwidth, ±26 V supply rails.

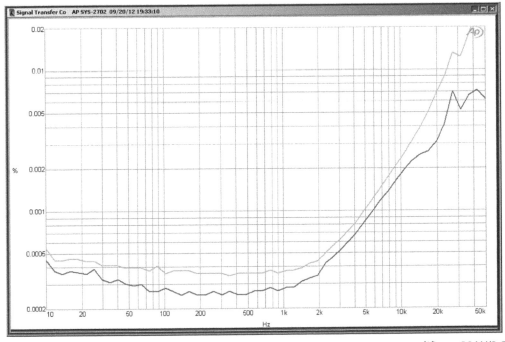

Figure 12.22. Distortion performance of EF-VAS, CFP output, inclusive compensation power amplifier at 20 W/8 Ω. 30 kHz and 80 kHz measurement bandwidth, ±26 V supply rails.

when inclusive compensation is switched in. THD at 10 kHz drops by more than two times, and at 20 kHz by 1.6 times. This is admittedly less impressive than the improvements shown in Figure 12.14 for Amplifier 3, but still very much worth having, given that the extra component cost (one small ceramic capacitor and one resistor) is trivial. It certainly proves that output-inclusive compensation will work with a CFP stage, though I must say I had no reason to think otherwise.

Figure 12.22 shows the distortion behaviour with 80 kHz and 30 kHz measurement bandwidths, the latter demonstrating that with inclusive compensation in use, the THD is down in the noise up to 2 kHz.

Conclusions

This chapter takes a detailed look at several variations on the Blameless amplifier. It updates their known performance by the use of the Audio Precision SYS-2702. It demonstrates clearly that the use of a simple VAS rather than an EF-VAS is false economy, and that output-inclusive compensation will work effectively in a variety of situations.

References

1. Signal Transfer Company, http://www.signaltransfer.freeuk.com/compact.htm (accessed Sept. 2012).

2. Groner, S, *Comments on Audio Power Amplifier Design Handbook*, 2011, pp. 34−44, SG Acoustics, http://www.sg-acoustics.ch/analogue_audio/power_amplifiers/index.html (accessed Aug. 2012).

Compensation and Stability

'Stability,' said the Controller, 'stability.'
        Aldous Huxley, *Brave New World*

Soon ... you'll attain the stability you strive for, in the only way that it's granted...
        Jefferson Airplane, 'Crown of Creation'

Compensation and Stability

The compensation of an amplifier is the tailoring of its open-loop gain and phase characteristics so that it is dependably stable when the global feedback loop is closed. The basic theory of feedback and stability can be found in many textbooks, and I only give a very quick overview here.

The distortion performance of an amplifier is determined not only by open-loop linearity, but also the negative feedback factor applied when the loop is closed; in practical circumstances, doubling the NFB factor halves the distortion. We have seen that this results in the distortion from a Blameless amplifier consisting almost entirely of crossover artefacts, because of their high-order and hence high frequency. Audio amplifiers using more advanced compensation are rather rare; some of these techniques are described here.

It must be said straight away that 'compensation' is a thoroughly misleading word to describe the subject of this chapter. It implies that one problematic influence is being balanced out by another opposing force; nothing like that is happening. Here it means the process of tailoring the open-loop gain and phase of an amplifier so that it is satisfactorily stable when the global feedback loop is closed. The derivation of the word is historical, going back to the days when all servomechanisms were mechanical, and usually included an impressive Watt governor pirouetting on top of the machinery.

An amplifier requires compensation because its basic open-loop gain is still high at frequencies where the internal phase-shifts are reaching 180°. This turns negative feedback into positive at high frequencies, and causes oscillation, which in audio amplifiers can be very destructive. The way to prevent this is to ensure that the loop gain (the open-loop gain minus the feedback factor) falls to below unity before the phase-shift reaches 180°; oscillation therefore cannot develop. Compensation is vital to make an amplifier stable, but there is much more to it than just establishing bulletproof stability. The exact way in which compensation is applied has a very important influence on the closed-loop distortion.

This chapter concentrates on applying compensation to the classical three-stage amplifier architecture with transconductance input, transimpedance VAS, and unity-gain output stage. Two-stage amplifiers with transconductance input and unity-gain output have been built but are ill-suited to power-amp impedances. Four-stage amplifiers are described in Chapter 4; the best-known is probably the design by Otala[1] which has the low open-loop gain of 52 dB (due to the dogged use of local feedback) and only 20 dB of global feedback.

There are two popular ways to describe how safe an amplifier is from instability. The *gain margin* is the ratio by which the open-loop gain has dropped below unity when the phase-shift has reached the critical 180°. The *phase margin* is the amount by which the phase is less than 180° when the loop-gain is unity.

The phase margin is more commonly used as it has a more direct relation with frequency response and transient behaviour; this is shown in Figure 13.1. Note that for the 60° case a barely-visible hump in the frequency

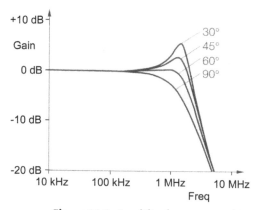
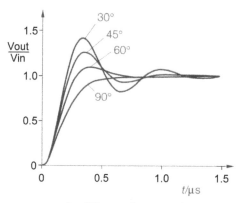

Figure 13.1. Amplifier frequency and transient response for different phase margins.

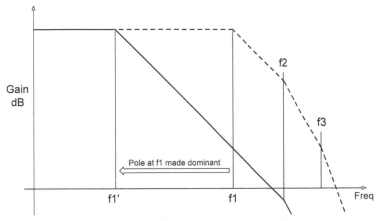

Figure 13.2. Stabilising an amplifier with dominant-pole compensation. Dotted line is frequency response before compensation, solid line is after.

response gives an obvious transient overshoot. Solid-state audio power amplifiers are not expected to show frequency response peaking or overshoot, and so a phase margin of much less than 90° is not normally acceptable. Ringing into capacitive loads is usually due to interaction with the series output inductor and has nothing to do with amplifier stability. Valve amplifiers are not likely to have phase margins as good as 90° because of phase-shift in their output transformers.

Amplifiers are sometimes described as 'unconditionally stable', which is intended to mean, if anything, that they are stable into any load, reactive or otherwise, they are likely to encounter. The phrase is borrowed from Control Theory, where it means something totally different. It means that the amplifier remains stable even if the open-loop gain is reduced. This might sound like something that can be taken for granted, and with dominant pole compensation that is true, but with other forms of compensation, it is definitely not so. For any amplifier the open-loop gain is effectively reduced during clipping, current-limiting, slew-limiting, and at power-up and power-down; this is when a conditionally stable amplifier may show oscillation. The first misuse of 'unconditionally stable' I have located was by Bailey in *Wireless World* in 1966.

Dominant Pole Compensation

Dominant pole compensation is the simplest and most reliable kind. Simply take the lowest pole (roll-off point) f1, and make it dominant; in other words so much lower in frequency than the next pole f2 that the total loop-gain falls below unity before enough phase-

shift accumulates to cause HF oscillation, as shown in Figure 13.2. The dotted line shows the open-loop gain without compensation. There are poles at f1, f2, and f3 (and there may be many more higher in frequency) which cause the response to cross the unity loop-gain line with a phase of 270°, ensuring instability. If, however, pole f1 is made dominant by moving it down to f1', the gain has gone below the unity line before f2 is reached. If *pole-splitting* occurs (this is described in more detail below) then f2 moves up in frequency as f1 moves down, increasing the stabilising effect.

With a single pole, the open-loop gain falls at 6 dB/octave, corresponding to a constant 90° phase shift. Thus the phase margin will always be 90°, giving good stability. If there are two poles in the forward path, then at a high enough frequency, the open-loop gain will have a slope of 12 dB/octave, with a constant 180° phase shift; the phase margin is thus zero; in theory, the amplifier would tremble on the brink of instability, but in practice oscillation can be confidently expected. If the open-loop gain slope is between 6 and 12 dB/octave, the amplifier will show intermediate phase margins and safety of stability.

Maximal Negative Feedback

Looking at Figure 13.2, you can see that there is an enormous area of unusable gain between the compensated and uncompensated responses. It is instructive to look at the various approaches to making use of this, even if they are not all applicable to audio amplifiers. One of the oldest methods is to give the compensated response a slope intermediate between 6 and 12 dB/octave. You may be wondering why anyone would want to fool around with intermediate gain slopes if 6 dB/octave gives the very healthy and desirable phase margin of 90°. The answer is that early valves were expensive, power-hungry, non-linear, and unreliable. These were major problems in the field where negative feedback was invented — stringing together large numbers of telephone repeaters to achieve long-distance operation. The need for acceptable linearity with cascaded amplifiers was the spur for the invention of negative feedback by Harold Black in 1927.[2] There was a very great incentive to get as much linearising feedback factor as possible, and if it meant complicated feedback networks sprinkled with inductors, that was very acceptable if it minimised the number of valves. The great Hendrik Bode published the fundamental work in this field in 1945.[3] Bode showed that the phase margin was proportional to the slope of the open-loop gain when it hit the unity loop-gain line, thus:

$$phase\ margin = 180 - (15 \cdot slope) \qquad \text{Equation 13.1}$$

Where:

slope is in dB/octave.
phase margin is in degrees.

Applying this equation gives us 90° for a 6 dB/octave slope and 0° for 12 dB/octave, as expected.

In telephone repeaters a 30° phase margin was considered adequate, and this corresponds to a 10 dB/octave slope. If the calculations are done to preserve a 30° phase margin across the whole frequency band, you get not a straight line at 10 dB/octave, running from f1 to the unity-gain intercept at f4, but the rather strange shape shown in Figure 13.3. The roughly triangular area between the maximal-feedback line and the 10 dB/octave asymptote is called the Bode Fillet.[4] (No fish were harmed in the making of this graph.) Making use of this fillet means that the maximum gain can be maintained for another octave upwards, from f1 to f2 without affecting stability. The fillet blends back into the straight line at f3, which is well above the unity-gain crossover point at f4. In audio use the Bode Fillet is likely to be acting around 1 kHz, where there is usually already enough feedback to put the distortion below the noise; what we really want is more feedback in the 10 kHz–20 kHz region. It will not be considered further. Bode also proposed introducing a flat section of frequency response between f5 and f7, to improve the stability margins; this is usually called the Bode Step.[5]

A phase margin of 30° may be all right for controlling oil refineries, but in an audio amplifier would frighten me considerably. It is ironic that in process control the relevant time-constants can usually be determined quite accurately, while the details of real amplifier responses, particularly in the output stage, are rather more obscure; see Chapter 3 for a demonstration of this. A more reasonable phase margin of 60° would use a slope of 8 dB/octave.

Be aware that Figure 13.3 is an approximation using asymptotes, and in real life putting sharp corners into frequency response is not possible. Implementing slopes that are not multiples of 6 dB/octave is also going to present problems.

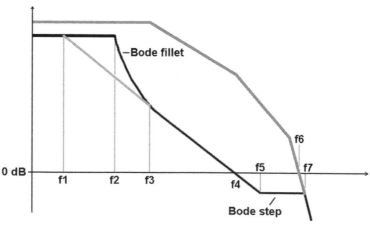

Figure 13.3. Bode's maximal feedback. The Bode fillet on top of a 10 dB/octave line, the Bode step, and the open-loop response.

You may be wondering why I have wandered into this elevated area of feedback theory. It seems unlikely that anyone is going to insert a Bode Step into the frequency response of a power amplifier. The answer is that it is of fundamental importance in showing just how much feedback you can get away with, in theory at least. Some of the more advanced forms of compensation are approximations to Bode's idea of maximal feedback. The use of some sort of Bode Fillet, if it can be done without undue circuit complexity, looks like a possibility. Most if not all advanced compensation schemes for audio power amplifiers retain a final slope of 6 dB/octave to maximise the phase margin.

Dominant Pole Miller Compensation

The Miller effect describes how connecting a capacitance between the input and output of an inverting amplifier makes it behave like a much larger capacitor; the greater the amplifier gain, the greater the capacitance shown.[6] It is very useful for amplifier compensation because a relatively low dominant pole frequency can be implemented while keeping the currents flowing in the capacitor small. Figure 13.4 shows how it works in the usual three-stage amplifier. Ccompen is the added compensation capacitor, in parallel with Cbc, the existing collector-base capacitance in the VAS transistor. Cbc varies with the voltage on the VAS transistor and this can cause serious distortion if not dealt with in some way. This issue is dealt with in detail in Chapter 7.

Figure 13.5a shows in more detail the Miller method of creating a dominant pole. The collector pole of Q3 is lowered by adding the external Miller-capacitance Cdom to that which unavoidably exists as the internal Cbc of the VAS transistor. However, there are some other beneficial effects; Cdom causes *pole-splitting,*

in which the pole at Q1 collector is pushed up in frequency as f1 is moved down — most desirable for stability. How effective this is when the input pair is loaded with a current-mirror rather than resistor loads is currently uncertain. Simultaneously the local NFB through Cdom linearises the VAS, so the unused open-loop gain is not just thrown away but is instead pressed into use as purely local feedback at high frequencies. Negative feedback is applied globally at low frequencies, but is smoothly transferred by Cdom to be local solely to the VAS as frequency increases.

Assuming that input-stage transconductance is set to a plausible 5 mA/V, and stability considerations set the maximal 20 kHz open-loop gain to 50 dB, then from Equations 5.1–5.3 in Chapter 5, Cdom must be 100 pF. This is a practical real-life value, and it is very commonly used.

The peak current that flows in and out of this capacitor for an output of 20 Vrms at 20 kHz, is 447 μA. Since the input stage must sink Cdom current while the VAS collector load sources it, and likewise the input stage must source it while the VAS sinks it, there are four possible ways in which the slew-rate may be limited by inadequate current capacity; if the input stage is properly designed, then the usual limiting factor is VAS current-sourcing. In this example a peak current of less than 0.5 mA should be easy to deal with, and the maximum frequency for unslewed output will be comfortably above 20 kHz.

Dominant Pole Miller Compensation at High Gains

In this book it is almost everywhere assumed that the closed-loop gain of a power amplifier is desired to be in the region of 20 to 25 times (usually 23 times), and the discussions of compensation and stability reflect that.

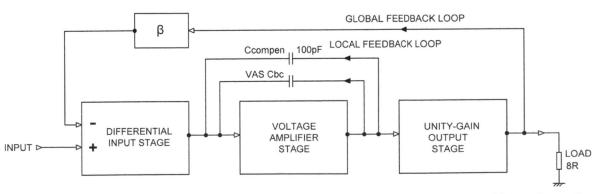

Figure 13.4. The feedback paths in a conventional Miller-compensated amplifier. Note the presence of the non-linear Cbc.

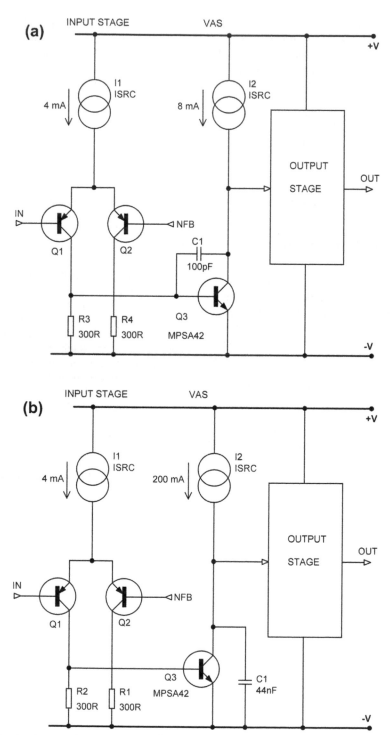

Figure 13.5. (a) Miller dominant-pole compensation by connecting C1 from collector to base; (b) Shunt dominant-pole compensation by connecting C1 from collector to ground. Note much increased VAS standing current.

There is an implicit assumption that normal moving-coil loudspeakers are to be driven. This, however, is not always the case. If electrostatic speakers are to be driven directly by a power amplifier, rather than via a step-up transformer, much higher gains are needed.

In an actual example an output of 180 Vrms into a small capacitance was required. This could be provided by an amplifier with a gain of 23 times; the necessary input level is high at 7.8 Vrms, but that is do-able by an opamp so long as it is not expected to have any headroom. If it is directly connected to the power amplifier, there is no problem as the latter will always clip first.

However, there are some snags to this approach which intrude when you get to the details of design:

1. A full output of 180 Vrms at 22 kHz requires a slew-rate of 35 V/usec, which is pushing the limits of the conventional three-stage amplifier as usually implemented.
2. If a low-impedance feedback network is used to give the best noise performance (the usual network in this book is 2k2−100 Ω), then at full output the power dissipated in the 2k2 upper feedback resistor is a wholly excessive 15 Watts.

Both of these difficulties can be surmounted by running the power amplifier at a higher closed-loop gain. The reduced amount of negative feedback means that the Miller compensation capacitor can be much smaller for the same stability, and the potential slew-rate problems disappear. For the same reason the upper feedback resistor will be much higher in value, while keeping the lower feedback resistor at a suitable value for low noise, such as 100 Ω.

The power amplifier was therefore given a closed-loop gain of 158 times (+44 dB), 6.9 times greater than the usual 23 times (+27 dB). The amount of Miller compensation required for HF stability is reduced by the same factor, so a dominant pole Miller capacitance of 100 pF can be reduced to 15 pF, returning the NFB factor to its original figure; this really does work in practice. The maximum slew-rate is increased by 6.9 times for the same quiescent currents in the amplifier, and the upper feedback resistor becomes 15.7 kΩ, which now dissipates a more reasonable 2.0 W at full output. The input for full output is reduced to 1.1 Vrms, which should eliminate the pre-amplifying opamp stage.

Dominant Pole Shunt Compensation

Figure 13.5b shows another method of dominant pole compensation that is much less satisfactory − the addition of capacitance to ground from the VAS collector. This is usually called shunt or lag compensation, but is sometimes known as parallel compensation. As Peter Baxandall[7] put it, 'The technique is in all respects sub-optimal', which for Peter was strong language indeed.

We have already seen in Chapter 7 that loading the VAS collector resistively to ground is a very poor option for reducing LF open-loop gain, and a similar argument shows that capacitive loading to ground for compensation purposes is an even worse idea. To reduce open-loop gain at 20 kHz to 50 dB as before, the shunt capacitor Clag must be 43.6 nF, which is a whole different order of things from a Miller capacitor of 100 pF. The current in and out of Clag at 20 V rms, 20 kHz, is 155 mA peak, which is going to require some serious electronics to provide it. The Miller version only requires a much more practical 447 μA. This important result is yielded by simple calculation, confirmed by SPICE simulation. The input stage no longer constrains the slew-rate limits, which now depends entirely on the VAS as that stage is both sourcing and sinking the capacitor current.

A VAS working under these conditions will have poor linearity. The Ic variations in the VAS, caused by the heavy extra loading, produce more distortion and there is no local NFB through a Miller capacitor to correct it. To make matters worse, the dominant pole P1 will probably need to be set to a lower frequency than for the Miller case, to maintain the same stability margins, as there is now no pole-splitting action to increase the frequency of the pole at the input-stage collector. Hence Clag may have to be even larger than 43 nF, requiring yet higher peak currents. The bad effects of adding much smaller shunt capacitances than this (say, 1 nF) to a VAS collector are illustrated in Figure 13.24 on p. 348. The use of a very small capacitor (say, 33 pF) from the VAS collector to ground is often useful in suppressing output stage parasitics, but this has nothing to do with the amplifier compensation. This handy fix is discussed in detail later in this chapter.

Another very serious disadvantage of shunt compensation is that the HF open-loop gain after compensation is not fixed by the value of the compensation capacitor, as it is with the Miller method. In shunt compensation the gain depends on both the capacitance and the current gain (beta) of the VAS transistor.[8] Since transistor beta is variable between different samples of the same transistor type, and also dependent on operating conditions such as the collector current, this introduces very unwelcome uncertainty about HF stability.

The theory of shunt compensation is dealt with in detail by Huijsing in[9] Takahashi[10] produced a fascinating paper on shunt compensation, showing one way of generating the enormous compensation currents required for good slew-rates. The only thing missing is an explanation of why shunt compensation was chosen in the first place.

Output-inclusive Compensation

Looking at Figure 13.4, we note that the output stage has unity gain, and it has occurred to many people that the open-loop gain and the feedback factor at various frequencies would be unchanged if the Miller capacitor Ccompen were driven from the amplifier output, as in Figure 13.6, creating a semi-local feedback loop enclosing the output stage. 'Semi-local' means that it encompasses two stages, but not all three — that is the function of the global feedback loop. The inclusion of the output stage in the Miller loop has always been seen as a highly desirable goal because it promises that crossover distortion from the output stage can be much reduced by giving it all the benefit of the open-gain inside the Miller loop, which does not fall off with frequency until the much smaller Cbc takes effect.

Figure 13.6 shows the form of output-inclusive compensation that is usually advocated, and it has a major drawback — it does not work. I have tried it many times and the result was always intractable high-frequency instability. On a closer examination this is not surprising.

The Problems of Output Inclusion

Using local feedback to linearise the VAS is reliably successful because it is working in a small local loop with no extra stages that can give extra phase-shift beyond that inherent in the Ccompen dominant pole. Experience shows that you can insert a cascode or a small-signal emitter-follower into this loop, but a slow output stage with all sorts of complexities in its frequency response is a very different matter. Published information on this is very scanty, but Bob Widlar stated in 1988[11] that output stage behaviour must be well controlled up to 100 MHz for the technique to be reliable; this would appear to be flat-out impossible for discrete power stages, made up of devices with varying betas, and driving a variety of loads.

Trying to evaluate what sort of output stage behaviour, in particular, frequency response, is required to make this form of inclusive compensation workably stable quickly runs into major difficulties. The devices in a typical Class-B output stage work in voltage and current conditions that vary wildly over a cycle that covers the full output voltage swing into a load. Consequently the transconductances and frequency responses of those devices, and the response of the output stage overall, also vary by a large amounts.

The output stage also has to drive loads that vary widely in both impedance modulus and phase angle. To some extent, the correct use of Zobel networks and output inductors reduces the phase angle problem, but load modulus still has a direct effect on the magnitude of currents flowing in the output stage devices, and has corresponding effects on their transconductance and frequency response.

Input-inclusive Compensation

An alternative form of inclusive compensation that has been proposed is enclosing the input stage and the

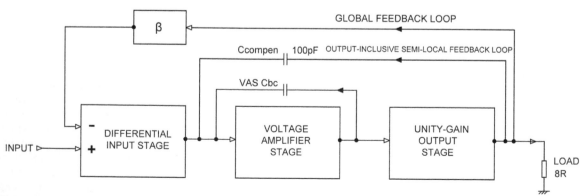

Figure 13.6. The feedback paths in an output-inclusive compensated amplifier. Note that the collector-base capacitance Cbc is still very much present.

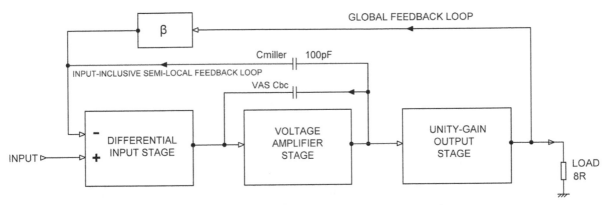

Figure 13.7. The feedback paths in an input-inclusive compensated amplifier.

VAS in an inner feedback loop, leaving just the output stage forlornly outside. This approach, shown in Figure 13.7, has been advocated many times, one of its proponents being the late John Linsley-Hood, for example in.[12] A typical circuit is shown in Figure 13.8. It was his contention that this configuration reduced the likelihood of input-device overload (i.e., slew-limiting) on fast transients because current flow into and out of the compensation capacitor was no longer limited by the maximum output current of the input pair (essentially the value of the tail current source).

This scheme is only going to be stable if the phase-shift through the input stage is very low, and this is now actually less likely because there is less Miller feedback to reduce the input impedance of the VAS, therefore less of a pole-splitting effect, and so there is more likely to be a significant pole at the output of the input stage. There is still some local feedback around the VAS, but what there is goes through the signal-dependent capacitance Cbc.

My experience with this configuration was that it was unstable, and any supposed advantages it might have had were therefore irrelevant. I corresponded with

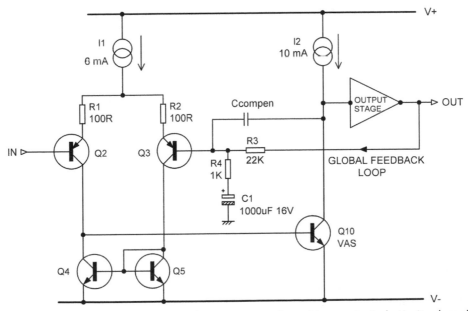

Figure 13.8. Attempt to implement input-inclusive compensation, with some typical circuit values shown.

JLH on this matter in 1994, hoping to find exactly how it was supposed to work, but no consensus on the matter could be reached.

A similar input-inclusive compensation configuration was put forward by Marshall Leach,[13] the intention being not the reduction of distortion, but to decrease the liability of oscillation provoked by capacitive loading on the output, and so avoid the need for an output inductor.

Attempting to include the input stage in the inner loop to reduce its distortion seems to me to be missing the point. The fact of the matter is that the linearity of the input stage can be improved almost as much as you like, either by further increasing the tail current, and increasing the emitter degeneration resistors to maintain the transconductance, or by using more a slightly more complex input stage. Any improvements in the slew-rate that might be achieved would be of little importance, as obtaining a more than adequate slew-rate with the three-stage amplifier architecture is completely straightforward; see Chapter 15.

If any stage needs more feedback around, it is the output stage, as this will reduce its intractable crossover distortion. I think that trying to create a semi-local loop around the input stage and VAS is heading off in wholly the wrong direction.

Stable Output-inclusive Compensation: The History

Considering the problems described earlier, it is clear that trying to include the output stage in the VAS compensation loop over its full bandwidth is impractical. What can be done, however, is to include it over the bandwidth that affects audio signals, but revert to purely local VAS compensation at higher frequencies where the extra phase-shifts of the output stage are evident. The method described here was suggested to me by the late Peter Baxandall, in a document I received in 1995,[14] commenting on some work I had done on the subject in 1994. This document has now been published.[15] He sent me six pages of theoretical analysis, but did not make it completely clear if he had personally evaluated it on a real amplifier. However, he said that he had 'devoted much thought and experiment to the problem' which implies that he had. He did not say he had invented the technique, and recent research has shown that it is completely disclosed by a US patent granted to Kunio Seki in 1979[16] called '*Multistage Amplifier Circuit*', a title which is a classic example of obfuscation if I ever saw one, and assigned to Hitachi. The intended application was IC power amplifiers. Peter never mentioned this patent to me and I do not know if he was aware of it.

A possibly accidental use of inclusive compensation was described in a paper by Gunderson in 1984,[17] where the Miller capacitor is fed from a cascode device that is driven from the amplifier output. He described this as an OFICC (Output-Following Intermediate Cascode Circuit). This paper seems to have received little attention, possibly because of the rather tangled explanation of how the OFICC is supposed to work. At no point is the enclosure of the output stage in the Miller feedback loop explicitly mentioned. What appears to be the same principle appeared in the Rotel RB-1090-3 power amplifier (introduced in 2001).

The TDA7293 is a monolithic power amplifier IC that, according to its data sheet (Dec. 1999), uses output inclusive compensation. The internal circuitry appears not to have been published.

The general method of output-inclusive compensation was discussed in a DIYaudio forum[18] in March 2007; as usual, there appears to have been much simulation but no actual measurement of hardware. I reported my own experiments and measurements in *Linear Audio* in 2010.[19]

It is common to find out that ideas go back further than you might think, and important to realise that patents sometimes get granted even though there is prior art. The notion may well pre-date the Seki patent, but nothing earlier has been found so far. The method shown here requires a unity-gain output stage, so it is most unlikely to have originated in the valve era.

Stable Output-inclusive Compensation: Implementation

The basic technique is shown in Figure 13.9. At low frequencies C1 and C2 have little effect, and the whole of the open-loop gain is available for negative feedback around the global feedback loop. As frequency increases, semi-local feedback through Ri and C1 begins to smoothly roll off the open-loop gain, but the output stage is still included in this semi-local loop. At higher frequencies still, where it is not feasible to include the output stage in the semi-local loop, the impedance of C2 is becoming low compared with that of Ri, and the configuration smoothly changes again so that the local Miller loop gives dominant pole compensation in the usual way. If the series combination of C1 and C2 gives the same capacitance as a normal dominant-pole Miller capacitor, then stability should be unchanged.

Figure 13.10 is a conceptual diagram of a VAS compensated in this way, for simulation with a minimum of distracting complications. The current feed from the input stage is represented by Rin, which

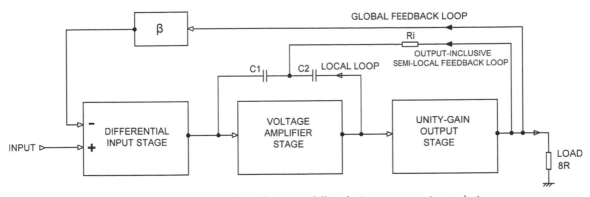

Figure 13.9. The basic principle of the Baxandall inclusive compensation technique.

delivers a constant current as the opamp inverting input is at virtual ground. The 'opamp' is in fact a VCVS (Voltage-Controlled Voltage Source) with a flat voltage gain of 10,000x, once more to keep things simple. Since there is no global feedback loop as in a complete amplifier, if the input is 1 Volt, the output signal will be measured in kilo-Volts, at least at low frequencies; this is not exactly realistic but the magnitude does not alter the basic mechanism being studied.

Figure 13.11 shows how the feedback current through C1 is sourced via Ri at low frequencies, and via C2 at high frequencies. The lower the value of Ri, the higher the frequency at which the transition between the two routes occurs. With C1 and C2 set at 220 pF, and Ri = 1K, this occurs at 723 kHz. At very high frequencies the effective Miller capacitance is 110 pF, a slight increase on the usual value of 100 pF; this is simply because 220 pF capacitors are readily available.

Below 100 kHz, almost all of the current through C1 is supplied through Ri, and very little through C2. This means that there is less loading on the VAS as C2 is effectively bootstrapped, and this has the potential to reduce VAS distortion. Note that the C1 current is only constant here because the stage is operating open-loop. When a global feedback loop is closed around it, the current through C1 will increase with frequency to keep the output voltage constant.

At low frequencies the amount of semi-local feedback is controlled by the impedance of C1; at 220 pF it is more than twice the size of the usual 100 pF Miller capacitor, and so the reduced open-loop gain means the feedback factor available is actually 6.5 dB *less* than normal over the audio band. This sounds like a bad thing, but the fact that the semi-local loop includes the output stage more than makes up for it. At 723 kHz the impedance of C1 has fallen to the point where it is equal to Ri. At 1.45 MHz the impedance of the parallel combination of C1 and C2 now reaches that of Ri, and the capacitors dominate, giving strictly local Miller compensation with approximately the usual capacitance (110 pF).

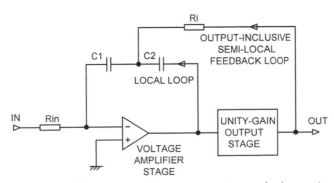

Figure 13.10. Conceptual diagram of stable output-inclusive compensation method, sometimes known as Transitional Miller Compensation, or TMC.

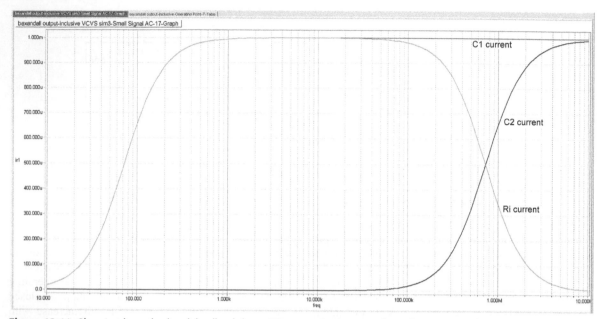

Figure 13.11. Showing how the local feedback loop (C2 current) takes over from the output-inclusive compensation (Ri current) as frequency increases.

Figure 13.12 attempts to illustrate this; it shows open-loop gain with the closed-loop gain (+27.2 dB) subtracted to give a plot of the feedback factor. Note that the kink in the plot only extends over an octave, and so in reality is a gentle transition between the two straight-line segments. This diagram is for Ri = 10 kΩ, as this raises the gain plateau above the X-axis and makes thing a bit clearer.

The dotted line shows normal Miller compensation; with a 110pF Miller capacitor the feedback factor reaches a plateau around 20 Hz as this is the maximum gain of input stage and VAS combined, without compensation.

Figure 13.13 shows the practical implementation of the inclusive technique to a Blameless power amplifier; some of the unaltered parts of the circuitry are omitted for greater clarity.

Figure 13.14 compares the distortion performance of the standard Blameless amplifier with the new output-inclusive version; the THD at 10 kHz has been reduced

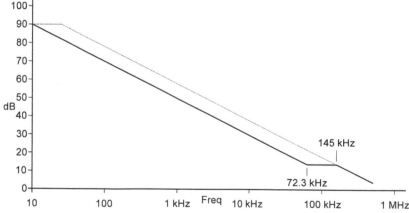

Figure 13.12. How the feedback factor varies with frequency. C1 = C2 = 220 pF, Ri = 10 KΩ. The dotted line shows the feedback factor with normal Miller compensation with the capacitor equal to half C1, C2.

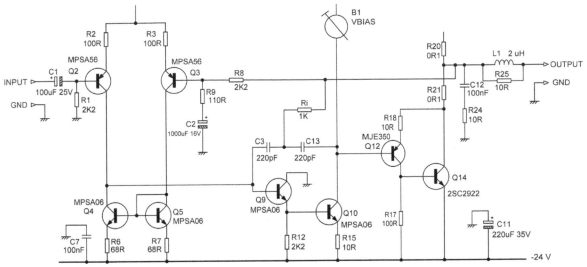

Figure 13.13. Practical implementation of output-inclusive compensation to a Blameless amplifier (circuit simplified for clarity).

from 0.0025% to 0.00074%, an improvement of at least three times; noise makes a significant contribution to the latter figure. There may be no official definition of 'ultra-low distortion', but I reckon anything less than 0.001%

(10 ppm) at 10 kHz qualifies. The THD measurement system was an Audio Precision SYS-2702.

Figure 13.15 shows another view of the output-inclusive distortion performance. Because the THD

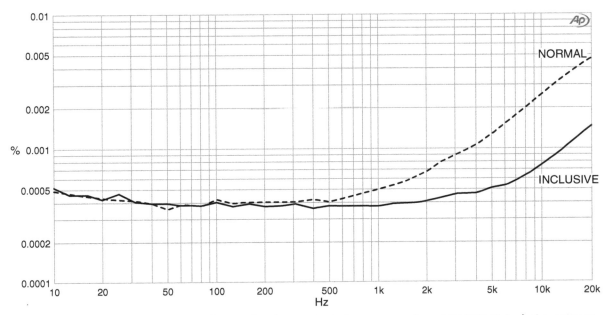

Figure 13.14. Distortion performance with normal and output-inclusive compensation, at 22 W/8 Ω. Inclusive compensation yields a THD + Noise figure of 0.00075% at 10 kHz, less than a third of the 0.0026% given by conventional compensation. Measurement bandwidth 80 kHz.

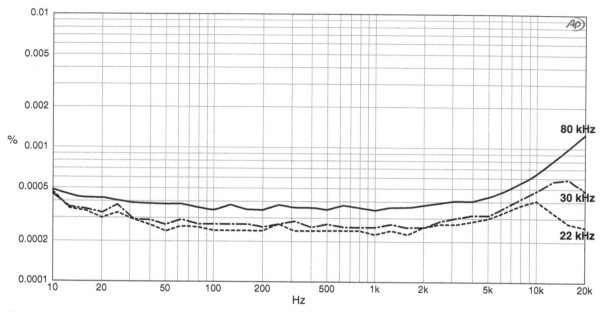

Figure 13.15. Distortion performance with output-inclusive compensation at 22 W/8 Ω. Bandwidths are 80 kHz, 30 kHz and 22 kHz.

levels are so low, noise is a significant part of the reading. Three measurement bandwidths are therefore shown. The 22 kHz and 30 kHz bandwidths eliminate most of the harmonics when the fundamental is 10 kHz or above, so they do not give much information in this region. It is, however, clear that a 10 kHz THD of 0.00074% is an overestimate, and 0.0006% (6 ppm) is probably more accurate.

This form of compensation is very effective, but it is still necessary to optimise the quiescent bias of the Class-B output stage. This is easier to do without the distortion-suppression of output-inclusion, so you might consider adding a jumper so that the connection via Ri can be broken for bias setting.

As Figure 13.15 shows, this amplifier pretty much takes us to the limits of THD analysis; it is impossible to read any distortion at all below 2 kHz. The 22 kHz trace is only just above the AP distortion output; this varies slightly with output voltage but is below 0.00025% (2.5 ppm) up to 10 kHz.

It is only right that I point out that a slight positional tweak of the output inductor was required to get the best THD figures. I attribute this to small amounts of uncorrected inductive distortion, where the half-wave currents couple into the input or feedback paths; it is an insidious cause of non-linearity. The amplifier I used for the tests was thought to be free from this,

but the reduced output stage distortion appears to have exposed some remaining vestiges of it. At 10 kHz the inductive coupling will be ten times greater than at 1 kHz, but in a conventional Blameless amplifier this effect is normally masked by crossover products.

The information here is not claimed to be a fully worked-out design such as you might put into quantity production. As with any unconventional compensation system, it would be highly desirable to check the HF stability at higher powers, with 4 Ω loads and below, and with highly reactive loads.

Since the distortion performance is decisively better than the standard Blameless Amplifier, I think a new name is appropriate. I call it an Inclusive Amplifier.

Experimenting with Output-inclusive Compensation

One of the great advantages of this approach is that it can be added to an existing amplifier for the cost of a few pence. Possibly one of the best bargains in audio! This does not mean, of course that it can be applied to any old amplifier; it can only work in a three-stage architecture, and the amplifier needs to be Blameless to begin with to get a real benefit.

Another advantage is that if you decide that inclusive compensation is not for you, the amplifier can be instantly converted back to standard Miller compensation by breaking the connection to Ri. There is in fact a continuum between conventional and output-inclusive compensation. As the value of Ri increases, the local/semi-local transition occurs at lower and lower frequencies, until conventional Miller compensation is reached when Ri is infinite.

The quickest and most effective way to experiment with this form of compensation is to start with an amplifier that is known to be Blameless. A very suitable design is the Load-Invariant amplifier produced by The Signal Transfer Company;[20] this has all the circuit features described above for Blameless performance, and the PCB layout is carefully optimised to eliminate inductive distortion. I need to declare an interest here; I am, with my colleague Gareth Connor, the technical management of The Signal Transfer Company.

Ultra-low Distortion Performance Comparisons

It is instructive to compare the output-inclusive method of compensation with other ways of achieving ultra-low distortion. The Halcro power amplifier is noted for its low distortion, achieved by applying error-correction techniques to the FET output stage; however, what appears to be a basic version, as disclosed in a patent,[21] uses 31 transistors. More information is given in a later patent.[22] Giovanni Stochino (for whose abilities I have great respect) has also used error-correction in a most ingenious form[23,24] but it requires two separate amplifiers to implement; the main amplifier uses 20 transistors and the auxiliary correction amplifier has 17, totalling 37. The output-inclusive compensation version of the Blameless amplifier uses only 13 transistors (excluding overload protection), and gives, I think, extraordinary results for such simple circuitry.

Two-pole Compensation

The compensation methods looked at so far are all examples of the dominant-pole technique. The gain roll-off slope is always 6 dB/octave, so it is bound to have that value when the loop-gain crosses the unity line, and the phase margin will be a solid 90°. However, as we saw back in Figure 13.2, this leaves a lot of gain unavailable to the global feedback loop, even if it is used locally to improve VAS linearity. The simplest way of exploiting this is two-pole compensation.

Two-pole compensation is well known as a technique for squeezing the best performance from an opamp[25,26,27] but it has rarely been applied to power amplifiers; a notable example is the LM12, an early integrated power amplifier designed by Bob Widlar.[8] This device also uses output-inclusive compensation, described above; the schematic can be found in the data sheet.[28]

An extra HF time constant C2 - R3 is inserted in the compensation path around the VAS, as in Figure 13.16a. This adds a second low-frequency pole and a zero at a higher frequency. The two poles give an open-loop gain curve that typically peaks, and then falls at almost 12 dB/octave as frequency increases, finally reverting asymptotically to a 6 dB/octave slope around 100 kHz as the zero takes effect, as in Figure 13.17. This is derived from simulation of a complete Blameless amplifier driving an 8 Ω load, not just a conceptual model. The reversion is arranged to happen well before the unity loop-gain line is reached, and so stability should be the same as for the conventional dominant-pole scheme, but with increased negative feedback over a large part of the operational frequency range. So long as the slope returns to 6 dB/octave before the unity loop-gain crossing occurs, stability should remain good, with a phase margin of 90°. As we saw in the section above on maximal feedback, it is not a problem if the slope does not reach exactly 6 dB/octave before the crossing occurs (and it cannot as it is only approaching it asymptotically) as a slightly reduced phase margin of, say, 85° is unlikely to be noticed.

Figure 13.17 shows the loop-gain, i.e., the open-loop gain with the +27 dB closed-loop gain subtracted, for both conventional Miller and two-pole compensation. For the latter I have used the values C1 = 1000 pF, C2 = 120 pF, and R1 = 2K2, which are employed in several examples in this book. The horizontal line at +30 dB shows that the usual amount of NFB is applied at 20 kHz. The time constants arranged so the 6 dB/octave slopes for conventional Miller and two-pole compensation above 100 kHz are equal. The closed-loop response is shown running along the 0 dB line, beginning to roll-off above 400 kHz as the amount of NFB available falls to nothing.

The midband open-loop gain peak at 900 Hz may look worrying, but I have so far failed to detect any resulting ill-effects in the closed-loop behaviour. Peter Baxandall pointed out to me, and demonstrated mathematically, that the open-loop gain peak has no repercussions at all in the closed-loop gain plot. It is not a question of a resonance being masked or heavily

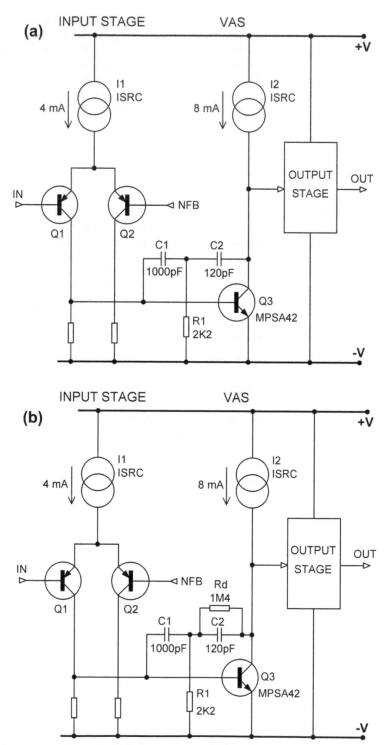

Figure 13.16. Two-pole compensation with realistic component values. The version at (b) has a resistor added to suppress the midband peak in the loop gain.

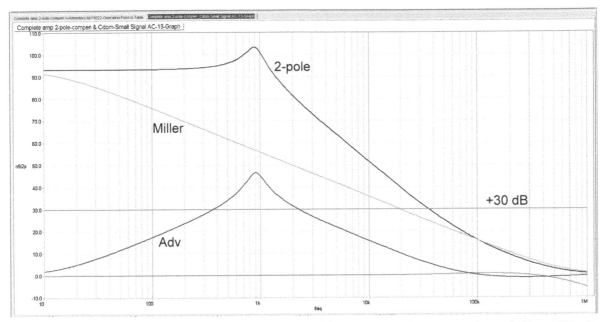

Figure 13.17. The loop-gain plot for a complete Blameless amplifier using two-pole compensation with C1 = 1000 pF, C2 = 120 pF, R1 = 2K2. The 'Adv' line shows the advantage in the amount of feedback available. The input pair emitter degeneration resistors were 100 Ω as usual.

suppressed — it simply does not exist any more. The full demonstration has now been published in.[29]

The trace labelled 'Adv' shows the advantage gained in terms of extra feedback. This reaches an impressive +46 dB just below 1 kHz — we have a powerful technique here. On either side of the peak, the advantage falls off at 6 dB/octave. What we are really interested in is the advantage over the range 1–20 kHz, where distortion normally increases as the amount of global NFB available falls. At 1 kHz we have an advantage of +45 dB, where it is not much needed, but this is down to 16 dB at 10 kHz and only 10 dB at 20 kHz. This is unfortunate as 10 kHz–20 kHz is the area where we most need more linearisation, but it is still at least three times better than the equivalent Miller dominant pole compensation. If two-pole compensation is applied correctly, the overall reduction in distortion is dramatic and extremely valuable. Crossover glitches on the THD residual visually almost disappear.

In 2010, an excellent paper was published by Dymond and Mellor[30] that gives for the first time a mathematical analysis of two-pole compensation. Dymond points out that it greatly reduces the signal current required from the input stage over much of the audio band; this is of course equivalent to saying that the open-loop gain is greater. The paper gives equations

for the open-loop response, etc.; particularly useful is the formula for the zero frequency at which the response blends back into the 6 dB/octave slope.

$$f_z = \frac{1}{2\pi R1(C1 + C2)} \qquad \text{Equation 13.2}$$

This shows us that the ultimate stability at high frequencies depends on the sum of C1 and C2, as expected. The sum of C1 and C2 should have the same value as it would for stable single-pole compensation, usually 100 pF. C1 should be significantly larger than C2; at least twice as big. R1 is usually in the region 1 kΩ–10 kΩ. The zero frequency for the values used here works out at 64.6 kHz.

The practical results are excellent with no obvious reduction in stability. See Figure 13.18 for the happy result of applying this technique to a Blameless Class-B amplifier.

Factors Affecting the Two-pole Loop-gain Response

Figure 13.19 shows how the loop-gain response is affected by the value of R1. As it increases, the 12 dB/octave region moves to a lower frequency and

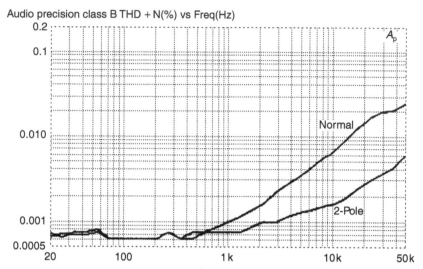

Figure 13.18. Distortion reduction with two-pole compensation.

becomes less extensive, while the 6 dB/octave region becomes larger. A slope closer to 6 dB/octave at the unity crossing may improve stability, but it will reduce the feedback in the 10 kHz–20 kHz octave where it is most needed. A value for R1 of the order of 2.2 kΩ is required if a significant improvement in distortion performance is to be had.

Figure 13.20 shows how the loop-gain response is affected by the ratio of C1/C2. Since the value of their series combination is already fixed at 100 pF by the need for HF stability, this is the only parameter left that can be altered. As the ratio moves from 21 to 2 times, the peak and the roll-off move upward in frequency, and the stability at HF may be reduced.

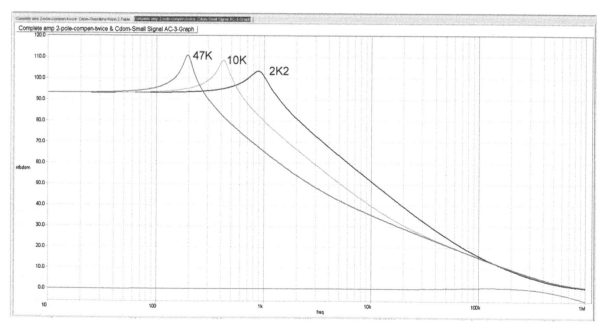

Figure 13.19. How the loop-gain response varies with the value of R1. C1 = 1000 pF, C2 = 120 pF.

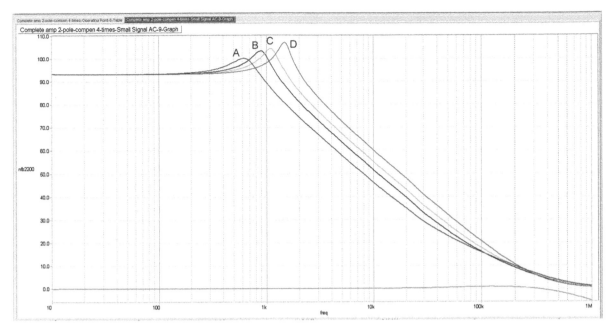

Figure 13.20. How the loop-gain response varies with the C1/C2 ratio. C1/C2 = (A) 21; (B) 8.3; (C) 5.8; (D) 2.0. R1 = 2.2 kΩ.

To a first approximation, the response is unchanged by swapping the values of C1 and C2. It is, however, advantageous to make C2 the smaller capacitor. At HF, C2 has low impedance and allows R1 to directly load the VAS collector to ground, which could worsen VAS linearity. To illustrate this, assume there is a 10 Vrms 10 kHz signal at the power amplifier output. If C1 = 1000 pF, C2 = 120 pF and R1 = 2.2 kΩ, then the current in C1 is 10.4 uA rms and that in C2 is a tiny 1.24 uA rms. Swapping the values still gives 10.4 uA rms through C1 but the current through C2 is now 86.7 uA rms, an increase of about seventy times. In reference,[27] C2 is larger than C1 but it is driven from an opamp output.

In tests on an experimental amplifier based on the Load Invariant design but not fully optimised, I started off with C1 at 100 pF and C2 at 1000 pF; the THD at 10 kHz was 0.0043% (25 W/8 Ω). Swapping the capacitor values dropped it to 0.00317%, due to reduced VAS loading.

Effect of Two-pole Compensation on the Closed-loop Gain

Despite the large amounts of feedback that are commonly employed in amplifiers, the open-loop response does affect the closed-loop response. This is usually only significant at high frequencies where the feedback is less. As explained above, the midband peak in the loop-gain has no effect on the closed-loop frequency response, but that does not mean that the use of two-pole compensation has no effect on it. Because the loop-gain plot does not cross the unity-gain line at exactly 6 dB/octave, there is likely to be some very mild peaking at ultrasonic frequencies; this can be seen in an exaggerated form in Figure 13.1. What actually happens is given in Figure 13.21, for the compensation values used in Figure 13.16. The standard values of C1 = 1000 pF, C2 = 120 pF and R1 = 2.2 kΩ give rise to a +0.70 dB peak at 127 kHz. The response variations at 20 kHz are minimal, not exceeding 0.15 dB.

Although all the responses appear to be plunging rapidly southwards, this is because of the greatly magnified scale of ±1 dB; in all cases the ultimate roll-off is at the usual 6 dB/octave.

Similarly, Figure 13.22 shows how the changes in capacitor ratio explored in Figure 13.20 affect the closed-loop response. The ratio 8.3 corresponds to C1 = 1000 pF, C2 = 120 pF and R1 = 2.2 kΩ. It appears that these values are a good compromise as they give useful amounts of extra feedback without excessive peaking of the closed-loop response.

The peak in the closed-loop response obviously has an effect on the transient response. Figure 13.23, for

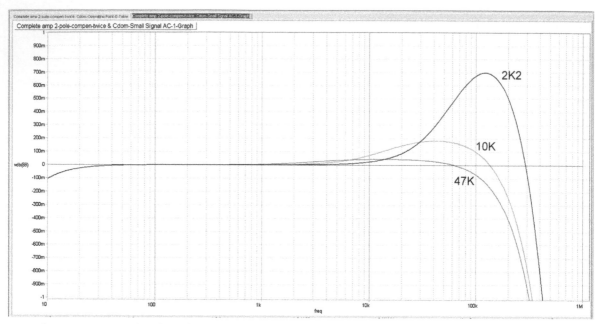

Figure 13.21. How the closed-loop gain response varies with the value of R1. C1 = 1000 pF, C2 = 120 pF.

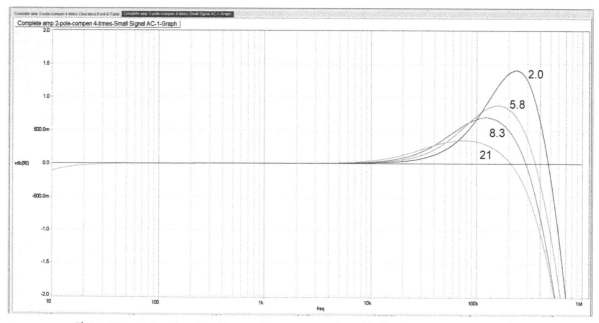

Figure 13.22. How the closed-loop gain response varies with the C1/C2 ratio. R1 = 2.2 kΩ.

C1 = 1000 pF, C2 = 120 pF and R1 = 2.2 kΩ shows that it induces a small amount of overshoot: 8% of the peak-to-peak amplitude of the square wave. Component choices that increase the frequency response peaking naturally increase the overshoot, and vice versa. The C1/C2 ratio of 2.0 in Figure 13.20 gives an overshoot of 16%.

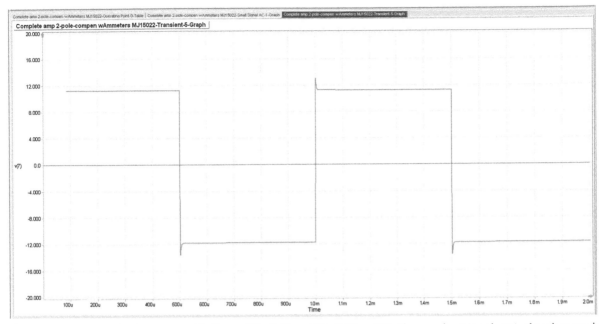

Figure 13.23. The transient response with C1 = 1000 pF, C2 = 120 pF, R1 = 2K2. The overshoot is only 8% of peak-to-peak amplitude.

The transient response of Figure 13.23 was obtained using a square-wave input with rise and fall times of only 100 nsec. This affects the overshoot; increasing the rise and fall times to 1 usec reduces the overshoot to 7%. No power amplifier is going to experience this sort of input except under test, for in real operation there will always be some bandwidth-limiting circuitry upstream. This can reduce the gain peaking and the overshoot markedly. For example, if a simple first-order RC lowpass filter with a -3 dB frequency of 234 kHz is used (e.g., 1 kΩ and 680 pF), the overall frequency response is flat to within 0.2 dB up to 100 kHz, and the overshoot is reduced to 4%. (Power amplifiers must not be driven directly from a 1 kΩ source impedance as this will degrade linearity; buffering is required. See Chapter 6.) In control theory this is called pre-filtering. The concept of a pre-filter may strike you as a heinous bodge, but it is respectable enough to be used in satellite control systems.[31] A more complex pre-filter could eliminate the overshoot entirely, but it is not clear there is anything to be gained by doing this. In practice the circuitry upstream may have multiple HF roll-offs.

Like the other simulations, those for closed-loop performance were done with a complete amplifier circuit, in other words, with a full scale output stage driving 8 Ω.

Eliminating the Two-pole Midband Loop-gain Peak

The midband peak in the open-loop gain has nothing to do with the closed-loop peaking around 100 kHz that we have just looked at, and, as noted earlier, it has no ill-effects. If you cannot stop worrying and love the peak, then it can be eliminated by a method suggested to me by Peter Baxandall: adding a damping resistor Rd across C2, as in Figure 13.16b. Trial and error, I beg your pardon, manual optimisation, with the complete amplifier simulation showed that the peak just disappears when R = 1.4 MΩ, as in Figure 13.24. There is very little effect on the response away from the peak frequency. If lower values of damping resistor are used, the length of the 12 dB/octave response shrinks, and the feedback advantage is much reduced. There appears to be no valid reason to do this; it is not likely to increase overall stability as the response above 100 kHz is not changed.

An alternative method of eliminating the midband peak is to put a very small capacitor across C1 and C2, i.e., from VAS base to VAS collector. With the two-pole components as above, the optimal value of this capacitor in simulation is only 1.5 pF, which seems to indicate that in real life stray capacitances may be enough to eliminate the peak by themselves. The damping resistor is marginally cheaper and will have tighter tolerances than a capacitor.

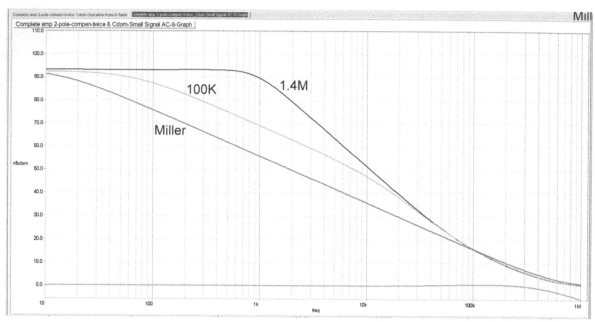

Figure 13.24. The loop-gain plot with the peak just flattened (1.4 MΩ), and with a higher degree of damping (100 KΩ). Conventional Miller compensation also shown.

Two-pole Compensation and PSRR

Elsewhere (Chapter 26) it is described how the Power-Supply Rejection Ratio (PSRR) of the amplifier, for the rail on which the VAS transistor sits, is intimately connected with the way that compensation is applied around that transistor. The basic problem is that the VAS has its ground reference connected to one supply rail (always the negative one in this book) and it is only the global negative feedback that keeps the supply rail ripple from reaching the output at unacceptable levels. Dymond and Mellor[30] describe how the PSRR under two-pole compensation can be much improved by connecting the resistor R1 (see Figure 13.14) to ground instead of the negative rail.

Two-pole Compensation: Summary

Two-pole compensation is an attractive technique, as it can be simply applied to an existing design by adding two inexpensive components; adding/removing the shunt resistor allows instant comparison between the two kinds of compensation. It is, however, only sensible to be cautious about any technique that increases the NFB factor, however it does it; power amplifiers face varying conditions and it is difficult to be sure that a design will always be stable under all circumstances. The

danger is not so much that the compensation method will not work — negative feedback has been studied intensively over the last 80 years and the mathematics is sound. The worry is that alternative compensation may inadvertently set up the conditions for parasitic oscillation in the output stage, probably by affecting the impedance at the VAS collector; this is not susceptible to mathematical analysis, and makes designers rather conservative about compensation. If you do decide to use unconventional compensation, then you need to allow plenty of time for assessing HF stability.

Amplifiers with two-pole compensation are only conditionally stable; when the open-loop gain is effectively reduced by clipping, current-limiting, slew-limiting, or at power-up/down they may show instability. I have not at the time of writing put a two-pole amplifier into quantity production.

Combining Two-pole and Output-inclusive Compensation

We have seen that both output-inclusive compensation and two-pole compensation are powerful techniques for reducing amplifier distortion, and it is an obvious idea that they should be combined. Reference[27] combines them in an opamp application, but the intention is apparently to provide enough drive for C2 rather than reduce opamp output stage distortion.

The LM12 power amp IC uses both methods.[28] The Yamaha A-720 amplifier (1986) appears to have some very strange compensation arrangements that appear to combine two-pole and output-inclusive compensation, using a transformer-coupled path for the compensation signal. Yamaha refers to this technology as ZDR (Zero Distortion Rule) circuitry, which is somewhat less than informative. Their literature suggests that they regarded it as a form of error-correction, though since output-inclusive compensation can be regarded as correcting output-stage errors, the boundary here is a little blurred. The relevant patents appear to be[32] and.[33]

The Yamaha A-520 and Yamaha A-1020 (1986) also have ZDR. A further complication with this series of amplifiers is that the larger models, the Yamaha A-720 and the A-1020 (1986) also have a feature called 'Auto Class-A power' which allows the output bias to be increased to give Class-A operation at low levels only. This is unrelated to the ZDR circuitry. ZDR was also applied in other Yamaha amplifier series, including the M-40 (1984), M-50 (1982), M-60 (1984), M-70 (1982), M-80 (1979), M-85 (1986) and A-1000 (1984).

Other Forms of Compensation

In this chapter so far we have examined dominant-pole compensation, output-inclusive compensation, and two-pole compensation in some detail. Of these, only two-pole compensation makes use of gain slopes greater than 6 dB/octave; it was described at the start of the chapter how the use of steeper slopes well away from the unit-gain intersection frequency allows the use of higher levels of negative feedback without imperilling HF stability.

There are other ways to achieve steeper gain slopes, such as the use of lead-lag networks (series R and C) across the emitter degeneration resistors or the collector loads of a differential pair. This can be seen in the Otala design in Chapter 4; this is a four-stage amplifier with lead-lag applied to the emitters of the first two differential pairs. Oscar Bonello[34] gives design rules for achieving an average 9 dB/octave slope by this method, and claims that practical amplifiers based on this were built and showed low distortion.

Stability and VAS-collector-to-ground Capacitance

In the search for HF stability, a capacitor from the VAS collector to ground can be a very present help in time of trouble; see C1 in Figure 13.25. I will be the first to admit that this is a strictly empirical modification that looks a bit suspect, but the fact is that it works. It is especially useful if there are stability issues with capacitive loads. Note that the shunt capacitor is very small in value, often 10 pF; the largest value I have so far used is 33 pF. The value is not critical.

The basic function of this component is the suppression of parasitic oscillation in the output stage. The exact theoretical mechanism is not fully known, but the key point appears to be that the impedance seen at the VAS collector is prevented from becoming inductive at very high frequencies.

This expedient is *not* the same as Lag Compensation, which is roundly condemned earlier in this chapter. C1 does not replace the dominant-pole capacitor, which remains at its original value, and C1 is orders of magnitude smaller in value than a typical lag capacitor. Obviously if C1 is too big, there may be effects on both linearity and maximum slew-rate; if it needs to be larger than, say, 47 pF, there may be something wrong with the output stage or output network design.

Figure 13.26 shows that small values of shunt capacitor C1 can be added without significantly affecting a good distortion performance. The amplifier used was one of my more recent commercial designs (2008).

With this sort of measure, it is always worth enquiring as to how far it can be taken before things go wrong; this will help you avoid picking a value that initially appears OK but is actually poised on the brink of disaster. The results of this enquiry are shown in Figure 13.27.

With C1= 280 pF, the HF distortion is slightly worse, but only above 20 kHz where its effect is less important. 480 pF causes a sharp increase of distortion at 40 kHz, characteristic of slew-rate-limiting, but linearity is not much worse from 10 kHz to 20 kHz. With 1000 pF, twice as large, we predictably get slew-limiting at 20 kHz, i.e., half the frequency. Output power was 180 W into 8 Ω (38 Vrms), a large voltage swing on the VAS being chosen to bring out the possibility of slew-limiting.

I think these results confirm that small values of shunt capacitor can be used to improve stability without affecting the distortion performance.

Nested Feedback Loops

Nested feedback is a way to apply more NFB around the output stage without increasing the global feedback factor. If an extra voltage gain stage is bolted on before the output stage, then a local feedback loop can be closed around these two stages. This NFB around

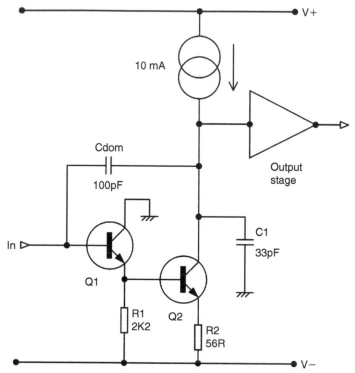

Figure 13.25. Adding a small shunt capacitor C1 from the VAS collector to ground can be very helpful in obtaining dependable HF stability.

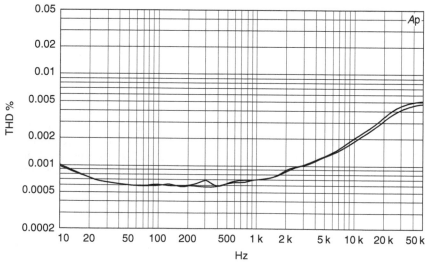

Figure 13.26. Demonstrating that adding C1 need not compromise a good distortion performance. Lower trace C1 = 10 pF, upper trace C1 = 37 pF. Power 180 W into 8 Ω.

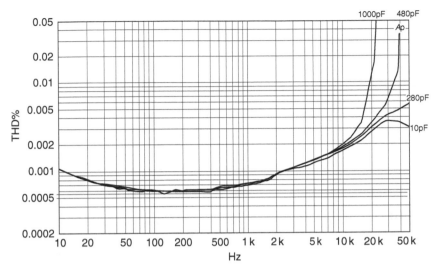

Figure 13.27. What happens to amplifier distortion when the over-large values of C1 shown are used. Power 180 W into 8 Ω.

the composite output bloc reduces output stage distortion and increases frequency response, to make it safe to include in the global NFB loop.

Suppose that bloc A1 (Figure 13.28a) is a Distortionless small-signal amplifier providing all the open-loop gain and so including the dominant pole. A3 is a unity-gain output stage with its own main pole at 1 MHz and distortion of 1% under given conditions; this 1 MHz pole puts a firm limit on the amount of global NFB that can be safely applied. Figure 13.28b shows a nested-feedback version; an extra gain-bloc A2 has been added, with local feedback around the output stage. A2 has the modest gain of 20 dB so there is a good chance of stability when this loop is closed to bring the gain of A3 + A2 back to unity. A2 now experiences 20 dB of NFB, bringing the distortion down to 0.1%, and raising the main pole to 10 MHz, which should allow the application of 20 dB more global NFB around the overall loop that includes A1. We have thus decreased the distortion that exists before global NFB is applied, and simultaneously increased the amount of NFB that can be safely used, promising that the final linearity could be very good indeed. For another theoretical example, see Pernici et al.[35]

Real-life examples of this technique in power amps are not easy to find (see the Pioneer A-8 in the next section), but it is widely used in opamps. Many of us were long puzzled by the way that the much-loved 5534 maintained such low THD up to high frequencies. Contemplation of its enigmatic entrails appears to reveal

a three-gain-stage design with an inner Miller loop around the third stage, and an outer Miller loop around the second and third stages; global NFB is then applied externally around the whole lot. Nested Miller compensation has reached its apotheosis in some CMOS opamps – the present record appears to be three nested Miller loops plus the global NFB applied by the user; do not try this one at home. More details on the theory of nested feedback can be found in Scott and Spears;[35] the treatment is wholly mathematical.

The problem is how to apply nested feedback to the usual three-stage amplifier structure. With three stages there can only be one inner feedback loop, and it is highly desirable to close it around the VAS and the output stage. This is equivalent to output inclusive compensation, as described earlier.

Nested Differentiating Feedback Loops

One implementation of nested feedback loops is the concept of Nested Differentiating Feedback Loops (NDFL), introduced by Edward Cherry in 1982. The original JAES paper[36] is tough going mathematically. A somewhat more readable account was published in *Electronics Today International* in 1983,[37] including a practical design for a 60W NDFL amplifier, though I cannot help thinking that Cherry lost 99% of his audience when he launched suddenly into complex algebra and Laplace variables. The relevant US patent is 4,243,943.[38]

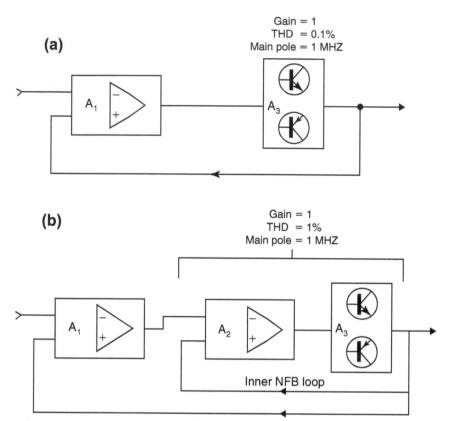

Figure 13.28. (a) Normal single-loop global negative feedback; (b) nested feedback.

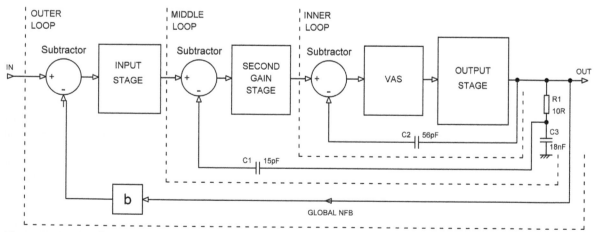

Figure 13.29. The Cherry NDFL system applied to a four-stage amplifier to give three nested feedback loops, as used in the Pioneer A-5 and A-6.

Pioneer launched NDFL in their A-5 and A-6 integrated amplifiers (1981). Cherry's initiation of the technology is acknowledged in the service manual; it appears Pioneer bought a licence to use Cherry's patent. The circuitry is not dissimilar to the ETI design, having an unusual four-stage architecture with a differential input stage, an extra single-ended gain stage, the VAS, and the output stage. That allows two inner feedback loops. The A-5 and A-6 also have a non-switching feature (see Chapter 4).

The basic idea is shown in Figure 13.29. C2 closes an inner feedback loop around the VAS and output stage. This loop is equivalent to output inclusive compensation, as described earlier in this chapter, and is not normally stable unless the loop transitions to enclose the VAS only at HF. There is no such arrangement here, so presumably the compensation of earlier stages is supposed to make the inner loop stable, though it is not easy to see how that could work.

The middle feedback loop via C1 encloses the second stage and the inner loop. It is an enigmatic feature of NDFL that C1 is fed from the middle of the Zobel network; this would appear at first sight to add destabilising phase lag; I wonder if there is some covert two-pole compensation there. Finally, the global feedback encloses the input stage and the middle loop. It is the use of C1 and C2 that gives rise to the 'differentiating' part of 'Nested Differentiating Feedback Loops', though the effect on the forward paths is of course is to make them integrating.

Distortion of the A-5 was specified as less than 0.009% 20 Hz−20 kHz at 35 W/8 Ω, which doesn't seem like an astounding leap forward.

References

1. Otala, M, An Audio Power Amplifier for Ultimate Quality Requirements, *IEEE Trans on Audio and Electroacoustics*, Vol AU-21, No. 6, Dec. 1973.

2. Black, H S, Stabilized Feed-Back Amplifiers, *Electrical Engineering*, 53, Jan. 1934, pp. 114−120.

3. Bode, H W, *Network Analysis and Feedback Amplifier Design*, New York: Van Nostrand Company Inc, 1945, p. 455.

4. Roddam, T, The Bode Fillet, *Wireless World*, Feb. 1964, p101.

5. Bode, H W, *Network Analysis and Feedback Amplifier Design*, New York: Van Nostrand Company Inc, 1945, p. 472.

6. Miller, J, M, *Dependence of the Input Impedance of a Three-Electrode Vacuum Tube Upon the Load in the Plate Circuit*. Scientific Papers of the Bureau of Standards, 15(351): 367−385, 1920.

7. Baxandall, P, Audio Power Amplifier Design: Part 4, *Wireless World*, July 1978, p. 76.

8. Dostal, J, *Operational Amplifiers* 2nd Edition, Oxford: Butterworth-Heinemann, 1993, p. 345.

9. Huijsing, J H, *Operational Amplifiers*, Dordrecht: Kluwer Academic Publishers, 2001, p. 211 onwards.

10. Takahashi, S et al., Design and Construction of High Slew-Rate Amplifiers, paper presented at AES 60th Convention, Preprint No. 1348 (A-4) 1978.

11. Widlar, B and Yamatake, M, A Monolithic Power Op-Amp, *IEEE J Solid-State Circuits*, 23, No 2, April 1988.

12. Linsley-Hood, J, Solid-State Audio Power, *Electronics World +Wireless World*, Nov. 1989, p. 047.

13. Leach, M, Feedforward Compensation of the Amplifier Output Stage for Improved Stability with Capacitive Loads, *IEE Trans on Consumer Electronics*, 34(2), May 1988, pp. 334−338. (So far as I can see, this is a misuse of the word feedforward. What is actually described is semi-local feedback around the input stage, so that the output stage actually sees *less* HF feedback. The aim is to avoid using an output inductor; I feel this is a mistake. The amplifier is divided into only two stages for analysis.)

14. P Baxandall, Private communication, 1995.

15. Baxandall, P and Self, D, *Baxandall and Self On Audio Power*, Linear Audio Publishing, Sept. 2011 ISBN 978-94-90929-03-9 p. 115 (inclusive compen).

16. Seki, K, *Multistage Amplifier Circuit*, US Patent 4,145,666, 20 Mar. 1979.

17. Gunderson, A, Topology to Linearise Miller-Effect Compensated Amplifiers (OFICC = Output Following Intermediate Cascode Circuit) *JAES*, June 1984, p. 430.

18. diyAudio bulletin board: http://www.diyaudio.com/forums/solid-state/171159-bob-cordells-power-amplifier-book-92.html (That page and many following) (accessed June 2012).

19. Self, D, Inclusive Compensation and Ultra-Low Distortion Amplifiers, *Linear Audio*, Volume 0, Sept. 2010, p. 6.

20. The Signal Transfer Company http://www.signaltransfer.freeuk.com/ (accessed June 2012).

21. Candy, B H, *Amplifier Having Ultra-low Distortion*, US patent No. 5,892,398 (1999).

22. Candy, B H, *Low Distortion Amplifier*, US patent No. 6,600,367 (2003).

23. Stochino, G, Audio Design Leaps Forward? *Electronics World*, Oct. 1994, p. 818.

24. Stochino, G and Porra, S, *Audio Power Amplifier Apparatus*, US patent, No. 7,564,304. (2009).

25. National Semi, *Fast Compensation Extends Power Bandwidth*, Linear Brief 4, Santa Clara, CA: NatSem Linear Apps Handbook, 1991, p. 1186.

26. Feucht, D, *Handbook of Analog Circuit Design, San Diego:* Academic Press, 1990, p. 264.

27. Dobkin, R C, *Fast Compensation Extends Power Bandwidth*, National Semiconductor Linear Brief 4; 1969.

28. Anon, LM12CL 80W Operational Amplifier data sheet. Texas Instruments, 2008, p. 14.

29. Baxandall, P and Self, D, *Baxandall and Self on Audio Power*, Linear Audio Publishing, Sept. 2011 ISBN 978-94-90929-03-9, p. 112 (2-pole).

30. Dymond, H and Mellor, P. Analysis of Two-Pole Compensation in Linear Audio Amplifiers, AES Convention paper, 129th Convention, San Francisco, Nov. 2010.

31. Lurie, B and Enright, P, *Classical Feedback Control*, Boca Raton, FL: CRC Press, Taylor & Francis Group, 2012, p. 111.

32. Iwamatsu, M, *Amplifier with Distortion Cancellation*, US patent 4,476,442, Oct. 1984.

33. Yokoyama, K, *Power Amplifier Circuit for an Audio Circuit* US patent 4,785,257, Nov 1988.

34. Bonello, O, Advanced Negative Feedback Design for High Performance Amplifiers, paper presented at 67th AES Convention, New York, Oct. 1980, Preprint 1706 (D-5).

35. Scott, J, Spears, G, and Cherry, E M, On the Advantages of Nested Feedback *Loops JAES* 39, March 1991, p. 115.

36. Cherry, E, Nested Differentiating Feedback Loops in Simple Audio Power Amplifiers, *JAES* 30(5), May 1982, p. 295.

37. Cherry, E, Designing NDFL Amps, *Electronics Today International*, April and May 1983.

38. Cherry, E, *Feedback Systems*, US Patent 4,243,943, Jan. 1981.

Chapter 14

Output Networks and Load Effects

Poverty is the load of some, and wealth is the load of others, perhaps the greater load of the two. It may weigh them to perdition.

Augustine of Hippo (354–430 AD)

Output Networks

The usual output networks for a power amplifier are shown in Figure 14.1, with typical values. They comprise a shunt Zobel network, for stability into inductive loads, and a series output inductor/damping resistor for stability into capacitive loads.

Amplifier Output Impedance

The main effect of output impedance is usually thought to be its effect on the Damping Factor. This is wrong, as explained in Chapter 1. Despite this demonstration of its irrelevance, I will refer to the Damping Factor here, to show how an apparently impressive figure dwindles as more parts of the speaker-cable system are included.

Figure 14.1 shows a simplified amplifier with Zobel network and series output inductor, plus simple models of the connecting cable and speaker load. The output impedance of a solid-state amplifier is very low if even a modest amount of global NFB is used.

I measured a Blameless Class-B amplifier similar to Figure 14.9 with the usual NFB factor of 29 dB at 20 kHz, increasing at 6 dB/octave as frequency falls. Figure 14.2 shows the output impedance at point B before the output inductor, measured by injecting a 10 mA signal current into the output via a 600 Ω resistance.

The low-frequency output impedance is approximately 9 mΩ (an 8 Ω Damping Factor of 890). To put this into perspective, one metre of thick 32/02 equipment cable (32 strands of 0.2 mm diameter) has a resistance of 16.9 mΩ. The internal cabling resistance in an amplifier can equal or exceed the output impedance of the amplifier itself at LF. Cable resistance is looked at in more detail in Chapter 25.

Output impedance rises at 6 dB/octave above 3 kHz, as global NFB falls off, reaching 36 mΩ at 20 kHz. The 3 kHz break frequency does not correspond with the amplifier dominant pole frequency, which is much lower at around 10 Hz.

The closed-loop output impedance of any amplifier is set by the open-loop output impedance and the negative feedback factor. The output impedance is not simply the output impedance of the output stage alone, because the latter is driven from the VAS, so there is a significant and frequency-varying source impedance at point A in Figure 14.1.

When the standard EF and CFP stages are driven from a zero-impedance source, in both cases the raw

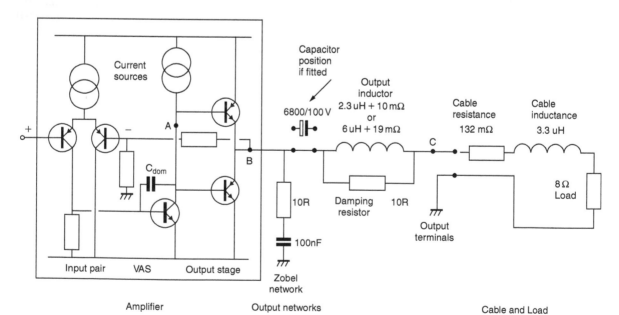

Figure 14.1. The amplifier-cable-speaker system. Simplified amplifier with Zobel network and damped output inductor, and a resistive load. Cable resistance and inductance values are typical for a 5 m length.

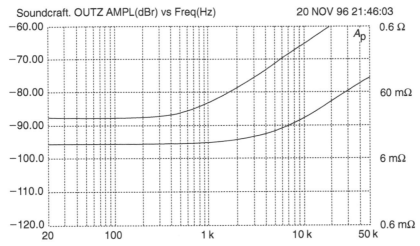

Figure 14.2. Output impedance of a Blameless amplifier, with and without 6 µH output inductor. Adding the inductor (upper trace) increases both the flat LF output impedance, due to its series resistance, and the rising HF impedance.

output impedance is in the region of 150–180 mΩ. This assumes the emitter resistors Re are 0.1 Ω. Increasing Re to 0.22 Ω increases output impedance to the range 230–280 mΩ, showing that these resistors in fact make up most of the output impedance. The output devices and drivers have little influence.

If the average open-loop output impedance is 200 mΩ, and the NFB factor at 20 kHz is 29 dB, or 28 times, we would expect a closed-loop output impedance of approximately 200/28, which is 7 mΩ. Since it is actually about 33 mΩ at this frequency, there is clearly more going on than simple theory implies. In a real amplifier the output stage is not driven from a zero impedance, but a fairly high one that falls proportionally with frequency; for my Blameless Class-B design, it falls from 3 kΩ at 1 kHz to about 220 Ω at 20 kHz. A 220 Ω source impedance produces an open-loop output impedance of about 1 Ω, which when reduced by a factor of 28 when global feedback is applied, gives 35 mΩ. This is close to the value measured at 20 kHz at point B in Figure 14.1.

All of these measured closed-loop output impedances are very low compared with the other impedances in the amp-cable-speaker system. It would appear they can in most cases be ignored.

Figure 14.2 was produced using an output inductor of approximately 6 µH, at the high end of the permissible range. This is limited by the HF roll-off into the lowest load resistance to be driven. The 6 µH inductor is a substantial component comprising 20 turns of 1.5 mm diameter copper wire, wound in a 1 in. diameter coil, with a DC resistance of 19 mΩ. This small extra

resistance raises the flat section of the impedance plot to 24 mΩ, and in fact dominates the LF output impedance as measured at the amplifier terminals (point C). It also sharply reduces the notional Damping Factor from 890 to 330.

Naturally the inductance of the coil pushes the rising portion of the impedance curve higher. The output impedance now starts to rise from 700 Hz, still at 6 dB per octave, reaching 0.6 Ω at 20 kHz. See Figure 14.2.

Minimising Amplifier Output Impedance

This issue is worth considering, not because it optimises speaker dynamics, which it does not, but because it minimises frequency response variations due to varying speaker impedance. There is also, of course, specmanship to be considered.

It is clear from Figure 14.2 that the output impedance of a generic amplifier will very probably be less than the inductor resistance, so the latter should be attended to first. Determine the minimum output inductance for stability with capacitive loads, because lower inductance means fewer turns of wire and less resistance. Some guidance on this is given in the next section. Note, however, that the inductance of the usual single-layer coil varies with the square of the number of turns, so halving the inductance only reduces the turns, and hence the series resistance, by root-two. The coil wire must be as thick as the cost/quality tradeoffs allow.

It is also desirable to minimise the resistance of the amplifier internal wiring, and to carefully consider any extra resistance introduced by output relays, speaker switching, etc. When these factors have been reduced as far as cost and practicality allow, it is likely that the output impedance of the actual amplifier will still be the smallest component of the total.

Zobel Networks

All power amplifiers except for the most rudimentary kinds include a Zobel network in their arrangements for stability. This simple but somewhat enigmatic network comprises a resistor and capacitor in series from the amplifier output rail to ground. It is always fitted on the inside (i.e., upstream) of the output inductor, though a few designs have a second Zobel network after the output inductor; the thinking behind this latter approach is obscure. The resistor approximates to the expected load impedance, and is usually between 4.7 and 10 Ω. The capacitor is almost invariably 100 nF, and these convenient values and their constancy in the face of changing amplifier design might lead one to suppose that they are not critical; in fact, experiment suggests that the real reason is that the traditional values are just about right.

The function of the Zobel network (sometimes also called a Boucherot cell) is rarely discussed, but is usually said to prevent too inductive a reactance being presented to the amplifier output by a loudspeaker voice-coil, the implication being that this could cause HF instability. It is intuitively easy to see why a capacitive load on an amplifier with a finite output resistance could cause HF instability by introducing extra lagging phase-shift into the global NFB loop, but it is less clear why an inductive load should be a problem; if a capacitive load reduces stability margins, then it seems reasonable that an inductive one would increase them.

At this point I felt some experiments were called for, and so I removed the standard 10 Ω/0.1 μF Zobel from a Blameless Class-B amplifier with CFP output and the usual NFB factor of 32 dB at 20 kHz. With an 8 Ω resistive load, the THD performance and stability were unchanged. However, when a 0.47 mH inductor was added in series, to roughly simulate a single-unit loudspeaker, there was evidence of local VHF instability in the output stage; there was certainly no Nyquist instability of the global NFB loop.

I also attempted to reduce the loading placed on the output by the Zobel network. However, increasing the series resistance to 22 Ω still gave some evidence of stability problems, and I was forced to the depressing conclusion that the standard values are just about

right. In fact, with the standard 10 Ω/0.1 μF network the extra loading placed on the amplifier at HF is not great; for a 1 V output at 10 kHz the Zobel network draws 6.3 mA, rising to 12.4 mA at 20 kHz, compared with 125 mA drawn at all frequencies by an 8 Ω resistor. These currents can be simply scaled up for realistic output levels, and this allows the Zobel resistor power rating to be determined. Thus an amplifier capable of 20 V rms output must have a Zobel resistor capable of sustaining 248 mA rms at 20 kHz, dissipating 0.62 W; a 1 W component could be chosen.

In fact, the greatest stress is placed on the Zobel resistor by HF instability, as amplifier oscillation is often in the range 50–500 kHz. It should therefore be chosen to withstand this for at least a short time, as otherwise faultfinding becomes rather fraught; ratings in the range 3 to 5 W are usual.

To conclude this section, there seems no doubt that a Zobel network is required with any load that is even mildly inductive. The resistor can be of an ordinary wire-wound type, rated to 5 W or more; this should prevent its burn-out under HF instability. A wire-wound resistor may reduce the effectiveness of the Zobel at VHF, but seems to work well in practice; the Zobel still gives effective stabilisation with inductive loads.

Output Inductors

Only in the simplest kinds of power amplifier is it usual for the output stage to be connected directly to the external load. Direct connection is generally only feasible for amplifiers with low feedback factors, which have large safety margins against Nyquist instability caused by reactive loads.

When the stability of amplifiers into various loads is discussed, the phrase 'unconditional stability' is usually bandied about by people who are under the impression it means 'stable with any load you can think up'. Its original meaning, which comes from control theory, is quite different. In a normal dominant pole compensated amplifier, reducing the loop gain (e.g., by reducing the amount of NFB) simply makes it more stable; this is the true meaning of 'unconditional stability'. If, however, you have a complicated compensation scheme, it is not hard to come up with an amplifier that becomes unstable when the loop gain is reduced, and this is called 'conditional stability'.

For many years designers have been wary of what might happen when a capacitive load is connected to their amplifiers; a fear that dates back to the introduction of the first practical electrostatic loudspeaker from Quad

Acoustics, which was crudely emulated by adding a 2 µF capacitor in parallel to the usual 8 Ω resistive test load. The real load impedance presented by an electrostatic speaker is far more complex than this, largely as a result of the step-up transformer required to develop the appropriate drive voltages, but a 2 µF capacitor alone can cause instability in an amplifier unless precautions are taken.

When a shunt capacitor is placed across a resistive load in this way, and no output inductor is fitted, it is usually found that the value with the most destabilising effect is nearer 100 nF than 2 µF.

The most effective precaution against this form of instability is a small air-cored inductor in series with the amplifier output. This isolates the amplifier from the shunt capacitance, without causing significant losses at audio frequencies. The value is normally in the region 1−7 µH, the upper limit being set by the need to avoid significant HF roll-off into a 4 Ω load. If 2 Ω loads are contemplated, then this limit must be halved.

It is usual to test amplifier transient response with a square wave while the output is loaded with 8 Ω and 2 µF in parallel to simulate an electrostatic loudspeaker, as this is often regarded as the most demanding condition. However, there is an inductor in the amplifier output, and when there is significant capacitance in the load they resonate together, giving a peak in the frequency response at the HF end, and overshoot and ringing on fast edges.

This test therefore does not actually examine amplifier response at all, for the damped ringing that is almost universally seen during these capacitive loading tests is due to the output inductor resonating with the test load capacitance, and has nothing whatever to do with amplifier stability. The ringing is usually around 40 kHz or so, and this is much too slow to be blamed on any normally compensated amplifier. The output network adds ringing to the transient response even if the amplifier itself is perfect.

It is good practice to put a low-value damping resistor across the inductor; this reduces the Q of the output LC combination on capacitive loading, and thus reduces overshoot and ringing.

If a power amplifier is deliberately provoked by shorting out the output inductor and applying a capacitive load, then the oscillation is usually around 100−500 kHz, and can be destructive of the output transistors if allowed to persist. It is nothing like the neat ringing seen in typical capacitive load tests. In this case there is no such thing as 'nicely damped ringing' because damped oscillation at 500 kHz probably means you are one bare step away from oscillatory disaster.

Attempts to test this on a standard Blameless amplifier were frustrated because it is actually rather resistant to capacitance-induced oscillation, probably because the level of global feedback is fairly modest. 100 nF directly across the output induced damped ringing at 420 kHz, while 470 nF gave ringing at 300 kHz, and 2 µF at 125 kHz.

While the 8Ω/2 µF test described above actually reveals nothing about amplifier transient response, it is embedded in tradition, and it is too optimistic to expect its doubtful nature to be universally recognised. Minimising output ringing is of some commercial importance; several factors affect it, and can be manipulated to tidy up the overshoot and avoid deterring potential customers:

- The output inductance value. Increasing the inductance with all other components held constant reduces the overshoot and the amount of response peaking, but the peak moves downward in frequency so the rising response begins to invade the audio band. See Figures 14.3 and 14.4.

- The value of the damping resistor across the output coil. Reducing its value reduces the Q of the output LC tuned circuit, and so reduces overshoot and ringing. The resistor is usually 10 Ω, and can be a conventional wirewound type without problems due to self-inductance; 10 Ω reduces the overshoot from 58% without damping to 48%, and much reduces ringing. Response peaking is reduced with only a slight effect on frequency. See Figures 14.5 and 14.6. The damping resistor can in fact be reduced to as low as 1 Ω, providing the amplifier stability into capacitance remains dependable, and this reduces the transient overshoot further from 48% to 19%, and eliminates ringing altogether; there is just a single overshoot. Whether this is more visually appealing to the potential customer is an interesting point.

- The load capacitance value. Increasing this with the shunt resistor held at 8 Ω gives more overshoot and lower frequency ringing that decays more slowly. The response peaking is both sharper and lower in frequency, which is not a good combination. However, this component is part of the standard test load and is outside the designer's control. See Figures 14.7 and 14.8.

- In actual fact, by far the most important factor affecting overshoot and ringing is the rise-time of the applied square wave. This is yet another rather important audio fact that seems to be almost unknown. Figure 14.9 shows how the overshoot given by the circuit in Figure 14.1 plus load capacitance is

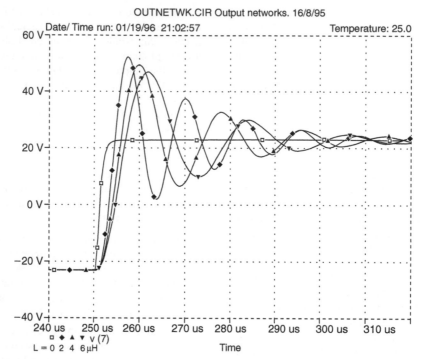

Figure 14.3. Transient response with varying output inductance; increasing L reduces ringing frequency without much effect on overshoot. Input rise-time 1 μsec.

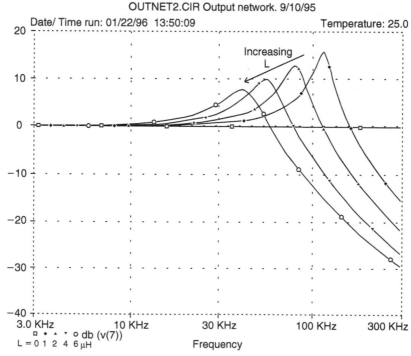

Figure 14.4. Increasing the output inductance reduces frequency response peaking and lowers its frequency.

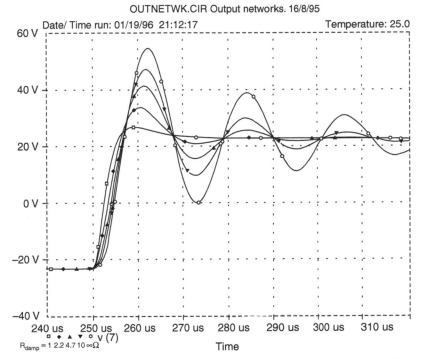

Figure 14.5. The effect of varying the damping resistance on transient response. 1 Ω almost eliminates overshoot.

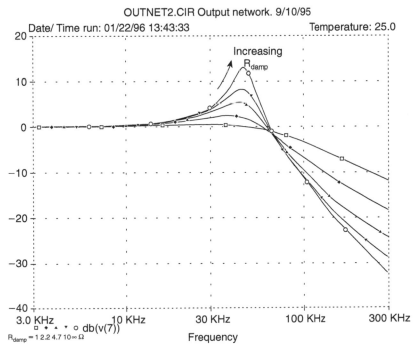

Figure 14.6. The effect of varying damping resistance on frequency response. Lower values reduce the peaking around 40 kHz.

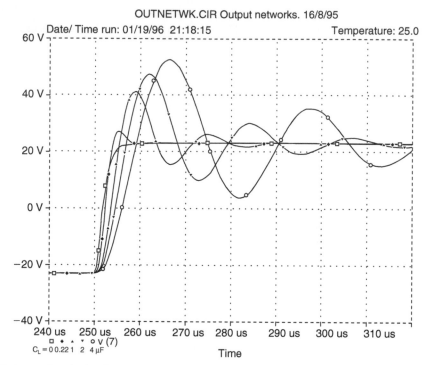

Figure 14.7. Increasing the load capacitance increases the transient overshoot, while lowering its frequency.

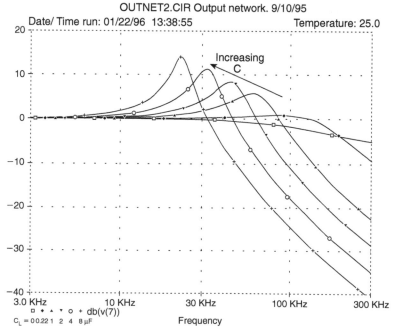

Figure 14.8. Increasing the load capacitance increases frequency response peaking and lowers its frequency.

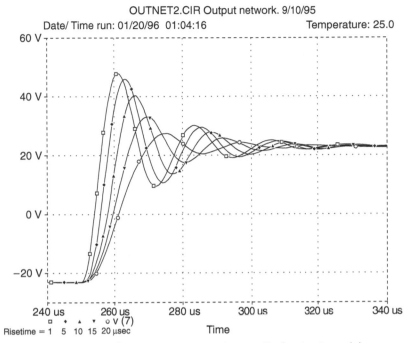

Figure 14.9. The most important factor in the transient response is actually the rise-time of the square-wave input, especially for overshoot percentage. The ringing frequency is unaffected.

51% for a 1 μsec rise-time, but only 12% for a 20 μsec rise-time. It is clear that the 'transient response' measured in this test may depend critically on the details of the testgear and the amplifier slew-rate, and can be manipulated to give the result you want.

An output inductor should be air-cored to eliminate the possibility of extra distortion due to the saturation of magnetic materials. Wire thick enough to handle the output current will be self-supporting and no former is required. Ferrite-based VHF chokes give stable operation, but their linearity must be considered dubious. In the 1970s there was a fashion for using one of the big power-supply electrolytics as a coil-former, but this is a really terrible idea. The magnetic characteristics of the capacitor are unknown, and its life-time may be reduced by heat dissipated in the coil winding resistance.

The resistance of an air-cored 6 μH coil made from 20 turns of 1.5 mm diameter wire (this is quite a substantial component 3 cm in diameter and 6 cm long) is enough to cause a measurable power loss into a 4 Ω load, and to dominate the output impedance as measured at the amplifier terminals. The coil wire should therefore be as thick as your cost/quality tradeoffs allow.

The power rating for the damping resistor is assessed as follows. For a resistive 8 Ω load the voltage across the output inductor increases slowly with frequency, and the damping resistor dissipation only reaches 1.2 mW at 20 kHz for 1 V rms output. This assumes a normal 10 Ω damping resistor; if the value is reduced to 1 Ω to eliminate ringing into capacitive loads, as described above, then the dissipation is ten times as great at 12 mW.

A much greater potential dissipation occurs when the load is the traditional 8 Ω/2 μF combination. The voltage across the output inductor peaks as it resonates with the load capacitance, and the power dissipated in a 10 Ω damping resistor at resonance is 0.6 W for 1 V rms. This is, however, at an ultrasonic frequency (around 50 kHz with a 7 μH inductor) and is a fairly sharp peak, so there is little chance of musical signals causing high dissipation in the resistor in normal use. However, as for the Zobel network, some allowance must be made for sinewave testing and oscillatory faults, so the damping resistor is commonly rated at between 1 and 5W. An ordinary wire-wound component works well with no apparent problems due to self-inductance.

An alternative method of stabilisation is to put in a small series resistor instead of the inductor; this approach has been used by at least one English manufacturer. I found that with 100 nF loading, a 0.1 Ω wire-wound output resistor completely removed ringing on the amplifier output. This is cheaper than an inductor, but obviously less efficient as 100 mΩ of extra resistance have been introduced instead of the 10 mΩ of the new 2.3 μH inductor. The so-called 'damping factor' relative to 8 Ω with a 0.1 Ω series resistor cannot exceed 80. A more important objection is that the 4 Ω output power is significantly reduced − a 200 W/4 Ω amplifier is reduced to a 190 W unit, which does not look so good in the specs, even though the reduction in perceived loudness is negligible.

An example of this approach was the Rotel RA-820B integrated amplifier (released 1985) which in series with the output a 0.22 Ω 2 W resistor, and also a 3.15 Amp fuse. Ignoring the fuse resistance, the 'damping factor' could not have exceeded 36.

Designing the Output Inductor: Single-layer Coils

As mentioned above, the output inductor for my earlier amplifier designs started out at 20 turns and approximately 6 μH, with the aim of erring on the side of safety as regards stability. This gives roll-off into a resistive 4 Ω load of −3 dB at 106 kHz, −1 dB at 53 kHz, and −0.15 dB at 20 kHz. Clearly this is about the maximum inductance usable without introducing an unwanted droop at the top of the audio range. If 2 Ω loads or worse are a possibility, the inductance will need to be halved- and you will also want to make sure that the coil is wound with some pretty thick wire, to reduce resistive losses. My 6 uH inductor was close-wound (with adjacent turns touching) and made from copper wire 1.5 mm diameter, which is more than strong enough to make the coil self-supporting. It was quite a hefty component.

After a lot of extensive testing the physical size of my 'standard output inductor' was cut in half, so it only had 10 turns, 2.3 μH inductance, and 10.1 mΩ DC resistance. Note that the inductance dropped to less than half; inductance is in general proportional to the square of the number of turns, but for short coils like this, the end effects are significant. This inductor has proved adequate for stability with various types of amplifier, and a very wide range of loads. It does now look more like an 'average' amplifier output inductor, rather than an oversized one.

It may surprise you that there is no universally accepted exact formula for the inductance of a single-layer coil. An attempt at an exact calculation is

ferociously complicated, partly due to the end effects. The calculation is much simpler for a section of an infinitely long coil, but the price of copper being what it is, you don't see many of those around these days.

A good approximation is Wheeler's formula[1,2] shown as Equation 14.1 which is accurate to 1% when the coil diameter/length ratio is less than 3. That is more than good enough for our purposes; the answer will be only 4% low for diameter/length = 5.

$$L = 1000 \frac{r^2 N^2}{228r + 254l} \qquad \text{Equation 14.1}$$

where:

L is inductance in uH	N is the number of turns
r is coil radius in metres	l is coil length in metres

If the coil is close-wound, the coil length l is not an independent quantity but is determined by the number of turns N and the diameter of the wire chosen, as in Equation 14.2. If both lead-out wires are soldered into the PCB, which is almost always the case, then the number of turns has to be an integer, obtained by rounding up rather than down to ensure the inductance value is high rather than low.

$$N = \frac{w}{2\pi r} \qquad \text{Equation 14.2}$$

where:

r is coil radius in metres	w is wire length (not coil length) in metres
	N is number of turns

From this we calculate the coil length:

$$l = bN \qquad \text{Equation 14.3}$$

where:

    b is the diameter of the wire in metres
    N is number of turns

It is best to set up these equations on a spreadsheet; the design procedure for a 2 uH inductor then goes like this

1. Select the wire diameter you want to use. We will be generous and use 1.5 mm diameter wire.

2. Make a guess at a suitable coil radius (10 mm) and Goal-Seek by altering the wire length w until you get the desired inductance L, the new wire length w, and the turns N, using Equation 14.2. Our answers are 2.002 uH, 0.727 metres, and 11.56 turns.
3. Make N an integer by rounding up, giving 12 turns. Plug that into the equations.
4. Check the new larger value of inductance; we now have 2.101 uH, near enough to cause no anxiety. Better too much than too little.
5. Calculate the coil length from Equation 14.3, as 18.0 mm.
6. Check the diameter/length ratio gives an acceptable coil efficiency. As described below, this basically means not letting the length exceed the diameter. Here diameter/length = 1.11 and we get 95% of the maximum possible inductance for the wire used.
7. Using the new wire length w, the wire diameter b, and the resistivity of the metal used (almost always copper) check that the series resistance is acceptable. First calculate the cross-sectional area a of the wire from Equation 14.4, then plug it into Equation 14.5 to get the resistance.

$$a = \pi \left(\frac{b}{2}\right)^2 \qquad \text{Equation 14.4}$$

where:

b is the diameter of the wire in metres	a is cross-sectional are of wire in metre^2

We will add 5 mm to each end of the coil for the lead-out wires, making the total wire length 0.727 + 0.010 metres.

$$R = \frac{\rho w}{a} \qquad \text{Equation 14.5}$$

where:

w is wire length (not coil length) in metres
a is cross-sectional are of wire in metre^2
ρ (rho) = resistivity of the metal used (1.72 x 10^{-8} Ohm-metres for copper)

(Be aware that the resistivity of copper varies slightly according to the level of impurities in it, and can vary from 1.71 x 10^{-8} Ohm-metres to 1.8 x 10^{-8}. Fortunately the coil resistances we are likely to encounter are so low that this makes no practical difference.)

This gives us a resistance of 7.17 milliOhms. As described earlier in this chapter, that is a small fraction of the typical total resistance of coil, internal wiring, and external loudspeaker cables.

Output coils are usually close-wound, to minimise the space taken up. Spreading out the turns not only occupies more room but also reduces the inductance for the same length of wire and so makes the coil less cost-effective.

Amplifier coils come in various shapes and sizes, and it may well have occurred to you that there is an optimal coil configuration which gives the maximum inductance for a given length of wire, or, to put it another way, the minimal length of wire for a given inductance. There is, and it occurs when the ratio of diameter/length is 2.22. Getting somewhere near this optimal point is of some importance, not only because of the high cost of copper, but because it minimises the series resistance. The relation between diameter/length ratio and the normalised inductance is shown in Figure 14.10 which was produced by using the Wheeler formula with a constant 1.9 metre of wire, and designing a set of coils with varying radii, which naturally causes the number of turns and the length to vary as one turn will consist of varying lengths of wire.

The inductance maximum is a very broad peak and does not fall below 99% so long as diameter/length is between 1.5 and 3; in fact, so long as the length does not exceed the diameter, the coil will be better than 93% efficient, which will be good enough most of the time. What is definitely to be avoided is the use of long thin coils; it is not uncommon to see output inductors that are four or more times as long as their diameter, so diameter/length is 0.25 or less, and they are working on the far left of Figure 14.10. A coil four times as long as its diameter has only 60% of the inductance of the optimal shape, and to increase its inductance to the equivalent of 100% would require 66% more copper. Figure 14.11 shows an optimal coil compared with one four times longer than its diameter.

Coils with diameter much greater than their length lose far less efficiency, but are likely to extend inconveniently far above the PCB, lack mechanical strength, and are thus vulnerable to knocks during production.

There may of course be other considerations than maximising the inductor efficiency on this basis. For example, staying within a height limit or minimising the PCB footprint.

Unless you have an unusually-proportioned coil, wire of 1 mm diameter or greater should be self-supporting. 1.5 mm wire has excellent strength; if you treat a length of wire as a structural beam, its stiffness is proportional to the fourth power of its diameter, so the 1.5 mm wire is 5.1 times stiffer. The 1.5 mm wire also gives a satisfyingly low series resistance, but the copper cost is correspondingly greater, and it is hard to solder into a PCB

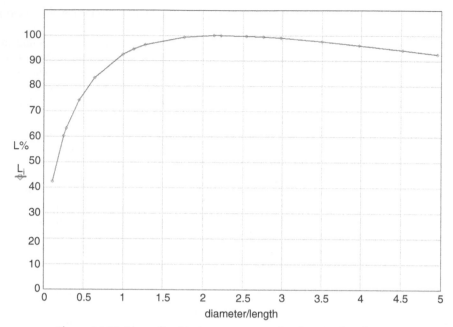

Figure 14.10. Normalised inductance versus the diameter/length ratio.

because it rapidly conducts heat away from the joint. A high-power soldering-iron is recommended.

There should be no metal plates close to the coil because the shorted-turn effect will reduce the effective inductance.

Designing the Output Inductor: Multi-layer Coils

If space is tight and there is not enough room for a single-layer coil that gives adequate inductance, then a multi-layer coil will give substantially more inductance for a given PCB footprint area.

Multi-layer coils usually have an odd number of layers. Take the case of a 3-layer coil; the winding goes from start to finish, goes back to the start for the second layer, then from start to finish again for the third layer so the lead-out wires are at opposite ends and give suitable mechanical support. This change in winding direction means that the turns in each layer do not sit as closely together as if both layers were wound in the same direction.

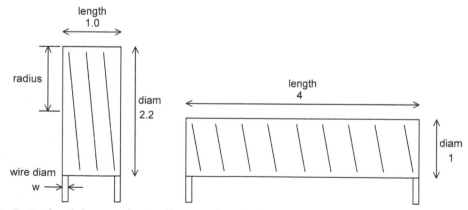

Figure 14.11. Optimal coil shape, and an inefficient coil with a diameter/length ratio of 0.25, using the same length of wire. The latter has only 60% of the inductance of the optimal shape.

As with single layer coils, there are proportions which give the maximum efficiency. This is called a Brooks coil, and an example is shown in Figure 14.12. The coil windings have a square cross-section, with c = length and radius = 3c/2. To put it another way the inner diameter is equal to twice the height (or length) of the coil winding. For this special case a suitably accurate approximation for the inductance is delightfully simple, as in Equation 14.6.

$$L = 1.6694rN^2 \qquad \text{Equation 14.6}$$

Where:

    r is coil radius in metres
    N is the number of turns
    L is inductance in uH

The square cross-section for the coil is efficient because it brings all the turns close together, increasing the coupling between them and hence the total inductance of the coil. A circular cross-section might be slightly better, but it would be harder to wind, and, as mentioned above, in practice the turns cannot fit together in total intimacy anyway.

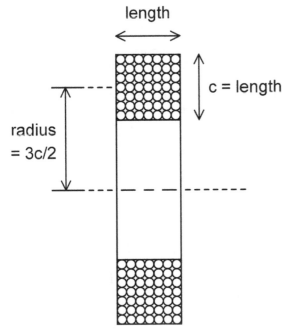

Figure 14.12. The dimensions of a multi-layer coil; the optimal proportions (a Brooks coil) are shown. In a practical output inductor there will be fewer turns and they will not lie together as neatly as shown.

Good approximation equations exist for multi-layer coils with all the dimensions arbitrary, but they are much more complicated, and rarely required because a coil has to deviate quite significantly from the Brooks proportions before L is much reduced.

Let's compare a multi-layer coil with the single-layer coil designed in the previous section, which had coil radius of 10 mm, a length of 18 mm, and 12 turns of 1.5 mm diameter wire. If we use the same radius, we find we need 10.85 turns to get exactly 2.0 uH. Obviously we have to round that up to 11 turns, giving 2.06 uH, and at first it looks as if we have gained very little over the single-layer coil.

But consider that this is a multi-layer coil; the obvious way to make it is with three layers of 4 turns each, (which would give 2.45 uH) though it may be possible for your coil-winder to put only 3 turns on one layer. This is near enough to a square winding to have very little effect on the calculation of inductance. We therefore get a coil length of only 4 times the wire diameter, i.e., 6 mm. That is a third of the length of the single-layer coil, and the coil diameter has only increased by 3 mm. The multi-layer coil will save significant PCB area, though little if any weight of copper. It is likely to be slightly more expensive to wind than a single-layer coil.

Crosstalk in Amplifier Output Inductors

When designing a stereo power amplifier, the issue of interchannel crosstalk is always a concern. Now that amplifiers with up to seven channels for home theatre are becoming more common, the crosstalk issue is that much more important, if only because the channels are likely to be more closely packed. Here I deal with one aspect of it. Almost all power amplifiers have output coils to stabilise them against capacitive reactances, and a question often raised is whether inductive coupling between the two is likely to degrade crosstalk. It is sometimes suggested that the coils — which are usually in solenoid form, with length and diameter of the same order — should be mounted with their axes at right angles rather than parallel, to minimise coupling. But does this really work?

I think I am pretty safe in saying there is no published work on this, so it was time to make some. The coil coupling could no doubt be calculated (though not by me) but as often in the glorious pursuit of electronics, it was quicker to measure it.

The coils I used were both single-layer close-wound with 14 turns of 1 mm diameter copper wire, overall length 22 mm and diameter 20 mm. This has an inductance of about 2 μH, and is pretty much an 'average' output coil, suitable for stabilising amplifiers up to about 150 W/8 Ω. Different coils will give somewhat different results, but extrapolation to whatever component you are using should be straightforward; for example, twice the turns on both coils means four times the coupling.

Figure 14.13a shows the situation in a stereo power amplifier. The field radiated due to the current in Coil A is picked up by Coil B and a crosstalk voltage added to the output signal at B.

Figure 14.13b shows the experimental set-up. Coil A is driven from a signal generator with a source impedance of 50 Ω, set to 5 Vrms. Virtually all of this is dropped across the source resistance, so Coil A is effectively driven with a constant current of 100 mA rms.

Figure 14.14 shows the first configuration measured; the coils are coaxial with varying spacing between them. A spacing of zero, with the ends touching was, as expected, the worst case for coupling, and the crosstalk at 20 kHz was taken as the 0 dB reference, marked 'CAL' on the plots that follow. This yielded 2.4 mV rms across Coil B. Since 100 mA rms in Coil A corresponds to 800 mVrms across an 8 Ω, load, this gives a voltage crosstalk figure from channel to channel of 800/2.4 = −50 dB at 20 kHz. It carries on deteriorating above 20 kHz but no one can hear it. All crosstalk figures given below are at 20 kHz.

The crosstalk rises at 6 dB/octave, because the voltage induced in Coil B is proportional to the rate of change of flux, and the magnitude of peak flux is fixed. This is clearly not the same as conventional transformer action, where the frequency response is flat. In a transformer the primary inductance is much greater than the circuit series impedance, so the magnetic flux that couples with the secondary halves as the input frequency doubles, and the voltage induced in the secondary is constant.

The coils were then separated 10 mm at a time, and with each increment the crosstalk dropped by 10 dB, as seen in Figure 14.15. At a 110 mm spacing, which is quite practical for most designs, the crosstalk had fallen by 47 dB from the reference case, giving an overall crosstalk figure of −54 and −47 dB = −101 dB total. This is a very low level, and at the very top of the audio band. At 1 kHz, where the ear is much more sensitive, the crosstalk will be some 25 dB less, which brings it down to −126 dB total which I can say with some confidence is not going to be a problem. This is obtained with what looks like the least favourable orientation of coils. The crosstalk is −32 dB at 50 mm

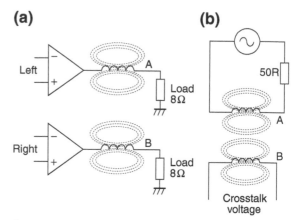

Figure 14.13. (a) The coupling of output coils in a stereo power amplifier; (b) the experimental circuit the 'transmitting' Coil A is driven with an effectively constant current, and the voltage across the 'receiving' Coil B measured.

spacing, and this figure will be used to compare the configurations.

The next configuration tested was that of Figure 14.16, where the coils have parallel axes but are displaced to the side. The results are shown in Figure 14.17; the crosstalk is now −38 dB at 50 mm. With each 10 mm spacing increment the crosstalk dropped by 7 dB. This set-up is worse than the crossed-axis version but better than the coaxial one.

The final configurations had the axes of the coils at 90°; the crossed-axis condition. The base position is with the corners of the coils touching; see Figure 14.18. When the coils are touching, crosstalk almost vanishes as there is a cancellation null. With the coils so close, this is a very sharp null and exploiting it in quantity production is quite impractical. The slightest deformation of either coil ruins the effect. Moving the Coil A away from B again gives the results in Figure 14.19. The crosstalk is now −43 dB

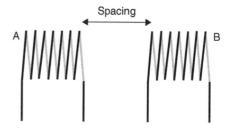

Figure 14.14. The physical coil configuration for the measurement of coaxial coils.

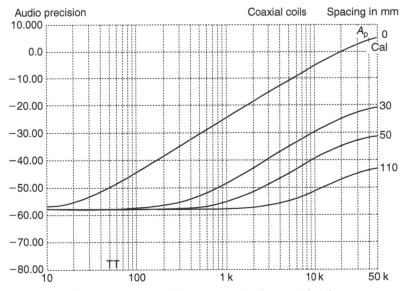

Figure 14.15. Crosstalk versus spacing for coaxial coils.

at 50 mm, only an improvement of 11 dB over the coaxial case; turning coils around is clearly not as effective as might be supposed. This time, with each 10 mm spacing increment the crosstalk dropped by 8 dB rather than 10 dB.

The obvious next step is to try combining distance with cancellation as in Figure 14.20. This can give a good performance even if a large spacing is not possible. Figure 14.21 shows that careful coil positioning can give crosstalk better than −60 dB (−114 dB total) across the audio band, although the spacing is only 20 mm. The other curves show the degradation

of performance when the coil is misaligned by moving it bodily sideways by 1, 2, 3 and 4 mm; just a 2 mm error has worsened crosstalk by 20 dB at 20 kHz. Obviously in practice the coil PCB hole will not move − but it is very possible that coils will be bent slightly sideways in production.

Figure 14.22 gives the same results for a 50 mm spacing, which can usually be managed in a stereo design. The null position once more just gives the noise floor across the band, and a 2 mm misalignment now only worsens things by about 5 dB. This is definitely the best arrangement if the spacing is limited.

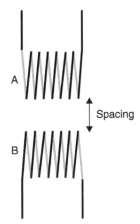

Figure 14.16. The coil configuration for non-coaxial parallel-axis coils.

Coil Crosstalk Conclusions

Coil orientation can help. Simply turning one coil through 90° gives an improvement of only 11 dB, but if it is aligned to cancel out the coupling, there is a big improvement. See how −38 dB in Figure 14.17 becomes −61 dB in Figure 14.22 at 20 kHz. On a typical stereo amplifier PCB, the coils are likely to be parallel − probably just for the sake of appearance −but their spacing is unlikely to be less than 50 mm unless the output components have been deliberately grouped together. As with capacitive crosstalk, physical distance is cheaper than anything else, and if the results are not good enough, use more of it. In this case the overall crosstalk at 20 kHz will be −54 plus −38 dB = −92 dB total, which is probably already well below other forms of interchannel crosstalk.

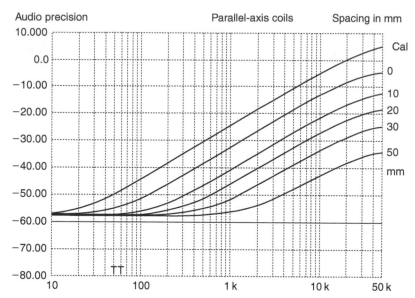

Figure 14.17. Crosstalk versus spacing for non-coaxial parallel-axis coils.

A quick quarter-turn of the coil improves this to at least −114 dB. It should do.

Coil Placement Issues

We have just looked at the issue of inter-coil crosstalk, and we noted earlier in the chapter that the coil should be kept away from metal plates that will reduce the inductance by shorted-turn effect. Another consideration, dealt with in Chapter 11 in the section on magnetic distortion, is that the coil should be distanced from ferrous metals to avoid the introduction of distortion, though the effects of this are less serious than you might think. A most important point is that the

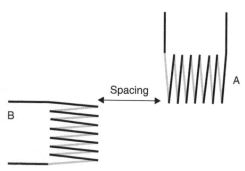

Figure 14.18. The coil configuration for crossed-axis measurements.

coil is sensitive to external magnetic fields, and so half-wave-rectified Class-B supply currents must be kept away; this can be tricky as the coil is usually placed near the output stage.

Cable Impedance Effects

Looking at the amplifier-cable-load system as a whole, the amplifier and cable impedances have the following effects with an 8 Ω resistive load:

- A constant amplitude loss due to the cable resistance forming a potential divider with the 8 Ω load. The resistive component from the amplifier output is usually negligible.
- A high-frequency roll-off due to the cable inductance forming an LR lowpass filter with the 8 Ω load. The amplifier's output inductor (to give stability with capacitive loads) adds directly to this to make up the total series inductance. The shunt capacitance of any normal speaker cable is trivially small, and can have no significant effect on frequency response or anything else.

The main factors in speaker cable selection are therefore series resistance and inductance. If these parameters are below 100 mΩ and 3 μH, any effects will be imperceptible. This can be met by 13 A mains cable, especially if all three conductors are used.

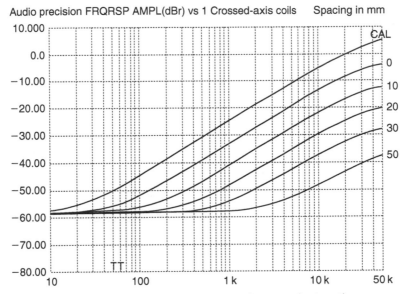

Figure 14.19. Crosstalk versus spacing for crossed-axis coils.

If the amplifier is connected to a typical loudspeaker rather than a pure resistance, the further effects are:

- The frequency response of the voltage at the loudspeaker terminals shows small humps and dips as the uneven speaker impedance loads the series combination of amplifier output impedance and cable resistance.
- The variable loading affects the amplifier distortion performance. HF crossover distortion reduces as load resistance increases above 8 Ω; even 68 Ω loading increases HF distortion above the unloaded

condition. For heavier loading than 8 Ω, crossover may continue to increase, but this is usually masked by the onset of Large Signal Non-linearity; see Chapter 10.

Severe dips in impedance may activate the overload protection circuitry unexpectedly. Signal amplitudes are higher at LF so impedance dips here are potentially more likely to draw enough current to trigger protection.

Reactive Loads and Speaker Simulation

Amplifiers are almost universally designed and tested running into a purely resistive load, although they actually spend their working lives driving loudspeakers, which contain both important reactive components and also electromechanical resonances. At first sight this is a nonsensical situation; however, testing into resistive loads is neither naïve nor an attempt to avoid the issue of real loads; there is in fact little alternative.

Loudspeakers vary greatly in their design and construction, and this is reflected in variations in the impedance they present to the amplifier on test. It would be necessary to specify a *standard speaker* for the results from different amplifiers to be comparable. Second, loudspeakers have a notable tendency to turn electricity into sound, and the sinewave testing of a 200 W amplifier would be a demanding experience for all those in earshot; soundproof chambers are not easy or cheap to construct. Third, such a standard test

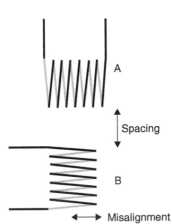

Figure 14.20. The coil configuration for crossed-axis with cancellation.

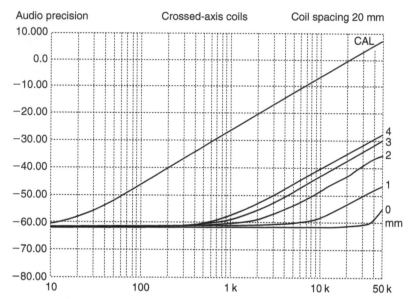

Figure 14.21. Crosstalk versus alignment for crossed-axis coils spaced at 20 mm, using cancellation.

speaker would have to be capable of enormous power-handling if it were to be able to sustain long-term testing at high power; loudspeakers are always rated with the peak/average ratio of speech and music firmly in mind, and the lower signal levels at high frequencies are also exploited when choosing tweeter power ratings. A final objection is that loudspeakers are not noted for perfect linearity, especially at the LF end, and if the

amplifier does not have a very low output impedance, this speaker non-linearity may confuse the measurement of distortion. Amplifier testing would demand a completely different sort of loudspeaker from that used for actually listening to music; the market for it would be very, very small, so it would be expensive.

A most ingenious solution to this problem was put forward by Dymond and Mellor.[3] The test load consists

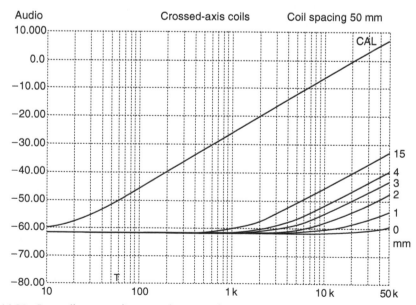

Figure 14.22. Crosstalk versus alignment for crossed-axis coils spaced at 50 mm, using cancellation.

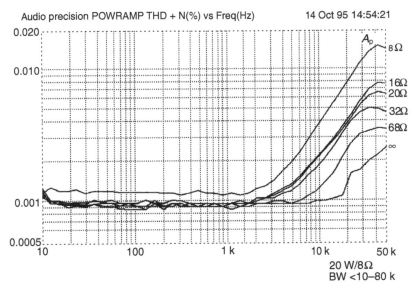

Figure 14.23. The reduction of HF THD as resistive amplifier loading is made lighter than 8 Ω.

not of passive or electro-mechanical components, but of another power amplifier. Their design can be set to emulate linear complex loads with modulus from 4 Ω to 50 Ω, with load angles from -60 to +60 degrees by altering the feedback arrangements.

Resistive Loads

Amplifiers are normally developed through 8 Ω and 4 Ω testing, though intermediate values such as 5.66 Ω (the geometric mean of 8 and 4) are rarely explored, considering how often they occur in real use. This is probably legitimate in that if an amplifier works well at 8 Ω and 4 Ω, it is most unlikely to give trouble at intermediate loadings. In practice, few nominal 8 Ω speakers have impedance dips that go below 5 Ω, and design to 4 Ω gives a safety margin, if not a large one.

The most common elaboration on a simple resistive load is the addition of 2 µF in parallel with 8 Ω to roughly simulate an electrostatic loudspeaker; this is in fact not a particularly reactive load, for the impedance of a 2 µF capacitor only becomes equal to the resistance at 9.95 kHz, so most of the audio band is left undisturbed by phase shift. This load is in fact a worse approximation to a moving-coil speaker than is a pure resistance.

Modelling Real Loudspeaker Loading

The impedance curve of a real loudspeaker may be complex, with multiple humps and dips representing various features of the speaker. The resonance in the bass driver unit will give a significant hump in LF impedance, with associated phase changes. Reflex (ported enclosure) designs have a characteristic double-hump in the LF, with the middle dip corresponding to the port tuning. The HF region is highly variable, and depends in a complicated fashion on the number of drive units, and their interactions with the crossover components.

Connection of an amplifier to a typical speaker impedance rather than a resistance has several consequences:

- The frequency response, measured in terms of the voltage across the loudspeaker terminals, shows small humps and bumps due to the uneven impedance loading the series combination of amplifier output impedance and connecting cable resistance.

- Severe dips in impedance may activate the overload protection circuitry prematurely. This has to be looked at in terms of probability, because a high amplitude in a narrow frequency band may not occur very often, and if it does, it may be so brief that the distortion generated is not perceptible. Amplitudes are higher at LF and so impedance dips here are potentially more serious.

- The variable loading affects the distortion performance.

Figure 14.23 shows how the HF crossover distortion varies with load resistance for loads lighter than those usually considered. Even 68 Ω, loading increases HF distortion.

Figure 14.24 shows an electrical model of a single full-range loudspeaker unit. While a single-driver design is unlikely to be encountered in hi-fi applications, many PA, disco and sound reinforcement applications use full-range drive units, for which this is a good model. Rc and Lc represent the resistance and inductance of the voice-coil. Lr and Cr model the electromechanical resonance of the cone mass with the suspension compliance and air-spring of the enclosure, with Rr setting the damping; these last three components have no physical existence, but give the same impedance characteristics as the real resonance.

The input impedance magnitude this network presents to an amplifier is shown in Figure 14.25. The peak at 70 Hz is due to the cone resonance; without the sealed enclosure, the restoring force on the cone would be less and the free-air resonance would be at a lower frequency. The rising impedance above 1 kHz is due to the voice-coil inductance Lc.

When the electrical model of a single-unit load replaces the standard 8 Ω resistive load, something remarkable happens; HF distortion virtually disappears, as shown in Figure 14.26. This is because a Blameless amplifier driving 8 Ω only exhibits crossover distortion, increasing with frequency as the NFB factor falls, and the magnitude of this depends on the current drawn from the output stage; with an inductive load, this current falls at high frequencies.

Most hi-fi amplifiers will be driving two-way or three-way loudspeaker systems, and four-way designs are not unknown. This complicates the impedance characteristic, which in a typical two-way speaker looks something like Figure 14.27, though the rise

above 10 kHz is often absent. The bass resonance remains at 70 Hz as before, but there are two drive units, and hence two resonances. There is also the considerable complication of a crossover network to direct the HF to the tweeter and the LF to the low-frequency unit, and this adds several extra variables to the situation. In a bass reflex design the bass resonance hump may be supplemented by another LF resonant peak due to the port tuning. An attempt at a representative load simulator for a two-way infinite-baffle loudspeaker system is shown in Figure 14.28. This assumes a simple crossover network without compensation for rising tweeter coil impedance, and is partially based on a network proposed by Ken Kantnor in Atkinson.[4]

Some loudspeaker crossover designs include their own Zobel networks, typically placed across the tweeter unit, to compensate for the HF rise in impedance due to the voice-coil inductance. If these Zobels are placed there to terminate the crossover circuitry in a roughly resistive load, then the loudspeaker designer has every right to do it; electroacoustic design is quite difficult enough without adding extra restrictions. However, if they are incorporated simply to make the impedance curve look tidier, and allow a claim that the load has been made easier for the amplifier to drive, then this seems misguided. The actual effect is the opposite; a typical amplifier has no difficulty driving an inductive reactance, and the HF crossover distortion can be greatly reduced when driving a load with an impedance that rises above the nominal value at HF.

This is only an introduction to the huge subject of real amplifier loads. More detailed information is given in Benjamin.[5]

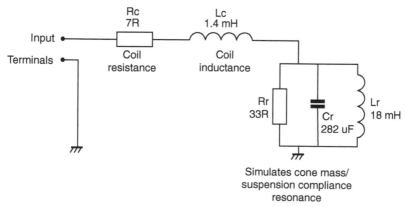

Figure 14.24. Electrical model of a single speaker unit in a sealed enclosure.

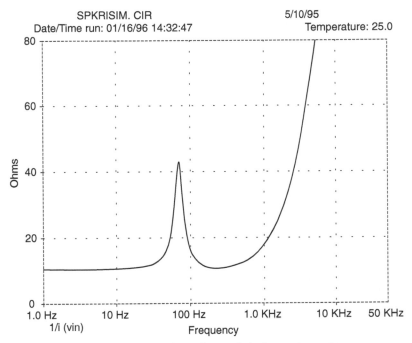

Figure 14.25. Input impedance of single speaker unit.

Loudspeaker Loads and Output Stages

There is a common assumption that any reactive load is more difficult for an amplifier to drive than a purely resistive one; however, it is devoutly to be wished that people would say what they mean by 'difficult'. It could mean that stability margins are reduced, or that the stresses on the output devices are increased. Both problems can exist, but I suspect that this belief is rooted in anthropomorphic thinking. It is easy to assume that if a signal is more complex to contemplate, it is harder for an amplifier to handle. This is not, however, true; it is not necessary to understand the laws of physics to obey them. Everything does anyway.

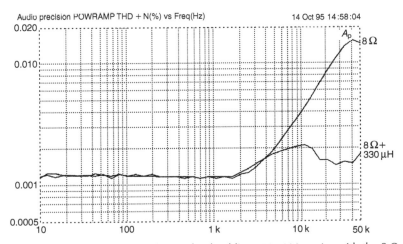

Figure 14.26. The reduction of HF THD with an inductive load; adding 330 μH in series with the 8 Ω reduces the 20 kHz THD by more than four times.

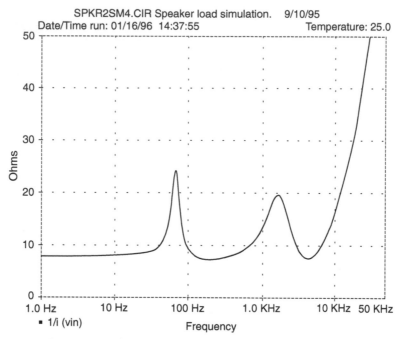

Figure 14.27. The impedance plot of the 2-way speaker model.

When solid-state amplifiers show instability, it is always at ultrasonic frequencies, assuming we are not grappling with some historical curiosity that has AC coupling in the forward signal path. It never occurs in the middle of the audio band although many loudspeakers have major convulsions in their impedance curves in this region. Reactive loading can and does imperil stability at high frequencies unless precautions are taken, usually in the form of an output inductor. It does not cause oscillation or ringing mid-band.

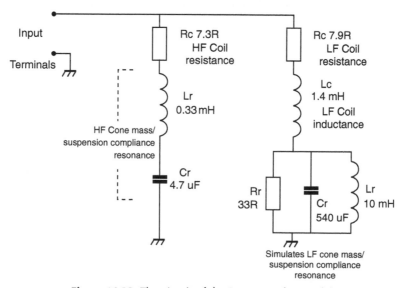

Figure 14.28. The circuit of the 2-way speaker model.

Reactive loads do increase output device stresses. In particular, peak power dissipation is increased by the altered voltage/current phase relationships in a reactive load.

Single-speaker Load

Considering a single speaker unit with the equivalent circuit of Figure 14.24, the impedance magnitude never falls below the 8 Ω nominal value, and is much greater in some regions; this suggests the overall amplifier power dissipation would be less than for an 8 Ω resistive load.

Unfortunately this is not so; the voltage/current phase relationship brought about by the reactive load is a critical factor. When a pure resistance is driven, the voltage across the output device falls as the current through it rises, and they never reach a maximum at the same time. See Figure 14.29, for Class-B with an 8 Ω resistive load. The instantaneous power is the product of instantaneous current and voltage drop, and in Class-B has a characteristic two-horned shape, peaking twice at 77 W during its conducting half-cycle.

When the single-speaker load is driven at 50 Hz, the impedance is a mix of resistive and inductive, at 8.12+3.9 j Ω. Therefore the current phase-lags the voltage, altering the instantaneous product of voltage and power to that shown in Figure 14.30. The average dissipation over the Class-B half-cycle is slightly reduced, but the peak instantaneous power increases by 30% due to the voltage/current phase shift. This could have serious results on amplifier reliability if not considered at the design stage. Note that this impedance is equivalent *at 50 Hz only* to 8.5 Ω in series with 10.8 mH. Trying to drive this replacement load at any other frequency, or with a non-sine waveform, would give completely wrong results. Not every writer on this topic appears to appreciate this.

Similarly, if the single-speaker load is driven at 200 Hz, on the other side of the resonance peak, the impedance is a combination of resistive and capacitive at 8.4 − 3.4 j Ω and the current leads the voltage. This gives much the same result as Figure 14.30, except that the peak power now occurs in the first part of the half-cycle. The equivalent load *at 200 Hz only* is 10.8 Ω in parallel with 35 μF.

When designing output stages, there are four electrical quantities to accommodate within the output device ratings: peak current, average current, peak power and average power. (Junction temperatures must of course also be considered at some point.) The

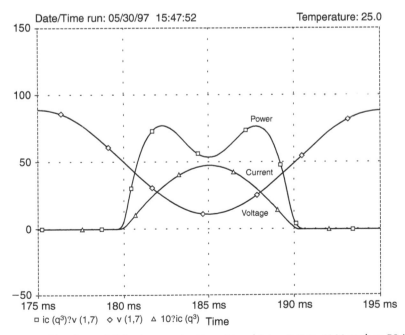

Figure 14.29. Instantaneous Vce, Ic, and Pdiss in an output transistor driving 8 Ω to 40 V peak at 50 H from ±50 V rails. Device dissipation peaks twice at 77 watts in each half-cycle.

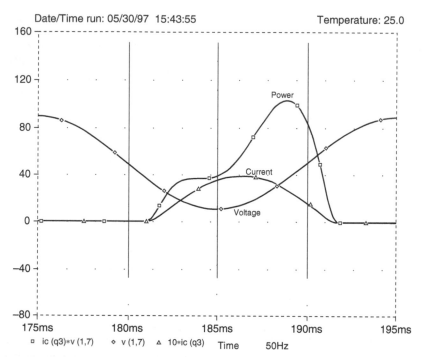

Figure 14.30. As Figure 14.29, but driving 50 Hz into the single-speaker load. At this frequency the load is partly inductive so current lags voltage and the instantaneous power curve is asymmetrical, peaking higher at 110 watts towards the end of the half-cycle.

critical quantities for semiconductor safety in amplifiers are usually the peak instantaneous values; for heatsink design average power is what counts, while, for the power supply, average current is the significant quantity.

To determine the effect of real speaker loads on device stress, I simulated an EF output stage driving a single-speaker load with a 40 V peak sinewave, powered from ±50 V rails. The load was as Figure 14.24 except for a reduction in the voice-coil inductance to 0.1 mH; the resulting impedance curve is shown in Figure 14.31. Transient simulations over many cycles were done for 42 spot frequencies from 20 Hz to 20 kHz, and the peak and average quantities recorded and plotted. Many cycles must be simulated as the bass resonance in the impedance model takes time to reach steady state when a sinewave is abruptly applied; not everyone writing on this topic appears to have appreciated this point.

Steady sinewave excitation was used as a practical approach to simulation and testing, and does not claim to be a good approximation to music or speech. Arbitrary non-cyclic transients could be investigated by the same method, but the number of waveform possibilities is infinite. It would also be necessary to be careful about the initial conditions.

Figures 14.31, 14.32 and 14.33 are the distilled results of a very large number of simulations. Figure 14.38 shows that the gentle foothills of the impedance peak at bass resonance actually increase the peak instantaneous power stress on the output devices by 30%, despite the reduced current drawn.

The most dangerous regions for the amplifier are the sides of a resonance hump where the phase shift is the greatest. Peak dissipation only falls below that for an 8 Ω resistor (shown dotted) around the actual resonance peak, where it drops quickly to a quarter of the resistive case.

Likewise, the increase in impedance at the HF end of the spectrum, where voice-coil inductance is significant, causes a more serious rise in peak dissipation to 50% more than the resistive case. The conclusion is that, for peak power, the phase angle is far more important than the impedance magnitude.

The effects on the average power dissipation, and on the peak and average device current in Figure 14.33, are more benign. With this type of load network, all three quantities are reduced when the speaker impedance increases, the voltage/current phase shifts having no effect on the current.

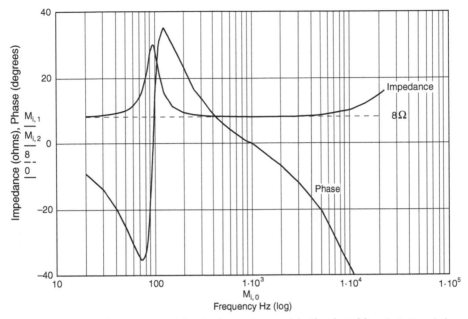

Figure 14.31. Impedance curve of the single-speaker model. The dotted line is 8 Ω resistive.

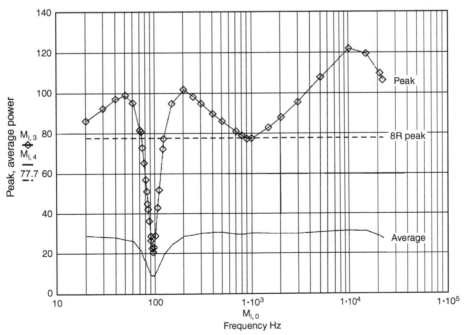

Figure 14.32. Peak and average output device power dissipation driving the single-unit speaker impedance as Figure 14.33. The dotted line is peak power for 8 Ω resistive.

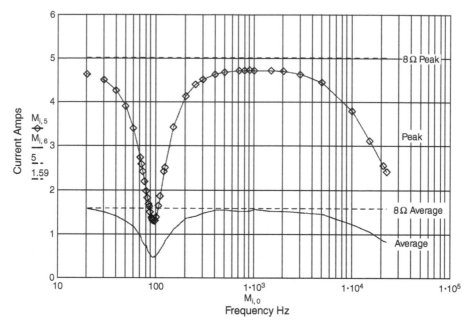

Figure 14.33. Peak and average output device current driving the single-unit speaker impedance. Dotted lines are peak and average current into 8 Ω.

Two-way Speaker Loads

The impedance plot for the simulated two-way speaker load of Figure 14.34 is shown in Figure 14.34 at 59 spot frequencies. The curve is more complex and shows a dip below the nominal impedance as well as peaks above; this is typical of multi-speaker designs. An impedance dip causes the maximum output device stress as it combines increased current demand with phase shifts that increase peak instantaneous dissipation.

In Figure 14.35 the impedance rise at bass resonance again causes increased peak power dissipation due to phase shifts; the other three quantities are reduced. In the HF region, there is an impedance dip at 6 kHz which nearly doubles peak power dissipation on its lower slopes, the effect being greater because both phase-shift and increased current demand are acting. The actual bottom of the dip sharply reduces peak power where the phase angle passes through zero, giving the notch effect at the top of the peak.

Average power (Figure 14.35) and peak and average current (Figure 14.36) are all increased by the impedance dip, but to a more modest extent.

Peak power would appear to be the critical quantity. Power device ratings often allow the power and second-breakdown limits (and sometimes the bondwire current limit also) to be exceeded for brief periods. If you

attempt to exploit these areas in an audio application, you are living very dangerously, as the longest excursion specified is usually 5 msec, and a half-cycle at 20 Hz lasts for 25 msec.

From this it can be concluded that a truly 'difficult' load impedance is one with lots of small humps and dips giving significant phase shifts and increased peak dissipation across most of the audio band. Impedance dips cause more stress than peaks, as might be expected. Low impedances at the high-frequency end (above 5 kHz) are particularly undesirable as they will increase amplifier crossover distortion.

Enhanced Loudspeaker Currents

When amplifier current capability and loudspeaker loading are discussed, it is often said that it is possible to devise special waveforms that cause a loudspeaker to draw more transient current than would at first appear to be possible. This is perfectly true. The issue was raised by Otala et al.,[6] and expanded on in Otala and Huttunen.[7] The effect was also demonstrated by Cordell.[8]

The effect may be demonstrated with the electrical analogue of a single speaker unit as shown in Figure 14.24. Rc is the resistance of the voice-coil and

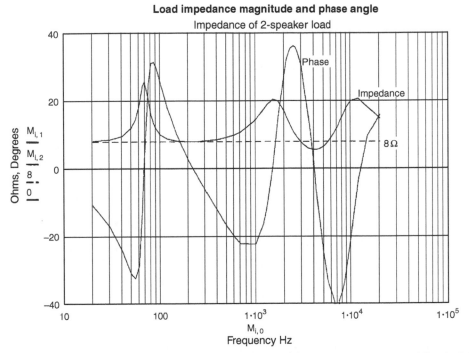

Figure 14.34. Impedance curve of model of the two-unit speaker model in Figure 14.28. Dotted line is 8 Ω resistive.

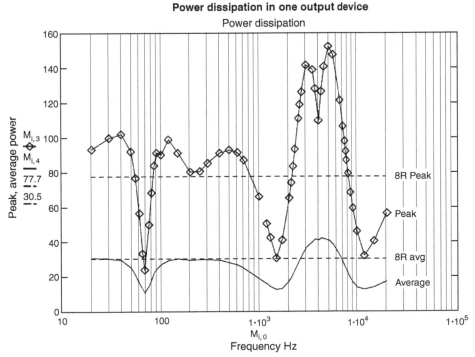

Figure 14.35. Peak and average output device power dissipation driving the two-way speaker model. Dotted lines are peak and average for 8 Ω.

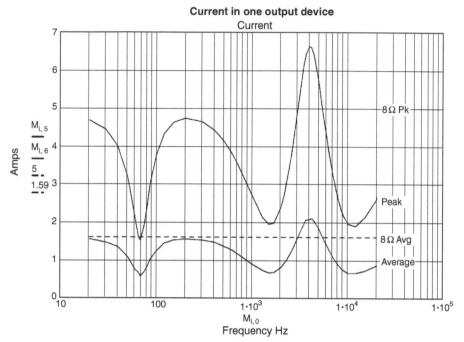

Figure 14.36. Peak and average output device current driving two-way speaker impedance as Figure 14.28 Dotted lines are peak and average for 8 Ω.

Lc its inductance. Lr and Cr model the cone resonance, with Rr controlling its damping. These three components simulate the impedance characteristics of the real electromechanical resonance. The voice-coil inductance is 0.29 mH, and its resistance 6.8 Ω, typical for a 10 inch bass unit of 8 Ω nominal impedance. Measurements on this circuit cannot show an impedance below 6.8 Ω at any frequency, and it is easy to assume that the current demands can therefore never exceed those of a 6.8 Ω resistance. This is not so.

The secret of getting unexpectedly high currents to flow is to make use of the energy stored in the circuit reactances. This is done by applying an asymmetrical waveform with transitions carefully timed to match the speaker resonance. Figure 14.37 shows PSpice simulation of the currents drawn by the circuit of Figure 14.24. The rectangular waveform is the current in a reference 8 Ω resistance driven with the same waveform. A ±10V output limit is used here for simplicity but this could obviously be much higher, depending on the amplifier rail voltages.

At the start of the waveform at A, current flows freely into Cr, reducing to B as the capacitance charges. Current is also slowly building up in Lr, causing the total current drawn to increase again to C. A positive transition to the

opposite output voltage then takes the system to point D; this is not the same state as at A because energy has been stored in Lr during the long negative period.

A carefully timed transition is then made at E, at the lowest point in this part of the curve. The current change is the same amplitude as at D, but it starts off from a point where the current is already negative, so the final peak goes much lower to 2.96 amps, 2.4 times greater than that drawn by the 8 Ω resistor. I call this the Current Timing Factor, or CTF.

Otala and Huttunen[7] show that the use of multi-way loudspeakers, and more complex electrical models, allows many more degrees of freedom in maximising the peak current. They quote a worst case CTF of 6.6 times. An amplifier driving 50 W into 8 Ω must supply a peak current into an 8 Ω resistance of 3.53 A; amplifiers are usually designed to drive 4 Ω or lower to allow for impedance dips and this means the peak current capability must be at least 7.1 amps. However, a CTF of implies that the peak capability should be at least 23 A. This peak current need only be delivered for less than a millisecond, but it could complicate the design of protection circuitry.

The vital features of the provocative waveform are the fast transitions and their asymmetrical timing. The

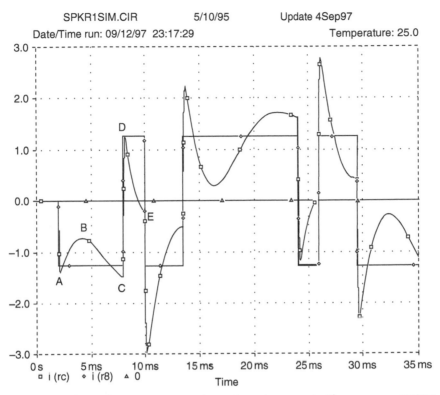

Figure 14.37. An asymmetrical waveform to generate enhanced speaker currents. The sequence ABCDE generates a negative current spike; to the right, the inverse sequence produces a positive spike. The rectangular waveform is the current through an 8 Ω resistive load.

optimal transition timing for high currents varies with the speaker parameters. The waveform in Figure 14.37 uses ramped transitions lasting 10 μsec; if these transitions are made longer, the peak currents are reduced. There is little change up to 100 μsec, but with transitions lengthened to 500 μsec, the CTF is reduced from 2.4 to 2.1.

Without doing an exhaustive survey, it is impossible to know how many power amplifiers can supply six times the nominal peak current required. I suspect there are not many. Is this therefore a neglected cause of real audible impairment? I think not, because:

1. Music signals do not contain high-level rectangular waveforms, nor trapezoidal approximations to them. A useful investigation would be a statistical evaluation of how often (if ever) waveforms giving significant peak current enhancement occur. As an informal test, I spent some time staring at a digital scope connected to general-purpose rock music, and saw nothing resembling the test waveform. Whether the asymmetrical timings were present is not easy to say; however, the large-amplitude vertical edges were definitely not.

2. If an amplifier does not have a huge current-peak capability, then the overload protection circuitry will hopefully operate. If this is of a non-latching type that works cleanly, the only result will be rare and very brief periods of clipping distortion when the loudspeaker encounters a particularly unlucky waveform. Such infrequent transient distortion is known to be inaudible and this may explain why the current enhancement effect has attracted relatively little attention so far.

Amplifier Stability

Amplifier stability can be one of the more challenging areas of design. Instability can refer to unwanted oscillations at either HF or LF, but the latter is very rare in solid-state amplifiers, though still very much an issue for valve designers. Instability has to be taken very seriously, because it may not only destroy the amplifier that hosts it, but also damage the loudspeakers.

Instability at middle frequencies such as 1 kHz is virtually impossible unless you have a very eccentric design with roll-offs and phase-shifts in the middle of the audio band.

HF Instability

HF instability is probably the most difficult problem that may confront the amplifier designer, and there are several reasons for this:

1. The most daunting feature of HF oscillation is that under some circumstances it can cause the destruction of the amplifier in relatively short order. It is often most inadvisable to let the amplifier sit there oscillating while you ponder its shortcomings.

   BJT amplifiers will suffer overheating because of conduction overlap in the output devices; it takes time to clear the charge carriers out of the device junctions. Some designs deal with this better than others, but it is still true that subjecting a BJT design to prolonged sinewave testing above 20 kHz should be done with great caution. Internal oscillations may of course have much higher frequencies than this, and in some cases the output devices may be heated to destruction in a few seconds. The resistor in the Zobel network will probably also catch fire.

   FET amplifiers are less vulnerable to this overlap effect, due to their different conduction mechanism, but show a much greater tendency to parasitic oscillation at high frequencies, which can be equally destructive. Under high-amplitude oscillation, plastic-package FETs may fail explosively; this is usually a prompt failure within a second or so and leaves very little time to hit the off switch.

2. Various sub-sections of the amplifier may go into oscillation on their own account, even if the global feedback loop is stable against Nyquist oscillation. Even a single device may go into parasitic oscillation (e.g., emitter-followers fed from inappropriate source impedances) and this is usually at a sufficiently high frequency that it either does not fight its way through to the amplifier output, or does not register on a 20 MHz scope. The presence of this last kind of parasitic is usually revealed by excessive and unexpected non-linearity.

3. Another problem with HF oscillation is that it cannot in general be modelled theoretically. The exception to this is global Nyquist oscillation (i.e., oscillation around the main feedback loop because the phase-shift has become too great before the loop gain has dropped below unity) which can be avoided by calculation, simulation, and design. The forward-path gain and the dominant pole frequency are both easy to calculate, though the higher pole frequencies that cause phase-shift to accumulate are usually completely mysterious; to the best of my knowledge virtually no work has been done on the frequency response of audio amplifier output stages. Design for Nyquist stability therefore reduces to deciding what feedback factor at 20 kHz will give reliable stability with various resistive and reactive loads, and then apportioning the open-loop gain between the transconductance of the input stage and the trans-resistance of the VAS.

The other HF oscillations, however, such as parasitics and other more obscure oscillatory misbehaviour, seem to depend on various unknown or partly known second-order effects that are difficult or impossible to deal with quantitatively and are quite reasonably left out of simulator device models. This means we are reduced to something not much better than trial-and-error when faced with a tricky problem.

The CFP output stage has two transistors connected together in a very tight 100% local feedback loop, and there is a clear possibility of oscillation inside this loop. When it happens, this tends to be benign, at a relatively high frequency (say, 2–10 MHz) with a clear association with one polarity of half-cycle.

LF Instability

Amplifier instability at LF (motorboating) is largely a thing of the past now that amplifiers are almost invariably designed with DC-coupling throughout the forward and feedback paths. The theoretical basis for it is exactly as for HF Nyquist oscillation; when enough phase-shift accumulates at a given frequency, there will be oscillation, and it does not matter if that frequency is 1 Hz or 1 MHz. It can be as destructive of bass drivers as HF oscillation is of tweeters, especially with bass reflex designs that impose no cone loading at subsonic frequencies.

At LF things are actually easier, because all the relevant time-constants are known, or can at least be pinned down to a range of values based on electrolytic capacitor tolerances, and so the system is designable, which is far from the case at high frequencies. The techniques for dealing with almost any number of LF poles and zeros were well known in the valve era, when AC coupling between stages was usually unavoidable, because of the large DC voltage difference between the anode of one stage and the grid of the next.

The likeliest cause of LF instability is probably a mis-designed multi-pole DC servo; see Chapter 23 for more on this. Oscillation at LF is very unlikely to be provoked by awkward load impedances. This is not true at HF, where a capacitive load can cause serious instability. However, this problem at least is easily handled by adding an output inductor.

References

1. Langford-Smith, F (ed.) *Radio Designer's Handbook* 4th edition, (revised 1967) Newnes 1997 p. 432.

2. Wheeler, H A, Simple Inductance Formulas for Radio Coils, *Proc. I.R.E.*, Vol 16, Oct.1928, p. 1398.

3. Dymond, H and Mellor, P, An Active Load and Test Method for Evaluating the Efficiency of Audio Power Amplifiers *JAES* Vol 58, No 5, May 2010, p. 394

4. Atkinson, J, *Review of Krell KSA-50S Power Amplifier Stereophile* Aug 1995, p. 168.

5. Benjamin, E, Audio Power Amplifiers for Loudspeaker Loads *JAES* Vol 42, Sept 1994, p. 670.

6. Otala, M, et al., *Input Current Requirements of High-Quality Loudspeaker Systems* AES preprint #1987 (D7) for 73rd Convention, March 1983.

7. Otala, M and Huttunen, P, *Peak Current Requirement of Commercial Loudspeaker Systems* JAES, June 1987, p. 455. See Ch. 12, p. 294.

8. Cordell, R, Interface Intermodulation in Amplifiers, *Wireless World*, Feb 1983, p. 32.

Speed and Slew-rate

...the speed was power, and the speed was joy, and the speed was pure beauty.

Richard Bach

Speed and Slew-rate in Audio Amplifiers

It seems self-evident that a fast amplifier is a better thing to have than a slow one, but — what is a fast amplifier? Closed-loop bandwidth is not a promising yardstick; it is virtually certain that any power amplifier employing negative feedback will have a basic closed-loop frequency response handsomely in excess of any possible aural requirements, even if the overall system bandwidth is defined at a lower value by earlier filtering.

There is always a lot of loose talk about the importance of an amplifier's open-loop bandwidth, much of it depressingly ill-informed. I demonstrated[1] that the frequency of the dominant pole P1 that sets the open-loop bandwidth is a variable and rather shifty quantity that depends on transistor beta and other ill-defined parameters. (I also showed how it can be cynically manipulated to make it higher by reducing open-loop gain below P1.) While P1 may vary, the actual gain at HF (say, 20 kHz) is thankfully a much more dependable parameter that is set only by frequency, input stage transconductance, and the value of Cdom. It is this which is the meaningful figure in describing the amount of NFB that an amplifier enjoys.

The most meaningful definition of an amplifier's 'speed' is its maximal slew-rate. The minimum slew-rate for a 100 W/8 Ω amplifier to cleanly reproduce a 20 kHz sinewave is easily calculated as 5.0 V/usec; so 10 V/μsec is adequate for 400 W/8 Ω, a power level that takes us somewhat out of the realms of domestic hi-fi. A safety-margin is desirable, and if we make this a bare factor of two, then it could be logically argued that 20 V/μsec is enough for any hi-fi application; there is in fact a less obvious but substantial safety-margin already built in, as 20 kHz signals at maximum level are mercifully rare in music; the amplitude distribution falls off rapidly at higher frequencies.

Firm recommendations on slew-rate are not common; Peter Baxandall made measurements of the slew-rate produced by vinyl disc signals, and concluded that they could be reproduced by an amplifier with a slew-limit corresponding to maximum output at 2.2 kHz. For the 100 W amplifier, this corresponds to 0.55 V/μsec.[2]

Nelson Pass made similar tests, with a moving-magnet (MM) cartridge, and quoted a not dissimilar maximum of 1 V/μsec at 100 W. A moving-coil (MC) cartridge doubled this to 2 V/μsec, and Pass reported[3] that the absolute maximum, with a moving-magnet (MM) cartridge, and possible with a combination of direct-cut discs and MC cartridges was 5 V/μsec at 100 W. This is comfortably below the 20 V/μsec figure arrived at theoretically above; Pass concluded that even if a generous 10:1 factor of safety were adopted, 50 V/μsec would be the highest speed ever required from a 100 W amplifier.

However, in the real world we must also consider the Numbers Game; if all else is equal, then the faster amplifier is the more saleable. As an example of this, it has been recently reported in the hi-fi press that a particular 50 W/8 Ω amplifier has been upgraded from 20 V/μsec to 40 V/usec[4] and this is clearly expected to elicit a positive response from intending purchasers. This report is exceptional, for equipment reviews in the hi-fi press do not usually include slew-rate measurements. It is therefore difficult to get a handle on the state of the art, but a trawl through the accumulated data of years shows that the most highly specified equipment usually plumps for 50 V/μsec — slew-rates always being quoted in suspiciously round numbers. There was one isolated claim of 200 V/μsec, but I must admit to doubts about the reality of this.

A high slew-rate can be a bad thing. If an amplifier succumbs to HF instability, the amplitude of oscillation is normally limited by the slew-rate, so faster slewing means more oscillation, with possibly fatal results for the output devices. There was at least one Japanese amplifier that was well known for suffering from this problem. The slew-rate should be high enough to give a good safety-margin in reproducing 20 kHz, but making it much greater than it needs to be is dangerous.

The Class-B amplifier shown in Figure 15.1 is that already described in Chapter 12; the same component numbers have been preserved. This generic circuit has many advantages, though an inherently good slew performance is not necessarily one of them; however, it remains the basis for the overwhelming majority of amplifiers so it seems the obvious place to start. I have glibly stated that its slew-rate calculated at 40 V/μsec, which by the above arguments is more than adequate. However, let us assume that a major improvement in slew-rate is required to counter the propaganda of the Other Amplifier Company down the road, and examine how it might be done. As in so many areas of life, things will prove much more complicated than expected.

The Basics of Amplifier Slew-limiting

At the simplest level, the slew-rate S of a conventional amplifier configuration like Figure 15.1 depends on

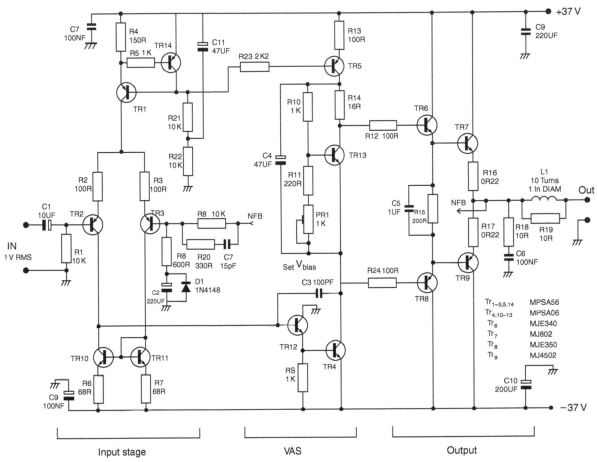

Figure 15.1. The Class-B amplifier. At the simplest level, the maximum slew-rate is defined by current sources, TR1, TR5, and the value of C3.

how fast the current I can be moved in and out of Cdom (C3). There is the convenient relation in Equation 15.1:

$$\text{Slew-rate} = \frac{I}{Cdom} \text{ V/}\mu\text{sec} \qquad \text{Equation 15.1}$$

for I in μA, Cdom in pF

The maximum output frequency for a given slew-rate S and voltage Vpk is:

$$\text{Freq max} = \frac{S}{2\pi Vpk} \qquad \text{Equation 15.2}$$

So, for example, with a slew-rate of 20 V/μsec the maximum freq at which 35 V rms can be sustained is

64 kHz, and if Cdom is 100 pF, then the input stage must be able to source and sink 2 mA peak. Likewise, a sinewave of given amplitude and frequency has a maximum slew-rate (at the zero-crossing) of:

$$\begin{aligned} S \;=\; dV/dt \;&=\; \varpi Vpk \\ &=\; 2\pi f Vpk \end{aligned} \qquad \text{Equation 15.3}$$

where f is frequency.

For Figure 15.1, our slew-rate equation yields 4000/100, or about 40 V/μsec, as quoted above, if we assume (as all textbooks do) that the only current limitation is the tail-source of the input pair. If this differential pair has a current-mirror collector load— and there are pressing reasons why it should — then almost the full tail-current is available to service Cdom. This seems

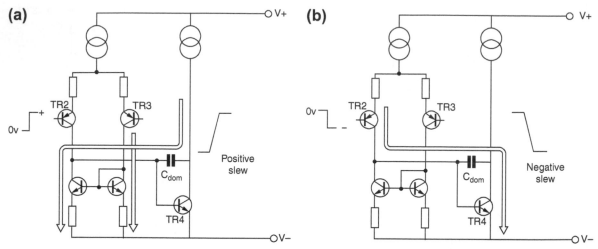

Figure 15.2. (a) The current path for positive slewing. At the limit all of the slewing current has to pass through the current-mirror, TR2 being cut off. (b) The current path at negative slew limit. TR2 is saturated and the current-mirror is cut off.

very simple — to increase slew-rate increase the tail-current. But …

The tail-current is not the only limit on the slew current in Cdom. (This point was touched on by me in.[5]) Figure 15.2 shows the current paths for positive and negative slew-limit, and it can be seen at once that the positive current can only be supplied by the VAS current-source load. This will reduce the maximum positive rate, causing slew asymmetry, if the VAS current-source cannot supply as much current as the tail source. In contrast, for negative slewing TR4 can turn on as much as required to sink the Cdom current, and the VAS collector load is not involved.

In most designs the VAS current-source value does not appear to be an issue, as the VAS is run at a higher current than the input stage to ensure enough pull-up current for the top half of the output stage; however, it will transpire that the VAS source can still cause problems.

Slew-rate Measurement Techniques

Directly measuring the edge-slopes of fast square waves from a scope screen is not easy, and without a delayed timebase, it is virtually impossible. A much easier (and far more accurate) method is to pass the amplifier output through a suitably-scaled differentiator circuit; slew-rate then becomes simple amplitude, which is much easier to read from a graticule. The circuit in Figure 15.3 gives a handy 100 mV output for each V/μsec of slew; the RC time-constant must be very short for reasonable accuracy. The differentiator was driven directly by the amplifier, and *not* via an output inductor. Be aware that this circuit needs to be coupled to the scope by a proper ×10 probe; the capacitance of plain screened cable gives serious under-readings. We are dealing here with sub-microsecond pulse techniques, so bear in mind that waveform artefacts such as ringing are as likely to be due to test cabling as to the amplifier.

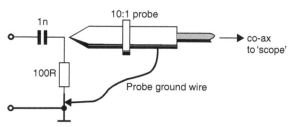

Figure 15.3. A simple (but very useful) differentiator. A local probe ground is essential for accuracy to exceed ±10%.

Applying a fast-edged square wave to an amplifier does not guarantee that it will show its slew-rate limits. If the error voltage so generated is not enough to saturate the input stage, then the output will be an exponential response, without non-linear effects. For most of the tests described here, the amplifier had to be driven hard to ensure that the true slew-limits were revealed; this is due to the heavy degeneration that reduces the transconductance of the input pair. Degeneration increases the error voltage required for saturation, but does not directly alter slew limits.

Running a slew test on a standard Blameless amplifier with an 8 Ω load sharply highlights the inadequacies of simple theory. The differentiator revealed asymmetrical slew-rates of +21 V/μsec up and −48 V/μsec down, which is both a letdown and a puzzle considering that the simple theory promises 40 V/μsec. To get results worse than theory predicts is merely the common lot of the engineer; to simultaneously get results that are *better* is grounds for the gravest suspicions.

Improving the Slew-rate

Looking again at Figure 15.1, the VAS current-source value is apparently already bigger than required to source the current Cdom requires when the input stage is sinking hard, so we confidently decrease R4 to 100 Ω (to match R13) in a plausible attempt to accelerate slewing. With considerable disappointment we discover that the slew-rate only changes to +21 V/μsec, −62 V/μsec; the negative rate still exceeds the new theoretical value of 60 V/μsec. Just what is wrong here? Honesty compels us to use the lower of the two figures in our ads (doesn't it?) and so the priority is to find out why the positive slewing is so feeble.

At first it seems unlikely that the VAS current source is the culprit, as with equal-value R4 and R13, the source should be able to supply all the input stage can sink. Nonetheless, we can test this cherished belief by increasing the VAS source current while leaving the tail-current at its original value. We find that R4 = 150 Ω, R13 = 68 Ω gives +23 V/usec, −48 V/μsec, and this small but definite increase in positive rate shows clearly there is something non-obvious going on in the VAS source. (This straightforward method of slew acceleration by increasing standing currents means a significant increase in dissipation for the VAS and its current source. We are in danger of exceeding the capabilities of the TO92 package, leading to a cost increase. The problem is less in the input stage, as dissipation is split between at least three devices.)

Simulating Slew-limiting

When circuits turn truculent, it's time to simplify and simulate. The circuit was reduced to a *model* amplifier by replacing the Class-B output stage with a small-signal Class-A emitter follower; this was then subjected to some brutally thorough PSPICE simulation, which revealed the various mechanisms described below.

Figure 15.4 shows the positive-going slew of this model amplifier, with both the actual output voltage and its differential, the latter suitably scaled by dividing by 10^6 so it can be read directly in V/μsec from the same plot. Figure 15.5 shows the same for the negative-going slew. The plots are done for a series of changes to the resistors R4, R13 that set the standing currents.

Several points need to be made about these plots; first, the slew-rates shown for the lower R4, R13 values are not obtainable in the real amplifier with output stage, for reasons that will emerge. Note that almost imperceptible wobbles in the output voltage put large spikes on the plot of the slew-rate, and it is unlikely that these are being simulated accurately, if only because circuit strays are neglected. To get valid slew-rates, read the flat portions of the differential plots.

Using this method, the first insight into slew-rate asymmetry was obtained. At audio frequencies, a constant current-source provides a fairly constant current and that is the end of the matter, making it the usual choice for the VAS collector load; as a result, its collector is exposed to the full output swing and the full slew-rate. When an amplifier slews rapidly, there is a transient feedthrough from the collector to the base (see Figure 15.6) via the collector-base capacitance. If the base voltage is not tightly fixed, then fast positive slewing drives the base voltage upwards, reducing the voltage on the emitter and hence the output current. Conversely, for negative slew, the current-source output briefly increases; see Erdi.[6] In other words, fast positive slewing itself reduces the current available to implement it.

Having discovered this hidden constraint, the role of isolation resistor R23 feeding TR6 base immediately looks suspect. Simulation confirms that its presence worsens the feedthrough effect by increasing the impedance of the reference voltage fed to TR6 base. As is usual, the input-stage tail-source TR1 is biased from the same voltage as TR6; this minor economy complicates things significantly, as the tail current also varies

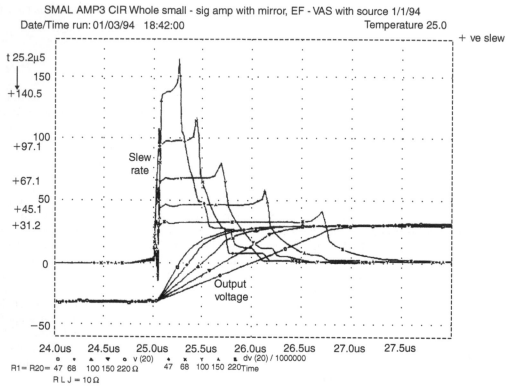

Figure 15.4. Positive slewing of simulated model amplifier. The lower traces show the amplifier output slewing from −30 to +30 V while the upper traces are the scaled differentiation.

during fast transients, reducing for positive slew, and increasing for negative.

Slewing Limitations in Real Life

Bias isolation resistors are not unique to the amplifier of Figure 15.1; they are very commonly used. For an example taken at random, see Meyer.[7] My own purpose in adding R23 was not to isolate the two current sources from each other at AC (something it utterly fails to do) but to aid fault-finding. Without this resistor, if the current in either source drops to zero (e.g., if TR1 fails open-circuit), then the reference voltage collapses, turning off both sources, and it can be time-consuming to determine which has died and which has merely come out in sympathy. Accepting this, we return to the original Figure 15.1 values and replace R23 with a link; the measured slew-rates at once improve from +21, −48 to +24, −48 (from here on the V/μsec is omitted). This is already slightly faster than our first attempt at acceleration, without the thermal penalties of increasing the VAS standing current.

The original amplifier used an active tail-source, with feedback control by TR14; this was a mere whim, and a pair of diodes gave identical THD results. It seems likely that reconfiguring the two current-sources so that the VAS source is the active one would make it more resistant to feedthrough, as the current-control loop is now around TR5 rather than TR1, with feedback applied directly to the quantity showing unwanted variations (see Figure 15.7). There is indeed some improvement, from +24, −48 to +28, −48.

This change seems to work best when the VAS current is increased, and R4 = 100 Ω, R13 = 68 Ω now gives us +37, −52, which is definite progress on the positive slewing. The negative rate has also slightly increased, indicating that the tail-current is still being increased by feedthrough effect. It seems desirable to minimise this transient feedthrough, as it works against us just at the wrong time. One possibility would be a cascode transistor to shield TR5 collector from rapid voltage changes; this would require more biasing components and would reduce the positive output swing, albeit only slightly.

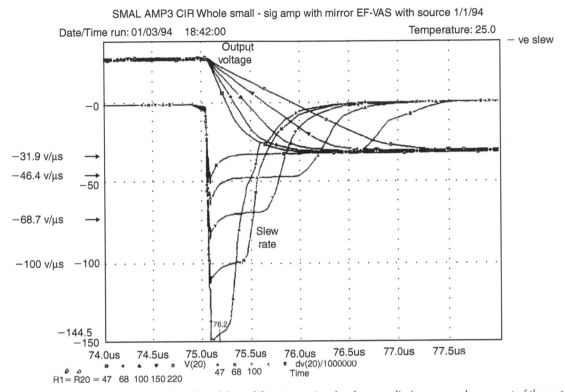

SMAL AMP3 CIR Whole small - sig amp with mirror EF-VAS with source 1/1/94

Date/Time run: 01/03/94   18:42:00   Temperature: 25.0

Figure 15.5. Negative slewing of simulated model amplifier. Increasing the slew-rate limit causes a larger part of the output transient to become exponential, as the input pair spends less time saturated. Thus the differential trace has a shorter flat period.

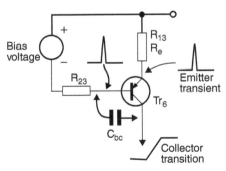

Figure 15.6. One reason why simple theory fails. Fast positive edges on the collector of the VAS source TR6 couple through the internal Cbc to momentarily reduce standing current.

Since it is the VAS current-source feedthrough capacitance that causes so much grief, can we turn it against itself, so that an abrupt voltage transition increases the current available to sustain it, rather than reducing it? Oh, yes we can, for if a small capacitance

Cs is added between the TR5 collector (carrying the full voltage swing) and the sensing point A of the active tail source, then as the VAS collector swings upward, the base of TR14 is also driven positive, tending to turn it off and hence increasing the bias applied to VAS source TR5 via R21. This technique is highly effective, but it smacks of positive feedback and should be used with caution; Cs must be kept small. I found 7.5 pF to be the highest value usable without degrading the amplifier's HF stability.

With R4 = 100 Ω, R13 = 68 Ω adding Cs = 6 pF takes us from +37, −52 to +42, −43; and the slew asymmetry that has dogged this circuit from the start has been corrected. Fine adjustment of this capacitance is needed if good slew-symmetry is demanded.

Some Additional Complications

Some other unsuspected effects were uncovered in the pursuit of speed; it is not widely known that slew-rate is affected both by output loading and the output stage

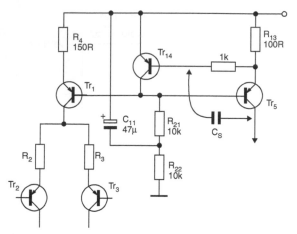

Figure 15.7. A modified biasing system that makes TR5 current the controlled variable, and reduces the feedthrough effect.

operating class. For example, above we have noted that R4 = 100 Ω, R13 = 68 Ω yields +37, −52 for Class-B and an 8 Ω load. With a 4 Ω loading, this changes to +34, −58, and again the loss in positive speed is the most significant. If the output stage is biased into Class-A (for an 8 Ω load) then we get +35, −50. The explanation is that the output stage, despite the cascading of drivers and output devices, draws significant current from the VAS stage. The drivers draw enough base current in the 4 Ω case to divert extra current from Cdom and current is in shortest supply during positive slew. The effect in Class-A is more severe because the output device currents are always high, the drivers requiring more base current even when quiescent, and again this will be siphoned off from the VAS collector.

Speeding-up this amplifier would be easier if the Miller capacitor Cdom was smaller. Does it really need to be that big? Well, yes, because if we want the NFB factor to be reasonably low for dependable HF stability, the HF loop gain must be limited. Open-loop gain above the dominant pole frequency P1 is the product of input stage gm with the value of Cdom, and the gm is already as low as it can reasonably be made by emitter degeneration. Emitter resistors R2, R3 at 100 Ω are large enough to mildly compromise the input offset voltage, because the tail-current splits in two through a pair of resistors that are unlikely to be matched to better than 1%, and noise performance is also impaired by this extra resistance in the input pair emitters. Thus, for a given NFB factor at 20 kHz, Cdom is fixed.

Despite these objections, the approach was tested by changing the distribution of open-loop gain between the input stage and the VAS. R2, R3 were increased from 100 Ω to 220 Ω, and Cdom reduced to 66 pF; this does not give exactly the same NFB factor, but in essence we have halved the transconductance of the input stage, while doubling the gain of the VAS. This gain-doubling allows Cdom to be reduced to 66 pF without reduction of stability margins.

With R4 = 100 Ω, R13 = 68 Ω as before, the slew-rate is increased to +50, −50 with Cs = 6 pF to maintain slewing symmetry. This is a 25% increase in speed rather than the 50% that might be expected from simple theory, and indicates that other restrictions on speed still exist; in fact PSPICE showed there are several.

One of these restrictions is as follows; when slewing positively, TR4 and TR12 must be turned off as fast as possible, by pulling current out of Cdom. The input pair therefore causes TR10 to be turned on by an increasing voltage across TR11 and R7. As TR10 turns on, its emitter voltage rises due to R6, while at the same time the collector voltage must be pulled down to near the -ve rail to turn off Q4. In the limit TR10 runs out of Vce, and is unable to pull current out of Cdom fast enough. Once more it is the positive rate that suffers. The simplest way to reduce this problem is to reduce the resistors R6, R7 that degenerate the current-mirror. This risks HF distortion variations due to input-pair Ic imbalance (values down to 12 Ω have given acceptable results for this), but perhaps more importantly significantly degrades the amplifier noise performance; see Chapter 6.

Another way to reduce the value needed for Cdom is to lower the loop-gain by increasing the feedback network attenuation, or in other words, to run the amplifier at a higher closed-loop gain. This might be no bad thing; the current 'standard' of 1 V for full output is (I suspect) due to a desire for low closed-loop gain in order to maximise the NFB factor, so reducing distortion. I recall JLH advocating this strategy back in 1974. However, we must take the world as we find it, and so I have left closed-loop gain alone. We could of course attenuate the input signal so it can be amplified more, though I have an uneasy feeling about this sort of thing; amplifying in a pre-amp then attenuating in the power amp implies a headroom bottleneck, if such a curdled metaphor is permissible. It might be worth exploring this approach; this amplifier has good open-loop linearity and I do not think excessive THD would be a problem.

Having previously spent some effort on minimising distortion, we do not wish to compromise the THD of a Blameless amplifier. Mercifully, none of the modifications set out here have any significant effect on overall THD, though there may be minor variations around 10–20 kHz.

On Asymmetrical Slew-rates

There is an assumption about that there is some indefinable advantage in having symmetrical slew-rates, i.e., the same in both positive and negative directions. I can see no merit in this; if there is a healthy safety-margin in each direction, slewing will never occur so its exact rate is of little interest.

However, I can see an advantage to deliberately making the slew-rates asymmetrical. If the amplifier should burst into oscillation, its amplitude is typically limited by slew-limiting. This in itself is a good argument for not designing in excessive slew-rates; there is plenty of anecdotal evidence that commercial amplifiers designed for fast slewing are more subject to destructive instability. If the slew-rates are asymmetrical, then oscillation will cause a shift in the average DC output level away from 0 V, and the DC-offset protection

circuitry will kick in to disconnect the offending amplifier from the loudspeakers.

Further Improvements and other Configurations

The results I have obtained in my attempts to improve slewing are not exactly stunning at first sight; however, they do have the merit of being as grittily realistic as I can make them. I set out in the belief that enhancing slew-rate would be fairly simple; the very reverse has proved to be the case. It may well be that other VAS configurations, such as the push-pull VAS, will prove more amenable to design for rapid slew-rates; however, such topologies appear to have other disadvantages to overcome.

Stochino, in a fascinating paper,[8] has presented a topology, which, although a good deal more complex than the conventional arrangement, claims to make slew-rates up to 400 V/μsec achievable.

Finally, it is notable that a fast slew-rate has not always been seen as a good thing. In the early 1970s it was sometimes recommended that VAS slew-rate be deliberately restricted to avoid excessive currents from overlap conduction in slow output transistors.[9]

References

1. Self, D, Distortion in Power Amplifiers, Part 1, *Electronics World+WW*, Aug 1993, p. 631.

2. Baxandall, P, Audio Power Amplifier Design, *Wireless World*, Jan. 1978, p. 56.

3. Pass, N, Linearity, Slew Rates, Damping, Stasis and... *Hi-Fi News and RR*, Sept. 1983, p. 36.

4. Hughes, J, Arcam Alpha5/Alpha6 Amplifier Review, *Audiophile*, Jan. 1994, p. 37.

5. Self, D, Distortion in Power Amplifiers, Part 7, *Electronics World+WW*, Feb. 1994, p. 138.

6. Erdi, G A, 300 v/uS Monolithic Voltage Follower, *IEEE J. of Solid-State Circuits*, Dec. 1979, p. 1062.

7. Meyer, D, Assembling a Universal Tiger, *Popular Electronics*, Oct. 1970.

8. Stochino, G, Ultra-Fast Amplifier Electronics, *World+WW*, Oct. 1995, p. 835.

9. Alves, J, Power bandwidth limitations in Audio Amplifiers, *IEEE Transactions on Broadcast & TV,* May 1973, pp. 79–86.

Power Dissipation in Amplifiers

Crowds without company, and dissipation without pleasure.

Edward Gibbon, referring to London

Output Stage Conditions

There are several important considerations in designing an output stage that is up to the job. The average power dissipated determines how big the heatsink must be to keep output device junction temperatures down to safe levels. The peak power dissipated in the output devices must also be considered as the effects of this are not averaged by the heatsink mass. Audio waveforms have large low-frequency components, too slow for peak currents and powers to be allowed to exceed the DC limits on the data sheet. The peak collector current must be examined to make sure is within the limits for each device. These quantities determine the type and number of output devices.

For a resistive load, the peak power is fixed and easily calculable. With a reactive load, the peak power excursions are less easy to determine but even more important because they are increased by the changed voltage/phase relationships in the output device. Thus for a given load impedance modulus the peak power needs to be plotted against load phase angle as well as output fraction to give a complete picture.

The average power drawn from the rails is also a vital prerequisite for the power-supply design; since the rail voltage is substantially constant, this can be easily converted into a current demand, which allows the sizing of reservoir capacitors, rectifiers, and mains transformers.

The Mathematical Approach

Most textbooks, when dealing with Class-B power amplifier efficiency, use a purely mathematical method to determine the dissipation in the output devices; the product of voltage and current is integrated over a half-cycle. To make it mathematically tractable, the situation is highly idealised, assuming an exact 50% conduction period, no losses in emitter resistors or drivers, and so on. A classic example is.[1]

The result is Figure 16.1, which plots dissipation against the fraction of the maximum output voltage used. We see the familiar information that for Class-B maximum device dissipation occurs at 64% of maximum output voltage, representing 42% of the maximum output power. These specific numbers result from the use of a sine waveform, and other waveforms give different values. In most amplifier types the power dissipation varies strongly with output signal amplitude as it goes from zero to maximum, so the information is best presented as a graph of dissipation against the fraction of the available rail-to-rail output swing. I have called this parameter the voltage fraction.

Figure 16.1 was calculated using ±50 V rails and an 8 Ω load. The theoretical maximum power output is 156 W, and the output stage dissipation peaks at 64 W. This is divided equally between the output devices, of which there must be at least two, so each device must cope with a maximum dissipation of 32 W.

The mathematics is relatively straightforward, and can be found in many references such as.[1] However, solving the same problem for Class-AB, where the device conduction period varies with signal amplitude, is considerably more complicated due to the varying integration limits, while Class-G is even worse.

Dissipation by Simulation

The mathematical difficulties can be circumvented by simulation. This is not only easier, but also more accurate, as all losses and circuit imperfections are included, and the power dissipations in every part of the circuit, including power drawn from the supply rails, are made available by a single simulation run.

It is an obvious choice to use a sine waveform in the transient simulations, as this allows a reality check against the mathematical results. Reactive loads are easily handled, so long as it is appreciated that the simulation often has to be run for 10 or more cycles to allow the conditions in the load to reach a steady state. In the past, some people have failed to appreciate this, with most regrettable results. All simulations were run with ±50 V rails and an 8 Ω resistive or reactive load. The output emitter resistors were 0.1 Ω. The voltage fraction is increased from 0 in steps of 0.05, stopping at 0.95 to avoid clipping effects. The X-axis may be linear or logarithmic; here it is linear.

With some simulator packages this can be rather labour-intensive. The computed peak and averaged power dissipations at the end of the cycle often have to be read out from a cursor and recorded by hand into an application such as Mathcad.

Power Partition Diagrams

The graph in Figure 16.1 gives only one quantity, the amplifier output stage dissipation. There is a more informative graph format that I call a Power Partition

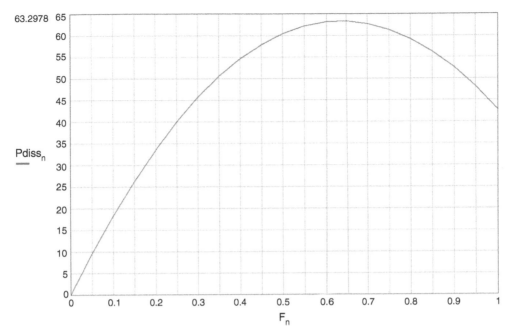

Figure 16.1. Output stage dissipation for mathematically ideal Class-B with sinewave drive, against output voltage. Maximum dissipation is at 64% of maximum voltage output, and is equivalent to 42% of maximum power output.

Diagram, which shows how the input power drawn is partitioned between amplifier dissipation, useful power in the load, and losses in drivers, emitter resistors, etc. I have called the upper output transistor (connected to the V+ rail) the source, and the lower (connected to the V- rail) the sink.

Class-B: CFP and EF Power Partition

Figure 16.2 shows the Power Partition Diagram (PPD) for a Class-B Complementary Feedback Pair (CFP) stage, which significantly has a very low quiescent current compared with the Emitter-Follower (EF) version. Line 1 plots the dissipation in the sink device, while Line 2 is source plus sink dissipation. Line 3 is source plus sink plus load dissipation. The topmost line 4 is the total power drawn from the power supply, and so the narrow region between 3 and 4 is the power dissipated in the rest of the circuit — mainly the drivers and output emitter resistors. This power loss increases with output drive, but remains small compared with the other powers examined. Note that power is represented by the *vertical* distances between the lines — the areas between them have no special significance.

Figure 16.2 shows that the power drawn from the supply increases proportionally to the drive voltage

fraction. This is partitioned between the load power (the curved region between lines 2 and 3), and the output devices; note how the peak in their power dissipation accommodates the curve of the load power as it increases with the square of the voltage fraction.

Figure 16.3 shows the Power Partition Diagram (PPD) for a Class-B Emitter-Follower (EF) output stage as shown in Figure 9.4b. The important difference is that the quiescent current of an EF output stage is much larger than for the CFP version (here it is 150 mA) so there is now greater power dissipation around zero output. At higher levels the curves are unchanged, so it is an important finding that there is absolutely no need for greater heatsinking than for the CFP case. If it is adequate for worst-case dissipation with a CFP output, it is adequate for an EF output.

Class-AB Power Partition

Figure 16.4 shows Class-AB, with the bias increased so that Class-A operation and linearity are maintained up to 5 Wrms into 8 Ω. The quiescent current is now 370 mA, so the quiescent power dissipation is significantly greater for voltage fractions below 0.1. The device dissipation is still greatest at a drive fraction of around 0.6, so once again no extra cooling is required to deal with the

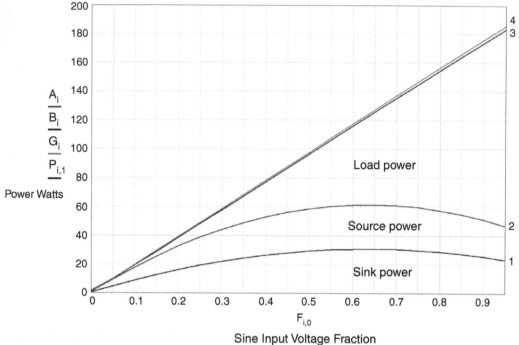

Figure 16.2. Class-B CFP Power Partition Diagram for 8 Ohm resistive load.

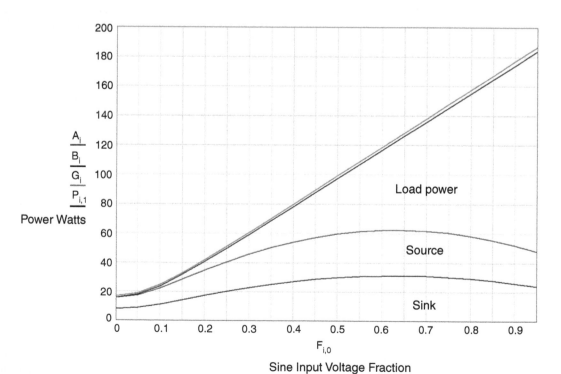

Figure 16.3. Class-B EF Power Partition into 8 Ohm resistive load.

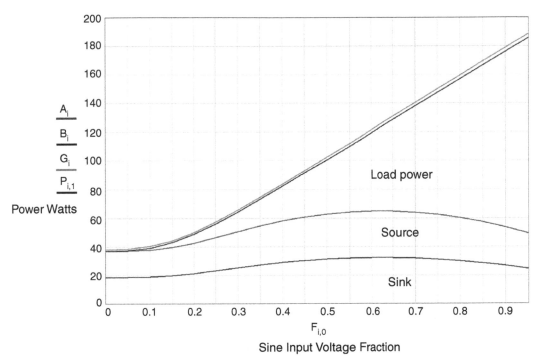

Figure 16.4. Class-AB sinewave drive. If the range of class-A operation is extended by increasing the bias, the area below level = 0.1 advances upwards and to the right.

increased quiescent dissipation. It is a good question at what level of AB biasing more heatsinking *is* required; perhaps someone else would like to work it out, as my lack of enthusiasm for the extra distortion generated by Class-AB, compared with optimally biased Class-B, is well known.

Class-A Power Partition

A push-pull Class-A amplifier draws a large standing current, and the picture in Figure 16.5 looks totally different. The power drawn from the supply is constant, but as the output increases, dissipation transfers from the output devices to the load, so minimum amplifier heating is at maximum output. The significant point is that amplifier dissipation is only meaningfully reduced at a voltage fraction of 0.5 or more, i.e., only 6 dB from clipping. If the amplifier is run at a more realistic level of, say, 15 dB below clipping, the sinewave efficiency is a pitiful 1.8%. Compared with other classes, an enormous amount of energy is wasted internally. Truly 'dissipation without pleasure' as per Mr Gibbon at the head of this chapter.

Single-ended constant-current or resistive-load versions of Class-A have even lower efficiency, worse

linearity, and no countervailing advantages. If Class-B had just been invented, it would be hailed as a breakthrough in efficiency.

Class XD Power Partition: Constant-current and Push-pull

This is a new kind of output stage that I devised. The principle is crossover displacement; a constant or varying current is injected into the output to displace the crossover point away from 0 V, so there is a region of pure Class-A operation, after which the output stage moves into Class-B with just the usual crossover artefacts, *not* the big steps in gain found in Class-AB. It is described in detail in Chapter 18. The first amplifier of its type was the Cambridge Audio 840A (2006), and for marketing motives crossover displacement was called Class XD. The PPD for the constant-current displacer version is shown in Figure 16.6.

Here the displacer sinks 1 Amp from the output node, giving Class-A operation up to about 5 Wrms/8 Ω, and it is in parallel with the sink output device. The displacer dissipation is large, about equal to that of the source, but the sink dissipation is much lower; in fact, it does not

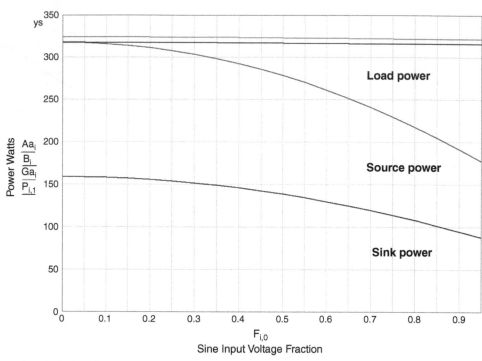

Figure 16.5. Class-A push-pull. Almost all the power drawn is dissipated in the amplifier, except at the largest outputs.

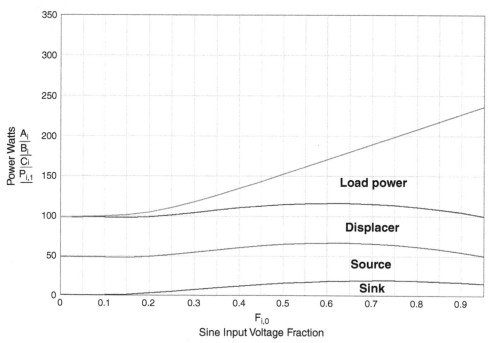

Figure 16.6. Class-XD with constant-current displacer. There is significant dissipation in the source and displacer, but much less in the sink.

conduct at all until the limit of the Class-A region is reached as the voltage fraction increases. Below that power, the output voltage is solely controlled by the source.

Figure 16.7 shows the PPD for Class-XD with a push-pull displacer. Now the displacer current varies synchronously with the signal from zero at maximum positive output voltage, to 2 Amps at maximum negative output voltage. This mode of operation reduces the dissipation in source, sink, and displacer, and is overall more efficient as it can be seen that less power is drawn from the supply. Distortion is also reduced as the varying displacer current reduces the current variations in the rest of the output stage.

The sink transistor now starts to conduct later as the voltage fraction increases, and there is a larger Class-A region than with a constant-current displacer.

Class-G Power Partition

As described in Chapter 19, Class-G aims to reduce amplifier power dissipation by exploiting the high peak-mean ratio of music. At low outputs power is drawn from a pair of low voltage rails; for the relatively infrequent excursions into high power, higher rails are drawn from. We will start with the lower rails at ±15 V,

30% of the higher ±50 V rails; I call this Class G-30%, though it could be more concisely written as G-30 or even G3.

This gives the discontinuous PPD in Figure 16.8. Line 1 is the dissipation in the low-voltage inner sink device, which is kept low by the small voltages across it. Line 2 adds the dissipation in the high-voltage outer sink; this is zero below the rail-switching threshold. Above this are added the identical (due to symmetry) dissipations in the inner and outer source devices, as Lines 3 and 4. Line 5 adds the power in the load, and 6 is the total power drawn, as before. Comparing this with the Class-B diagram in Figure 16.2, the power drawn from the supply, and the amplifier dissipation at low outputs, are much reduced; above the threshold these quantities are only slightly less than for Class-B. Class-G-30% does not show its power-saving abilities well under sinewave drive, because a sine function spends a lot of its time at high values. The sinewave results are not a basis for evaluating the real-life efficiency of Class-G, or deciding on the lower supply voltage. (The upper supply voltage is determined by the maximum power output required.) This issue is dealt with later in this chapter.

If we raise the lower supply rails to ±30 V, 60% of the higher ±50 V rails, we have Class G-60%,

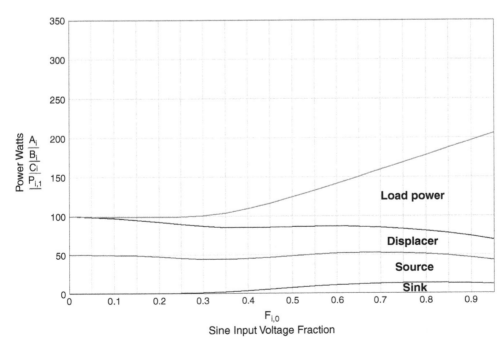

Figure 16.7. Class-XD with push-pull displacer. Dissipation in source, sink, and displacer is reduced, giving greater efficiency.

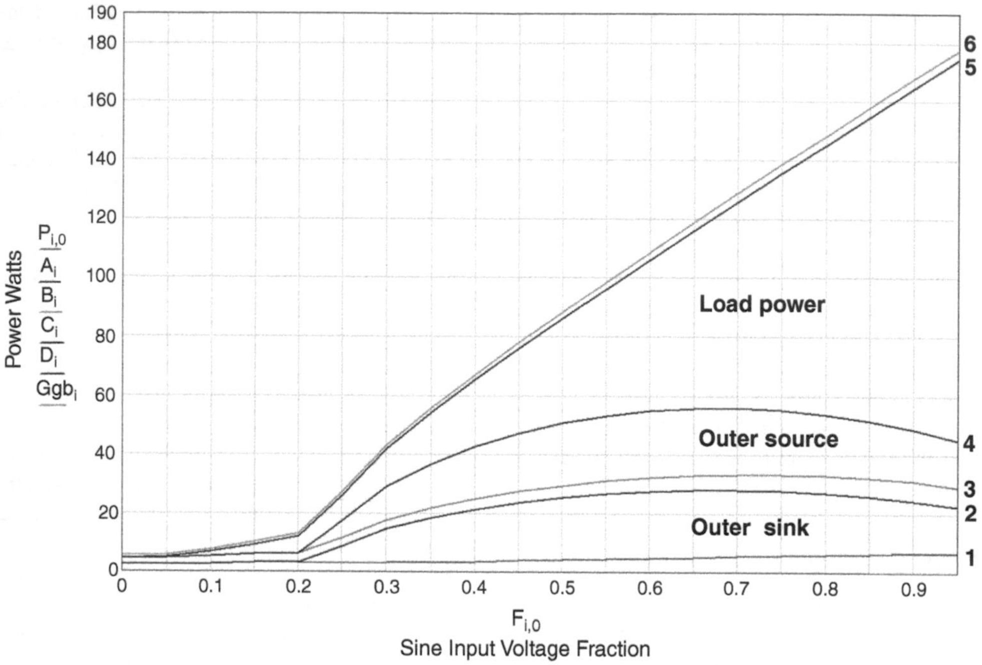

Figure 16.8. Class-G 30% (lower supply rails are 30% of upper rails) with sinewave drive. Compare Figure 16.2; amplifier dissipation at low levels is much reduced.

(or G-60) and the Power Partition Diagram is as Figure 16.9. It is clear that the amplifier dissipation is increased at low powers but much reduced at high powers. This is for a sine input; as we shall see later, the position with real musical waveforms is reversed, with Class G-30% the more efficient of the two.

Assessments of Class-G efficiency when more than two rails are involved, and for other cases, were made by Bortoni and others.[2]

Class-B EF with Reactive Loads

The simulation method outlined above is also suitable for reactive loads. It is, however, necessary to run the simulation not just for one cycle, but sometimes for as many as twenty, to ensure that steady-state conditions have been reached. The diagrams below are for steady-state 200 Hz sinewave drive; the frequency must be defined so the load impedance can be set by suitable component values, but otherwise makes no difference.

Figure 16.10 shows what happens in a Class-B EF output stage when driving a 60-degree capacitive-reactive load with a modulus of 8 Ω. Comparing it with Figure 16.2, the power drawn from the supply is essentially unchanged, and is still proportional to output voltage fraction. The larger vertical distances at the bottom show that more power is being dissipated in the output devices and correspondingly less reaches the load, because the phase shift causes the periods of high voltage across and high current through the output devices to overlap more. The amplifier must now dispose of 95 Watts of heat worst-case, rather than 60 W. Average device dissipation no longer peaks, but increases monotonically up to maximum output. 45° phase angles are common when loud-speakers are driven, and it is generally accepted that an amplifier should be able to provide full voltage swing into such a load.

When the load is purely reactive, with a phase angle of 90°, it can dissipate no power and so all that delivered to it is re-absorbed and dissipated in the amplifier. Figure 16.11 shows that the worst-case device dissipation is much greater at 185 W, absorbing all the power drawn from the supply, and therefore necessarily increasing proportionally with output level; there is no maximum at medium levels. This is a very severe test for a power amplifier. This case is instructive but wholly unrealistic, as no assembly of moving-coil speaker elements can ever present a purely reactive impedance; 60° loads are normally the most reactive catered for.

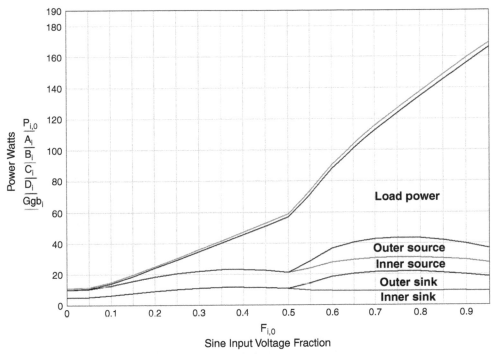

Figure 16.9. Class-G 60% with lower supply rails set at 60% of upper rails, with sinewave drive. Compare Figure 16.4; amplifier dissipation at low levels is much reduced.

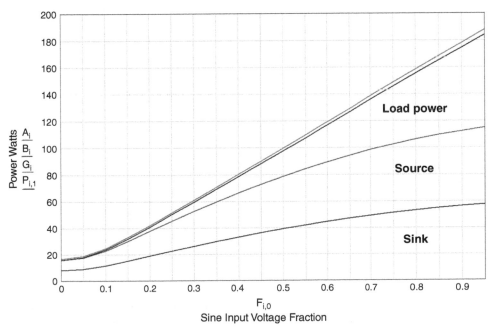

Figure 16.10. Class-B EF with 45-degree reactive load (11.3 Ω in parallel with 71 uF, impedance modulus 8 Ω at 200 Hz). The amplifier dissipation is increased, and the power delivered to the load decreased.

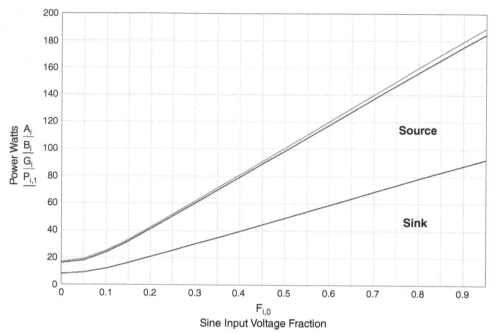

Figure 16.11. Class-B EF with-90 degree reactive load (99.5 uF capacitor, impedance modulus 8 Ω at 200 Hz. All the supply power is now being absorbed by the amplifier, and none by the load. The efficiency is zero.

Table 16.1 shows the worst-case cycle-averaged dissipation for various load angles, showing how the position of maximum dissipation moves towards full output as the angle increases.

This is best displayed in 3D as Figure 16.12, which plots power vertically; the slight hump at the front (for a non-reactive load) disappearing as the load becomes more reactive. The dissipation hump is of little practical significance. An audio amplifier will almost certainly be required to drive 45° loads, and these cause higher power dissipations than resistive loads driven at any level. The dissipation at 45°, or possibly 60°, is the design condition to be met. We are still speaking here of sinewave

drive, which does not much resemble music in its demands. However, it must be taken into account because this is usually how amplifiers are tested.

Figure 16.13 shows the same plot for peak power dissipation in the output devices, which increases monotonically with both output fraction and load angle.

Figure 16.14 summarises all this data for design purposes. It shows worst-case peak and average power in one output device against load reactance, with sinewave drive. Peak powers are taken at 0.95 of full output, average power at whatever output fraction gives maximum dissipation. Therefore to design an amplifier to cope with 45-degree loads, note that average power is increased by 1.4 times, and peak power by 2.7 times, over the resistive case. This can mean that it is necessary to increase the number of output devices simply to cope with the much greater peak power.

Table 16.1. Worst-case cycle-average dissipation for various load angles

Load deg	Max Pdiss	Voltage fraction
0	60 W	0.64
10	63 W	0.65
20	67 W	0.70
30	70 W	0.75
45	95 W	0.95
60	115 W	1.00
90	185 W	1.00

Conclusions on Reactive Loads

1. Amplifier power consumption and average supply current drawn do not vary with load phase angle if the impedance modulus remains constant.
2. The average current in the output devices is not altered so long as the impedance modulus remains constant. This follows from (1).

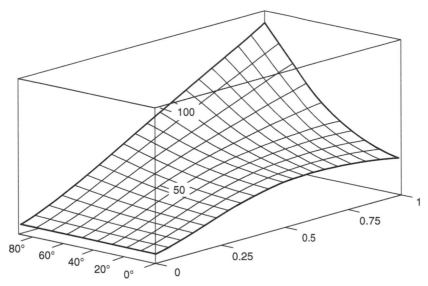

Figure 16.12. The average power dissipation (vertical axis) against load angle (left horizontal axis) and output fraction (right horizontal axis).

3. The peak current in the output devices is not altered so long as the impedance modulus remains constant.
4. Average device dissipation also increases as the load angle increases. A 45° load increases average dissipation by 1.4 times, and a 60° load by 1.8 times. See Figure 16.12.
5. Peak device dissipation increases more rapidly than average dissipation as the load angle increases.

A 45° load increases power peaks by 2.7 times, and a 60° load by 3.4 times. See Figure 16.12.

Considering simple reactive loads like those above gives an essential insight into the extra stresses they impose on semiconductors, but is still some way removed from real signals and real loudspeaker loads, where the impedance modulus varies along with the phase, due to electromechanical resonances or crossover dips. Single

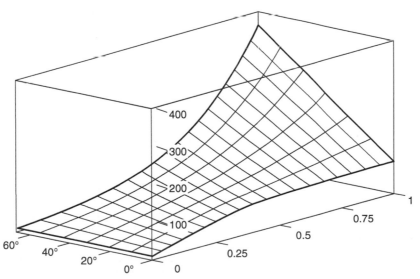

Figure 16.13. Peak power dissipation plotted as in Figure 16.12. The vertical scale must accommodate much higher power levels than Figure 16.12.

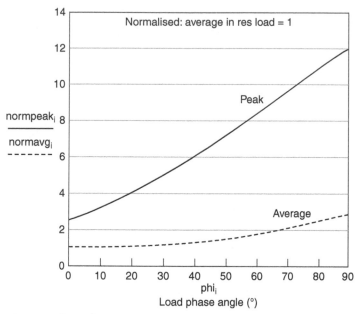

Figure 16.14. Peak power increases faster than worst-case average power as the load becomes more reactive and its phase angle increases. Class-B EF as before.

and two-unit loudspeaker models are examined in Chapter 14, where the maximum phase angle found is 40°. To summarise, the results are:

1. Amplifier power consumption and average supply current drawn vary with frequency due to impedance modulus changes.
2. Average current in the output devices increases by a maximum of 1.3 times.
3. Peak device current increases by a maximum of 1.3 times at the modulus minima.
4. Average device dissipation increases by a maximum of 1.4 times.
5. Peak device dissipation increases by a maximum of 2 times, mostly due to phase shift rather than impedance dips.

These numbers come from two specific simulation models that attempted to represent 'average' speakers, and the circuit conditions for more demanding outcomes could easily have been set up. Benjamin gives an excellent account of real speaker loading in [3]; 21 models were tested and the worst load angle found was 67°. Eliminating the two most extreme cases reduced this to 60°.

The most severe effect of reactive loads is the increase in peak power, followed by the increase in average power. Both are a strong function of load phase, and so the maximum load angle to be driven is a vital part of any power amplifier specification because of its great effect on the number of output devices required and the heatsink design, and hence on the amplifier cost. It is likely that a failure to appreciate just how quickly peak power increases with load angle is the root cause of many amplifier failures.

The Peak-to-Mean Ratio of Music

So far we have seen how the power consumed by amplifiers of various classes was partitioned between internal dissipation and the power delivered to the load. This was determined for the usual sinewave case.

The snag with this approach is that a sinewave does not remotely resemble real speech or music in its characteristics, and in many ways is almost as far from it as you could get. In particular, it is well known that music has a large Peak-to-Mean Ratio (PMR), though the actual value of this ratio in dB is a very vague quantity. Signal statistics for music are in surprisingly short supply.

Very roughly, general-purpose rock music has a PMR of 10 dB to 30 dB, while classical orchestral material (which makes very little use of fuzzboxes and the like) is 20 to 30 dB. The muzak you endure in lifts is limited in PMR to 3 to 10 dB, and compressed bass material in live PA systems is similar. It is clear that the power dissipation in PA bass amplifiers is going to be radically different from that in hi-fi amplifiers reproducing orchestral material at the same peak level. The PMR of a sinewave is 4.0 dB, so results from this are only relevant to lifts ...

Recognising that music actually has a Peak-to-Mean Ratio is a start, but it is actually not much help as it reduces the statistics of signal levels to a single number. This does not give enough information for the estimation of power dissipation with real signals.

To calculate the actual power dissipations, two things are needed: (1) a plot of the instantaneous power dissipation against level; and (2) a description of how much time the signal spends at each level. The latter is formally called the Probability Density Function (PDF) of the signal; more of this later.

The Instantaneous Power Partition Diagram (IPDD) is obtained by running the output stage simulation with a sawtooth input and no per-cycle averaging. The instantaneous power dissipation can therefore be read

out for any input voltage fraction simply by running the cursor up the sawtooth. Figure 16.15 is the IPDD for the Class-B CFP case, where the quiescent current is very small. This looks very like the averaged-sinewave Power Partition Diagram in Figure 16.2, but with the device dissipation maximum at 50% voltage rather than 64% for the sinewave case. The instantaneous powers are much higher, as they are not averaged over a cycle, and there is only one device-power area at the bottom as only one device conducts at a time.

Output device dissipation at the moment when the signal is halfway between rail and ground (input fraction 50%) is 76 W, and the power in the load is 75 W. This totals to 151 W, on the lower of the two straight lines, while the power drawn from the supply is shown as 153 by the upper straight line. The 2 W difference represents losses in the driver transistors and the output emitter resistors.

All the IPDDs for various output stages look very similar in shape to the averaged-sine PPDs earlier in this chapter, but the peak values on the Y-axis are higher. The IPDD can be combined with any PDF to give a much more realistic picture of how power

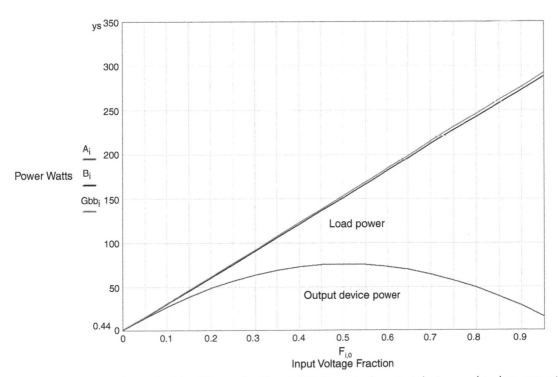

Figure 16.15. Instantaneous Power Partition Diagram for Class-B CFP. Power in the output devices peaks when output is at half the rail voltage.

dissipation changes as the level of a given type of signal is altered.

The Probability Density Function (PDF)

The most difficult part of the process above is obtaining the Probability Density Function. For repetitive waveforms the PDF can be calculated,[4] but music and speech need a statistical approach.

It is often assumed that musical levels have a Gaussian (normal) probability distribution, as the sum of many random variables, but positive statements on this are hard to find. Benjamin[5] says: 'music can be represented accurately as a Gaussian distribution' while Raab[6] states 'music and mixed sounds typically have Gaussian PDFs'. It appears likely this assumption is true for multi-part music which can be regarded as a summation of a many random processes; whatever the PDF of each component, the result is always Gaussian (the Central Limit Theorem).

If the distribution is Gaussian, its mean is clearly zero, as there is no DC component, which leaves the variance (i.e., width of the bell-curve) as the only parameter left to determine. The Gaussian distribution tails off to infinity, implying that enormous levels can occur, though very rarely. In reality, the headroom is fixed, and I have dealt with this by setting variance so the maximum value (O dB) occurs 1% of the time. This is realistic as music very often requires judicious limiting of occasional peaks to optimise the dynamic range.

The PDF presents some conceptual difficulties, as it shows a density rather than a probability. If a signal level ranges between 0 and 100%, then clearly it might be expected to spend some of its time around 50%. However, the probability that it will be at exactly 50.000% is zero, because a single level value has zero extent. Hence the PDF at x is the probability that the signal variable is in the interval (x,x+dx), where dx is the usual calculus infinitesimal.

The Cumulative Distribution Function (CDF)

If the probability that the instantaneous voltage will be *above* — not *at* — a given level is plotted against that level, then we get a *cumulative distribution function* or CDF. This is important as it is easier to measure than the PDF. If the variable is x, then the PDF is often called P(x) and the CDF called F(x). These are related by:

$$P(x) = \frac{d}{dx}F(x)$$ Equation 16.1

or to put it the other way round:

$$F(x) = \int_0^x P(a)\,da$$ Equation 16.2

where a is a dummy variable. The integration in Equation 16.2 starts at zero because we are not going to get signal levels below zero.

Generating a CDF by integrating a given PDF is straightforward, but going the other way, determining the PDF from the CDF, can be troublesome as the differentiation accentuates noise on the data.

Figure 16.16 shows the calculated PDF of a sinewave. As with every PDF, the total area under the curve is one, because the signal must be at some level or other all of the time. However, the function blows up (i.e., heads off to infinity) at each end because the peaks of the wave are 'flat', and so the signal dwells there for infinitely longer than on the slopes where things are changing. However, these 'flat' bits are infinitely small in time extent, and so the area under the curve is still unity. This shows you why PDFs are not always the easiest things to handle. The CDF for a sinewave is shown in Figure 16.17; the probability of exceeding the level on the axis falls slowly at first, but then accelerates to zero as the rounded peaks are reached.

Measuring the PDF

But is all music Gaussian? I was not satisfied that this had been conclusively established from just two brief references, so I decided to measure some musical PDFs myself. In essence this is simple. The first thing to decide is the length of time over which to examine the signal. For most contemporary music the obvious answer is 'one track', a complete

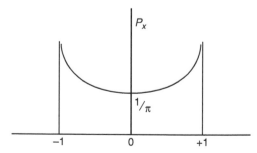

Figure 16.16. The Probability Density Function (PDF) of a sinewave. The peaks at each end go towards infinity.

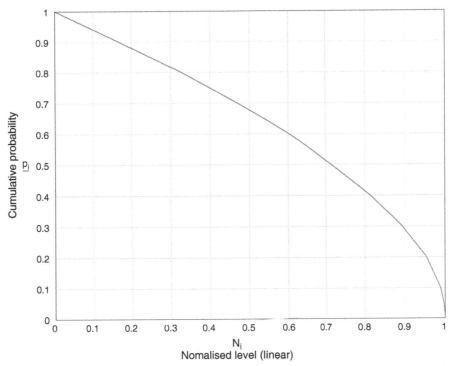

Figure 16.17. Cumulative Distribution Function (CDF) of a sinewave. Drawn with measured data from the circuit of Figure 16.4 as a reality check.

composition lasting typically between three and eight minutes.

Very simple circuitry can be used to determine a CDF, and hence the PDF, though the process is protracted. A comparator was driven by the signal to be measured, with its switching point set by a variable voltage that covered the range of signal amplitude. The comparator output is an irregular rectangular waveform which is averaged by a long time-constant then applied to a moving-coil meter. This allows reading of a changing signal, though not with any great accuracy. The meter deflection is proportional to the amount of time that the signal exceeds the variable threshold, and so gives the CDF. Full details of the circuitry I used are given in.[7]

The circuit only measures one polarity of the waveform, so signal symmetry is assumed. This is safe unless you plan to do a lot of work with solo instruments or single acapella voices; the human vocal waveform is notably asymmetrical.

The hardware is very simple but only yields one data point at a time, for each level. Since something like twenty points are required for a good graph, this gets pretty tedious. It is wise to pick a short track.

The CDF thus obtained for Alannah Myles' 'Black Velvet' is Figure 16.18, and the PDF derived from it is Figure 16.19, complete with some rather implausible ups and downs produced by differentiating data that is accurate to ±1% at best.

Several rock tracks were measured, and also short classical works by Albinoni and Bach. The results are surprisingly similar to Figure 16.18; see the composite CDF in Figure 16.20.

This is good news because we can use a single PDF to evaluate amplifiers faced with varying musical styles. However, I decided the method needed a reality check, by deriving the PDF in a completely different way.

A DSP system offers the possibility of determining as many data points as you want on one playing of the music specimen. In this case a very simple 56001 program sorted the audio samples into 65 amplitude bins. The result for 30 seconds of disco music was somewhere between triangular and Gaussian. The difference between them is small, and either can be used. The triangular PDF simplifies the mathematics, but if like me you use Mathcad to do the work, it is very simple to plug in whatever distribution seems appropriate.

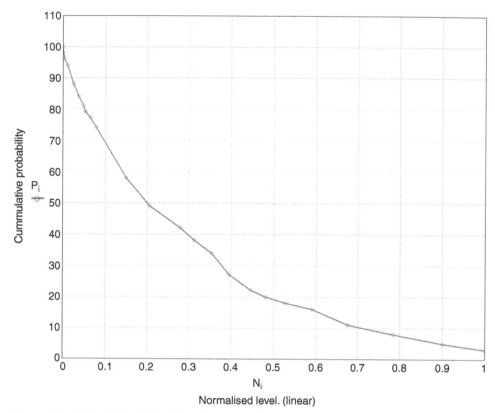

Figure 16.18. The CDF obtained from Alannah Myles' 'Black Velvet' by the comparator method.

Deriving the Actual Power Dissipation

Having found the PDF, it is combined with the Instantaneous Power Partition Diagram. In this case the IPDD is divided into twenty steps of voltage fraction, and each one multiplied by the probability the signal is in that region. The summation of these products yields a single number: the average power dissipation in Watts for a real signal that just reaches clipping for 1% of the time.

An obvious extension of the idea is to plot the average power derived as above, against signal level on the X-axis. This gives an immediate insight into how amplifier dissipation varies with the volume setting. Figure 16.21 shows how level changes affect the PDF. Line 1 is full volume, just reaching maximum level at the right. Line 2 is half volume (-6 dB) and so hits the X-axis at 0.5; it is above Line 1 to the left as the probability of lower levels must be higher to maintain unity area under it. This process continues as volume is reduced, until at zero volume the zero-level probability is 1 and all other levels have zero probability.

Having generated twenty PDF functions, the powers that result for each one are plotted with the volume setting (not the output fraction) as the X-axis. The results for some common amplifier classes are as follows.

Actual Power Dissipation for Class-B CFP

Figure 16.22 shows the instantaneous power plot for Class-B CFP combined with a triangular PDF, showing how the load and device power vary with volume setting. A signal with triangular PDF spends most of its time at low values, below 0.5 output fraction, and so there is no longer a dissipation maximum around half output. Device dissipation at bottom increases monotonically with volume. Load power increases with a square law, which is a reassuring check on all these calculations.

Figure 16.23 is Figure 16.22 replotted with a log X-axis, which is more applicable to human hearing. Domestic amplifiers are rarely operated on

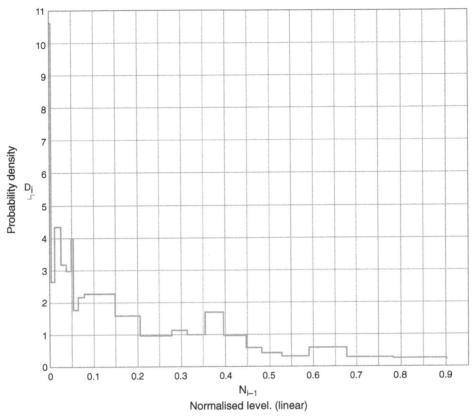

Figure 16.19. The PDF derived from the CDF of Alannah Myles' 'Black Velvet'

the edge of clipping; a realistic operating point is more like −15 or −10 dB. The plot reveals that the efficiency at −15 dB is only 16%, with much more power dissipated in the output devices than reaches the load.

Actual Power Dissipation for Class-AB

Figure 16.24 shows the dB plot for Class-AB, biased so Class-A operation is maintained up to 5W RMS output. The quiescent current is now 370 mA, so there is greater quiescent dissipation at zero volume. There is also substantial conduction overlap, and so sink and source would be different if the plot only considered voltage excursions in one direction away from 0 V. The positive and negative half-cycles therefore have to be averaged to get the final plot. The total device dissipation is unchanged but the boundary between the source and sink areas is halfway, as in Figure 16.24.

Actual Power Dissipation for Class-A Push-pull

I have stuck with the same +/−50 V rails for ease of comparison, and this yields a very powerful Class-A amplifier. The power drawn from the supply is constant, and as output increases, dissipation transfers from the output devices to the load, giving minimum amplifier heating at maximum output. The result for sinewave drive is bad enough (see Figure 16.5 above) but Figure 16.25 reveals that with real signals, almost all the energy supplied is wasted internally, even at maximum volume. Class-A has always been stigmatised as inefficient; this shows that under realistic conditions it is hopelessly inefficient, so much so that it grates on my sense of engineering aesthetics. At typical listening volumes of, say, −15 dB, the efficiency barely reaches 1%, compared with 16% for Class-B. Given the low-distortion capabilities of a Blameless Class-B amplifier, especially when fitted with output-inclusive compensation (see Chapter 13), it is questionable if the use of Class-A is ever justified, even in its most efficient

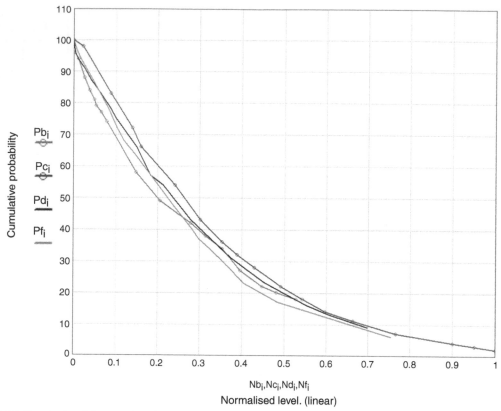

Figure 16.20. The CDFs of three rock and one classical tracks, showing only small differences. The classical track (Bach) is marked with round data points.

push-pull form. Certainly it seems that there is no place for single-ended Class-A which has even lower efficiency; half or a quarter of that of push-pull.

Actual Power Dissipation for Class-G

In Figure 16.26 the lower rails are +/-15 V, 30% of the higher +/−50 V rails; I call this Class-G-30%.

The lower area is the power in the inner devices (i.e., those on all the time) and the larger area just above is that in the outer devices (those only activated when running from the higher rails); this is zero below the rail-switching threshold at a volume of 0.2. The total device dissipation is reduced from 48 W in Class-B to 40 W, which is not a good return for twice as many power transistors. This is because the lower rail voltage is poorly chosen for signals with a triangular PDF.

If the lower rails are increased to +/−30 V, this becomes Class-G-60% as in Figure 16.27. Here the

low-dissipation region now extends up to a voltage fraction of 0.5, but inner device dissipation is higher due to the increased lower rail voltages. The overall result is that total device power is reduced from 48 W in Class-B to 34 W, which is a definite improvement. It is not suggested that 60% is the optimum lower-rail voltage in all cases. The efficiency of Class-G amplifiers depends very much on signal statistics.

Actual Power Dissipation with Reactive Loads

The disadvantage of using instantaneous power is that it ignores signal and circuit history, and so cannot give meaningful information with reactive loads. The peak dissipations that these give rise to with real signals are difficult to simulate; it would be necessary to drive the circuit with stored music signals for many cycles; and that would only cover a few seconds of a CD or concert. The anomalous speaker currents examined in[8]

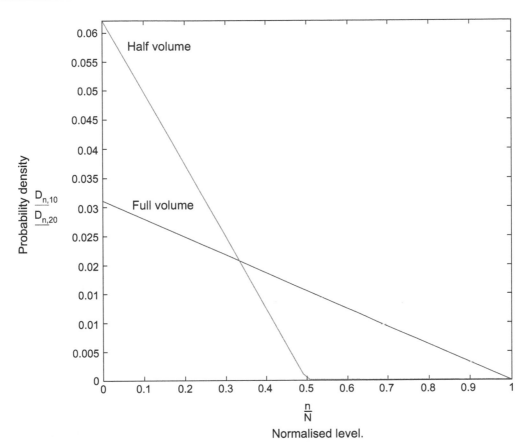

Figure 16.21. The triangular PDF, and how level changes affect it. Line 1 is full volume, and Line 2 half-volume.

show how significant history effects can be with some waveforms.

Dissipation Summary

Tables 16.2 and 16.3 summarise how a triangular-PDF signal, rather than a sinewave, reduces average power dissipation and the power drawn from the supply.

These economies are significant; the power amplifier market is highly competitive, and it is desirable to at least partially exploit the cost savings in heatsinks and PSU components made possible by designing for real signals rather than sinewaves. In particular, Class-G shows valuable economies in device dissipation and PSU capacity, though to get the former, the lower supply voltage must be carefully chosen. However, it must not be forgotten that amplifiers are tested with sinewaves and must be able to sustain such testing for a reasonable period.

Table 16.2. Device dissipation, worst-case volume

	Sinewave power	PDF power	Factor
Class-B CFP	64 W	48 W	0.75
Class-AB	64 W	55 W	0.78
A push-pull	324 W	324 W	—
Class-G-30%	43 W	40 W	0.93
Class-G-60%	56 W	34 W	0.61

Table 16.3. Power drawn, worst-case. (Always max. output)

	Sinewave power	PDF power	Factor
Class-B CFP	186 W	97 W	0.52
Class-AB	188 W	105 W	0.58
A push-pull	324 W	324 W	—
Class-G-30%	177 W	93 W	0.52
Class-G-60%	169 W	81 W	0.48

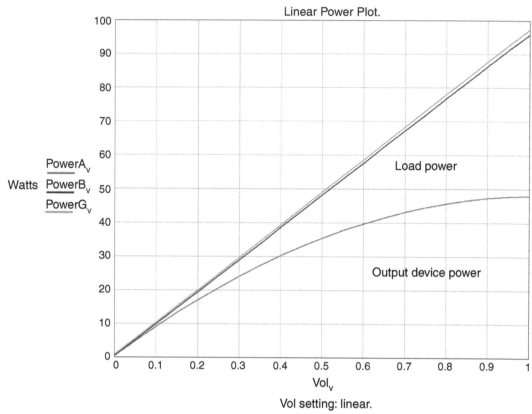

Figure 16.22. Class-B CFP Power Partition versus level. The Class-B IPPD has been combined with the triangular PDF. Device dissipation (lower area) now increases monotonically with volume.

Consideration of a triangular-PDF signal is unlikely to reduce the number of power devices required as real signals give no corresponding reduction in peak device power or peak device current over a sinewave.

A Power Amplifier Design Procedure

Armed with the information above, we can set out a design procedure for a power amplifier which will allow us to specify the transformer, the number of power devices, heatsink size, and so on. The method outlined below is best implemented as a mathematical model on a spreadsheet, as the basic method is to calculate the output power from the transformer secondary voltage, and then go back to tweak the transformer as required, probably requiring several iterations. I have used for many years a rather extensive mathematical model which gives pretty much all the information

needed for a basic design, down to the projected lifetime of the reservoir capacitors.

This design procedure is valid for Class-B amplifiers, and also Class-AB, so long as the quiescent current is not set so high that it requires the heatsink size to be increased over that for Class-B. For other classes, the power dissipation in the output devices must be determined from the graphs given earlier in the chapter. There is much more detail on power supply design in Chapter 26.

It goes like this:

1. Estimate the required transformer secondary voltage in Vrms.
2. Calculate the voltage on the reservoir capacitors by taking the peak transformer voltage ($\sqrt{2}$ times the rms voltage) and subtracting 2 V to allow for losses in the bridge rectifier; the current goes through two diodes in the bridge so two 1 V drops are subtracted. This voltage is greater than the usual 600 mV we

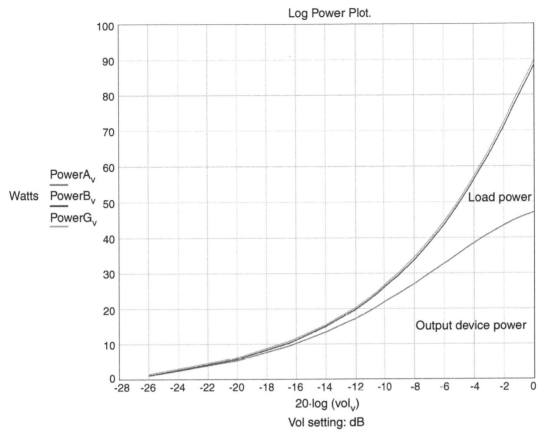

Figure 16.23. Class-B CFP plotted with volume on a more useful log (dB) X-axis. The shape looks quite different from Figure 16.22.

associate with a diode because of the current flows in large peaks.

3. Not all the supply voltage is available to the load. There will be voltage drops in the output emitter resistors, which can be easily calculated from a potential divider model. If 50 V is applied to a divider with 0.1 Ω at the top and 8 Ω at the bottom, only 49.4 V peak is going to reach the load. The loss is obviously greater with loads of 4 Ω or less, and these need to be considered at the same time.

A further limit on the output voltage is how closely the amplifier can drive its output to the supply rails; this depends on the small-signal part of the amplifier and is not usually easy to calculate precisely. 2 V top and bottom can be used as a place-holder, but simulation or measurement of a similar amplifier is necessary to get any accuracy; there is more on this in Chapter 17 on Class-A amplifiers, where you need to squeeze out

every Volt you can. If you are using an NPN VAS, then clipping will usually occur earlier on negative peaks; this is covered in Chapter 7 on the single-ended VAS.

4. We now have the maximum peak voltage that can reach the load. From that we calculate the maximum rms voltage by dividing by $\sqrt{2}$, and from that the output power using $P = V^2/R$. If the power comes out too low or high compared with the spec, we go back to Step 1 and adjust the transformer secondary voltage. It is usual to aim for 5% or so more power than the bare spec figure to allow for odd losses in cable resistance, etc., and because it never looks good when an amplifier fails to reach its rated power on review.

We also find the peak output current simply by using Ohm's law. That gives us an idea of how many output devices are required to handle the peak current. Note that we have so far ignored the ripple

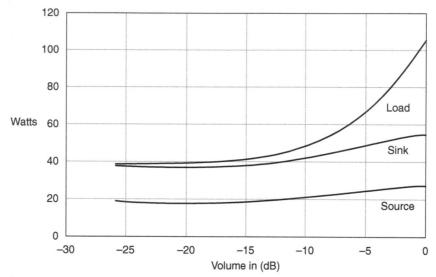

Figure 16.24. Class-AB Power Partition Diagram, stage biased to give Class-A up to 5W. Averaged over whole cycle.

on the reservoir capacitors, although it is the troughs of the ripple that determine when clipping starts.

5. From the peak output current we determine the average current drawn from each supply rail. For a Class-B amplifier and sine wave this is simply the peak current divided by π (3.1416). Other amplifier types or waveforms would require a different constant. This enables us to specify the current rating of the transformer, and select

the bridge rectifier, bearing in mind that both have to handle the current drawn from *both* supply rails, and that twice over again for a stereo amplifier.

6. Knowing the average current drawn from each rail, we can calculate the ripple on the reservoir capacitors, making an initial guess at the capacitance required. It is a rough but very usable approximation that the reservoirs charge linearly for 3 msec

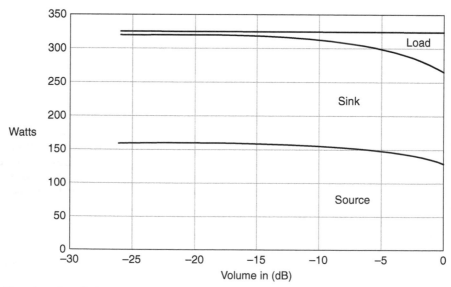

Figure 16.25. Class-A push-pull, for 150 Watts output. The internal dissipation completely dominates, even at maximum volume.

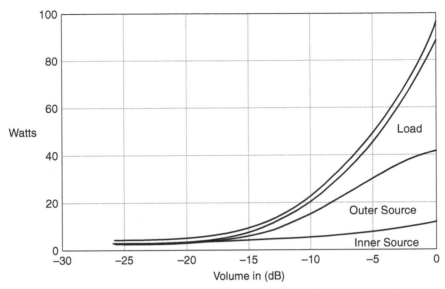

Figure 16.26. Class-G-30%. Low rail voltage is 30% of the high rail. Rail-switching occurs at about −15 dB relative to maximum output.

and then discharge linearly for 7 msec, so we can use this equation:

$$Vpk-pk = \frac{I \cdot \Delta t \cdot 1000}{C} \qquad \text{Equation 16.3}$$

where:

> Vpk-pk is the peak-to-peak ripple voltage on the reservoir capacitor
> I is the average current drawn from that supply rail in Amps
> Δt is the length of the capacitor discharge time, taken as 7 msec
> C is the size of the reservoir capacitor in uF
> The '1000' factor simply gets the decimal point in the right place

We now subtract the peak-peak ripple from the supply rail voltage determined in Step 2, to get the voltage at the bottom of the ripple trough, which determines when clipping occurs. We have to plug this correction into Step 2 to get a new and more realistic value for the output power calculated in Step 4. If you try to do this directly, your spreadsheet will probably complain bitterly that it is being asked to perform a circular calculation, but this is easily evaded by making a manual correction and going round the loop a few times until the values settle.

At this point you can experiment with different reservoir capacitances; more capacitance means a slightly greater amplifier output because the ripple amplitude is less. The usual levels of ripple on full load are around 1 to 3 V peak-to-peak.

The reservoir capacitor voltage rating can also be decided, bearing in mind that the transformer secondary voltage will rise when there is no load because there is no amplifier output. The rise is likely to be at least 5%, and you must also assume that the mains voltage may be 10% high, as do the safety regulations. It is wise to add a healthy margin on top of that.

7. At this point the ripple current through the reservoir capacitances can be found. This is the AC current that flows through the capacitor because there is a varying voltage across its terminals (the ripple). It can be calculated from this equation:

$$I = \frac{1.445 \cdot V_{pk-pk} \cdot \pi \cdot C}{10^5} \qquad \text{Equation 16.4}$$

where:

> I is the rms ripple current in Amps
> Vpk-pk is the peak-to-peak ripple voltage on the reservoir capacitor
> C is the size of the reservoir capacitor in uF
> The 10^5 factor on the bottom gets the decimal point in the right place.

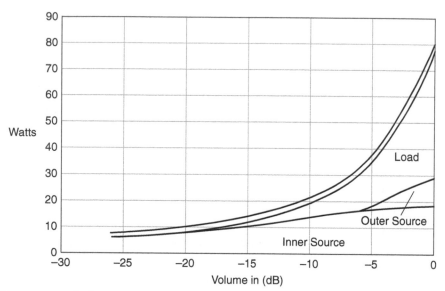

Figure 16.27. Class-G-60%. The lower rail voltage is now 60% of the high rail. This reduces both dissipation and power consumed, compared with Figure 16.14.

The reservoir capacitors are rated for maximum ripple current by the manufacturers, and exceeding this will overheat the capacitor and shorten its life drastically. However, in normally proportioned amplifier power supplies, the ripple current is usually a small fraction of the capacitor rating, often around 15%, and there are no problems.

If the ripple current does come out as uncomfortably high, more reservoir capacitance is needed. If you use two capacitors where one grew before, the capacitance is doubled and the ripple voltage halved. The same overall amount of ripple current therefore flows, but it is now divided between two capacitors.

8. Knowledge of the average current drawn from each supply rail gives the power dissipated as heat in each output device. Adding up the contributions from all the devices on the heatsink allows us to calculate its temperature rise above the ambient, so long as we know the heatsink thermal resistance in °C/W. From this we can derive the temperature rise above the heatsink of the output device package, by using the thermal resistance of the thermal washer that couples it to the heatsink. Finally we can calculate how much hotter the device junction gets compared with the package by using the junction-to-package thermal resistance which is given on the device data sheet. The maximum junction temperature for a power transistor is usually 150 °C, though some parts such as the MJ15022-15024 are rated at 200 °C. It is wise

to keep it below 100 °C for reliability. There is much more on transistor mounting and thermal resistances in Chapter 22 on thermal dynamics.

Knowing the average current drawn from each supply rail also allows the sizes of the mains fuse, transformer secondary fuses, and HT rail fuses to be chosen. The first two must make a generous allowance for current inrush at switch-on.

Design Procedure Results

I have successfully used the procedure outlined above for many amplifier designs that have gone into large-scale production. It is possible to get the mains transformer exactly right first time (I have done it) but the uncertainties in voltage regulation and so on usually mean that a small adjustment has to be made to get the power output greater than the spec figure by a small but reliable margin. The procedure also allows mechanical engineering to proceed at once as the size of heatsink required, and the number of output devices attached to it, are known at a very early stage in the design.

To put it roughly, a single pair of MT-200 output devices are usually adequate for hi-fi usage up to 100 W/8 Ω. For more power, or if you are expecting to drive low impedances, two pairs will be needed. I have successfully used three pairs of MT-200s for 250 W/8 Ω; toneburst power into 2 Ω was about 800 W.

References

1. Ed., *National Semiconductor Audio/Radio Handbook*, 1980 Section 4, p. 49.

2. Bortoni, R, Filhom, S and Seara, R, On the Design and Efficiency of Class A, B, AB, G and H Audio Power Amplifier Output Stages, *Journ Audio Eng Society*, 50(7/8), July/Aug. 2002.

3. Benjamin, E, Audio Power Amplifiers for Loudspeaker Loads, *Journ Audio Eng Society*, 42(9) Sept. 1994, p. 670.

4. Carlson, A, *Communication Systems*, 3rd edn, Maidenhead: McGraw-Hill, 1986, p. 144.

5. Benjamin, E, Audio Power Amplifiers for Loudspeaker Loads, *Journal of Audio Eng Society*, 42(9), Sept. 1994, p. 670.

6. Raab, F, Average Efficiency of Class-G Amplifiers, *IEEE Trans on Consumer Electronics*, Vol CE-22, No. 2, May 1986.

7. Self, D, *Self on Audio* 2nd edn, Oxford: Newnes, 2006, pp. 452–453.

8. Self, D, Loudspeaker Undercurrents, *Wireless World*, Feb. 1998, p. 98.

Class-A Power Amplifiers

If you can't stand the heat, get out of the kitchen.
                                    Harry S. Truman

An Introduction to Class-A

The two salient facts about Class-A amplifiers are that they are inefficient, and that they give the best possible distortion performance. They will never supplant Class-B amplifiers; but they will always be around.

The quiescent dissipation of the classic push-pull Class-A amplifier is equal to twice the maximum output power, making massive power outputs impractical, if only because of the discomfort engendered in the summer months. However, the nature of human hearing means that the power of an amplifier must be considerably increased to sound significantly louder. Doubling the sound pressure level (SPL) is not the same as doubling subjective loudness, the latter being measured in Sones rather than dB above threshold, and it appears that doubling subjective loudness requires nearer a 10 dB rather than 6 dB rise in SPL.[1] This implies amplifier power must be increased something like ten-fold, rather than merely quadrupled, to double subjective loudness. Thus a 40 W Class-B amplifier does not sound much larger than its 20 W Class-A cousin.

There is certainly an attractive simplicity and purity about Class-A. Most of the distortion mechanisms studied so far stem from Class-B, and we can thankfully forget crossover and switch-off phenomena (Distortions 3b, 3c), non-linear VAS loading (Distortion Four), injection of supply-rail signals (Distortion Five), induction from supply currents (Distortion Six), and erroneous feedback connections (Distortion Seven). Beta-mismatch in the output devices can also be ignored.

The great disadvantage of Class-A is its inefficiency. Most of the quiescent current sluices straight through the amplifier from one rail to the other without doing anything useful. Chapter 16 on amplifier dissipation demonstrates that the sinewave efficiency is very low unless the output is near the maximum, and the efficiency with musical signals is very low indeed (1%) in normal use. Almost all of the input power is pointlessly expended in what Edward Gibbon called 'dissipation without pleasure'.

Inevitably human ingenuity has been applied to creating compromises between Class-A and the much more efficient Class-B. The simplest is traditional Class-AB, which is simply Class-B with the bias turned up to give a region of Class-A around zero. As compromises go, this is not a happy one (see Chapters 9 and 10) because when the output stage leaves the Class-A region, there is a step-change in gain which generates significantly greater high-order distortion than optimally biased Class-B. My contribution to the party is Crossover Displacement (aka Class XD) which gives a Class-A region around zero without steps in gain on transitioning to Class-B; this is fully described in Chapter 18.

Another attempt to get the linearity of Class-A without the heat is the so-called non-switching amplifier. This is a Class-B stage with its output devices clamped to always pass a minimum current. However, it is not obvious that a sudden halt in current-change as opposed to complete turn-off makes a better crossover region. Those residual oscillograms that have been published seem to show that some kind of discontinuity still exists at crossover.[2] There is much more on non-switching amplifiers in Chapter 4.

One potential problem is the presence of maximum ripple on the supply-rails at zero signal output; the PSRR must be taken seriously if good noise and ripple figures are to be obtained. This problem is simply solved by the measures proposed for Class-B designs in Chapter 26.

Class-A Configurations and Efficiency

There is a canonical sequence of efficiency in Class-A amplifiers. The simplest version is single-ended and resistively loaded, as at Figure 17.1a. When it sinks output current, there is an inevitable voltage drop across the emitter resistance, limiting the negative output capability, and resulting in an efficiency of 12.5% (erroneously quoted in at least one textbook as 25%, apparently on the grounds that power not dissipated in silicon does not count). This would be of purely theoretical interest — and not much of that — except that at least one single-ended resistive design appeared. This was produced by Fuller Audio, and reportedly produced a 10 W output for a dissipation of 120 W, with the output swing predictably curtailed in one direction.[3]

A better method — Constant-current Class-A — is shown in Figure 17.1b. The current sunk by the lower constant-current source is no longer related to the voltage across it, and so the output voltage can approach the negative rail with a practicable quiescent current (hereafter shortened to Iq). Maximum efficiency is doubled to 25% at maximum output; for an example with 20 W output (and a big fan) see Nelson.[4] Some designs, such as those by Krell, have a current-source that is rapidly increased in output by a sort of noise gate sidechain, in response to signal level. With no further stimulus the quiescent current is reduced after

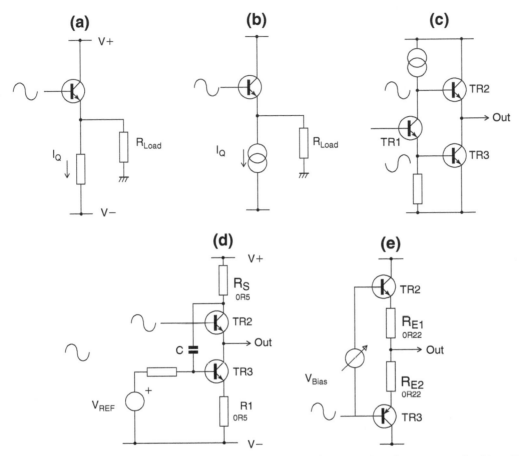

Figure 17.1. The canonical sequence of Class-A configurations. c, d and e are push-pull variants, and achieve 50% efficiency. e is simply a Class-B stage with higher Vbias.

a delay. They call this Sustained Plateau Bias, and an instance of its use is the Krell KSA-100S (1994), which had its bias altered in four discrete steps rather than continuously, the level being sustained for 20–30 seconds. There is a later version called Sustained Plateau Bias II which takes into account the loudspeaker impedance as well as the signal level. The basic version is described in US Patent 5, 331, 291, which indicates that the bias control system is driven by the output stage current. In any system that responds to signal level, there must be questions as to how quickly it can react to a sudden transient. Noise gates can be straight-forwardly designed to respond in tens of microseconds, but changing the value of a substantial quiescent current is likely to take longer.

Push-pull operation once more doubles full-power efficiency, getting us to a more practical 50%; most commercial Class-A amplifiers have been of this type.

Both output halves now swing from zero to twice the Iq, and least voltage corresponds with maximum current, reducing dissipation. There is also the intriguing prospect of cancelling the even-order harmonics generated by the output devices.

Push-pull action can be induced in several ways. Figures 17.1c, d show the lower constant current-source replaced by a voltage-controlled current source (VCIS). This can be driven directly by the amplifier forward path, as in Figure 17.1c,[5] or by a current-control negative feedback loop, as in Figure 17.1d.[6] The first of these methods has the drawback that the stage generates gain, phase-splitter TR1 doubling as the VAS; hence there is no circuit node that can be treated as the input to a unity-gain output stage, making the circuit hard to analyse, as VAS distortion cannot be separated from output stage non-linearity. There is also no guarantee that upper and lower output devices will be driven appropriately for Class-A; in the Linsley-Hood design,[5] the effective quiescent varies by more than 40% over the cycle.

The second push-pull method in Figure 17.1d is more dependable, and I have designed several versions that worked well. The disadvantage with the simple version shown is that a regulated supply is required to prevent rail ripple from disrupting the current-loop control. Designs of this type have a limited current-control range — in Figure 17.1d TR3 cannot be turned on any further once the upper device is fully off — so the lower VCIS will not be able to respond to an unforeseen increase in the output loading. In this event there is no way of resorting to Class-AB to keep the show going and the amplifier will show some form of asymmetrical hard clipping.

The best push-pull stage seems to be that in Figure 17.1e, which probably looks rather familiar. Like all the conventional Class-B stages examined in Chapters 9 and 10, this one will operate effectively in pure push-pull Class-A if the quiescent bias voltage is sufficiently increased; the increment over Class-B is typically 700 mV, depending on the value of the emitter resistors. For an early example of using a high-biased Class-B output stage to get Class-A, see Nelson-Jones.[7] This topology has the great advantage that, when confronted with an unexpectedly low load impedance, it will operate in Class-AB. The distortion performance will be inferior not only to Class-A but also to optimally biased Class-B, once above the AB transition level, but can still be made very low by proper design.

The push-pull concept has a maximum efficiency of 50%, but this is only achieved at maximum sinewave output; due to the high peak/average ratio of music, the true average efficiency probably does not exceed 10%, even at maximum volume before obvious clipping. In my book, *Self on Audio*,[8] I examined the efficiency of many kinds of amplifier when handling signals with a more realistic Probability Density Function, and it became clear that 10% was actually pretty optimistic. With realistic listening levels, say, 15 dB, with respect to full output, efficiency just about reached 1%. The output signal barely disturbs the current continuously sluicing through the amplifier from rail to rail. This is not a very elegant situation.

Other possibilities are signal-controlled variation of the Class-A amplifier rail voltages, either by a separate Class-B amplifier or by a modulated switch-mode supply. Both approaches are capable of high power output, but involve extensive extra circuitry, and present some daunting design problems.

A Class-B amplifier has a limited voltage output capability, but is flexible about load impedances; more current is simply turned on when required. However, Class-A has also a current limitation, after which it enters Class-AB, and so loses its *raison d'être*. The choice of quiescent value has a major effect on thermal design and parts cost; so Class-A design demands a very clear idea of what load impedance is to be driven in pure A before we begin. The calculations to determine the required Iq are straightforward, though lengthy if supply ripple, Vce(sat)s, and Re losses, etc. are all considered, so I just give the results here. (An unregulated supply with 10,000 μF reservoirs is assumed.)

A 20 W/8 Ω amplifier will require rails of approximately ±24 V and a quiescent of 1.15 A. If this is extended to give roughly the same voltage swing into 4 Ω, then the output power becomes 37 W, and to deliver this in Class-A the quiescent must increase to 2.16 A, almost doubling dissipation. If, however, full voltage swing into 6 Ω will do (which it will for many reputable speakers), then the quiescent only needs to increase to 1.5 A; from here on I assume a quiescent of 1.6 A to give a margin of safety.

Output Stages in Class-A

I consider here only the increased-bias Class-B topology, because it is probably the best approach, effectively solving the problems presented by the other methods. Figure 17.2 shows a SPICE simulation of the collector currents in the output devices versus output voltage, and also the sum of these currents. This sum of device currents is in principle constant in Class-A, though it need not be so for low THD; the output signal is the difference of device currents, and is not inherently related to the sum. However, a large deviation from this constant-sum condition means increased inefficiency, as the stage must be conducting more current than it needs to for some part of the cycle.

The constancy of this sum-of-currents is important because it shows that the voltage measured across Re1 and Re2 together is also effectively constant so long as the amplifier stays in Class-A. This in turn means that quiescent current can be simply set with a constant-voltage bias generator, in very much the same way as Class-B.

Figures 17.3, 17.4 and 17.5 show SPICE gain plots for open-loop output stages, with 8 Ω loading and 1.6 A quiescent; the circuitry is exactly as for the Class-B output stages in Chapter 9. The upper traces show Class-A gain, and the lower traces optimal-bias Class-B gain for comparison. Figure 17.3 show the gain plot for an

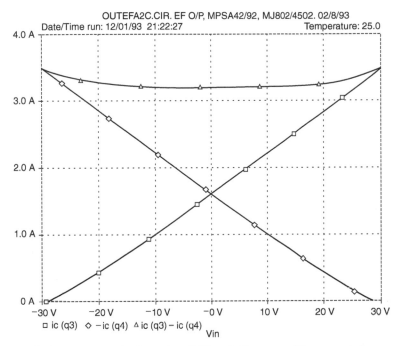

Figure 17.2. How output device current varies in push-pull Class-A. The sum of the currents is near-constant, simplifying biasing.

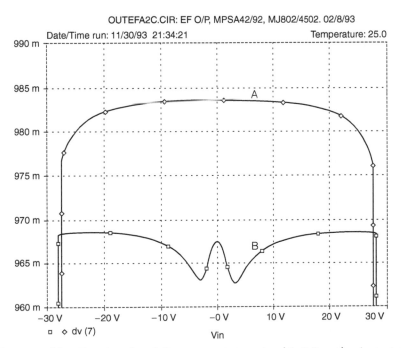

Figure 17.3. Gain linearity of the Class-A emitter-follower output stage. Load is 8 Ω, and quiescent current (Iq) is 1.6A.

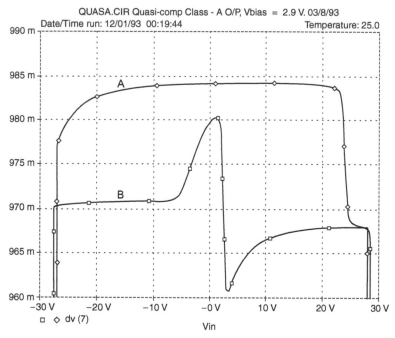

Figure 17.4. Gain linearity of the Class-A quasi-complementary output stage. Conditions as Figure 17.3.

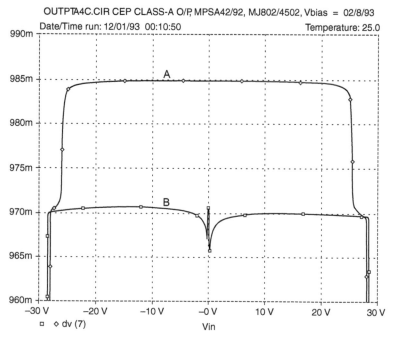

Figure 17.5. Gain linearity of the Class-A CFP output stage.

emitter-follower output, Figure 17.4 for a simple quasi-complementary stage, and Figure 17.5 for a CFP output.

We would expect Class-A stages to be more linear than B, and they are. (Harmonic and THD figures for the three configurations, at 20 V Pk, are shown in Table 17.1.) There is absolutely no gain wobble around 0 V, as in Class-B, and push-pull Class-A really can and does cancel even-order distortion.

It is at once clear that the emitter-follower has more gain variation, and therefore worse linearity, than the CFP, while the quasi-comp circuit shows an interesting mix of the two. The more curved side of the quasi gain plot is on the negative side, where the CFP half of the quasi circuit is passing most of the current; however, we know by comparing Figure 17.3 and Figure 17.5 that the CFP is the more linear structure. Therefore it appears that the shape of the gain curve is determined by the output half that is turning off, presumably because this shows the biggest gm changes. The CFP structure maintains gm better as current decreases, and so gives a flatter gain curve with less rounding of the extremes.

The gain behaviour of these stages is reflected in their harmonic generation; Table 17.1 reveals that the two symmetrical topologies give mostly odd-order harmonics, as expected. The asymmetry of the quasi-comp version causes a large increase in even-order harmonics, and this is reflected in the higher THD figure. Nonetheless, all the THD figures are still 2 to 3 times lower than for their Class-B equivalents.

This modest factor of improvement may seem a poor return for the extra dissipation of Class-A, but not so. The crucial point about the distortion from a Class-A output stage is not just that it is low in magnitude, but that it is low-order, and so benefits much more from the typical NFB factor that falls with frequency than does high-order crossover distortion.

Table 17.1. Harmonic levels generated by different output stages

Harmonic	Emitter-Follower (%)	Quasi-comp (%)	CFP output (%)
Second	0.00012	0.0118	0.00095
Third	0.0095	0.0064	0.0025
Fourth	0.00006	0.0011	0.00012
Fifth	0.00080	0.00058	0.00029
THD	0.0095	0.0135	0.0027

Note: THD is calculated from the first nine harmonics, though levels above the fifth are very small.

The choice of Class-A output topology is now simple. For best performance, use the CFP; apart from greater basic linearity, the effects of output device temperature on Iq are servo-ed out by local feedback, as in Class-B. For utmost economy, use the quasi-complementary with two NPN devices; these need only a low Vce(max) for a typical Class-A amp, so here is an opportunity to recoup some of the money spent on heatsinking. The rules here are somewhat different from Class-B; the simple quasi-complementary configuration gives first-class results with moderate NFB, and adding a Baxandall diode to simulate a complementary emitter-follower stage gives little improvement in linearity. See, however, Nelson-Jones[7] for an example of its use.

It is sometimes assumed that the different mode of operation of Class-A makes it inherently short-circuit proof. This may be true with some configurations, but the high-biased type studied here will continue delivering current in time-honoured Class-B fashion until it bursts, and overload protection seems to be no less essential.

Quiescent Current Control Systems

Unlike Class-B, precise control of quiescent current is not required to optimise distortion; for good linearity there just has to be enough of it. However, the Iq must be under some control to prevent thermal runaway, particularly if the emitter-follower output is used. A badly designed quiescent controller can ruin the linearity, and careful design is required. There is also the point that a precisely held standing-current is considered the mark of a well-bred Class-A amplifier; a quiescent that lurches around like a drunken sailor does not inspire confidence.

Straightforward thermal compensation with a Vbe-multiplier bias generator thermally coupled to the main heatsink can certainly be made to work,[9] and will prevent thermal runaway so long as the heatsink is of adequate size. At least one design — which in other respects appears to be a sincere homage to this chapter — has been published in which the bias generator is thermally coupled not to the main heatsink, but to the driver heatsinks, which are small PCB-mounting types for TO-220.[10] I am not at all convinced that this will give proper control of the quiescent current.

However, this sort of approach misses a golden opportunity. Unlike Class-B, the use of Class-A offers the possibility of tightly controlling Iq by negative feedback. This is profoundly ironic because now that we can precisely control Iq, it is no longer critical. Nevertheless, it seems churlish to ignore the opportunity, and so feedback quiescent control will be examined.

There are two basic methods of feedback current-control. In the first, the current in one output device is monitored, either by measuring the voltage across one emitter-resistor (Rs in Figure 17.6a), or by a collector sensing resistor, as in Figure 17.6b; the second method monitors the sum of the device currents, which as described above, is constant in Class-A.

The first method, as implemented in Figure 17.6a,[7] compares the Vbe of TR4 with the voltage across Rs, with filtering by RF, CF. If quiescent is excessive, then TR4 conducts more, turning on TR5 and reducing the bias voltage between points A and B. In Figure 17.6b, which uses the VCIS approach, the voltage across collector sensing resistor Rs is compared with Vref by TR4, the value of Vref being chosen to allow for TR4 Vbe.[11] Filtering is once more by RF, CF.

For either Figure 17.6a or b, the current being monitored contains large amounts of signal, and must be low-pass filtered before being used for control purposes. This is awkward as it adds one more time-constant to worry about if the amplifier is driven into asymmetrical clipping. In the case of collector-sensing, there are unavoidable losses in the extra sense resistor. It is also my experience that imperfect filtering causes a serious rise in LF distortion.

The Better Way is to monitor current in *both* emitter resistors; as explained above, the voltage across both is very nearly constant, and, in practice, filtering is

unnecessary. An example of this approach is shown in Figure 17.6c, based on a concept originated by Nelson Pass.[12] Here TR4 compares its own Vbe with the voltage between X and B; excessive quiescent turns on TR4 and reduces the bias directly. Diode D is not essential to the concept, but usefully increases the current-feedback loop-gain; omitting it more than doubles Iq variation with TR4 temperature in the Pass circuit.

The trouble with this method is that TR3 Vbe directly affects the bias setting, but is outside the current-control loop. A multiple of Vbe is established between X and B, when what we really want to control is the voltage between X and Y. The temperature variations of TR4 and TR3 Vbe partly cancel, but only partly. This method is best used with a CFP or quasi output so that the difference between Y and B depends only on the driver temperature, which can be kept low. The *reference* is TR4 Vbe, which is itself temperature-dependent; even if it is kept away from the hot bits it will react to ambient temperature changes, and this explains the poor performance of the Pass method for global temp changes (Table 17.2).

A Novel Quiescent Current Controller

To solve this problem, I would like to introduce the novel control method in Figure 17.7. We need to compare the floating voltage between X and Y with a fixed reference, which sounds like a requirement for two differential amplifiers. This can be reduced to one

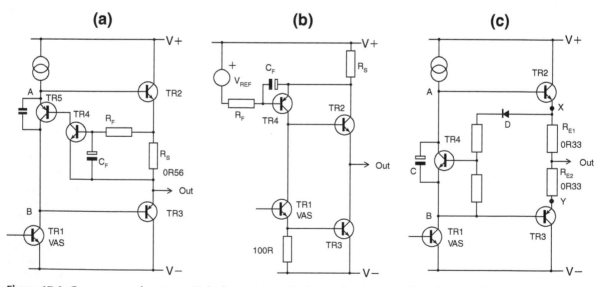

Figure 17.6. Current-control systems. Only that at (c) avoids the need to low-pass filter the control signal; capacitor C simply provides feedforward to speed up signal transfer to TR2.

Table 17.2. Iq change per degree C

	Changing TR7 temp only (%)	Changing global temp (%)
Quasi + Vbe-multiplier	+0.112	−0.43
Pass: as Figure 17.6c	+0.0257	−14.1
Pass: no diode D	+0.0675	−10.7
New system:	+0.006	−0.038

is that TR13 is part of an NFB loop that establishes a voltage at A that will keep the bias voltage between A and B constant. This comes to the same thing as maintaining a constant Vbias across TR5. As might be imagined, this loop does not shine at transferring signals quickly, and this duty is done by feed forward capacitor C4. Without it, the loop (rather surprisingly) works correctly, but HF oscillation at some part of the cycle is almost certain. With C4 in place, the current-loop does not need to move quickly, since it is not required to transfer the signal but rather to maintain a DC level.

The experimental study of Iq stability is not easy because of the inaccessibility of junction temperatures. Professional SPICE implementations like PSpice allow both the global circuit temperature and the temperature of individual devices to be manipulated; this is another aspect where simulators shine. The exact relationships of component temperatures in an amplifier is hard to predict, so I show here only the results of changing the global temperature of all devices, and changing the junction temp of TR7 alone (Figure 17.7) with different current-controllers. TR7 will be one of the hottest

by sitting the reference Vref on point Y; this is a very low-impedance point and can easily swallow a reference current of 1 mA or so. A simple differential pair TR15, 16 then compares the reference voltage with that at point Y; excess quiescent turns on TR16, causing TR13 to conduct more and reducing the bias voltage.

The circuitry looks enigmatic because the high-impedance of the TR13 collector would seem to prevent the signal from reaching the upper half of the output stage; this is in essence true, but the vital point

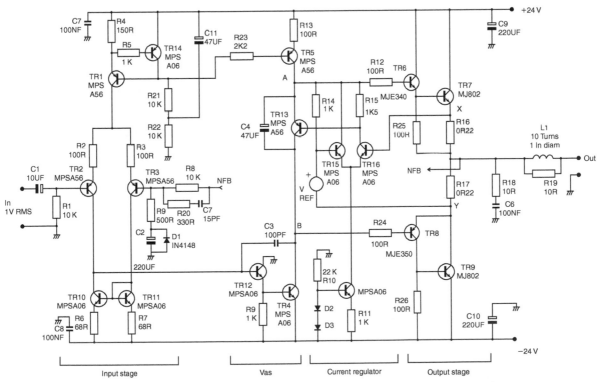

Figure 17.7. A Blameless 20W Class-A power amplifier, using the novel current-control system.

transistors and unlike TR9 it is not in a local NFB loop, which would greatly reduce its thermal effects.

A Class-A Design

A design example of a Blameless 20 W/8 Ω Class-A power amplifier is shown in Figure 17.7. This is as close as possible in operating parameters to the previous Class-B design, to aid comparison; in particular, the NFB factor remains 30 dB at 20 kHz. The front-end is as for the Class-B version, which should not be surprising as it does exactly the same job, input Distortion One being unaffected by output topology. As before, the input pair uses a high tail current, so that R2, 3 can be introduced to linearise the transfer characteristic and set the transconductance. Distortion Two (VAS) is dealt with as in Chapter 7, the added emitter-follower TR12 preventing the non-linear Cbc of TR4 from causing trouble. There is no need to worry about Distortion Four (non-linear loading by output stage) as the input impedance of a Class-A output, while not constant, does not have the sharp variations shown by Class-B.

Figure 17.7 uses a standard quasi output. This may be replaced by a CFP stage without problems. In both cases the distortion is extremely low, but gratifyingly the CFP proves even better than the quasi, confirming the simulation results for output stages in isolation.

The operation of the current regulator TR13, 15, 16 has already been described. The reference used is a National LM385/1.2. Its output voltage is fixed at 1.223 V nominal; this is reduced to approximately 0.6 V by a

1 kΩ−1 kΩ divider (not shown). Using this band-gap reference, a 1.6 Amp Iq is held to within ±2 mA from a second or two after switch-on. Looking at Table 17.2, there seems no doubt that the new system is effective.

As before, a simple unregulated power supply with 10,000 μF reservoirs was used, and despite the higher prevailing ripple, no PSRR difficulties were encountered once the usual decoupling precautions were taken.

The closed-loop distortion performance (with conventional compensation) is shown in Figure 17.8 for the quasi-comp output stage, and in Figure 17.9 for a CFP output version. The THD residual is pure noise for almost all of the audio spectrum, and only above 10 kHz do small amounts of third-harmonic appear. The expected source is the input pair, but this so far remains unconfirmed.

The distortion generated by the Class-B and Class-A design examples is summarised in Table 17.3, which shows a pleasing reduction as various measures are taken to deal with it. As a final fling, two-pole compensation was applied to the most linear (CFP) of the Class-A versions, reducing distortion to a rather small 0.0012% at 20 kHz, at some cost in slew-rate (see Figure 17.10). While this may not be the fabled Distortionless Amplifier, it must be a near relation.

The Trimodal Amplifier

I present here my own contribution to global warming in the shape of an improved Class-A amplifier; it is believed to be unique in that it not only copes with

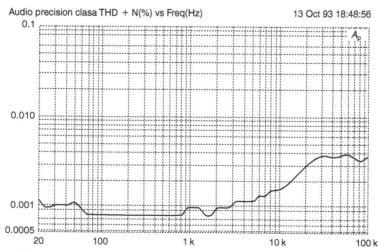

Figure 17.8. Class-A amplifier THD performance with quasi-comp output stage. The steps in the LF portion of the trace are measurement artefacts.

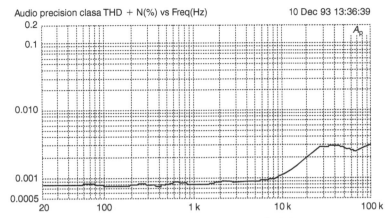

Figure 17.9. Class-A distortion performance with CFP output stage.

load impedance dips by means of the most linear form of Class-AB possible, but will also operate as a Blameless Class-B engine. The power output in pure Class-A is 20 to 30 W into 8 Ω, depending on the supply-rails chosen.

This amplifier uses a Complementary Feedback Pair (CFP) output stage for best possible linearity, and some incremental improvements have been made to noise, slew-rate and maximum DC offset. The circuit naturally bears a very close resemblance to a Blameless Class-B amplifier, and so it was decided to retain the Class-B Vbe-multiplier, and use it as a safety-circuit to prevent catastrophe if the relatively complex Class-A current-regulator failed. From this the idea arose of making the amplifier instantly switchable between Class-A and Class-B modes, which gives two kinds of amplifier for the price of one, and permits of some interesting listening tests. Now you really can do an A/B comparison.

Table 17.3. THD levels from different types of amplifier

	1 kHz (%)	10 kHz (%)	20 kHz (%)	Power (W)
Class-B EF	<0.0006	0.0060	0.0120	50
Class-B CFP	<0.0006	0.0022	0.0040	50
Class-B EF 2-pole	<0.0006	0.0015	0.0026	50
Class-A quasi	<0.0006	0.0017	0.0030	20
Class-A CFP	<0.0006	0.0010	0.0018	20
Class-A CFP 2-pole	<0.0006	0.0010	0.0012	20

Note: All for 8 Ω loads and 80 kHz bandwidth. Single-pole compensation unless otherwise stated.

In the Class-B mode, the amplifier has the usual negligible quiescent dissipation. In Class-A, the thermal dissipation is naturally considerable, as true Class-A operation is extended down to 6 Ω resistive loads for the full output voltage swing, by suitable choice of the quiescent current; with heavier loading, the amplifier gracefully enters Class-AB, in which it will give full output down to 3 Ω before the Safe-Operating-Area (SOAR) limiting begins to act. Output into 2 Ω is severely curtailed, as it must be with one output pair, and this kind of load is definitely not recommended.

In short, the amplifier allows a choice between:

1. being very linear all the time (Blameless Class-B) and:
2. being ultra-linear most of the time (Class-A) with occasional excursions into Class-AB. The AB mode is still extremely linear by current standards, though inherently it can never be quite as good as properly handled Class-B. Since there are three classes of operation, I have decided to call the design a Trimodal power amplifier.

It is impossible to be sure that you have read all the literature; however, to the best of my knowledge this is the first ever Trimodal amplifier.

As previously said, designing a low-distortion Class-A amplifier is in general a good deal simpler than the same exercise for Class-B, as all the difficulties of arranging the best possible crossover between the output devices disappear. Because of this it is hard to define exactly what 'Blameless' means for a Class-A amplifier. In Class-B the situation is quite different, and 'Blameless' has a very specific meaning; when each of the eight or more distortion mechanisms has been minimised in effect, there always remains the

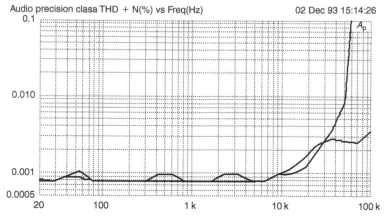

Figure 17.10. Distortion performance for CFP output stage with 2-pole compensation. The THD drops to 0.0012% at 20 kHz, but the extra VAS loading has compromised the positive-going slew capability.

crossover distortion inherent in Class-B, and there appears to be no way to reduce it without departing radically from what might be called the generic Lin amplifier configuration. Therefore the Blameless state appears to represent some sort of theoretical limit for Class-B, but not for Class-A.

However, Class-B considerations cannot be ignored, even in a design intended to be Class-A only, because if the amplifier does find itself driving a lower load impedance than expected, it will move into Class-AB, and then all the additional Class-B requirements are just as significant as for a Class-B design proper. Class-AB can never give distortion as low as optimally biased Class-B, but it can be made to approach it reasonably closely, if the extra distortion mechanisms are correctly handled.

In a Class-A amplifier, certain sacrifices are made in the name of quality, and so it is reasonable not to be satisfied with anything less than the best possible linearity. The amplifier described here therefore uses the Complementary Feedback Pair (CFP) type of output stage, which has the lowest distortion due to the local feedback loops wrapped around the output devices. It also has the advantage of better output efficiency than the emitter-follower (EF) version, and inherently superior quiescent current stability. It will shortly be seen that these are both important for this design.

Half-serious thought was given to labelling the Class-A mode 'Distortionless' as the THD is completely unmeasurable across most of the audio band. However, detectable distortion products do exist above 10 kHz, so this provocative idea was regretfully abandoned.

It seemed appropriate to take another look at the Class-A design, to see if it could be inched a few steps nearer perfection. The result is a slight improvement

in efficiency, and a 2 dB improvement in noise performance. In addition, the expected range of output DC offset has been reduced from ±50 mV to ±15 mV, still without any adjustment.

Load Impedance and Operating Mode

The amplifier is 4 Ω capable in both A/AB and B operating modes, though it is the nature of things that the distortion performance is not quite so good. All solid-state amplifiers (without qualification, as far as I am aware) are much happier with an 8 Ω load, both in terms of linearity and efficiency; loudspeaker designers please note. With a 4 Ω load, Class-B operation gives better THD than Class-A/AB, because the latter will always be in AB mode, and therefore generating extra output stage distortion through gm-doubling. (Which should really be called gain-deficit-halving, but somehow I do not see this term catching on.) These not entirely obvious relationships are summarised in Table 17.4.

Table 17.4. Distortion and dissipation for different output stage classes

Load (Ω)	Mode	Distortion	Dissipation
8	A/AB	Very low	High
4	A/AB	High	High
8	B	Low	Low
4	B	Medium	Medium

Note: 'High distortion' in the context of this sort of amplifier means about 0.002% THD at 1 kHz and 0.01% at 10 kHz.

Figure 17.11 attempts to show diagrammatically just how power, load resistance, and operating mode are related. The rails have been set to ±20 V, which just allows 20 W into 8 Ω in Class-A. The curves are lines of constant power (i.e., V × I in the load), the upper horizontal line represents maximum voltage output, allowing for Vce(sat)s, and the sloping line on the right is the SOAR protection locus; the output can never move outside this area in either mode. The intersection between the load resistance lines sloping up from the origin and the ultimate limits of voltage-clip and SOAR protection define which of the curved constant-power lines is reached.

In A/AB mode, the operating point must be left of the vertical push-pull current-limit line (at 3 A, twice the quiescent current) for Class-A. If we move to the right of this limit along one of the impedance lines, the output devices will begin turning off for part of the cycle; this is the AB operation zone. In Class-B mode, the 3 A line has no significance and the amplifier remains in optimal Class-B until clipping or SOAR limiting occurs. Note that the diagram axes represent instantaneous power in the load, but the curves show sinewave RMS power, and that is the reason for the apparent factor-of-two discrepancy between them.

Efficiency

Concern for efficiency in Class-A may seem paradoxical, but one way of looking at it is that Class-A Watts are precious things, wrought in great heat and dissipation, and so for a given quiescent power it makes sense to ensure that the amplifier approaches its limited theoretical efficiency as closely as possible. I was confirmed in this course by reading of another recent design[13] which seems to throw efficiency to the winds by using a hybrid BJT/FET cascode output stage. The voltage losses inherent in this arrangement demand ±50 V rails and six-fold output devices for a 100 W Class-A capability; such rail voltages would give 156 W from a 100% efficient amplifier.

The voltage efficiency of a power amplifier is the fraction of the supply-rail voltage which can actually be delivered as peak-to-peak voltage swing into a specified load; efficiency is invariably less into 4 Ω due to the greater resistive voltage drops with increased current.

The Blameless Class-B amplifier has in general a voltage efficiency of 91.7% for positive swings, and 92.5% for negative, into 8 Ω. Amplifiers are not in general completely symmetrical, and so two figures need to be quoted; alternatively the lower of the two

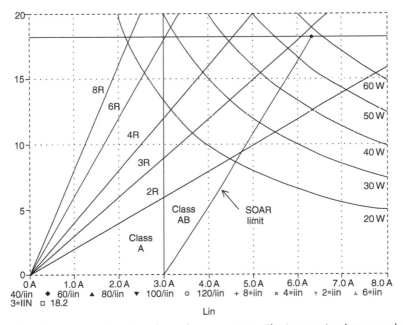

Figure 17.11. The relationships between load, mode, and power output. The intersection between the sloping load resistance lines and the ultimate limits of voltage-clipping and SOAR protection define which of the curved constant-power lines is reached. In A/AB mode, the operating point must be to the left of the vertical push-pull current-limit line for true Class-A.

can be given as this defines the maximum undistorted sinewave. These figures above are for an emitter-follower output stage, and a CFP output does better, the positive and negative efficiencies being 94.0% and 94.7%, respectively. The EF version gives a lower output swing because it has two more Vbe drops in series to be accommodated between the supply-rails; the CFP is always more voltage-efficient, and so selecting it over the EF for the current Class-A design is the first step in maximising efficiency.

Figure 17.12 shows the basic CFP output stage, together with its two biasing elements. In Class-A the quiescent current is rigidly controlled by negative feedback; this is possible because in Class-A across both emitter resistors Re is constant throughout the cycle. In Class-B, this is not the case, and we must rely on 'thermal feedback' from the output stage, though to be strictly accurate, this is not feedback at all, but a kind of feedforward (see Chapter 22). Another big advantage of the CFP configuration is that Iq depends only on

driver temperature, and this is important in the Class-B mode, where true feedback control of quiescent current is not possible, especially if low-value Res such as 0.1 Ω, are chosen, rather than the more usual 0.22 Ω; the motivation for doing this will soon become clear.

The voltage efficiency for the quasi-complementary Class-A circuit in Figure 17.7 into 8 Ω is 89.8% positive and 92.2% negative. Converting this to the CFP output stage increases this to 92.9% positive and 93.6% negative. Note that a Class-A quiescent current (Iq) of 1.5 A is assumed throughout; this allows 31 W into 8 Ω in push-pull, if the supply-rails are adequately high. However, the assumption that loudspeaker impedance never drops below 8 Ω is distinctly doubtful, to put it mildly, and so as before this design allows for full Class-A output voltage swing into loads down to 6 Ω.

So how else can we improve efficiency? The addition of extra and higher supply-rails for the small-signal section of the amplifier surprisingly does not give a significant increase in output; examination of Figure 17.13

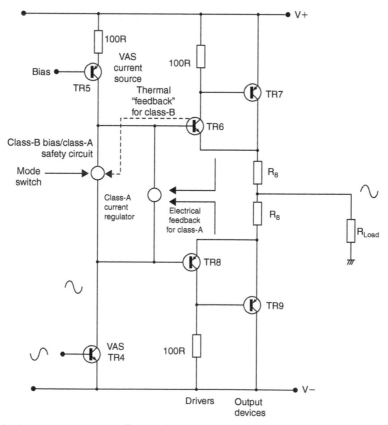

Figure 17.12. The basic CFP output stage, equally suited to operating Classes B, AB and A, depending on the magnitude of Vbias. The emitter resistors Re may be from 0.1 to 0.47 Ω.

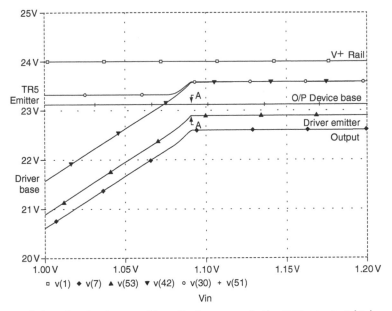

Figure 17.13. PSpice simulation showing how positive clipping occurs in the CFP output. A higher sub-rail for the VAS cannot increase the output swing, as the limit is set by the minimum driver Vce, and not the VAS output swing.

shows why. In this region, the output device TR6 base is at a virtually constant 880 mV from the V+ rail, and as TR7 driver base rises, it passes this level, and keeps going up; clipping has not yet occurred. The driver emitter follows the driver base up, until the voltage difference between this emitter and the output base (i.e., the driver Vce) becomes too small to allow further conduction; this choke point is indicated by the arrows A–A. At this point the driver base is forced to level off, although it is still about 500 mV below the level of V+. Note also how the voltage between V+ and the TR5 emitter collapses. Thus a higher rail will give no extra voltage swing, which I must admit came as something of a surprise. Higher sub-rails for small-signal sections only come into their own in FET amplifiers, where the high Vgs for FET conduction (5 V or more) makes their use almost mandatory.

The efficiency figures given so far are all greater for negative rather than positive voltage swings. The approach to the rail for negative clipping is slightly closer because there is no equivalent to the 0.6 V bias established across R13; however, this advantage is absorbed by the need to lose a little voltage in the RC filtering of the V- supply to the current-mirror and VAS. This is essential if really good ripple/hum performance is to be obtained (see Chapter 26).

In the quest for efficiency, an obvious variable is the value of the output emitter resistors Re. The performance of the current-regulator described, especially when combined with a CFP output stage, is more than good enough to allow these resistors to be reduced while retaining first-class Iq stability. I took 0.1 Ω as the lowest practicable value, and even this is comparable with PCB track resistance, so some care in the exact details of physical layout is essential; in particular, the emitter resistors must be treated as four-terminal components to exclude unwanted voltage drops in the tracks leading to the resistor pads.

If Re is reduced from 0.22 Ω to 0.1 Ω, then voltage efficiency improves from 92.9%/93.6%, to 94.2%/95.0%. Is this improvement worth having? Well, the voltage-limited power output into 8 Ω is increased from 31.2 to 32.2 W with ±24 V rails, at zero cost, but it would be idle to pretend that the resulting increase in SPL is highly significant; it does, however, provide the philosophical satisfaction that as much Class-A power as possible is being produced for a given dissipation; a delicate pleasure.

The linearity of the CFP output stage in Class-A is very slightly worse with 0.1 Ω emitter resistors, though the difference is small and only detectable open-loop; the simulated THD for 20 V pk–pk into 8 Ω is only increased from 0.0027% to 0.0029%. This is probably due simply to the slightly lower total resistance seen by the output stage.

However, at the same time, reducing the emitter resistors to 0R1 provides much lower distortion when

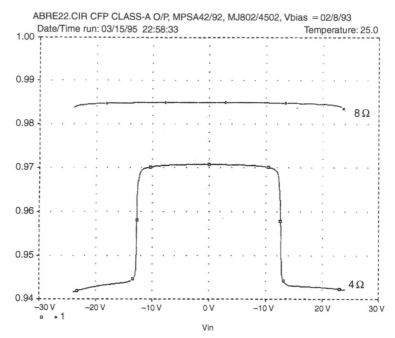

ABRE22.CIR CFP CLASS-A O/P, MPSA42/92, MJ802/4502, Vbias = 02/8/93

Figure 17.14. CFP output stage linearity with Re = 0R22. Upper trace is Class-A into 8 Ω, lower is Class-AB operation into 4 Ω, showing step changes in gain of 0.024 units.

the amplifier runs out of Class-A; it halves the size of the step gain changes inherent in Class-AB, and so effectively reduces distortion into 4 Ω loads. See Figures 17.14 and 17.15 for output linearity simulations; the measured results from a real and Blameless Trimodal amplifier shown in Figure 17.16, where it can be clearly seen that THD has been halved by this simple change. To the best of my knowledge, this is a new result; if you must work in Class-AB, then keep the emitter resistors as low as possible, to minimise the gain changes.

Having considered the linearity of Class-A and Class-AB, we must not neglect what effect this radical Re change has on Class-B linearity. The answer is, not very much; see Figure 17.17, where crossover distortion seems to be slightly higher with Re = 0.2 Ω than for either 0.1 or 0.4 Ω. Whether this is a consistent effect (for CFP stages anyway) remains to be seen.

The detailed mechanisms of bias control and mode-switching are described in the next section.

On Trimodal Biasing

Figure 17.18 shows a simplified rendering of the Trimodal biasing system; the full version appears in Figure 17.19. The voltage between points A and B is determined by one of two controller systems, only one

of which can be in command at a time. Since both are basically shunt voltage regulators sitting between A and B, the result is that the lowest voltage wins. The novel Class-A current-controller introduced earlier is used here, adapted for 0.1 Ω emitter resistors, mainly by reducing the reference voltage to 300 mV, which gives a quiescent current (Iq) of 1.5 A when established across the total emitter resistance of 0.2 Ω.

In parallel with the current-controller is the Vbe-multiplier TR13. In Class-B mode, the current-controller is disabled, and critical biasing for minimal crossover distortion is provided in the usual way by adjusting preset PR1 to set the voltage across TR13. In Class-A/AB mode, the voltage TR13 attempts to establish is increased (by shorting out PR1) to a value greater than that required for Class-A. The current-controller therefore takes charge of the voltage between X and Y, and unless it fails, TR13 does not conduct. Points A, B, X, and Y are the same circuit nodes as in the simple Class-A design (see Figure 17.6c).

Class-A/AB Mode

In Class-A/AB mode, the current-controller (TR14, 15, 16 in Figure 17.18) is active and TR13 is off, as TR20 has shorted out PR1. TR15, 16 form a simple differential

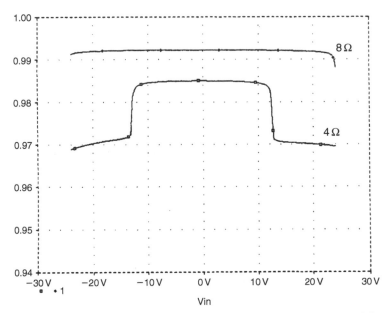

Figure 17.15. CFP output linearity with Re = 0R1, re-biased to keep Iq at 1.5 A. There is slightly poorer linearity in the flat topped Class-A region than for Re = 0R22, but the 4 Ω AB steps are halved in size at 0.01 2 units. Note that both gains are now closer to unity; same scale as Figure 17.14.

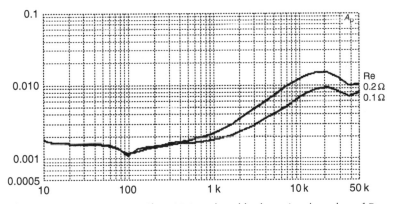

Figure 17.16. Distortion in Class-AB is reduced by lowering the value of Re.

amplifier that compares the reference voltage across R31 with the Vbias voltage across output emitter resistors R16 and R17; as explained above, in Class-A this voltage remains constant despite delivery of current into the load. If the voltage across R16, 17 tends to rise, then TR16 conducts more, turning TR14 more on and reducing the voltage between A and B. TR14, 15 and 16 all move up and down with the amplifier output, and so a tail current-source (TR17) is used.

I am very aware that the current-controller is more complex than the simple Vbe-multiplier used in most Class-B designs. There is an obvious risk that an assembly error could cause a massive current that would prompt the output devices to lay down their lives to save the rail fuses. The tail-source TR17 is particularly vulnerable because any fault that extinguishes the tail-current removes the drive to TR14, the controller is disabled, and the current in the output stage will be very large. In Figure 17.18, the Vbe-multiplier TR13 acts as a safety-circuit which limits Vbias to about 600 mV rather than the normal 300 mV, even if the current-controller is completely

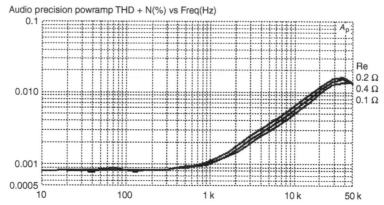

Figure 17.17. Proving that emitter resistors matter much less in Class-B. Output was 20 W in 8 Ω, with optimal bias. Interestingly, the bias does not need adjusting as the value of Re changes.

non-functional and TR14 is fully off. This gives a *quiescent* of 3.0 A, and I can testify this is a survivable experience for the output devices in the short term; however, they may eventually fail from overheating if the condition is allowed to persist.

There are some important points about the current-controller. The entire tail-current for the error-amplifier, determined by TR17, is siphoned off from the VAS current source TR5, and must be taken into account when ensuring that the upper output half gets enough drive current.

There must be enough tail-current available to turn on TR14, remembering that most of TR16 collector-current flows through R15, to keep the pair roughly balanced. If you feel moved to alter the VAS current, remember also that the base current for driver TR6 is higher in Class-A than Class-B, so the positive slew-rate is slightly reduced in going from Class-A to Class-B.

The original Class-A amplifier used a National LM385/1.2, its output voltage fixed at 1.223 V nominal; this was reduced to approximately 0.6 V by a 1 kΩ − 1 kΩ potential divider. The circuit also worked well with Vref provided by a silicon diode, 0.6 V being an appropriate Vbias drop across two 0.22 Ω output emitter resistors. This is simple, and retains the immunity of Iq to heatsink and output device temperatures, but it does sacrifice the total immunity to ambient temperature that a band-gap reference gives.

The LM385/1.2 is the lowest voltage band-gap reference commonly available; however, the voltages shown in Figure 17.18 reveal a difficulty with the new lower Vbias value and the CFP stage; points A and Y are now only 960 mV apart, which does not give the reference room to work in if it is powered from node A, as

in the original circuit. The solution is to power the reference from the V+ rail, via R42 and R43. The mid-point of these two resistors is bootstrapped from the amplifier output rail by C5, keeping the voltage across R43 effectively constant. Alternatively, a current-source could be used, but this might reduce positive headroom. Since there is no longer a strict upper limit on the reference voltage, a more easily obtainable 2.56 V device could be used, providing R30 is suitably increased to 5k to maintain Vref at 300 mV across R31.

In practical use, Iq stability is very good, staying within 1% for long periods. The most obvious limitation on stability is differential heating of TR15, 16 due to heat radiation from the main heatsink. TR14 should also be sited with this in mind, as heating it will increase its beta and slightly imbalance TR15, 16.

Class-B Mode

In Class-B mode, the current-controller is disabled, by turning off tail-source TR17 so TR14 is firmly off, and critical biasing for minimal crossover distortion is provided as usual by Vbe-multiplier TR13. With 0.1 Ω, emitter resistors Vbias (between X and Y) is approximately 10 mV. I would emphasise that in Class-B this design, if constructed correctly, will be as Blameless as a purpose-built Class-B amplifier. No compromises have been made in adding the mode-switching.

As in the previous Class-B design, the addition of R14 to the Vbe-multiplier compensates against drift of the VAS current-source TR5. To make an old but much-neglected point, the preset should always be in the bottom arm of the Vbe divider R10, 11, because when presets fail, it is usually by the wiper going

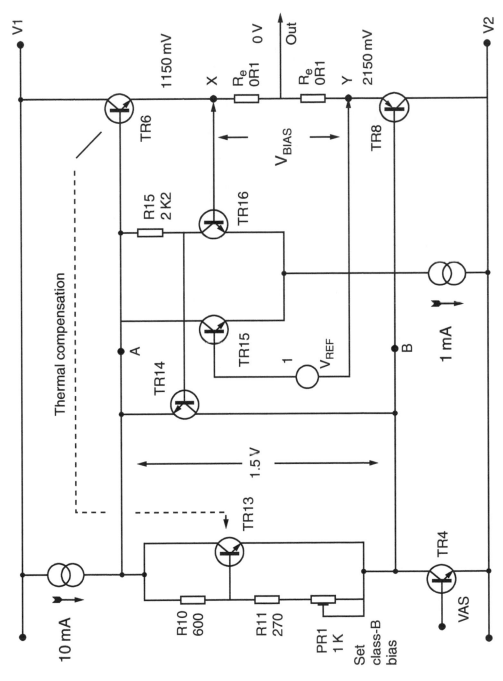

Figure 17.18. The simplified current-controller in action, showing typical DC voltages in Class-A. Points A, B, X and Y are in Figure 17.6 on p. 430.

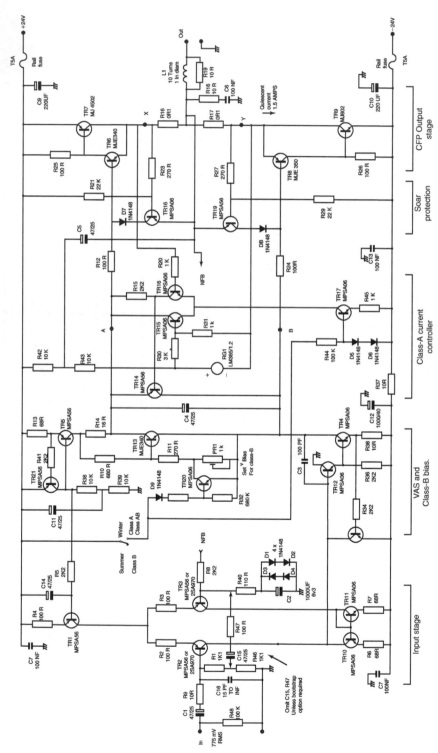

Figure 17.19. The complete circuit diagram of Trimodal amplifier, including the optional bootstrapping components, R47 and C15.

open; in the bottom arm this gives minimum Vbias, but in the upper it would give maximum.

In Class-B, temperature compensation for changes in driver dissipation remains vital. Thermal runaway with the CFP is most unlikely, but accurate quiescent setting is the only way to minimise crossover distortion. TR13 is therefore mounted on the same small heatsink as driver TR6. This is often called thermal feedback, but it is no such thing as TR13 in no way controls the temperature of TR6; 'thermal feedforward' would be a more accurate term.

The Mode-switching System

The dual nature of the biasing system means Class-A/Class-B switching can be implemented fairly simply. A Class-A amplifier is an uneasy companion in hot weather, and so I have been unable to resist the temptation to subtitle the mode switch *Summer/Winter,* by analogy with a car air intake.

The switchover is DC-controlled, as it is not desirable to have more signal than necessary running around inside the box, possibly compromising interchannel crosstalk. In Class-A/AB mode, SW1 is closed, so TR17 is biased normally by D5, 6, and TR20 is held on via R33, shorting out preset PR1 and setting TR13 to safety mode, maintaining a maximum Vbias limit of 600 mV. For Class-B, SW1 is opened, turning off TR17 and therefore TR15, 16 and 14. TR20 also ceases to conduct, protected against reverse-bias by D9, and reduces the voltage set by TR13 to a suitable level for Class-B. The two control pins of a stereo amplifier can be connected together, and the switching performed with a single-pole switch, without interaction or increased crosstalk.

The mode-switching affects the current flowing in the output devices, but not the output voltage, which is controlled by the global feedback loop, and so it is completely silent in operation. The mode may be freely switched while the amplifier is handling audio, which allows some interesting A/B listening tests.

It may be questioned why it is necessary to explicitly disable the current controller in Class-B; TR13 is establishing a lower voltage than the current controller which latter subsystem will therefore turn TR14 off as it strives futilely to increase Vbias. This is true for 8 Ω loads, but 4 Ω impedances increase the currents flowing in R16, 17 so they are transiently greater than the Class-A Iq, and the controller will therefore intermittently take control in an attempt to reduce the average current to 1.5 A. Disabling the controller by turning off TR17 via R44 prevents this.

If the Class-A controller is enabled, but the preset PR1 is left in circuit (e.g., by shorting TR20 base-emitter) we have a test mode which allows suitably cautious testing; Iq is zero with the preset fully down, as TR13 over-rides the current-controller, but increases steadily as PR1 is advanced, until it suddenly locks at the desired quiescent current. If the current-controller is faulty, then Iq continues to increase to the defined maximum of 3.0 A.

Thermal Design

Class-A amplifiers are hot almost by definition, and careful thermal design is needed if they are to be reliable, and not take the varnish off the Sheraton. The designer has one good card to play; since the internal dissipation of the amplifier is maximal with no signal, simply turning on the prototype and leaving it to idle for several hours will give an excellent idea of worst-case component temperatures. In Class-B the power dissipation is very program-dependent, and estimates of actual device temperatures in realistic use are notoriously hard to make.

Table 17.5 shows the output power available in the various modes, with typical transformer regulation, etc.; the output mode diagram in Figure 17.11 shows exactly how the amplifier changes mode from A to AB with decreasing load resistance. Remember that in this context 'high distortion' means 0.002% at 1 kHz. This diagram was produced in the analysis section of PSpice simply by typing in equations, and without actually simulating anything at all.

The most important thermal decision is the size of the heatsink; it is going to be expensive, so there is a powerful incentive to make it no bigger than necessary. I have ruled out fan cooling as it tends to make concern for ultra-low electrical noise look rather foolish; let us rather spend the cost of the fan on extra cooling fins and convect in ghostly silence. The exact thermal design calculations are simple but tedious, with many parameters to enter; the perfect job for a spreadsheet.

Table 17.5. Power capability

	Load resistance			
	8 Ω	**6 Ω**	**4 Ω**	**Distortion**
Class-A	20 W	27 W	15 W	Low
Class-AB	n/a	n/a	39 W	High
Class-B	21 W	28 W	39 W	Medium

Table 17.6. Temperature rises resulting in a 100°C junction temperature

	Thermal resistance °C/W	Heat flow (W)	Temp rise (°C)	Temp (°C)
Juncn to TO3 Case	0.7	36	25	100 Junction
Case to Sink	0.23	36	8	75 TO 3 case
Sink to air	0.65	72	47	67 Heatsink
Total			80	20 Ambient

The final answer is the margin between the predicted junction temperatures and the rated maximum. Once power output and impedance range are decided, the heatsink thermal resistance to ambient is the main variable to manipulate; and this is a compromise between coolness and cost, for high junction temperatures always reduce semiconductor reliability. Looking at it very roughly in Table 17.6.

Table 17.6 shows that the transistor junctions will be 80° above ambient, i.e., at around 100°C; the rated junction maximum is 200°C, but it really is not wise to get anywhere close to this very real limit. Note the Case-Sink thermal washers were high-efficiency material, and standard versions have a slightly higher thermal resistance.

The heatsinks used in the prototype had a thermal resistance of 0.65°C/W per channel. This is a substantial chunk of metal, and since aluminium is basically congealed electricity, it's bound to be expensive.

A Complete Trimodal Amplifier Circuit

The complete Class-A amplifier is shown in Figure 17.19, complete with optional input bootstrapping. It may look a little complex, but we have only added four low-cost transistors to realise a high-accuracy Class-A quiescent controller, and one more for mode-switching. Since the biasing system has been described above, only the remaining amplifier subsystems are dealt with here.

The input stage follows my design methodology by using a high tail current to maximise transconductance, and then linearising by adding input degeneration resistors R2, 3 to reduce the final transconductance to a suitable level. Current-mirror TR10, 11 forces the collector currents of the two input devices TR2, 3 to be equal, balancing the input stage to prevent the generation of

second-harmonic distortion. The mirror is degenerated by R6, 7 to eliminate the effects of Vbe mismatches in TR10, 11. With some misgivings I added the input network R9, C15, which is definitely not intended to define the system bandwidth, unless fed from a buffer stage; with practical values the HF roll-off could vary widely with the source impedance driving the amplifier. It is intended rather to give the possibility of dealing with RF interference without having to cut tracks. R9 could be increased for bandwidth definition if the source impedance is known, fixed, and taken into account when choosing R9; bear in mind that any value over 47 Ω will measurably degrade the noise performance. The values given roll off above 150 MHz to keep out UHF.

The input-stage tail current is increased from 4 to 6 mA, and the VAS standing current from 6 to 10 mA over the original circuit. This increases maximum positive and negative slew-rates from +21, −48 V/μsec to +37, −52 V/μsec; as described in Chapter 15, this amplifier architecture is bound to slew asymmetrically. One reason is feedthrough in the VAS current source; in the original circuit an unexpected slew-rate limit was set by fast edges coupling through the current-source c-b capacitance to reduce the bias voltage during positive slewing. This effect is minimised here by using the negative feedback type of current source bias generator, with VAS collector current chosen as the controlled variable. TR21 senses the voltage across R13, and if it attempts to exceed Vbe, turns on further to pull up the bases of TR1 and TR5. C11 filters the DC supply to this circuit and prevents ripple injection from the V+ rail. R5, C14 provide decoupling to prevent TR5 from disturbing the tail-current while controlling the VAS current.

The input tail-current increase also slightly improves input-stage linearity, as it raises the basic transistor gm and allows R2, 3 to apply more local NFB.

The VAS is linearised by beta-enhancing stage TR12, which increases the amount of local NFB through Miller dominant-pole capacitor C3 (i.e., Cdom). R36 has been increased to 2k2 to minimise power dissipation, as there seems no significant effect on linearity or slewing. Do not omit it altogether, or linearity will be affected and slewing much compromised.

As described in Chapter 26, the simplest way to prevent ripple from entering the VAS via the V- rail is old-fashioned RC decoupling, with a small R and a big C. We have some 200 mV in hand (see p. 437) in the negative direction, compared with the positive, and expending this as the voltage-drop through the RC decoupling will give symmetrical clipping. R37 and C12 perform this function; the low rail voltages in this design allow the 1000 µF C12 to be a fairly compact component.

The output stage is of the Complementary Feedback Pair (CFP) type, which as previously described, gives the best linearity and quiescent stability, due to the two local negative feedback loops around driver and output device. Quiescent stability is particularly important with R16, 17 at 0.1Ω, and this low value might be rather dicey in a double emitter-follower (EF) output stage. The CFP voltage efficiency is also higher than the EF version. R25, R26 define a suitable quiescent collector current for the drivers TR6, 8, and pull charge carriers from the output device bases when they are turning off. The lower driver is now a BD136; this has a higher fT than the MJE350, and seems to be more immune to odd parasitics at negative clipping.

The new lower values for the output emitter resistors R16, 17 halve the distortion in Class-AB. This is equally effective when in Class-A with too low a load impedance, or in Class-B but with Iq maladjusted too high. It is now true in the latter case that too much Iq really is better than too little − but not much better, and AB still comes a poor third in linearity to Classes A and B.

SOAR (Safe Operating ARea) protection is given by the networks around TR18, TR19. This is a single-slope SOAR system that is simpler than two-slope SOAR, and therefore somewhat less efficient in terms of getting the limiting characteristic close to the true SOAR of the output transistor. In this application, with low rail voltages, maximum utilisation of the transistor SOAR is not really an issue; the important thing is to observe maximum junction temperatures in the A/AB mode.

The global negative feedback factor is 32 dB at 20 kHz, and this should give a good margin of safety against Nyquist-type oscillation. Global NFB increases at 6dB/octave with decreasing frequency to a plateau of around 64 dB, the corner being at a rather ill-defined 300 Hz; this is then maintained down to 10 Hz. It is fortunate that magnitude and frequency here are non-critical, as they depend on transistor beta and other doubtful parameters.

It is often stated in hi-fi magazines that semiconductor amplifiers sound better after hours or days of warm-up. If this is true (which it certainly is not in most cases) it represents truly spectacular design incompetence. This sort of accusation is applied with particular venom to Class-A designs, because it is obvious that the large heatsinks required take time to reach final temperature, so I thought it important to state that in Class-A this design stabilises its electrical operating conditions in less than a second, giving the full intended performance. No 'warm-up time' beyond this

is required; obviously the heatsinks take time to reach thermal equilibrium, but as described above, measures have been taken to ensure that component temperature has no significant effect on operating conditions or performance.

The Power Supply

A suitable unregulated power supply is that shown in Figure 9.2; a transformer secondary voltage of 20−0−20 Vrms and reservoirs totalling 20,000 µF per rail will give approximately ±24V. This supply must be designed for continuous operation at maximum current, so the bridge rectifier must be properly heat-sunk, and careful consideration given to the ripple-current ratings of the reservoirs. This is one reason why reservoir capacitance has been doubled to 20,000 µF per rail, over the 10,000 µF that was adequate for the Class-B design; the ripple voltage is halved, which improves voltage efficiency as it is the ripple troughs that determine clipping onset, but in addition the ripple current, although unchanged in total value, is now split between two components. (The capacitance was *not* increased to reduce ripple injection, which is dealt with far more efficiently and economically by making the PSRR high.) Do not omit the secondary fuses; even in these modern times rectifiers do fail, and transformers are horribly expensive.

The Performance

The performance of a properly designated Class-A amplifier challenges the ability of even the Audio Precision measurement system. To give some perspective on this, Figure 17.20 shows the distortion of the AP oscillator driving the analyser section directly for various bandwidths. There appear to be internal mode changes at 2 kHz and 20 kHz, causing step increases in oscillator distortion content; these are just visible in the THD plots for Class-A mode.

Figure 17.21 shows Class-B distortion for 20 W into 8 and 4 Ω, while Figure 17.22 shows the same in Class-A/AB. Figure 17.23 shows distortion in Class-A for varying measurement bandwidths. The lower bandwidths misleadingly ignore the HF distortion, but give a much clearer view of the excellent linearity below 10 kHz. Figure 17.24 gives a direct comparison of Class-A and Class-B. The HF rise for B is due to high-order crossover distortion being poorly linearised by negative feedback that falls with frequency.

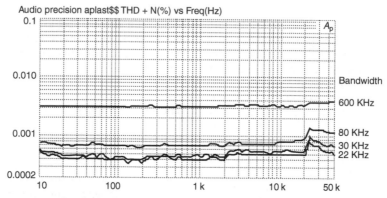

Figure 17.20. The distortion in the AP-1 system at various measurement bandwidths.

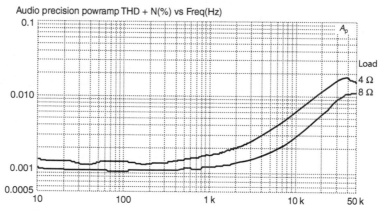

Figure 17.21. Distortion in Class-B (Summer) mode. Distortion into 4 Ω is always worse. Power was 20 Win 8 Ω and 40W in 4 Ω, bandwidth 80 kHz.

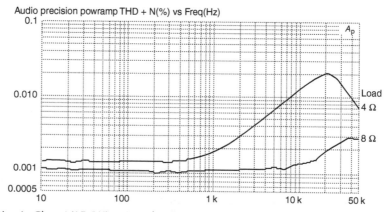

Figure 17.22. Distortion in Class-A/AB (Winter) mode, same power and bandwidth of Figure 17.21. The amplifier is in AB mode for the 4 Ω case, and so distortion is higher than for Class-B into 4 Ω. At 80 kHz bandwidth, the Class-A plot below 10 kHz merely shows the noise floor.

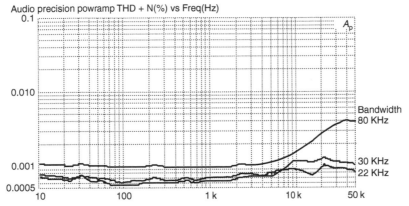

Figure 17.23. Distortion in Class-A only (20 W/8 Ω) for varying measurement bandwidths. The lower bandwidths ignore HF distortion, but give a much clearer view of the excellent linearity below 10 kHz.

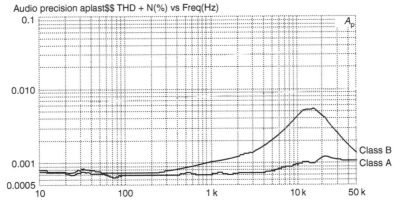

Figure 17.24. Direct comparison of Classes A and B (20 W/8 Ω) at 30 kHz bandwidth. The HF rise for B is due to the inability of negative feedback that falls with frequency to linearise the high-order crossover distortion in the output stage.

Further Possibilities

One interesting extension of the ideas presented here is the Adaptive Tri-modal Amplifier. This would switch into Class-B on detecting device or heatsink over-temperature, and would be a unique example of an amplifier that changed mode to suit the operating conditions. The thermal protection would need to be latching; flipping from Class-A to Class-B every few minutes would subject the output devices to unnecessary thermal cycling.

References

1. Moore, B J, *An Introduction to the Psychology of Hearing*, New York: Academic Press, 1982, pp. 48—50.

2. Tanaka, S A, New Biasing Circuit for Class-B Operation, *JAES*, Jan/Feb 1981, p. 27.

3. Fuller, S, Private communication.

4. Pass, N, Build A Class-A Amplifier, *Audio*, Feb 1977, p. 28 (Constant-current).

5. Linsley-Hood, J, Simple Class-A Amplifier, *Wireless World*, April 1969, p. 148.

6. Self, D, High-Performance Preamplifier, *Wireless World*, Feb 1979, p. 41.

7. Nelson-Jones, L, Ultra-Low Distortion Class-A Amplifier, *Wireless World*, March 1970, p. 98.

8. Self, D, *Self On Audio*, 2nd edn, Oxford: Newnes, 2006, p. 459.

9. Giffard, T, Class-A Power Amplifier, *Elektor*, Nov. 1991, p. 37.

10. Simpson, L and Smith, P, 20W Class-A Amplifier Module, *Everyday Practical Electronics*, Oct. 2008, p. 32.

11. Linsley-Hood, J, High-Quality Headphone Amp, *HiFi News and RR*, Jan. 1979, p. 81.

12. Pass, N, The Pass/A40 Power Amplifier, *The Audio Amateur*, 1978, p. 4 (Push-pull).

13. Thagard, N, Build a 100W Class-A Mono Amp, *Audio*, Jan. 95, p. 43.

Chapter 18

Class XD: Crossover Displacement

Class XD™ is a new output stage technology I have devised which abolishes crossover distortion up to a certain power level, without any accompanying compromises. 'XD' is derived from the phrase 'Crossover Displacement' and while it is certainly a novel technology, it is not a basic 'Class' like Class-A, Class-B, or Class-C. Of such is marketing. By my classification system (see Chapter 4), it is Class-B·A, as it essentially consists of a Class-B amplifier connected in parallel with a Class-A circuit. The technology is covered by British patent GB2424137B and is proprietary to Cambridge Audio. At the time of writing it has so far been used in the Azur 840A, 840W and 851A power amplifiers, for the first two of which I did all the electronic design. 'Class XD' is a trademark of Audio Partnership PLC, and it should be pointed out that the use of the Class XD concept and its trademark is restricted; I have permission from Cambridge Audio to use the term and describe the circuitry but no licence to use the technology is implied or granted by the publication of this description.

Having held various posts in companies concerned with audio power amplifiers, I have frequently had to deal with enthusiastic inventors who feel they have come up with an output stage technology that overcomes the crossover distortion problems of conventional Class-B, and who are anxious to sell the idea to me. Two stick in the mind. There was the consortium that took out extensive worldwide patents on an idea that had been disclosed in *Wireless World* a quarter-century before, and which didn't work properly anyway. Then there was the chap who offered me an error-correcting output stage that 'only requires another 140 transistors'. I would have liked to have seen that circuit diagram, but not enough to pay money to do so.

In the light of this sort of thing, anyone is entitled to be sceptical about New And Improved amplifier output stages. However, Class XD is different; it really does work, doing what it claims with total reliability and minimal extra circuitry, as I shall now demonstrate.

One of the main themes of this book is the difficulty of dealing with crossover distortion in a Class-B output stage. I have described various methods of attack, such as the use of multiple output devices to reduce the current changes in each output transistor, the use of two-pole compensation to increase the global negative feedback factor, and the use of output-inclusive compensation to apply more feedback around the output stage (see Chapter 13). These methods usefully reduce the amount of crossover distortion but do not eliminate it. As a result, one of the great divides in amplifier technology is still between efficient but imperfect Class-B and beautifully linear but disheartimgly inefficient Class-A. As I demonstrated in my book, *Self on Audio*, a Class-A amplifier may theoretically be 50% efficient with a maximum sinewave output, but when it reproduces a real music signal this falls to 1 or 2%.[1] For those that care at all about the economic utilisation of energy, a Class-A amplifier is not an attractive proposition.

Class-B linearity can be very good. The Blameless amplifier design methodology, especially in its Load-Invariant form, yields less than 0.001% THD at 1 kHz. This is in its simplest form without multiple output devices or advanced compensation. The limitation is that a Class-B amplifier inherently generates crossover distortion, and most inconveniently does so at the zero-crossing, so it is always present, no matter how low the signal amplitude. At one unique setting of quiescent conditions, the distortion produced is at a minimum, and this characterises optimal Class-B; but at no value can it be made to disappear. It is inherent in the classical Class-B operation of a pair of output transistors.

Given these two alternatives, there has always been a desire for a compromise between the efficiency of Class-B and the linearity of Class-A. The most obvious approach is to turn up the quiescent current of a Class-B stage, to create an area of Class-A operation, with both output transistors conducting, around the zero-crossing. This area widens as the quiescent current increases, until ultimately it encompasses the entire voltage output range of the amplifier, and we have created a pure Class-A design where both output transistors are conducting all the time. There is thus a range of quiescent current between Class-B and Class-A, and this mode of operation is called Class-AB. It is certainly a compromise between Class-A and Class-B, but not a good compromise, as it introduces extra distortion of its own.

This appears when the signal exceeds the limits of the Class-A region. The THD worsens abruptly due to the sudden gain-changes when the output transistors turn on and off, and linearity is inferior not only to Class-A but also to optimally-biased Class-B. This effect is often called 'gm-doubling', and is dealt with in detail in Chapters 9 and 10. Class-AB distortion can be made very low by proper design, such as using the lowest practicable emitter resistors, but it remains at least twice as high as for the equivalent Class-B situation. The bias control of a Class-B amplifier does NOT give a straightforward trade-off between power dissipation and linearity at all levels, despite the constant repetition this misguided notion receives in some parts of the audio press. To demonstrate this, Figure 18.1 shows THD plotted against output level for Classes AB and B.

What we really want is an amplifier that would give Class-A performance up to the transition level, with Class-B after that, rather than the unsatisfactory Class-AB. This would abolish the abrupt AB gain changes that generate the extra distortion.

The Crossover Displacement Principle

When we consider Class-B, it is clear that it would be better if the crossover region were anywhere else rather than where it is. If we can displace the crossover point away from its zero-crossing position, then the amplifier output will not traverse it until the output reaches a certain voltage level. Below this level the performance is pure Class-A; above it the performance is optimal Class-B, the only difference being that crossover discontinuities on the THD residual are no longer evenly spaced. The harmonic structure of the crossover distortion produced is not significantly changed, as explained in more detail below.

The central idea of the Crossover Displacement principle is the injection of an extra current, either fixed or varying with the signal, into the output point of a conventional Class-B amplifier. Figure 18.2a shows a conventional Class-B EF output stage; in Figure 18.2b there is added a black box I have called the Displacer which

draws a controlled current from the output node and sinks it into the negative rail; sourcing current from the positive rail and injecting it into the output would be equally valid. The displacer current may be constant, in which case the displacer is simply a constant-current source, or it may vary with the signal. The latter improves both efficiency and linearity. In either case the Displacer is working in Class-A and never turns off.

The displacement current does not directly alter the output-voltage because the output stage has an inherently low output impedance, which is further reduced by the global negative feedback. What it does do is alter the pattern of current flowing in the output devices. The displacement current in the version shown here is sunk to V- from the output. This is arbitrary as the direction of displacement makes no difference. The extra current therefore flows through Re1, and the extra voltage drop across it means the output voltage must go some way negative before the current through Re1 stops and that in Re2 starts. In other words, the crossover point when Q2 hands over to Q4 has been moved to a point negative of the 0 V rail; I refer to this as the 'transition point' between Class-A and Class-B. For output levels below transition both Q2 and Q4 are conducting and no crossover distortion is generated. The resulting change in the incremental gain of the output stage is shown in Figure 18.3. Here the crossover region is moved 8 V negative of ground by a constant 1 Amp displacement current; if the Displacer had been connected to the positive rail, the

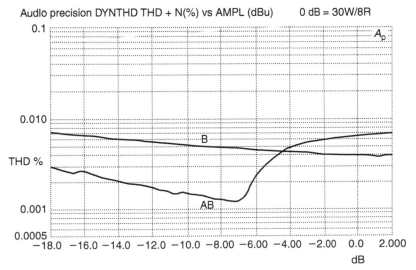

Figure 18.1. THD vs level for Class-B and Class-AB (0 dB is 30 W into 8 Ω).

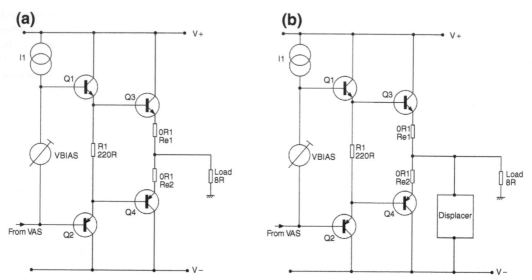

Figure 18.2. (a) A conventional Class-B output stage with drivers and bias voltage source; (b) adding a displacer system that draws current from the output and sinks it into the negative rail.

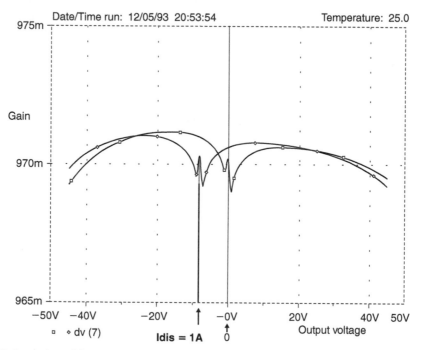

Figure 18.3. SPICE simulation of the output stage gain variation with and without a constant 1 Amp of displacement current. The central peak is moved left from 0V to −8V.

crossover region would have been pulled upwards. Note that the vertical scale is very much exaggerated, and that the crossover region has been moved but remains the same shape − the existing linearity has not been compromised. My classification scheme describes this as Class-B•A, as the Class-B output stage and the Class-A Displacer are in parallel; the B comes first because that is the stage that controls the output voltage.

The amount of crossover displacement is in no way critical. Changes in the displacement current only alter the output power level at which crossover occurs, and do not affect the basic linearity of the amplifier, nor the accuracy of output stage bias setting. Almost any amount of crossover displacement is better than none.

I should emphasise here that crossover displacement in no way renders output stage bias adjustment unnecessary; if it is wrong, the same distortion will occur, though only above a certain output level. This should not be regarded as making the adjustment less critical. Getting it right costs no more than getting it roughly right, so there is really nothing to be gained by compromising on this.

We now have before us the intriguing prospect of a power amplifier with three output devices, which, if nothing else, is novel. The operation of the output stage is inherently asymmetrical, and indeed this is the whole point, but it should not cause alarm. Circuit symmetry is often touted as being a prerequisite for either low distortion or respectable operation in general, but this has no real foundation. A perfectly symmetrical circuit may have no even-order distortion, but it may still have frightening amounts of odd-order non-linearity, such as a cubic characteristic. Odd-order harmonics are normally considered more dissonant than even-order, so circuit symmetry in itself is not enough.

In a conventional optimal Class-B amplifier, the crossover events are evenly spaced in time. In the crossover-displacement amplifier, the crossover events are asymmetrical in time and put energy into both even and odd harmonics when operating above the transition point. However, since both even and odd exist already in conventional amplifiers, there is no cause for concern. As always, the real answer is to reduce the distortion, of whatever order, to so far below the noise floor that it could not possibly be audible and you never need to fret about it.

Crossover Displacement Realisation

There are several ways in which a suitable displacement current can be drawn from the main amplifier output node. The simplest method is resistive crossover displacement. Connect a suitable power resistor between the output rail and a supply rail, as shown in Figure 18.4a, and the crossover point will be displaced. In this and all the following examples, the crossover point is displaced negatively by sinking a current into the negative rail.

The resistive method suffers from poor efficiency, as the resistance acts as another load on the amplifier output, effectively in parallel with the normal load. It also threatens ripple-rejection problems as R is connected directly to a supply rail, which in most cases is unregulated and carrying substantial 100 Hz ripple. A regulated supply to the resistor could be used, but this would be relatively expensive and even less efficient due to the voltage drop in the regulator.

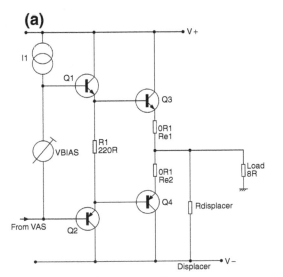
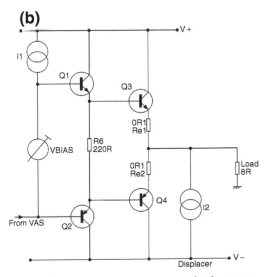

Figure 18.4. (a) The concept of resistive crossover displacement; (b) constant-current crossover displacement.

The resistive system is inefficient because the displacement of the crossover region occurs when the output is negative of ground, but when the output is positive, the resistor is still connected and a greater current is drawn from it as the voltage across it increases. This increasing current is of no use in the displacement process and simply results in increased power dissipation in the positive output half-cycles.

This method has the other drawback that the distortion performance of the basic amplifier will be worsened because of the heavier loading it sees, the resistor being connected to ground as far as AC signals are concerned.

A superior solution is constant-current displacement, as shown in Figure 18.4b; here a constant-current source is connected between the output and negative rail. Efficiency is better as no output power is dissipated due to the high dynamic impedance of the current source. The output of the current source does not need to be controlled to very fine limits. Long-term variations in the current only affect the degree to which the crossover region is displaced, and this is not a critical parameter. Noise or ripple on the displacement current is greatly attenuated by the very low impedance of the basic power amplifier and its global negative feedback, so sophisticated current-control circuitry is not required. The efficiency of this configuration is greater, because the output current of the displacer does not increase as the output moves more positive. The voltage across the current source increases, so its dissipation is still increased, but by a lesser amount than for the resistor. Likewise, the upper output transistor Q3 is passing less current on positive excursions so its power dissipation is less.

Having moved from a simple resistor displacer to a constant-current source, the obvious next step is to move from a constant current to a voltage-controlled current source (VCIS) whose output is modulated by the signal to further improve efficiency. The most straightforward way to do this is to make the displacement current proportional to the output voltage. Thus, if the displacement current is 1 Amp with the output quiescent at 0 V, it is set to increase to 2 Amps with the output fully negative, and to reduce to zero with the output fully positive. The displacer current is set by the equation:

$$I_\mathrm{d} = I_\mathrm{q}\left(1 - \frac{V_\mathrm{out}}{V_\mathrm{rail}}\right) \qquad \text{Equation 18.1}$$

where Iq is the quiescent displacement current (i.e. with the output at 0 V) and Vrail is the bottom rail voltage,

which must be inserted as a negative number to make the arithmetic work. It is not essential for the displacement current to swing from zero to twice the quiescent value; it could be modulated to a lesser extent, and there is in fact a continuum of possible solutions from constant-current displacement to the full push-pull case.

Depending on the design of the VCIS, a scaling factor X is required to drive it correctly; see Figure 18.5 Since a signal polarity inversion is also necessary to get the correct mode of operation, active controlling circuitry is necessary.

The use of push-pull displacement is analogous to the use of push-pull current sources in Class-A amplifiers, where there is a well-known canonical sequence of increasing efficiency, which is fully described in Chapter 17. This begins with a resistive load giving only 12.5% efficiency at full power, moves to a constant-current source with high dynamic impedance giving 25%, and finally to a push-pull controlled current-source, giving 50% efficiency. In the push-pull case the sink transistor acts in a sense as a negative resistance, though it is more usefully regarded as a driven source (VCIS) than a pure negative resistance, as the current does not depend on rail voltage. In each of these moves, the efficiency doubles. These efficiency figures are ideal, ignoring circuit losses; note that Class-A efficiency is very seriously reduced at output powers less than the maximum. In the same way, there is a canonical sequence of sophistication in crossover displacers, though the differences in efficiency are smaller.

The push-pull displacement approach has another benefit; it also reduces distortion when operating above transition in the Class-B mode. This is because the push-pull system acts to reduce the current swings in the output devices, the displacement current varying in the correct sense to do this. This is equivalent to a decrease in output stage loading; this is the exact inverse of what occurs with resistive displacement, which increases output loading. Lighter loading is known to make the current crossover between the output devices more gradual, and so reduces the size of the gain-wobble that causes crossover distortion; this is described in Chapters 9 and 10. In addition, the crossover region is spread over more of the output voltage range, so the distortion harmonics generated are lower-order and receive more linearisation from a negative feedback factor that falls with frequency. Large-Signal Non-linearity (typically experienced with loads of 4 Ω and less) is also somewhat reduced. In push-pull displacement operation, the accuracy of the

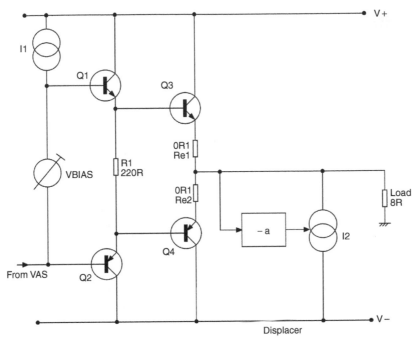

Figure 18.5. The concept of push-pull crossover displacement. The control circuitry implements a scaling factor of -a in the signal to the controlled current-source.

current variation does not have to be high to get the full reduction of the distortion, because of the low output impedance of the main amplifier, which maintains control of the output voltage. The global feedback around this amplifier is effective in reducing the inherently low output impedance of the output stage in the usual way, being unaffected by the addition of the displacer.

While the constant-current displacement method is simple and effective, the push-pull version of crossover displacement is to be preferred for the best linearity and efficiency; the extra control circuitry required is simple and works at low power so it adds minimally to total amplifier cost.

Circuit Techniques for Crossover Displacement

The constant-current displacer is the simplest practical displacement technique, the resistive version being discarded for the reasons given above.

A practical circuit for a constant-current displacer is shown in Figure 18.6a; for clarity the Class-B output stage is omitted. The displacement current typically chosen will be in the region of 0.5 to 1 A, and therefore a driver transistor Q5 is used, exactly as drivers are used in the main amplifier, so the control circuitry can work at low power levels. The power device Q6 is going to get hot, so its Vbe must be excluded from having a direct effect on current stability. Therefore the CFP (complementary feedback pair) structure shown is used, so the effect of Vbe variations is reduced by the negative feedback around the local loop Q5–Q6. The bias for the constant current is shown as a Zener diode D1; if greater accuracy is required, a low voltage reference IC such as the LM385 could be used instead, but there is no real need to do so. The voltage across R1 should not be large enough to limit the output swing; but on the other hand, if it is small compared with the Vbe of Q5, then the current value may drift excessively with temperature, as Q5 warms up.

Power transistor Q6 dissipates significant heat; clearly the greater the crossover displacement required, the greater the displacement current, and the greater the dissipation. Q6 is therefore normally mounted on the same heatsink as the amplifier output devices. This provides the intriguing sight of a power amplifier with an odd number of output transistors, which might conceivably be exploited for marketing purposes.

The push-pull controller drives the displacer so that, as the output rail goes positive, the displacer supplies less current. The basic problem is to apply a scaled and inverted version of the output voltage to the

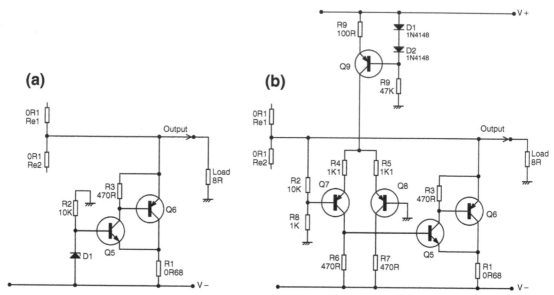

Figure 18.6. (a) constant-current displacer with complementary feedback pair structure; (b) push-pull displacer with differential pair controller.

displacer. The signal must also have its reference transferred to the negative rail, which can be assumed to carry mains ripple and distorted signal components. Transferring the reference is done by using the high-impedance (like a current-source) output from a bipolar transistor collector. As before, a driver transistor Q5 is used to drive the Displacer Q6 so the control circuitry can work at low power levels. This not only minimises total current consumption but also reduces the effect of Vbe changes due to device heating. See Figure 18.6b.

The controller is simply a differential pair of transistors with one input grounded and the other driven by the main amplifier output voltage, scaled down appropriately by R2, R8. The differential pair has heavy local feedback applied by the addition of the emitter resistors R4, R5, in order to minimise distortion and achieve an accurate gain. The drive to the VCIS displacer is taken from collector load R6, to give the required phase inversion. R7 is present simply to equalise the dissipation in the differential pair transistors to maintain balance.

The tail of the differential pair is fed by the 6 mA constant-current source Q9. This gives good common-mode rejection, which prevents the significant ripple voltages on the supply rails from interfering with the control signal. Since half of the standing current through the differential pair flows through R6, the value of the tail current-source sets the quiescent

displacement current. The stability of the current generated by this source therefore sets the stability of the quiescent (no-signal) value of the displacement current. Figure 18.6b shows a simple current-source biased by a pair of silicon diodes. This has proven to work well in practice but more sophisticated current-sources using negative feedback could be used if greater stability is required. However, even if the tail current-source is perfect, the value of the displacement current still depends on the temperature of Q5. More sophisticated circuitry could be used to remove this dependency; for example, the voltage across R1 could be sensed by an opamp instead of by Q5. The opamp used would need to be able to work with a common-mode voltage down to the negative rail, or an extra supply-rail would have to be provided.

A further possible refinement is the addition of a safety resistor in the differential pair tail to limit the amount of current flowing in the event of component failure. Such a resistor has no effect on normal operation, but it must be employed with care as its presence may mean that the circuit will not start working until the supply rails have risen to a large fraction of the working value. This is a serious drawback as it is wise to test power amplifiers by slowly raising the rail voltages from zero, and the lower the voltage at which they start working, the safer this procedure is.

A Complete Crossover Displacement Power Amplifier Circuit

Figure 18.7 shows the practical circuit of a push-pull crossover displacement amplifier. The Class-B amplifier is based on the Load Invariant design and follows the Blameless design philosophy described elsewhere in this book. Conventional dominant-pole compensation is used. The design uses the following robust techniques described in this book to bring the distortion down to the irreducible minimum generated by a Class-B output stage.

1. The local negative feedback in the input differential pair Q1, Q2 is increased by running it at a high collector current, and then defining the stage transconductance and linearising it by local negative feedback introduced by the emitter resistors R10, R11.
2. The crucial collector-current balance between the two halves of the input differential pair is enforced by the use of a degenerated current-mirror Q3, Q4.
3. The local negative feedback around the voltage-amplifier transistor Q10 is increased by adding the emitter-follower Q11 inside the Miller Cdom loop.
4. The output stage uses a Complementary Feedback Pair (CFP) configuration to establish local negative feedback around the output devices. This increases linearity and also minimises the effect of output junction temperatures on the bias conditions. The

bias generator Q15 has its temperature coefficient increased by the addition of D3, R28, R29 to improve the accuracy of the thermal compensation; see Chapter 22 for more details. Q15 is thermally coupled to one of the drivers and preferably mounted on top of it; for this reason, Q15 is a MJE340 simply so the packages are the same.

The circuit shown is capable of at least 50 W without modification. Powers above 100 W into 8 Ω will require two parallel power transistors in the main amplifier output stage. The displacer transistor does not necessarily require doubling; it depends on the degree of crossover displacement desired.

The displacer control circuitry is essentially the same as Figure 18.6b. A push-on link can be connected across R6 so that the crossover displacement action can be manually disabled to simplify testing and fault-finding.

Note that overload protection circuitry has been omitted from the diagram for simplicity.

The Measured Performance

The measurements shown here demonstrate how cross-over displacement not only deals with crossover distortion, but also reduces distortion in general when the push-pull variant is employed. Tests were done with an amplifier similar to that shown in Figure 18.7.

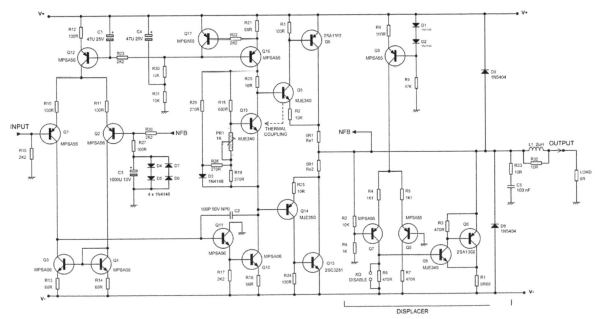

Figure 18.7. Complete circuit of a Blameless amplifier using push-pull crossover displacement.

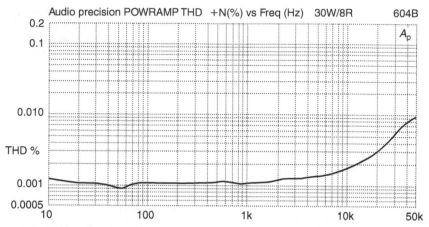

Figure 18.8. THD vs frequency for a standard Blameless Class-B amplifier at 30 W/8 Ω (604b).

Figure 18.8 shows THD vs frequency for a standard Blameless Class-B amplifier giving 30 W into 8 Ω. The distortion shown only emerges from the noise floor at 2 kHz, and is here wholly due to crossover artefacts; the bias is optimal and this is essentially as good as Class-B gets. The distortion only gets really clear of the noise floor at 10 kHz, so this frequency was used for all the THD/amplitude tests below. This frequency provides a demanding test for an audio power amplifier. In all these tests the measurement bandwidth was 80 kHz. This may filter out some ultrasonic harmonics, but is essential to reduce the noise bandwidth; it is also a standard setting on many distortion analysers.

Figure 18.9 plots distortion vs amplitude at 10 kHz, over the power range 200 mW−20 W; this covers the levels at which most listening is done. (0 dB on the graph is 30 W into 8 Ω.) Trace B is the result for the Blameless Class-B amplifier measured in Figure 18.8; the THD percentage increases as the power is reduced, partly because of the nature of crossover distortion, but more because the noise level becomes proportionally greater as level is reduced. Trace A shows the result for a Class-A amplifier of my design (see Chapter 17) which is pretty much distortion-free at 10 kHz, and once more simply shows the increasing relative noise level as power reduces;

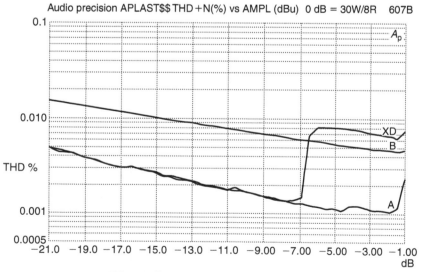

Figure 18.9. THD+N vs power out for Class-A, Class-B, and Class-B with constant-current crossover displacement (Trace XD). Tested at 10 kHz to get enough distortion to measure; 0 dB = 30 W into 8 Ω (607B).

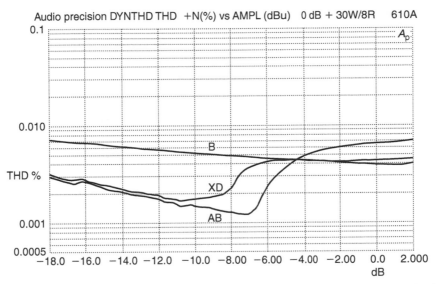

Figure 18.10. THD vs power out for Class-B, Class-AB, and Class-B with constant-current crossover displacement added. (Trace XD). At 10 kHz, 0 dB = 30 W/8 Ω (610A).

Trace A is lower than Trace B because of the complete absence of crossover distortion. Trace XD demonstrates how a constant-current crossover displacement amplifier has the same superb linearity as Class-A up to an output of -7 dB, but distortion then rises to the Class-B level as the output swing begins to traverse the displaced crossover region. (In fact, it slightly exceeds Class-B in this case, as the data was acquired before the prototype was fully optimised.)

A similar THD/amplitude plot in Figure 18.10 compares Class-B with Class-AB and constant-current crossover displacement; emphasising that XD is superior to AB. Here the transition point from Class-A to Class-B is at -8 dB and the Class-AB case was biased so that gm-doubling began at −7 dB. At this point the experimental amplifier was fully optimised and the constant-current crossover displacement distortion above the transition point is now the same as Class-B.

Figure 18.11 demonstrates that push-pull crossover displacement gives markedly lower distortion than constant-current crossover displacement. The transition points are not quite the same (−8 dB for push-pull versus −11 dB for constant-current) but this has no significant effect on the distortion produced. The salient point is that at −2 dB, for example, the THD is very significantly lowered from 0.0036% to 0.0022% by the use of the push-pull method, because of the way it reduces the magnitude of the current changes in the output transistors of the main amplifier.

Figure 18.12 is the same as Figure 18.11 with a THD vs level plot for Class-AB added, underlining the point that Class-AB gives significantly greater distortion above its transition point (say, at −4 dB) than Class-B. As before, constant-current crossover displacement gives slightly less distortion than Class-B, and push-pull crossover displacement gives markedly less and is clearly the best mode of operation.

Figure 18.13 returns to the THD/frequency format, and is included to confirm that XD push-pull gives lower THD over the whole of the upper audio band from 1 kHz to 20 kHz. Below 1 kHz the noise floor dominates.

The Effect of Loading Changes

When a new amplifier concept is considered, it is essential to consider its behaviour into real loads, which deviate significantly from the classical 8 Ω resistance.

Firstly, what is the effect of changing the load resistance, for example, by using a 4 Ω load? The signal currents in the output stage are doubled, so the voltage by which the crossover region is displaced is halved. Half the voltage across half the resistance means the output power is halved, so the volume at the transition point has been reduced by 3 dB. In terms of SPL and human perception, the reduction is not very significant. One of the concerns facing conventional Class-B amplifiers driving 4 Ω or less is the onset of Large Signal

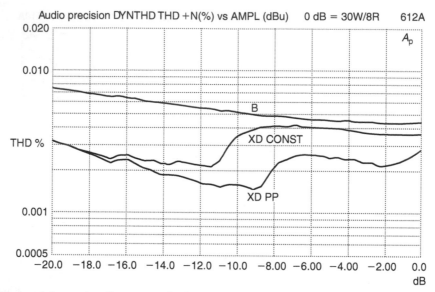

Figure 18.11. This shows that push-pull crossover displacement (XD PP) gives much lower distortion than constant-current crossover displacement (XD CONST). At 10 kHz, 0 dB = 30 W/8 Ω.(612A).

Non-linearity caused by increased output device currents and consequent fall-off of beta in the driver transistors. The use of push-pull displacement reduces LSN in same way that crossover distortion is reduced – by reducing the range of current variation in the output stage.

Secondly, what about reactive loads? In particular, we must scrutinise the way that the push-pull displacer is driven by output voltage rather than device currents. In a conventional Class-B amplifier, adding a reactive element to the load alters the phase relationship between the output voltage and the crossover events; this is because voltage and current are now out of phase, and crossover is a current-domain phenomenon. It has never been suggested this presents any sort of psychoacoustic problem. Putting reactive loading on a crossover-displacement amplifier moves its crossover

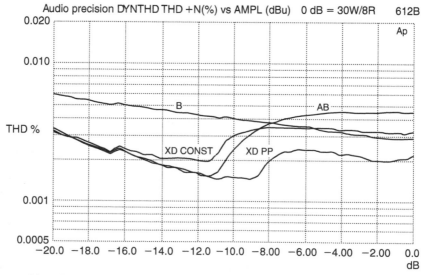

Figure 18.12. Adding the Class-AB plot shows that it is clearly the least linear mode above −8 dBu (612B).

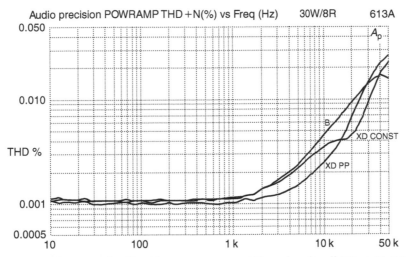

Figure 18.13. THD vs frequency for Class-B, constant-current XD, and push-pull XD at 30 W/8 Ω (613A).

events (if the power level is above transition − otherwise they are not generated at all) in time with respect to the voltage output in exactly the same way, and there is absolutely no reason to suppose that this is any cause for concern.

Stability into reactive loads is not affected by the addition of the displacement system. As previously described, the displacer system has only a very minor effect on the output signal, its effects being confined to making the output stage operate in a more advantageous way, and so it has virtually no effect on the characteristics of the forward amplifier path, and therefore no effect on stability margins.

The Efficiency of Crossover Displacement

The crossover displacement technique obviously increases the total power dissipated in the output stage, and the efficiency is therefore somewhat worse than Class-B. The dissipation in the upper transistor is increased by the displacement current flowing through it, while that in the lower transistor is unchanged. There is also the additional dissipation in the displacer itself, which is likely to be mounted on the same heatsink as the main output devices.

When using techniques such as Class-AB or crossover displacement that give a limited power output in Class-A, you must decide at the start just how much Class-A power you are prepared to pay for in terms of extra heat liberated, and just what load impedance you intend to drive in that mode. For example, assume that 5 Watts of Class-A operation into 8 Ω is required from

an amplifier with a full output of 50 W. The crossover point must therefore be displaced by the peak voltage corresponding to 5 W, which is 8.9 V, and this will require a displacement current of about 1.1 Amps. It is well established that it takes about a 10 dB increase in sound intensity to double subjective loudness,[2] and 10 dB is a power ratio of ten times. Therefore if there is only a doubling in loudness between the transition point into Class-B and full output, the amplifier will be in the Class-A region most of the time; this seems like an eminently reasonable approach.

Table 18.1 shows the calculated efficiency for the various types of amplifier. The calculations were not based on the usual simplistic theory that ignore voltage drops in emitter resistors, transistor saturation voltages and so on, but emerged from a lengthy series of SPICE simulations of complete output stage circuits. The effects of transistor non-linearity and so on are fully taken into account. The efficiency results are therefore slightly worse than simple theory predicts. I think they are as accurate as extensive calculation can make

Table 18.1. Efficiency of amplifier types

Class	Full output (%)	Half power (%)	1/10 power (%)
Class-B	74	54	23
Class XD push-pull	66	46	14
Class XD constant	57	39	11
Class-A	43	23	4

them. For comparison, the 'classical' calculations for Class-B give a full power efficiency of 78%, but the more detailed simulations show that it is only 74% when typical losses are included.

The output stages were simulated using ±50V rails, giving a maximum power of about 135W into 8 Ω. Displacement currents were set to give a transition from Class-A to Class-B at 5W. All emitter resistors were 0.1 Ω.

Here we have demonstrated that there is some penalty in efficiency when crossover displacement is used, but it is far more economical than Class-A. The push-pull XD mode is clearly better than constant-current operation. As I mentioned before, if real musical signals are used rather than sinewaves, the Class-A amplifier comes off a *lot* worse, with efficiency reduced to 1 or 2%.

Other Methods of Push-pull Displacement Control

This chapter describes in detail push-pull crossover displacement implemented by controlling the displacer from the amplifier voltage output. This has the great merit of simplicity, but its method of operation requires design assumptions to be made about the minimum load impedances to be driven. If the load is of higher impedance than expected, which can often occur in loudspeaker loads because of voice-coil inductance, the displacer current may be increased more than is necessary for the desired amount of crossover displacement, leading to unnecessary power dissipation. This is because the voltage-control method is an open-loop or feedforward system.

This situation could be avoided by using a current-controlled system which senses the current flowing in the main amplifier output devices and turns on enough displacer current to give the amount of crossover displacement desired. This is a second negative feedback loop operating at the full signal frequency, and experience has shown that high frequency instability can be a serious problem with this sort of approach. Nonetheless I feel the concept is worthy of further investigation.

Summary: Advantages and Disadvantages

Crossover displacement provides a genuine way to compromise between the linearity of Class-A and the efficiency of Class-B. I think it is now firmly established

that the conventional use of Class-AB, by simply turning up the bias, is not such a compromise because it introduces extra distortion. An important merit of crossover displacement technology is that it is robust and completely dependable. During the development of the Cambridge Audio Azur 840A and 840W power amplifiers, neither of the crossover displacement systems required so much as a change in a resistor value.

Here I summarise the pros and cons.
Advantages:

- Crossover distortion is moved away from the central point where an amplifier output spends most of its time; in normal use, an amplifier can almost always run in Class-A.
- Push-pull displacement also reduces both crossover distortion and LSN when in Class-B operation.
- It is simple. Only 5 extra transistors are used, of which 3 are small-signal and of very low cost.
- There are no extra pre-sets or adjustments.
- It does not affect HF stability.
- No extra overload-protection circuitry is needed, as the displacer is inherently current-limited.
- The technology is versatile. It can be attached to almost any kind of Class-B amplifier.

Disadvantages:

- There is some extra power dissipation, but far less than the use of Class-A.
- There is some extra cost in circuitry, but not much. Only one more power transistor is required.

Crossover displacement has now been in use in the up-market Cambridge Audio amplifiers for some five years, at the time of writing (2012). A new application of it, the Azur 851A Integrated Class XD amplifier, has recently been released and the technology is clearly giving long-term satisfaction. This book contains many examples of new amplifier technologies that failed to catch on and disappeared quite quickly, as they proved to give no real benefit apart from a short-term marketing advantage. I hope I have demonstrated in this chapter that crossover displacement does provide real technical benefits. It is one of the very few successful attempts to reduce the effect of crossover distortion in a Class-B output stage.

References

1. Self, D, *Self on Audio*, Oxford: Newnes/Elsevier, 2006, p. 459.

2. Moore, B J, *An Introduction to the Psychology of Hearing*, New York: Academic Press, 1982, pp. 48–50.

Class-G Power Amplifiers

Most types of audio power amplifier are less efficient than Class-B; for example, Class-AB is markedly less efficient at the low end of its power capability, while it is clear from Chapter 17 that Class-A wastes virtually all the energy put into it. Building amplifiers with higher efficiency is more difficult. Class-D, using ultrasonic pulse-width modulation, promises high efficiency and indeed delivers it, but it is undeniably a difficult technology, and its linearity is still a long way short of Class-B. The practical efficiency of Class-D rests on details of circuit design and device characteristics. The apparently unavoidable LC output filter — second order at least — can only give a flat response into one load impedance, and its magnetics are neither cheap nor easy to design. There are likely to be some daunting EMC difficulties with emissions. Class-D is not an attractive proposition for high-quality domestic amplifiers that must work with separate speakers of unknown impedance characteristics.

There is, however, the Class-G method. Power is drawn from either high- or low-voltage rails as the signal level demands. This technology has taken a long time to come to fruition, but is now used in very-high-power amplifiers for large PA systems, where the savings in power dissipation are important, and is also making its presence felt in home theatre systems; if you have seven or eight power amplifiers instead of two, their losses are rather more significant. Class-G is firmly established in powered subwoofers, and even in ADSL telephone-line drivers. Given the current concern for economy in energy consumption, Class-G may well become more popular in mainstream areas where its efficiency can be used as a marketing point. It is a technology whose time has come.

The Principles of Class-G

Music has a large peak-to-mean level ratio. For most of the time the power output is a long way below the peak levels, and this makes possible the improved efficiency of Class-G. Even rudimentary statistics for this ratio for various genres of music are surprisingly hard to find, but it is widely accepted that the range between 10 dB for compressed rock, and 30 dB for classical material, covers most circumstances.

If a signal spends most of its time at low power, then while this is true, a low-power amplifier will be much more efficient. For most of the time lower output levels are supplied from the lowest-voltage rails, with a low voltage drop between rail and output, and correspondingly low dissipation. The most popular Class-G configurations have two or three pairs of supply rails, two being usual for hi-fi, while three is more common in high-power PA amplifiers.

When the relatively rare high-power peaks do occur, they must be handled by some mechanism that can draw high power, causing high internal dissipation, but which only does so for brief periods. These infrequent peaks above the transition level are supplied from the high-voltage pair of rails. Clearly the switching between rails is the heart of the matter, and anyone who has ever done any circuit design will immediately start thinking about how easy or otherwise it will be to make this happen cleanly with a high-current 20 kHz signal.

There are two main ways to arrange the dual-rail system: series and parallel (i.e. shunt). This chapter deals only with the series configuration, as it seems to have had the greatest application to hi-fi. The parallel version is more often used in high-power PA amplifiers.

Hitachi introduced Class-G, and first applied it to the HMA-8300 power amplifier in 1977. This was quite a powerful machine giving 200 W/8 Ω. They called the principle 'Dynaharmony', a name which singularly failed to catch on.

Introducing Series Class-G

A series configuration Class-G output stage using two rail voltages is shown in Figure 19.1. The so-called inner devices are those that work in Class-B; those that perform the rail-switching on signal peaks are called the outer devices — by me, anyway. In this design study the EF type of output stage is chosen because of its greater robustness against local HF instability, though the CFP configuration could be used instead for inner, outer, or both sets of output devices, given suitable care. For maximum power efficiency the inner stage normally runs in Class-B, though there is absolutely no reason why it could not be run in Class-AB or even Class-A; there will be more discussion of these intriguing possibilities later. If the inner power devices are in Class-B, and the outer ones conduct for much less than 50% of a cycle, being effectively in Class-C, then according to the classification scheme I have proposed,[1] this should be denoted Class-B + C. The plus sign indicates the series rather than shunt connection of the outer and inner power devices. This basic configuration was developed by Hitachi to reduce amplifier heat dissipation.[2,3] Musical signals spend most of their time at low levels, having a high peak/mean ratio, and power dissipation is

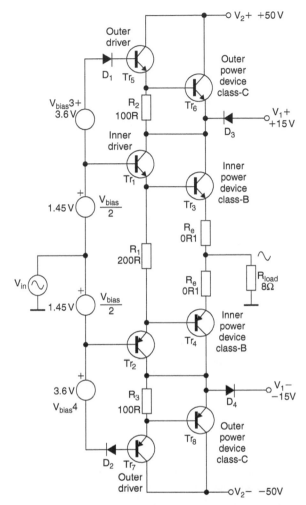

Figure 19.1. A series Class-G output stage, alternatively Class-B+C. Voltages and component values are typical. The inner stage is Class-B EF. Biasing by my method.

greatly reduced by drawing from the lower ± V1 supply rails at these times.

The inner stage TR3, 4 operates in normal Class-B. TR1, 2 are the usual drivers and R1 is their shared emitter resistor. The usual temperature-compensated Vbias generator is required, shown here theoretically split in half to maintain circuit symmetry when the stage is SPICE simulated; since the inner power devices work in Class-B, it is their temperature which must be tracked to maintain quiescent conditions. Power from the lower supply is drawn through D3 and D4, often called the commutating diodes, to emphasise their rail-switching action. The word 'commutation' avoids confusion with the usual Class-B crossover at

zero volts. I have called the level at which rail-switching occurs the transition level.

When a positive-going instantaneous signal exceeds low rail +V1, D1 conducts, TR5 and TR6 turn on and D3 turns off, so the entire output current is now taken from the high-voltage +V2 rail, with the voltage drop and hence power dissipation shared between TR3 and TR6. Negative-going signals are handled in exactly the same way. Figure 19.2 shows how the collector voltages of the inner power devices retreat away from the output rail as it approaches the lower supply level.

Class-G is commonly said to have worse linearity than Class-B, the blame usually being loaded onto the diodes and problems with their commutation. As usual, received wisdom is only half of the story, if that, and there are other linearity problems that are not due to sluggish diodes, as will be revealed shortly. It is inherent in the Class-G principle that if switching glitches do occur, they only happen at moderate power or above, and are well displaced away from the critical crossover region where the amplifier spends most of its time. A Class-G amplifier has a low-power region of true Class-B linearity, just as a Class-AB amplifier has a low-power region of true Class-A performance.

Efficiency of Class-G

The standard mathematical derivation of Class-B efficiency with sinewave drive uses straightforward integration over a half-cycle to calculate internal dissipation against voltage fraction, i.e., the fraction of possible output voltage swing. As is well known, in Class-B the maximum heat dissipation is about 40% of maximum output power, at an output voltage fraction of 63%, which also delivers 40% of the maximum output power to the load.

The mathematics is simple because the waveforms do not vary in shape with output level. Every possible idealisation is assumed, such as zero quiescent current, no emitter resistors, no Vce(sat) losses, and so on. In Class-G, on the other hand, the waveforms are a strong function of output level, requiring variable limits of integration, and so on, and it all gets very unwieldy.

The SPICE simulation method described by me in Chapter 16 is much simpler, if somewhat laborious, and can use any input waveform, yielding a Power Partition Diagram (PPD), which shows how the power drawn from the supply is distributed between output device dissipation and useful power in the load.

No one disputes that sinewaves are poor simulations of music for this purpose, and their main advantage is

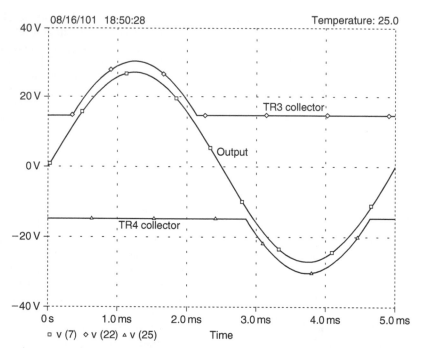

Figure 19.2. The output of a Class-G stage and the voltages on the collectors of the inner output devices.

that they allow direct comparison with the purely mathematical approach. However, since the whole point of Class-G is power saving, and the waveform used has a strong effect on the results, I have concentrated here on the PPD of an amplifier with real musical signals, or at any rate, their statistical representation. The triangular Probability Distribution Function (PDF) approach is described in Chapter 16.

Figure 19.3 shows the triangular PDF PPD for conventional Class-B EF, while Figure 19.4 is that for Class-G with $\pm V2 = 50$ V and $\pm V1 = 15$ V, i.e., with the ratio of V1/V2 set to 30%. The PPD plots power dissipated in all four output devices, the load, and the total drawn from the supply rails. It shows how the input power is partitioned between the load and the output devices. The total sums to slightly less than the input power, the remainder being accounted for as usual by losses in the drivers and Res. Note that in Class-G power dissipation is shared, though not very equally, between the inner and outer devices, and this helps with efficient utilisation of the silicon.

In Figure 19.4 the lower area represents the power dissipated in the inner devices and the larger area just above represents that in the outer devices; there is only one area for each because in Class-B and Class-G only one side of the amplifier conducts at a time. Outer device dissipation is zero below the rail-switching

threshold at -15 dB below maximum output. The total device dissipation at full output power is reduced from 48 W in Class-B to 40 W, which may not appear at first to be a very good return for doubling the power transistors and drivers.

Figure 19.5 shows the same PPD but with $\pm V2 = 50$ V and $\pm V1 = 30$ V, i.e. with V1/V2 set to 60%. The low-voltage region now extends up to -6dB ref full power, but the inner device dissipation is higher due to the higher V1 rail voltages. The result is that total device power at full output is reduced from 48 W in Class-B to 34 W, which is a definite improvement. The efficiency figure is highly sensitive to the way the ratio of rail voltages compares with the signal characteristics. Domestic hi-fi amplifiers are not operated at full volume all the time, and in real life the lower option for the V1 voltage is likely to give lower general dissipation. I do not suggest that V1/V2 = 30% is the optimum lower-rail voltage for all situations, but it looks about right for most domestic hi-fi.

Practicalities

In my time I have wrestled with many 'new and improved' output stages that proved to be anything but improved. When faced with a new and intriguing

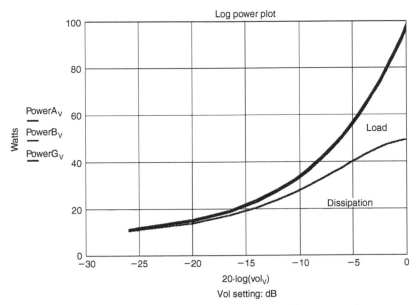

Figure 19.3. Power partition diagram for a conventional Class-B amplifier handling a typical music signal with a triangular Probability Density Function. X-axis is volume.

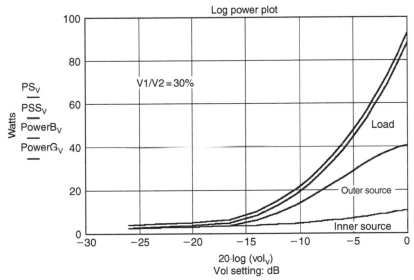

Figure 19.4. Power partition diagram for Class-G with V1/V2 = 30%. Signal has a triangular PDF. X-axis is volume; outer devices dissipate nothing until −15 dB is reached.

possibility, I believe the first thing to do is sketch out a plausible circuit such as Figure 19.1 and see if it works in SPICE simulation. It duly did.

The next stage is to build it, power it from low supply rails to minimise the size of any explosions, and see if it

works for real at 1 kHz. This is a bigger step than it looks.

SPICE simulation is incredibly useful but it is no substitute for testing a real prototype. It is easy to design clever and complex output stages that work

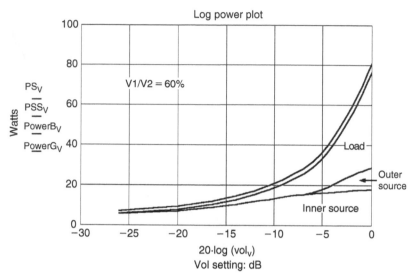

Figure 19.5. Power partition diagram for Class-G with V1/V2 = 60%. Triangular PDF. Compared with Figure 19.4, the inner devices dissipate more and the outer devices almost nothing except at maximum volume.

beautifully in simulation but in reality prove impossible to stabilise at high frequencies. Some of the more interesting output-triple configurations seem to suffer from this.

The final step — and again it is a bigger one than it appears — is to prove real operation at 20 kHz and above. Again it is perfectly possible to come up with a circuit configuration that either just does not work at 20 kHz, due to limitations on power transistor speeds, or is provoked into oscillation or other misbehaviour that is not set off by a 1 kHz testing.

Only when these vital questions are resolved, is it time to start considering circuit details, and assessing just how good the amplifier performance is likely to be.

The Biasing Requirements

The output stage bias requirements are more complex than for Class-B. Two extra bias generators Vbias3, Vbias4 are required to make TR6 turn on before TR3 runs out of collector voltage. These extra bias voltages are not critical, but must not fall too low, or become much too high. Should these bias voltages be set too low, so the outer devices turn on too late, then the Vce across TR3 becomes too low, and its current sourcing capability is reduced. When evaluating this issue, bear in mind the lowest impedance load the amplifier is planned to drive, and the currents this will draw from the output devices. Fixed Zener diodes of normal

commercial tolerance are quite accurate and stable enough for setting Vbias3 and Vbias4.

Alternatively, if the bias voltage is set too high, then the outer transistors will turn on too early, and the heat dissipation in the inner power devices becomes greater than it need be for correct operation. The latter case is rather less of a problem, so if in doubt this bias should be chosen to be on the high side rather than the low.

The original Hitachi circuit[1] put Zeners in series with the signal path to the inner drivers to set the output quiescent bias, their voltage being subtracted from the main bias generator which was set at 10 V or so, a much higher voltage than usual (see Figure 19.6). SPICE simulation showed me that the presence of Zener diodes in the forward path to the inner power devices gave poor linearity, which is not exactly a surprise. There is also the problem that the quiescent conditions will be affected by changes in the Zener voltage. The 10V bias generator, if it is the usual Vbe multiplier, will have much too high a temperature coefficient for proper thermal tracking.

I therefore rearrange the biasing as in Figure 19.1. The amplifier forward path now goes directly to the inner devices, and the two extra bias voltages are in the path to the outer devices; since these do not control the output directly, the linearity of this path is of lesser importance. The Zeners are out of the forward path and the bias generator can be the standard sort. It must be thermally coupled to the inner power devices; the outer ones have no effect on the quiescent conditions.

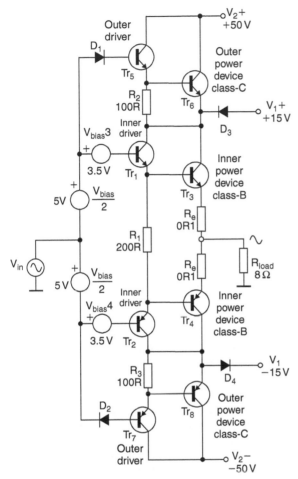

Figure 19.6. The original Hitachi Class-G biasing system, with inner device bias derived by subtracting Vbias3, 4 from the main bias generator.

The Linearity Issues of Series Class-G

Series Class-G has often had its linearity called into question because of difficulties with supply-rail commutation. Diodes D3 and D4 must be power devices capable of handling a dozen amps or more, and conventional silicon rectifier diodes that can handle such currents take a long time to turn off, due to their stored charge carriers. This has the following unhappy effect: when the voltage on the cathode of D3 rises above V1, the diode tries to turn off abruptly, but its charge carriers sustain a brief but large reverse current as they are swept from its junction. This current is supplied by TR6, attempting as an emitter-follower to keep its emitter up to the right voltage. So far all is well.

However, when the diode current ceases, TR6 is still conducting heavily, due to its own charge-carrier storage. The extra current it turned on to feed D3 in reverse now goes through TR3 collector, which accepts it because of TR3's low Vce, and passes it onto the load via TR3 emitter and Re.

This process is readily demonstrated by a SPICE commutation transient simulation; see Figures 19.7 and 19.8. Note there are only two of these events per cycle — not four, as they only occur when the diodes turn off. In the original Hitachi design this problem was reportedly tackled by using fast transistors and relatively fast gold-doped diodes, but according to Sampei et al.,[2] this was only partially successful.

It is now simple to eradicate this problem. Schottky power diodes are readily available, as they were not in 1976, and are much faster due to their lack of minority carriers and charge storage. They have the added advantage of a low forward voltage drop at large currents of 10A or more. The main snag is a relatively low reverse withstand voltage, but fortunately in Class-G usage the commutating diodes are only exposed at worst to the difference between V2 and V1, and this only when the amplifier is in its low power domain of operation. Another good point about Schottky power diodes is that they do appear to be robust; I have subjected 50 amp Motorola devices to 60 amps-plus repeatedly without a single failure. This is a good sign. The spikes disappear completely from the SPICE plot if the commutating diodes are Schottky rectifiers. Motorola MBR5025L diodes capable of 50A and 25 PIV were used in simulation.

The Static Linearity

SPICE simulation shows in Figure 19.9 that the static linearity (i.e. that ignoring dynamic effects like diode charge-storage) is distinctly poorer than for Class-B. There is the usual Class-B gain-wobble around the crossover region, exactly the same size and shape as for conventional Class-B, but also there are now gain-steps at ±16V. The result with the inner devices biased into push-pull Class-A is also shown, and proves that the gain-steps are not in any way connected with crossover distortion. Since this is a DC analysis, the gain-steps cannot be due to diode switching-speed or other dynamic phenomena, and Early Effect was immediately suspected. (Early Effect is the increase in collector current when the collector voltage increases, even though the Vbe remains constant.) When unexpected distortion appears in a SPICE simulation of this

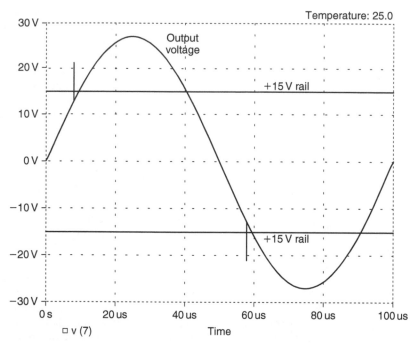

Figure 19.7. Spikes due to charge storage of conventional diodes, simulated at 10 kHz. They only occur when the diodes turn off, so there are only two per cycle. These spikes disappear completely when Schottky diodes are used in the SPICE model.

kind, and effects due to finite transistor beta and associated base currents seem unlikely, a most useful diagnostic technique is to switch off the simulation of Early Effect for each transistor in turn. In SPICE transistor models the Early Effect can be totally disabled by setting the parameter VAF to a much higher value than the default of 100, such as 50,000. This experiment demonstrated in short order that the gain-steps were caused wholly by Early Effect acting on both inner drivers and inner output devices. The gain-steps are completely abolished with Early Effect disabled. When TR6 begins to act, TR3 Vce is no longer decreasing as the output moves positive, but substantially constant as the emitter of Q6 moves upwards at the same rate as the emitter of Q3. This has the effect of a sudden change in gain, which naturally degrades the linearity.

This effect appears to occur in drivers and output devices to the same extent. It can be easily eliminated in the drivers by powering them from the outer rather than the inner supply rails. This prevents the sudden changes in the rate in which driver Vce varies. The improvement in linearity is seen in Figure 19.10, where the gain-steps have been halved in size. The resulting circuit is shown in Figure 19.11. Driver

power dissipation is naturally increased by the increased driver Vce, but this is such a small fraction of the power consumed that the overall efficiency is not significantly reduced. It is obviously not practical to apply the same method to the output devices, because then the low-voltage rail would never be used and the amplifier is no longer working in Class-G. The small-signal stages naturally have to work from the outer rails to be able to generate the full voltage swing to drive the output stage.

We have now eliminated the commutating diode glitches, and halved the size of the unwanted gain-steps in the output stage. With these improvements made, it is practical to proceed with the design of a Class-G amplifier with midband THD below 0.002%.

Practical Class-G Design

The Class-G amplifier design expounded here uses very similar small-signal circuitry to the Blameless Class-B power amplifier, as it is known to generate very little distortion of its own. If the specified supply voltages of ±50 V and ±15 V are used, the maximum power output is about 120 W into 8 Ω, and the rail-switching transition occurs at 28 W.

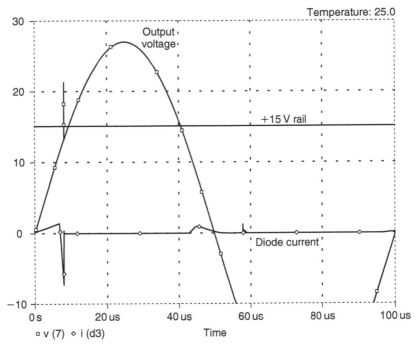

Figure 19.8. A close-up of the diode transient. Diode current rises as output moves away from zero, then reverses abruptly as charge carriers are swept out by reverse-biasing. The spike on the output voltage is aligned with the sudden stop of the diode reverse current.

This design incorporates various techniques described in this book, and closely follows the Blameless Class-B amp though some features derive from the Trimodal (Chapter 17) and Load Invariant (Chapter 10) amplifiers. A notable example is the low-noise feedback network, complete with its option of input bootstrapping to give a high impedance when required. Single-slope VI limiting is incorporated for overload protection; this is implemented by Q12, 13. Figure 19.12 shows the circuit.

As usual, in my amplifiers the global NFB factor is a modest 30 dB at 20 kHz.

Controlling Small-Signal Distortion

The distortion from the small-signal stages is kept low by the same methods as for the other amplifier designs in this book, and so this is only dealt with briefly here. The input stage differential pair Q1, 2 is given local feedback by R5 and R7 to delay the onset of third-harmonic Distortion One. Internal Re variations in these devices are minimised by using an unusually high tail current of 6 mA. Q3, 4 are a degenerated current-mirror that enforces accurate balance of the

Q1, 2 collector currents, preventing the production of second-harmonic distortion. The input resistance (R3 + R4) and feedback resistance R16 are made equal and made unusually low, so that base current mismatches stemming from input device beta variations give a minimal DC offset. Vbe mismatches in Q1 and Q2 remain, but these are much smaller than the effects of Ib. Even if Q1 and Q2 are high-voltage types with relatively low beta, the DC offset voltage at the output should be kept to less than ±50mV. This is adequate for all but the most demanding applications. This low-impedance technique eliminates the need for balance presets or DC servo systems, which is most convenient.

A lower value for R16 implies a proportionally lower value for R15 to keep the gain the same, and this reduction in the total impedance seen by Q2 improves noise performance markedly. However, the low value of R3 plus R4 at 2k2 gives an input impedance which is not high enough for many applications.

There is no problem if the amplifier is to have an additional input stage, such as a balanced line receiver. Proper choice of opamp will allow the stage to drive a 2k2 load impedance without generating additional distortion. Be aware that adding such a stage – even if it is properly designed and the best available opamps are used – will

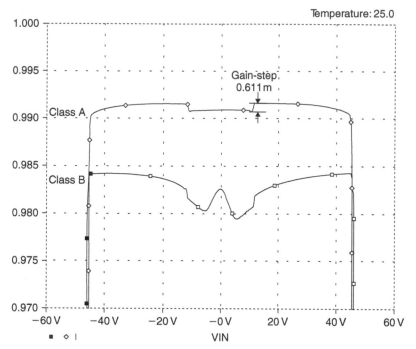

Figure 19.9. SPICE simulation shows variations in the incremental gain of an EF-type Class-G series output stage. The gain-steps at transition (at ±16V) are due to Early effect in the transistors. The Class-A trace is the top one, with Class-B optima below. Here the inner driver collectors are connected to the switched inner rails, i.e. the inner power device collectors, as in Figure 19.1.

degrade the signal-to-noise ratio significantly. This is because the noise generated by the power amplifier itself is so very low — equivalent to the Johnson noise of a resistor of a few hundred ohms — that almost anything you do upstream will degrade it seriously.

If there is no separate input stage, then other steps must be taken. What we need at the input of the power amplifier is a low DC resistance, but a high AC resistance; in other words, we need either a 50 Henry choke or recourse to some form of bootstrapping. There is to my mind no doubt about the way to go here, so bootstrapping it is. The signal at Q2 base is almost exactly the same as the input, so if the mid-point of R3 and R4 is driven by C3, so far as input signals are concerned, R3 has a high AC impedance. When I first used this arrangement I had doubts about its high-frequency stability, and so added resistor R9 to give some isolation between the bases of Q1 and Q2. In the event I have had no trouble with instability, and no reports of any from the many constructors of the Trimodal and Load-Invariant designs, which incorporate this option.

The presence of R9 limits the bootstrapping factor, as the signal at R3-R4 junction is thereby a little smaller than at Q2 base, but it is adequate. With R9 set to 100R, the AC input impedance is raised to 13 k, which should be high enough for almost all purposes. Higher values than this mean that an input buffer stage is required.

The value of C8 shown (1000 μF) gives an LF roll-off in conjunction with R15 that is -3 dB at 1.4 Hz. The purpose is not impossibly extended sub-bass, but the avoidance of a low-frequency rise in distortion due to non-linearity effects in C8. If a 100 μF capacitor is used here, the THD at 10 Hz worsens from <0.0006% to 0.0011%, and I regard this as unacceptable aesthetically — if not perhaps audibly. This is not the place to define the low-frequency bandwidth of the system — this must be done earlier in the signal chain, where it can be properly implemented with more accurate non-electrolytic capacitors. The protection diodes D1 to D4 prevent damage to C2 if the amplifier suffers a fault that makes it saturate in either direction; it looks like an extremely dubious place to put diodes, but since they normally have no AC or DC voltage across them, no measurable or detectable distortion is generated.

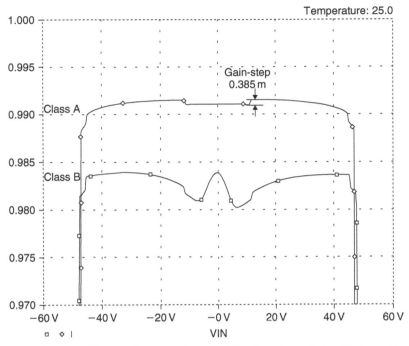

Figure 19.10. Connecting the inner driver collectors to the outer V2 rails reduces Early effect non-linearities in them, and halves the transition gain-steps.

The Voltage-Amplifier-Stage (VAS) Q11 is enhanced by emitter-follower Q10 inside the Miller-compensation loop, so that the local negative feedback that linearises the VAS is increased. This effectively eliminates VAS non-linearity. Thus increasing the local feedback also reduces the VAS collector impedance, so a VAS-buffer to prevent Distortion Four (loading of VAS collector by the non-linear input impedance of the output stage) is not required. Miller capacitor Cdom is relatively big at 100pF, to swamp transistor internal capacitances and circuit strays, and make the design predictable. The slew-rate calculates as 40 V/μsec use in each direction. VAS collector load Q7 is a standard current source.

Almost all the THD from a Blameless amplifier derives from crossover distortion, so keeping the quiescent conditions optimal to minimise this is essential. The bias generator for an EF output stage, whether in Class-B or Class-G, is required to cancel out the Vbe variations of four junctions in series: those of the two drivers and the two output devices. This sounds difficult, because the dissipation in the two types of devices is quite different, but the problem is easier than it looks. In the EF type of output stage, the driver dissipation is almost constant as power output varies, and so the

problem is reduced to tracking the two output device junctions. The bias generator Q8 is a standard Vbe-multiplier, with R23 chosen to minimise variations in the quiescent conditions when the supply rails change. The bias generator should be in contact with the top of one of the inner output devices, and not the heatsink itself. This position gives much faster and less attenuated thermal feedback to Q8. The VAS collector circuit incorporates not only bias generator Q8 but also the two Zeners D8, D9 which determine how early rail-switching occurs as the inner device emitters approach the inner (lower) voltage rails.

The output stage was selected as an EF (emitter-follower) type as this is known to be less prone to parasitic or local oscillations than the CFP configuration, and since this design was to some extent heading into the unknown, it seemed wise to be cautious where possible. R20 and R32 are the usual shared emitter resistor for the inner drivers. The outer drivers Q16 and Q17 have their own emitter resistors R33 and R36, which have their usual role of establishing a reasonable current in the drivers as they turn on, to increase driver transconductance, and also in speeding up turn-off of the outer output devices by providing a route for charge-carriers to leave the output device bases.

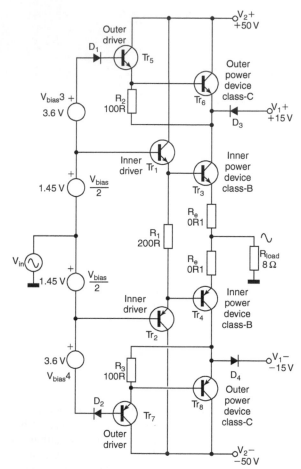

Figure 19.11. A Class-G output stage with the drivers powered from the outer supply rails.

As explained above, the inner driver collectors are connected to the outer rails to minimise the gain-steps caused by the abrupt change in collector voltage when rail transition occurs.

Deciding the size of heatsink required for this amplifier is not easy, mainly because the heat dissipated by a Class-G amplifier depends very much on the rail voltages chosen and the signal statistics. A Class-B design giving 120 W into 8 Ω would need a heatsink with thermal resistance of the order of 1°C/W (per channel); a good starting point for a Class-G version giving the same power would be about half the size, i.e., 2°C/W. The Schottky commutating diodes do not require much heatsinking, as they conduct only intermittently and have a low forward voltage drop. It is usually convenient to mount them on the main heatsink, even if this does

mean that most of the time they are being heated rather than cooled.

C15 and R19 make up the usual Zobel network. The coil L1, damped by R39, isolates the amplifier from load capacitance. Using 10 to 20 turns at 1 inch diameter should work well; the value of inductance for stability is not all that critical.

The Performance

Figure 19.13 shows the THD at 20 W and 50 W (into 8 Ω) and I think this demonstrates at once that the design is a practical competitor for Class-B amplifiers. Compare these results with the upper trace of Figure 19.14, taken from a Blameless Class-B amplifier at 50 W, 8 Ω. Note the lower trace of Figure 19.14 is for 30 kHz bandwidth, used to demonstrate the lack of distortion below 1 kHz; the THD data above 10 kHz is in this case meaningless as all the harmonics are filtered out. All the Class-G plots here are taken at 80 kHz to make sure any high-order glitching is properly measured.

Figure 19.15 shows the actual THD residual at 50 W output power. The glitches from the gain-steps are more jagged than the crossover disturbances, as would be expected from the output stage gain plot in Figures 19.9, 19.11. Figure 19.16 confirms that at 20 W, below transition, the residual is indistinguishable from that of a Blameless Class-B amplifier; in this region, where the amplifier is likely to spend most of its time, there are no compromises on quality.

Figure 19.17 shows THD versus level, demonstrating how THD increases around 28 W as transition begins. The steps at about 10 W are nothing to do with the amplifier − they are artefacts due to internal range-switching in the measuring system.

Figure 19.18 shows for real the benefits of powering the inner drivers from the outer supply rails. In SPICE simulation (see above) the gain-steps were roughly halved in size by this modification, and Figure 19.18 duly confirms that the THD is halved in the HF region, the only area where it is sufficiently clear of the noise floor to be measured with any confidence.

Deriving a New Kind of Amplifier: Class-A + C

A conventional Class-B power amplifier can be almost instantly converted to push-pull Class-A simply by increasing the bias voltage to make the required quiescent current flow. This is the only real circuit change, though naturally major increases in heatsinking and

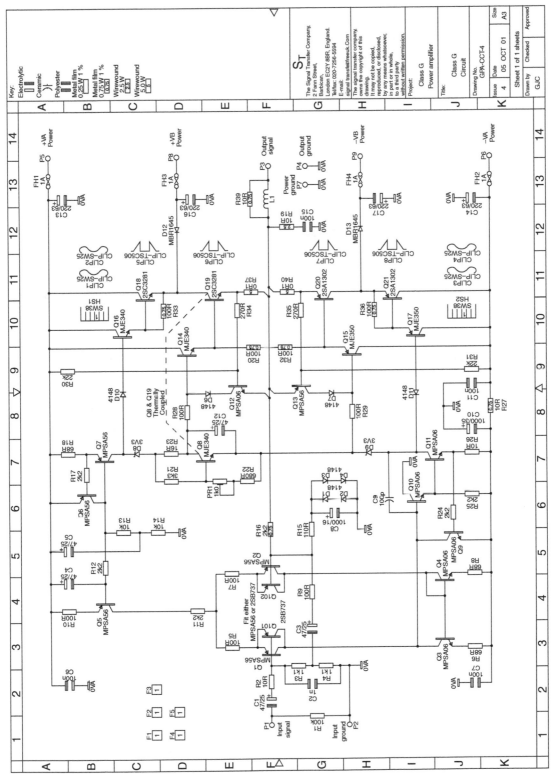

Figure 19.12. The full circuit diagram of the Class-G amplifier.

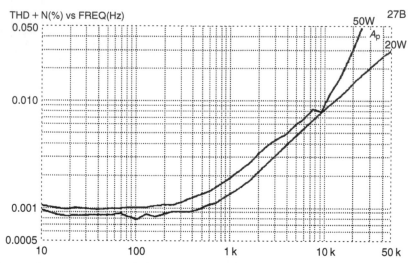

Figure 19.13. THD versus frequency, at 20 W (below transition) and 50 W into an 8 Ω load. The joggle around 8 kHz is due to a cancellation of harmonics from crossover and transition. 80 kHz bandwidth.

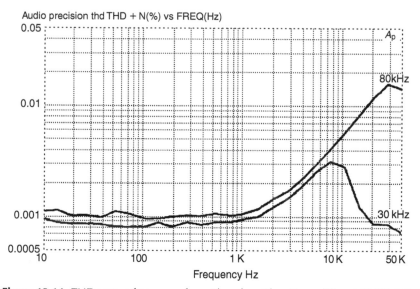

Figure 19.14. THD versus frequency for a Blameless Class-B amplifier at 50 W, 8 Ω.

power-supply capability are required for practical use. Exactly the same principle applies to the Class-G amplifier. In the book *Self on Audio*,[4] I suggested a new and much more flexible system for classifying amplifier types and here it comes in very handy. Describing Class-G operation as Class-B + C immediately indicates that only a bias increase is required to transform it into Class-A + C, and a new type of amplifier is born. This amplifier configuration combines the superb

linearity of classic Class-A up to the transition level, with only minor distortion artefacts occurring at higher levels, as demonstrated for Class-B + C above. Using Class-A means that the simple Vbe-multiplier bias generator can be replaced with precise negative feedback control of the quiescent current, as implemented in the Trimodal amplifier in Chapter 17. There is no reason why an amplifier could not be configured as a Class-G Trimodal, i.e., with the low-voltage

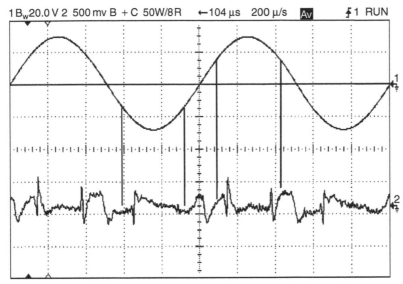

Figure 19.15. The THD residual waveform at 50 W into 8 Ω. This residual may look rough, but in fact it had to be averaged eight times to dig the glitches and crossover out of the noise; THD is only 0.0012%. The vertical lines show where transition occurs.

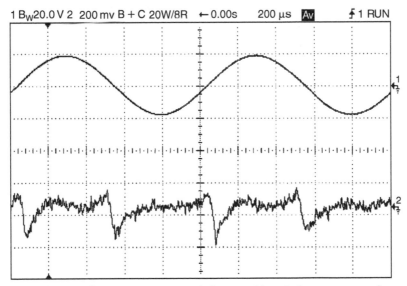

Figure 19.16. The THD residua waveform at 20 W into 8 Ω, below transition. Only crossover artefacts are visible as there is no rail-switching.

section manually switchable between Class-A and Class-B. That would indeed be an interesting machine.

In Figure 19.19 is shown the THD plot for such a Class-A + C amplifier working at 20 W and 30 W into 8 Ω. At 20 W, the distortion is very low indeed, no higher than a pure Class-A amplifier. At 30 W, the transition gain-steps appear, but the THD remains very well controlled, and no higher than a Blameless Class-B design. Note that as in Class-B, when the THD does start to rise, it only does so at 6 dB/octave. The quiescent current was set to 1.5 A.

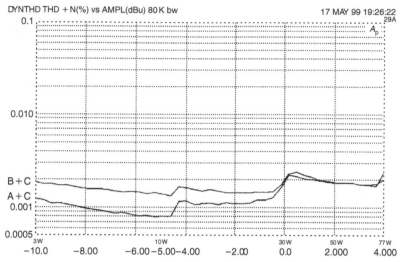

Figure 19.17. THD versus level, showing how THD increases around 28 W as transition begins. Class-A + C is the lower and Class-B + C the upper trace.

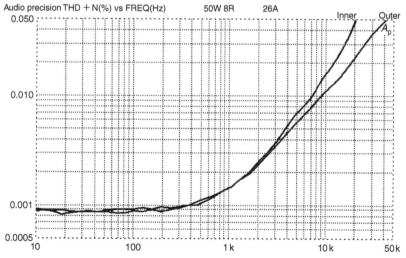

Figure 19.18. THD plot of real amplifier driving 50 W into 8 Ω. Rails were ±40 and ±25 V. Distortion at HF is halved by connecting the inner drivers to the outer supply rails rather than the inner rails.

Figure 19.20 reveals the THD residual during Class-A + C operation. There are absolutely no crossover artefacts, and the small disturbances that do occur happen at such a high signal level that I really do think it is safe to assume they could never be audible. Figure 19.21 shows the complete absence of artefacts on the residual when this new type of amplifier is working below transition; it gives pure Class-A linearity. Finally, Figure 19.22 gives the THD when the amplifier is driving the full 50 W into 8 Ω; as before, the Class-A + C THD plot is hard to distinguish from Class-B, but there is the immense advantage that there is no crossover distortion at low levels, and no critical bias settings.

I modestly suggest that this might be the lowest distortion Class-G amplifier so far.

Class-G with Two-pole Compensation

I have previously shown elsewhere in this book that amplifier distortion can be very simply reduced by changes to the compensation, which means a scheme

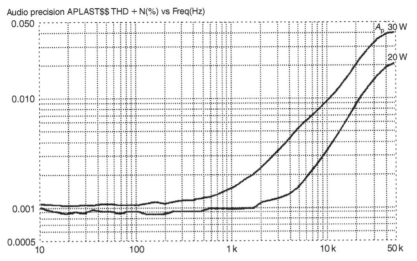

Figure 19.19. The THD plot of the Class-A + C amplifier (30 W and 20 W into 8 Ω). Inner drivers powered from outer rails.

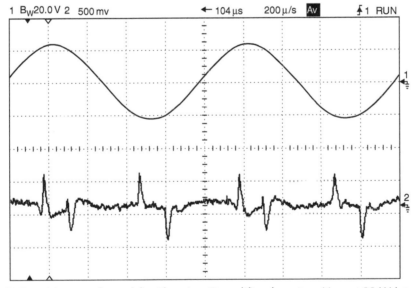

Figure 19.20. The THD residua waveform of the Class-A + C amplifier above transition, at 30 W into 8 Ω. Switching artefacts are visible but not crossover distortion.

more sophisticated than the near-universal dominant pole method. It must be borne in mind that any departure from the conventional 6 dB/octave-all-the-way compensation scheme is likely to be a move away from unconditional stability. (I am using this phrase in its proper meaning; in Control Theory, unconditional stability means that increasing open-loop gain above a threshold causes instability, but the system is stable for all lower values. Conditional stability means that lower open-loop gains can also be unstable.)

A conditionally stable amplifier may well be docile and stable into any conceivable reactive load when in normal operation, but show the cloven hoof of oscillation at power-up and power-down, or when clipping. This is because under these conditions the effective open-loop gain is reduced.

Class-G distortion artefacts are reduced by normal dominant pole feedback in much the same way as crossover non-linearities, i.e., not all that effectively, because the artefacts take up a very small part of the cycle and

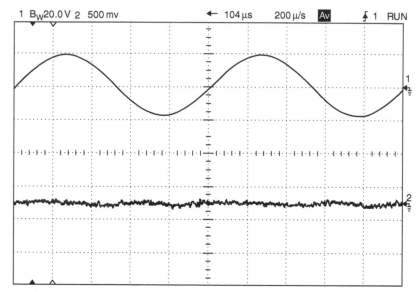

Figure 19.21. The THD residual waveform plot of the Class-A + C amplifier (20 W into 8 Ω)

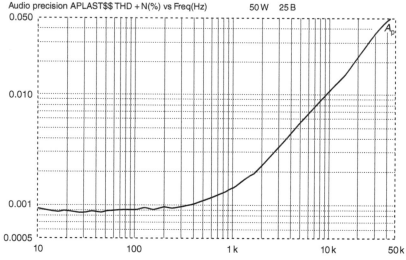

Figure 19.22. The THD plot of the Class-A + C amplifier (50 W into 8 Ω). Inner drivers powered from outer rails.

are therefore composed of high-order harmonics. Therefore a compensation system that increases the feedback factor at high audio frequencies will be effective on switching artefacts, in the same way that it is for crossover distortion. The simplest way to implement two-pole circuit compensation is shown in Figure 19.23. Further details are given in Chapter 13.

The results of two-pole compensation for Class-B + C are shown in Figure 19.24; comparing it with Figure 19.13 (the normal Miller compensated Class-

B + C amplifier) the above-transition (30 W) THD at 10 kHz has dropped from 0.008% to 0.005%; the sub-transition (20 W) THD at 10 kHz has fallen from 0.007% to 0.003%. Comparisons have to be done at 10 kHz or thereabouts to ensure there is enough to measure.

Now comparing the two-pole Class-B + C amplifier with Figure 19.19 (the Class-A + C amplifier), the above-transition (30 W) THD at 10 kHz of the former is lower at 0.005% compared with 0.008%. As I have

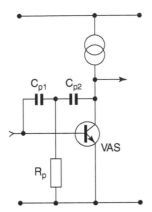

Figure 19.23. The circuit modification for two-pole compensation.

demonstrated before, proper use of two-pole compensation can give you a Class-B amplifier that is hard to distinguish from Class-A — at least until you put your hand on the heatsink.

Class-G with Output-inclusive Compensation

We have seen in Chapters 12 and 13 that output-inclusive compensation, in which the output stage is included in the VAS compensation loop, can give a dramatic reduction in amplifier distortion. This distortion is mainly due to crossover artefacts in the output

stage, and it seems very likely that it would be equally effective against the rail-switching glitches.

Class-G Mode Indication

Having designed and built a Class-G amplifier, the question arises, does it do what it says on the tin? How often does it actually draw power from the upper supply rails? This question can be answered by the simple indicator circuit presented here. It illuminates an LED whenever the amplifier switches over to draw power from the higher (outer) supply rails $\pm V2$, and has a fast-attack slow-decay characteristic so that even short excursions into the high-power domain are clearly signalled.

The first question is how to define exactly what constitutes entering the high-power mode. I have taken this to happen when the commutating diodes become reverse-biased; there is no doubt then that all the power is being drawn from the upper supply rails through the outer power devices.

The circuit is shown in Figure 19.25; the component numbers follow on from those of the main diagram in Figure 19.12. Its operation is as follows. When the amplifier is drawing power from the lower $\pm V1$ rails, the commutating diode D12 is conducting, and its cathode is therefore about 0.2 V negative of its anode. (Remember it is a Schottky diode with low forward voltage drop.) Q22 is therefore firmly off; the diode D14 is simply a precaution to prevent Q22 exploding

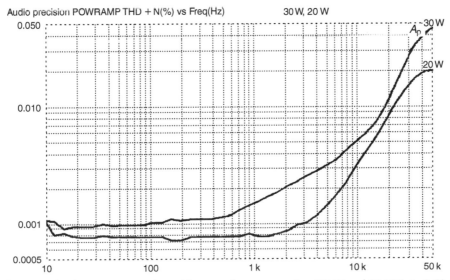

Figure 19.24. The THD plot for B + C operation with two-pole compensation (20 W and 30 W into 8 Ω). Compare with Figures 19.13 (B + C) and 19.19 (A + C).

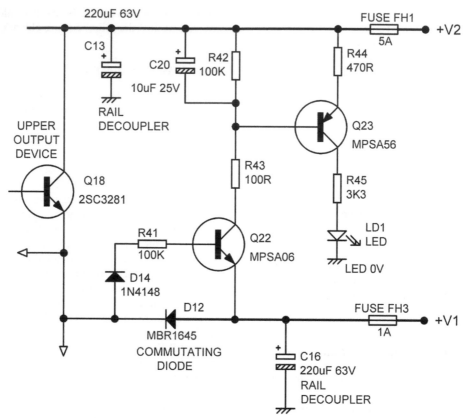

Figure 19.25. Class-G mode indicator. When current is drawn from the supply ±V2 rails D12 is reverse-biased, Q22 and Q23 turn on, and the LED illuminates.

if D12 should fail open-circuit. When the amplifier output moves sufficiently positive, the amplifier enters the high-power mode and Q18 turns on. The commutating diode D12 becomes reverse-biased, D14 conducts, and Q22 is turned on via R41 which limits the base current. The collector current of Q22 rapidly charges C20 through R43, which again limits the current flowing to an amount that will not inconvenience Q22. Q23 then turns on and current flows through dropper resistor R45 to illuminate LED LD1. R42 makes the discharge time more predictable as most of the discharge current is flowing through a resistor rather than a Q23 base, with the beta-variations that this implies.

I usually mount the indicating LED directly on the amplifier PCB, where it acts as a useful indication that all is working properly; not everyone wants a display of flashing lights while they are soaking up lute music. The front panel is also not a good place to put the LED if you are planning double-blind listening tests to determine the merits or otherwise of Class-G.

The value for C20 given here (10 uF) provides a decay time of approx 500 msec, to give a good clear indication even on fast transients. Reducing this to 2.2 uF gives much snappier operation if preferred. The LED 0 V connection must be made to a 'dirty' part of the grounding system and not mixed up with clean signal grounds. A separate wire back to the junction of the power supply reservoir capacitors is all that is required. Failure to get this right is likely to lead to crunching noises on the audio as the LED goes on and off.

The indicator presented here is unipolar − it only responds to positive excursions into the high-voltage region. A fully comprehensive monitoring system would have a similar circuit working on the negative supply side. If it is built purely as the complement of the positive monitor, it would be a quite separate circuit driving its own LED. Alternatively, it would be simple to connect the two monitor circuits together so that both activated the same LED.

This Class-G mode indicator is included in the Signal Transfer Class-G amplifier.[5]

Further Variations on Class-G

This by no means exhausts the possible variations that can be played on Class-G. For example, it is not necessary for the outer devices to operate synchronously with the inner devices. So long as they turn on in time, they can turn off much later without penalty except in terms of increased dissipation. In so-called syllabic Class-G, the outer devices turn on fast but then typically remain on for 100 msec or so to prevent glitching; see Funada and Akiya[6] for one version. Given the good results now obtainable with straight Class-G, this no longer seems a promising route to explore.

In the Fifth edition of this book, I confidently predicted: 'With the unstoppable advance of multi-channel amplifiers and powered sub-woofers, Class-G is at last coming into its own. It has recently even appeared in a Texas ADSL driver IC'.[7] Actually, as far as I can see that has not happened. Class-D seems to be significantly more popular for these applications, despite its mediocre performance in every area apart from efficiency. This seems to me to be an opportunity missed.

References

1. Self, D, *Self On Audio*, Oxford: Newnes, 2006, p. 347.

2. Sampei, T et al., Highest Efficiency and Super Quality Audio Amplifier Using MOS Power FETs in Class-G Operation, *IEEE Trans on Consumer Electronics*, Vol CE–24, #3 Aug. 1978, p. 300.

3. Feldman, L, Class-G, High Efficiency Hi-Fi Amplifier, *Radio Electronics*, Aug. 1976, p. 47.

4. Self, D, *Self On Audio*, Oxford: Newnes, 2006, p. 405, ISBN 0–7506–4765–5, p. 405.

5. The Signal Transfer Company, http://www.signaltransfer.freeuk.com/ (accessed Oct. 2012).

6. Funada, S and Akiya, H, A Study of High-Efficiency Audio Power Amplifiers Using a Voltage Switching Method, *JAES* 32(10), Oct. 1984, p. 755.

7. Wilson, J, *Zero Overhead Class-G Drivers Improve Power Efficiency in ADSL Line Cards*. Texas Instruments.

Class-D Power Amplifiers

Since the first edition of this book, Class-D amplifiers have increased enormously in popularity. This is because Class-D gives the highest efficiency of any of the amplifier classes, although the performance, particularly in terms of linearity, is not so good. They are used in active subwoofers, mobile phones, low-end home theatre systems, and sound reinforcement applications. Their penetration into general hi-fi has been small due to justified concerns over distortion performance and frequency response variations with different loudspeakers. The rapid rate of innovation means that this section of the book is much more of a snapshot of a fast-moving scene than the rest of the material. I do not want to keep repeating 'At the time of writing' as each example is introduced, so I hope you will take that as read.

Class-D amplifiers have very little in common with all the other amplifier technologies described in this book, such as Class-A, Class-B, Class-G, and Class XD. All of those have the same basic structure and use the same kinds of components, differing only in how the output stage is implemented. Class-D is quite different, and a thorough exploration of it would probably be a book of at least the size of this one. I have no plans to write it. For reason of space this chapter can only be a concise account of the most important points.

The fields of application for Class-D amplifiers can be broadly divided into two areas: low and high power outputs. The low power field reaches from a few milliwatts (for digital hearing aids) to around 5 W, while the high-power applications go from 80 W to 1400 W. At present there seems to be something of a gap in the middle, for reasons that will emerge.

The low-power area includes applications such as mobile phones, personal stereos, and laptop computer audio. These products are portable, and battery-driven, so power economy is very important. A major application of Class-D is the production of useful amounts of audio power from a single low-voltage supply rail. A good example is the National Semiconductor LM4671, a single-channel amplifier IC that gives 2.1 W into a 4 Ω speaker from a 5 V supply rail, using a 300 kHz switching frequency. This is a very low supply voltage by conventional power amplifier standards, and requires an H-bridge output structure, of which more later.

The high-power applications include PA amplifiers, home theatre systems, and big sub-woofers. These are all energised from the mains supply, so power economy is not such a high priority. Here Class-D is used because it keeps dissipation and therefore power

supply and heatsink size to a minimum, leading to a smaller and neater product. High-power Class-D amplifiers are also used in car audio systems, with power capabilities of 1000 W or more into 2 Ω; here minimising the power drain is of rather greater importance, as the capabilities of the engine-driven alternator that provides the 12 V supply are finite.

There is a middle ground between these two areas, where an amplifier is powered from the mains but of no great output power – say, a stereo unit with an output of 30 W into 8 Ω per channel. The heatsinks will be small, and eliminating them altogether will not be a great cost saving. The power supply will almost certainly be a conventional toroid-and-bridge-rectifier arrangement, and the cost savings by reducing its size by using Class-D will not be large. In this area the advantage gained by accepting the limitations of Class-D are not at present enough to justify it.

Class-D amplifiers normally come as single ICs or as chip sets with separate output stages. Since the circuitry inside these ICs is complex, and not disclosed in detail, they are not very instructive to those planning to design their own discrete Class-D amplifier.

A Bit of History

The history of the Class-D amplifier goes back, as is so often the case with technology, much further than you might think, though very little progress seems to have happened in the valve era, the combination of high switching frequencies and output transformers presenting some unenticing difficulties.

One of the earliest discussions of Class-D techniques was an article called 'Modulated Pulse AF Amplifiers' by D. R. Birt in *Wireless World* in 1963.[1] He takes us even further back by referencing a 1930 patent by B. D. Bedford, using thyratrons.[2] The Birt article did not give a complete working circuit, and the first published design for an audio Class-D amplifier that I am aware of appeared in *Wireless World* in April 1965;[3] it used six germanium transistors and gave 2 W rms into a 15 Ω loudspeaker. The second harmonic distortion alone at a quarter of a Watt output was 1%. This article was followed by two more articles analysing Class-D efficiency in 1967.[4]

The first commercial appearance of Class-D in the UK was the Sinclair X-10 module, designed by Gordon Edge, which ambitiously claimed an output of 10 W; remarkably this appeared in December 1964, *before* the *WW* articles mentioned above. It was followed by the X-20 in 1965, alleging an equally

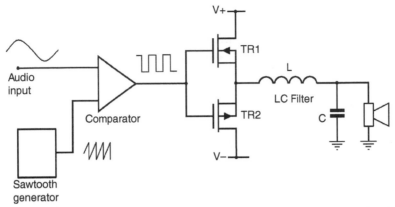

Figure 20.1. A basic Class-D amplifier with PWM comparator, FET output stage, and second-order LC output filter.

ambitious 20 W. I resurrected one of the latter in 1976, when my example proved to yield about 3 W into 8 Ω. The THD was about 5% and the rudimentary output filter did very little to keep the low switching frequency out of the load. Nonetheless, the X-20 was generally held to have a more predictable performance than the unlamented X-10.

The biggest problem of the technology at that time was that bipolar transistors of suitable power-handling capacity were too slow for the switching frequencies required; this caused serious losses that undermined the whole point of Class-D, and also produced unappealingly high levels of distortion. It was not until power FETs, with their very fast switching times, appeared that Class-D began to become a really practical proposition.

Basic Principles

Amplifiers working in Class-D differ radically from the more familiar Classes of A, B and G. In Class-D there are no output devices operating in the linear mode. Instead they are switched on and off at an ultrasonic frequency, the output being connected alternately to each supply rail. When the mark-space ratio of the input signal is varied, the average output voltage varies with it, the averaging being done by a low-pass output filter, or by the loudspeaker inductance alone. This is called Pulse Width Modulation (PWM), and is the commonest implementation, but other forms of modulation are possible and are described later. Since the output is periodically connected to the supply rails, the averaged output signal is also directly proportional to the supply voltage; there is no inherent power supply rejection (PSRR) at all with this sort of output

stage, unlike the Class-B output stage. The use of negative feedback helps with this, but its application to Class-D presents difficulties, as described later. The switching frequencies used range from 50 kHz to 1 MHz. A higher frequency makes the output filter simpler and smaller, but tends to increase switching losses and distortion.

The classic method of generating the drive signal is to use a differential comparator. One input is driven by the incoming audio signal, and the other by a sawtooth waveform at the required switching frequency. A basic Class-D amplifier is shown in Figure 20.1, and the PWM process is illustrated in Figure 20.2.

Clearly the sawtooth needs to be very linear (i.e., with constant slope) to prevent distortion being introduced at this stage. Assuming it is, the PWM process may be theoretically free from distortion, but this assumes zero switching times and no glitching of any

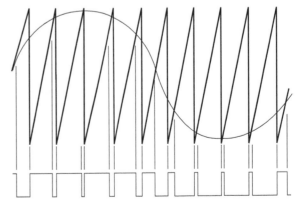

Figure 20.2. The PWM process as performed by a differential comparator.

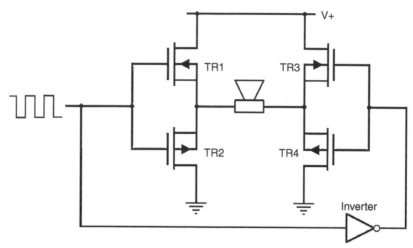

Figure 20.3. The H-bridge output configuration. The output filter is not shown.

kind in the output devices. These assumptions are very far from true in the real world, and Class-D output stages in general produce an averaged output with considerable distortion.

The PWM signal can also be generated directly from a digital input, such as SPDIF, by suitable DSP algorithms. This is not done by directly emulating the triangle-and-comparator method just described, as the time resolution available with practical clock frequencies is only a few hundredths of a switching interval, and this generates a lot of quantization distortion. To reduce this, DSP noise shaping techniques are used which transfer the quantisation artefacts into frequencies above the audible range.

When the aim is to produce as much audio power as possible from a low-voltage supply such as 5 V, the H-bridge configuration is employed, as shown in Figure 20.3. It allows twice the voltage-swing across the load, and therefore theoretically four times the output power, and also permits the amplifier to run from one supply rail without the need for bulky output capacitors of doubtful linearity. This method is also called the Bridge-Tied Load, or BTL.

The use of two amplifier outputs requires a somewhat more complex output filter. If the simple 2-pole filter of Figure 20.4a is used, the switching frequency is kept out of the loudspeaker, but the wiring to it will carry a large common-mode signal from OUT−. A balanced filter is therefore commonly used, in either the Figure 20.4b or Figure 20.4c versions. Figure 20.4d illustrates a four-pole output filter − note that you can save a capacitor. This is only used in quality applications because inductors are never cheap.

Class-D Technology

The theory of Class-D has an elegant simplicity about it; but in real life complications quickly begin to intrude.

While power FETs have a near-infinite input resistance at the gate, they require substantial current to drive them at high frequencies, because of the large device capacitances, and the gate drive circuitry is a non-trivial part of the amplifier. Power FETs, unlike bipolars, require several volts on the gate to turn them on. This means that the gate drive voltage needed for the high-side FET TR1 in Figure 20.1 is actually above the positive voltage rail. In many designs a bootstrap supply driven from the output is used to power the gate-drive circuitry. Since this supply will not be available until the high-side FET is working, special arrangements are needed at start-up. There can also be problems when the average output level is near one of the rails, as this means that one polarity of pulse will be very narrow, and the bootstrap supply may have trouble handling this. For this reason, a separate higher supply rail from the power supply is sometimes provided instead.

The more powerful amplifiers usually have external Schottky diodes connected from output to the supply rails for clamping flyback pulses generated by the inductive load. These are not merely to protect the output stage from damage, but to improve efficiency, as described in the section below.

The application of negative feedback to reduce distortion and improve supply rail rejection is complicated by the switched nature of the output waveform. Feedback can be taken from after the output filter, or alternatively taken from before it and passed through

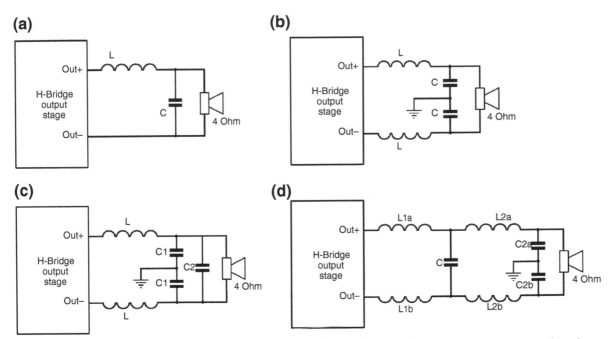

Figure 20.4. Filter arrangements for the H-bridge output. 4a is the simplest but allows a common-mode signal on the speaker cabling. 4b and 4c are the most usual versions. 4d is a 4-pole filter.

an opamp active filter to remove the switching frequency. In either case the filtering adds phase shift and limits the amount of negative feedback that can be applied while still retaining Nyquist stability.

Other enhancements that are common are selectable input gain, and facilities for synchronising the switching frequencies of multiple amplifiers to avoid audible heterodyne tones. Figure 20.5 shows a Class-D amplifier including these features.

Perhaps the gravest problem with Class-D is that you either have to use proprietary and therefore single-sourced parts, or design and build it yourself from standard components. The latter is a very serious undertaking; do not underestimate the amount of background research, the length of the design investigations and the protracted periods of optimisation that will be required before you have a reliable product with reasonable performance. In most cases the only realistic option is to use proprietary parts, which, as always, carry with them the risk that the manufacturer will suddenly disappear, leaving you well and truly in the lurch. If you are lucky, you may be able to do a last-time-buy that will give you enough time to do some high-speed redesign, but you may not be lucky, and this is the sort of thing that sinks companies.

A recent example is Tripath Technologies, who called their approach to Class-D by the name 'Class T',[5] though this was just a trademark rather than an actual class of operation. While the company had some success, their chipsets being used by Sony, Panasonic and Blaupunkt,[6] financial difficulties caused Tripath to file for Chapter 11 bankruptcy protection in February 2007, and their parts are no longer available.

Output Filters

The purpose of the output filter is to prevent radiation of switching frequencies for amplifiers that have external speaker cables, and also to improve efficiency. The inductance of a loudspeaker coil alone will in general be low enough to allow some of the switching frequency energy to pass through it to ground, causing significant losses. While some low-power integrated applications have no output filter at all, most Class-D amplifiers have a second-order LC filter between the amplifier output and the loudspeaker. In some cases a fourth-order filter is used, as in Figure 20.4d. The Butterworth alignment is usually chosen to give maximal flatness of frequency response.

As described in the chapter on real speaker loads, a loudspeaker, even a single-element one, is a long way from being a resistive load. It is therefore rather surprising that at least one manufacturer provides filter design equations that assume just that. When a Class-D amplifier is to be used with separate

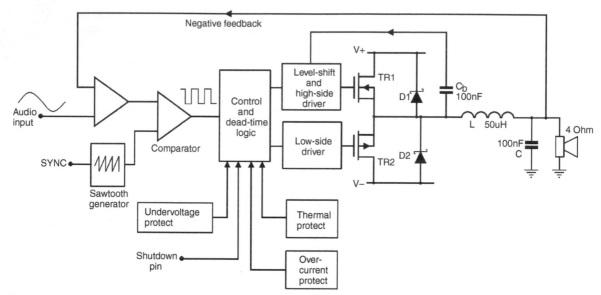

Figure 20.5. Showing the main features of a practical Class-D amplifier including Schottky clamp diodes, bootstrap supply, and one form of negative feedback.

loudspeakers of unknown impedance characteristic, the filter design can only proceed on the basis of plausible assumptions, and there are bound to be some variations in frequency response. This is a major drawback of Class-D amplifiers. Filter-loudspeaker interaction can only be dealt with properly when the loudspeaker is of a known and fixed type, as occurs in mobile phones and active sub-woofers.

Output filter capacitors need to be selected for low equivalent series resistance (ESR) and equivalent series inductance if they are to be effective. Both parameters must be included in simulation models if the results are to be realistic.

The inductor values required are typically in the region 10 μH to 50 μH, which is much larger than the 1 μH to 2 μH air-cored coils used to ensure stability with capacitive loads in Class-B amplifiers. It is therefore necessary to use ferrite-cored inductors, and care must be taken that they do not saturate at maximum output. It is important to include the series resistance, the inter-turn capacitance, and the stray capacitance to ground of filter inductors in simulation models.

The series resistance of the filter inductors, and the difficulty of applying negative feedback to the output stage, mean that the output impedance of a Class-D amplifier is usually much higher than that of a conventional design, even if the reactive impedances of the filter are not considered. It can become so high that it really does affect the damping of a loudspeaker unit.

The issue of distortion generated by the output filter inductors was investigated by Knott et al.[7] who pointed out that the filter capacitors can also produce distortion. Inductors alone were shown to give distortion levels from 0.03% to 0.20% in a 200 W Class-D amplifier.

Negative Feedback in Class-D

The thoughtful application of negative feedback to a conventional Class-B amplifier allows its distortion to be reduced to very low levels. In fact, the negative feedback loop is essential because without it the correct DC conditions in the amplifier are not established. This is not required for a typical Class-D amplifier, and negative feedback is in a sense optional. On the other hand, Class-D output stages produce significant distortion, and have effectively no supply rail rejection (PSRR), so some corrective negative feedback would be extremely welcome.

Unfortunately applying negative feedback presents some serious problems. If the feedback is taken directly from the output stage, it is still in the form of a high-frequency PWM signal, and must be averaged to obtain an audio frequency signal that can be compared with the input signal in the usual way. This method can do nothing to correct distortion generated in the output filter, which, as described in the previous section, can be substantial. Even though the feedback is taken from before the filter, it suffers some phase shift as a result of the PWM process; on average, the PWM waveform will not change until a quarter of a switching interval has passed, and this effectively delays the signal by a fixed amount dependent on the switching frequency.

If negative feedback is taken from after the output low-pass filter, it has suffered there extra phase-shift which means that the amount of feedback that can be applied without inducing HF instability is limited. The use of a higher switching frequency allows higher output filter cut-off frequencies, and so less phase-shift and larger permissible amounts of negative feedback. This will reduce distortion but increase output stage losses. Class-D is all about compromise.

Protection

All the implementations of Class-D on the market have internal protection systems to prevent excessive output currents and device temperatures.

In the published circuitry, DC offset protection is conspicuous by its absence. It is understandable that there is little enthusiasm for adding output relays to personal stereos — they might consume more power than the amplifier. However, it is surprising that they also appear to be absent from 500 W designs where

relay size and power consumption are minor issues. Are such amplifiers really that reliable?

Most Class-D systems also have undervoltage protection. If the supply voltage falls too low, then there may not be enough gate-drive voltage to turn the output FETs fully on, and they will dissipate excessive power. A lock-out circuit prevents operation below a certain voltage. A shut-down facility is almost always provided; this inhibits any switching in the output stage and allows power consumption to be very low indeed in the stand-by mode.

Efficiency

The efficiency of a Class-D amplifier is of the first importance because, quite frankly, it is the only advantage it has. At the most elementary level of theory, the efficiency is always 100%, at all output levels. In practice, of course, the mathematical idealisations do not hold, and the real-life efficiency of most implementations is between 80% and 90% over most of the power output range. At very low powers, the efficiency falls off steeply, as there are fixed losses that continue to dissipate power in the amplifier when there is no audio output at all (see Figure 20.6). When the peak-mean ratio of typical signals is considered, they spend

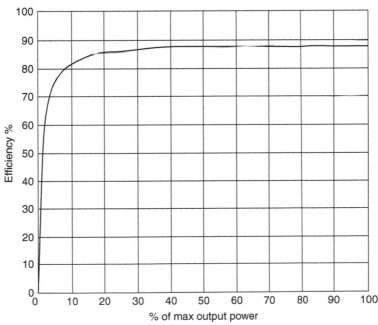

Figure 20.6. A typical efficiency curve for a Class-D amplifier driving a 4 Ω load at 1 kHz.

a small proportion of their time on the right-hand side of the plot and the overall efficiency is not likely to exceed 80%.

The losses in the output stage are due to several mechanisms. The most important are:

1. The output FETs have a non-zero resistance even when they are turned hard on. This is typically in the range 100–200 mΩ, and can double as the device temperature increases from 0 to 150°C, the latter being the usual maximum operating temperature. This resistance causes I^2R losses.

2. The output devices have non-zero times for switching on and off. In the period when the FET is turning on or turning off, it has an intermediate value of resistance which again causes I^2R losses. It is essential to minimise the stray inductance in the drain and source circuits, as this not only extends the switching times but also causes voltage transients at turn-off that can overstress the FETs.

3. The flyback pulses generated by an inductive load can cause conduction of the parasitic diodes that are part of the FET construction. These diodes have relatively long reverse recovery times and more current will flow than is necessary. To prevent this, many Class-D designs have Schottky clamp diodes connected between the output line and supply rails, as in Figure 20.5. These turn on at a lower voltage than the parasitic FET diodes and deal with the flyback pulses. They also have much faster reverse recovery times.

4. Last, and perhaps most dangerous, is the phenomenon known as 'shoot-through'. This refers to the situation when one FET has not stopped conducting before the other starts. This gives rise to an almost direct short between the supply rails; its duration is very brief, but large amounts of unwanted heating can occur, and efficiency will suffer badly. To prevent this, the gate-drive to the FET that is about to be turned on is slightly delayed, by a 'dead-time' circuit. The introduction of dead-time increases distortion, as the average output voltage is no longer proportional to the pulse width, and so only the minimum is applied. A 40 nsec dead-time is sufficient to create more than 2% THD in a 1 kHz sinewave.

Because the amount of dead-time used is measured in nanoseconds, and has to steer a very narrow course between excessive distortion and unacceptable shoot-through, setting it correctly presents formidable difficulties. If it is to be controlled to within a nanosecond, a digital system to do this would need at least a 1 GHz clock. While clock frequencies of 1 GHz and greater are common place in PC processors, this has only been achieved by a considerable investment in research time and money. Making chips this fast to be sold in much smaller numbers is a challenge in economics.

The problem of determining the correct amount of dead-time (which may vary between different samples of output devices) can be addressed by using adaptive dead-time control. If shoot-through pulses occur, they can be detected by monitoring the voltage across a very low value resistor, and readily distinguished from normal pulses as they are very narrow and of high amplitude. It is therefore possible to use a control loop so that the dead-time is maintained at just enough to stop shoot-through, continuously adjusted to cope with varying conditions such as output device temperature. This seems like a very good idea.

Alternative Modulation Systems

What might be called 'classic' Class-D uses pulse-width modulation (PWM) with a constant frequency. Another technique, called a self-oscillating loop, dispenses with an oscillator to provide the switching frequency; instead there is a feedback path taken from before the output filter which feeds an integrator; the integrator defines the oscillation frequency, and a second input to its summing point allows the input signal to modulate both the mark-space ratio and the switching frequency.

Another way to create the required waveform is a delta-sigma modulator,[8] as used in ADC and DAC chips. Their output is a form of pulse density modulation. A first-order delta-sigma modulator has a feedback loop closed around an integrator and a one-bit quantiser (i.e. a comparator). Higher-order delta-sigma modulators have more than one integrator in the loop, and this allows them to be configured for noise-shaping, with the quantisation distortion reduced in the audio band and transferred to ultrasonic frequencies. Higher-order modulators can be unstable if incorrectly designed.

The Tripath chips used a digitisation loop similar to a higher order delta-sigma (or sigma-delta) modulator, and it was claimed that this allowed a higher loop gain, and therefore lower distortion, at high audio frequencies than would be possible in a conventional single-pole amplifier.

Class-D Examples

This chapter has inevitably been mostly theoretical, but you can get a feel for current Class-D technology by examining some of the many examples out there. At the time of writing, these are all worth looking at:

Bang & Olufsen sell a series of Class-D modules called ICEpower.[9] This is one of the few Class-D solutions with which I have some personal experience. A company with which I was associated used them in an active sub-woofer; as usual, the loudspeaker itself was very inefficient because of the need to get deep bass out of a reasonably small box. A considerable amount of amplifier power was therefore required, but only below 200 Hz. This is an ideal application for a Class-D amplifier as the loudspeaker unit is a fixed and known quantity, and the output filtering requirements are relaxed because there is no need to worry about keeping the switching frequency out of the tweeter. The bought-in amplifier module incorporated a universal-mains switch-mode power supply. I was not involved in detailed performance testing, but I can say that the modules were reliable. And that is no small thing in a power amplifier. THD is well controlled at lower powers. As an example, the ICEpower 80AM2 module, rated at 80 W/4 Ω stereo or 160 W/8 Ω bridged mono, has THD below 0.01% up to 40 W, but … after that, it rises steeply, exceeding 1% at 55 W, and hitting 10% at 70 W. These are the manufacturer's figures.

Texas make a series of Purepath single-chip Class-D amplifiers.[10] These range from the TAS5261 (315 W/4 Ω mono)[11] and TAS5162 (210 W/4 Ω stereo)[12] to the TAS5132 (20 W/4 Ω stereo)[13] at the other end of the power spectrum. Typical THD is 0.2% at maximum output. All the power output powers are spec'd at 10% THD, which is usual in this part of the market. The assumption is that anyone so dead to all sense of decency as to play gangsta rap on his mobile will be quite happy with 10% THD; one manufacturer notes that 30% THD is unacceptable, which I am relieved to hear.

National Semi also make Class-D amplifier chips, an example being the LM4675 (2006) in the 'Boomer' series. This is a mono low power (2.6 W/4 Ω) device

intended for mobile phones and other portable devices. It is billed as requiring no output filter with inductive loads, such as loudspeakers, and to accomplish this, it uses spread-spectrum PWM. The switching frequency varies around a 300 kHz centre frequency, spreading the HF energy over a wide bandwidth. The cycle-to-cycle variation of the switching period is said not to affect the efficiency. It comes in a 9-bump micro-SMD package or a Leadless Leadframe Package (LLP). The THD is 0.03% at 1 W/8 Ω. Once again, the output power is specified at 10% THD.

Another example of Class-D that makes use of spread-spectrum PWM is the Maxim MAX9744 20 W/4 Ω stereo amplifier.[14] It can be switched to fixed-frequency PWM and has a built-in 64-step volume control.

Microsemi make the AudioMAX™ LX1721 (mono) and LX1722 (stereo) Class-D controller chips,[15] which contain all the circuitry except the output devices and filter components. They drive two pairs of external power FETs in an H-bridge, and come in a 44-Pin QSOP Package.

This is by no means a comprehensive survey but I hope it is representative. It is notable that in every case the distortion performance is much worse than a Blameless Class-B amplifier, which will give less than 0.0005% THD at 1 kHz, right up to the point at which hard clipping begins. Despite the progress that has been made, Class-D is still about efficiency, not performance.

Further Development

Work on the many issues of Class-D continues apace. To get an insight into the wide range of research activities, see the *Proceedings of the AES 27th International Conference*. 'Efficient Audio Power Amplification', in 2005[16] and *Proceedings of the AES 37th International Conference*, 'Class-D Audio Amplification', in 2009.[17] To date, these are the only AES conferences that were devoted to Class-D.

References

1. Birt D R, Modulated Pulse AF Amplifiers, *Wireless World*, Feb. 1963, p.76.

2. B D Bedford, Patent 389,855, Granted 25 Sept. 1931.

3. Turnbull and Townsend, A Feedback Pulse Width Modulated Audio Amplifier, *Wireless World*, Apr. 1965, p. 160.

4. Turnbull and Townsend, Efficiency Considerations in a Class-D Amplifier, Parts 1 and 2, *Wireless World*, May, June 1967.

5. Hickman I, Is Class-T Hi-fi? *Electronics World*, April 2000, p. 274.

6. Wikipedia http://en.wikipedia.org/wiki/Class_T_amplifier (accessed Oct. 2012).

7. Knott et al., Modelling Distortion Effects in Class-D Amplifier Filter Inductors, paper presented at 128th AES Convention, London, May 2010, Paper 7997 P6-2.

8. Wikipedia http://en.wikipedia.org/wiki/Delta-sigma_modulation (accessed Oct. 2012).

9. Bang & Olufsen http://www.icepower.bang-olufsen.com/ (accessed Oct. 2012).

10. Texas: http://www.ti.com/general/docs/lit/getliterature.tsp?literatureNumber=slyt312&fileType=pdf (accessed Oct. 2012).

11. Texas http://www.ti.com/corp/docs/landing/tas5261/indextracking.htm (accessed Oct. 2012).

12. Texas http://www.ti.com/lit/ds/symlink/tas5162.pdf (accessed Oct. 2012).

13. Texas http://www.ti.com/general/docs/lit/getliterature.tsp?genericPartNumber=tas5132&fileType=pdf (accessed Oct. 2012).

14. Maxim http://www.maximintegrated.com/datasheet/index.mvp/id/5178 (accessed Oct. 2012).

15. Microsemi http://www2.microsemi.com/datasheets/lx1721_22.pdf (accessed Oct. 2012).

16. AES, *Proceedings of the AES 27th International Conference*. Efficient Audio Power Amplification, Copenhagen, Denmark, September 2–4 2005 (17 papers).

17. AES, Class-D Audio Amplification, in *Proceedings of the AES 37th International Conference*. Copenhagen, Denmark, August 28–30 2009 (18 papers).

Chapter 21

FET Output Stages

The Characteristics of Power FETS

A field-effect transistor (FET) is essentially a voltage-controlled device. So are bipolar junction transistors (BJTs), despite a wrong-headed minority that persists in regarding them as current-controlled. They really are not, even if BJT base currents do happen to be non-negligible.

The power FETs normally used are enhancement devices — in other words, with no voltage between gate and source, they remain off. In contrast, the junction FETs found in small-signal circuitry are depletion devices, requiring the gate to be taken negative of the source (for the most common N-channel devices) to reduce the drain current to usable proportions. (Please note that the standard information on FET operation is in many textbooks and will not be repeated here.)

Power FETs have large internal capacitances, both from gate to drain, and from gate to source. The gate-source capacitance is effectively bootstrapped by the source-follower configuration, but the gate-drain capacitance, which can easily total 2000 pF, remains to be driven by the previous stage. There is an obvious danger that this will compromise the amplifier slew-rate if the VAS is not designed to cope.

FETs tend to have much larger bandwidths than BJT output devices. My own experience is that this tends to manifest itself as a greater propensity for parasitic oscillation rather than anything useful, but the tempting prospect of higher global NFB factors due to a higher output stage pole remains. The current state of knowledge does not yet permit a definitive judgement on this.

A great deal has been said on the thermal coefficients of the Vbias voltage. It is certainly true that the temp coefficient at high drain currents is negative — in other words, drain current falls with increasing temperature — but, on the other hand, the coefficient reverses sign at low drain currents, and this implies that precise quiescent-current setting will be very difficult. A negative-temperature coefficient provides good protection against thermal runaway, but this should never be a problem anyway.

FET versus BJT Output Stages

On beginning any power amplifier design, one of the first decisions that must be made is whether to use BJTs or FETs in the output stage. This decision may of course already have been taken for you by the marketing department, as the general mood of the marketplace is that if FETs are more expensive, they must be better. If, however, you are lucky enough to have this crucial decision left to you, then FETs normally disqualify themselves on the same grounds of price. If the extra cost is not translated into either better performance and/or a higher sustainable price for the product, then it appears to be foolish to choose anything other than BJTs.

Power MOSFETS are often hailed as the solution to all amplifier problems, but they have their own drawbacks, not the least being low transconductance, poor linearity, and a high on-resistance that makes output efficiency mediocre. The high-frequency response may be better, implying that the second pole P2 of the amplifier response will be higher, allowing the dominant pole P1 to be raised with the same stability margin, and so in turn giving more NFB to reduce distortion. However, we would need this extra feedback (if it proves available in practice) to correct the worse open-loop distortion, and even then the overall linearity would almost certainly be worse. To complicate matters, the compensation cannot necessarily be lighter because the higher output-resistance makes more likely the lowering of the output pole by capacitive loading.

The extended FET frequency response is, like so many electronic swords, two-edged if not worse, and the HF capabilities mean that rigorous care must be taken to prevent parasitic oscillation, as this is often promptly followed by an explosion of disconcerting violence. FETs should at least give freedom from switch-off troubles (Distortion 3c) as they do not suffer from BJT charge-storage effects.

Advantages of FETs

1. For a simple complementary FET output stage, drivers are not required. This is somewhat negated by the need for gate-protection Zener diodes.
2. There is no second-breakdown failure mechanism. This may simplify the design of overload protection systems, especially when arranging for them to cope with highly reactive loads.
3. There are no charge-storage effects to cause switch-off distortion.

Disadvantages of FETs

1. Linearity is very poor by comparison with a BJT degenerated to give the same transconductance. The Class-B conduction characteristics do not cross over smoothly, and there is no equivalent to the optimal Class-B bias condition that is very obvious with a BJT output stage.

2. The Vgs required for conduction is usually of the order of 4–6 V, which is much greater than the 0.6–0.8 V required by a BJT for base drive. This greatly reduces the voltage efficiency of the output stage unless the preceding small-signal stages are run from separate and higher-voltage supply rails.

3. Power FETs draw negligible DC current through their gate connections, but they have high internal capacitances which must be charged and discharged rapidly for high-frequency operation. This often requires extra complications in the driver circuitry to provide these large currents at low distortion.

4. The minimum channel resistance of the FET, known as Rds(on), is high and gives a further reduction in efficiency compared with BJT outputs.

5. Power FETs are liable to parasitic oscillation. In severe cases, a plastic-package device will literally explode. This is normally controllable in the simple complementary FET output stage by adding gate-stopper resistors, but is a serious disincentive to trying radical experiments in output stage circuit design.

6. Some commentators claim that FET parameters are predictable; I find this hard to understand as they are notorious for being anything but. From one manufacturer's data (Harris), the Vgs for the IRF240 FET varies between 2.0 and 4.0 V for an Id of 250 μA; this is a range of two to one. In contrast, the Vbe/Ic relation in bipolars is fixed by a mathematical equation for a given transistor type, and is much more reliable. Nobody uses FETs in log converters.

7. Since the Vgs spreads are high, this will complicate placing devices in parallel for greater power capability. Paralleled BJT stages rarely require current-sharing resistors of greater than 0.1 Ω, but for the FET case they may need to be a good deal larger, reducing efficiency further.

8. At the time of writing, there is a significant economic penalty in using FETs. Taking an amplifier of given power output, the cost of the output semiconductors is increased by between 1.5 and 2 times with FETs.

IGBTs

Insulated-Gate Bipolar Transistors (IGBTs) represent a relatively new option for the amplifier designer. They have been held up as combining the best features of FETs and BJTs. In my view, this is a dubious proposition as I find the advantages of FETs for audio to be heavily outweighed by the drawbacks, and if IGBTs do have any special advantages, they have not so far emerged.

An IGBT consists of an FET driving a bipolar transistor,[1] both fabricated in the same junction structure. They got off to a bad start as this structure originally contained a parasitic transistor which acted as an auto-destruct system, turning the device fully on when triggered by high currents. This little difficulty appears to have been overcome, and they are now much used for electric motor switching and pulse applications.

Notwithstanding this, at least one IGBT design has been published for construction.[2] The output stage of this 90 W/8 Ω design is a hybrid (described below) with BJT drivers and IGBT output devices, arranged to give gain in the output stage, which is unusual. Interestingly, there are very few circuit changes from a 60 W/8 Ω version using HEXFETs for output devices, and also having gain in the output stage. The design appears to work reasonably well. There are also a few amateur designs circulating on the internet.

I have no personal information on the linearity of IGBTs, but the combination of FETs and BJTs does not sound promising. At present, the use of IGBTs in an amplifier is still somewhat a step into the unknown. They appear to have all the problems of MOSFETs relating to complicated temperature coefficients, complicating the maintenance of the desired bias conditions. On the whole, they do not seem to be a promising direction to pursue for audio purposes.

Power FET Output Stages

Three types of FET output stage are shown in Figure 21.1, and Figures 21.2–21.5 show SPICE gain plots, using 2SK135/2SJ50 devices. Most FET amplifiers use the simple source-follower configuration in Figure 21.1a; the large-signal gain plot at Figure 21.2 shows that the gain for a given load is lower (0.83 rather than 0.97 for bipolar, at 8 Ω) because of low gm, and this, with the high on-resistance, reduces output efficiency seriously. Open-loop distortion is markedly higher; however, LSN does not increase with heavier loading, there being no equivalent of 'Bipolar Gain-Droop'. The crossover region has sharper and larger gain deviations than a bipolar stage, and generally looks pretty nasty; Figure 21.3 shows the impossibility of finding a correct Vbias setting.

Figure 21.1b shows a hybrid (i.e., bipolar/FET) quasi-complementary output stage, first described by me.[3] This topology is intended to maximise economy rather than performance, once the decision has been made (presumably for marketing reasons) to use FETs,

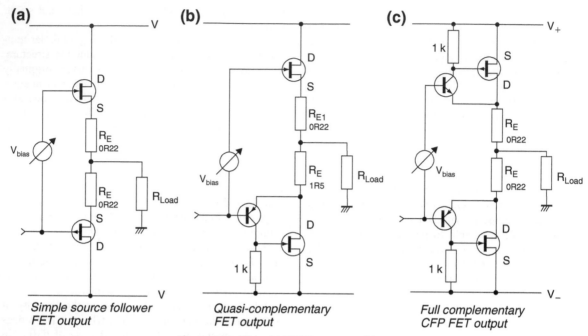

Figure 21.1. Three MOSFET output architectures.

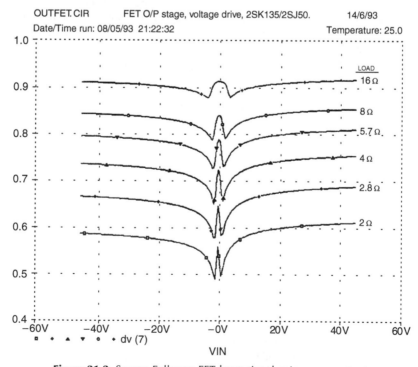

Figure 21.2. Source-Follower FET large-signal gain versus output.

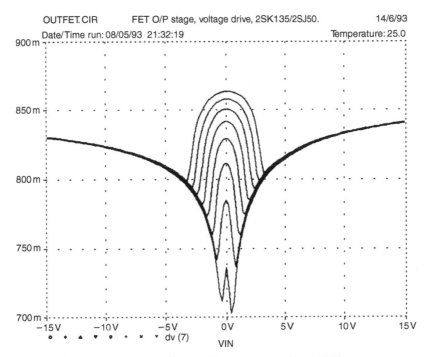

Figure 21.3. Source-Follower FET crossover region ±15 V range.

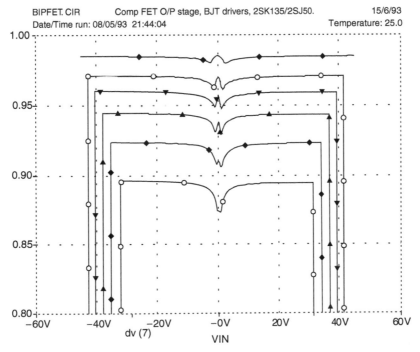

Figure 21.4. Complementary Bipolar-FET gain versus output.

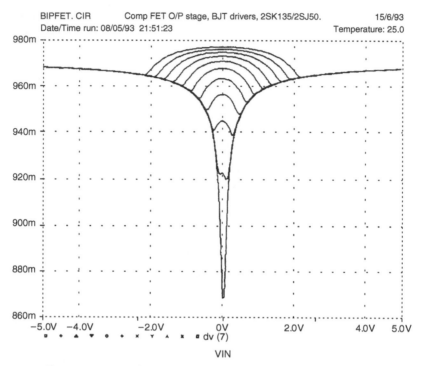

BIPFET. CIR   Comp FET O/P stage, BJT drivers, 2SK135/2SJ50.   15/6/93
Date/Time run: 08/05/93 21:51:23   Temperature: 25.0

Figure 21.5. Complementary BJT-FET crossover region ±15 V range.

by making both output devices cheap N-channel devices; complementary MOSFET pairs remain relatively rare and expensive. The basic configuration is badly asymmetrical, the hybrid lower half having a higher and more constant gain than the source-follower upper half. Increasing the value of Re2 gives a reasonable match between the gains of the two halves, but leaves a daunting crossover discontinuity. To the best of my knowledge, this idea has not caught on, which is probably a good thing.

The hybrid full-complementary stage in Figure 21.1c was conceived[4] to maximise FET performance by linearising the output devices with local feedback and reducing Iq variations due to the low power dissipation of the bipolar drivers. It is very linear, with no gain-droop at heavier loadings (Figure 21.4), and promises freedom from switch-off distortions; however, as shown, it is rather inefficient in voltage-swing. The crossover region in Figure 21.5 still has some unpleasant sharp corners, but the total crossover gain deviation (0.96–0.97 at 8 Ω) is much smaller than for the quasi-hybrid (0.78–0.90) and so less high-order harmonic energy is generated.

Table 21.1 summarises the SPICE curves for 4 Ω and 8 Ω loadings. Each was subjected to Fourier analysis to calculate THD% results for a ±40 V input. The BJT results from earlier chapters are included for comparison.

Power FETs and Bipolars: the Linearity Competition

There has been much debate as to whether power FETs or bipolar junction transistors (BJTs) are superior in power amplifier output stages, e.g., Hawtin.[5] As the debate rages, or at any rate flickers, it has often been flatly stated that power FETs are more linear than BJTs, usually in tones that suggest that only the truly benighted are unaware of this.

In audio electronics it is a good rule of thumb that if an apparent fact is repeated times without number, but also without any supporting data, it needs to be looked at very carefully indeed. I therefore present my own view of the situation here.

I suggest that it is now well established that power FETs when used in conventional Class-B output stages are a good deal less linear than BJTs. The gain-deviations around the crossover region are far more severe for FETs than the relatively modest wobbles of correctly biased BJTs, and the shape of the FET

Table 21.1. THD percentages and average gains for various types of output stage, for 8 and 4 ohm loading

	Emitter-Follower	CFP	Quasi Simple	Quasi Bax	Triple Type 1	Simple MOSFET	Quasi MOSFET	Hybrid MOSFET
8Ω (%)	0.031	0.014	0.069	0.050	0.13	0.47	0.44	0.052
Gain	0.97	0.97	0.97	0.96	0.97	0.83	0.84	0.97
4Ω (%)	0.042	0.030	0.079	0.083	0.60	0.84	0.072	0.072
Gain	0.94	0.94	0.94	0.94	0.92	0.72	0.73	0.94

gain-plot is inherently jagged, due to the way in which two square law devices overlap. The incremental gain range of a simple FET output stage is 0.84−0.79 (range 0.05) and this is actually much greater than for the bipolar stages examined earlier; the EF stage gives 0.965−0.972 into 8 Ω (range 0.007) and the CFP gives 0.967−0.970 (range 0.003). The smaller ranges of gain-variation are reflected in the much lower THD figures when PSpice data is subjected to Fourier analysis.

However, the most important difference may be that the bipolar gain variations are gentle wobbles, while all FET plots seem to have abrupt changes that are much harder to linearise with NFB that must decline with rising frequency. The basically exponential Ic/Vbe characteristics of two BJTs approach much more closely the ideal of conjugate (i.e., always adding up to 1) mathematical functions, and this is the root cause of the much lower crossover distortion.

A close-up examination of the way in which the two types of device begin conducting as their input voltages increase shows that FETs move abruptly into the square law part of their characteristic, while the exponential behaviour of bipolars actually gives a much slower and smoother start to conduction.

Similarly, recent work[6] shows that less conventional approaches, such as the CC-CE configuration of Mr Bengt Olsson,[7] also suffer from the non-conjugate nature of FETs, and show sharp changes in gain. Gevel[8] shows that this holds for both versions of the stage proposed by Olsson, using both N- and P-channel drivers. There are always sharp gain-changes.

FETs in Class-A Stages

It occurred to me that the idea that FETs are more linear was based not on Class-B power-amplifier applications, but on the behaviour of a single device in Class-A. It might be argued that the roughly square law nature of an FET's Id/Vgs law is intuitively more 'linear' than

the exponential Ic/Vbe law of a BJT, but it is a bit difficult to know quite how to define 'linear' in this context. Certainly a square law device will generate predominantly low-order harmonics, but this says nothing about the relative amounts produced.

In truth, the BJT/FET contest is a comparison between apples and aardvarks, the main problem being that the raw transconductance (gm) of a BJT is far higher than for any power FET. Figure 21.6 illustrates the conceptual test circuit; both a TO3 BJT (MJ802) and a power-FET (IRF240) have an increasing DC voltage Vin applied to their base/gate, and the resulting collector and drain currents from PSpice simulation are plotted in Figure 21.7. Voffset is used to increase the voltage applied to FET M1 by 3.0 V because nothing much happens below Vgs = 4 V, and it is helpful to have the curves on roughly the same axis. Curve A, for the BJT, goes almost vertically skywards, as a result of its far higher gm. To make the comparison meaningful, a small amount of local negative feedback is added to Q1 by Re, and as this emitter degeneration is increased from 0.01 Ω to 0.1 Ω, the Ic curves become closer in slope to the Id curve.

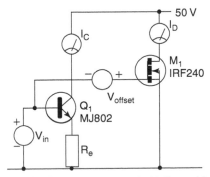

Figure 21.6. The linearity test circuit. Voffset adds 3 V to the DC level applied to the FET gate, purely to keep the current curves helpfully adjacent on a graph.

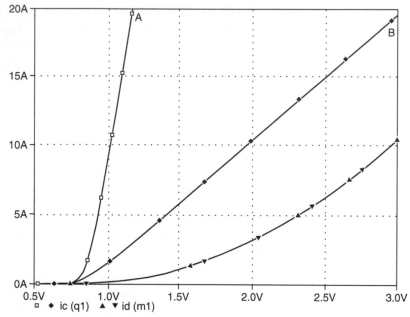

Figure 21.7. Graph of Ic and Id for the BJT and the FET. Curve A shows Ic for the BJT alone, while Curve B shows the result for Re = 0.1 Ω. The curved line is the Id result for a power FET without any degeneration.

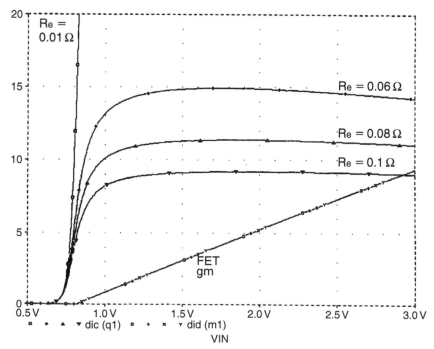

Figure 21.8. Graph of transconductance versus input voltage for BJT and FET. The near-horizontal lines are BJT gm for various Re values.

Because of the curved nature of the FET Id plot, it is not possible to pick an Re value that allows very close gm equivalence; Re = 0.1 Ω was chosen as a reasonable approximation; see Curve B. However, the important point is that I think no one could argue that the FET Id characteristic is more linear than Curve B.

This is made clearer by Figure 21.8, which directly plots transconductance against input voltage. There is no question that FET transconductance increases in a beautifully linear manner — but this 'linearity' is what results in a square law Id increase. The near-constant gm lines for the BJT are a much more promising basis for the design of a linear amplifier.

To forestall any objections that this comparison is all nonsense because a BJT is a current-operated device, I add here a reminder that this is completely untrue. The BJT is a voltage-operated device, and the base current that flows is merely an inconvenient side-effect of the collector current induced by said base voltage. This is why beta varies more than most BJT parameters; the base current is an unavoidable error rather than the basis of transistor operation.

The PSpice simulation shown here was checked against manufacturers' curves for the devices, and the agreement was very good — almost unnervingly so. It therefore seems reasonable to rely on simulator output for these kinds of studies — it is certainly infinitely quicker than doing the real measurements, and the comprehensive power FET component libraries that are part of PSpice allow the testing to be generalised over a huge number of component types without actually buying them.

To conclude, I think it is probably irrelevant to simply compare a naked BJT with a naked FET. Perhaps the vital point is that a bipolar device has much more raw transconductance gain to begin with, and this can be handily converted into better linearity by local feedback, i.e., adding a little emitter degeneration. If the transconductance is thus brought down roughly to FET levels, the bipolar has far superior large-signal linearity. I must admit to a sneaking feeling that if practical power BJTs had come along after FETs, they would have been seized upon with glee as a major step forward in power amplification.

References

1. Langdon, S, Audio amplifier designs using IGBTs, MOSFETs, and BJTs, Toshiba Application Note X3504, Vol. 1, Mar. 1991.

2. Build Your Own High-end Audio Equipment, *Elektor Electronics*, 1995, p. 57.

3. Self, D, Sound MOSFET Design, *Electronics* and Wireless *World*, Sept. 1990, p. 760.

4. Self, D, MOSFET Audio Output, Letter, Electronics and Wireless *World*, May 1989, p. 524 (see also note 2 above).

5. Hawtin, V, Letters, *Electronics World*, Dec. 1994, p. 1037.

6. Sclf, D, Two-Stage Amplifiers and the Olsson Output Stage, *Electronics World*, Sept. 1995, p. 762.

7. Olsson, B, Better Audio from Non-Complements? *Electronics World*, Dec. 1994, p. 988.

8. M Gevel, Private communication, Jan. 1995.

Chapter 22

Thermal Compensation and Thermal Dynamics

Why Quiescent Conditions are Critical

In earlier sections of this book we looked closely at the distortion produced by amplifier output stages, and it emerged that a well-designed Class-B amplifier with proper precautions taken against the easily fixed sources of non-linearity, but using basically conventional circuitry, can produce startlingly low levels of THD. The distortion that actually is generated is mainly due to the difficulty of reducing high-order crossover non-linearities with a global negative-feedback factor that declines with frequency; for 8 Ω loads, this is the major source of distortion, and unfortunately crossover distortion is generally regarded as the most pernicious of non-linearities. For convenience, I have chosen to call such an amplifier, with its small signal stages freed from unnecessary distortions, but still producing the crossover distortion inherent in Class-B, a Blameless amplifier (see Chapter 4).

Chapter 9 suggests that the amount of crossover distortion produced by the output stage is largely fixed for a given configuration and devices, so the best we can do is ensure the output stage runs at optimal quiescent conditions to minimise distortion.

Since it is our only option, it is therefore particularly important to minimise the output- stage gain irregularities around the crossover point by holding the quiescent conditions at their optimal value. This conclusion is reinforced by the finding that for a Blameless amplifier increasing quiescent current to move into Class-AB makes the distortion worse, not better, as gm-doubling artefacts are generated. In other words, the quiescent setting will only be correct over a relatively narrow band, and THD measurements show that too much quiescent current is as bad (or at any rate very little better) than too little.

The initial quiescent setting is simple, given a THD analyser to get a good view of the residual distortion; simply increase the bias setting from minimum until the sharp crossover spikes on the residual merge into the noise. Advancing the pre-set further produces edges on the residual that move apart from the crossover point as bias increases; this is gm-doubling at work, and is a sign that the bias must be reduced again.

It is easy to attain this optimal setting, but keeping it under varying operating conditions is a much greater problem because quiescent current (Iq) depends on the maintenance of an accurate voltage-drop Vq across emitter resistors Re of tiny value, by means of hot transistors with varying Vbe drops. It's surprising it works as well as it does.

Some kinds of amplifier (e.g., Class-A or current-dumping types) manage to evade the problem altogether, but in general the solution is some form of thermal compensation, the output-stage bias voltage being set by a temperature-sensor (usually a Vbe-multiplier transistor) coupled as closely as possible to the power devices.

There are inherent inaccuracies and thermal lags in this sort of arrangement, leading to program-dependency of Iq. A sudden period of high power dissipation will begin with the Iq increasing above the optimum, as the junctions will heat up very quickly. Eventually the thermal mass of the heatsink will respond, and the bias voltage will be reduced. When the power dissipation falls again, the bias voltage will now be too low to match the cooling junctions and the amplifier will be under-biased, producing crossover spikes that may persist for some minutes. This is very well illustrated in an important paper by Sato et al.[1]

The output device, thermal coupler (if any) and the heatsink should so far as possible form a symmetrical system. If one end of a heatsink gets significantly hotter than the other, the output device junctions will be at different temperatures, and compensating them with a single thermal sensor will be much more difficult.

As explained in Chapter 9, crossover distortion is reduced when more devices are used in parallel. This has another beneficial effect, as it means that the junction temperature rises are less, as the dissipated power is divided among more devices, and temperature compensation becomes that much easier.

Accuracy Required of Thermal Compensation

Quiescent stability depends on two main factors. The first is the stability of the Vbias generator in the face of external perturbations, such as supply voltage variations. The second and more important is the effect of temperature changes in the drivers and output devices, and the accuracy with which Vbias can cancel them out.

Vbias must cancel out temperature-induced changes in the voltage across the transistor base-emitter junctions, so that Vq remains constant. From the limited viewpoint of thermal compensation (and given a fixed Re) this is very much the same as the traditional criterion that the quiescent current must remain constant, and no relaxation in exactitude of setting is permissible.

I have reached some conclusions on how accurate the Vbias setting must be to attain minimal distortion. The two major types of output stage, the EF and the CFP, are quite different in their behaviour and bias requirements, and this complicates matters considerably. The results are approximate, depending partly on visual

Table 22.1. Vbias tolerance for 8 Ω

		EF output	CFP output
Crossover spikes obvious	Underbias	2.25 V	1.242 V
Spikes just visible	Underbias	2.29	1.258
Optimal residual	Optimal	2.38	1.283
gm-doubling just visible	Overbias	2.50	1.291
gm-doubling obvious	Overbias	2.76	1.330

assessment of a noisy residual signal, and may change slightly with transistor type, etc. Nonetheless, Table 22.1 gives a much-needed starting point for the study of thermal compensation.

From these results, we can take the permissible error band for the EF stage as about ±100 mV, and for the CFP as about ±10 mV. This goes some way to explaining why the EF stage can give satisfactory quiescent stability despite its dependence on the Vbe of hot power transistors.

Returning to the PSpice simulator, and taking Re = 0R1, a quick check on how the various transistor junction temperatures affect Vq yields:

• The EF output stage has a Vq of 42 mV, with a Vq sensitivity of −2 mV/°C to driver temperature, and −2 mV/°C to output junction temperature. No surprises here.

• The CFP stage has a much smaller Vq. (3.1 mV) Vq sensitivity −2 mV/°C to driver temperature, and only −0.1 mV/°C to output device temperature. This confirms that local NFB in the stage makes Vq relatively independent of output device temperature, which is just as well as Table 22.1 shows it needs to be about ten times more accurate.

The CFP output devices are about 20 times less sensitive to junction temperature, but the Vq across Re is something like 10 times less; hence the actual relationship between output junction temperature and crossover distortion is not so very different for the two configurations, indicating that, as regards temperature stability, the CFP may only be twice as good as the EF, and not vastly better, which is perhaps the common assumption. In fact, as will be described, the CFP may show poorer thermal performance in practice.

In real life, with a continuously varying power output, the situation is complicated by the different dissipation characteristics of the drivers as output varies. See Figure 22.1, which shows that the CFP driver dissipation is more variable with output, but on average runs cooler. For both configurations, driver temperature is equally important, but the EF driver dissipation does not vary much with output power, though the initial drift at switch-on is greater as the standing dissipation is higher. This, combined with the two-times-greater sensitivity to output device temperature and the greater self-heating of the EF output devices,

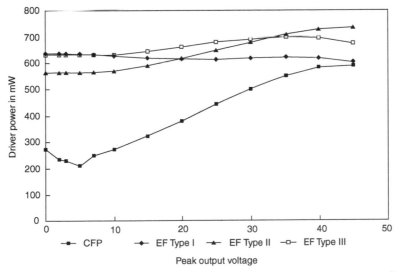

Figure 22.1. Driver dissipation versus output level. In all variations on the EF configuration, power dissipation varies little with output; CFP driver power, however, varies by a factor of two or more.

may be the real reason why most designers have a general feeling that the EF version has inferior quiescent stability. The truth as to which type of stage is more thermally stable is much more complex, and depends on several design choices and assumptions.

Having assimilated this, we can speculate on the ideal thermal compensation system for the two output configurations. The EF stage has Vq set by the subtraction of four dissimilar base-emitter junctions from Vbias, all having an equal say, and so all four junction temperatures ought to be factored into the final result. This would certainly be comprehensive, but four temperature-sensors per channel is perhaps overdoing it. For the CFP stage, we can ignore the output device temperatures and only sense the drivers, which simplifies things and works well in practice.

If we can assume that the drivers and outputs come in complementary pairs with similar Vbe behaviour, then symmetry prevails and we need only consider one half of the output stage, so long as Vbias is halved to suit. This assumes the audio signal is symmetrical over time-scales of seconds to minutes, so that equal dissipations and temperature rises occur in the top and bottom halves of the output stage. This seems a pretty safe bet, but, as one example, the unaccompanied human voice has positive and negative peak values that may differ by up to 8 dB, so prolonged acapella performances have at least the potential to mislead any compensator that assumes symmetry. One amplifier that does use separate sensors for the upper and lower output sections is the Adcom GFA-565 (introduced 1991).

For the EF configuration, both drivers and outputs have an equal influence on the quiescent Vq, but the output devices normally get much hotter than the drivers, and their dissipation varies much more with output level. In this case the sensor goes on or near one of the output devices, thermally close to the output junction. It has been shown experimentally that the top of the TO3 can is the best place to put it; see Figure 22.4. Recent experiments have confirmed that this holds true also for the TO3P package (a large flat plastic package like an overgrown TO220, and nothing at all like TO3) which can easily get 20° hotter on its upper plastic surface than does the underlying heatsink.

Since the first edition of this book, the TO3 has all but disappeared, being replaced in high-power applications by the MT200 package. This is a large flat plastic format like a wider version of the TO3P package. Once again, the top of the package gets hotter than the adjacent heatsink.

In the CFP the drivers have most effect and the output devices, although still hot, have only one-twentieth the

influence on the quiescent conditions. Driver dissipation is also much more variable, so now the correct place to put the thermal sensor is as near to the driver junction as you can get it.

Schemes for the direct servo control of quiescent current have been mooted,[2] but all suffer from the difficulty that the quantity we wish to control is not directly available for measurement, as except in the complete absence of signal, it is swamped by Class-B output currents. Nevertheless, several designs have been put forward in which the quiescent current is controlled by measuring the average current in the output stage. While this could be made to apparently work with steady test signals, it seems inevitable that there would be serious bias errors with a dynamic music signal.

In contrast, the quiescent current of a Class-A amplifier is easily measured, allowing very precise feedback control; ironically its value is not critical for distortion performance.

So, how accurately must quiescent current be held? This is not easy to answer, not least because it is the wrong question. Chapter 9 established that the crucial parameter is not quiescent current (hereafter Iq) as such, but rather the quiescent voltage-drop Vq across the two emitter resistors Re. This takes a little swallowing — after all, people have been worrying about quiescent current for 30 years or more — but it is actually good news, as the value of Re does not complicate the picture. The voltage across the output stage inputs (Vbias) is no less critical, for once Re is chosen, Vq and Iq vary proportionally. The two main types of output stage, the Emitter-Follower (EF) and the Complementary-Feedback Pair (CFP) are shown in Figure 22.2. Their Vq tolerances are quite different.

From the measurements shown above, the permissible error band for Vq in the EF stage is ±100 mV, and for the CFP is ±10 mV. These tolerances are not defined for all time; I only claim that they are realistic and reasonable. In terms of total Vbias, the EF needs 2.93 V ±100 mV, and the CFP 1.30 V ±10 mV. Vbias must be higher in the EF as four Vbe's are subtracted from it to get Vq, while in the CFP only two driver Vbe's are subtracted.

The CFP stage appears to be more demanding of Vbias compensation than EF, needing 1% rather than 3.5% accuracy, but things are not so simple. Vq stability in the EF stage depends primarily on the hot output devices, as EF driver dissipation varies only slightly with power output. Vq in the CFP depends almost entirely on driver junction temperature, as the effect of output device temperature is reduced by the local

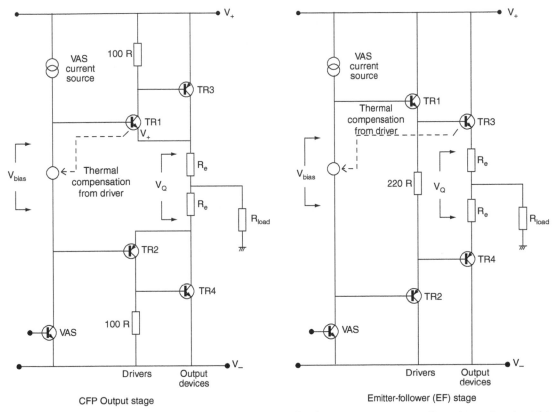

Figure 22.2. The Emitter-Follower (EF) and Complementary-Feedback Pair (CFP) output configurations, showing Vbias and Vq.

negative feedback; however, CFP driver dissipation varies strongly with power output so the superiority of this configuration cannot be taken for granted. Driver heatsinks are much smaller than those for output devices, so the CFP Vq time-constants promise to be some ten times shorter.

From the statements made above, it would appear that with an EF type output stage, it is essential to put the sensor on an output device, or as a second best onto the main heatsink, particularly as EF driver dissipation varies only slightly with power output, and therefore apparently gives little indication of what the output device dissipation is. In fact, this is not the complete story. An EF-type amplifier with satisfactory bias stability *can* be made by putting the sensor on one of the driver heatsinks. I used such a driver sensor in a recent project, as a result of some mechanical constraints that made putting a sensor on the main heatsink very difficult, and the amplifier is still on the market and selling well. The approach can be made to work satisfactorily; unfortunately I have no space to explore it further here.

Basic Thermal Compensation

In Class-B, the usual method for reducing quiescent variations is so-called 'thermal feedback'. Vbias is generated by a thermal sensor with a negative temperature-coefficient, usually a Vbe-multiplier transistor mounted on the main heatsink. This system has proved entirely workable over the last 30-odd years, and usually prevents any possibility of thermal runaway. However, it suffers from thermal losses and delays between output devices and temperature sensor that make maintenance of optimal bias rather questionable, and in practice quiescent conditions are a function of recent signal and thermal history. Thus the crossover linearity of most power amplifiers is intimately bound up with their thermal dynamics, and it is surprising this area has not been examined more closely; Sato et al.[1] is one of the few serious papers on the subject, though the conclusions it reaches appear to be unworkable, depending on calculating power dissipation from amplifier output voltage without considering load impedance.

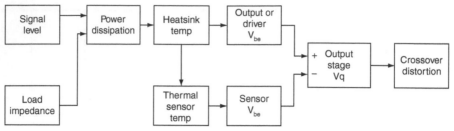

Figure 22.3. Thermal signal flow of a typical power amplifier, showing that there is no thermal feedback to the bias generator. There is instead feedforward of driver junction temperature, so that the sensor Vbe will hopefully match the driver Vbe.

As is almost routine in audio design, things are not as they appear. So-called 'thermal feedback' is not feedback at all — this implies the thermal sensor is in some way controlling the output stage temperature; it is not. It is really a form of approximate *feedforward* compensation, as shown in Figure 22.3. The quiescent current (Iq) of a Class-B design causes a very small dissipation compared with the signal, and so there is no meaningful feedback path returning from Iq to the left of the diagram. (This might be less true of Class-AB, where quiescent dissipation may be significant.) Instead this system aspires to make the sensor junction temperature mimic the driver or output junction temperature, though it can never do this promptly or exactly because of the thermal resistances and thermal capacities that lie between driver and sensor temperatures in Figure 22.3. It does not place either junction temperature or quiescent current under direct feedback control, but merely aims to cancel out the errors. Hereafter I simply call this *thermal compensation*.

Assessing the Bias Errors

The temperature error must be converted to mV error in Vq, for comparison with the tolerance bands suggested above. In the CFP stage this is straightforward; both driver Vbe and the halved Vbias voltage decrease by 2 mV per °C, so temperature error converts to voltage error by multiplying by 0.002. Only half of each output stage will be modelled, exploiting symmetry, so most of this chapter deals in half-Vq errors, etc. To minimise confusion this use of half-amplifiers is adhered to throughout, except at the final stage when the calculated Vq error is doubled before comparison with the tolerance bands quoted above.

The EF error conversion is more subtle. The EF Vbias generator must establish four times Vbe plus Vq, so the Vbe of the temperature-sensing transistor is multiplied by about 4.5 times, and so decreases at 9 mV/°C. The CFP Vbias generator only multiplies 2.1 times, decreasing at 4 mV/°C. The corresponding values for a half-amplifier are 4.5 and 2 mV/°C.

However, the EF drivers are at near-constant temperature, so after two driver Vbes have been subtracted from Vbias, the remaining voltage decreases faster with temperature than does output device Vbe. This runs counter to the tendency to under-compensation caused by thermal attenuation between output junctions and thermal sensor; in effect the compensator has thermal *gain,* and this has the potential to reduce long-term Vq errors. I suspect this is the real reason why the EF stage, despite looking unpromising, can in practice give acceptable quiescent stability.

Thermal Simulation

Designing an output stage requires some appreciation of how effective the thermal compensation will be, in terms of how much delay and attenuation the 'thermal signal' suffers between the critical junctions and the Vbias generator.

We need to predict the thermal behaviour of a heatsink assembly over time, allowing for things like metals of dissimilar thermal conductivity, and the very slow propagation of heat through a mass compared with near-instant changes in electrical dissipation. Practical measurements are very time-consuming, requiring special equipment such as multi-point thermocouple recorders. A theoretical approach would be very useful.

For very simple models, such as heat flow down a uniform rod, we can derive analytical solutions to the partial differential equations that describe the situation; the answer is an equation directly relating temperature to position along-the-rod and time. However, even slight complications (such as a non-uniform rod) involve rapidly increasing mathematical complexities, and anyone who is not already deterred should consult Carslaw and Jaeger;[3] this will deter them.

To avoid direct confrontation with higher mathematics, finite-element and relaxation methods were developed; the snag is that Finite-Element-Analysis is a rather specialised taste, and so commercial FEA software is expensive.

I therefore cast about for another method, and found I already had the wherewithal to solve problems of thermal dynamics; the use of electrical analogues is the key. If the thermal problem can be stated in terms of lumped electrical elements, then a circuit simulator of the SPICE type can handle it, and as a bonus has extensive capabilities for graphical display of the output. The work here was done with PSpice. A more common use of electrical analogues is in the electro-mechanical domain of loudspeakers; see Murphy[4] for a virtuoso example.

The simulation approach treats temperature as voltage, and thermal energy as electric charge, making thermal resistance analogous to electrical resistance, and thermal capacity to electrical capacitance. Thermal capacity is a measure of how much heat is required to raise the temperature of a mass by 1°C. (And if anyone can work out what the thermal equivalent of an inductor is, I would be interested to know: possibly Glauber's salt might be useful.) With the right choice of units the simulator output will be in Volts, with a one-to-one correspondence with degrees Celsius, and Amps, similarly representing Watts of heat flow; see Table 22.2. It is then simple to produce graphs of temperature against time.

Since heat flow is represented by current, the inputs to the simulated system are current sources. A voltage source would force large chunks of metal to change temperature instantly, which is clearly wrong. The ambient is modelled by a voltage source, as it can absorb any amount of heat without changing temperature.

Table 22.2. The relation between real thermal units and the electrical units used in simulation

	Reality	Simulation
Temperature	°C	Volts
Heat quantity	Joules (Watt-seconds)	Coulombs (Amp-seconds)
Heat flow rate	Watts	Amps
Thermal resistance	°C/Watt	Ohms
Thermal capacity	°C/Joule	Farads
Heat source	Dissipative element, e.g., transistor	Current source
Ambient	Medium-sized planet	Voltage source

Modelling the EF Output Stage

The major characteristic of emitter-follower (EF) output stages is that the output device junction temperatures are directly involved in setting Iq. This junction temperature is not accessible to a thermal compensation system, and measuring the heatsink temperature instead provides a poor approximation, attenuated by the thermal resistance from junction to heatsink mass, and heavily time-averaged by heatsink thermal inertia. This can cause serious production problems in initial setting up; any drift of Iq will be very slow as a lot of metal must warm up.

For EF outputs, the bias generator must attempt to establish an output bias voltage that is a summation of four driver and output Vbes. These do not vary in the same way. It seems at first a bit of a mystery how the EF stage, which still seems to be the most popular output topology, works as well as it does. The probable answer is Figure 22.1, which shows how driver dissipation (averaged over a cycle) varies with peak output level for the three kinds of EF output described in Chapter 9, and for the CFP configuration. The SPICE simulations used to generate this graph used a triangle waveform, to give a slightly closer approximation to the peak-average ratio of real waveforms. The rails were ±50 V, and the load 8 Ω.

It is clear that the driver dissipation for the EF types is relatively constant with power output, while the CFP driver dissipation, although generally lower, varies strongly. This is a consequence of the different operation of these two kinds of output. In general, the drivers of an EF output remain conducting to some degree for most or all of a cycle, although the output devices are certainly off half the time. In the CFP, however, the drivers turn off almost in synchrony with the outputs, dissipating an amount of power that varies much more with output. This implies that EF drivers will work at roughly the same temperature, and can be neglected in arranging thermal compensation; the temperature-dependent element is usually attached to the main heatsink, in an attempt to compensate for the junction temperature of the outputs alone. The Type I EF output keeps its drivers at the most constant temperature; this may (or may not) have something to do with why it is the most popular of the EF types.

(The above does not apply to integrated Darlington outputs, with drivers and assorted emitter resistors combined in one ill-conceived package, as the driver sections are directly heated by the output junctions. This would seem to work directly against quiescent stability, and why these compound devices are ever used in audio amplifiers remains a mystery to me.)

The drawback with most EF thermal compensation schemes is the slow response of the heatsink mass to thermal transients, and the obvious solution is to find some way of getting the sensor closer to one of the output junctions (symmetry of dissipation is assumed). If TO3 devices are used, then the flange on which the actual transistor is mounted is as close as we can get without a hacksaw. This is, however, clamped to the heatsink, and almost inaccessible, though it might be possible to hold a sensor under one of the mounting bolts. A simpler solution is to mount the sensor on the top of the TO3 can. This is probably not as accurate an estimate of junction temperature as the flange would give, but measurement shows the top gets much hotter much faster than the heatsink mass, so while it may appear unconventional, it is probably the best sensor position for an EF output stage. Figure 22.4 shows the results of an experiment designed to test this. A TO3 device was mounted on a thick aluminium L-section thermal coupler in turn clamped to a heatsink; this construction is representative of many designs. Dissipation equivalent to 100 W/8 Ω was suddenly initiated, and the temperature of the various parts monitored with thermocouples. The graph clearly shows that the top of the TO3 responds much faster, and with a larger temperature change, though after the first two minutes the temperatures are all increasing at the same rate. The whole assembly took more than an hour to asymptote to thermal equilibrium.

Figure 22.5 shows a TO3 output device mounted on a thermal coupling bar, with a silicone thermal washer giving electrical isolation. The coupler is linked to the heatsink proper via a second conformal material; this need not be electrically insulating so highly efficient materials like graphite foil can be used. This is representative of many amplifier designs, though a good number have the power devices mounted directly on the heatsink; the results hardly differ. A simple thermal-analogue model of Figure 22.5 is shown in Figure 22.6; the situation is radically simplified by treating each mass in the system as being at a uniform temperature, i.e., isothermal, and therefore representable by one capacity each. The boundaries between parts of the system are modelled, but the thermal capacity of each mass is concentrated at a notional point. In assuming this, we give capacity elements zero thermal resistance, e.g., both sides of the thermal coupler will always be at the same temperature. Similarly, elements such as the thermal washer are assumed to have zero heat capacity, because they are very thin and have negligible mass compared with other elements in the system. Thus the parts of the thermal system can be conveniently divided into two categories: pure thermal resistances and pure thermal capacities. Often this gives adequate results; if not, more sub-division will be needed. Heat losses from parts other than the heatsink are neglected.

Real output stages have at least two power transistors; the simplifying assumption is made that power dissipation will be symmetrical over anything but the extreme short term, and so one device can be studied by slicing the output stage, heatsink, etc., in half.

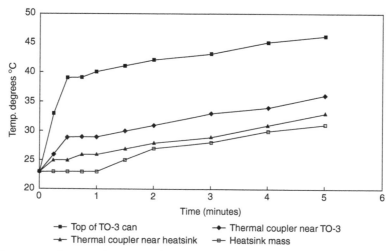

Figure 22.4. Thermal response of a TO3 device on a large heatsink when power is suddenly applied. The top of the TO3 can responds most rapidly.

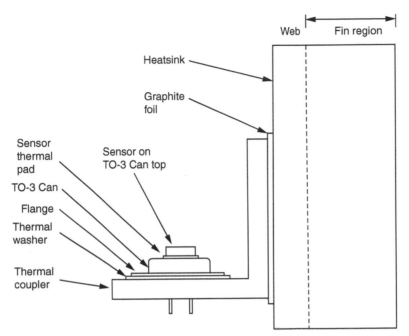

Figure 22.5. A TO3 power transistor attached to a heatsink by a thermal coupler. Thermal sensor is shown on can top; more usual position would be on thermal coupler.

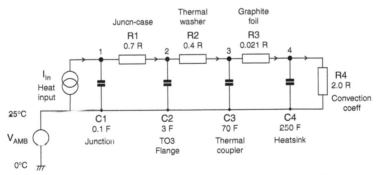

Figure 22.6. A thermal/electrical model of Figure 22.5, for half of one channel only. Node 1 is junction temperature, node 2 flange temperature, and so on. Vamb sets the baseline to 25°C. Arrows show heat flow.

VIt is convenient to read off the results directly in °C, rather than temperature rise above ambient, so Figure 22.6 represents ambient temperature with a voltage source Vamb that offsets the baseline (node 10) 25°C from simulator ground, which is inherently at 0°C (0 V).

Values of the notional components in Figure 22.6 have to be filled in with a mixture of calculation and manufacturer's data. The thermal resistance R1 from junction to case comes straight from the data book, as does the resistance R2 of the TO3 thermal washer; also R4, the convection coefficient of the heatsink itself, otherwise known as its thermal resistance to ambient. This is always assumed to be constant with temperature, which it very nearly is. Here R4 is 1°C/W, so this is doubled to 2 as we cut the stage in half to exploit symmetry.

R3 is the thermal resistance of the graphite foil; this is cut to size from a sheet and the only data is the bulk thermal resistance of 3.85 W/mK, so R3 must be calculated. Thickness is 0.2 mm, and the rectangle area in this example was 38 × 65 mm. We must be careful to convert all lengths to metres.

$$\text{Heat flow/}^\circ\text{C} = \frac{3.85 \times \text{Area}}{\text{Thickness}}$$

$$= \frac{3.83 \times (0.038 \times 0.065)}{0.0002}$$

$$= 47.3\text{W/}^\circ\text{C}$$

Equation 22.1

$$\text{So thermal resistance} = \frac{1}{47.3}$$

$$= 0.021^\circ\text{C/W}$$

Thermal resistance is the reciprocal of heat flow per degree, so R3 is 0.021°C/W, which just goes to show how efficient thermal washers can be if they do not have to be electrical insulators as well.

In general, all the thermal capacities will have to be calculated, sometimes from rather inadequate data, thus:

Thermal capacity = Density × Volume × Specific heat

A power transistor has its own internal structure, and its own internal thermal model (Figure 22.7). This represents the silicon die itself, the solder that fixes it to the copper header, and part of the steel flange the header is welded to. I am indebted to Motorola for the parameters, from an MJ15023 TO3 device.[5] The time-constants

are all extremely short compared with heatsinks, and it is unnecessary to simulate in such detail here.

The thermal model of the TO3 junction is therefore reduced to lumped component C1, estimated at 0.1J/°C; with a heat input of 1 W and no losses, its temperature would increase linearly by 10°C/sec. The capacity C2 for the transistor package was calculated from the volume of the TO3 flange (representing most of the mass) using the specific heat of mild steel. The thermal coupler is known to be aluminium alloy (not pure aluminium, which is too soft to be useful) and the calculated capacity of 70J/°C should be reliable. A similar calculation gives 250J/°C for the larger mass of the aluminium heatsink. Our simplifying assumptions are rather sweeping here, because we are dealing with a substantial chunk of finned metal which will never be truly isothermal.

The derived parameters for both output TO3s and TO-225 AA drivers are summarised in Table 22.3. The drivers are assumed to be mounted onto small individual heatsinks with an isolating thermal washer; the data is for the popular Redpoint SW38-1 vertical heatsink.

Figures 22.8 and 22.9 show the result of a step-function in heat generation in the output transistor; 20 W dissipation is initiated, corresponding approximately to the dissipation in one half of the output stage following a sudden demand for full sinewave power from a quiescent 200 W amplifier. The junction

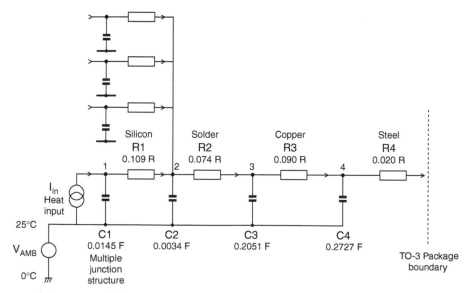

Figure 22.7. Internal thermal model for a TO3 transistor. All the heat is liberated in the junction structure, shown as N multiples of C1 to represent a typical interdigitated power transistor structure.

Table 22.3. The parameters for an output stage with TO3 outputs and TO-225AA drivers

			Output device	Driver
C1	Junction capacity	J/°C	0.1	0.05
R1	Junction-case resistance	°C/W	0.7	6.25
C2	Transistor package capacity	J/°C	3.0	0.077
R2	Thermal washer res	°C/W	0.4	6.9
C3	Coupler capacity	J/°C	70	–
R3	Coupler-heatsink res	°C/W	0.021	–
C4	Heatsink capacity	J/°C	250	20.6
R4	Heatsink convective res	°C/W	2.0	10.0

temperature V(1) takes off near-vertically, due to its small mass and the substantial thermal resistance between it and the TO3 flange; the flange temperature V(2) shows a similar but smaller step as R2 is also significant. In contrast, the thermal coupler, which is so efficiently bonded to the heatsink by graphite foil that they might almost be one piece of metal, begins a slow exponential rise that will take a very long time to asymptote. Since after the effects of C1 and C2 have died away, the junction temp is offset by a constant amount from the temp of C3 and C4, V(1) also shows a slow rise. Note the X-axis of Figure 22.9 must be in kilo-seconds, because of the relatively enormous thermal capacity of the heatsink.

This shows that a temperature sensor mounted on the main heatsink can never give accurate bias compensation for junction temperature, even if it is assumed to be isothermal with the heatsink; in practice, there will be some sensor cooling which will make the sensor temperature slightly under-read the heatsink temperature V(4). Initially the temperature error V(1)−V(4) increases rapidly as the TO3 junction heats, reaching 13° in about 200 ms. The error then increases much more slowly, taking 6 seconds to reach the effective final value of 22°. If we ignore the thermal-gain effect mentioned above, the long-term Vq error is +44 mV, i.e., Vq is too high. When this is doubled to allow for both halves of the output stage, we get +88 mV,

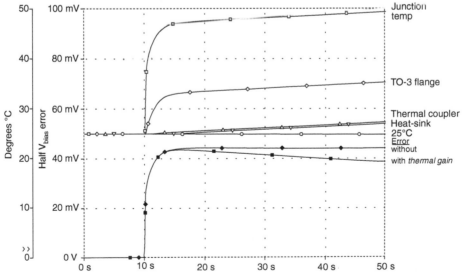

Figure 22.8. Results for Figure 22.6, with step heat input of 20 W to junction initiated at Time = 10 sec. Upper plot shows temperatures, lower the Vbias error for half of output stage.

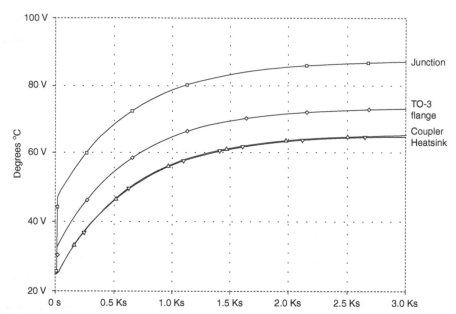

Figure 22.9. The long-term version of Figure 22.8, showing that it takes over 40 minutes for the heatsink to get within 1° of final temperature.

which uses up nearly all of the ±100 mV error band, without any other inaccuracies. (Hereafter all Vbias/Vq error figures quoted have been doubled and so apply to a complete output stage.) Including the thermal gain actually makes little difference over a 10-sec timescale; the lower Vq-error trace in Figure 22.8 slowly decays as the main heatsink warms up, but the effect is too slow to be useful.

The amplifier Vq and Iq will therefore rise under power, as the hot output device Vbe voltages fall, but the cooler bias generator on the main heatsink reduces its voltage by an insufficient amount to compensate.

Figure 22.9 shows the long-term response of the system. At least 2500 sec pass before the heatsink is within a degree of final temperature.

In the past I have recommended that EF output stages should have the thermal sensor mounted on the top of the TO3 can, despite the mechanical difficulties. This is not easy to simulate as no data is available for the thermal resistance between junction and can top. There must be an additional thermal path from junction to can, as the top very definitely gets *hotter* than the flange measured at the very base of the can. In view of the relatively low temperatures, this path is probably due to internal convection rather than radiation.

A similar situation arises with TO3P packages for the top plastic surface can get at least 20° hotter than the heatsink just under the device. Recent work has shown

that this also applies to the MT-200 and the TO-264 plastic packages.

Using the real thermocouple data from Figure 22.4, I have estimated the parameters of the thermal paths to the TO3 top. This gives Figure 22.10, where the values of elements R20, R21, C5 should be treated with caution, though the temperature results in Figure 22.11 match reality fairly well; the can top (V20) gets hotter faster than any other accessible point. R20 simulates the heating path from the junction to the TO3 can and R21 the can-to-flange cooling path, C5 being can thermal capacity.

Figure 22.10 includes approximate representation of the cooling of the sensor transistor, which now matters. R22 is the thermal pad between the TO3 top and the sensor, C6 the sensor thermal capacity, and R23 is the convective cooling of the sensor, its value being taken as twice the data-sheet free-air thermal resistance as only one face is exposed. The sensor transistor is assumed to be isothermal, and not significantly heated by its own standing current.

Placing the sensor on top of the TO3 would be expected to reduce the steady-state bias error dramatically. In fact, it overdoes it, as after factoring in the thermal-gain of a Vbe-multiplier in an EF stage, the bottom-most trace of Figure 22.11 shows that the bias is over-compensated; after the initial positive transient error, Vbias falls too low, giving an error of −30 mV,

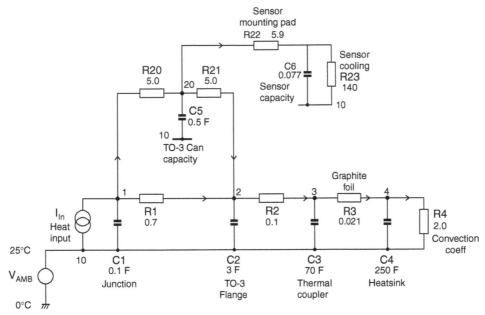

Figure 22.10. Model of EF output stage with thermal paths to TO3 can top modelled by R20, R21. C5 simulates can capacity. R23 models sensor convection cooling; node 21 is sensor temperature.

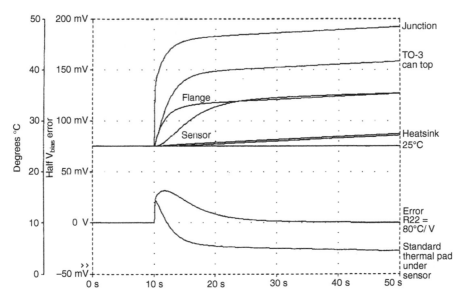

Figure 22.11. The simulation results for Figure 22.10; lower plot shows Vbias errors for normal thermal pad under sensor, and 80°C/W semi-insulator. The latter has near-zero long-term error.

slowly worsening as the main heatsink warms up. If thermal-gain had been ignored, the simulated error would have apparently fallen from +44 (Figure 22.8) to +27 mV; apparently a useful improvement, but actually illusory.

Since the new sensor position over-compensates for thermal errors, there should be an intermediate arrangement giving near-zero long-term error. I found this condition occurs if R22 is increased to 80°C/W, requiring some sort of semi-insulating material rather

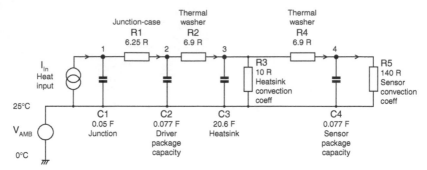

Figure 22.12. Model of a CFP stage. Driver transistor is mounted on a small heatsink, with sensor transistor on the other side. Sensor dynamics and cooling are modelled by R4, C4 and R5.

than a thermal pad, and gives the upper error trace in the lower half of Figure 22.11. This peaks at +30 mV after 2 seconds, and then decays to nothing over the next 20 seconds. This is much superior to the persistent error in Figure 22.8, so this new technique may be useful, but bear in mind that it slows the sensor response.

It has been suggested that a sensor position that needs long wires to connect to the rest of the circuitry could make HF stability uncertain. I have had no trouble with wires up to 20 cm in length, and I think this is not surprising because the result is presumably an increase in the capacitance to ground at the VAS collector of a few pF. Since the effect of such capacitance is, perhaps counter-intuitively but quite definitely, to increase HF stability (see Chapter 13 for more details), this seems to be something of a non-problem.

Modelling the CFP Output Stage

In the CFP configuration, the output devices are inside a local feedback loop, and play no significant part in setting Vq, which is dominated by thermal changes in the driver Vbes. Such stages are virtually immune to thermal runaway; I have found that assaulting the output devices with a powerful heat gun induces only very small Iq changes. Thermal compensation is mechanically simpler as the Vbe-multiplier transistor is usually mounted on one of the driver heatsinks, where it aspires to mimic the driver junction temperature.

It is now practical to make the bias transistor of the same type as the drivers, which may help to give the best matching of Vbe,[6] though given the differences in Ic, how important this is in practice is uncertain. It definitely avoids the difficulty of trying to attach a small-signal (probably TO92) transistor package to a heatsink.

Since it is the driver junctions that count, output device temperatures are here neglected. The thermal

parameters for a TO225AA driver (e.g., MJE340/350) on the SW38-1 vertical heatsink are shown in Table 22.3; the drivers are on individual heatsinks so their thermal resistance is used directly, without doubling.

In the simulation circuit (Figure 22.12) V(3) is the heatsink temperature; the sensor transistor (also MJE340) is mounted on this sink with thermal washer R4, and has thermal capacity C4. R5 is convective cooling of the sensor. In this case the resulting differences in Figure 22.13 between sink V(3) and sensor V(4) are very small.

We might expect the CFP delay errors to be much shorter than in the EF; however, simulation with a heat step-input suitably scaled down to 0.5 W (Figure 22.13) shows changes in temperature error V(1)−V(4) that appear rather paradoxical; the error reaches 5° in 1.8 seconds, levelling out at 6.5° after about 6 sec. This is markedly slower than the EF case, and gives a total bias error of +13 mV, which, after doubling to +26 mV, is well outside the CFP error band of ±10 mV.

The initial transients are slowed down by the much smaller step heat input, which takes longer to warm things up. The *final* temperature, however, is reached in 500 rather than 3000 sec, and the timescale is now in hundreds rather than thousands of seconds. The heat input is smaller, but the driver heatsink capacity is also smaller, and the overall time-constant is less.

It is notable that both timescales are much longer than musical dynamics.

The Integrated Absolute Error Criterion

Since the thermal sensor is more or less remote from the junction whose gyrations in temperature will hopefully be cancelled out, heat losses and thermal resistances cause the temperature change reaching the sensor

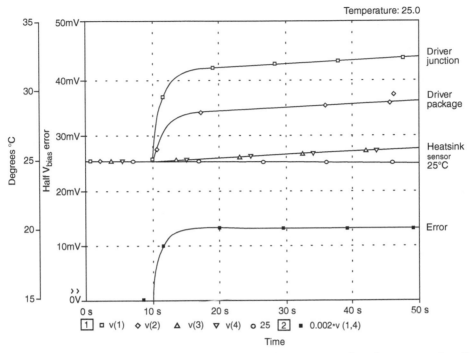

Figure 22.13. Simulation results for CFP stage, with step heat input of 0.5 W. Heatsink and sensor are virtually isothermal, but there is a persistent error as driver is always hotter than heatsink due to R1, R2.

to be generally too little and too late for complete compensation.

In this section, all the voltages and errors here are for one-half of an output stage, using symmetry to reduce the work involved. These 'half-amplifiers' are used throughout this chapter, for consistency, and the error voltages are only doubled to represent reality (a complete output stage) when they are compared against the tolerance bands previously quoted.

We are faced with errors that vary not only in magnitude, but also in their persistence over time; judgement is required as to whether a prolonged small error is better than a large error which quickly fades away.

The same issue faces most servomechanisms, and I borrow from Control Theory the concept of an *Error Criterion* which combines magnitude and time into one number.[7,8] The most popular criterion is the Integrated Absolute Error (IAE) which is computed by integrating the absolute-value of the error over a specified period after giving the system a suitably provocative stimulus; the absolute-value prevents positive and negative errors cancelling over time. Another common criterion is the Integrated Square Error (ISE) which solves the polarity problem by squaring the error before

integration – this also penalises large errors much more than small ones. It is not immediately obvious which of these is most applicable to bias control and the psychoacoustics of crossover distortion that changes with time, so I have chosen the popular IAE.

One difficulty is that the IAE error criterion for bias voltage tends to accumulate over time, due to the integration process, so any constant bias error quickly comes to dominate the IAE result. In this case, the IAE is little more than a counter-intuitive way of stating the constant error, and must be quoted over a specified integration time to mean anything at all. This is why the IAE concept was not introduced earlier in this chapter.

Much more useful results are obtained when the IAE is applied to a situation where the error decays to a very small value after the initial transient, and stays there. This can sometimes be arranged in amplifiers, as I hope to show. In an ideal system where the error decayed to zero without overshoot, the IAE would asymptote to a constant value after the initial transient. In real life, residual errors make the IAE vary slightly with time, so for consistency all the IAE values given here are for 30 sec after the step-input.

Improved Thermal Compensation: the Emitter-follower Stage

It was shown above that the basic emitter-follower (EF) stage with the sensor on the main heatsink has significant thermal attenuation error and therefore under-compensates temperature changes. (The Vq error is +44 mV, the positive sign showing it is too high. If the sensor is on the TO3 can top, it over-compensates instead; Vq error is then −30 mV.)

If an intermediate configuration is contrived by putting a layer of controlled thermal resistance (80°C/W) between the TO3 top and the sensor, then the 50-sec timescale component of the error can be reduced to near-zero. This is the top error trace in the bottom half of Figure 22.14; the lower trace shows the wholly misleading result if sensor heat losses are neglected in this configuration.

Despite this medium-term accuracy, if the heat input stimulus remains constant over the very long term (several kilo-seconds), there still remains a very slow drift towards over-compensation due to the slow heating of the main heatsink (Figure 22.15).

This long-term drift is a result of the large thermal inertia of the main heatsink and since it takes 1500 sec (25 minutes) to go from zero to −32 mV is of doubtful relevance to the time-scales of music and signal level changes. On doubling to −64 mV, it remains within the EF Vq tolerance of ±100 mV. On the shorter 50-sec timescale, the half-amplifier error remains within a ±1 mV window from 5 sec to 60 sec after the step-input.

For the EF stage, a very long-term drift component will always exist so long as the output device junction temperature is kept down by means of a main heatsink that is essentially a weighty chunk of finned metal.

The EF system stimulus is a 20 W step as before, being roughly worst-case for a 200 W amplifier. Using the 80°C/W thermal semi-insulator described above gives the upper error trace in Figure 22.16, and an IAE of 254 mV-sec after 30 sec. This is relatively large because of the extra time-delay caused by the combination of an increased R22 with the unchanged sensor thermal capacity C6. Once more, this figure is for a half-amplifier, as are all IAEs in this chapter.

Up to now I have assumed that the temperature coefficient of a Vbe-multiplier bias generator is rigidly fixed at

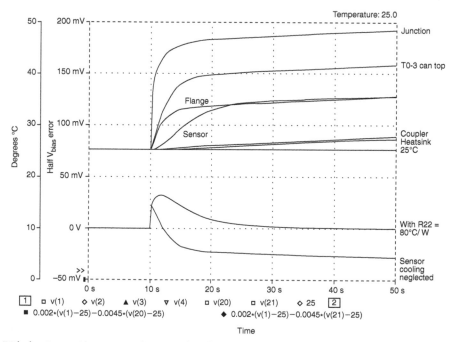

Figure 22.14. EF behaviour with semi-insulating pad under sensor on TO3 can top. The sensor in the upper temperature plot rises more slowly than the flange, but much faster than the main heatsink or coupler. In lower Vq-error section, upper trace is for an 80 C/W thermal resistance under the sensor, giving near-zero error. Bottom trace shows serious effect of ignoring sensor-cooling in TO3-top version.

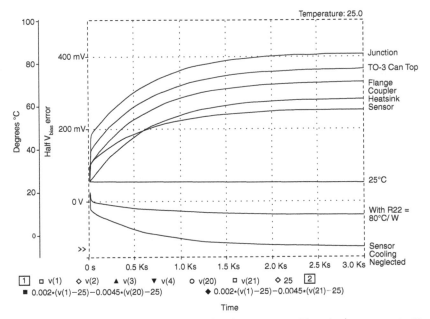

Figure 22.15. Over a long timescale, the lower plot shows that the Vq error, although almost zero in Figure 22.14, slowly drifts into over-compensation as the heatsink temperature (upper plot) reaches asymptote.

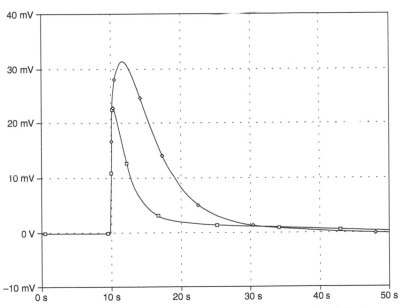

Figure 22.16. The transient error for the semi-insulating pad and the low-tempco version. The latter responds much faster, with a lower peak error, and gives less than half the Integrated Absolute Error (IAE).

−2 mV/°C times the Vbe-multiplication factor, which is about 4.5× for EF and 2× for CFP. The reason for the extra thermal gain displayed by the EF was set out on p. 510.

The above figures are for both halves of the output stage, so the half-amplifier value for EF is −4.5 mV/°C, and for CFP −2 mV/°C. However, if we boldly assume that the Vbias generator can have its thermal coefficient varied at will, the insulator and its aggra-vated time-lag can be eliminated.

If a thermal pad of standard material is once more used between the sensor and the TO3 top, the optimal Vbias coefficient for minimum error over the first 40 seconds proves -2.8 mV/°C, which is usefully less than -4.5. The resulting 30-sec IAE is 102 mV-sec, more than a two times improvement; see the lower trace in Figure 22.16, for comparison with the semi-insulator method described above.

From here on, I am assuming that a variable-temperature-coefficient (tempco) bias generator can be made when required; the details of how to do it are not given here. It is an extremely useful device, as thermal attenuation can then be countered by increasing the thermal gain; it does not, however, help with the problem of thermal delay.

In the second EF example above, the desired tempco is -2.8 mV/°C, while an EF output stage plus has an actual tempco of -4.5 mV/°C. (This inherent thermal gain in the EF was explained on p. 510.) In this case we need a bias generator that has a *smaller* tempco than the standard circuit. The conventional EF with its temp sensor on the relatively cool main heatsink would require a *larger* tempco than standard.

A potential complication is that amplifiers should also be reasonably immune to changes in ambient temperature, quite apart from changes due to dissipa-tion in the power devices. The standard tempco gives a close approach to this automatically, as the Vbe-multiplication factor is naturally almost the same as the number of junctions being biased. However, this will no longer be true if the tempco is significantly different from standard, so it is necessary to think about a bias generator that has one tempco for power-device temperature changes, and another for ambient changes. This sounds rather daunting, but is actually fairly simple.

Improved Compensation for the CFP Output Stage

As revealed earlier, the Complementary-Feedback Pair (CFP) output stage has a much smaller bias tolerance of ±10 mV for a whole amplifier, and surprisingly long time-constants. A standard CFP stage therefore has larger relative errors than the conventional Emitter-Follower (EF) stage with a thermal sensor on the main heatsink; this is the opposite of conventional wisdom. Moving the sensor to the top of the TO3 can was shown to improve the EF performance markedly,

so we shall attempt an analogous improvement with driver compensation.

The standard CFP thermal compensation arrange-ments have the sensor mounted on the driver heatsink, so that it senses the heatsink temperature rather than that of the driver itself. (See Figure 22.17a for the mechanical arrangement, and Figure 22.18 for the thermal model.) As in the EF, this gives a constant long-term error due to the sustained temperature differ-ence between the driver junction and the heatsink mass; see the upper traces in Figure 22.20 on p. 524, plotted for different bias tempcos. The CFP stimulus is a 0.5 W step, as before. This constant error cannot be properly dealt with by choosing a tempco that gives a bias error passing through a zero in the first 50 seconds, as was done for the EF case with a TO3 top sensor, as the heatsink thermal inertia causes it to pass through zero very quickly and head rapidly South in the direction of ever-increasing negative error. This is because it has allowed for thermal attenuation but has not decreased thermal delay. It is therefore pointless to compute an IAE for this configuration.

A Better Sensor Position

By analogy with the TO3 and TO3P transistor packages examined earlier, it will be found that driver packages such as TO225AA on a heatsink get hotter faster on their exposed plastic face than any other accessible point. It looks as if a faster response will result from putting the sensor on top of the driver rather than on the other side of the sink as usual. With the Redpoint SW 38-1 heatsink, this is fairly easy as the spring-clips used to secure one plastic package will hold a stack of two TO225AAs with only a little physical persuasion. A standard thermal pad is used between the top of the driver and the metal face of the sensor, giving the sand-wich shown in Figure 22.17b. The thermal model is shown in Figure 22.19. This scheme greatly reduces both thermal attenuation and thermal delay (lower traces in Figure 22.20) giving an error that falls within a ±1 mV window after about 15.5 seconds, when the tempco is set to -3.8 mV/°C. The IAE computes to 52 mV, as shown in Figure 22.21, which demonstrates how the IAE criterion tends to grow without limit unless the error subsides to zero. This value is a distinct improvement on the 112 mV IAE which is the best that could be got from the EF output.

The effective delay is much less because the long heatsink time-constant is now partly decoupled from the bias compensation system.

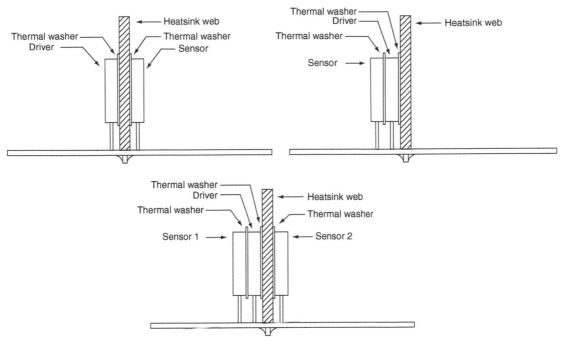

Figure 22.17. (a) The sensor transistor on the driver heatsink; (b) an improved version, with the sensor mounted on top of the driver itself, is more accurate; (c) using two sensors to construct a junction-estimator.

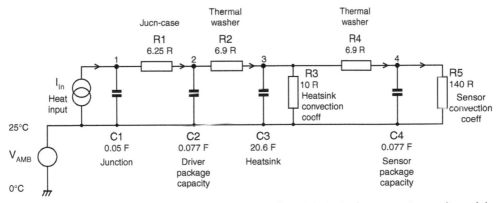

Figure 22.18. Thermal circuit of normal CFP sensor mounting on heatsink. R3 is the convective cooling of the heatsink, while R5 models heat losses from the sensor body itself.

A Junction-temperature Estimator

It appears that we have reached the limit of what can be done, as it is hard to get one transistor closer to another than they are in Figure 22.17b. It is, however, possible to get better performance, not by moving the sensor position, but by using more of the available information to make a better estimate of the true driver junction temperature. Such 'estimator' subsystems are widely used in servo control systems where some vital variable is inaccessible, or only knowable after such a time-delay as to render the data useless.[9] It is often almost as useful to have a model system, usually just an abstract set of gains and time-constants, which gives an estimate of what the current value of the unknown variable must be, or at any rate, *ought* to be.

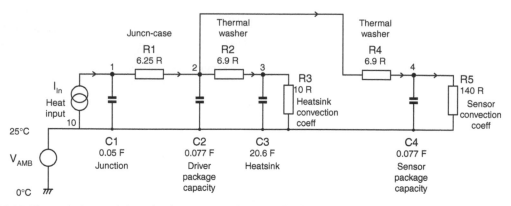

Figure 22.19. Thermal circuit of driver-back mounting of sensor. The large heatsink time-constant R2−C2 is no longer in the direct thermal path to the sensor, so the compensation is faster and more accurate.

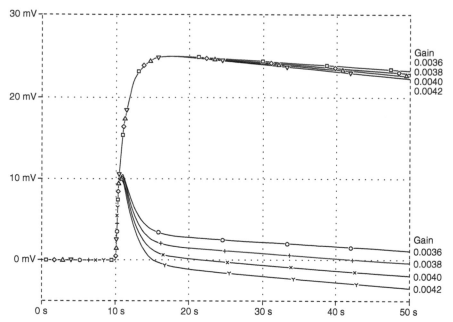

Figure 22.20. The Vq errors for normal and improved sensor mounting, with various tempcos. The improved method can have its tempco adjusted to give near-zero error over this timescale. Not so for the usual method.

The situation here is similar, and the first approach makes a better guess at the junction temperature V(1) by using the known temperature drop between the package and the heatsink. The simplifying assumption is made that the driver package (not including the junction) is isothermal, so it is modelled by one temperature value V(2).

If two sensors are used, one placed on the heatsink as usual, and the other on top of the driver package, as described above (Figure 22.17c), then things get interesting. Looking at Figure 22.19, it can be seen that the difference between the driver junction temperature and the heatsink is due to R1 and R2; the value of R1 is known, but not the heat flow through it. Neglecting small incidental losses, the temperature drop through R1 is proportional to the drop through R2. Since C2 is much smaller than C3, this should remain reasonably true even if there are large thermal transients. Thus, measuring the difference between V(2) and V(3) allows a reasonable estimate of the difference between V(1) and V(2);

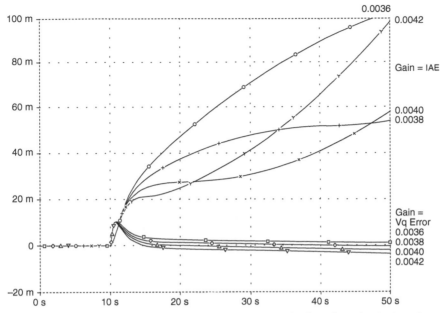

Figure 22.21. The Vq error and IAE for the improved sensor mounting method on driver back. Error is much smaller, due both to lower thermal attenuation and to less delay. Best IAE is 52 mV-sec (with gain = 0.0038); twice as good as the best EF version.

when this difference is added to the known V(2), we get a rather good estimation of the inaccessible V(1). This system is shown conceptually in Figure 22.22, which gives only the basic method of operation; the details of the real circuitry must wait until we have decided exactly what we want it to do.

We can only measure V(2) and V(3) by applying thermal sensors to them, as in Figure 22.17c, so we actually have as data the sensor temperatures V(4) and V(5). These are converted to bias voltage and subtracted, thus estimating the temperature drop across R1. The computation is done by Voltage-Controlled-Voltage-Source E1, which in PSpice can have any equation assigned to define its behaviour. Such definable VCVSs are very handy as little 'analogue computers' that do calculations as part of the simulation model. The result is then multiplied by a scaling factor called *estgain* which is incorporated into the defining equation for E1, and is adjusted to give the minimum error; in other words, the variable-tempco bias approach is used to allow for the difference in resistance between R1 and R2.

The results are shown in Figure 22.23, where an *estgain* of 1.10 gives the minimum IAE of 25 mV-sec. The transient error falls within a ±1 mV window after about 5 sec. This is a major improvement, at what promises to be little cost.

A Junction Estimator with Dynamics

The remaining problem with the junction-estimator scheme is still its relatively slow initial response; nothing can happen before heat flows through R6 into C5, in Figure 22.22. It will take even longer for C4 to respond, due to the inertia of C3, so we must find a way to speed up the dynamics of the junction-estimator.

The first obvious possibility is the addition of phase-advance to the forward bias-compensation path. This effectively gives a high gain initially, to get things moving, which decays back over a carefully set time to the original gain value that gave near-zero error over the 50-sec timescale. The conceptual circuit in Figure 22.24 shows the phase-advance circuitry added to the compensation path; the signal is attenuated 100× by R50 and R51, and then scaled back up to the same level by VCVS E2, which is defined to give a gain of 110 times incorporating estimated gain = 1.10. C causes fast changes to bypass the attenuation, and its value in conjunction with R50, R51 sets the degree of phase-advance or lead. The slow behaviour of the circuit is thus unchanged, but transients pass through C and are greatly amplified by comparison with steady-state signals.

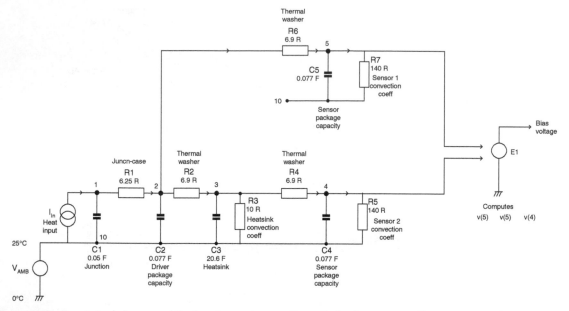

Figure 22.22. Conceptual diagram of the junction-estimator. Controlled-voltage-source El acts as an analogue computer performing the scaling and subtraction of the two sensor temperatures V(4) and V(5), to derive the bias voltage.

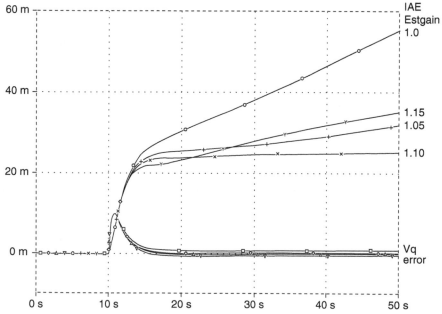

Figure 22.23. Simulation results for the junction-estimator, for various values of estgain. The optimal IAE is halved to 25 mV-sec; compare with Figure 22.21.

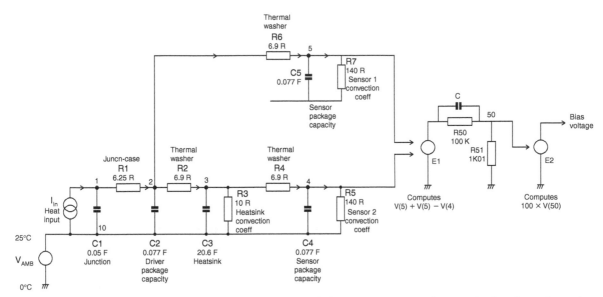

Figure 22.24. The conceptual circuit of a junction-estimator with dynamics. C gives higher gain for fast thermal transients and greatly reduces the effects of delay.

The result on the initial error transient of varying C around its optimal value can be seen in the expanded view of Figure 22.25. The initial rise in Vq error is pulled down to less than a third of its value if C is made 10 μF; with a lower C value the initial peak is still larger than it need be, while a higher value introduces some serious undershoot that causes the IAE to rise again, as seen in the upper traces in Figure 22.26. The big difference between no phase-advance, and a situation where it is even approximately correct, is very clear.

With C set to 10 μF, the transient error falls into a ±1 mV window after only 0.6 sec, which is more than twenty times faster than the first improved CFP version (sensor put on driver) and gives a nicely reduced IAE of 7.3 mV-sec at 50 sec. The real-life circuitry to do this has not been designed in detail, but presents no obvious difficulties. The result should be the most accurately bias-compensated Class-B amplifier ever conceived.

Conclusions about the Simulations

Some of the results of these simulations and tests were rather unexpected. I thought that the CFP would show relatively smaller bias errors than the EF, but it is the EF that stays within its much wider tolerance bands, with either heatsink or TO3-top mounted sensors. The thermal-gain effect in the EF stage seems to be

the root cause of this, and this in turn is a consequence of the near-constant driver dissipation in the EF configuration.

However, the cumulative bias errors of the EF stage can only be reduced to a certain extent, as the system is never free from the influence of the main heatsink with its substantial thermal inertia. In contrast, the CFP stage gives much more freedom for sensor placement and gives scope for more sophisticated approaches that reduce the errors considerably.

Hopefully it is clear that it is no longer necessary to accept 'Vbe-multiplier on the heatsink' as the only option for the crucial task of Vbias compensation. The alternatives presented promise greatly superior compensation accuracy.

Power Transistors with Integral Temperature Sensors

For a very long time it was obvious that all attempts at estimating device junction temperature would come a poor second to having a sensor built right in to the junction structure. At last such power transistors appeared when Sanken introduced the SAP series of transistors with integral sensing diodes. These were Darlington devices; these are usually not good for bias stability as the driver transistor is heated up by the adjacent output device, but in this case the integral diodes were intended to compensate both driver and output

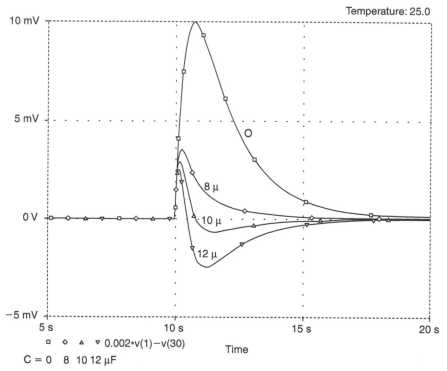

Figure 22.25. The initial transient errors for different values of C. Too high a value causes undershoot.

Vbe changes. The SAP transistors had one diode built into the NPN part, and five diodes built into the PNP part, designed so that 2.5 mA through the diodes gave good matching with a transistor quiescent Ic of 40 mA. They also had an integral 0R22 emitter resistor which proved not to be a good idea as it was more electrically fragile than the transistors themselves.

The SAP series has now been replaced by the STD03N and the STD03P which have the same diode structure but no internal emitter resistor; see Figure 22.27a. A Darlington output stage needs four Vbe drops, plus a few mV across the emitter resistors, so six diodes' worth of bias may appear excessive; the answer is that the five diodes in the PNP device are Schottkys with a lower voltage drop. The diodes are part of the main transistor die and so have the fastest response possible. The main drawback of this approach is that it permits little flexibility in circuit design. It has also been pointed out that if thermal compensation is *too* fast, thermal distortion might be introduced as the bias changed during a cycle.

Quite recently ON Semiconductor (Motorola that was) introduced some very interesting output pairs with integral diodes, under the name ThermalTrak. They are the NJL4281 (NPN) and NJL4302 (PNP)

pair, with a Vceo of 350V, and the NJL0281 (NPN) and NJL0302 (PNP) pair with a Vceo of 260V. It has to be said that the published data was not very helpful, and it took a remarkable collaboration on the DIYaudio forum in 2008 to determine their construction and characteristics.[10]

It emerged that they differ from the Sanken devices in that a single silicon diode of the MUR120 type is mounted on the copper lead frame and is electrically isolated from the transistors. The electrical isolation gives much greater freedom of electrical design, but at a cost. The thermal response is inevitably much slower, as the lead frame is coupled to the heatsink through a low-thermal-resistance washer, and so it follows the heatsink temperature much more than it follows the junction temperature. It appears a sensor mounted on the top of the package will still respond more rapidly than the internal diode.

Bob Cordell has made some measurements[11] which indicate that the transistor Vbe temperature coefficient is −2.14 mV/ °C while the diode has −1.7 mV/ °C, and the best voltage matching occurs when the diode current is a quarter of the transistor Ic, so it would appear that applying these devices to get the best

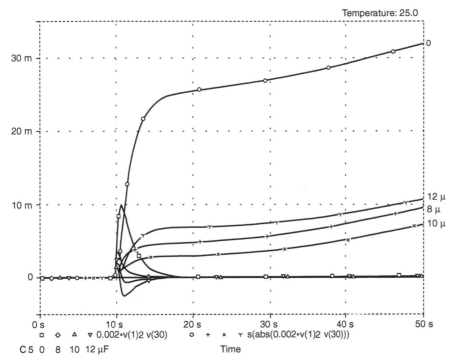

Figure 22.26. The IAE for different values of C. 10 μF is clearly best for minimum integrated error (IAE = 7.3 mV−sec) but even a rough value is a great improvement.

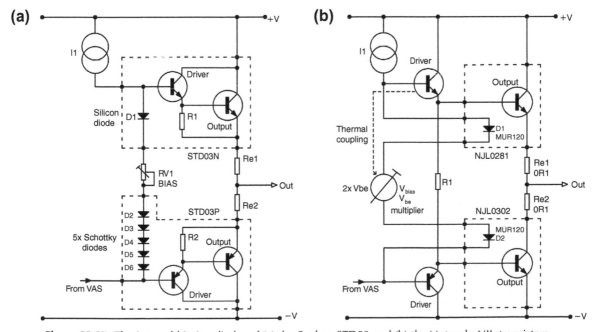

Figure 22.27. The internal biasing diodes of (a) the Sanken STD03 and (b) the Motorola NJL transistors.

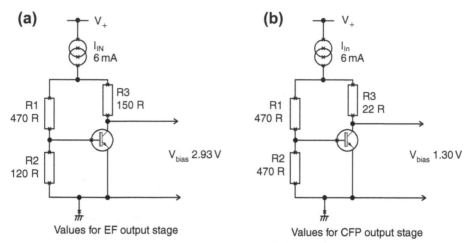

Figure 22.28. The classical Vbe-multiplier bias generator. Two versions are shown: for biasing EF (a) and (b) CFP output stages. The EF requires more than twice the bias voltage for optimal crossover performance.

compensation is going to take a bit of thought. The drivers are now separate devices and will have their own temperature, so the internal diodes can only compensate the output transistors. One possible configuration is shown in Figure 22.27b, where a conventional Vbe multiplier generating two Vbe drops is thermally coupled to the drivers only; current source I1 will need to be set for the best diode-output-device matching. The Motorola devices have beautifully flat beta/Ic curves and so should give low Large Signal Distortion (see Chapter 9). Some tests on the ThermalTrak NJL0281 and NJL0302 are described at the end of this chapter.

Variable-tempco Bias Generators

The standard Vbe-multiplier bias generator has a temperature coefficient that is fixed by the multiplication factor used, and so ultimately by the value of Vbias required. However, at many points in this chapter it has been assumed that it is possible to make a bias generator with an arbitrary temperature coefficient. This section shows how to do it.

Figure 22.28 shows two versions of the usual Vbe-multiplier bias generator. Here the lower rails are shown as grounded to simplify the results. The first version in Figure 22.28a is designed for an EF output stage, where the voltage Vbias to be generated is (4 × Vbe) + Vq, which totals +2.93 V. Recall that Vq is the small quiescent voltage across the emitter-resistors Re; it is *this* quantity we are aiming to keep constant, rather than the quiescent current, as is usually assumed. The optimal Vq for an EF stage is in the region of 50 mV.

The second bias generator in Figure 22.28b is intended for a CFP output stage, for which the required Vbias is less at (2 × Vbe) + Vq, or approximately 1.30 V in total. Note that the optimal Vq is also much smaller for the CFP type of output stage, being about 5 mV.

It is assumed that Vbias is trimmed by varying R2, which will in practice be a pre-set with a series end-stop resistor to limit the maximum Vbias setting. It is important that this is the case, because a pre-set normally fails by the wiper becoming disconnected, and if it is in the R2 position, the bias will default to minimum. In the R1 position an open-circuit pre-set will give maximum bias, which may blow fuses or damage the output stage. The adjustment range provided should be no greater than that required to take up production tolerances; it is, however, hard to predict just how big that will be, so the range is normally made wide for pre-production manufacture, and then tightened in the light of experience.

The EF version of the bias generator has a higher Vbias, so there is a larger Vbe-multiplication factor to generate it. This is reflected in the higher temperature coefficient (hereafter shortened to 'tempco'). See Table 22.4.

Table 22.4. Temperature coefficients for different Vbias voltages

	Vbias volts	R1 Ω	R2 Ω	R3 Ω	Tempco mV/°C
EF	.93	120R	470R	22R	−9.3
CFP	1.30	470R	470R	150R	−3.6

Creating a Higher Tempco

A higher (i.e., more negative) tempco than normal may be useful to compensate for the inability to sense the actual output junction temperatures. Often the thermal losses to the temperature sensor are the major source of steady-state Vbias error, and to reduce this, a tempco is required that is larger than the standard value given by: 'Vbe-multiplication factor times −2 mV/°C'. Many approaches are possible, but the problem is complicated because in the CFP case the bias generator has to work within two rails only 1.3 V apart. Additional circuitry outside this voltage band can be accommodated by bootstrapping, as in the Trimodal amplifier biasing system in Chapter 10, but this does add to the component count.

A simple new idea is shown in Figure 22.29. The aim is to increase the multiplication factor (and hence the negative tempco) required to give the same Vbias. The diagram shows a voltage source V1 inserted in the R2 arm. To keep Vbias the same, R2 is reduced. Since the multiplication factor $(R1 + R2)/R2$ is increased, the tempco is similarly increased. In Table 22.5, a CFP bias circuit has its tempco varied by increasing V1 in 100 mV steps; in each case the value of R2 is then reduced to bring Vbias back to the desired value, and the tempco is increased.

Table 22.5. Varying tempco by varying V1

V1 mV	Vbias V	R2 Ω	Tempco mV/C
0	1.287	470	−3.6
100	1.304	390	−4.0
200	1.287	330	−4.4
300	1.286	260	−5.0
400	1.285	190	−6.9

A practical circuit is shown in Figure 22.30, using a 2.56 V bandgap reference to generate the extra voltage across R4. This reference has to work outside the bias generator rails, so its power-feed resistors R7, R8 are bootstrapped by C from the amplifier output, as in the Trimodal amplifier design.

Ambient Temperature Changes

Power amplifiers must be reasonably immune to ambient temperature changes, as well as changes due to dissipation in power devices. The standard compensation system deals with this pretty well, as the Vbe-multiplication factor is inherently almost the same as the number of junctions being biased. This is no longer true if the tempco is significantly modified. Ideally we require a bias generator that has one increased tempco for power-device temperature changes only, and another standard tempco for

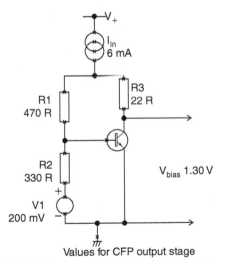

Figure 22.29. Principle of a Vbe multiplier with increased tempco. Adding voltage source V1 means the voltage-multiplication factor must be increased to get the same Vbias. The tempco is therefore also increased, here to −4.4 mV/°C.

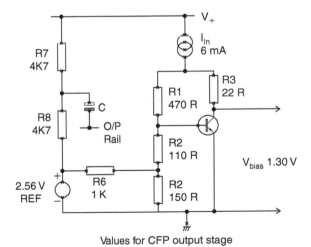

Figure 22.30. Shows a practical version of a Vbe multiplier with increased tempco. The extra voltage source is derived from the bandgap reference by R6, R4. Tempco is increased to −5.3 mV/°C.

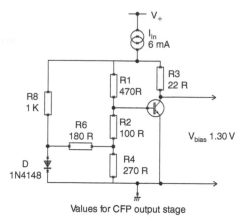

Figure 22.31. Practical Vbe multiplier with increased tempco, and also improved correction for ambient temperature changes, by using diode D to derive the extra voltage.

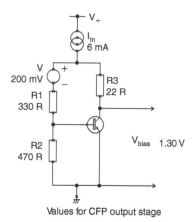

Figure 22.32. The principle of a Vbe multiplier with reduced tempco. The values shown give −3.1 mV/°C.

ambient changes affecting all components. One approach to this is Figure 22.31, where V1 is derived via R6, R4 from a silicon diode rather than a bandgap reference, giving a voltage reducing with temperature. The tempco for temperature changes to Q1 only is −4.0 mV/°C, while the tempco for global temperature changes to *both* Q1 and D1 is lower at −3.3 mV/°C. Ambient temperatures vary much less than output device junction temperatures, which may easily range over 100°C.

Creating a Lower Tempco

Earlier in this chapter I showed that an EF output stage has 'thermal gain' in that the thermal changes in Vq make it appear that the tempco of the Vbias generator is higher than it really is. This is because the bias generator is set up to compensate for four base-emitter junctions, but in the EF output configuration the drivers have a roughly constant power dissipation with changing output power, and therefore do not change much in junction temperature. The full effect of the higher tempco is thus felt by the output junctions, and if the sensor is placed on the power device itself rather than the main heatsink, to reduce thermal delay, then the amplifier can be seriously over-compensated for temperature. In other words, after a burst of power, Vq will become too low rather than too high, and crossover distortion will appear. We now need a Vbias generator with a *lower* tempco than the standard circuit.

The principle is exactly analogous to the method of increasing the tempco. In Figure 22.32, a voltage source is inserted in the upper leg of potential divider

R1, R2; the required Vbe-multiplication factor for the same Vbias is reduced, and so therefore is the tempco.

Table 22.6 shows how this works as V1 is increased in 100 mV steps. R1 has been varied to keep Vbias constant, in order to demonstrate the symmetry of resistor values with Table 22.5; in reality, R2 would be the variable element, for the safety reasons described above.

Current Compensation

Both bias generators in Figure 22.28 are fitted with a current-compensation resistor R3. The Vbe multiplier is a very simple shunt regulator, with low loop gain, and hence shows a significant series resistance. In other words, the Vbias generated shows unwanted variations in voltage with changes in the standing current through it. R3 is added to give first-order cancellation of Vbias variations caused by these current changes. It subtracts a correction voltage proportional to this current. Rather than complete cancellation, this gives a peaking of the output voltage at a specified current, so that current changes around this peak value cause

Table 22.6. Creating a lower tempco by varying V1

V1 mV	Vbias V	R1 Ω	Tempco mV/C
0	1.287	470	−3.6
100	1.304	390	−3.3
200	1.287	330	−3.1
300	1.286	260	−2.8
400	1.285	190	−2.5

only minor voltage variations. This peaking philosophy is widely used in IC bias circuitry.

R3 should never be omitted, as without it mains voltage fluctuations can seriously affect Vq. Table 22.4 shows that the optimal value for peaking at 6 mA depends strongly on the Vbe multiplication factor.

Figure 11.18 in Chapter 11 demonstrates the application of this method to the Class-B amplifier. The graph in Figure 11.19 shows the variation of Vbias with current for different values of R3. The slope of the uncompensated (R3 = 0) curve at 6 mA is approximately 20 Ω, and this linear term is cancelled by setting R14 to 18 Ω in Figure 11.18.

The current through the bias generator will vary because the VAS current source is not a perfect circuit element. Biasing this current source with the usual pair of silicon diodes does not make it wholly immune to supply-rail variations. I measured a generic amplifier (essentially the original Class-B Blameless design) and varied the incoming mains from 212 V to 263 V, a range of 20%. This in these uncertain times is perfectly plausible for a power amplifier travelling around Europe. The VAS current-source output varied from 9.38 mA to 10.12 mA, which is a 7.3% range. Thanks to the current-compensating resistor in the bias generator, the resulting change in quiescent voltage Vq across the two Res is only from 1.1 mV (264 V mains) to 1.5 mV (212 V mains). This is a very small absolute change of 0.4 mV, and within the Vq tolerance bands. The ratio of change is greater, because Vbias has had a large fixed quantity (the device Vbes) subtracted from it, so the residue varies much more. Vq variation could be further suppressed by making the VAS current source more stable against supply variations.

The finite ability of even the current-compensated bias generator to cope with changing standing current makes a bootstrapped VAS collector load much less attractive than the current-source version; from the above data, it appears that Vq variations will be at least three times greater.

A quite different approach reduces Vbias variations by increasing the loop gain in the Vbe multiplier. Figure 22.33 shows the circuit of a two-transistor version that reduces the basic resistance slope from 20 to 1.7 Ω. The first transistor is the sensor. An advantage is that Vbias variations will be smaller for all values of VAS current, and no optimisation of a resistor value is required. A drawback is slightly greater complexity in an area where reliability is vital. Figure 22.34 compares the two-transistor configuration with the standard version (without R3). Multi-transistor feedback loops raise the possibility of instability and oscillation, and

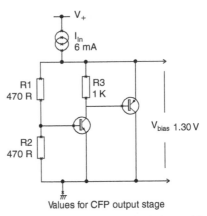
Values for CFP output stage

Figure 22.33. Circuit of a two-transistor Vbe multiplier. The increased loop gain holds Vbias more constant against current changes.

this must be carefully guarded against, as it is unlikely to improve amplifier reliability.

This section of the Thermal Dynamics chapter describes simple Vbias generators with tempcos ranging from −2.5 to −6.9 mV/°C. It is hoped that this, in combination with the techniques described earlier, will enable the design of Class-B amplifiers with greater bias accuracy, and therefore less afflicted by crossover distortion. However, there is another factor which causes quiescent conditions to vary, and which must be considered when setting the current-compensation. This is dealt with in the next section.

Early Effect in Output Stages

There is another factor that affects the accuracy with which quiescent conditions can be maintained. If you take a typical power amplifier (with an unregulated power supply) and power it from a variable-voltage transformer, you are very likely to find that Vq varies with the mains voltage applied. This at first seems to indicate that the apparently straightforward business of compensating the bias generator for changes in standing current has fallen somewhat short of success. However, even if this compensation appears to be correct, and the constant-current source feeding the bias generator and VAS is made absolutely stable, the quiescent conditions are still likely to vary. At first this seems utterly mysterious, but the true reason is that the transistors in the output stage are reacting directly to the change in their collector-emitter voltage (Vce). As Vce increases, so does the Vq and the quiescent current. This is called Early effect. It is a narrowing of the base-collector

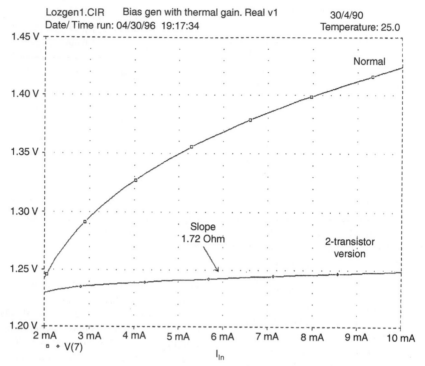

Figure 22.34. The two-transistor configuration gives a consistently lower series resistance, and hence Vbias variation with current, compared with the standard version without R3.

region as Vce increases, which will cause an increase in the collector current Ic even if Vbe and Ib are held constant. In a practical EF output stage, the result is a significant variation in quiescent conditions when the supply voltage is varied over a range such as ±10%.

Table 22.7 shows the effect as demonstrated by SPICE simulation, using MJE340/50 for drivers and MJ15022/23 as output devices, with fixed bias voltage of 2.550 V, which gave optimal crossover in this case. It is immediately obvious that (as usual) things are more complicated than they at first appear. The Vq increases with rail voltage, which matches reality. However, the way in which this occurs is rather

unexpected. The Vbes of the drivers Q1 and Q2 reduce with increasing Vce, as expected. However, the output devices Q3 and Q4 show a Vbe that increases — but by a lesser amount, so that after subtracting all the Vbe drops from the fixed bias voltage the aggregate effect is that Vq, and hence quiescent current Iq, both increase. Note that the various voltages have been summed as a check that they really do add up to 2.550 V in each case.

Table 22.8 has the results of real Vbe measurements. These are not easy to do, because any increase in Iq increases the heating in the various transistors, which will cause their Vbes to drift. This happens to such an

Table 22.7. SPICE Vbe changes with supply rail voltage (MJE340/50 and MJ15022/3). All devices held at 25°C

±rail V	Vq mV	Q1 Vbe mV	Q3 Vbe mV	Q2 Vbe mV	Q4 Vbe mV	Sum V
10	7.8	609	633	654	646	2.550
20	13	602	640	647	648	2.550
30	18	597	643	641	649	2.550
40	23	593	647	637	650	2.550
50	28	589	649	634	650	2.550

Table 22.8. Real Vbe changes with supply rail voltage (2SC4382, 2SA1668 drivers and 2SC2922, 2SA1216 output)

±rail V	Vq mV	Q1 Vbe mV	Q3 Vbe mV	Q2 Vbe mV	Q4 Vbe mV	Sum V
40	1.0	554	568	541	537	2.201
45	1.0	544	556	533	542	2.176
50	1.0	534	563	538	536	2.172
55	1.0	533	549	538	540	2.161
60	1.0	527	552	536	535	2.151
65	1.0	525	540	536	539	2.141
70	1.0	517	539	537	539	2.133

extent that sensible measurements are impossible. The measurement technique was therefore slightly altered. The amplifier was powered up on the minimum rail voltage, with its Vq set to 1.0 mV only. This is far too low for good linearity, but minimises heating while at the same time ensuring that the output devices are actually conducting. The various voltages were measured, the rail voltage increased by 5 V, and then the bias control turned down as quickly as possible to get Vq back to 1.0 mV, and the process is repeated. The results are inevitably less tidy as the real Vbes are prone to wander around by a millivolt or so, but it is clear that in reality, as in SPICE, most of the Early effect is in the drivers, and there is a general reduction in aggregate Vbe as rail voltage increases. The sum of Vbes is no longer constant as Vq has been constrained to be constant instead.

It may seem at this point as if the whole business of quiescent control is just too hopelessly complicated. Not so. The cure for the Early effect problem is to overcompensate for VAS standing current changes, by making the value of resistor R3 described above larger than usual. The best and probably the only practical way to find the right value is the empirical method. Wind the HT up and down on the prototype design with a variable transformer and adjust the value of R3 until the Vq change is at a minimum. (Unfortunately this interacts with the bias setting, so there is a bit of twiddling to do — however, for a given design you only need to find the optimal value for R3 once.) The resistance value will be a good deal larger than that required to merely compensate for changes in the output of the VAS current source; in one design, with a negative feedback biased current source, R3 = 16 Ω gave optimal rejection of current source changes, but 100 Ω was the value required to give minimal change in Vq as the variable transformer was wound up and down over a mains

range from 80% to 110%. The results (for a different amplifier, but with the same output stage configuration) are shown in Table 22.9. It is a good question as to how much this effect will vary between different specimens of the same output transistor type; right now I have no answers on that.

If R3 is as high as 100 Ω, the extra voltage drop across R3 will be between 600 mV and 1V, depending on the VAS standing current, and this may reduce the positive output swing slightly. This simple method assumes that the supply-rail rejection of the VAS current source and its biasing circuitry is predictable and stable; with the circuits normally used, this seems to be the case, but some further study in this area is required. A potential problem is that the current source biasing circuitry is likely to include RC filtering to prevent rail ripple getting in, and this could introduce a delay so that rapid mains variations are not properly compensated.

Thermal Dynamics by Experiment

One of the main difficulties in the study of amplifier thermal dynamics is that some of the crucial quantities, such as transistor junction temperatures, are not directly measurable; this is why simulation is so important in this

Table 22.9. Changes in Vq with mains voltage

Mains voltage (%)	Vq in mV with R3 = 16Ω	Vq in mV R3 = 100Ω
110	18.2	14.6
100	14.4	14.4
90	10.6	14.3
80	8.0	14.2

field. However, some insight into the way that bias conditions are altering can be obtained by observing changes in the THD residual as viewed on a scope or recorded against time. This does not of course tell you anything about how the various contributions to the bias state are varying — you just get the single result as a THD figure.

At the end of the day, what really matters is the crossover distortion produced by the output stage, and measuring this gets to the heart of the matter. One method I have used with success works as follows. The amplifier under study is deliberately underbiased by a modest amount. I choose a bias setting that gives about 0.02% THD with a peak responding measurement mode. This creates crossover spikes that are clear of the rest of the THD residual, to ensure the analyser is reading these spikes and ignoring noise and other distortions at a lower level. The Audio Precision System-1 and SYS-2702 have a mode that plots a quantity against time (it has to be said that the way to do this is not at all obvious from the AP screen menus — essentially 'time' is treated as an external stimulus — but it *is* in the manual) and this effectively gives that most desirable of plots — crossover conditions against time. In both cases below the amplifier was turned on with the input signal already present, so that the power dissipation stabilised within a second or so.

It takes a significant time for the testgear to take a THD reading, and this limits the time resolution of the results.

Crossover Distortion Against Time: Some Results

The first test amplifier examined has a standard EF output stage. The drivers have their own small heatsinks and have no thermal coupling with the main output device heatsink. The most important feature is that the bias sensor transistor is not mounted on the main heatsink, as is usual, but on the back of one of the output devices, as I recommended above. This puts the bias sensor much closer thermally to the output device junction. A significant feature of this test amplifier is its relatively high supply rails. This means that even under no load, there is a drift in the bias conditions due to the drivers heating up to their working temperature. This drift can be reduced by increasing the size of the driver heatsinks, but not eliminated. Figure 22.35 shows the THD plot taken over 10 minutes, starting from cold and initiating some serious power dissipation at $t = 0$. The crossover distortion drops at once; Figure 22.1 at the start of this chapter shows that driver dissipation is not much affected by output level, so this appears be due to the output device junctions

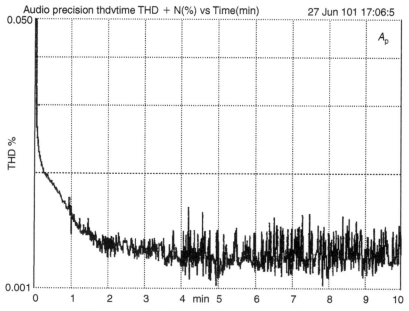

Figure 22.35. Peak THD vs time over 10 minutes.

heating up and increasing Vq. There is then a slower reduction until the THD reading stabilises at about 3 minutes.

The second amplifier structure examined is more complex. It is a triple-EF design with drivers and output devices mounted on a large heatsink with considerable thermal inertia. The pre-drivers are TO220 devices mounted separately without heatsinks. It may seem perverse to mount the drivers on the same heatsink as the outputs, because some of the time they are being heated up rather than cooled down, which is exactly the opposite of what is required to minimise Vbe changes. However, they need a heatsink of some sort, and given the mechanical complications of providing a separate thermally isolated heatsink just for the drivers, they usually end up on the main heatsink. All that can be done (as in this case) is to mount them on the heatsink in the area that stays coolest in operation. Once more the bias sensor transistor is not mounted on the main heatsink, but on the back of one of the output devices. See Figure 22.36 for the electrical circuit and thermal coupling paths.

The results are quite different. Figure 22.37 shows at A the THD plot taken over 10 minutes, again starting from cold and initiating dissipation at $t = 0$. Initially THD falls rapidly, as before, as the output device junctions heat. It then commences a slow rise over 2 minutes, indicative of falling bias, and this represents the time lag in heating the sensor transistor. After this

there is a much slower drift downwards, at about the same rate as the main heatsink is warming up. There are clearly at least three mechanisms operating with very different time-constants. The final time-constant is very long, and the immediate suspicion is that it must be related to the slow warming of the main heatsink. Nothing else appears to be changing over this sort of timescale. In fact, this long-term increase in bias is caused by cooling of the bias sensor compared with the output device it is mounted on. This effect was theoretically predicted above, and it is pleasing to see that it really exists, although it does nothing but further complicate the quest for optimal Class-B operation. As the main heatsink gets hotter, the heat losses from the sensor become more significant, and its temperature is lower than it should be. Therefore the bias voltage generated is too high, and this effect grows over time as the heatsink warms up.

Knowledge of how the long-term drift occurs leads at once to a strategy for reducing it. Adding thermal insulation to cover the sensor transistor, in the form of a simple pad of plastic foam, gives plot B, with the long-term variation reduced. Plot C reduces it still further by more elaborate insulation; a rectangular block of foam with a cut-out for the sensor transistor. This is about as far as it is possible to go with sensor insulation; the long-term variation is reduced to about 40% of what it was. While this technique certainly appears to improve bias control, bear in mind that it is

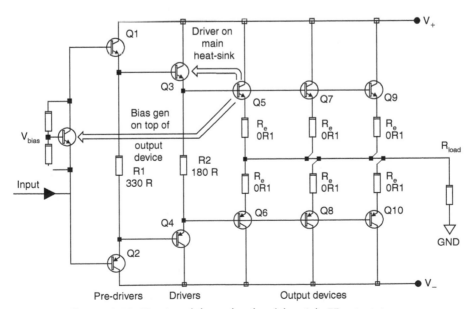

Figure 22.36. Circuit and thermal paths of the triple-EF output stage.

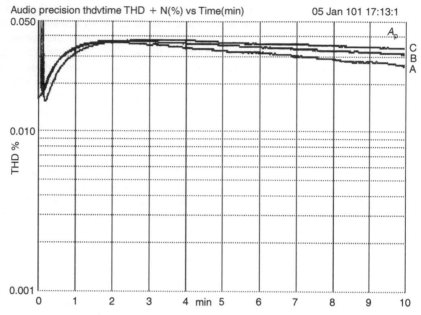

Figure 22.37. Peak THD versus time over 10 sec.

being tested with a steady sinewave. Music is noted for not being at the same level all the time, and its variations are much faster than the slow effect we are examining. It is very doubtful if elaborate efforts to reduce sensor cooling are worthwhile. I must admit this is the first time I have applied thermal lagging to an amplifier output stage.

More Measurements: Conventional and ThermalTrak

I recently revisited this technique to investigate the ThermalTrak transistors described earlier in this chapter. I used the NJL0281 (NPN) and NJL0302 (PNP) pair. The test amplifier was essentially the Load-Invariant design with the output stage converted from CFP to EF Type 2. The maximum power was 20 W/8R but the amplifier was run at 8.3 W output to maximise the dissipation; it is always a good idea to keep the power level of experimental amplifiers low if possible as they are much more tolerant of misplaced probes and slipping screwdrivers.

The main heatsink was deliberately small to speed up heating and cooling; it got warm but not hot in 200 seconds; measuring it with the Mk1 fingertip, I would say about 40°C. The driver heatsinks were made large compared with the power level, to minimise driver heating and keep the thermal situation simpler. They

stayed at ambient temperature. The first three tests used conventional compensation with a simple Vbe-multiplier to establish some basis for comparisons. The ThermalTrak diodes were not used.

The first test, in Figure 22.38, had the Vbe-multiplier sensor transistor mounted on top of one of the drivers. Because these stayed cold, the amplifier was definitely undercompensated, with the bias level steadily increasing, and therefore the crossover distortion steadily decreasing. Note the initial transient lasting about 20 seconds. I assume — but I do *not* know for certain at this point — that this is due to output device junction heating while a temperature gradient junction-case-heatsink is being established.

Undercompensated so crossover distortion decreases as bias increases with time.

Next the sensor was moved to a conventional position on the main heatsink, about 2 cm from one of the output devices; see Figure 22.39. The long-term result is much better; but somewhat overcompensated as the sensor Vbe is being multiplied 4 times but the drivers are staying cold. The bias therefore slowly decreases over 200 seconds. The initial transient appears unchanged.

In the third test the sensor was mounted on top of one of the output devices, with a silicone washer between them; see Figure 22.40. This is the sensor position which was recommended earlier in this chapter. The initial transient is now faster (10 sec rather than 20 sec) but no

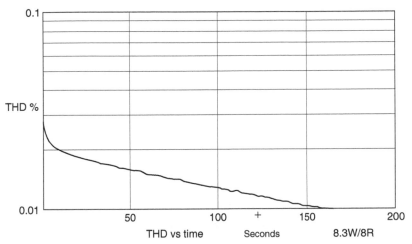

Figure 22.38. Sensor transistor mounted on the driver heatsink. Undercompensated so crossover distortion decreases as bias increases with time.

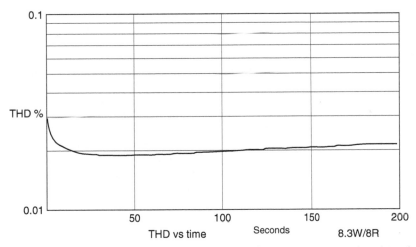

Figure 22.39. Sensor transistor mounted on the main heatsink, somewhat overcompensated, so bias decreases with time.

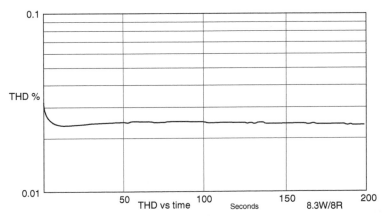

Figure 22.40. Sensor transistor mounted on top of one of the output devices. Good compensation over 200 seconds.

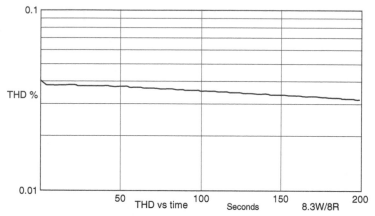

Figure 22.41. ThermalTrak diodes running at 9.2 mA in series with the sensor Vbe-multiplier; clearly somewhat undercompensated.

smaller, which is what might expected. The long-term compensation is also good, though that is more luck than judgement — the amount of thermal coupling to the sensor happens to be about right. This is intended to represent the best that conventional compensation can do.

Now we make use of the ThermalTrak diodes for the first time. They were simply put in series with the sensor Vbe-multiplier, as shown in Figure 22.27b above. The Vbe-multiplier was turned down in voltage as it is now only compensating for the driver Vbes, and the sensor was moved back to one of the driver heatsinks. Everything there remained cold, so we should be able to see clearly how the ThermalTrak diodes compensate their associated transistors. The VAS current was 9.2 mA.

The results are seen in Figure 22.41. The initial transient is now both smaller and faster, but over 200 seconds the amplifier is somewhat undercompensated. This was expected, because as stated earlier in this chapter, it appears that the transistor Vbe temperature coefficient is -2.14 mV/°C while the diode has -1.7 mV/°C, at a current of 25 mA.[10] The initial transient is a lot faster but still measured in seconds.

In the ThermalTrak discussions on the diyAudio Forums, several people suggested that the discrepancy in temperature coefficients could be corrected by running the diodes at a much lower current; diode tempcos increase slowly as current is reduced (by approx. 0.2 mV/°C per decade decrease of current).

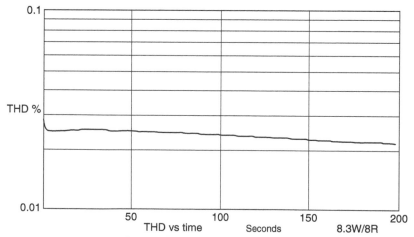

Figure 22.42. ThermalTrak diodes running at 800 uA to increase their tempco; a puzzlingly small improvement, and still definitely undercompensated.

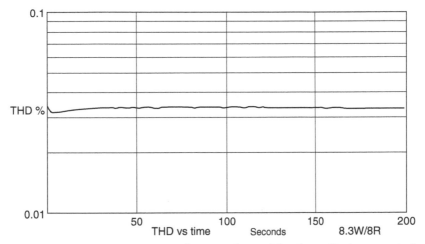

Figure 22.43. Sensor transistor mounted on top of one of the ThermalTrak output devices.

The diode tempco needs to be increased by 0.44 mV/°C, so the diode current must be reduced by 2.2 decades, or 158 times. That gives us a diode current of only 158 uA (no, not a typo − that's just how the numbers work out). This is much too low for a VAS operating current so the diodes were moved to a circuit that replicated the diode voltage at much greater current in the VAS collector. The ThermalTrak diodes were actually fed with 800 uA, rather than 158 uA, for practical reasons, so we do not expect perfect compensation, but it should be much better. This is where theory and practice diverge, because the results seen in Figure 22.42 are better, but only slightly so; the THD change over 200 seconds is 0.006% rather than the 0.008% in the previous test.

For the next test I decided to radically increase the ThermalTrak diode tempco rather than just tweaking it. The diodes are once more passing about 9.2 mA but their Vf is now multiplied by a factor of two. The long-term compensation as seen in Figure 22.43 is now quite good, though our theory says it should be seriously over-compensated as 2×1.7 mV/°C $= 3.4$ mV/°C, much more than 2.14 mV/°C. There is also a rather worrying dip in the first 20 seconds − clearly the internal diodes are not following the transistor junction temperatures rapidly. As noted earlier in this chapter, the position of the sense diode on the lead frame means it is much closer in temperature to the heatsink than the junction.

The final two tests were intended to investigate the greater speed of compensation response that the internal

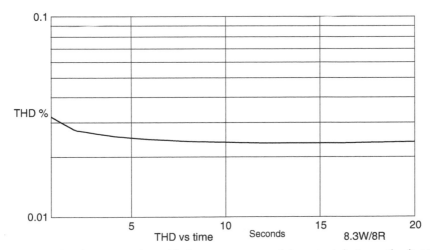

Figure 22.44. Conventional compensation with the sensor on top of the output device − the first 20 seconds.

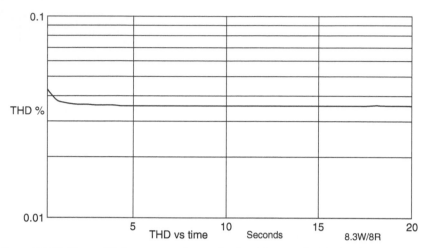

Figure 22.45. ThermalTrak compensation – the first 20 seconds. Diode current 9.2 mA.

diodes should give. Figure 22.44 zooms in on the first 20 seconds of the response for conventional compensation with the sensor on top of the output device, as shown in Figure 22.40 above. Figure 22.45 similarly shows the first 20 seconds of the first ThermalTrak experiment as in Figure 22.41. It is clear that the latter has a faster initial transient. For conventional compensation the time-constant (time to reach 63% of final value) is about 2.5 seconds, while for the ThermalTrak case it is about 0.5 seconds. I must admit I was expecting the ThermalTrak response to be faster than that, and I suspect there is more here than meets the eye.

You will have gathered that this is work in progress, but it clearly shows now that the bias problem is more complex than it looks. I cannot at present demonstrate the perfect biasing system, any more than I can demonstrate the perfect amplifier.

While the ThermalTrak transistors do not allow direct sensing of the junction temperature, the addition of the diode does give access to more information about the thermal situation in an output device. With suitable circuitry that information might be used to give an indirect estimate of the junction temperature.

References

1. Sato, T et al., *Amplifier Transient Crossover Distortion Resulting from Temperature Change of Output Power Transistors,* AES Preprint. AES Preprint 1896 for 72nd Convention, Oct. 1982.

2. Brown, I, Opto-Bias Basis for Better Power Amps, *Electronics World*, Feb. 1992, p. 107.

3. Carslaw, H S and Jaeger, J C, *Conduction of Heat in Solids*, Oxford: Oxford University Press, 1959.

4. Murphy, D, Axisymmetric Model of a Moving-Coil Loudspeaker, *JAES*, Sept. 1993, p. 679.

5. Motorola, Toulouse, Private communication, Dec. 2008.

6. Evans, J, Audio Amplifier Bias Current, Letters, *Electronics & Wireless World*, Jan. 1991, p. 53.

7. Chen, C-T, *Analog and Digital Control System Design*, Saunders-HBJ, 1993, p. 346.

8. Harriot, P, *Process Control*, New York: McGraw-Hill, 1964, pp. 100–102.

9. Liptak, B (ed.), *Instrument Engineer's Handbook-Process Control*, Oxford: Butterworth-Heinemann, 1995, p. 66.

10. http://www.diyaudio.com/forums/solid-state/71534-semi-thermaltrak.html (accessed 21 May 2012).

11. Cordell, B, DIYAudio Forums, Biasing and Thermal Compensation of Thermal Trak Transistors, available at: http://www.diyaudio.com/. Page 5.htm.

The Design of DC Servos

In the section of this book dealing with input stages I have gone to some lengths to demonstrate that a plain unassisted amplifier — if designed with care — can provide DC offset voltages at the output which are low enough for most practical purposes, without needing either an offset-nulling preset or a DC servo system. For example, the Trimodal amplifier can be expected not to exceed ±15mV at the output. However, there may be premium applications where this is not good enough. In this case the choice is between manual adjustment and DC servo technology. As precision opamps have got cheaper, the use of DC servos has increased.

DC Offset Trimming

Pre-set adjustment to null the offset voltage has the advantage that it is simple in principle and most unlikely to cause any degradation of audio performance. In servicing, the offset should not need renulling unless one of relatively few components are changed; the input devices have the most effect, because the new parts are unlikely to have exactly the same beta, but the feedback resistors also have some influence as the input stage base currents flow through them.

The disadvantages are that an extra adjustment is required in production, and since this is a set-and-forget pre-set, it can have no effect on DC offsets that may accumulate due to input stage thermal drift or component ageing.

Figure 23.1 shows one simple way to add a DC trim control to an amplifier, by injecting a small current of whatever polarity is required into the feedback point. Since the trim circuit is powered from the main HT rails, which are assumed to be unregulated, careful precautions against the injection of noise, ripple and

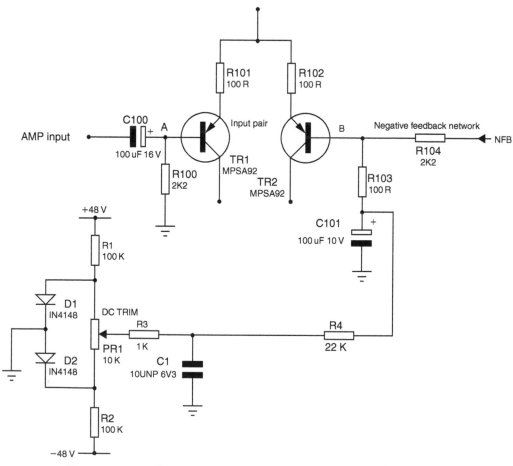

Figure 23.1. DC offset trim with injection into the negative feedback network.

DC fluctuations must be taken. The diodes D1, D2 set up a stable voltage across the potentiometer. They do of course have a thermal coefficient, but this is not likely to be significant over the normal temperature range. R3 and C1 form a low-pass filter to reduce noise and ripple, and the trimming current is injected through R4. This resistor has a relatively high value to minimise its effect on the closed-loop gain, and to give a powerful filtering action in conjunction with the large value of C101, to remove any remaining noise and ripple. Note that the trim current is injected at the bottom of R103, and not into the actual feedback point at B, as this would feed any disturbances on C1 directly into the amplifier path. From the point of view of the amplifier, R4 is simply a resistance to ground in parallel with R101, so its effect on the gain can be easily taken into account if required. This DC trim circuit should not degrade the noise performance of the amplifier when it is added, even though the amplifier itself is unusually quiet due to the low impedance of the feedback network.

So long as the input is properly AC-coupled (DC-blocked) the trim current can also be fed into the input at point A, but the possible effect on the noise and hum performance is less predictable as the impedance feeding the amplifier input is not known.

DC Offset Control by Servo-loop

A DC servo system (presumably so-called to emphasise that it does not get directly involved in the main feedback loop) provides continuous active nulling of the amplifier offset by creating another feedback path that has a high gain at DC and very low frequencies, but limited control of the output DC level. This second path uses an opamp, usually configured as an integrator, to perform the feedback subtraction in which the output DC level is compared with ground. It is straightforward to select an opamp whose input offset specification is much better than the discrete input stage, because DC precision is where opamp technology can really excel. For example, both the Analog Devices AD711JN and OPA134 offer a maximum offset of ± 2 mV at 25 °C, rising to 3 mV over the full commercial temperature range. Performance of an order of magnitude better than this is available, e.g., the OPA627, but the price goes up by an order of magnitude too. FET input opamps are normally used to avoid bias current offsets with high-value resistors.

An unwelcome complication is the need to provide ± 15 V (or thereabouts) supply rails for the opamp, if it does not already exist. It is absolutely essential that this supply is not liable to drop-out if the main amplifier

reproduces a huge transient that pulls down the main supply rails. If it does drop out sufficiently to disrupt the operation of the servo, disturbances will be fed into the main amplifier, possibly causing VLF oscillation. This may not damage the amplifier, but is likely to have devastating results for the loudspeakers connected to it.

Advantages of DC Servos

1. The output opamp DC offset of the amplifier can be made almost as low as desired. The technology of DC precision is mature and well understood.
2. The correction process is continuous and automatic, unlike the DC trimming approach. Thermal drift and component ageing are dealt with, and there is only one part on which the accuracy of offset nulling depends – the servo opamp, which should not significantly change its characteristics over time.
3. The low-frequency roll-off of the amplifier can be made very low without using huge capacitors. It can also be made more accurate, as the frequency is now set by a non-electrolytic capacitor.
4. The use of electrolytics in the signal path can be avoided, and this will impress some people.
5. The noise performance of the power amplifier can be improved because lower value resistances can be used in the feedback network, yielding a very quiet amplifier indeed.

Points 3, 4, and 5 are all closely related, so they are dealt with at greater length below.

Basic Servo Configurations

Figure 23.2a shows a conventional feedback network, as used in the Load-Invariant amplifier in this book. The usual large capacitor C is present at the bottom of the feedback network; its function is to improve offset accuracy by reducing the closed-loop gain to unity at DC. Figure 23.2b shows a power amplifier with a DC servo added, in the form of a long-time-constant integrator feeding into the feedback point. C is no longer required, as the servo can do all the work of maintaining the DC conditions, though sometimes it might be a good idea to retain it to keep the DC loop gain of the servo system high, and so improve its accuracy; if you do, check carefully for LF stability, as you have introduced another time-constant. Note that the output of the integrator is at ground as far as audio frequencies are concerned, and so the addition of R3 puts it effectively in parallel with R2 and causes a small increase in closed-loop gain which must be taken into account.

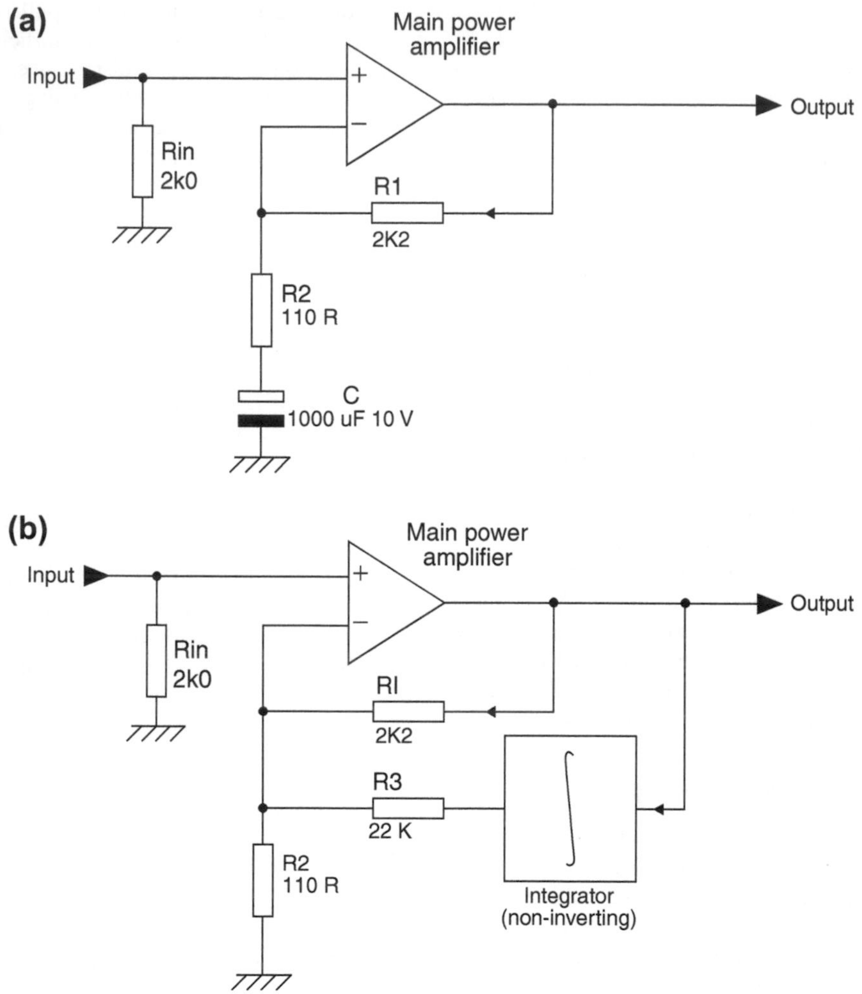

Figure 23.2. Power amplifiers without and with a DC servo in the feedback path.

It had better be said at once that if the integrator constant is suitably long, a negligible amount of the audio signal passes through it, and the noise and distortion of the main amplifier should not be degraded in any way; more on this later.

As with manual trimming, there are many ways to implement a DC servo. This method shown in Figure 23.2b works very well, and I have used it many times. One important point is that the integrator block must be non-inverting for the servo feedback to be in the correct phase. The standard shunt-feedback integrator is of course inverting, so something needs to be done about that. Several non-inverting integrators are examined below.

Injection of the servo signal into the input is possible, and in this case a standard inverting integrator can be used. However, as for manual trimming, using the input gives a greater degree of uncertainty in the operating conditions as the source impedance is unknown. If there is no DC blocking on the input, the DC servo will probably not work correctly as the input voltage will be controlled by the low impedance of a pre-amp output. If there is DC blocking, then the blocking capacitor may introduce an extra pole into the servo response, which, if nothing else, complicates things considerably.

Injection of the servo correction into the amplifier forward path is not a good idea as the amplifier has its own priorities — in particular, keeping the input pair exactly balanced. If, for example, you feed the servo output into the current-mirror at the bottom of the input pair, the main amplifier can only accommodate its control demands by unbalancing the input pair collector currents, and this will have dire effects on the high-frequency distortion performance.

Noise, Component Values, and the Roll-off

When you design an amplifier feedback network, there is a big incentive to keep the Johnson noise down by making the resistor values as low as possible. In the simple feedback network shown in Figure 23.2a, the source impedance seen by the input stage of the amplifier is effectively that of R2; if the rest of the amplifier has been thoughtfully designed, then this will be a significant contributor to the overall noise level. Since the Johnson noise voltage varies as the square root of the resistance, minor changes (such as allowing for the fact that R1 is effectively in parallel in R2) are irrelevant. Because of the low value of R2, the feedback capacitor C tends to be large as its RC time-constant with R2 (not R1 + R2) is what sets the LF roll-off. If R2 is low, then C is big, and practical values of C put a limit on how far R2 can be reduced. Hence there is a trade-off between low-frequency response and noise performance, controlled by the physical size of C.

When a DC servo is fitted, it is usual to let it do all the work, by removing capacitor C from the bottom arm of the negative feedback network. The components defining the LF roll-off are now transferred to the servo, which will use high-value resistors and small non-electrolytic capacitors. The value of R2 is no longer directly involved in setting the LF roll-off and there is the possibility that its resistance can be further reduced to minimise its noise contribution, while at the same time the LF response is extended to whatever frequency is thought desirable. The limit of this approach to noise reduction is set by how much power it is desirable to dissipate in R1.

There is a temptation to fall for the techno-fallacy that if it can be done, it should be done. A greatly extended LF range (say, below 0.5 Hz) exposes the amplifier to some interesting new problems of DC drift. A design with its lower point set at 0.1 Hz is likely to have its output wavering up and down by tens of milliVolts, as a result of air currents differentially cooling the input pair, introducing variations that are slow but still too fast for the servo to correct. Whether these perturbations are likely to cause subtle intermodulations in speaker units is a moot point; it is certain that it does not look good on an oscilloscope, and could cause reviewers to raise their eyebrows. Note that unsteady air currents can exist even in a closed box due to convection from internal heating.

A cascode input stage reduces this problem by greatly lowering the voltage drop across the input transistors, and hence their dissipation, package temperature, and vulnerability to air currents. While it has been speculated that an enormously extended LF range benefits reproduction by reducing phase distortion at the bottom of the audio spectrum, there seems to be no hard evidence for this, and in practical terms there is no real incentive to extend the LF bandwidth greatly beyond what is actually necessary.

Non-inverting Integrators

The obvious way to build a non-inverting integrator is to use a standard inverting integrator followed by an inverter. The first opamp must have good DC accuracy as it is here the amplifier DC level is compared with 0 V. The second opamp is wholly inside the servo loop so its DC accuracy is not important. This arrangement is shown in Figure 23.3. It is not a popular approach because it is perfectly possible to make a non-inverting integrator with one opamp. It does, however, have the advantage of being conceptually simple; it is very easy to calculate. The frequency response of the integrator is needed to calculate the low-frequency response of the whole system.

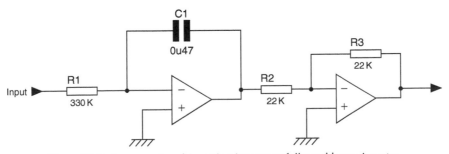

Figure 23.3. A conventional inverting integrator followed by an inverter.

The component values shown in Figure 23.3 give unity gain at 1 Hz.

The 2C Integrator

Figure 23.4 shows a non-inverting integrator that has often been used in DC servo applications, having the great advantage of requiring one opamp. It does, however, use two capacitors; if you are aiming for a really low roll-off, these can become quite large for non-electrolytics and will be correspondingly expensive. Despite the presence of two RC time-constants, this circuit is still a simple integrator with a standard −6dB/octave frequency response.

At the input is a simple RC lag, with the usual exponential time response to step changes; its deviation from being an integrator is compensated for by the RC lead network in the feedback network. A good question is, what happens if the two RC time-constants are not identical? Does the circuit go haywire? Fortunately not. A mismatch only causes gain errors at very low frequencies, and these are unlikely to be large enough to be a problem. An RC mismatch of ±10% leads to an error of ±0.3 dB at 1.0 Hz, and this error has almost reached its asymptote of ±0.8 dB at 0.1 Hz.

The frequency domain response of Figure 23.4 is:

$$A = \frac{1}{j\varpi RC}$$ Equation 23.1

where $\omega = 2\pi f$ exactly as for the simple integrator of Figure 23.3. The values shown give unity gain at 1 Hz.

The 1C Integrator

Figure 23.5 displays an apparently superior non-inverting integrator circuit that requires only one opamp and one capacitor. This is sometimes called a Deboo integrator after its inventor Gordon Deboo. How it works is by no means immediately obvious, but work it does. R1 and C1 form a simple lag circuit at the input. By itself, this naturally does not give the desired integrator response of a steadily rising or falling capacitor voltage as a result of a step input; instead it gives the familiar exponential response, because as the capacitor voltage rises, the voltage across R1 falls, and the rate of capacitor charging is reduced. In this circuit, however, as the capacitor voltage rises, the output of the opamp rises at twice the rate, due to the gain set up by R3 and R4, and so the increasing current flowing into C1 through R2 exactly compensates for the decreasing current flowing through R1, and the voltage on C1 rises linearly, as though it were being charged from a constant current source. This is in fact the case, because the circuit can be viewed as equivalent to a Howland current source driving into a capacitor.

As for the previous circuit, doubts may be entertained as to what happens when the compensation is less than perfect. For example, here it depends on R1 and R2 being the same value, and also the equality of R3 and R4, to set a gain of exactly two. Note that R3 and R4 can be high value resistors. Stray capacitances are dealt with by the addition of C2, which in most cases will be found to be essential for the HF stability of this configuration; this extra capacitor somewhat

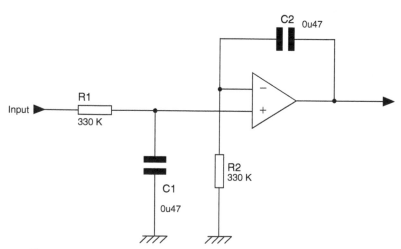

Figure 23.4. A non-inverting integrator that requires only one op-amp.

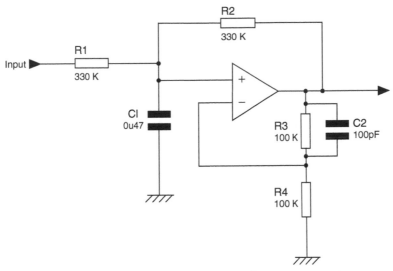

Figure 23.5. A non-inverting integrator that requires only one opamp and one capacitor.

detracts from the economy of the circuit, but it will be a small ceramic type and of much less cost than the non-electrolytic capacitor used to set the integrator time-constant.

The frequency domain response is now different:

$$A = \frac{1}{j\varpi\frac{R}{2}C}$$

Equation 23.2

where $R = R1 = R2$.

The R/2 term appears because C1 is now being charged through two equal resistors R1 and R2. The values shown therefore give unity gain at 2 Hz.

Choice of Integrator

The 1C integrator is clearly the most economical, because big non-electrolytic capacitors are relatively expensive, and I have used it successfully in several applications where it was appropriate. However, it does have some non-obvious disadvantages. If there is a significant power amplifier offset to be servo-ed out, the accuracy with which it is done depends rather critically on the matching of the two resistor pairs R1-R2 and R3-R4 in Figure 23.5. This holds even if a perfect servo opamp with zero input offset voltage of its own is assumed.

A significant offset typically occurs when the bases of the two transistors in the power amplifier input

differential pair are fed from resistances of very different values. Looking at Figure 23.2b, R2 connects the inverting input to ground with a low resistance of 110 Ω, the value being kept as low as possible to minimise Johnson noise. A resistor Rin is connected from the non-inverting input to ground, to define the DC conditions, but this is typically much larger, so as not to load unduly the signal source. It is usually of the order of 2 kΩ if there is some opamp circuitry (such as a balanced input amplifier) upstream, as this is high enough to avoid excessively loading an opamp and so introducing distortion. However, it could be a good deal higher at 10 kΩ or more if the amplifier is intended to be driven directly from the outside world. Even if we assume exactly equal base currents, the much higher value of Rin will give a positive offset of tens of millivolts at the non-inverting input. This would not be the case in Figure 23.2a, which has a capacitor at the bottom of the NFB network, as it often possible to make Rin = R1 and so aim for offsets that are equal at each input and so cancel out.

In a respectable power amplifier the collector currents of the input pair should be almost exactly equal to minimise distortion (see Chapter 6 on input stages), but this does not mean that the base currents, or what would in an opamp spec sheet be called the input bias currents, are equal, as the input devices will have differing betas.

To take a real example, an amplifier as in Figure 23.2b with Rin = 2k2 and R2 = 110 Ω gave an offset of +26 mV on the non-inverting input; if the input transistors had had the minimum beta on their spec sheet, it could have been several times greater. Using this value in a SPICE simulation using a 1C servo circuit as per Figure 23.5, with R3 = 2k2 and zero opamp offset gave a highly satisfactory offset of +37 uV at the power amp output. But ... this simulation had both resistor pairs R1-R2 and R3-R4 set to be exactly correct. If R3 was set just 1% high, the power amp output offset leapt up to +29 mV; when it was 1% low, the output offset was −31 mV. Deviations of 1% in the values of R1, R2 gave similar errors.

This is a very good illustration of the caution you need to apply to simulator results; it is not obvious on inspecting the circuit that its operation depends crucially on perfectly matched resistors. The simulator answer is absolutely correct, but not applicable to the real world of imperfect components.

Another disadvantage of the 1-C circuit is that when you use a real opamp, as opposed to a simulated perfect one, its own input offset appears doubled at the power amplifier output, due to the gain of two set up by R3 and R4. If you do not have access to the opamp offset-null pins, there is no easy way to add a DC trimming network, as connecting even high-value resistors such as 10 M disturbs the balance of this circuit and stops it working properly.

Having examined the quite serious limitations of the 1C non-inverting integrator, let's go back to the 2C version and see if that is more tractable.

Firstly, the 2C circuit does not require accurately matched resistors to work properly. In Figure 23.4, R1-R2 mismatch has a negligible effect on the DC accuracy, and only a microscopic effect on the AC

response below 1 Hz. The opamp offset is not multiplied by two.

Secondly, an important point is that it is now possible to add a DC trimming network without disrupting the integrator's operation. This can be used to null to zero the small offset (typically 1−2 mV) that remains when a servo is added. If the opamp used has offset-null pins, then these should be used with whatever nulling circuitry the manufacturer recommends, but for economy it is often the case that the servo is one half of a dual opamp with no offset-null pins, the other half typically being used for over-temperature detection. If this is so, a DC trimming network can be added to the 2C circuit − unlike the 1C version. Figure 23.6 shows a network that can be added to allow nulling to less than a millivolt. Its range of adjustment is limited to only ±5 mV at the power amp output. The component values are those used in a production amplifier − the negative feedback network had the values as shown in Figure 23.2b − note, however, that the integrator resistors R1, R2 have been changed to 180 K and the servo injection resistor Rinj has been changed from 22 k to 2k2 to obtain the desired LF roll-off frequency.

Figure 23.6 does not include any filter components to prevent noise or hum on the ±15 V rails from entering the servo; details on how to do this are given in Figure 23.1. It should be pointed out that while this sort of external nulling is usually quite satisfactory, it will not perform as well over a wide temperature range as using the official offset-null pins if they are available.

It can be concluded that in most cases the 2C integrator is superior to the 1C version. It is true that two capacitors are needed instead of one, but in many

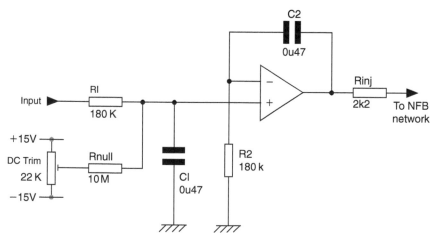

Figure 23.6. Adding a DC trimming network to the 2C servo integrator.

cases the price is well worth paying for better and more predictable servo performance.

Choice of Opamps

All of these integrator circuits use high resistor values to keep the size of the capacitors down. It is essential to use FET-input opamps, with their near-zero bias and offset currents. Bipolar opamps have many fine properties, but they are not useful here. You will need a reasonably high quality FET opamp to beat non-servo power amplifiers, which can be designed so their output offset does not exceed ±15 mV offset at the output.

Table 23.1 presents some prime candidates, giving the maximum ± offset voltages, and the relative cost at the time of writing.

In many designs a dual opamp is the best choice, the remaining section being used as a comparator driven by an over-temperature detection device such as a thermistor. If the opamp is accurate enough to do its job as a servo, it will almost certainly be good enough for temperature detection.

Servo Authority

The phrase 'servo authority' refers to the amount of control that the DC servo system has over the output DC level of the amplifier. It is, I hope, clear that the correct approach is to design a good input stage that gives a reasonably small DC offset unaided, and then add the servo system to correct the last few dozen millivolts, rather than to throw together something that needs to be hauled into correct operation by brute-force servo action.

In the latter case, the servo must have high authority in order to do its job, and if the servo opamp dies and its output hits one of its rails, the amplifier will follow suit. The DC offset protection should come into action to prevent disaster, but it is still an unhappy situation.

Table 23.1. Opamp specs compared

Type	Offset at 25°C (mV)	Offset over −40 to +85°C (mV)	Relative cost
TLO51	1.5	2.5	1.00
OPA134	2	3	1.34
AD711JN	2	3	1.48
OPA627AP	0.28	0.5	16.0

Note that the TL051 looks like quite a bargain, and going for a serious improvement on this with the OPA627AP will cost you deep in the purse.

However, if the input stage is well designed, so the servo is only called upon to make fine adjustments, it is possible to limit the servo authority, by proportioning the circuit values so that R3 in Figure 23.2 is relatively high. Then, even if the opamp fails, the amplifier offset will be modest. In many cases it is possible for the amplifier to continue to function without any ill effects on the loudspeakers. This might be valuable in sound reinforcement applications and the like.

Calculating the effects of opamp failure in the circuit of Figure 23.2 is straightforward. The system appears as a shunt-feedback amplifier where R3 is the input resistance and R1 is the feedback resistance. Thus if the opamp is working from ±15 rails, then, ignoring saturation effects, the main amplifier output will be displaced by ±1.5 V.

When limiting servo authority, it is of course essential to allow enough adjustment to deal with any combination of component tolerances that may happen along. Do not limit it too much.

Design of LF Roll-off Point

Calculating the frequency response of the servo-controlled system is surprisingly easy. The −3 dB point will occur where the feedback through the normal network and the integrating servo path are equal in amplitude; it is −3dB rather than −6dB because the two signals are displaced in phase by 90°. This is exactly the same as the −3 dB point obtained with a RC circuit, which happens at the frequency where the impedance of the R and C are equal in magnitude, though displaced in phase by 90°.

As a first step, decide what overall gain is required; this sets the ratio of R1 and R2. Next, determine how low R1 can conveniently be made, to minimise the noise contribution of R2. This establishes the actual values of R1 and R2. It is important to remember that the servo injection resistor R3, being connected to an effective AC ground at the servo opamp output, is effectively in parallel with R2 and has a small influence on the main amplifier gain. Thirdly, decide how low a −3dB point you require for the overall system, and what servo authority you are prepared to allow. I shall take 0.2 Hz as an example, to demonstrate how a servo system makes such a low value easy to attain. Using the values shown in Figure 23.2, the section above demonstrates that the servo authority is more than enough to deal with any possible offset errors, while not being capable of igniting the loudspeakers if the worst happens. R3 is therefore 22 k, which is ten times R2, so at the −3 dB point the integrator output must be ten times the main amplifier output; in other words, it must have a gain of ten at 0.2 Hz.

The next step is to choose the integrator type; the one opamp, one capacitor version of Figure 23.5 is clearly the most economical so we will use that. The frequency response equation given above is then used to set suitable values for R1, R2 and C1 in Figure 23.5. Non-electrolytic capacitors of 470 nF are reasonably priced and this gives a value for R1, R2 of 338 kΩ; the preferred value of 330 k is quite near enough.

Servo Overload

The final step is to check that the integrator will not be overdriven by the audio-frequency signals at the amplifier output, bearing in mind that the opamp will be running off lower supply rails that are half or less of the main amplifier rail voltages. Here I will assume the amplifier rails are ±45 V, i.e., three times the ±15 V opamp rails. Hence the integrator will clip with full amplifier output at the frequency where integrator gain is 1/3. The integrator we have just designed has a gain of ten times at 0.2 Hz and a slope of −6dB per octave, so its gain will have fallen to unity at 2 Hz, and to 1/3 at 6 Hz. Hence the integrator can handle any amplifier output down to maximum power at 6 Hz, which is somewhat below the realm of audio, and all should be well.

Servo Testing

One problem with servo designs of this type is that they are difficult to test; frequencies of 0.2 Hz and below are well outside the capabilities of normal audio test equipment. It is not too hard to find a function generator that will produce the range 0.1 to 1.0 Hz, but measuring levels to find the -3dB point is difficult. A storage oscilloscope will give approximate results if you have one; the accuracy is not usually high.

One possibility is the time-honoured method of measuring the tilt on a low-frequency square wave. Accuracy is still limited, but you can use an ordinary oscilloscope. Even very low frequency roll-offs put an easily visible tilt on a 20 Hz square wave, and this should be fast enough to give reasonable synchronisation on a non-storage oscilloscope. Table 23.2 is a rough guide.

Performance Issues

The advantages of using a DC servo have been listed above, without mentioning any disadvantages, apart

Table 23.2. Tilt on 20 Hz square wave with different LF roll-off frequencies

−3dB point in Hz	Tilt (%)
0.15	2.5
0.23	3.5
0.32	5.0
0.50	7.4
0.70	10.5
1.0	15.2
1.4	20
2.1	28

Note that the tilt is expressed as a percentage of the zero to peak voltage, not peak-to-peak.

from the obvious one that more parts are required and a little power is needed to run the opamp. It could easily be imagined that another and serious drawback is that the presence of an opamp in the negative feedback network of an amplifier could degrade both the noise and distortion performance. However, this is not the case. When the system in Figure 23.2b is tested with a load-invariant amplifier, and an OPA134 opamp as a servo, there is no measurable effect on either quantity.

The distortion performance is unaffected because the servo integrator passes very little signal at audio frequencies. The noise performance is preserved because the integrators are very quiet due to their falling frequency response, and with the long integration constants used here, they are working at a noise-gain of unity at audio frequencies. Both parameters benefit from the fact that the servo feedback path via R3 has one-tenth of the gain of the main feedback path through R1.

Multipole Servos

All the servos shown above use an integrator and therefore have a single pole. It is possible to make servos that have more than one pole, and they have been used in some designs, though the motivation for doing it is somewhat unclear. The usual arrangement has a single opamp non-inverting integrator followed by a simple RC lag network that feeds into the feedback point. Naturally, once you have more than one pole in a system, there is the possibility of an under-damped response and gain peaking, so this approach demands careful design, not least because measuring gain peaking at 0.1 Hz is not that easy.

Amplifier and Loudspeaker Protection

And what price peace of mind?
　　　'Carrying No Cross' by the band UK

Categories of Amplifier Protection

Properly equipped power amplifiers require protection systems not only to protect them from adverse external conditions, such as a short-circuited output or blocked ventilation, but also to protect the loudspeakers from amplifier faults. The various hazards to be dealt with are:

1. Overload of the output. The protection of solid-state amplifiers against overload is largely a matter of safeguarding them from load impedances that are too low and endanger the output devices; the most common and most severe condition being a short across the output. Highly reactive loads with a moderate impedance can also cause excessive peaks in power dissipation. The word 'overload' here must be distinguished from the casual and erroneous use of the word to mean excessive signal levels that cause clipping and audible distortion.
2. Excessive output voltage excursions. These occur when the drive to a load with some inductance is abruptly cut off, usually by the overload protection system. This can destroy output devices unless they are protected by clamping or catching diodes.
3. DC offset faults. When a solid-state amplifier fails, the output often jams hard against one supply rail. The resulting DC flowing through loudspeakers can damage them very rapidly, so an offset-detect system is used to open the output circuit and so protect the loudspeakers from the amplifier.
4. Overheating. Failure of ventilation or excessive ambient temperature can cause output device failure. Usually a temperature-sensing device is placed on a heatsink and this triggers shutdown when the temperature becomes too high.
5. Output transient suppression. Many amplifiers, even those running from dual supply rails, generate size-able thumps or bangs when they are powered up. A turn-on delay ensures that the output relay is not closed until the internal conditions have had time to settle. Some designs generate annoying turn-off transients when the power is removed, and in bad cases these might be large enough to damage sensitive loudspeakers. Mains-fail detection allows the opening of the output circuit before the supply rails have decayed enough to induce transients.
6. Clipping detection. Solid-state amplifiers are unlikely to be damaged by sustained clipping, though it could in some cases put excessive stress on the power supply. The main danger is to delicate tweeters, and some amplifiers have clip-detect systems that merely signal infrequent clipping, but trigger shutdown if it is heavy and prolonged.
7. Internal protection. All amplifiers require internal fusing to minimise the consequences of a component failure, such as a short-circuit bridge rectifier or damaged output devices that draw fault currents from the supply rails. These in effect protect the amplifier from itself, and maintain safety in the event of a mains wiring fault. Fuses are usually fitted in both the DC supply rails and the transformer secondaries. Special 'anti-domino-effect' current-limiting circuitry may be built into the power amplifier itself to prevent the failure of one device from initiating a train of destruction.

Another consideration is safety testing. In Europe, this includes sequentially shorting every transistor collector-to-emitter to check that in no case does a hazard arise. This includes resistors bursting into flames, and, to prevent this, special fusible resistors are used at strategic points in the circuit. Under excessive current, these go open-circuit in a controlled way without the emission of flame. They are usually metal-film types with a special flame-retardant coating; if the resistor can affect the signal, be sure to check that it really is metal-film, as other types can introduce distortion. Fusible resistors are more expensive than the standard metal-film types and are therefore only fitted at critical points.

Semiconductor Failure Modes

Solid-state output devices have several main failure modes, including excess current, excess power dissipation, and excess voltage. These are specified in manufacturer's data sheets as Absolute Maximum Ratings, usually defined by some form of words such as 'exceeding these ratings even momentarily may cause degradation of performance and/or reduction in operating lifetime'. For semiconductor power devices, ratings are usually plotted as a Safe Operating Area (SOA) which encloses all the permissible combinations of voltage and current. Sometimes there are extra little areas, notably those associated with second-breakdown in BJTs, with time limits (usually in microseconds) on how long you can linger there before something awful happens.

It is of course also possible to damage the base-emitter junction of a BJT by exceeding its current or reverse voltage ratings, but this is unlikely in power amplifier applications. In contrast, the insulated gate of an FET is more vulnerable and Zener clamping of gate to source is usually considered mandatory, especially since FET amplifiers often have separate higher supply-rails for their small-signal sections.

BJTs have an additional important failure mode known as second breakdown, which basically appears as a reduction in permissible power dissipation at high voltages, due to local instability in current flow. The details of this mechanism may be found in any textbook on transistor physics.

Excessive current usually causes failure when the I^2R heating in the bond wires becomes too great and they fuse. This places a maximum on the current-handling of the device no matter how low the voltage across it, and hence the power dissipation. In a TO3 package only the emitter bond wire is vulnerable, as the collector connection is made through the transistor substrate and flange. If this wire fails with high excess current, then on some occasions the jet of vaporised metal will drill a neat hole through the top of the TO3 can — an event which can prove utterly mystifying to those not in the know.

Any solid-state device will fail from excess dissipation, as the internal heating will raise the junction temperatures to levels where permanent degradation occurs.

Excess emitter-collector or source-drain voltage will also cause failure. This failure mode does not usually require protection as such, because designing against it should be fairly easy. With a resistive load the maximum voltage is defined by the power supply-rails, and when the amplifier output is hard against one rail, the voltage across the device that is turned off will be the sum of the two rails, assuming a DC-coupled design. If devices with a Vce(max.) greater than this are selected, there should be no possibility of failure. However, practical amplifiers will be faced with reactive load impedances, and this can double the Vce seen by the output devices. It is therefore necessary to select a device that can withstand at least twice the sum of the HT rail voltages, and allow for a further safety margin on top of this. Even greater voltages may be generated by abrupt current changes in inductive loads, and these may go outside the supply-rail range causing device failure by reverse biasing. This possibility is usually dealt with by the addition of *catching* diodes to the circuit (see below) and does not in itself affect the output device specification required.

Power semiconductors have another failure mode initiated by repeated severe temperature changes. This is usually known as *thermal cycling* and results from stresses set up in the silicon by the differing expansion coefficients of the device chip and the header it is bonded to. This constitutes the only real wear-out mechanism that semiconductors are subject to. The average lifetime of a device subjected to temperature variations ΔT can be approximately predicted by

$$N = 10^7 \cdot e^{-0.05 \, \Delta T} \qquad \text{Equation 24.1}$$

where $N =$ cycles to failure, and ΔT is the temperature change.

This shows clearly that the only way open to the designer to minimise the risk of failure is to reduce the temperature range or the number of temperature cycles. Reducing the junction temperature range requires increasing heatsink size or improving the thermal coupling to it. Thermal coupling can be quickly improved by using high-efficiency thermal washers, assuming their increased fragility is acceptable in production, and this is much more cost-effective than increasing the weight of heatsink. The number of cycles can only be minimised by leaving equipment (such as Class-A amplifiers) powered long-term, which has distinct disadvantages in terms of energy consumption and possibly safety.

Overload Protection

Solid-state output devices are much less tolerant of overload conditions than valves, and often fail virtually instantaneously. Some failure modes (such as overheating) take place slowly enough for human intervention, but this can never be relied upon. Overload protection is therefore always an important issue, except perhaps for specialised applications such as amplifiers built into powered loudspeakers, where there are no external connections and no possibility of inadvertent short-circuits. However, even here, protection against shorted voice-coils may be desirable.

Driven by necessity, workable protection systems were devised relatively early in the history of solid-state amplifiers; see Bailey,[1] Becker,[2] and Motorola.[3] Part of the problem is defining what constitutes adequate current delivery into a load. Otala[4] has shown that a complex impedance, i.e., containing energy-storage elements, can be made to draw surprisingly large currents if specially optimised pulse waveforms are used that catch the load at the worst part of the cycle; however, such waveforms do not occur in real life.

Verifying that overload protection works as intended over the wide range of voltages, currents, and load impedances possible is not a light task. Peter Baxandall introduced a most ingenious method of causing an amplifier to plot its own limiting lines.[5]

Overload Protection by Fuses

The use of fuses in series with the output line for overload protection is no longer considered acceptable, as it is virtually impossible to design a fuse that will blow fast enough to protect a semiconductor device, and yet be sufficiently resistant to transients and turn-on surges. There are also the obvious objections that the fuse must be replaced every time the protection is brought into action, and there is every chance it will be replaced by a higher value fuse which will leave the amplifier completely vulnerable. Fuses can react only to the current flowing through them, and are unable to take account of other important factors such as the voltage drop across the device protected.

Series output fuses are sometimes advocated as a cheap means of DC offset protection, but they are not dependable in this role. Placing a fuse in series with the output will cause low-frequency distortion due to cyclic thermal changes in the fuse resistance. The distortion problem can, in theory at least, be side-stepped by placing the fuse inside the global feedback loop; however, what will the amplifier do when its feedback is abruptly removed when the fuse blows? (See also the section on DC offset protection below.)

One way of so enclosing fuses that I have seen advocated is to use them instead of output emitter-resistors Re; I have no personal experience of this technique, but since it appears to add extra time-dependent thermal uncertainties (due to the exact fuse resistance being dependent upon its immediate thermal history) to a part of the amplifier where they already cause major difficulties, I do not see this as a promising path to take. There is the major difficulty that the failure of only one fuse will generate a maximal DC offset, so we may have dealt with the overload, but there is now a major DC offset to protect the loudspeaker from. The other fuse may blow as a consequence of the large DC current flow, but sizing a fuse to protect properly against both overload and DC offset may prove impossible.

Amplifier circuitry should always include fuses in each HT line. These are not intended to protect the output devices, but to minimise the damage when the output devices have already failed. They can and should therefore be of the slow-blow type, and rated with a good safety margin, so that they are entirely reliable; a fuse operated anywhere near its nominal fusing current has a short lifetime, due to heating and oxidation of the fuse wire. HT fuses cannot save the output devices, but they do protect the HT wiring and the bridge rectifier, and prevent fire. There should be separate DC fuses for each channel, as this gives better protection than one fuse of twice the size, and allows one channel to keep working in an emergency.

Similarly, the mains transformer secondaries should also be fused. If this is omitted, a failure of the rectifier will inevitably cause the mains transformer to burn out, and this could produce a safety hazard. The secondary fuses can be very conservatively rated to make them reliable, as the mains transformer should be able to withstand a very large fault current for a short time. The fuses must be of the slow-blow type to withstand the current surge into the reservoir capacitors at switch-on.

The final fuse to consider is the mains fuse. The two functions of this are to disconnect the live line if it becomes shorted to chassis, and to protect against gross faults such as a short between live and neutral. This fuse must also be of the slow-blow type, to cope with the transformer turn-on current surge as well as charging the reservoirs. In the UK, there will be an additional fuse in the moulded mains plug. This does not apply to mains connectors in other countries and so a mains fuse built into the amplifier itself is absolutely essential.

Electronic Overload Protection

There are various approaches possible to overload protection. The commonest form (called electronic protection here to distinguish it from fuse methods) uses transistors to detect the current and voltage conditions in the output devices, and shunts away the base drive from the latter when the conditions become excessive. This is cheap and easy to implement (at least in principle) and, since it is essentially a clamping method, requires no resetting. Normal output is resumed as soon as the fault conditions are removed. The disadvantage is that a protection scheme that makes good use of the device SOA may allow substantial dissipation for as long as the fault persists undetected, and while this should not cause short-term failure if the protection has been correctly designed, the high temperatures generated may impair long-term reliability. In my recent designs a microcontroller detects when SOA limiting is happening and opens the output relay if it persists.

An alternative approach does not limit at all on a cycle-by-cycle basis, but simply drops out the DC

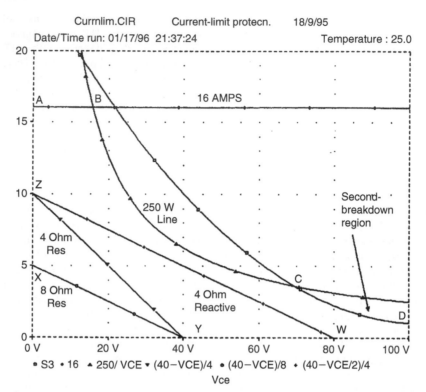

Figure 24.1. The Safe Operating Area (SOA) of a typical TO3 high-power transistor, in this case, the Motorola MJ15024.

protection relay when overload is detected. This will clearly only work if the output stage can survive uncontrolled overload dissipation for long enough for the circuitry to act and the mechanical parts of the relay to move.

In either case the output relay may either be opened for a few seconds delay, after which it resets, or stay latched open until the protection circuit is reset. This is normally done by cycling the mains power on and off, to avoid the expense of a reset button that would rarely be used.

If the equipment is essentially operated unattended, so that an overload condition may persist for some time, the self-resetting system will subject the output semiconductors to severe temperature changes, which may shorten their operational lifetime.

Plotting the Protection Locus

The standard method of representing the conditions experienced by output devices, of whatever technology, is to draw loadlines onto a diagram of the component's SOA, to determine where they cross the limits of the

area. This is shown in Figure 24.1, for an amplifier with ±40 V HT rails, which would give 100 W into 8 Ω and 200 W into 4 Ω, ignoring losses; the power transistor is a Motorola MJ15024. You do not need to fix the HT voltage before drawing most of the diagram; the position of the SOA limits is fixed by the device characteristics. The line AB represents the maximum current rating of 16 A, and the reciprocal curve BC the maximum power dissipation of 250 W. The maximum Vce is 250 V, and is far off the diagram to the right. Line CD defines the second-breakdown region, effectively an extra area removed from the high-voltage end of the power-limited region. Second-breakdown is an instability phenomenon that takes a little time to develop, so the manufacturer's data often allows brief excursions into the region between the second-breakdown line and the power limit. The nearer these excursions go towards the power limit, the briefer they must be if the device is to survive, and trying to exploit this latitude in amplifiers is living dangerously, because the permitted times are very short (usually tens of microseconds) compared with the duration of audio waveforms, where a large bass signal could easily have half-cycle durations lasting for 10 mSec or more. There is more on time-dependent overload protection below.

The resistive loadline XY represents an 8 Ω load, and as a point moves along it, the co-ordinates show the instantaneous voltage across the output device and the current through it. At point X, the current is maximal at 5.0 A with zero voltage across the device, as Vce(sat)s and the like can be ignored without significant error. The power dissipated in the device is zero, and what matters is that point X is well below the current-limit line AB. This represents conditions at clipping.

At the other end, at Y, the loadline has hit the X-axis and so the device current is zero, with one rail voltage (40 V) across it. This represents the normal quiescent state of an amplifier, with zero volts at its output, and zero device dissipation once more. So long as Y is well to the left of the maximum-voltage line, all is well. Note that while you do not need to decide the HT voltage when drawing the SOA for the device, you must do so before the loadlines are drawn, as all lines for purely resistive loads intersect the X-axis at a voltage representing one of the HT rails.

Intermediate points along XY represent instantaneous output voltages between 0 V and clipping; voltage and current co-exist and so there is significant device dissipation. If the line cuts the maximum-power rating curve BC, the dissipation is too great and the device will fail.

Different load resistances are represented by lines of differing slope; ZY is for a 4 Ω load. The point Y must be common to both lines, for the current is zero and the rail voltage unchanged no matter what load is connected to a quiescent amplifier. Point Z is, however, at twice the current, and there is clearly a greater chance of this low-resistance line intersecting the power limit BC. Resistive loads cannot reach the second-breakdown region with these rails.

Unwelcome complications are presented by reactive loading. Maximum current no longer coincides with the maximum voltage, and vice versa. A typical reactive load turns the line XY into an ellipse, which gets much nearer to the SOA limit. The width (actually the minor axis, to be mathematical) of the ellipse is determined by the amount of reactance involved, and since this is another independent variable, the diagram could soon become over-complex. The solution is to take the worst-case scenario for all possible reactive loads of the form R + jX, and instead of trying to draw hundreds of ellipses, to simply show the envelope made up of all their closest approaches to the SOA limit. This is another straight line, drawn from the same maximum current point Z to a point W at twice the rail voltage.

There is clearly a much greater chance that the ZW line will hit the power-limit or second-breakdown lines than the 4 Ω resistive line ZY, and the power devices must have an SOA large enough to give a clear safety margin between its boundary and the reactive envelope line for the lowest rated load impedance. The protection locus must fit into this gap, so it must be large enough to allow for circuit tolerances.

The final step is to plot the protection locus on the diagram. This locus, which may be a straight line, a series of lines, or an arbitrary curve, represents the maximum possible combinations of current and voltage that the protection circuitry permits to exist in the output device. Most amplifiers use some form of VI limiting, in which the permitted current reduces as the voltage across the device increases, putting a rough limit on device power dissipation. When this relationship between current and voltage is plotted, it forms the protection locus.

This locus must always be above and to the right of the reactive envelope line for the lowest rated load, or the power output will be restricted by the protection circuitry operating prematurely. It must also always be to the left and below the SOA limit, or it will allow forbidden combinations of voltage and current that will cause device failure.

Simple Current-limiting

The simplest form of overload protection is shown in Figure 24.2, with both upper and lower sections shown. For positive output excursions, R1 samples the voltage drop across emitter-resistor Re1, and when it exceeds the Vbe of approximately 0.6 V, TR1 conducts and shunts current away from TR2 base. The component values in Figure 24.2 give a 5.5 A constant-current regime as shown in Figure 24.3, which was simulated using a model like that in Figure 24.9 below. The loadlines shown represent 8 Ω and 4 Ω resistive, and 4 Ω worst-case reactive (ZW). The current-limit line is exactly horizontal, though it would probably show a slight slope if the simulation were extended to include more of the real amplifier, such as real current sources, etc.

The value of Re1 is usually determined by the requirements of efficiency or quiescent stability, and so the threshold of current-limiting is set by R1 and R2. This circuit can only operate at a finite speed, and so R1 must be large enough to limit TR1 base current to a safe value. 100 Ω seems sufficient in practice. Re1 is usually the output emitter resistor, as well as current sensor, and so does double duty.

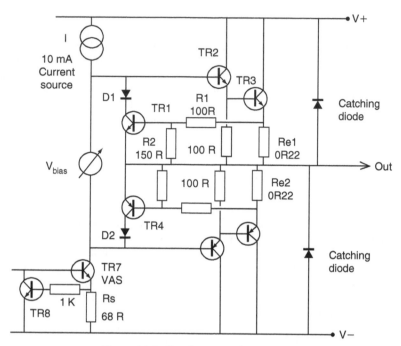

Figure 24.2. Simple current-limit circuit.

The current drawn by TR1 in shunting away TR2 base drive is inherently limited by I, the constant-current load of the VAS. There is no such limit on TR4, which can draw large and indeterminate currents through VAS transistor TR7. If this is a TO-92 device, it will probably fail. It is therefore essential to limit the VAS current in some way, and a common approach is shown in Figure 24.2. There is now a secondary layer of current-limiting, with TR8 protecting TR7 in the same way that TR1 protects TR2, 3. The addition of Rs to sense the VAS current does not significantly affect the VAS operation, and does not constitute local negative feedback. This is because the input to TR7 is a current from the input stage, and not a voltage; the development of a voltage across Rs does not affect the value of this current, as it is effectively being supplied from a constant-current source.

It has to be faced that this arrangement often shows signs of HF instability when current-limiting, and this can prove difficult or impossible to eradicate completely. (This applies to single and double-slope VI limiting also.) The basic cause appears to be that under limiting conditions there are two feedback systems active, each opposing the other. The global voltage feedback is attempting to bring the output to the demanded voltage level, while the overload protection must be able to override this to safeguard the output devices. HF oscillation is often a danger to BJT output devices, but in this case it does not seem to adversely affect survivability. Extensive tests have shown that in a conventional BJT output stage, the oscillation seems to reduce rather than increase the average current through the output devices, and it is arguable that it does more good than harm. It has to be said, however, that the exact oscillation mechanism remains obscure despite several investigations, and the state of our knowledge in this area is far from complete.

The diodes D1, D2 in the collectors of TR1, TR4 prevent them conducting in the wrong half-cycle if the Re voltage drops are large enough to make the collector voltage go negative. Under some circumstances you may be able to omit them, but the cost saving is negligible.

The *loadline* for an output short-circuit on the SOA plot is a vertical line, starting upwards from Y, the HT rail voltage on the X-axis, and representing the fact that current increases indefinitely without any reduction of the voltage drop across the output devices. An example is shown in Figure 24.3 for ±40 V rails. When the short-circuit line is prolonged upwards, it hits the 5.5 A limiting locus at 40 V and 5.5 A; at 220 W this is just inside the power-limit section of the

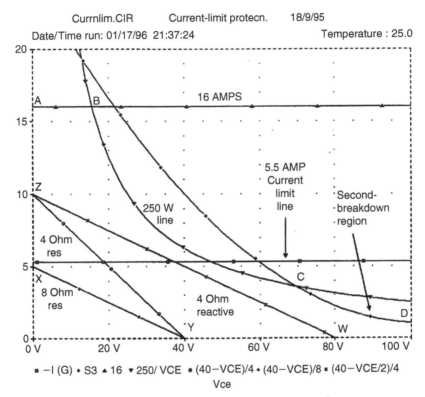

Figure 24.3. Current-limiting with ±40 VHT rails.

SOA. The devices are therefore safe against short-circuits; however, the 4 Ω resistive loadline also intersects the 5.5 A line, at Vce = 18 V and Ic = 5.5 A, limiting the 4 Ω output capability to 12 V peak. This gives 18 W rather than 200 W in the load, despite the fact that full 4 Ω output would in fact be perfectly safe. The full 8 Ω output of 100 W is possible as the whole of XY lies below 5.5A.

With 4 Ω reactive loads, the situation is worse. The line ZW cuts the 5.5 A line at 38 V, leaving only 2 V for output, and limiting the power to a feeble 0.5 W.

The other drawback of constant current protection is that if the HT rails were increased only slightly, to ±46 V, the intersection of a vertical line from Y the X-axis centre would hit the power-limit line, and the amplifier would no longer be short-circuit-proof unless the current limit was reduced.

Single-slope VI Limiting

Simple current-limiting makes very poor use of the device SOA; single-slope VI limiting is greatly superior

because it uses more of the available information to determine if the output devices are endangered. The Vce as well as the current is taken into account. An early exposition of this technique was by A. R. Bailey.[1]

The most popular circuit arrangement is seen in Figure 24.4, where R3 has been added to reduce the current-limit threshold as Vce increases. This simple summation of voltage and current seems crude at first sight, but Figure 24.5 shows it to be an enormous improvement over simply limiting the current.

The protection locus has now a variable slope, making it much easier to fit between reactive load lines and the SOA boundary; the slope is set by R3. In Figure 24.5, Locus 1 is for R3 = 15 kΩ, and Locus 2 for 10 kΩ. If Locus 2 is chosen, the short-circuit current is reduced to 2 A, while still allowing the full 4 Ω resistive output.

Current capability at Vce = 20 V is increased
from 5.5 A to 7.5 A.

An important fact about this circuit is that you should always make provision for the base-emitter resistor R2.

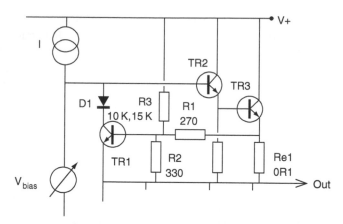

Figure 24.4. Single-slope VI limiter circuit.

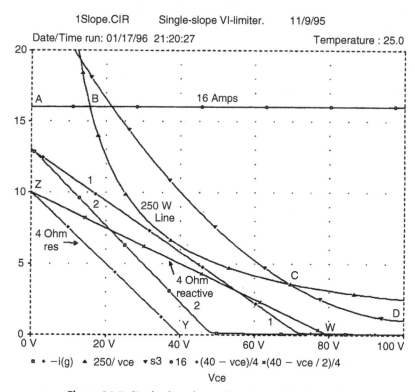

Figure 24.5. Single-slope locus plotted on MJ15024 SOA.

While it may be possible in the initial design to come up with values for R1, R3, and Re that make R2 unnecessary, it is highly likely that the limiting characteristics will need to be adjusted during development. The absence of a position for R2 on the PCB makes this much more difficult, and may require a board iteration.

Dual-slope VI limiting

The motivation for more complex forms of protection than single-slope VI limiting is usually the saving of money, by exploiting more of the output device SOA. In a typical amplifier required to give 165 W into 8 Ω and 250 W into 4 Ω (assuming realistic losses), the

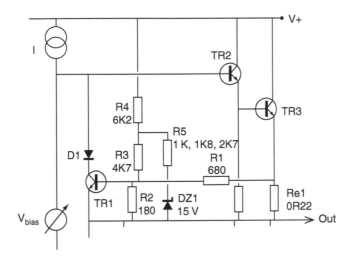

Figure 24.6. Dual-slope locus plotted on MJ15024 SOA.

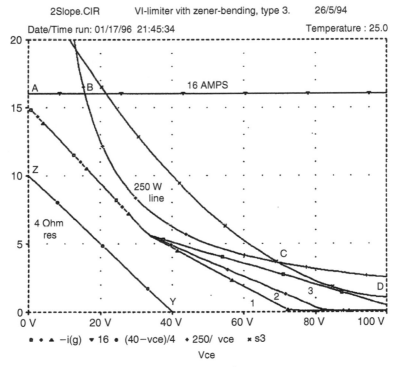

Figure 24.7. Dual-slope VI limiter circuit.

number of device pairs in the output stage can be reduced from three to two by the use of dual-slope protection, and the cost saving is significant. The single-slope limiting line is made dual-slope by introducing a breakpoint in the locus so it is made of two straight-line sections as in Figure 24.6, allowing it to

be moved closer to the curved SOA limit; the current delivery possible at low device voltages is further increased.

A dual-slope system is shown in Figure 24.7. The action of the Vce component on sensing transistor TR1 is reduced when Vce is high enough for Zener diode DZ1 to conduct. The series combination of R4 and R3 is chosen to give the required initial slope with low Vce (i.e., the left-hand slope) but as the voltage increases, the Zener conducts and diverts current through R5, whose value controls the right-hand slope of the protection locus. Locii 1, 2 and 3 are for R5 = 2k7, 1k8 and 1 kΩ, respectively.

Current capability at Vce = 20 V is further increased from 7.5 A to 9.5 A.

Time-dependent VI Limiting

It was noted earlier that bipolar transistors usually have safe-operating areas that are specified to allow very brief excursions into the second-breakdown region, but the time allowed is far too short to be of much use with a low-frequency audio signal, and therefore attempting to exploit this is to live dangerously.

Another time-domain consideration, however, is that the thermal mass of the output device die, header, etc., have an averaging effect on temperature rise due to rapidly changing rates of power dissipation. So long as the peak dissipation is below the permitted maximum, VI-limiting that lets through short peaks of power will allow more output on transients without putting the output devices in greater danger.

The example shown in Figure 24.8 shows a classic approach to time-dependent limiting, based on that of the Phase Linear 400 power amplifier. The famous Phase Linear 700 used a very similar circuit with some value changes.

The static VI-limiting characteristic is set by R1, R2, R3 and R4. D2 implements single-slope VI-limiting; as the output rail moves more positive, a greater amount of current is diverted away from Q1 base and a greater amount of current is allowed to pass through Re. C1 delays the turn-on of Q1, allowing moderate peaks to pass, but the network R1-C1-R2 is shunted by R5, so that large overloads are acted on promptly. The function of C2, and R6, C3 is more obscure; they are too small to have any significant effect on the time-response, and it seems unlikely that they would have prevented HF instability during limiting. As before, D1 prevents Q1 conducting when it is reverse-biased during the negative half-cycle.

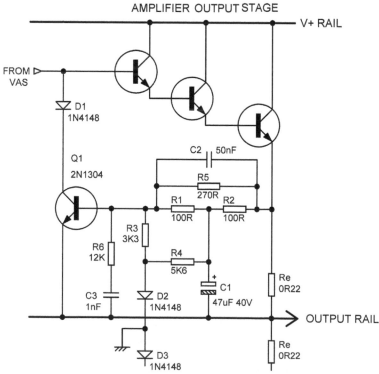

Figure 24.8. Time-dependent VI-limiting. The main delay is introduced by R2 and C1.

Alternative VI-limiter Implementations

They are many other ways to implement VI-limiting. Figure 24.9 shows a floating VI-limiter which senses the current in both upper and lower emitter-resistors Re; signal currents only flow in one of these at a time in a Class-B amplifier. When sufficient voltage appears across the base-emitter of Q1, it conducts, after a time delay due to C1. This turns on Q2 via D3, and overload signals from the other stereo channel are diode OR-ed in at this point by D4. Q2 then opens the output relay; this is the whole of the overload protection as there is no instantaneous VI-limiting. The voltage-sensing is performed by R7, D1 which shunt current away from Q1 base as the amplifier output goes positive, and R8, D2 which shunt current away from Q1 emitter as the amplifier output goes negative, giving single-slope limiting. The circuit is economical because only one transistor is used for VI-sensing, and it also demonstrates how a VI-limiter can be arranged to sense the average of the collector current in output stages with paralleled devices, by use of R1, R3 and R2, R4.

This kind of protection (with many variations in detail) has been extensively used in Japanese amplifiers.

VI Limiting and Temperature Effects

The component values for the VI limiters, of whatever type, are most conveniently determined by use of a SPICE simulator and a certain amount of cut-and-try. However, when the values settled on are put into practice, the results are often disappointing, with the amplifier distortion performance being degraded by the VI limiters starting to act when they should in theory be firmly off. The effect is demonstrated in Figure 24.10, where it can be seen that the distortion roughly triples when it rises above the noise.

VI limiters of the straightforward kind with the simple circuitry shown above depend on a voltage exceeding the sense transistor Vbe to make them operate. They are therefore somewhat temperature-sensitive, and this can be a real problem. The overload protection circuitry will almost always be close to the output devices being protected, and therefore near the heatsink. It is therefore very likely to get hotter than other parts of the amplifier, and the VI limiters will begin to act much earlier than expected. A SPICE simulator, unless told otherwise, will default all its component temperatures to a nice comfy 25°C, and will not give warning of the problem.

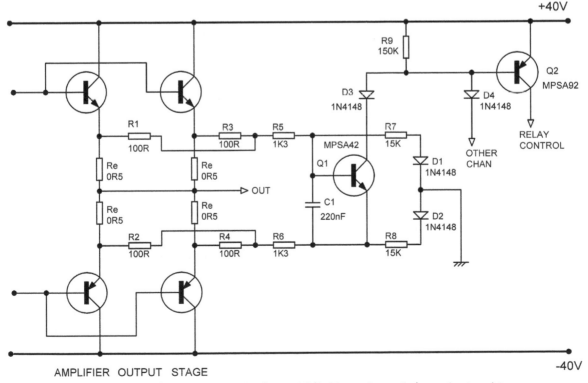

AMPLIFIER OUTPUT STAGE

Figure 24.9. An alternative way to implement VI-limiting, using a single sensing transistor.

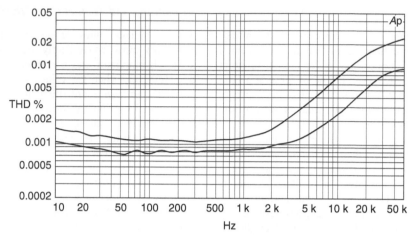

Figure 24.10. The effect on distortion of a prematurely active VI limiter; the lower trace has the limiter disabled. Output power 150 W/8 Ω.

Taking this temperature into account, and also ensuring that the VI limiters are firmly off in normal operation, so distortion performance is not degraded, means that the VI limiter component values differ significantly from those derived from room-temperature simulations. This implies that if the circuitry is designed not to come on early when the unit is hot, it will operate late when cold, if a fault occurs just after it has been switched on. The output stage must have enough capability to handle this without damage. Some commercial designs, such as the Hitachi HA-3700 and HA-4700 (introduced July 1980) have used thermistors to compensate the VI limiter for temperature changes.

It is of course always possible to design more complex VI limiting circuitry that is less temperature-sensitive. For example, the single sensing transistor could be replaced by a differential pair which would be much less temperature-sensitive.

More complex systems have been conceived, for example, by Crown, who have patented several sophisticated systems that are effectively analogue computers, but this approach has not been widely adopted. In systems designed to deal with fault conditions, simplicity, and therefore hopefully reliability, is a great virtue.

Simulating Overload Protection Systems

The calculations for protection circuitry can be time-consuming. Simulation is quicker; Figure 24.11 shows a conceptual model of a dual-slope VI limiter, which allows the simulated protection locus to be directly compared with the loadline and the SOA. The amplifier output stage is reduced to one half (the positive or upper half) by assuming symmetry, and the combination of the actual output device and the load represented by voltage-controlled current-source G. The output current from controlled-source G is the same as the output device current in reality, and passes through current-sense resistor Re1.

The 6 mA current-source I models the current from the previous stage that TR1 must shunt away from the output device. Usually this is an accurate model because the VAS collector load will indeed be a current-source.

The feedback loop is closed by making the voltage at the collector of TR1 control the current flowing through G and hence Re1.

In this version of VI-protection, the device voltage is sensed by R4 and the current thus engendered is added to that from R1 at the base of TR1. This may seem a crude way of approximating a constant power curve, and indeed it is, but it provides very effective protection for low- and medium-powered amplifiers.

Vin models the positive supply-rail, and exercises the simulation through the possible output voltage range. In reality, the emitter of TR1 and Re1 would be connected to the amplifier output, which would be move up and down to vary the voltage across the output devices, and hence the voltage applied across R1, R2. Here it is easier to alter the voltage source V, as the only part of the circuit connected to HT+. V+ is fixed at a suitable HT voltage, e.g., +50 V.

The simulation only produces the protection locus, and the other lines making up the SOA plot are added

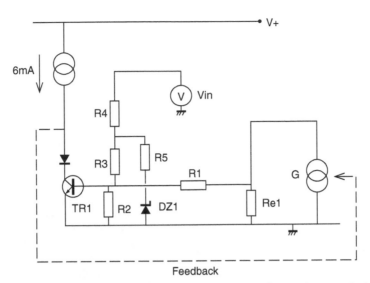

Figure 24.11. A conceptual model of an overload protection circuit that implements dual-slope limiting.

at the display stage. Ic(max.) is drawn by plotting a constant to give a horizontal line at 16 A. P(max.) is drawn as a line of a constant power, by using the equation 250/Vce to give a 250 W line. In PSpice there seems to be no way to draw a strictly vertical line to represent Vce(max.), but in the case of the MJ15024, this is 250 V, and is for most practical purposes off the right-hand end of the graph anyway. The second-breakdown region is more difficult to show, for in the manufacturer's data the region is shown as bounded by a non-linear curve. The voltage/current co-ordinates of the boundary were read from manufacturer's data, and approximately modelled by fitting a second order polynomial. In this case it is:

$$I = 24.96 - 0.463 \cdot V_{ce} + 0.00224 \cdot V_{ce}^2 \quad \text{Equation 24.2}$$

This is only valid for the portion that extends below the 250 W constant-power line, at the bottom right of the diagram.

As previously mentioned, simulation results for protection circuitry must be carefully checked against reality because of temperature effects.

Testing the Overload Protection

One of the more nerve-wracking aspects of amplifier testing is the verification of the overload protection system. This best done by slowly reducing the test load resistance from the rated value to one which is expected to trigger the overload, rather than wading straight in with a crowbar across the output terminals. If an amplifier has a rated load of 8 Ω, then the protection might be expected to act at 2 Ω, or perhaps 1 Ω, if the design is intended to deal authoritatively with deep dips in loudspeaker impedance.

Obviously there needs to be some way of monitoring that the VI limiters are beginning to act, and this may be as simple as observing the output waveform on an oscilloscope; when the peaks of the sinewave are starting to get clipped, then limiting is occurring. You need to make sure you are not seeing voltage clipping because the supply rails have been dragged down. When you are sure that the VI limiters are coming in as expected, then, and not before, is it time to start applying short-circuits to the output. This approach minimises the likelihood of output device damage. Blown output transistors are time-consuming to replace, with often quite a bit of dismantling involved, and there is also the likelihood of collateral damage to drivers and so on which has to be checked for and diagnosed; it is *much* more time-efficient to take a gradualist approach to overloading the output stage. I have recently pursued this method for four different commercial amplifiers, and in each case the verification procedure was completed without destroying a single transistor, a record of which I am mildly proud.

Complete verification includes overload testing with a 10% high mains supply voltage, and possibly at elevated ambient temperatures.

Speaker Short-circuit Detection

Some amplifiers test the speaker outputs for short-circuits before unmuting and connecting the power amplifiers to them. This usually entails using a change-over contact configuration for the output muting relay, with the external load connected to the moving contact and the power amplifier output connected to the normally-open contact. When the amplifier is muted, a very small current, too small to cause audible clicks, is passed through the normally-closed contact and into the loudspeaker load. The resulting voltage is applied to a comparator and, if it is too low, the amplifier is inhibited from unmuting. This prevents the output devices from being unnecessarily stressed by a short circuit. Since the test is only made when the amplifier is muted, the normal overload protection system must still be provided to cope with short circuits that occur while the amplifier is unmuted.

An interesting failure mode with this scheme can occur if the test current is made too small. Loudspeakers can also turn sound into electricity instead of vice versa, and I know of one design that would refuse to start up in a noisy environment.

Catching Diodes

These are reverse-biased power diodes connected between the supply-rails and the output of the amplifier, to allow it to absorb transients generated by fast current-changes into an inductive load. They are also known as clamp diodes or clamping diodes. All moving-coil loud-speakers present an inductive impedance over some frequencies.

When an amplifier attempts to rapidly change the current flowing in an inductive load, the inductance can generate voltage spikes that drive the amplifier output outside its HT rail voltages; in other words, if the HT voltage is ± 50 V, then the output might be forced by the inductive back-EMF to 80 V or more, with the likelihood of failure of the reverse-biased output devices. Catching diodes prevent this by conducting and clamping the output so it cannot move more than about 1 V outside the HT rails. These diodes are presumably so-called because they catch the output line if it attempts to move outside the rails.

So how can the output rail move outside the supply rails, no matter how fast the voltage change applied to a reactive load, if it is firmly held by the amplifier negative feedback loop? The answer is that a flyback pulse typically occurs when the amplifier is suddenly *disconnected* from a reactive load, rather than when there is

a sudden change in the signal. This happens when VI limiters cut in, turning off the half of the output stage that was until then driving the load. Now neither of the output devices are conducting, and this is when the voltage spike occurs and the clamp diodes justify their cost.

This sounds like a sharp crack of high amplitude; it is not a nice noise, but sometimes can be difficult to identify as it tends to happen during signal peaks. It can usually be more easily diagnosed by looking at an oscilloscope, as the sudden voltage excursion is much steeper than the signal waveforms. The only way to avoid these noises — for the catching diodes only limit the spike amplitude rather than suppressing it altogether — is to make sure that the output stage is big enough for its task, so the VI limiters can be designed with a big margin between normal use with likely loads, and the fault conditions that make it essential for them to act.

The diode current rating should be not less than 2 A, and the PIV 200 V or greater, and at least twice the sum of the HT rails. I usually specify 400 PIV 3 A diodes, and they never seem to fail.

DC-offset Protection

In some respects, any DC-coupled power amplifier is an accident waiting to happen. If the amplifier suffers a fault that causes its output to sit at a significant distance away from ground, then a large current is likely to flow through the loudspeaker system. This may cause damage either by driving the loudspeaker cones beyond their mechanical limits or by causing excessive thermal dissipation in the voice-coils, the latter probably being the most likely. In either case the financial loss is likely to be serious. There is also a safety issue, in that overheating of voice-coils or cross-over components could cause a fire.

Since most power amplifiers consist of one global feedback loop, there are many possible component failures that could produce a DC offset at the output, and in most cases this will result in the output sitting at one of the HT rail voltages. The only way to save the loud-speaker system from damage is to remove this DC output as quickly as possible. The DC protection system must be functionally quite separate from the power amplifier itself or the same fault may disable both.

There are several possible ways to provide DC protection:

1. By fusing in the output line, the assumption being that a DC fault will give a sustained current flow that

will blow the fuse when music-type current demands will not.

2. By means of a relay in the output line, which opens when a DC offset is detected.

3. By triggering a crowbar that shunts the output line to ground, and blows the HT fuses. The crowbar device is usually a triac, as the direction of offset to be dealt with is unpredictable.

4. By shutting down the power supply when a DC fault is detected. This can be done simply by an inhibit input if a switched-mode PSU is used. Conventional supplies are less easy.

5. The use of MOV surge suppressors across the speaker terminals has sometimes been suggested. This is a truly awful idea. It would be virtually impossible to select a component that could discriminate between clipping and a fault that jammed the output against a supply rail. Partial conduction would cause high distortion, and full conduction under DC fault conditions would, as for any crowbar method, be likely to cause further damage. This dreadful notion is not considered further.

DC-offset Protection by Fuses

Fuses in series with the output line are sometimes recommended for DC offset protection, but their only merit is cheapness. It may be true that they have a slightly better chance of saving expensive loudspeakers than the HT fuses, but there are at least three snags:

1. Selection of the correct fuse size is not at all easy. If the fuse rating is small and fast enough to provide some real loudspeaker protection, then it is likely to be liable to nuisance blowing on large bass transients. A good visual warning is given by behaviour of the fuse wire; if this can be seen sagging on transients, then it is going to fail sooner rather than later. At least one writer on DIY Class-A amplifiers gave up on the problem, and coolly left the tricky business of fuse selection to the constructor!

2. Fuses running within sight of their nominal rated current generate distortion at LF due to cyclic changes in their resistance caused by I^2R heating; the THD would be expected to rise rapidly as frequency falls, and Greiner[6] states that harmonic and intermodulation distortion near the burn-out point can reach 4%. It should be possible to eradicate this by including the fuse inside the global feedback network, for the distortion will be generated at low frequencies where the feedback factor is at its greatest, but there are problems with amplifier behaviour after the fuse has blown.

3. In my tests, the distortion generated was fairly pure third harmonic. Figure 24.12 shows the THD measured before and after a T1A (slow-blow) fuse in series with an 8 Ω load at 25 W. Below 100 Hz the distortion completely swamps that produced by the amplifier, reaching 0.007% at 20 Hz. The distortion rises at rather less than 6 dB/octave as frequency falls. The fuse in this test is running close to its rating, as increasing the power to 30 W caused it to blow.

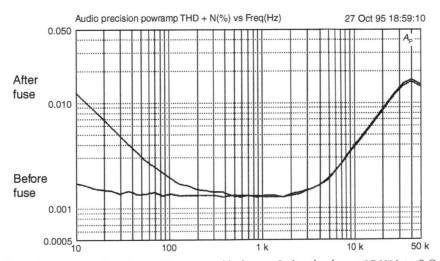

Figure 24.12. Fuse distortion. THD measured before and after the fuse at 25 W into 8 Ω.

4. Fuses obviously have significant resistance (otherwise they would not blow) so putting one in series with the output will degrade the theoretical damping factor. However, whether this is of any audible significance is very doubtful.

Despite these problems, some commercial designs have relied on a series output fuse for offset protection, and in some instances for overload protection as well. One example is the Sanyo DC-401, rated at 30 W/8 Ω with a 2 Amp slow-blow fuse in the output line. Another is the Rotel RA-820B (introduced 1985) rated at 25 W/8 Ω with a series 3.15 Amp fuse of unspecified time-characteristics; this amplifier also had a 0.22 Ω series output resistor for reasons that are unknown, but probably not unrelated to HF stability. As late as 2003, the Rotel RA-01 was introduced with no overload or DC-offset protection apart from a 4 Amp slow-blow (T type) fuse, with some negative feedback around it, in the output line. Strangely, output relays were fitted but were only used for thump suppression and A/B speaker switching; the Rotel RA-02 introduced in 2002 had the same arrangement. As a final example, take the Nikko TRM-750 (introduced in 1978) with transistor current limiting and output-relay offset protection, but still including a 5 Amp fuse of unspecified characteristics in the output line, outside the amplifier feedback loop.

Note that the HT rail fuses, as opposed to fuses in the output line, are intended only to minimise amplifier damage in the event of output device failure. They must not be relied upon for speaker protection against DC offset faults. Often when one HT fuse is caused to blow, the other also does so, but this cannot be relied upon, and obviously asymmetrical HT fuse blowing will in itself give rise to a large DC offset.

Relay DC-offset Protection and Muting Control

Relay protection against DC offsets has the merit that, given careful relay selection and control-circuitry design, it is virtually foolproof. The relay should be of the normally-open type so that if the protection fails, it will be to a safe condition.

The first problem is to detect the fault condition as soon as possible. This is usually done by low-pass filtering the audio output, to remove all signal frequencies, before the resulting DC level is passed to a comparator that trips when a set threshold is exceeded. This is commonly in the range of 1−2 V, well outside any possible DC-offsets associated with normal operation; these will almost certainly be below 100 mV. Any low-pass filter must introduce some delay between the

appearance of the DC fault and the comparator tripping, but with sensible design this will be too brief to endanger normal loudspeakers. There are other ways of tackling the fault-detection problem, for example, by detecting when the global negative feedback has failed, but the filtering approach appears to be the simplest method and is generally satisfactory. First-order filtering seems to be quite adequate, though at first sight a second-order active filter would give a faster response time for the same discrimination against false-triggering on bass transients. In general, there is much to be said for keeping protection circuitry as simple and reliable as possible.

Let us now examine DC offset detection circuitry in more detail. The problem falls neatly into two halves − distinguishing between acceptable large AC signals of up to 30 Vrms or more, and DC offsets which may only be a volt or so before stern action is desired, and applying the result to a circuit which can detect both positive and negative transgressions. To perform the first task, relatively straightforward lowpass filtering is often adequate, but the bidirectional detection can tackled in many ways, and sometimes presents a few unexpected problems.

At this point we might consider how quickly the DC offset protection must operate to be effective. Clearly there will always be some delay, as we are discriminating against normal high-amplitude bass information, but otherwise the quicker the better if the loudspeaker is to be saved. My experience of deliberately setting fire to loudspeaker elements is limited (and I hasten to point out that I have so far never set fire to one accidentally) but here is one test I can report.

I once had the entertaining task of determining just how long a speaker element − the LF unit, obviously, as the tweeter was protected by the crossover from any DC − could sustain an amplifier DC fault. The tests, which were conducted outdoors to avoid triggering the fire alarms, showed that a well-designed and conservatively rated loudspeaker could be turned into smouldering potential landfill in less than a second. The loudspeaker unit in question was a high-quality LF unit with the relatively small diameter of 5 in., made by a respected manufacturer. The test involved applying +40 V to it, as if its accompanying amplifier had failed. The cone and voice-coil assembly shot out of the magnetic gap as if propelled by explosives, and then burst into flames in less than a second. All we could really conclude as the smoke cleared was that a second was way too long a reaction time for a protection system.

Filtering for DC Protection

A good DC protection filter is that which discriminates best between powerful low frequency signals and a genuine amplifier problem. It is easy to make the filter time-constant so long that it will never be false-triggered by a thumping great bass note, but then its time-domain response will be so slow that your precious loudspeakers will be history before the amplifier reacts to protect them.

The simplest possible filter is a single-pole circuit that requires only one RC time constant; in many cases this is quite good enough, but some more sophisticated approaches are also described here.

The Single RC Filter

The time-constant needs to be long enough to filter out the lowest frequency anticipated, at the full voltage output of the amplifier. The ability to sustain 10 Hz at the onset of clipping is usually adequate for audio, but if you are designing subsonic amplifiers to drive vibration tables, you will need to go a bit lower. Figure 24.11a shows the single-pole filter with typical values of 47 kΩ and 47 µF that give a −3 dB point at 0.07 Hz. This is appropriate for low to medium amplifier powers, when feeding a later bidirectional detector that will trigger on an offset of the order of a Volt. The value of R1 is set by the current demands of this later stage — these can be significant, as we will see in the next section. The value of C1 is then determined by the required -3 dB frequency, and this means that it will be an electrolytic. It is important to remember at this point that DC offsets may arrive with either polarity, and may persist for long periods before someone notices there is a problem, so C1 needs to be either a non-polar electrolytic or constructed from two ordinary electrolytics connected back-to-back in the time-honoured fashion. Both methods are effective so it comes down to the fine details of the economics of component sourcing. Some amplifiers remove the supply from the power amplifier sections, so the offset does not persist, and this precaution may seem unnecessary; however, there is no point in trying to save fractions of a penny by possibly compromising the reliability of something as important as the DC offset protection. C1 should have a voltage rating at least equal to the supply rails of the amplifier concerned.

The single-pole filter in Figure 24.13 is −3 dB at 0.07 Hz. To evaluate it, it was fed from a power amplifier giving 55 V peak, and the filter output connected to a bidirectional detector that had trip points at ±2.0 V.

This set-up triggered at 2.0 Hz when a 55 V peak signal starting at 50 Hz was slowly reduced in frequency. This corresponds to a filter attenuation of −28.8 dB at 2.0 Hz, and this frequency was used as the criterion for bass rejection thereafter. When a fault was simulated so the input to the filter shot up to +55 V, and stayed there, the detector gave a DC offset indication after 78 msec.

This circuit is easily adapted to stereo usage by having two resistors feeding into it, as in Figure 24.13b. If the resistors remain the same value, then the resistance seen by C is halved, and its capacitance must be doubled to maintain the same roll-off frequency. The incoming DC offset is also halved, so the detector sensitivity must be doubled if it is to trigger from the same level of offset on one of the stereo amplifier outputs. You could also object that a positive offset on one channel might be cancelled out by a negative offset on the other; this seems laughably unlikely until you recall that bridged amplifiers are driven with input signals that are in anti-phase, so a DC error in the drive circuitry could present just this situation. More sophisticated circuits provide two independent inputs that do not interact, avoiding this problem. More on this later, in the section on detectors.

The Dual RC Filter

The thinking behind the use of more complicated filtering is that a faster response roll-off will give better discrimination against high-amplitude bass events, so a higher −3 dB frequency can be used with (hopefully) a quicker response in the time domain.

The simplest method is to cascade two single-pole RC filters, as shown in Figure 24.14. This obviously gives a rather soggy roll-off, but has the merit of not introducing any more semiconductors that might fail. The non-standard capacitor values shown give the same attenuation of −28.8 dB at 2.0 Hz as the previous circuit. The only real snag to this scheme is that it does not work. The time to react was 114 msec, half as long again as the simple filter above. However, I have seen it used in several designs, so you might come across it.

The Second-order Active Filter

Some amplifier designs use an active filter to separate the bass from the breakdowns. This obviously allows a nice sharp roll-off, and gives the freedom to set the filter damping factors and so on. But does it deliver? I

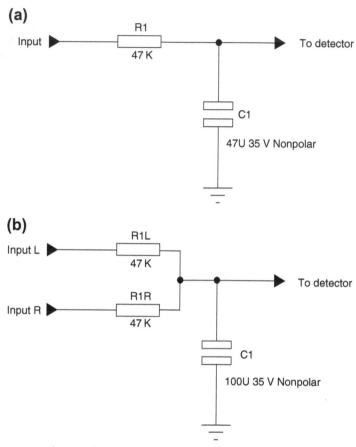

Figure 24.13. Mono (a) and stereo (b) single-pole filters for offset protection. The −3 dB point is 0.07 Hz.

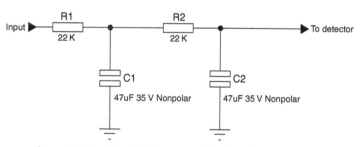

Figure 24.14. A dual RC low-pass filter for offset protection.

tested the circuit of Figure 24.15, a Sallen-and-Key configuration, which, with the values shown, gives a second-order Butterworth (maximal flatness) characteristic, with a −3 dB point at 0.23 Hz; due to the increased filter slope the attenuation is once more −28.8 dB at 2.0 Hz. The reaction time is 109 msec, which is better than the dual RC filter but yet somewhat inferior to the single-pole filter of Figure 24.13a. Most disappointing. The Bessel filter characteristic is noted

for a better response in the time domain, at the expense of a sharp roll-off, so I tried that. The component values in Figure 24.15 are now $R1 = R2 = 35$ kΩ, $C1 = 13.3$ μF, and $C2 = 10$ μF. The reaction time is actually worse, at 131 msec, which was rather a surprise.

Building active filters usually means using opamps. Putting an opamp into the system creates a need for low-voltage supplies within the power amplifier, which is highly inconvenient if they do not exist

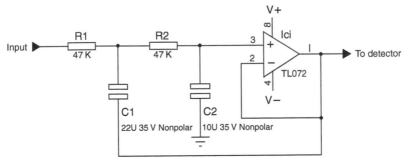

Figure 24.15. A second-order Sallen-and-Key filter input for offset protection.

already. Most protection designs use discrete transistors throughout, and one of the advantages of the Sallen-and-Key configuration is that it can be realised using a simple emitter-follower.

An important consideration is that opamps have a limited common-mode voltage capability, and they will not appreciate having the full power amplifier supply rail applied to them directly. It will be necessary to scale down the incoming voltages and allow for this when setting the detector thresholds.

The conclusion seems inescapable that, for once, the simplest circuit is the best; the single-pole filter is the way to go.

Bidirectional DC Detection

There are many, many ways to construct circuits that will respond to both positive and negative signals of a defined level, and here some of the more common and more useful ones are examined.

The Conventional Two-transistor Circuit

The circuit in Figure 24.16 is probably the most common approach to bidirectional detection. When the input exceeds +0.6 V, Q1 turns on and the output voltage falls while Q2 stays off. When the input goes negative, Q2 operates in common-base mode, and conducts, Q1 remaining off as its base-emitter junction is reverse-biased. In either case, current is drawn through R10 and the output voltage drops to signal an offset. There is a certain elegance in the way that the conducting base-emitter junction protects its neighbour from excess reverse bias, but this circuit has one great disadvantage. Since Q2 operates in common-base mode, it has near unity current gain, as opposed to Q1, which is in common-emitter

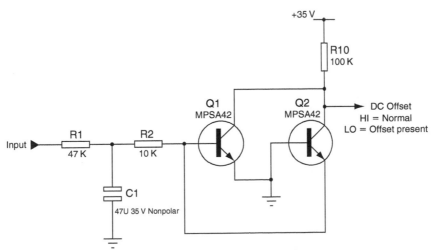

Figure 24.16. A common bidirectional detect circuit, giving very different thresholds for positive and negative inputs.

mode and therefore has current gain equal to the device beta.

This makes the two thresholds very asymmetrical. When the detector is driven from a single-pole RC filter with R1 = 47k, the positive threshold is +1.05 V but the negative threshold is −5.5 V. To reduce this asymmetry, R1 needs to be kept low, which leads to inconveniently large values of C1.

The One-transistor Version

Figure 24.17 shows a variation on this theme, saving a transistor by adding diodes and resistors. With current component pricing, the economic benefit is trivial, but it is still a circuit that has seen a great deal of use in Japanese amplifier designs. For positive inputs, D1, Q1 and D2 conduct. For negative inputs, D3 and Q1 conduct, the latter getting its base current through R2. As for the previous circuit, the current-gain difference between common-base and common-emitter modes of transistor operation gives asymmetrical thresholds; slightly less so because of the effect of D2 during positive inputs.

The Differential Detector

The interesting circuit of Figure 24.18, which has also seen use in Japanese hi-fi equipment, is based on a differential pair. This removes the objection to all the other circuits here, which is that it takes 0.6 V on the base to turn on a transistor directly, and so the detection

thresholds will be that or more, due to extra diodes and so on. In this circuit the differential pair Q1, Q2 cancels out the 0.6 V Vbe-drop, and sensitivity can be much higher; under what conditions this is actually necessary is a moot point. There is no low-pass filter as such; instead the same effect is achieved by high-pass filtering the signal, to remove DC and information. The result is then subtracted from the unfiltered signal by the differential pair, so only the DC and low frequency signals remain.

It works like this: for positive inputs, Q2 turns on more and Q1 less, so the voltage on Q2 collector falls and Q3 is turned on via D2, and passes an offset signal to the rest of the system. For negative inputs, Q1 turns on more and Q2 less, so the voltage on Q1 collector falls and Q3 is turned on via D1. The thresholds depend on the gain of the pair, set by the ratio of R5, R6 to R8, R9, and whatever voltage is set up on Q3 emitter. The circuit gives excellent threshold symmetry.

The Self Detector

Figure 24.19 shows my own version of a bidirectional detector. This has two advantages; it is symmetrical in its thresholds, and can be handily converted to an economical stereo or multichannel form without any loss of sensitivity. The only downside is that the thresholds are relatively high at about ±2.1 V with the component values shown. This is actually quite low enough to protect loudspeakers, and in any case, your typical

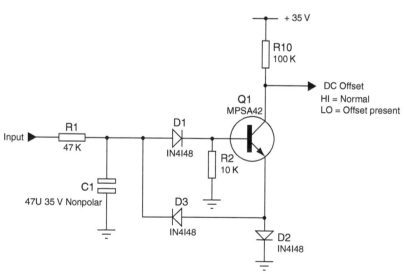

Figure 24.17. Another implementation of the same principle, saving a transistor but retaining the problem of asymmetrical thresholds.

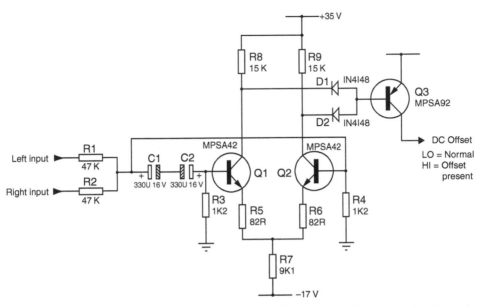

Figure 24.18. The differential detector, which can have very low thresholds. It uses a high-pass rather than a low-pass filter.

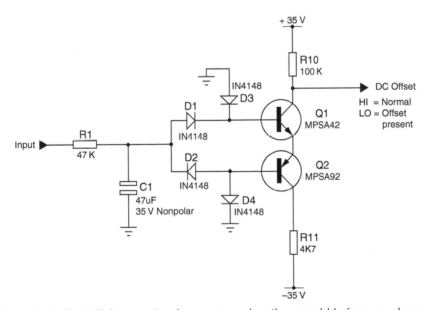

Figure 24.19. The Self detector. Good symmetry and easily expandable for more channels.

serious amplifier fault smacks the output hard against the supply rails, and detecting this is not very hard. The exactness of the threshold symmetry depends on the properties of the transistors used, but is more than good enough to eliminate any problems. It works well with transistors such as MPSA42/MPS92 which are

designed for high-voltage applications and therefore have low beta.

For positive inputs, D1, Q1, and Q2 conduct, with D4 supplying the base current for Q2. With a negative input, D2, Q2, and Q1 conduct, D3 now supplying base current for Q1. In each case there are two diode drops and two

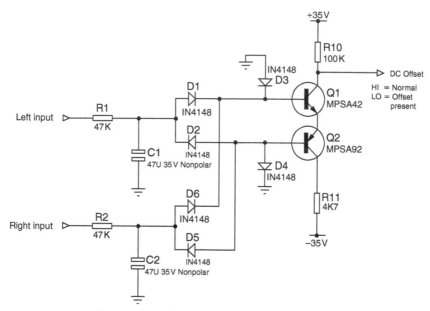

Figure 24.20. The stereo version of Figure 24.19.

Vbe drops in series, which, if each one was a nominal 0.6 V, would give thresholds of ±2.4 V; in practice, the diode supplying the much smaller base current has a lesser voltage across it, and the real thresholds come to ±2.1 V. Note that R11 is very definitely required to limit the current flowing through Q1, and Q2 when the input goes negative; R10 inherently limits it for positive inputs.

Figure 24.20 shows the stereo version, which uses separate filters for each channel, and two more diodes. The operation is exactly as before for each channel, and so the thresholds are unchanged. Equal-value positive and negative offsets on the two inputs do not cancel, and an offset is always clearly signalled. This circuit can be extended to cover more than two channels by just adding more RC filters (R1, C1) and diodes (D1, D2).

Having paid for a DC protection relay, it is only sensible to use it for system muting as well, to prevent thuds and bangs from the upstream parts of the audio system from reaching the speakers at power-up and power-down. Most power amplifiers, being dual-rail (i.e: DC-coupled) do not generate enormous thumps themselves, but they cannot be guaranteed to be completely silent, and will probably produce an audible turn-on thud.

An amplifier relay-control system should:

- Leave the relay de-energised when muted. At power-up, there should be a delay of at least 1 sec before the relay closes. This can be increased if required.

- Drop out the relay as fast as is possible at power-down, to stop the dying moans of the pre-amp, etc. from reaching the outside world. See the section on mains-fail detection below for more details.

- Drop out the relay as fast as is possible when a DC offset of more than 1−2 V, in either direction, is detected at the output of either power amp channel; the exact threshold is not critical. This is normally done by low-pass filtering the output (47 k and 47 μF works OK) and applying it to some sort of absolute-value circuit to detect offsets in either direction. The resulting signal is then OR-ed in some way with the muting signal mentioned above.

Do not forget that the contacts of a relay have a much lower current rating for breaking DC rather than AC. This is an issue that does not seem to have attracted the attention it deserves.

A block diagram of a relay control system meeting the above requirements is shown in Figure 24.21, which includes over-temperature protection. Any of the three *inhibit* signals can override the turn-on delay and pull out the relay.

Output Relay Selection

Obviously the relay contacts must be rated to pass the load current at maximum output (at a 10% high mains voltage) into the lowest specified impedance, and also

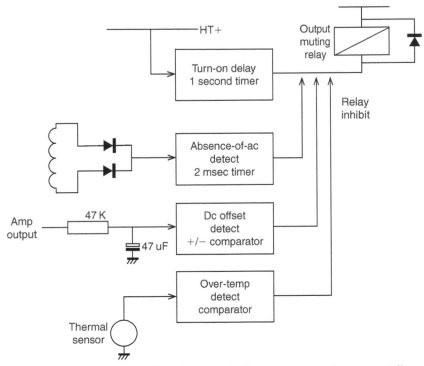

Figure 24.21. Output relay control combining DC offset protection and power-on/off muting.

whatever short-circuit currents the overload protection system may permit. Bear in mind that you cannot increase the current rating of a relay by connecting sets of contacts in parallel unless you are sure that the relay will never be called upon to break a current in excess of that which could be handled by one set of contacts alone. This is because in practice one set of contacts will always break (or make) before the other.

The most demanding requirement is the breaking of large DC currents during offset faults. Since this requires an arc between the contacts to be broken, the DC current rating of a relay is usually much less than its AC rating. In the worst case the arc will persist even when the contacts are fully open, destroying them and possibly generating enough heat to melt the relay insulation. The relay may be thus destroyed but there is no guarantee that the circuit will be broken.

Distortion from Output Relays

Relays remain the only simple and effective method of disconnecting an amplifier from its load. The contacts can carry substantial currents, and it has been questioned whether they can introduce non-linearities.

My experience is that silver-based contacts in good condition show effectively perfect linearity. Take a typical relay intended by its manufacturer for output-switching applications, with 'silver alloy' contacts — whatever that means — rated at 10 A. Figure 24.22 shows THD before and after the relay contacts while driving an 8 Ω load to 91 W, giving a current of 3.4 Arms. There is no significant difference; the only reason that the lines do not fall exactly on top of each other is because of the minor bias changes that Class-B is heir to. This apparently perfect linearity can be badly degraded if the contacts have been maltreated by allowing severe arcing — typically while trying and failing to break a severe DC fault.

Not everyone is convinced of this. If the contacts were non-linear for whatever reason, an effective way of dealing with it would be to include them in the amplifier feedback loop, as shown in Figure 24.23. R1 is the main feedback resistor, and R2 is a subsidiary feedback path that remains closed when the relay contacts open, and hopefully prevents the amplifier from going completely berserk. With the values shown, the normal gain is 15.4 times, and with the contacts open, it is 151 times. There is a feedback factor of about ten to linearise any relay problems.

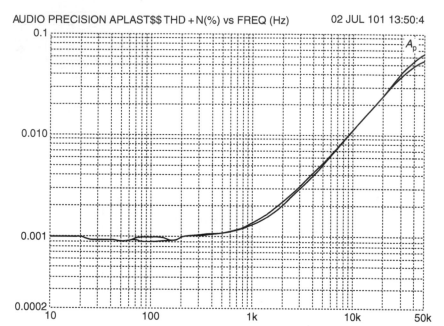

Figure 24.22. Demonstrating that relay contacts in themselves are completely distortion-free. Current through contacts was 3.4 Arms.

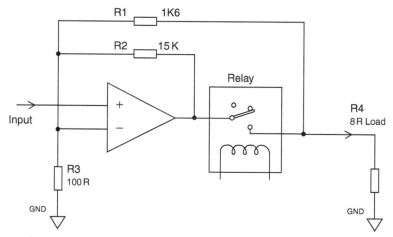

Figure 24.23. How to enclose relay contacts in the feedback loop. The gain shoots up when the relay contacts open, so muting the input signal is desirable.

The problem of course is that if there is to be a healthy amount of NFB wrapped around the relay contacts, R2 must be fairly high and so the closed-loop gain shoots up. If there is still an input signal, then the amplifier will be driven heavily into clipping. Some designs object to this, but even if the amplifier does not fail, it is likely to accumulate various DC offsets on its internal time-constants as a result of heavy clipping, and these could cause unwanted noises when the relay contacts close again. One solution to this is a muting circuit at the amplifier input that removes the signal entirely and prevents clipping. This need not be a sophisticated circuit, as huge amounts of muting are not required; −40 dB should be enough. It must, however, pass the signal cleanly when not muting.

An all-too-real form of non-linearity can occur if the relay is constructed so that its frame makes up part of the switched electrical circuit as well as the magnetic circuit. (This is not the case with the audio application relay discussed above.) A relay frame is made of soft iron, to prevent it becoming permanently magnetised, and this appears to present a non-linear resistance to a loudspeaker level signal, presumably due to magnetisation and saturation of the material. (It should be said at once that this is described by the manufacturer as a 'power relay' and is apparently not intended for audio use.) A typical example of this construction has massive contacts of silver/cadmium oxide, rated at 30 A AC, which in themselves appear to be linear. However, used as an amplifier output relay, this component generates more — much more — distortion than the power amplifier it is associated with.

The effect increases with increasing current; 4.0 Arms passing through the relay gives 0.0033% THD and 10 Arms gives 0.018%. The distortion level appears to increase with the square of the current. Experiment showed that the distortion was worst where the frame width was narrowest, and hence the current density greatest.

Figure 24.24 shows the effect at 200 Wrms/2 Ω (i.e., with 10 Amps rms through the load) before and after the relay. Trace A is the amplifier alone. This is

a Blameless amplifier and so THD is undetectable below 3 kHz, being submerged in the noise floor which sets a measurement limit of 0.0007%.

Trace B adds in the extra distortion from the relay. It seems to be frequency-dependent, but rises more slowly than the usual slope of 6dB/octave. Trace C shows the effect of closing the relay in the NFB loop using the circuit and component values of Figure 24.23; the THD drops to about a tenth, which is what simple NFB theory would predict. Note that from 10 kHz to 35 kHz, the distortion is now lower than before the relay was added; this is due to fortuitous cancellation of amplifier and relay distortion.

Figure 24.25 was obtained by sawing a 3 mm by 15 mm piece from a relay frame and wiring it in series with the amplifier output, by means of copper wires soldered at each end. As before, the level was 200 Wrms/2 Ω, i.e., 10 Amps rms. Trace A is the raw extra distortion; this is lower than shown in Figure 24.22 because the same current is passing through less of the frame material. Trace B is the result of enclosing the frame fragment in the NFB loop exactly as before. This removes all suspicion of interaction with coil or contacts and proves it is the actual frame material itself that is non-linear.

Wrapping feedback around the relay helps but, as usual, is not a complete cure. Soldering on extra wires

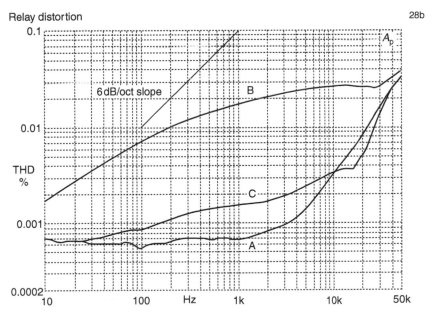

Figure 24.24. A is amplifier distortion alone, B total distortion with power relay in circuit. C shows that enclosing the relay in the feedback loop is not a complete cure.

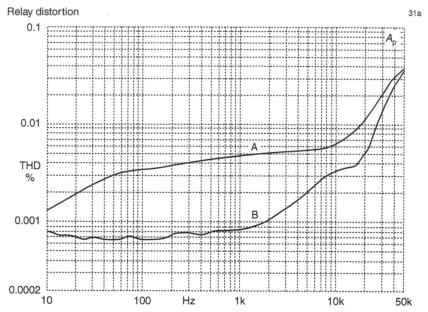

Figure 24.25. Trace A here is total distortion with a sample of the power relay frame material wired in circuit. B is the same, enclosed in the feedback loop as before.

to the frame to bypass as much frame material as possible is also useful, but it is awkward and there is the danger of interfering with proper relay operation. No doubt any warranties would be invalidated. Clearly it is best to avoid this sort of relay construction if you possibly can, but if high-current switching is required, more than an audio-intended relay can handle, the problem may have to be faced.

Output Crowbar DC Protection

Since relays are expensive and require control circuitry, and fuse protection is very doubtful, there has for at least two decades been interest in simpler and wholly solid-state solutions to the DC-protection problem. The circuit of Figure 24.26 places a triac across the output, the output signal being low-pass filtered by R and C. If sufficient DC voltage develops on C to fire the diac, it triggers the triac, shorting the amplifier output to ground.

While this approach has the merit of simplicity, in my experience it has proved wholly unsatisfactory. The triac needs to be very big indeed if it is to work more than once, because it must pass enough current to blow the HT rail fuses. If these fuses were omitted, the triac would have to dump the entire contents of a power-supply reservoir capacitor to ground through

a low total resistance, and the demands on it become quite unreasonable.

An output crowbar is also likely to destroy the output devices; the assumption behind this kamikaze crowbar system is presumably that the DC offset is most likely due to blown output devices, and a short across the output can do no more harm. This is quite wrong, because any fault in the small-signal part of the amplifier will also cause the output to saturate positive or negative, with the output devices in perfect working order. The operation of the crowbar under these

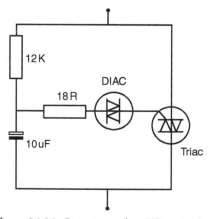

Figure 24.26. Output crowbar DC protection.

circumstances may destroy the output devices, for the overload protection may not be adequate to cope with such a very direct short-circuit.

Protection by Power-supply Shutdown

Conventional transformer power supplies can be shut down quickly by firing crowbar thyristors across the supply-rails; this overcomes one of the objections to output crowbars, as collateral damage to other parts of the circuit is unlikely, assuming of course you are correctly trying to blow the DC rail fuses, and not the transformer secondary fuses. The latter option would severely endanger the bridge rectifier, and the crowbar circuitry would have to handle enormous amounts of energy as it emptied the reservoir capacitors. Even blowing the DC fuses will require SCRs with a massive peak-current capability.

A conventional power supply can also be shut down by using relays to open the DC power supply rails. In a two-channel amplifier sharing one power supply and one offset-detect circuit, this is relatively straightforward, and one 2-make relay can do the job. If the two channels have separate power supplies, then they will also need separate offset-detect circuits and separate rail-switching relays to avoid the need for a 4-make relay, which is not a common component in high-current sizes.

Amplifiers have been designed with power switching relays between the transformer secondary and the rectifier. This is not a good plan as this circuit contains large charging-current pulses which are likely to cause excessive I^2R heating of the relay contacts.

If your amplifier is powered by a switch-mode supply, it may well have a logic input that gives the option of near-instant shutdown. This can be connected to a DC-detect low-pass filter, and the occurrence of a DC error then gives an apparently foolproof shutdown of everything.

There are (as usual) snags to this. Firstly, the high relative cost of switch-mode supplies means that almost certainly a single supply will be shared between two or more amplifier channels, and so both channels are lost if one fails. This is not a serious problem for domestic use, as few people are going to want to carry on listening to one channel of a stereo source. It may, however, be a disadvantage in sound-reinforcement applications. Secondly, and more worryingly, this provides very dubious protection against a fault in the supply itself. If such a fault causes one of the HT rails to collapse, then it may well also disable the shutdown facility, and all protection is lost.

Testing DC-offset protection

This is best done with an 8 Ω resistor load connected to the amplifier output and a suitable DC stimulus applied to its input. (You clearly don't want to try this with a loudspeaker connected.) This may be complicated by the presence of a DC servo which opposes your attempt to create a DC offset, but as described in the section on servo design, the authority of the servo should be limited so it can be easily overpowered by the external stimulus. It may be necessary to make sure that any electrolytic capacitors in the feedback path are not reverse-biased or subjected to excess forward voltages.

Thermal Protection

A properly designed amplifier will have adequate heat-sinking, and so on, and will not overheat in normal operation. Thermal protection is required to deal with two abnormal conditions:

1. The amplifier heatsinking is designed to be adequate for the reproduction of speech and music (which has a high peak-to-volume ratio, and therefore brings about relatively small dissipation) but cannot sustain long-term sinewave testing into the minimum specified load impedance without excessive junction temperatures. Heatsinking forms a large part of the cost of any amplifier, and so economics makes this a common state of affairs. Similar considerations apply to the rating of amplifier mains transformers, which are often designed to indefinitely supply only 70% of the current required for extended sinewave operation. Some form of thermal cut-out in the transformer itself then becomes essential (see Chapter 26).
2. The amplifier may be designed to withstand indefinite sinewave testing, but will still be vulnerable to having ventilation slots, etc. blocked, interfering either with natural convection or fan operation. In fan-cooled equipment there is the possibility of fan failure.

This section deals only with protecting the output semiconductors against excessive junction temperature; the thermal safeguarding of the mains transformer is dealt with in Chapter 26.

Output devices that are fully protected against excess current, voltage and power are by no means fully safeguarded. Most electronic overload protection systems allow the devices to dissipate much more power than in normal operation; this can and should be well inside the rated capabilities of the component itself, but this gives no assurance that the increased dissipation will not cause the heatsink to eventually reach such temperatures that the crucial junction temperatures are exceeded and the device fails. If no temperature protection is provided, this can occur after only a few minutes' drive into a short. Heatsink over-temperature may also occur if ventilation slots, etc. are blocked, or heatsink fins covered up.

The solution is a system that senses the heatsink temperature and intervenes when it reaches a pre-set maximum. This intervention may be in the form of:

1. Causing an existing muting/DC-protection relay to drop out, breaking the output path to the load. If such a relay is fitted, then it makes sense to use it.
2. Muting or attenuating the input signal so the amplifier is no longer dissipating significant power.
3. Removing the power supply to the amplifier sections. This normally implies using a bimetallic thermal switch to break the mains supply to the transformer primary, as anywhere downstream of here requires two lines to be broken simultaneously, e.g., the positive and negative HT rails.

Each of these actions may be either self-resetting or latching, requiring the user to initiate a reset. The possibility that a self-resetting system will cycle on and off for long periods, subjecting the output semiconductors to severe temperature changes, must be borne in mind. Such thermal cycling can greatly shorten the life of semiconductors. In an attempt to address this issue,

some IC power amplifiers mute and unmute very rapidly, almost on a per-cycle basis. The rationale behind this is that while the output devices never have time to cool down much, this is actually a good thing as sustained high temperatures are less damaging to device reliability than thermal cycling.

The two essential parts of a thermal protection system are the temperature-sensing element and whatever arrangement performs the intervention. While temperature can be approximately sensed by silicon diodes, transistor junctions, etc., these typically require some sort of set-up or calibration procedure, due to manufacturing tolerances. This is wholly impractical in production, for it requires the heatsink (which normally has substantial thermal inertia) to be brought up to the critical temperature before the circuit is adjusted. This not only takes considerable time, but also requires the output devices to reach a temperature at which they are somewhat endangered.

Until relatively recently I would have put thermistors into the same category, but they have now improved to the point where thermistor-based temperature protection systems can be made sufficiently accurate without worrying about calibration. They can also be obtained with a screw-on tab that makes mounting much easier.

Figure 24.27 shows the physical package and the resistance-temperature curves for a range of positive thermistors (i.e., parts where the resistance increases with temperature), made by Nichicon. A standard negative thermistor reduces its resistance as temperature increases. Figure 24.28 shows a comparator-based

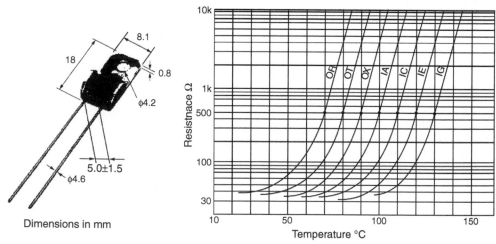

Figure 24.27. The physical package and the resistance-temperature curves of a suitable thermistor for over-temperature protection.

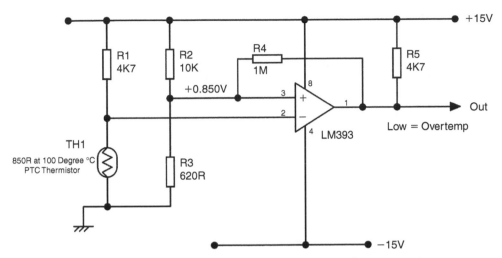

Figure 24.28. Over-temperature protection circuit using a positive thermistor and comparator.

over-temperature detector using a positive thermistor. This circuit uses an LM393 comparator; it is configured for the output to go low when the set temperature is exceeded. The thermistor is in the bottom arm of the potential divider R1, TH1, and the reference voltage is established by potential divider R2, R3 at +0.85 V. With the thermistor chosen, this represents a cut-out temperature of 100°C. R4 provides a little positive feedback to the reference divider, to create hysteresis around the switching point. The application of hysteresis is particularly necessary when comparators rather than opamps are used as they have no internal compensation and are more thus prone to oscillation. Good rail decoupling is also important.

The LM393 comparator has an open-collector output and so the pull-up resistor R5 is required.

If it would be more convenient for the output to go high when the set temperature is exceeded, the positions of the thermistor and reference divider could be swapped, but there is then a problem with the application of hysteresis as the resistance of the thermistor varies widely over the operating temperature range, and so the source resistance of the potential divider R1, TH1 also varies widely, and the amount of hysteresis is therefore temperature-dependent, which is not helpful to the design process.

The strong variation of thermistor resistance with temperature means that it is hard to significantly change the temperature cut-out setting by changing the reference voltage. For example, if the rightmost thermistor type in Figure 24.27 was used to implement cut-out at 150°C, it would be difficult to change that to 90°C

because at that temperature the thermistor resistance has flattened out at around 30 Ω, and shows little variation with temperature. It would be necessary to change the thermistor type, which is fine unless you already have 100,000 in stock.

If greater accuracy or more flexibility is required, the best method is the use of integrated temperature sensors that do not require any calibration. A good example is the National Semiconductor LM35DZ, a three-terminal device which outputs 10 mV for each degree Centigrade above Zero. Without any calibration procedure, the output voltage may be compared against a fixed reference. In this example, an opamp is used as a comparator, and this time the circuit is configured for the output to go high when the set temperature is exceeded. The resulting over-temperature signal is used to pull out the muting relay. An example of this is shown in Figure 24.29.

The output of the LM35 is applied to the non-inverting input of the opamp via R3, which in combination with R4 gives a little positive feedback which introduces hysteresis to the switching point and prevents dithering or oscillation of the output; the amount of hysteresis is constant because the output impedance of the LM35 is low and not temperature-dependent, unlike the thermistor sensor used in the previous circuit. When the output from the LM35 exceeds the reference voltage set up by the divider R1, R2, the opamp output switches from low to high, and this signal goes off to the rest of the control circuitry. The reference in this example is set to a voltage of +0.95 V, corresponding to a cut-out temperature of 95°C; because

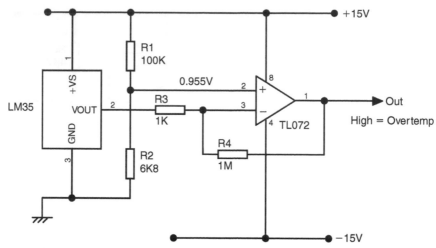

Figure 24.29. Over-temperature protection circuit using an LM35 temperature sensor and an opamp.

of the linear sensor output this can be easily altered over a very wide range by changing the value of R2. The assumption is made here that the opamp supply rails will be regulated by the usual methods, and, if so, this will be more than accurate enough for setting the reference voltage.

If the circuit is required to operate in the other sense (i.e., output low = overtemp) to interface with the rest of the protection system, then the opamp inputs can simply be interchanged, and the hysteresis must be applied to the reference divider instead.

The LM35 approach gives the most accurate and trouble-free temperature protection, in my experience. It allows a wide variation in setting without changing components. The downside is that IC temperature sensors are significantly more expensive than thermistors, etc.

Another pre-calibrated type of temperature sensor is the thermal switch, which usually operates on the principle of a bistable bimetallic element. These should not be confused with thermal fuses which are once-only components that open the circuit by melting an internal fusible alloy; they are in common use for transformer thermal protection, as they are cheaper than the self-resetting thermal switches. The trouble with thermal fuses is that they are relatively uncommon, and the chance of a blown thermal fuse being replaced with the correct component in the field is not high.

The physical positioning of the temperature sensor requires some thought. In an ideal world we would judge the danger to the output devices by assessing the actual junction temperature; since this is difficult to do,

the sensor must get as close as it can. It is shown elsewhere that the top of a TO-3 transistor can gets hotter than the flange, and as for quiescent biasing sensors, the top is the best place for the protection sensor. This does, however, present some mechanical problems in mounting. This approach may not be equally effective with plastic flat-pack devices such as TO-3P, for the outer surface is an insulator; however, it still gets hotter than the immediately adjacent heatsink, and this sensor position undoubtedly gives the fastest response.

Alternatively, the protection sensor can be mounted on the main heatsink, which is mechanically much simpler but imposes a considerable delay between the onset of device heating and the sensor reacting. For this reason, a heatsink-mounted sensor will normally need to be set to a lower trip temperature, usually in the region of 80°C, than if it is device-mounted. The more closely the sensor is mounted to the devices, the better they are protected. If two amplifiers share the same heatsink, the sensor should be placed between them; if it were placed at one end, the remote amplifier would suffer a long delay between the onset of excess heating and the sensor acting.

One well-known make of PA amplifiers implements (or used to) temperature protection by mounting a thermal switch in the live mains line on top of one of the TO3 cans in the output stage. This gains the advantage of fast response to dangerous temperatures, but there is the obvious objection that lethal mains voltages are brought right into the centre of the amplifier circuitry, where they are not normally expected, and this represents a real hazard to service personnel.

More sophisticated thermal protection systems attempt to estimate the actual temperature of the transistor die. This is sometimes called Junction Temperature Simulation or JTS. A notable example of this approach is the Crown protection system, as described in US patent 4,330,809; see Table 24.1. The power dissipation in an output device is estimated by multiplying the current through it and the voltage across it; multiple electrical time-constants are then used to simulate thermal capacities and delays.

Another approach patented by Pioneer is the measurement of output transistor junction temperature by measuring the Vbe voltage. A major difficulty is that the voltage changes to be measured are small, and, in the EF type of output stage, both base and emitter are typically going up and down at signal frequency; a DC-precise differential amplifier with very good common-mode rejection is therefore required. The method is described in US patent 5,383,083; see Table 24.1.

Output Transient Suppression

Start-up transients are suppressed by imposing a delay in energising the output mute relay after power has been applied, to prevent initial thumps and bangs from reaching the loudspeakers. There are many ways to delay the operation of a relay; here the requirements are for a delay of up to something like five seconds (too long a start-up delay will irritate the user, or make them fear that the amplifier is not going to start at all), which need not be implemented with any great accuracy.

Three similar methods are shown in Figure 24.30. The original amplifier was a switchable Class-A/Class-B design, based on the Trimodal amplifier of Chapter 17, which explains why the supply-rail voltages are relatively low. In a typical Class-B design with higher rails, it would be possible to run the relay coils in series rather than in parallel. The transistors types might also need to be changed as those shown have a Vceo of only 30 V.

In Figure 24.30a, the time delay is set by R1, C1 and the reference voltage of Zener diode D3. At switch-on, C1 charges through R1 until its voltage is two diode drops above the Zener voltage; Q1 and D2 then conduct, turning on Q2 and thus the relays and indicator LED. The use of two transistors is necessary to give enough current gain, and to give a reasonably snappy relay action. Even so, events are not instantaneous; with the values shown, the relay voltage takes about 35 msec to reach 90% of its final value. D1 is connected

Table 24.1. US patents details

Patent no.	Patent name	Date	Author	Comment
3,500,218	*Transistor Complementary Pair Amplifier with Active Current Limiting Means*	Mar. 1970	R. S. Burwen (assigned to Analog Devices)	Classic V-I limiting. The funny triangle and bar symbols are ordinary silicon diodes
3,526,846	*Protective Circuitry for High Fidelity Amplifier*	Sept. 1970	D. L. Campbell (assigned to McIntosh)	Another version of V-I limiting. Filed in 1967
4,330,809	*Thermal Protection for the Die of a Transistor*	May 1982	G. R. Stanley (assigned to Crown)	An analogue-computer method of calculating die temperature (Junction Temperature Simulation or JTS) by multiplying current and voltage, and using multiple electrical time-constants to simulate thermal capacities
4,611,180	*Grounded Bridge Amplifier Protection Through Transistor Thermo Protection*	Sept. 1986	G. R. Stanley (assigned to Crown)	JTS applied to bridged amplifiers
5,383,083	*Protective Apparatus for Power Transistor*	Jan. 1995	Shinoda (assigned to Pioneer)	Measuring output device junction temperature by sensing the Vbe
5,847,610	*Protection Circuit for an Audio Amplifier*	Dec. 1988	S. Fujita (assigned to Yamaha)	Describes a protection system that handles both over-current and DC-offset conditions. There seems to be nothing very new here
6,927,626	*Thermal Protection System for Output Stage of an Amplifier*	Aug. 2005	G. R. Stanley (assigned to Harman)	Elaborations and improvements to Patent 4,330,809

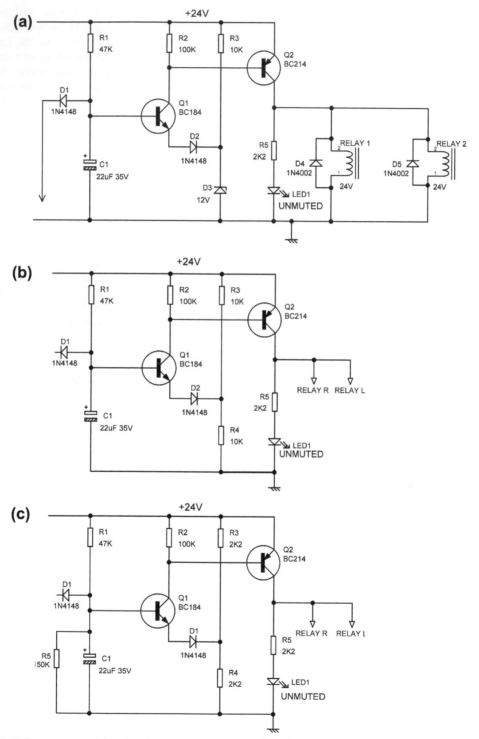

Figure 24.30. Relay power-up delay circuits: (a) uses a Zener as the reference, while (b) replaces this with a cheaper resistive divider, but there are snags, solved at (c).

to other parts of the protection circuitry, and when a DC-offset or over-temperature condition occurs, discharges C1 rapidly to ground.

Figure 24.30b shows a modified version which attempts to save money by using a potential divider R3, R4 to create the desired reference voltage. There are two snags to this apparently entirely reasonable plan. Firstly, the rate of change of the relay voltage is now much slower, due to the much higher impedance of the reference divider compared with the Zener diode, which reduces the gain of the circuit. With R3 = R4 = 10 KΩ, the relay voltage now takes about 300 msec to reach 90% of its final value, and this not only gives slow and slothful relay operation but also makes the LED illumination a noticeably leisurely business, which does not exactly give the desirable impression of snappily precise circuit operation. The slow turn-on of Q2 also means that a much greater amount of power is dissipated in Q2, as it is partially-on for much longer.

However, the circuit shown in Figure 24.30b has another, much more subtle drawback to snare the unwary. As C1 charges, the emitter of Q1 rises in voltage, for it is acting as a kind of emitter-follower. The circuit comes to rest with C1 fully charged and Q1 emitter only 0.6 V lower. This means that if there is an abrupt rise in the rail voltage, even of only a couple of Volts, Q1 may turn off, briefly dropping out the relays. This occurred on initial testing of the Class-A/Class-B design when it was switched from A to B mode and several Amps of quiescent current were abruptly switched off.

This problem was solved by adding R5 in Figure 24.30c, which prevents the voltage on C1 rising

to the level where rail dips can cause switch-off. The slow relay turn-on has also been fixed, by reducing the value of R3 and R4. With R3 = R4 = 4K7 the relay voltage takes about 130 msec to reach 90% of its final value, while with R3 = R4 = 2K2, it takes only 75 msec. This is fast enough to give satisfactorily prompt relay and LED behaviour. The downside is that the R3, R4 divider now draws more current from the supply-rail.

The other half of the transient suppression problem is addressed by muting the amplifier as fast as possible when the power is removed. Most audio circuitry will produce thumps and bangs at some point as the supply-rails collapse; if large reservoir capacitors are involved, this can be quite a lengthy process. It is therefore important to include a circuit that can detect loss of mains power quickly — before the voltage on the reservoirs in either the main amplifier or auxiliary opamp supplies can fall significantly.

The most common method of mains-fail detection is to feed a small capacitor via a rectifier from the transformer secondary. The capacitor is shunted with a resistance and the resulting RC time-constant sized so that the capacitor voltage falls much more quickly than that on any of the main power supply reservoirs. The principle is shown in Figure 24.31. In this case, the mains-fail circuit was driven from a transformer secondary that powered a +5V regulator powering a housekeeping PIC microcontroller. The mains-fail signal was fed directly to a PIC input port pin.

Note that half-wave rectification is used. R1, R2 and R3 are chosen so that Q1 stays on at least down to the minimum mains operating voltage (which is usually determined by the power supply regulators dropping out) but turns off promptly when the mains supply is

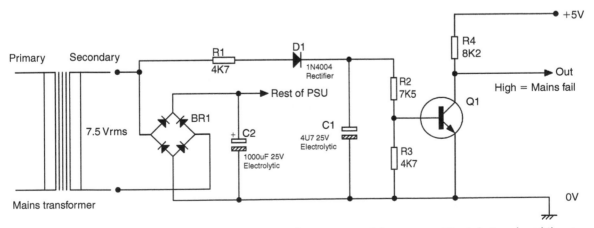

Figure 24.31. A simple mains-fail detection circuit, as used in a commercial power amplifier I designed a while ago.

removed. It is important to remember that there will be a substantial ripple voltage on the mains-fail capacitor C1, and since it will be a small electrolytic, you must check that the ripple current through it will not lead to excessive heating.

A more sophisticated technique for mains-fail detection is a 2 msec (or thereabouts) timer which is held reset by the AC on the mains transformer secondary, except for a brief period around the AC zero-crossing, which is not long enough for the timer to trigger. When the incoming AC disappears, the near-continuous reset is removed, the timer fires, and the relay is dropped out within 10 msec. This will be long before the various reservoir capacitors in the system can begin to discharge. However, if the mains switch contacts are generating RF that is in turn reproduced as a click by the pre-amp, then even this method may not be fast enough to mute it.

A simple way to implement this is shown in Figure 24.32. C4, D3, and Q3 implement the output relay turn-on delay. Q1 is always on except during zero-crossings, and keeps C3 discharged. If the mains fails, C3 charges rapidly via R5 and turns on Q2, which quickly discharges C4, turning off the output relays. Note that this circuit is designed to run off one of the power amplifier supply-rails and does not require an auxiliary power supply.

The Zeners across the output relays require a word of explanation. When a relay is driven by a transistor, a reversed diode across the coil is required to prevent the abrupt turn-off of current making the coil voltage reverse, driving the collector more negative. I have measured −120 V, enough to destructively exceed the Vceo of most transistors.

This protection conceals a lurking snag; relay drop-out time is hugely increased by the reversed diode, as it provides a path for coil current to circulate while the magnetic field decays. It takes roughly five times longer, which is really not what we need here. This is a good point to stop and consider exactly what we need to do: the aim is not 'suppress all back-EMF' but 'protect the transistor'. If the back-EMF is clamped to, say, −27V for a 24V relay by a suitable Zener diode in series with the reverse diode, the circulating current stops much sooner, and drop-out is almost as fast as for the non-suppressed relay. It is speeded up by a factor of about four on moving from conventional protection to Zener clamping. For relays of the usual size, a 500 mW Zener appears to be adequate. If this circuit is interfaced to a DC offset detector and an over-temperature circuit, via the input to Q2 collector, it implements the complete protection system in Figure 24.19.

The surprisingly complex and subtle subject of relay control is dealt with in much more detail in my book, *Self on Audio*.[7]

Very fast mains-fail detection does carry dangers. It can react to transient voltage drops on the mains that would otherwise be ignored by the equipment, enforcing the full power-up delay every time a glitch comes along. This will not be well received by the user. In one case, the prototype of a power amplifier of my design was being evaluated at home by one of my colleagues, and it showed just this behaviour, shutting down and restarting at fairly regular intervals. It was installed in an old house with an old refrigerator in the kitchen, and a little investigation showed that every time the rather large refrigerator motor started, it very briefly

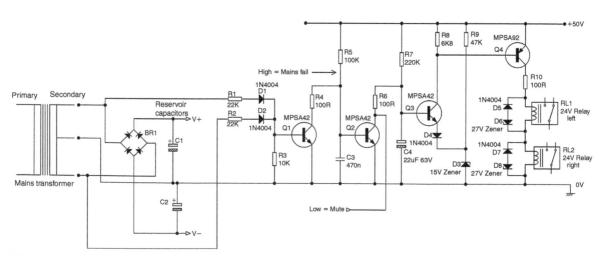

Figure 24.32. A more sophisticated mains-fail detection circuit, that gives a faster response to the removal of the power.

dragged the mains voltage down to less than half its normal value, due to the high resistance of the elderly house wiring. Fortunately the product in question did not emit noises the instant the power was removed, and reverting to a simpler and slower mains-fail detector as shown in Figure 24.28 solved the problem (and saved a few pennies in parts).

On the whole, it is best, if possible, to design the equipment so that it does not produce prompt noises on power-down, allowing the more tolerant (and cheaper and simpler) means of mains-fail detection to be used.

Clip Detection

Driving an amplifier with an excessive input signal so that its output stage is clipping is not in general dangerous to the amplifier, though it can cause excessive power to be delivered to the loudspeaker, and damage it mechanically or thermally. An exception to this is failure of the VAS transistor (see Chapter 7) which may be turned on excessively as the negative feedback loop attempts to pull the output lower than it can go. If the usual VAS current-limiting is applied to protect it when overload protection is operating, then this also protects against excessive VAS currents during clipping.

For a long time there was a belief that the extra high harmonics generated by heavy clipping put tweeters in particular in danger. This hypothesis appears to have been exploded by Montgomery Ross,[8] who shows that the apparent vulnerability of tweeters is in fact simply due to the increased power output associated with clipping, combined with the fact that the tweeter has a lower power rating than bass and middle speaker units because it receives much less power in normal usage

A clip detect circuit may simply drive an indicator, usually with some sort of pulse-stretching to make brief clipping clearly visible. It may, however, also provide a signal to a microcontroller so that prolonged episodes of clipping cause shut-down. It is important that both positive and negative clipping are detected, because a single unaccompanied voice, or a solo instrument, can have waveforms with considerable asymmetry in their peak values; up to 8 dB is often quoted.

Clip Detection by Rail-approach Sensing

The voltage variations in an unregulated power amplifier supply, due to mains voltage changes and varying current demands on the supply, mean that it is not

possible to accurately detect clipping by comparing the amplifier output with a fixed threshold. A far better method is to compare the output voltage with each supply rail. A thoroughly tested way to do this (I have been using it ever since I thought it up in 1975) is shown in Figure 24.33.

Normally the amplifier output is somewhere in the middle between the supply rails and both Q1 and Q2 are continuously conducting. Q2 collector is therefore pulled up to the positive supply rail; the CLIP signal is clamped at +5.1 V by R7 and Zener D2 so it can be applied to a port of a microcontroller such as a PIC.

If the output approaches the negative rail, then Q1 is no longer kept on via R1, R2, and the base drive to Q2 via R3, R4, R5 is removed. Q2 collector therefore drops to the negative rail, and the CLIP signal is clamped at −0.6 V by forward conduction of Zener D2; this small negative voltage is normally safe to apply to a port of a microcontroller without further clamping to reduce it (e.g., by a Schottky diode) but this is a point to check carefully.

If the output approaches the positive rail, then Q1 remains on but now D1 begins to conduct and pulls the junction of R4 and R5 positive, once more removing the base drive from Q2, which turns off, and its collector voltage drops to the negative rail again.

The resistor values R1 to R5 can be adjusted to match the clipping behaviour of the amplifier output stage. Initially D1 can be removed and the detection of negative clipping checked. D1 is then replaced and the relative values of R4 and R5 adjusted so that positive clipping is detected properly.

Note that this circuit detects the approach of the output to the rails rather than clipping as such, so it must trigger slightly before actual clipping. If it was set to trigger slightly *after* clipping, it would of course never operate. The circuit compensates for supply-rail variations due to both mains voltage changes and supply current drawn, but it cannot allow for the slightly earlier clipping that occurs with low-impedance loads due to the voltage drops in the output stage emitter resistors and the increased Vbe voltages of the output and driver devices.

Clip Detection by Input-output Comparison

A technique that takes everything into account is 'error sensing clip detection' which detects actual clipping as it senses when the output signal is no longer simply a scaled-up version of the input signal. This can

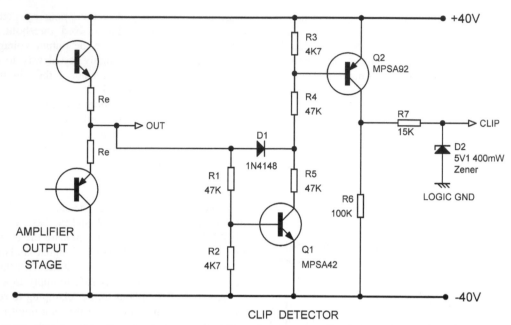

Figure 24.33. Clipping detection circuit that measures how far the output line is from the supply rails. This allows for supply-rail voltage changes but not for voltage drops in the output stage emitter resistors.

conveniently be done by monitoring the error voltage, i.e., the difference between the input signal and the feedback signal that is applied to the input pair of the amplifier. However, this detects several conditions as well as clipping; the operation of over-load protection, slew-limiting, a DC-offset fault, or even RF oscillation. A complicating factor is that an error voltage is always present, due to the finite open-loop gain of the amplifier, and this increases with frequency, due to the dominant-pole compensation that is usually employed.

An early instance of this approach was the 'Input-Output-Comparator' introduced by Crown in 1977 and used extensively since.

Amplifier Protection Patents

Table 24.1 earlier in this chapter shows some interesting US patents in the field of amplifier protection. This is a very small selection from the large number that exist. They may be accessed free of charge through Google Patents.

Powering Auxiliary Circuitry

Whenever it is necessary to power auxiliary circuitry, such as the relay control system described above, there

is an obvious incentive to use the main HT rails. A separate PSU requires a bridge rectifier, reservoir capacitor, fusing and an extra transformer winding, all of which will cost a significant amount of money.

The main disadvantage is that the HT rails are at an inconveniently high voltage for powering control circuitry. For low-current sections of this circuitry, such as relay timing, the problem is not serious as the same high-voltage small-signal transistors can be used as in the amplifier small-signal sections, and the power dissipation in collector loads, etc. can be controlled simply by making them higher in value. The biggest problem is the relay energising current; many relay types are not available with coil voltages higher than 24 V, and this is not easy to power from a 50 V HT rail without continuously wasting power in a big dropper resistor. This causes unwanted heating of the amplifier internals, and provides a place for service engineers to burn themselves.

One solution in a stereo amplifier is to run the two relays in series; the snag (and for sound reinforcement work, it may be a serious one) is that both relays must switch together, so if one channel fails with a DC offset, both are muted. In live work, independent relay control is much to be preferred, even though most of the relay control circuitry must be duplicated for each channel.

If the control circuitry is powered from the main HT rails, then its power must be taken off *before* the amplifier HT fuses. The control circuitry will then be able to mute the relays when appropriate, no matter what faults have occurred in the amplifiers themselves.

If there is additional signal circuitry in the complete amplifier it is not advisable to power it in this way, especially if it has high gain, e.g., a microphone preamplifier. When such signal circuits are powered in this way, it is usually by ±15 V regulators from the HT rails, with series dropper resistors to spread out some of the dissipation. However, bass transients in the power amplifiers can pull down the HT rails alarmingly, and if the regulators drop out, large disturbances will appear on the nominally regulated low-voltage rails, which can lead to very low-frequency oscillations extremely destructive of loudspeakers. In this case, the use of wholly separate clean rails run from an extra transformer secondary is strongly recommended. There will be no significant coupling between the supplies due to the use of a shared transformer primary.

References

1. Bailey, A, Output Transistor Protection in AF Amplifiers, *Wireless World*, June 1968, p. 154.

2. Becker, R, 'High-Power Audio Amplifier Design, *Wireless World*, Feb. 1972, p. 79.

3. Motorola, High Power Audio Amplifiers with Short Circuit Protection, Motorola Application Note AN-485 (1972).

4. Otala, M, Peak Current Requirement of Commercial Loudspeaker Systems, *JAES*, 35, June 1987, p. 455.

5. Baxandall, P, Technique for Displaying Current and Voltage Capability of Amplifiers, *JAES* 36, Jan/Feb 1988, p. 3.

6. Greiner, R, Amplifier-Loudspeaker Interfacing Loudspeakers, *JAES* 28(5), May 1980, pp. 310—315.

7. Self, D, *Self on Audio* 2nd edn, Oxford: Newnes 2006, p. 421.

8. Montgomery F. Ross (Rane Corp), An Investigation into How Amplifier Clipping is Said to Burn-Out Loudspeakers, and How Limiters Can Save Them, Paper Number: 2956, AES Convention: 89 (September 1990).

Chapter 25

Layout, Grounding, and Cooling

Audio Amplifier PCB Design

This section addresses the special printed circuit board (PCB) design problems presented by power amplifiers, particularly those operating in Class-B. All power amplifier systems contain the power-amp stages themselves, and usually associated control and protection circuitry; most also contain small-signal audio sections such as balanced input amplifiers, subsonic filters, output meters, and so on.

Other topics that are related to PCB design, such as grounding, safety, reliability, etc., are also dealt with.

The performance of an audio power amplifier depends on many factors, but in all cases the detailed design of the PCB is critical, because of the risk of inductive distortion due to crosstalk between the supply-rails and the signal circuitry; this can very easily be the ultimate limitation on amplifier linearity, and it is hard to over-emphasise its importance. The PCB design will to a great extent define both the distortion and crosstalk performance of the amplifier.

Apart from these performance considerations, the PCB design can have considerable influence on ease of manufacture, ease of testing and repair, and reliability. All of these issues are addressed below.

Successful audio PCB layout requires enough electronic knowledge to fully appreciate the points set out below, so that layout can proceed smoothly and effectively. It is common in many electronic fields for PCB design to be handed over to draughtspersons, who, while very skilled in the use of CAD, have little or no understanding of the details of circuit operation. In some fields, this works fine; in power amplifier design it will not, because basic parameters such as crosstalk and distortion are so strongly layout-dependent. At the very least, the PCB designer should understand the points set out below.

Crosstalk

All crosstalk has a transmitting end (which can be at any impedance) and a receiving end, usually either at high impedance or virtual-earth. Either way, it is sensitive to the injection of small currents. When interchannel crosstalk is being discussed, the transmitting and receiving channels are usually called the speaking and non-speaking channels, respectively.

Crosstalk comes in various forms:

- Capacitive crosstalk is due to the physical proximity of different circuits, and may be represented by a small notional capacitor joining the two circuits. It usually increases at the rate of 6 dB/octave, though

higher dB/octave rates are possible. Screening with any conductive material is a complete cure, but physical distance is usually cheaper.
- Resistive crosstalk usually occurs simply because ground tracks have a non-zero resistance. Copper is not a room-temperature superconductor. Resistive crosstalk is constant with frequency.
- Inductive crosstalk is rarely a problem in general audio design; it might occur if you have to mount two uncanned audio transformers close together, but otherwise you can usually forget it. The notable exception to this rule is the Class-B audio power amplifier, where the rail currents are halfwave sines that seriously degrade the distortion performance if they are allowed to couple into the input, feedback or output circuitry.

In most line-level audio circuitry the primary cause of crosstalk is unwanted capacitive coupling between different parts of a circuit, and in most cases this is defined solely by the PCB layout. Class-B power amplifiers, in contrast, should suffer very low or negligible levels of crosstalk from capacitive effects, as the circuit impedances tend to be low, and the physical separation large; a much greater problem is inductive coupling between the supply-rail currents and the signal circuitry. If coupling occurs to the same channel, it manifests itself as distortion, and can dominate amplifier non-linearity. If it occurs to the other (non-speaking) channel, it will appear as crosstalk of a distorted signal. In either case, it is thoroughly undesirable, and precautions must be taken to prevent it.

The PCB layout is only one component of this, as crosstalk must be both emitted and received. In general, the emission is greatest from internal wiring, due to its length and extent; wiring layout will probably be critical for best performance, and needs to be fixed by cable ties, etc. The receiving end is probably the input and feedback circuitry of the amplifier, which will be fixed on the PCB. Designing these sections for maximum immunity is critical to good performance.

Rail Induction Distortion

The supply-rails of a Class-B power-amp carry large and very distorted currents. As previously outlined, if these are allowed to crosstalk into the audio path by induction, the distortion performance will be severely degraded. This applies to PCB conductors just as much as cabling, and it is sadly true that it is easy to produce an amplifier PCB that is absolutely satisfactory in every respect but this one, and the only solution is another board iteration. The effect can be completely prevented but in the present state of knowledge I cannot give detailed guidelines to suit every constructional topology. The best approach is:

1. Minimise radiation from the supply rails by running the V+ and V- rails as close together as possible. Keep them away from the input stages of the amplifier, and the output connections; the best method is to bring the rails up to the output stage from one side, with the rest of the amplifier on the other side. Then run tracks from the output to power the rest of the amp; these carry no halfwave currents and should cause no problems.
2. Minimise pick-up of rail radiation by keeping the area of the input and feedback circuits to a minimum. These form loops with the audio ground and these loops must be as small in area as possible. This can often best be done by straddling the feed-back and input networks across the audio ground track, which is taken across the centre of the PCB from input ground to output ground.

Induction of distortion can also occur into the output and output-ground cabling, and even the output inductor. The latter presents a problem as it is usually difficult to change its orientation without a PCB update.

The Mounting of Output Devices

The most important decision is whether or not to mount the power output devices directly on the main amplifier PCB. There are strong arguments for doing so, but it is not always the best choice.

Advantages:

- The amplifier PCB can be constructed so as to form a complete operational unit that can be thoroughly tested before being fixed into the chassis. This makes testing much easier, as there is access from all sides; it also minimises the possibility of cosmetic damage (scratches, etc.) to the metalwork during testing.
- It is impossible to connect the power devices wrongly, providing you get the right devices in the right positions. This is important for such errors usually destroy both output devices and cause other domino-effect faults that are very time-consuming to correct.
- The output device connections can be very short. This seems to help stability of the output stage against HF parasitic oscillations.

Disadvantages:

- If the output devices require frequent changing (which obviously indicates something very wrong somewhere), then repeated resoldering will damage the PCB tracks. However, if the worst happens, the damaged track can usually be bridged out with short sections of wire, so the PCB need not be scrapped; make sure this is possible.
- The output devices will probably get fairly hot, even if run well within their ratings; a case temperature of 90°C is not unusual for a TO3 device. If the mounting method does not have a degree of resilience, then thermal expansion may set up stresses that push the pads off the PCB.
- The heatsink will be heavy, and so there must be a solid structural fixing between this and the PCB. Otherwise the assembly will flex when handled, putting stress on soldered connections.

Single and Double-sided PCBs

Single-sided PCBs used to be the usual choice for power amplifiers, because of their lower set-up charges and lower cost. It is not usually necessary to go double-sided for reasons of space or convoluted connectivity, because power amplifier components tend to be physically large, determining the PCB size, and in typical circuitry there are a large number of through-hole resistors, etc., that can be used for jumping tracks. However, the price differential between single- and double-sided plated-through-hole (PTH) is very much less than it used to be, and these are now the normal choice. It has to be said that single-sided PCBs now look rather old-fashioned.

Bear in mind that single-sided boards need thicker tracks to help ensure adhesion in case desoldering is necessary. Adding one or more ears (very short lengths of track leading nowhere) to pads with only one track running into them gives much better adhesion, and is highly recommended for the pads that are the most likely to need resoldering during maintenance, such as those for the drivers and output devices. This unfortunately is a very tedious manual task with many CAD systems, but if it is possible to define custom pad shapes, it becomes much easier.

The advantages of double-sided PTH for power amplifiers are as follows:

- No links are required.
- Double-sided PCBs may allow one side to be used primarily as a ground plane, minimising crosstalk and EMC problems.

- Much better pad adhesion on resoldering as the pads are retained by the through-hole plating.
- There is more total room for tracks, and so they can be wider, giving less volt-drop and so less PCB heating.
- The extra cost is small.

PCB Track Resistance and How to Reduce it

It is often the case that the resistance of tracks should be as low as possible, either to reduce crosstalk caused by shared resistance, to reduce resistive voltage losses, or simply to provide adequate capacity for the relatively large currents flowing in audio power amplifiers.

It is also useful to be able to calculate the resistance of a PCB track for the same reasons. This is slightly less than straightforward; given the smorgasbord of units that are in use in PCB technology, determining the cross-sectional area of the track can present some difficulty.

In the USA and the UK, and probably elsewhere, there is inevitably a mix of metric and imperial units on PCBs, as many important components come in dual-in-line packages which are derived from an inch grid; track widths and lengths are therefore very often in thousands of an inch, universally (in the UK at least) referred to as 'thou'. Conversely, the PCB dimensions and fixing-hole locations will almost certainly be metric as they interface with a world of metal fabrication and mechanical CAD that (once again, in the UK at least) went metric many years ago. Add to this the UK practice of quoting copper thickness in ounces (the weight of a square foot of copper) and all the ingredients for dimensional confusion are in place. In this section I have stuck with the way that the units are commonly used and have not converted them all to one system of units.

Given the copper thickness, multiplying by track width and length, gives the cross-sectional area. Since resistivity is always in metric units, it is best to

convert to metric at this point, so Table 25.1 gives area in square millimetres. This is then multiplied by the resistivity, not forgetting to convert the area to metres for consistency. This gives the 'resistance' column in the table, and it is then simple to treat this as part of a potential divider to calculate the usually unwanted voltage across the track.

For example, if the track in question is the ground return from an 8 Ω speaker load, this is the top half of a potential divider while the track is the bottom half, (I am of course ignoring here the fact loudspeakers are anything but resistive loads) and a quick calculation gives the fraction of the input voltage found along the track. This is expressed in the last column of Table 25.1 as attenuation in dB. This shows clearly that loudspeaker outputs should not have common return tracks if it can possibly be avoided, for the interchannel crosstalk will be dire. It is very clear from this table that relying on thicker copper on your PCB as means of reducing common-ground-path crosstalk is not very effective. Some alternative methods are described below.

PCB tracks have a limited current capability because excessive resistive heating will break down the adhesive holding the copper to the board substrate, and ultimately melt the copper. This is normally only a problem in power amplifiers and power supplies. It is useful to assess if you are likely to have problems before committing to a PCB design, and Table 25.2, based on MIL-standard 275, gives some guidance.

Note that Table 25.2 applies to tracks on the PCB surface only. Internal tracks in a multi-layer PCB experience much less cooling, and need to be about three times as thick for the temperature rise. This factor depends on laminate thickness, and so on, and you need to consult your PCB vendor.

Traditionally, overheated tracks could be detected visually because the solder mask on top of them would discolour to brown. I am not sure if this still

Table 25.1. Thickness of copper cladding and the calculation of track resistance

Weight oz	Thickness thou	Thickness micron	Width thou	Length inch	Area mm²	Resistance Ohm	Atten Ref 8 Ω dB
1	1.38	35	12	3	0.0107	0.123	−36.4
1	1.38	35	50	3	0.0444	0.029	−48.7
2	2.76	70	12	3	0.0213	0.061	−42.4
2	2.76	70	50	3	0.0889	0.015	−54.7
3	4.14	105	50	3	0.133	0.010	−58.1
4	5.52	140	50	3	0.178	0.0074	−60.7

Table 25.2. PCB track current capacity for a permitted temperature rise

Track temp rise	10 °C		20 °C		30 °C	
Copper weight Track width thou	1 oz	2 oz	1 oz	2 oz	1 oz	2 oz
10	1.0 A	1.4 A	1.2 A	1.6 A	1.5 A	2.2 A
15	1.2 A	1.6 A	1.3 A	2.4 A	1.6 A	3.0 A
20	1.3 A	2.1 A	1.7 A	3.0 A	2.4 A	3.6 A
25	1.7 A	2.5 A	2.2 A	3.3 A	2.8 A	4.0 A
30	1.9 A	3.0 A	2.5 A	4.0 A	3.2 A	5.0 A
50	2.6 A	4.0 A	3.6 A	6.0 A	4.4 A	7.3 A
75	3.5 A	5.7 A	4.5 A	7.8 A	6.0 A	10.0 A
100	4.2 A	6.9 A	6.0 A	9.9 A	7.5 A	12.5 A
200	7.0 A	11.5 A	10.0 A	16.0 A	13.0 A	20.5 A
250	8.3 A	12.3 A	12.3 A	20.0 A	15.0 A	24.5 A

applies with modern solder mask materials, as in recent years I have been quite successful in avoiding over-heated tracking.

The use of a double-sided PCB will allow extra parallel tracks to be used if the layout and topological considerations such as inductive distortion (see Distortion Six in Chapter 5) will allow, though double-sided is normally employed because it greatly simplifies layout, eliminates wire links, and gives much better pad adhesion during de-soldering because of the plated-through holes. Although making a double-sided PCB is a much more complex process than single-sided, this format is so commonly used that the extra cost for a given PCB area is only around 40%.

The standard thickness of PCB copper foil is known as one-ounce copper, which has a thickness of 1.4 thou (= 35 microns). Two-ounce copper is naturally twice as thick. The extra cost of specifying it is small, typically around 5% of the total cost, and this is a very simple way of halving track resistance; it can of course be applied very easily to an existing design without any fear of messing up a satisfactory layout. Four-ounce copper can also be obtained but is more rarely used and is therefore much more expensive. If heavier copper than two-ounce is required, the normal technique is to plate two-ounce up to three-ounce copper. The extra cost of this is surprisingly small and is in the region of 10% to 15%.

Another technique, historically used to reinforce tracks on a single-sided board, is to deposit solder on top of the tracks carrying power or the output signal. Since these are already relatively wide, it is feasible to

make long thin windows in the solder resist along the track, so that solder is deposited along it. A thickness of at least a millimetre can usually be obtained by the wave-solder process; it may be necessary to consider the direction of track relative to the direction the PCB goes through the machine. The result somewhat resembles corrugated iron, and does not really have a hi-tech look, though of course it is only visible when the amplifier is dismantled. While an extra millimetre of metal might be thought to reduce the track resistance to very low levels, it is not as effective as it appears because the resistivity of solder is high compared with that of copper — nine times higher if you compare copper with the 60/40 tin/lead solder that was used when this technique was in its heyday.[1] If one-ounce copper is used, then the resistance of one inch of the basic track is 0.041 Ω. A 1 mm layer of 60/40 solder (assumed here to be of uniform thickness, though in reality it takes up a rounded cross-section due to surface tension) will have a resistance of 0.013 Ω, the parallel combination of the two layers being 0.010 Ω, so the resistance of the path has been reduced by more than four times. The use of lead-free solder does not change the basic result. The obvious advantage of this technique is that it requires no extra manufacturing time; a downside is that you have to pay for the relatively large amounts of solder deposited on the PCB.

A more sophisticated method of providing a low resistance path is to replace tracks on the PCB by metal strips that run across the top surface, running vertically like walls so they obscure the minimum area of PCB. The metal is spaced from the PCB except where a leg extends down to be soldered into it; it is not good practice to rely solely on the solder resist for insulation. This obviously requires the fabrication of custom metal parts but they are relatively cheap if they are stamped out in flat shapes. Brass is a good material for this as it is obtainable in thin sheets, is non-magnetic (see Distortion Nine in Chapter 5), relatively resistant to corrosion, and solders readily. For the absolute minimum resistance copper can be used so long as it is plated with a non-corroding metal; gold looks very nice. I used this strip technique in the Cambridge Audio 840W to prevent any trace of inductive distortion, and to neatly distribute power to a large number of output devices. It is possible to make a very compact power distribution system by running two or more strips clamped together with a layer of insulation between them. If the clamping system involves drilling holes through the strips to insert plastic rivets, etc. it is essential to deburr the holes very carefully to prevent short-circuits.

A very straightforward method of reinforcing − or replacing altogether − a track is to use heavy gauge multi-strand cable terminating in the PCB at each end. This has the advantage that it can be easily added to an existing layout so long as you can fit in a couple of extra pads, and the cable can be fitted underneath the PCB for a better visual impression. Multi-strand cable comes in several standard formats, such as 7/02 (seven strands of 0.2 mm diameter wire), 16/02, 24/02, 32/02. The last is usually the biggest that is commonly used, but larger sizes are available. The next section gives more information on cable sizes and resistance.

Cable Resistance

Table 25.3 gives the resistance per metre of the common copper cable sizes in milli-Ohms. Such cables are often referred to as 'equipment wire' in catalogues. The last entry is an example of the larger sizes available, usually referred to as 'heavy current wire'. The version shown here drops the resistance per metre dramatically, but still has enough flexibility to make it relatively easy to handle.

With a well-designed amplifier, the output impedance, especially at low and medium frequencies, can be very low, to the point that a cable connection between the amplifier PCB and the output terminals has a significant effect on the so-called 'damping factor'. I have explained elsewhere that this is pretty much a meaningless specification, but it is still a specification, and if you can post a better number, some people

Table 25.3. Resistance per metre of common multi-strand copper cable formats

Cable size	Resistance per metre, mΩ
7/0.2	76.3
16/02	33.4
24/02	22.2
32/0.2	16.7
50/0.25	6.8

will be impressed. As 'damping factor' is defined as the ratio of the load impedance (usually taken as 8 Ω) to the amplifier output impedance, it changes a lot for small changes in amplifier output impedance. Suppose the total amplifier output impedance is 50 mΩ at a given frequency: the 'damping factor' is therefore 8/0.05 = 160. Reducing the overall output impedance by 10 mΩ, simply by using thicker output cabling, improves the spec to 8/0.4 = 200, which looks a lot better in print even if the improvement is in reality trivial. The effect of the amplifier output networks (output inductor, etc.) in conjunction with cable resistance on 'damping factor' is examined in detail in Chapter 14.

Cables are almost universally made of copper, for excellent reasons. It has the lowest resistance of any metal but silver. It is not wholly resistant to corrosion; a surface layer of dull copper oxide forms under normal conditions, but the effects are purely on the surface and do not have the damaging effects of rust on steel. In both single and multi-strand cables the copper is very often coated with tin to make it effectively immune to corrosion and to aid soldering. Being a heavy metal, it is unfortunately not that common in the earth's crust, and so is expensive compared with iron and steel. It is, however, cheap compared with silver.

Probably the silliest known way to reduce the resistance of a given size of cable is to use silver instead of copper. The conductivity of silver is the highest of any metal, but is only 6% better than copper. Silver is also subject to corrosion, but in this case a surface film of non-conducting silver sulphide forms as the metal reacts with hydrogen sulphide in the atmosphere. This why your teaspoons go black, and it can cause major problems with silver switch contacts working at audio line levels. Hydrogen sulphide can come from industrial pollution, but a major source is diesel engine exhaust, so virtually nowhere can be assumed to be free of it. Coating copper with tin to prevent corrosion is acceptable to everybody, but you are going to have problems persuading hi-fi enthusiasts that it might be a good idea to coat expensive silver with plebeian tin.

The price of both copper and silver varies due to economic and political factors, but at the time of writing silver was some 100 times more expensive by weight. Despite this, silver internal wiring has been used in some very expensive hi-fi amplifiers; silver has also been used for line level interconnects and loud-speaker cables. Output impedance-matching trans-formers wound with silver wire are not unknown in 'high-end' valve amplifiers. Since the technical advan-tages are negligible, such equipment has to be marketed solely on the basis of indefinable and indeed wholly imaginary subjective improvements.

It is worth noting that microwave components are often silver plated because the skin effect is very strong at such frequencies and minimising the resistance of the surface layer does bring real technical benefits.

Since silver is the most conductive metal there is, spending even more money on precious metals like gold and platinum cannot bring you a better conductor. Gold is a worse conductor than copper by a factor of 1.28, and platinum, which is even more expensive, is worse than copper by a factor of 6.2, but at least they are both highly resistant to corrosion. Don't even think about hoses filled with mercury; despite its high density it is 60 times less conductive than copper. RoHS compliance will present extreme difficulties, but given the insidiously poisonous nature of mercury, your medical problems may be even more pressing.

Power Supply PCB Layout

Power supply subsystems have special requirements due to the very high capacitor-charging currents involved:

- Tracks carrying the full supply-rail current must have generous widths. The board material used should have not less than Two-ounce copper. Four-ounce copper can be obtained but it is expensive and has long lead-times; not really recommended.
- Reservoir capacitors must have the incoming tracks going directly to the capacitor terminals; likewise the outgoing tracks to the regulator must leave from these terminals. In other words, do not run a tee off to the cap. Failure to observe this puts sharp pulses on the DC and tends to worsen the hum level.
- The tracks to and from the rectifiers carry charging pulses that have a considerably higher peak value than the DC output current. Conductor heating is therefore much greater due to the higher value of I^2R.

Heating is likely to be especially severe at PC-mount fuseholders. Wire links may also heat up and consideration should be given to two links in parallel; this sounds crude but actually works very effectively.

- Track heating can usually be detected simply by examining the state of the solder mask after several hours of full-load operation; the green mask materials currently in use discolour to brown on heating. If this occurs, then as a very rough rule the track is too hot. If the discoloration tends to dark brown or black, then the heating is serious and must definitely be reduced.
- If there are PCB tracks on the primary side of the mains transformer, and this has multiple taps for multi-country operation, then remember that some of these tracks will carry much greater currents at low voltage tappings; mains current drawn on 90 V input will be nearly 3 times that at 240 V.

Be sure to observe the standard safety spacings for creepage and clearance between mains tracks and other conductors. See Chapter 29 for the spacings required. (This applies to all track-track, track-PCB edge, and track-metal-fixings spacings.)

In general, PCB tracks carrying mains voltages should be avoided, as presenting an unacceptable safety risk to service personnel. If it must be done, then warnings must be displayed very clearly on both sides of the PCB. Mains-carrying tracks are unaccept-able in equipment intended to meet UL regulations in the USA, unless they are fully covered with insulating material that is non-flammable and can withstand at least 120°C (e.g., polycarbonate).

Power Amplifier PCB Layout Details

A simple unregulated supply is assumed:

- Power amplifiers have heavy currents flowing through the circuitry, and all of the requirements for power supply design also apply here. Thick tracks are essential, and 2-oz copper is highly desirable, espe-cially if the layout is cramped. If attempting to thicken tracks by laying solder on top, remember that ordinary 60:40 solder has a resistivity of about 6 times that of copper, so even a thick layer may not be very effective.
- The positive and negative rail reservoir caps will be joined together by a thick earth connection; this is called Reservoir Ground (RG). *Do not* attempt to use any point on this track as the audio-ground star-point, as it carries heavy charging pulses and will induce ripple into the signal. Instead take a thick tee from the centre of this track (through which the charging

pulses will not flow) and use the end of this as the star-point.

- Low-value resistors in the output stage are likely to get very hot in operation — possibly up to 200°C. They must be spaced out as much as possible and kept from contact with components such as electrolytic capacitors. Keep them away from sensitive devices such as the driver transistors and the bias-generator transistor.

- Vertical power resistors. The use of these in power amplifiers appears at first attractive, due to the small amount of PCB area they take up. However, the vertical construction means that any impact on the component, such as might be received in normal handling, puts a very great strain on the PCB pads, which are likely to be forced off the board. This may result in it being scrapped. Single-sided boards are particularly vulnerable, having much lower pad adhesion due to the absence of vias.

- Solderable metal clips to strengthen the vertical resistors are available in some ranges (e.g., Vitrohm) but this is not a complete solution, and the conclusion

must be that horizontal-format power resistors are preferable.

- Rail decoupler capacitors must have a separate ground return to the Reservoir Ground. This ground must *not* share any part of the audio ground system, and must *not* be returned to the star-point. See Figure 25.1.

- The exact layout of the feedback takeoff point is critical for proper operation. Usually the output stage has an *output rail* that connects the emitter power resistors together. This carries the full output current and must be substantial. Take a tee from this track for the output connection, and attach the feedback takeoff point to somewhere along this tee. *Do not* attach it to the track joining the emitter resistors.

- The input stages (usually a differential pair) should be at the other end of the circuitry from the output stage. Never run input tracks close to the output stage. Input stage ground, and the ground at the bottom of the feedback network must be the same track running back to the star-point. No decoupling capacitors, etc. may be connected to this track, but it seems to be

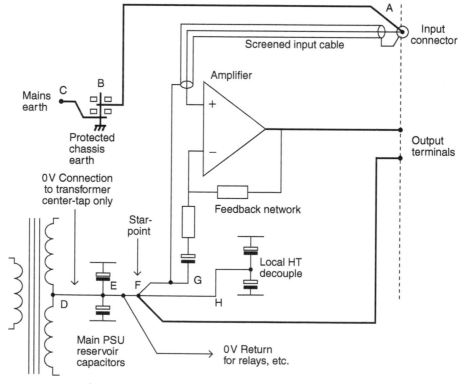

Figure 25.1. Grounding system for a typical power amplifier.

permissible to connect input bias resistors, etc. that pass only very small DC currents.

- Put the input transistors close together. The closer the temperature-match, the less the amplifier output DC offset due to Vbe mismatching. If they can both be hidden from 'seeing' the infra-red radiation from the heatsink (for example, by hiding them behind a large electrolytic), then DC drift is reduced.
- Most power amplifiers will have additional control circuitry for muting relays, thermal protection, etc. Grounds from this must take a separate path back to Reservoir Ground, and *not* the audio star-point.
- Unlike most audio boards, power amps will contain a mixture of sensitive circuitry and a high-current power-supply. Be careful to keep bridge-rectifier connections, etc., away from input circuitry.
- Mains/chassis ground will need to be connected to the power amplifier at some point. Do not do this at the transformer centre-tap as this is spaced away from the input ground voltage by the return charging pulses, and will create severe ground loop hum when the input ground is connected to mains ground through another piece of equipment. Connecting mains ground to the star-point is better, as the charging pulses are excluded, but the track resistance between input ground and star will carry any ground loop currents and induce a buzz. Connecting mains ground to the input ground gives maximal immunity against ground loops.
- If capacitors are installed the wrong way round, the results are likely to be explosive. Make every possible effort to put all capacitors in the same orientation to allow efficient visual checking. Mark polarity clearly on the PCB, positioned so it is still visible when the component is fitted.
- Drivers and the bias generator are likely to be fitted to small vertical heatsinks. Try to position them so that the transistor numbers are visible.
- All transistor positions should have emitter, base and collector or whatever marked on the top-print to aid fault-finding. TO3 devices need also to be identified on the copper side, as any screen-printing is covered up when the devices are installed.
- Any wire links should be numbered to make it easier to check they have all been fitted.

The Audio PCB Layout Sequence

PCB layout must be considered from an early stage of amplifier design. For example, if a front-facial layout shows the volume control immediately adjacent to a loudspeaker routing switch, then a satisfactory crosstalk performance will be difficult to obtain because of the relatively high impedance of the volume control wipers. Shielding metalwork may be required for satisfactory performance and this adds cost. In many cases the detailed electronic design has an effect on crosstalk, quite independently from physical layout.

1. Consider implications of fascia layout for PCB layout.
2. Circuitry designed to minimise crosstalk. At this stage try to look ahead to see how opamp halves, switch sections, etc. should be allocated to keep signals away from sensitive areas. Consider crosstalk at above PCB level; for example, when designing a module made up of two parallel double-sided PCBs, it is desirable to place signal circuitry on the inside faces of the boards, and power and grounds on the outside, to minimise crosstalk and maximise RF immunity.
3. Fascia components (pots, switches, etc.) placed to partly define available board area.
4. Other fixed components such as power devices, driver heatsinks, input and output connectors, and mounting holes placed. The area left remains for the purely electronic parts of the circuitry that do not have to align with metalwork, etc. and so may be moved about fairly freely.
5. Detailed layout of components in each circuit block, with consideration towards manufacturability.
6. Make efficient use of any spare PCB area to fatten grounds and high-current tracks as much as possible. It is not wise to fill in every spare corner of a prototype board with copper as this can be time-consuming (depending on the facilities of your PCB CAD system) and some of it will probably have to be undone to allow modifications. Ground tracks should always be as thick as practicable. Copper is free. (Once you've bought the laminate, that is.)

Miscellaneous points:

- On double-sided PCBs, copper areas should be solid on the component side, for minimum resistance and maximum screening, but will need to be cross-hatched on the solder side to prevent distortion of the PCB is flow-soldered. A common standard is 10 thou wide non-copper areas, i.e., mostly copper with small square holes; this is determined in the CAD package. If in doubt, consult those doing the flow-soldering.
- Do not bury component pads in large areas of solid copper, as this causes soldering difficulties due to the heat being conducted away.

- There is often a choice between running two tracks into a pad, or taking off a tee so that only one track reaches it. The former is better because it holds the pad more firmly to the board if desoldering is necessary. This is *particularly important* for components like transistors that are relatively likely to be replaced; for single-sided PCBs it is absolutely vital.
- If two parallel tracks are likely to crosstalk, then it is beneficial to run a grounded screening track between them. However, the improvement is likely to be disappointing, as electrostatic lines of force will curve over the top of the screen track.
- Jumper options must always be clearly labelled. Assume everyone loses the manual the moment they get it.
- Label pots and switches with their function on the screen-print layer, as this is a great help when testing. If possible, also label circuit blocks, e.g., *DC offset detect*. The labels must be bigger than component ident text to be clearly readable.

Amplifier Grounding

The grounding system of an amplifier must fulfil several requirements, among which are:

- The definition of a *star-point* as the reference for all signal voltages.
- In a stereo amplifier, grounds must be suitably segregated for good crosstalk performance. A few inches of wire as a shared ground to the output terminals will probably dominate the crosstalk behaviour.
- Unwanted AC currents entering the amplifier on the signal ground, due to external ground loops, must be diverted away from the critical signal grounds, i.e., the input ground and the ground for the feedback arm. Any voltage difference between these last two grounds appears directly in the output.
- Charging currents for the PSU reservoir capacitors must be kept out of all other grounds.

Ground is the point of reference for all signals, and it is vital that it is made solid and kept clean; every ground track and wire must be treated as a resistance across which signal currents will cause unwanted voltage-drops. The best method is to keep ground currents apart by means of a suitable connection topology, such as a separate ground return to the star-point for the local HT decoupling, but when this is not practical, it is necessary to make every ground track as thick as possible, and fattened up with copper at every possible point. It is vital

that the ground path has no necks or narrow sections, as it is no stronger than the weakest part. If the ground path changes board side, then a single via-hole may be insufficient, and several should be connected in parallel. Some CAD systems make this difficult, but there is usually a way to fool them.

Power amplifiers rarely use double-insulated construction and so the chassis and all metalwork must be permanently and solidly grounded for safety; this aspect of grounding is covered in Chapter 29. One result of permanent chassis grounding is that an amplifier with unbalanced inputs may appear susceptible to ground loops. One solution is to connect audio ground to chassis only through a 10 Ω resistor, which is large enough to prevent loop currents becoming significant. This is not very satisfactory as:

- The audio system as a whole may thus not be solidly grounded.
- If the resistor is burnt out due to misconnected speaker outputs, the audio circuitry is floating and could become a safety hazard.
- The RF rejection of the power amplifier is likely to be degraded. A 100 nF capacitor across the resistor may help.

A better approach is to put the audio-chassis ground connection at the input connector, so in Figure 25.1, ground loop currents must flow through A—B to the Protected Earth at B, and then to mains ground via B—C. They cannot flow through the audio path E—F. This topology is very resistant to ground loops, even with an unbalanced input; the limitation on system performance in the presence of a ground loop is now determined by the voltage-drop in the input cable ground, which is outside the control of the amplifier designer. A balanced input could in theory cancel out this voltage drop completely.

Figure 25.1 also shows how the other grounding requirements are met. The reservoir charging pulses are confined to the connection D—E, and do not flow E—F, as there is no other circuit path. E—F—H carries ripple, etc., from the local HT decouplers, but likewise cannot contaminate the crucial audio ground A—G.

Ground Loops: How they Work and How to Deal with them

A ground loop is created whenever two or more pieces of mains-powered equipment are connected together, so that mains-derived AC flows through shields and ground conductors, degrading the noise floor of the system. The effect is worst when two or more units are connected through mains ground as well as audio cabling, and this situation is what is normally meant by the term 'ground loop'. However, ground currents can also flow in systems that are not galvanically grounded; they are of lower magnitude but can still degrade the noise floor, so this scenario is also considered here.

The ground currents may either be inherent in the mains supply wiring (see 'Hum injection by mains grounding currents' below) or generated by one or more of the pieces of equipment that make up the audio system (see sections 'Hum injection by transformer stray magnetic fields' and 'Hum injection by transformer stray capacitance' below).

Once flowing in the ground wiring, these currents will give rise to voltage drops that introduce hum and buzzing noises. This may occur either in the audio interconnects, or inside the equipment itself if it is not well designed. See section 'Ground currents inside equipment', on p. 605.

Here I have used the word 'ground' for conductors and so on, while 'earth' is reserved for the damp crumbly stuff into which copper rods are thrust.

Hum Injection by Mains Grounding Currents

Figure 25.2 shows what happens when a so-called 'technical ground' such as a buried copper rod is attached to a grounding system which is already connected to 'mains ground' at the power distribution board. The latter is mandatory both legally and technically, so one might as well accept this and denote as the reference ground. In many cases this 'mains ground' is actually the neutral conductor, which is only grounded at the remote transformer substation. AB is the cable from substation to consumer, which serves many houses from connections tapped off along its length. There is substantial current flowing down the N + E conductor, so point B is often 1 V rms or more above earth. From B onwards, in the internal house wiring, neutral and ground are always separate (in the UK, anyway).

Two pieces of audio equipment are connected to this mains wiring at C and D, and joined to each other through an unbalanced cable F–G. Then an ill-advised connection is made to earth at D; the 1 V rms is now impressed on the path B–C–D, and substantial current is likely to flow through it, depending on the total resistance of this path. There will be a voltage drop from C to D, its magnitude depending on what fraction of the total BCDE resistance is made up by the section C–D. The earth wire C–D will be of at least 1.5 mm^2 cross-section, and so the extra connection FG down the audio cable is unlikely to reduce the interfering voltage much.

To get a feel for the magnitudes involved, take a plausible ground current of 1 A. The 1.5 mm^2 ground conductor will have a resistance of 0.012 Ω/m, so if the mains sockets at C and D are 1 m apart, the voltage C–D will be 12 mV rms. Almost all of this

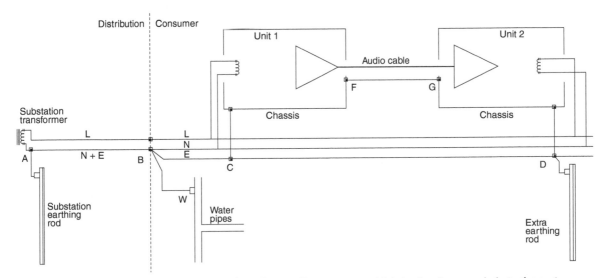

Figure 25.2. The pitfalls of adding a 'technical ground' to a system which is already grounded via the mains.

will appear between F and G, and will be indistinguishable from wanted signal to the input stage of Unit 2, so the hum will be severe, probably only 30 dB below the nominal signal level.

The best way to solve this problem is not to create it in the first place. If some ground current is unavoidable, then the use of balanced inputs (or ground-cancel outputs — it is not necessary to use both) should give at least 40 dB of rejection at audio frequencies.

Figure 25.2 also shows a third earthing point, which fortunately does not complicate the situation. Metal water pipes are bonded to the incoming mains ground for safety reasons, and since they are usually electrically connected to an incoming water supply, current flows through B—W in the same way as it does through the copper rod link D—E. This water-pipe current does not, however, flow through C—D and cannot cause a ground loop problem. It may, however, cause the pipes to generate an AC magnetic field which is picked up by other wiring.

Hum Injection by Transformer Stray Magnetic Fields

Figure 25.3 shows a thoroughly bad piece of physical layout which will cause ground currents to flow even if the system is correctly grounded to just one point.

Here Unit 1 has an external DC power supply; this makes it possible to use an inexpensive frame-type transformer which will have a large stray field. But note that the wire in the PSU which connects mains ground to the outgoing 0 V takes a half-turn around the transformer, and significant current will be induced into it, which will flow round the loop C—F—G—D, and give an unwanted voltage drop between F and G. In this case reinforcing the ground of the audio interconnection is likely to be of some help, as it directly reduces the fraction of the total loop voltage which is dropped between F and G.

It is difficult to put any magnitudes to this effect because it depends on many imponderables such as the build quality of the transformer and the exact physical arrangement of the ground cable in the PSU. If this cable is rerouted to the dotted position in the diagram, the transformer is no longer enclosed in a half-turn, and the effect will be much smaller.

Hum Injection by Transformer Stray Capacitance

It seems at first sight that the adoption of Class II (double-insulated) equipment throughout an audio system will give inherent immunity to ground loop problems. Life is not so simple, though it has to be said that when such problems do occur, they are likely to be much less severe. This mains transformer problem afflicts all Class II equipment to a certain extent.

Figure 25.4 shows two Class II units connected together by an unbalanced audio cable. The two mains transformers in the units have stray capacitance from both live and neutral to the secondary. If these capacitances were all identical, no current would flow, but in practice they are not, so 50 Hz currents are injected into the internal 0 V rail and flow through the resistance of F—G, adding hum to the signal. A balanced input or ground-cancelling output will remove or render negligible the ill-effects.

Reducing the resistance of the interconnect ground path is also useful — more so than with other types of ground loop, because the ground current is essentially fixed by the small stray capacitances, and so halving the resistance F—G will dependably halve the interfering voltage. There are limits to how far you can take this — while a simple balanced input will give 40 dB of rejection at low cost, increasing the cross-sectional area of copper in the ground of an audio cable by a factor of 100 times is not going to be either easy or cheap. Figure 25.4 shows equipment

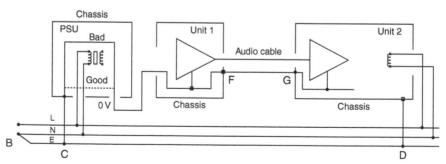

Figure 25.3. Poor cable layout in the PSU at left wraps a loop around the transformer and induces ground currents.

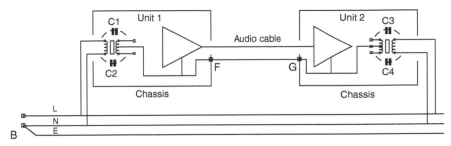

Figure 25.4. The injection of mains current into the ground wiring via transformer inter-winding capacitance.

with metal chassis connected to the 0 V (this is quite acceptable for safety approvals — what counts is the isolation between mains and everything else, not between low-voltage circuitry and touchable metal-work); note the chassis connection, however, has no relevance to the basic effect, which would still occur even if the equipment enclosure was completely non-conducting.

The magnitude of ground current varies with the details of transformer construction, and increases as the size of the transformer grows, as shown in Table 25.4. Therefore the more power a unit draws, the larger the ground current it can sustain. This is why many systems are subjectively hum-free until the connection of a powered sub-woofer, which is likely to have a larger transformer than other components of the system.

Ground Currents Inside Equipment

Once ground currents have been set flowing, they can degrade system performance in two locations: outside the system units, by flowing in the interconnect grounds, or inside the units, by flowing through internal PCB tracks, etc. The first problem can be dealt with effectively by the use of balanced inputs, but the internal effects of ground currents can be much more severe if the equipment is poorly designed.

Table 25.4. Typical ground currents for different sorts of equipment

Equipment type	Power consumption	Ground current
Turntable, CD, cassette deck	20 W or less	5 μA
Tuners, amplifiers, small TVs	20–100 W	100 μA
Big amplifiers, sub-woofers, large TVs	More than 100 W	1 mA

Figure 25.5 shows the situation. There is, for whatever reason, ground current flowing through the ground conductor CD, causing an interfering current to flow round the loop CFGD as before. Now, however, the internal design of Unit 2 is such that the ground current flowing through FG also flows through G-G' before it encounters the ground wire going to point D. G-G' is almost certain to be a PCB track with higher resistance than any of the cabling, and so the voltage drop across it can be relatively large, and the hum performance correspondingly poor. Exactly similar effects can occur at signal outputs; in this case the ground current is flowing through F-F'.

Balanced inputs will have no effect on this; they can cancel out the voltage drop along F—G, but if internal hum is introduced further down the internal signal path, there is nothing they can do about it.

The correct method of handling this is shown in Figure 25.6. The connection to mains ground is made right where the signal grounds leave and enter the units, and are made as solidly as possible. The ground current no longer flows through the internal circuitry. It does, however, still flow through the interconnection at FG, so either a balanced input or a ground-cancelling output will be required to deal with this.

Balanced Mains Power

There has been speculation in recent times as to whether a balanced mains supply is a good idea. This means that instead of live and neutral (230 V and 0 V) you have live and the other live (115 V—0—115 V) created by a centre-tapped transformer with the tap connected to neutral (see Figure 25.7).

It has been suggested that balanced mains has miraculous effects on sound quality, makes the sound stage ten-dimensional, etc. This is obviously nonsense. If a piece of gear is that fussy about its mains (and I do not believe any such gear exists), then dispose of it.

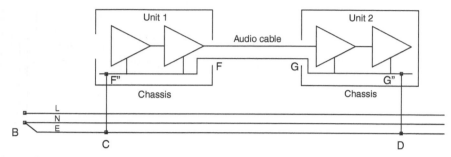

Figure 25.5. If ground current flows through the path F'FGG', then the relatively high resistance of the PCB tracks produces voltage drops between the internal circuit blocks.

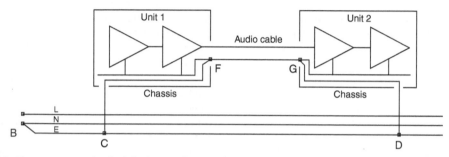

Figure 25.6. The correct method of dealing with ground currents; they are diverted away from internal circuitry.

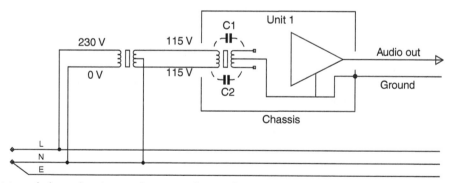

Figure 25.7. Using a balanced mains supply to cancel ground currents stemming from interwinding capacitance in the mains transformer. An expensive solution.

If there is severe RFI on the mains, an extra transformer in the path may tend to filter it out. However, a proper mains RFI filter will almost certainly be more effective — it is designed for the job, after all — and will definitely be much cheaper.

Where you might gain a real benefit is in a Class II (i.e., double-insulated) system with very feeble ground connections. Balanced mains would tend to cancel out the ground currents caused by transformer capacitance (see Figure 25.4 and above for more details on this)

and so reduce hum. The effectiveness of this will depend on C1 being equal to C2 in Figure 25.7, which is determined by the details of transformer construction in the unit being powered. I think that the effect would be small with well-designed equipment and reasonably heavy ground conductors in interconnects. Balanced audio connections are a much cheaper and better way of handling this problem, but if none of the equipment has them, then beefing up the ground conductors should give an improvement. If the results are not good enough, then as a last resort, balanced mains may be worth considering.

Finally, bear in mind that any transformer you add must be able to handle the maximum power drawn by the audio system at full throttle. This can mean a large and expensive component.

I would not be certain about the whole of Europe, but to the best of my knowledge it is the same as the UK, i.e., not balanced. The neutral line is at earth potential, give or take a volt, and the live is 230 V above this. The 3-phase 11 kV distribution to substations is often described as 'balanced' but this just means that power delivered by each phase is kept as near equal as possible for the most efficient use of the cables.

It has often occurred to me that balanced mains 115 V−0−115V would be a lot safer. Since I am one of those people put their hands inside live equipment a lot, I do have a kind of personal interest here.

Class I and Class II

Mains-powered equipment comes in two types: grounded and double insulated. These are officially called Class I and Class II, respectively.

Class I equipment has its external metalwork grounded. Safety against electric shock is provided by limiting the current the live connection can supply with a fuse. Therefore, if a fault causes a short-circuit between live and metalwork, the fuse blows and the metalwork remains at ground potential. A reasonably low resistance in the ground connection is essential to guarantee the fuse blows. A three-core mains lead is mandatory. Two-core IEC mains leads are designed so they cannot be plugged into three-pin Class I equipment. Class I mains transformers are tested to 1.5 kV rms.

Class II equipment is not grounded. Safety is maintained not by interrupting the supply in case of a fault, but by preventing the fault happening in the first place. Regulations require double insulation and a generally high standard of construction to prevent any possible connection between live and the chassis. A two-core IEC mains lead is mandatory; it is not permitted to sell a three-core lead with a Class II product. This would present no hazard in itself, but is presumably intended to prevent confusion as to what kind of product is in use. Class II mains transformers are tested to 3 kV rms, to give greater confidence against insulation breakdown.

Class II is often adopted in an attempt to avoid ground loops. Doing so eliminates the possibility of major problems, at the expense of throwing away all hope of fixing minor ones. There is no way to prevent capacitance currents from the mains transformer flowing through the ground connections. (See section 'Ground loops: how they work and how to deal with, them'.) It is also no longer possible to put a grounded electrostatic screen between the primary and secondary windings. This is serious as it deprives you of your best weapon against mains noise coming in and circuit RF emissions getting out. In Class II the external chassis may be metallic, and connected to signal 0 V as often as you like.

If a Class II system is not connected to ground at any point, then the capacitance between primaries and secondaries in the various mains transformers can cause its potential to rise well above ground. If it is touched by a grounded human, then current will flow, and this can sometimes be perceptible, though not directly, as a painful shock like static electricity. The usual complaint is that the front panel of the equipment is 'vibrating', or that it feels 'fuzzy'. The maximum permitted touch current (flowing to ground through the human body) permitted by current regulations is 700 μ A, but currents well below this are perceptible. It is recommended, though not required, that this limit be halved in the tropics where fingers are more likely to be damp. The current is measured through a 50 k resistance to ground.

When planning new equipment, remember that the larger the mains transformer, the greater the capacitance between primary and secondary, and the more likely this is to be a problem. To put the magnitudes into perspective, I measured a 500 VA toroid (intended for Class II usage and with no interwinding screen) and found 847 pF between the windings. At 50 Hz and 230 V this implies a maximum current of 63 μA flowing into the signal circuitry, the actual figure depending on precisely how the windings are arranged. A much larger 1500 VA toroidal transformer had 1.3 nF between the windings, but this was meant for Class I use and had a screen, which was left floating to get the figure above.

Warning

Please note that the legal requirements for electrical safety are always liable to change. This book does not attempt to give a complete guide to what is required for compliance. The information given here is correct at the time of writing, but it is the designer's responsibility to check for changes to compliance requirements. The information is given here in good faith but the author accepts no responsibility for loss or damage under any circumstances.

Cooling

All power amplifiers will have a heatsink that needs cooling, usually by free convection, and the mechanical design is often arranged around this requirement. There are three main approaches to the problem:

1. The heatsink is entirely internal, and relies on convected air entering the bottom of the enclosure, and leaving near the top (passive cooling).

   Advantages: The heatsink may be connected to any voltage, and this may eliminate the need for thermal washers between power device and sink. On the other hand, some sort of conformal material is still needed between transistor and heatsink. A thermal washer is much easier to handle than the traditional white oxide-filled silicone compound, so you will probably be using them anyway. With this form of construction there are no safety issues as to the heatsink temperatures.

   Disadvantages: This system is not suitable for large dissipations, due to the limited fin area possible inside a normal-sized box, and the relatively restricted convection path.
2. The heatsink is partly internal and partly external, as it forms one or more sides of the enclosure. Advantages and disadvantages are much as above; if any part of the heatsink can be touched, then the restrictions on temperature and voltage apply. Greater heat dissipation is possible.
3. The heatsink is primarily internal, but is fan-cooled (active cooling). Fans always create some noise, and this increases with the amount of air they are asked to move. Fan noise is very unwelcome in a domestic hi-fi environment; it seems pointless to strive for beautifully quiet electronics if there is a fan whirring away. Fan noise is of little importance in most PA applications, but could be an issue in small venues.
4. Fans allow maximal heat dissipation, but require an inlet filter to prevent the build-up of dust and fluff internally. Persuading people to clean such filters regularly is near-impossible.

The internal space in the enclosure will require some ventilation to prevent heat build-up; slots or small holes are desirable to keep foreign bodies out. Avoid openings on the top surface if you can as these will allow the entry of spilled liquids, and increase dust entry. BS415 is a good starting point for this sort of safety consideration, and this specifies that slots should be no more than 3 mm wide.

Reservoir electrolytics, unlike most capacitors, suffer significant internal heating due to ripple current. Electrolytic capacitor life is very sensitive to temperature, so mount them in the coolest position available, and if possible leave room for air to circulate between them to minimise the temperature rise.

Convection Cooling

Efficient passive heat removal requires extensive heatsinking with a free convective air flow, and this often indicates putting the heatsinks on the side of the amplifier; the front of the unit will carry at least the mains switch and power indicator light, while the back carries the in/out and mains connectors, leaving only the sides completely free.

It is important to realise that the buoyancy forces that drive natural convection are very small, and even small obstructions to flow can seriously reduce the rate of flow, and hence the cooling. If ventilation is by slots in the top and bottom of an amplifier case, then the air must be drawn under the unit, and then execute a sharp right-angle turn to go up through the bottom slots. This change of direction is a major impediment to air flow, and if you are planning to lose a lot of heat, then it feeds into the design of something so humble as the feet the unit stands on; the higher the better, for air flow. In one instance the amplifier feet were made 13 mm taller and all the internal amplifier temperatures dropped by 5°C. Standing a unit with low feet on a thick-pile carpet can be a really bad idea, but someone is bound to do it (and then drop their coat on top of it); hence the need for over-temperature cutouts if amplifiers are to be fully protected.

Heatsink Materials

Heatsinks in the audio business are almost always made of aluminium. It is cheaper than copper but of comparable conductivity, and has the very useful property that it can be extruded to form shapes with fins, and can be anodized in all sorts of colours. Aluminium is never used as a pure element because it is too soft to be practical; it is produced as an alloy with small amounts of copper and silicon. Even then it is still a relatively soft material, and there can be problems with tapped screw holes.

Copper heatsinks have become quite common in the specialised field of cooling overclocked computer processors, where performance is critical and people are prepared to pay for it. It has occasionally appeared in audio amplifiers. A copper heatsink may cost three times that of the same sized aluminium part because of the higher material and fabrication costs. Copper can't be extruded, so copper heatsinks must be machined, and the machining process is more demanding than for aluminium. Due to density and its abrasive nature, machining holes and other details in copper takes significantly longer and wears your tooling faster.

The cost of the copper raw material is about the same as aluminium by weight but it has three times the density of aluminium. Thus, to make a heatsink of a given size, the raw material cost is three times that of aluminium.

Silver is a better heatsink material than copper — it is in fact the only material that is better, being the element with the highest thermal conductivity — but the difference is only 7%, and the extra cost rules it out for all but the highest of high-end products. Silver heatsinks have been used for CPU cooling — so has silver-plated copper, the plating being purely for the visual effect. It would of course stop the copper oxidising, but silver is subject to blackening due to sulphide formation, so you may not be gaining much aesthetically. The surface of the silver can be lacquered to prevent chemical action, but this introduces an extra layer with much worse conductivity than metal.

Spending money on more precious metals is worse than pointless, for gold is a worse conductor than copper, and not much superior to aluminium. Platinum is very much worse, though I would not rule out the possibility that someone, somewhere, has made a heatsink out of it.

It is worth noting that aluminium is not the cheapest of metals, despite being relatively common, because it is produced from its ore by electrolytic refining, which uses huge amounts of electricity. This is why aluminium has been called 'congealed electricity'. Steel — which is cheap — is unfortunately the bottom metal in our conductivity table, being nine times worse than copper, and five times worse than aluminium. This is why you don't see steel heatsinks. I did know of one company that tried to use them, but they didn't get far with it. Apart from the problems of machining a harder material, steel rusts readily and so requires a protective coating which further impairs its thermal performance.

Table 25.5 gives the thermal conductivity of various substances, and also some bulk prices. These prices are of course subject to variation due to market conditions, and at the time of writing the trend is very much upwards due to the increasing demand for materials in the Far East. At the bottom of the table are included

Table 25.5. The thermal conductivity of various substances, some being more useful in amplifier design than others

Material	W/m-K	£/tonne 2005
Helium II	100000	
Diamond	2500	
Graphite	470	
Silver	429	816,670
Copper	401	2576
Gold	310	41,029,324
Aluminium	250	1044
Beryllium	177	
Tungsten	174	
Brass	109	
Platinum	70	21,233,329
Steel	46	400
Warth K381	3.8	
Warth K200	1.70	
Warth K177	0.79	

some bulk thermal conductivities for the Warth/Laird materials used in insulating thermal washers, which are described below.

Diamond makes a truly excellent heatsink, as it has extremely high thermal conductivity but is also an excellent electrical insulator. It has often been used for cooling exotic semiconductors such as microwave transistors and laser diodes. Diamond heatsinks are not of course big things with fins, but small flat pieces used to spread the heat out into cheaper materials. Flat bits of synthetic diamond for this purpose can be bought commercially with W/m-K values from 1000 to 1800. The drawbacks are the very high price and the very great difficulty of machining the material.

Discussing the use of diamond for amplifier heatsinks may appear frivolous, but given the remarkable advances in carbon chemistry in the last few years, it is not impossible that one day we may see big finned heatsinks grown from solid diamond. Which would, in all senses of the word, be truly cool.

Just to put things in perspective, the top entry in the table is for Helium II, which shows the highest thermal conductivity of any known substance as heat conduction in it occurs by an exceptional quantum-mechanical mechanism. Helium II is a liquid and only exists below 2.2 degrees K, within sight of Absolute Zero, and it must be regretfully concluded that it does not make a very practical heatsink for general use.

Heatsink Compounds

Heatsink compounds, in conjunction with the traditional mica washers, are still in common use by Far East manufacturers at the time of writing, though their use is definitely declining. The most common heatsink compound is the white-coloured paste, typically silicone oil filled with aluminium oxide, zinc oxide, or boron nitride. Some exotic brands of thermal interfaces (notably Arctic Silver) use micronised or pulverised silver.

While heatsink compound can provide good thermal performance, it is horribly messy to apply and creates problems when servicing equipment, as the force required to break the suction between the compound and heatsink is often enough to delaminate mica washers. Avoid it if you can.

Thermal Washers

These are typically made of thin silicone rubber loaded with highly thermally conductive but electrically insulating compounds such as aluminium oxide or boron nitride, and are usually reinforced by an inner weave of fibreglass to resist tearing. They are very much easier to apply than heatsink compounds.

The disadvantages are that because they are made very thin (0.25 mm at most, and sometimes much less) to increase thermal performance, they are vulnerable to being punctured by even very small pieces of metal contamination. Areas used for their assembly into equipment must be kept scrupulously clean of swarf and metal dust. Holes must also be carefully deburred to prevent cutting through the material. In general, the higher the performance of the washer, the more fragile it is and the more carefully it must be handled. This is because the increased proportion of particles of the thermally conductive compound in the silicone elastomer makes it harder and more brittle and crumbly. It is good practice to always design in the cheapest and toughest version of the thermal washers; these will also be the least efficient. If things do go a bit wrong thermally, you then have the option of switching to a more effective but costlier material to bring down device temperatures. In one power amplifier I was associated with, changing to more efficient thermal washers brought down some worryingly high junction temperatures by about 10°C, and we all slept better at night after that.

Many standard shapes are available for common semiconductor packages. For a TO-220 package and a mounting pressure of 50 psi, thermal resistance ranges from 1.5 to 3.4°C per Watt. Some of the bulk thermal conductivities for the Warth/Laird materials

are included at the bottom of Table 25.5, and it can be seen that there is quite a difference between the standard K177 and the high-performance K381. You can also see that even the best thermal washer materials have much less thermal conductivity than the worst of metals, and this is why thermal washers have to be so very thin.

A special type is the Phase Change thermal washer. These are made of a thermally conductive material that melts when it reaches its operating temperature, and therefore makes superior contact with the metal surface of the heatsink or semiconductor. A typical melting temperature is 65°C. They are made from a phase change compound coated on a fibreglass web, and can be handled just like ordinary thermal washers at room temperature. The downside is in servicing; the melting causes the thermal material to stick to the metal, and the washer cannot be easily removed, and certainly not re-used.

Sometimes you don't need electrical isolation, but you do need a conformal material between two surfaces to fill in the minor irregularities and so improve heat flow. In this case graphite foil, which is simply thin layers of graphite reinforced with a small amount of fibreglass to give it some (but not much) mechanical integrity, is highly effective. It is, however, fragile and needs careful handling. Like other thermal materials, tooling for custom shapes is relatively inexpensive. There is more on graphite foil in Chapter 22.

Fan Cooling

Fan cooling allows much more heat to be removed than relying on the very slow movements of natural convection, but it has disadvantages that mean you should think very hard before adopting it. A fan costs a significant amount of money, and introduces an electromechanical component that may well be the least reliable part of the unit. Fans create acoustic noise, and tend to fill up the enclosure with dust and dirt.

However, sometimes they are unavoidable. In hi-fi applications, big Class-A amplifiers will need fans to remove the prodigious heat output. In the sound reinforcement world, fans are useful because they mean amplifiers can be made lighter and more compact.

The fans used in audio applications are usually of the axial DC type, which are available in a wide range of diameters and airflow capabilities. The DC powering makes control of fan speed much simpler when it is required. The operating voltages generally available are nominally 5 V, 6 V, 12 V, 24 V, and 48 V, though most fans will operate over quite a wide range of reduced voltage. The 12 V versions are by far the most common and give the greatest choice of airflow/noise tradeoffs and so on, with the 24 V versions coming second.

If fan cooling is essential, there are certain things to be taken into consideration to make it as trouble-free as possible.

Firstly, the fan should be fitted so as to push air into the case rather than suck it out. Push operation means that all the air entering the case has passed through the filter, even if the case has a few leaks, which it almost certainly will have. The suction method may pull most of the air in through the filter, but some will be coming in through the leaks and will bring dirt and dust with it.

There are two schools of thought on the provision of air filters. The first regards an air filter as essential to keep dust, etc., out, and issues stern warnings that the filter must be cleaned regularly or dire things will come to pass, as the reduced air flow leads to overheating. The second recognises that in general people just don't clean air filters, no matter how much you threaten them, and so no filter is fitted; the obvious downside to this being that the dust brought in with the air soon causes the internals of the equipment to resemble the inside of a neglected vacuum cleaner, and eventually there will be problems.

Secondly, a DC fan motor uses current in quite hefty pulses and generates a corresponding magnetic field so keep it away from sensitive circuitry. The two wires to the fan must be twisted together or otherwise kept adjacent or otherwise the resulting loop will radiate nasty sharp spikes. Obviously the fan ground return must be kept quite separate from audio ground. A quick look at a particular 80 mm diameter 12 V fan shows that it draws 170 mA and produces pulses at 140 Hz; the frequency falls as the input voltage is reduced and the fan runs more slowly.

Thirdly, most DC fans will operate over quite a wide voltage range, though the rated voltage should not be exceeded. As the voltage is reduced, the fan turns more slowly and moves less air, but it also generates less acoustic noise. A large fan running below full speed may shift the same volume of air as a smaller fan at full throttle, but in most cases will make less noise about it. However, the large fan may of course be more expensive.

If you are planning to run a fan at the lower end of its voltage range to keep it quiet, then be very wary. As the bearings age and friction increases, the fan may stop turning altogether; it is difficult or impossible to predict how soon this is likely to be a problem.

Fan Control Systems

The simplest way to apply a fan is to select a model with enough flow rate for the most demanding conditions and have it running full speed continuously. Naturally this means that most of the time it is providing more cooling than is necessary and more noise than is desirable.

A fan can of course be controlled by a simple on/off thermostat, giving what in the world of control theory is called 'bang-bang' control, but this is crude and more audibly intrusive than having the fan running at full speed all the time.

Most fan applications have a proportional control system which aims to keep the fan running at a constant or at any rate very slowly varying speed which matches the power being dissipated by the heatsinks. This is usually a simple proportional servo circuit, with no PID complications, and the temperature control will not be very accurate as the loop gain has to be kept low. The heatsinks have a large thermal mass and the fan speed only affects their temperature slowly, so a high value of loop gain will lead to slow oscillations in fan speed, otherwise known as hunting. This is aurally very distracting and must be carefully avoided. Fortunately there is no need for great accuracy of temperature control so a low loop gain is not normally a problem. Figure 25.8 shows a simple proportional servo circuit to control a DC fan, showing how it interfaces to the power supply; the +15 V positive regulator is not shown. It assumes that supply rails of ±15 V or thereabouts are being derived to power opamps.

The LM35 temperature sensor IC is a very attractive component for this application. It puts out a voltage that is linearly dependent on temperature, so that its output at 25°C is +250 mV, and at 50°C is +500 mV. It uses very little power, is trimmed by the manufacturer at wafer level so it requires no calibration, and has a low output impedance. The downside is that it is relatively expensive, and in the TO-92 package, which is the most commonly available, is not easy to mount onto a heatsink surface without the use of either glue or a special clip. Thermistors are cheaper but have the disadvantage that they are strongly non-linear in their temperature-resistance relation, which means the loop gain of the servo depends on temperature, rather complicating the design of a stable control loop.

The servo circuit is essentially just an inverting amplifier with its gain set by the ratio of R4, R5. The set point is defined as 48°C by the reference divider R2, R3. Everything happens slowly in this circuit and decoupling this divider was not necessary. The fan

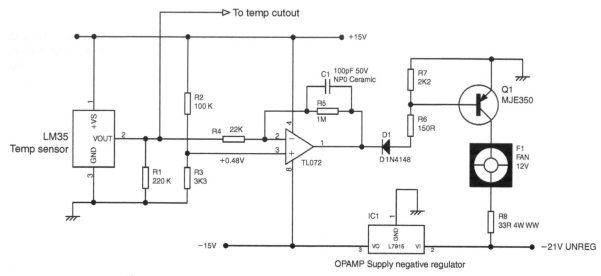

Figure 25.8. Fan control servo circuit using an LM35 temperature sensor.

drive circuitry may look a bit more complicated than necessary, but it was configured that way for a very good reason; I'm sure you never doubted that. The product in which this circuit was used was a powered mixer with a built-in DSP effects module for creating reverberation and so on. This was by today's standards a rather power-hungry device, and it took its +5 V supply via a regulator IC from the +15 V rail. In the interests of sharing the loading on the mains transformer secondaries, it was therefore desirable to run the fan between the unregulated −22 V rail and 0 V.

When the heatsink temperature exceeds the set point, the opamp output moves in the negative direction and turns on Q1 more via D1, R6, and R7. D1 prevents the base-emitter junction of Q1 from being reverse-biased when the opamp output is high. Q1 turns on more and the fan turns faster. The dropper resistor R8 looks like it simply wastes power, which in a sense it does. It is present solely because 12 V fans are much the most common, and fans running off 21 V do not exist. A 24 V model could have been used but it would have been more expensive and would never have run at its full capacity, so the resistive dropper solution was chosen. Running the fan off the regulated −15 V supply was not considered a good plan as it would greatly increase the dissipation of the −15 V regulator, and the pulsed current taken by the fan motor would have made the −15 V rail noisy.

Even with effective proportional control, the noise from a fan may be disturbing in the silences between the music. An excellent way to cure this is to arrange

for the fan to stop when the output of the power amplifiers falls below a certain threshold, on the basis that if the amplifier is no longer dissipating significant heat, the output device temperature cannot rise and there is no immediate need for further cooling. I used this philosophy with great success in a series of powered mixers intended for small venues, giving up to 250 W/8 Ω from two channels. In this application the need for relatively small and light heatsinks made fan cooling essential, but it was probable that fan noise would be intrusive − think about acoustic guitar music.

A somewhat more sophisticated version of this approach, which was adopted in the production versions, does not force the fan to immobility in the absence of signal, but instead shifts the set point upwards to a rather higher temperature. This recognises that in this sort of sound reinforcement application, ultimately cooling is more important than silence. Now, if the heatsink is on the hot side, the fan will continue to run when the music stops, though because of the proportional nature of the fan control it may well do so at less than full throttle. A circuit I have used for this approach can be seen in Figure 25.9.

The fan control servo itself is unchanged from Figure 25.8, apart from the lowering of the set point (without the music-sensing circuitry activated) to 35°C by reducing the value of R9. This ensures that the heatsink gets as much cooling as possible while the amplifier output masks the sound of a fan running quickly.

The output from the two channels of power amplification is summed at the base of Q2, and when there is

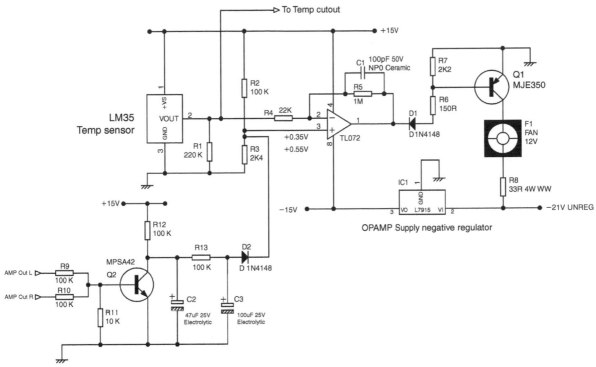

Figure 25.9. Improved servo circuit stops the fan when there is no audio output.

any significant volume, Q2 is held hard on for at least half the time. The voltage on the cathode of D2 is therefore only about 200 mV (the Vce(sat) of Q2) above 0 V, D2 is reverse-biased, and the set point voltage at the junction of R2, R3 is unaffected. When, however, the amplifier output fall below the threshold, Q2 now turns off and its collector voltage rises. After a delay set by R12, C2 and R13, C3, D2 conducts and raises the set point voltage of the servo; with the resistor values shown, it is increased from 35°C to 55°C. These temperatures can easily be modified. The time delays ensure that the fan set point is not being modulated at signal frequency or syllabic speed and prevent the fan turning on and off during very short pauses in the music.

It could be argued that what is really needed here is a maximum-selector circuit, such as a peak-rectifier driven by separate diodes, so that a positive input from one channel cannot be cancelled out by a negative input from the other. In practice this extra complication is quite unnecessary. The greatest amplitudes in music are always in the bass register, and bass instruments are usually panned to the centre to get the benefit of both channels of amplification, and so will almost

always be in phase; this is also important if you plan to cut a master for retro-vinyl, as large antiphase amplitudes will upset stylus tracking.

Fans do not stop turning immediately when the power is removed, but this is usually masked by the fadeout of the music. If the audio does stop abruptly, you may well hear the sound of the fan spooling down. It is not easy to think of a fix for this; fans have electronic commutation built in and so shorting their terminals does not induce dynamic braking.

Fan Failure Safety Measures

Most amplifiers have over-temperature sensors that initiate shutdown if the heatsink or power devices get too hot. This will eventually be triggered if the fan cooling fails, but heating things up to this extent is likely to stress the output devices and should be avoided if possible. To this end, fan-fail detectors are sometimes fitted. It is fairly simple to design a circuit that monitors the pulsating current drawn by a DC fan motor, and raises a warning signal if the pulses cease because the fan has stopped turning, because of bearing failure or the blades being choked with debris.

This gives some protection but does not of course detect the likeliest problem, which is the air filter becoming choked or the inlet port being blocked, so that insufficient air is being delivered although the fan is still flailing with all its might. A light vane placed in the air flow, its position being monitored by optoelectronic means, will give a warning against these conditions but is itself liable to jamming by dust and dirt and it is rarely worth the cost and complication. A better method of airflow detection is the use of two thermistors, arranged so they significantly self-heat by virtue of the current passing through them. If one thermistor is placed in the air flow and the other shielded from it, the first only will be cooled and the existence of a temperature difference, and hence a difference in electrical resistance, gives assurance that air is actually moving.

Heat Pipes

A heat pipe is a very effective method of transporting large quantities of heat from one place to another with a very small difference in temperature between the hot and cold zones. Typically it shifts heat from somewhere in the middle of a piece of equipment to an outside surface where it is more convenient to place a large finned heatsink. A heat pipe is simply a sealed length of pipe containing a small amount of fluid that boils at the hot end. The vapour moves to the cold end and condenses, and capillary action in a wick structure lining the pipe, or gravity, then returns the fluid to the hot end. Heat pipes are relatively expensive but their unique advantage is their great efficiency in transferring heat over relatively long distances. They are a much better heat conductor than the equivalent cross-section of solid copper.

Most heat pipes use either ammonia or water as their working fluid, the boiling temperature being controlled by the pressure set up in the sealed pipe at manufacture. Water has a useful range of 30 to 200°C which covers pretty much all electronic applications. To take a specific example, a 6 mm diameter heat pipe with a sintered metal wick transferring 10 Watts over a distance of 100 mm would have a thermal resistance of only 2.1°C /W. Doubling the distance only increases this to 2.5°C /W. If the heat was being transferred through a solid block of metal, doubling the distance would naturally double the thermal resistance.

Examples of commercial amplifiers using heat pipes include the Sony TA-F70 integrated amplifier (1979) with all its output devices mounted on a thermal coupler clamped around one end of a heatpipe which appears to be about 8 inches long; the majority of its length was finned, with the heat emitted inside the enclosure. They claimed that this form of construction reduced the length of cabling carrying large signal currents. Whether this implies they were thinking in terms of inductive distortion (Distortion Six) is unknown. However, they also claimed that the signal-to-noise ratio was improved, which does not seem likely.

Another example was the Luxman M-02 (1983) with a similar arrangement of heat pipes taking heat a couple of inches to internal finning; separate pipes were used for each channel. The publicity material referred to this as a 'liquid-cooled heatsink'.

One of the most enthusiastic exponents of heat pipes was Technics. Models using them include the SU-V7A (1982) the SU-V96, the SU V2X, and the SU-V303 / 505 / 707 series.

Heat pipes are now quite often used for cooling computer CPUs and high-end graphics processors; this is presumably bringing their price down. That does not mean that the introduction of heat pipes into audio amplifiers in significant numbers looks very likely. The technical advantages appear to be minimal.

Mechanical Layout and Design Considerations

The mechanical design adopted depends very much on the intended market, and production and tooling resources, but I offer below a few purely technical points that need to be taken into account:

Wiring Layout

There are several important points about the wiring for any power amplifier:

- Keep the + and − HT supply wires to the amplifiers close together. This minimises the generation of distorted magnetic fields which may otherwise couple into the signal wiring and degrade linearity. Sometimes it seems more effective to include the 0 V line in this cable run; if so it should be tightly braided to keep the wires in close proximity. For the same reason, if the power transistors are mounted off the PCB, the cabling to each device should be configured to minimise loop formation.
- The rectifier connections should go direct to the reservoir capacitor terminals, and then away again to the amplifiers. Common impedance in these connections superimposes charging pulses on the rail ripple waveform, which may degrade amplifier PSRR.

- Do not use the actual connection between the two reservoir capacitors as any form of star-point. It carries heavy capacitor-charging pulses that generate a significant voltage drop even if thick wire is used. As Figure 25.1 shows, the 'star-point' is tee-ed off from this connection. This is a star-point only insofar as the amplifier ground connections split off from here, so do not connect the input grounds to it, as distortion performance will suffer.

Semiconductor Installation

Driver transistors are usually in the TO-225AA (e.g., the MJE340, MJE350) or the TO-220 package. Power transistors are commonly in the TO-3P or MT-200 packages, though the venerable TO-3 all-metal package is still occasionally met with. For each package the effective cooling area (i.e., the metal part of the package) is given in Table 25.6, as this is needed to derive the temperature difference between the device and the heatsink, given the thermal conductivity of the thermal washer used and the power dissipated. The area lost to the mounting hole or holes has been allowed for in calculating effective area. A comparison is also made between the metal area and the total area, which gives some sort of notion how effective the package is. Thermal resistance from junction to case is also given; this is an average for different devices in the same package. The most popular packages are illustrated in Figure 25.10.

It is clear that in these terms the TO-225AA is not very area-efficient, but then it is not intended for high-power usage. The MT-200 is obviously the best plastic package, being beaten only by the TO-3 with its all-metal mounting flange — and all its mounting difficulties. The TO-264 package is important because it is used for the intriguing new five-leg

Table 25.6. The thermal parameters of common transistor packages

Package	Metal area mm^2	Body Area mm^2	Metal area (%)	Therm res juncn-case °C/W
TO-225AA	27	84	32	6.25
TO-220	91	154	59	1.66
TO-3P	313	533	58	0.83
TO-264	327	520	63	0.69
MT-200	626	779	80	0.63
TO-3	577	625	92	0.70

Onsemi ThermalTrak transistors with integral temperature-sensing diodes (see Chapter 22); the package is less efficient than the MT-200 but the thermal resistance junction-to-case is still quite respectable.

- TO-225AA driver transistor installation. These devices are usually mounted onto separate heatsinks that are light enough to be soldered into the PCB without further fixing. Silicone thermal washers ensure good thermal contact, and spring clips are used to hold the package firmly against the sink. Electrical isolation between device and heatsink is not normally essential, as the PCB need not make any connection to the heatsink fixing pads. If spring clips are not used, there is a single mounting hole for a bolt.
- TO-220 driver transistor installation. As for the TO-225AA package.
- TO-3P and TO-264 power transistor installation. These large flat plastic devices are best mounted on to the main heatsink with special spring clips, which are not only very rapid to install, but also generate less mechanical stress in the package than bolting the device down by its mounting hole. They also give a more uniform pressure onto the thermal washer material. Nonetheless, it is not always convenient to use the specially-shaped extrusions required to fit the spring clips, and screws are satisfactory so long as the appropriate torque setting is used. A good-sized washer should always be fitted under the screw head to spread the stress as much as possible. Being flat plastic devices, these transistors can be mounted directly onto the thick web of a heatsink, using blind tapped holes. Holes tapped into aluminium are not very durable and you have to be careful when installing or replacing devices.
- MT-200 power transistor installation. These even larger flat plastic devices have two mounting holes and once again can be mounted directly onto a large mass of metal. Once again, use good-sized washers under the screw heads. There is also an MT-100 package, which is somewhat smaller than the TO-3P, but it does not seem to have been so widely adopted.
- TO-3 power transistor installation. The rather dated TO-3 package is extremely efficient at heat transfer, but notably more awkward to mount, as they have to be bolted to a relatively thin piece of metal so that the base and emitter pins can protrude through the other side. Examples of TO-3 mounting are shown in Chapter 22 on Thermal Compensation and Thermal Dynamics.

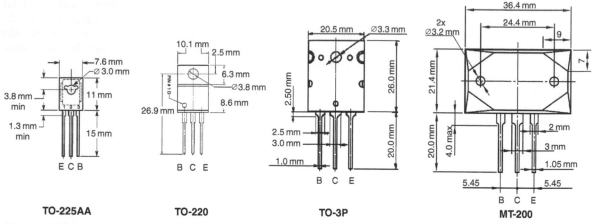

Figure 25.10. Dimensions of the most popular plastic transistor packages: TO-225AA, TO-220, TO-3P, and MT-200. Drawing is only approximately to scale.

When the first edition of this book was produced, TO3 packages were still fairly common in demanding power amplifier applications, but now they have pretty much fallen out of use, being replaced by plastic packages such as TO-3P and the larger MT-200 which offer much easier mounting. The section below has been allowed to stand as it will be useful to those rebuilding old equipment. There is also the point that audio fashion is an unpredictable thing, and for all I know there will be a sudden wave of TO-3 nostalgia. I understand some people still use valves …

My preference is for TO-3s to be mounted on an aluminium thermal-coupler which is bolted against the component side of the PCB. The TO-3 pins may then be soldered directly on the PCB solder side. The thermal-coupler is drilled with suitable holes to allow M3.5 fixing bolts to pass through the TO3 flange holes, through the flange, and then be secured on the other side of the PCB by nuts and crinkle washers which will ensure good contact with the PCB mounting pads. For reliability, the crinkle washers must cut through the solder-tinning into the underlying copper; a solder contact alone will creep under pressure and the contact force decay over time.

Insulating sleeves are essential around the fixing bolts where they pass through the thermal-coupler; nylon is a good material for these as it has a good high-temperature capability. Depending on the size of the holes drilled in the thermal-coupler for the two TO-3 package pins (and this should be as small as practicable to maximise the area for heat transfer), these are also likely to require insulation; silicone rubber sleeving carefully cut to length is very suitable.

An insulating thermal washer must be used between TO-3 and flange; these tend to be delicate and the bolts must not be over-tightened. If you have a torque-wrench, then 10 Nw/m is an approximate upper limit for M3.5 fixing bolts. *Do not* solder the two transistor pins to the PCB until the TO-3 is firmly and correctly mounted, fully bolted down, and checked for electrical isolation from the heatsink. Soldering these pins and *then* tightening the fixing bolts is likely to force the pads from the PCB. If this should happen, then it is quite in order to repair the relevant track or pad with a small length of stranded wire to the pin; 7/02 size is suitable for a very short run.

Alternatively, TO3s can be mounted off-PCB (e.g., if you already have a large heatsink with TO-3 drillings) with wires taken from the TO-3 pads on the PCB to the remote devices. These wires should be fastened together (two bunches of three is fine) to prevent loop formation; see above. I cannot give a maximum safe length for such cabling, but certainly 8 inches causes no HF stability problems in my experience. The emitter and collector wires should be substantial, e.g., 32/02, but the base connections can be as thin as 7/02. The cable routing will need to be carefully chosen to avoid Distortion Six − the inductive coupling of halfwave-rectified-sine pulses into sensitive parts of the circuitry; see Chapter 11 for more on this.

There are (or used to be) sockets for TO-3 transistors. These were retained by the two mounting bolts, and gripped the two pins with spring-loaded contacts, the idea being that device replacement was simpler. Don't touch them. The contacts are in general not adequate to handle large currents through the emitter pin. However, in my experience, there were worse problems than that; it was found that even in a clean domestic environment the contacts would corrode and become intermittent after a couple of years.

Reference

1. http://alasir.com/reference/solder_alloys/ (accessed Sept. 2012).

Power Supplies and PSRR

my power is made perfect …

2 Corinthians 12:9

Power Supply Technologies

There are three principal ways to power an amplifier:

1. a simple unregulated power supply consisting of transformer, rectifiers, and reservoir capacitors;
2. a linear regulated power supply;
3. a switch-mode power supply.

It is immediately obvious that the first and simplest option will be the most cost-effective, but at a first glance it seems likely to compromise noise and ripple performance, and possibly interchannel crosstalk. It is therefore worthwhile to examine the pros and cons of each technology in a little more detail. I am here dealing only with the main supply for the actual power amplifier rails. Many amplifiers now have some form of micro-controller to handle on/off switching by mains relays and other housekeeping functions; this is usually powered by a separate small standby transformer, which remains powered when the amplifier supply is switched off. The design of this is straightforward — or at least it was until the introduction of new initiatives to limit the amount of stand-by power that a piece of equipment is allowed to consume. The International Energy Agency is urging a one-Watt stand-by power limit for all energy-using products.

Simple Unregulated Power Supplies

Advantages:

- Simple, reliable, and cheap. (Relatively speaking — the traditional copper and iron mains transformer will probably be the most expensive component in the amplifier.)
- No possibility of instability or HF interference from switch-mode frequencies.
- The amplifier can deliver higher power on transient peaks, which is just what is required.

Disadvantages:

- The power into 4 Ω will not be twice that into 8 Ω, because the supply voltage will fall with increased current demand. On the other hand, the amplifier will always deliver the maximum possible power it can.
- Significant ripple is present on the DC output and so the PSRR of the amplifier will need careful attention; the problem is, however, not hard (if you read the second part of this chapter) and output hum levels below −100 dBu are easily attainable.
- The mains transformer will be relatively heavy and bulky.
- Transformer primary tappings must be changed for different countries and mains voltages.
- The absence of switch-mode technology does not mean total silence as regards RF emissions. The bridge rectifier will generate bursts of RF at a 100 Hz repetition rate as the diodes turn off. This worsens with increasing current drawn.

Linear Regulated Power Supplies

Advantages:

- A regulated supply-rail voltage means that the amplifier can be made to approximate more closely to a perfect voltage source, which would give twice the power into 4 Ω that it gives into 8 Ω. This is considered to have marketing advantages in some circles, though it is not clear why you would want to operate an amplifier on the verge of clipping. There are, however, still load-dependent losses in the output stage to consider. More on this later.
- A regulated supply-rail voltage to a power amplifier gives absolutely consistent audio power output in the face of mains voltage variation.
- Clipping behaviour will be cleaner, as the clipped peaks of the output waveform are not modulated by the ripple on the supply rails. Having said that, if your amplifier is clipping regularly, you might consider turning it down a bit.
- Can be designed so that virtually no ripple is present on the DC output (in other words, the ripple is below the white noise the regulator generates) allowing relaxation of amplifier supply-rail rejection requirements. However, you can only afford to be careless with the PSRR of the power amp if the regulators can maintain completely clean supply-rails in the face of sudden current demands. If not, there will be interchannel crosstalk unless there is a separate regulator for each channel. This means four for a stereo amplifier, making the overall system very expensive.
- The possibility exists of electronic shutdown in the event of an amplifier DC fault, so that an output relay can be dispensed with. However, this adds significant circuitry, and there is no guarantee that a failed output device will not cause a collateral failure in the regulators which leaves the speakers still in jeopardy.

Disadvantages:

- Complex and therefore potentially less reliable. The overall amplifier system is at least twice as complicated. The much higher component-count must reduce overall reliability, and getting it working in the first place will take longer and be more difficult. For an example, consider the circuit put forward by John Linsley-Hood in.[1] To regulate the positive and negative rails for the output stage, this PSU uses 16 transistors and a good number of further parts; a further 6 transistors are used to regulate the supplies to the small-signal stages. It is without question more complex and more expensive than most power amplifiers.
- If the power amplifier fails, due to an output device failure, then the regulator devices will probably also be destroyed, as protecting semiconductors with fuses is a very doubtful business; in fact, it is virtually impossible. The old joke about the transistors protecting the fuse is not at all funny to power-amplifier designers, because this is very often precisely what happens. Electronic overload protection for the regulator sections is therefore essential to avert the possibility of a domino-effect failure, and this adds further complications, as it will probably need to be some sort of foldback protection characteristic if the regulator transistors are to have a realistic prospect of survival.
- Comparatively expensive, requiring at least two more power semiconductors, with associated control circuitry and over-current protection. These power devices in turn need heatsinks and mounting hardware, checking for shorts in production, etc.
- Transformer tappings must still be changed for different mains voltages.
- IC voltage regulators are usually ruled out by the voltage and current requirements, so it must be a discrete design, and these are not simple to make bullet-proof. Cannot usually be bought in as an OEM item, except at uneconomically high cost.
- May show serious HF instability problems, either alone or in combination with the amplifiers powered. The regulator output impedance is likely to rise with frequency, and this can give rise to some really unpleasant sorts of HF instability. Some of my worst amplifier experiences have involved (very) conditional stability in such amplifiers.
- The amplifier can no longer deliver higher power on transient peaks.
- The overall power dissipation for a given output is considerably increased, due to the minimum voltage-drop though the regulator system.

- The response to transient current demands is likely to be slow, affecting slewing behaviour.

Switch-mode Power Supplies

Advantages:

- Has most of the advantages of linear regulated supplies, as listed above.
- Ripple can be considerably lower than for unregulated power supplies, though never as low as a good linear regulator design. 20 mV pk−pk is typical.
- There is no heavy mains transformer, giving a considerable saving in overall equipment weight. This can be important in PA equipment.
- Can be bought in as an OEM item; in fact, this is virtually compulsory in most cases as switch-mode design is a specialised job for experts.
- Can be arranged to shut down if amplifier develops a dangerous DC offset.
- Can be specified to operate properly, and give the same audio output without adjustment, over the entire possible worldwide mains-voltage range, which is normally taken as 90−260 V.

Disadvantages:

- Switch mode supplies are a prolific source of high-frequency interference. This can be extremely difficult to eradicate entirely from the audio output.
- The 100 Hz ripple output is significant, as noted above, and will require the usual PSRR precautions in the amplifiers.
- Much more complex and therefore less reliable than unregulated supplies. Dangerous if not properly cased, as high DC voltage is present.
- The response to transient current demands is likely to be relatively slow.
- Their design is very much a matter for specialists.

On perusing the above list, it seems clear to me that regulated supplies for power amplifiers are a Bad Thing. Not everyone agrees; see, for example, Linsley-Hood.[2] Unfortunately he did not adduce any evidence to support his case.

The usual claim − in fact, it is probably the closest thing to a Subjectivist consensus there is − is that linear regulated supplies give 'tighter bass' or 'firmer bass'; advocates of this position are always careful not to define 'tighter bass' too closely, so no one can disprove the notion. If the phrase means anything, it presumably refers to changes in the low-frequency transient response; however, since no such changes can be objectively detected, this appears to be simply untrue. If properly designed, all three approaches can give excellent sound, so it makes sense to go for the easiest solution; with the unregulated supply the main challenge is to keep the ripple out of the audio, which will be seen to be straightforward if tackled logically. The linear regulated approach presents instead the challenge of designing not one but two complex negative feedback systems, close-coupled in what can easily become a deadly embrace if one of the partners shows any HF instability. Before everyone runs off with the idea that I am irrevocably prejudiced against supply regulation, I will mention here that the first power amplifier system I ever designed did indeed have regulated power supplies, because at the time I was prepared to believe that it was the only way to achieve a good hum performance. Remarkably, considering that the only testgear I had was an old moving-coil test-meter, it all worked first time and without any misbehaviour I could detect. I still have it in the cellar. However, I did take away from the experience the conviction that if the power supplies were more complex than the amplifier, something was wrong with my design philosophy.

The generic amplifier designs examined in this book have excellent supply-rail rejection, and so a simple unregulated supply is perfectly adequate. The use of regulated supplies is definitely unnecessary, and I would recommend strongly against their use. At best, you have doubled the amount of high-power circuitry to be bought, built, and tested. At worst, you could have intractable HF stability problems, peculiar slew-limiting, and some expensive device failures.

A Devious Alternative to Regulated Power Supplies

In the list of the advantages of linear regulated supplies set out above, the one that seems to have most appeal to people is the first. It allows an amplifier to approximate more closely to a perfect voltage source, which would give exactly twice the power into 4 Ω that it gives into 8 Ω. In the not-always-rational world of hi-fi, this kind of amplifier behaviour is often considered a mark of solid merit, implying that there are huge output stages and heavyweight power supplies that can gracefully handle any kind of loudspeaker demand. I disagree, for the reasons set out above, but let's follow the train of thought for a bit, until it derails.

A regulated supply clearly gives a closer approach to this ideal than an unregulated supply whose voltage will

droop when driving the 4 Ω load. However, even if the regulated supply is as stiff as a girder of pure unbendium, there will still be load-dependent losses in the output stage which will make the 4 Ω output less than twice that into 8 Ω. Assume for the moment that we have an amplifier which gives 100 Watts into 8 Ω. There will be emitter resistors in the output stage, and the lowest value they are likely to have is 0.1 Ω. (There are good reasons why these resistors should be as low as practicable, because this improves linearity as well as efficiency; see Chapter 9.) These resistors are in series with the output and so form a potential divider with the load. Their presence alone, without considering other losses such as increased output device Vbes at higher currents, and the wiring resistance, will cause the 4 Ω output to be 195.1 Watts rather than 200 Watts. That perfect voltage source is not so easy to make after all.

However, to make a rather ambitious generalisation (and all generalisations are of course dangerous), it can be said that the power deficit from this cause is rather less than that due to unregulated supply rails drooping, which can cause twice the loss in terms of Watts. This factor depends very much on how big the mains transformer is, how big the reservoir capacitors are (because that affects the depth of the ripple troughs which is where clipping occur first), and so on — I said it was a generalisation. It is therefore perhaps worthwhile to look a little closer at the regulated supply issue.

I was once faced with this situation; the managing director wanted exact power doubling in a high-power design, but I was less than enthusiastic about trying to make heavy-current regulated power supplies work dependably. Time for some thought. If you accept that there is no problem in making a hum-free amplifier that runs from unregulated and ripply rails — which is emphatically true, as demonstrated in the second half of this chapter — then the function of the regulators is simply to keep part of the supply voltage away from the amplifiers. In effect, the output stage is a giant clipping circuit. So … why not do the clipping at the input of the amplifier, where it can be done with a couple of diodes, and go back to an unregulated power supply? The idea is shown in Figure 26.1. The electrical power previously wasted in the regulators is now absorbed by the output devices, perhaps necessitating a bit more heat-sinking, but all the complications of regulators disappear. As with a regulated supply, the clipping will be clean and uncontaminated by ripple — in fact probably cleaner because a small-signal clipping circuit will have no time-constants which may gather unwanted charges during overload. Now you may think that this

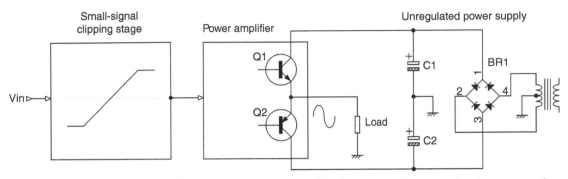

Figure 26.1. Putting a small-signal clipping circuit at the amplifier input to emulate a regulated power supply.

is cheating — the managing director certainly did, but even he was forced to admit that what I proposed was functionally identical to an amplifier with regulated supplies, and *much* cheaper. However, the idea of deliberately restricting amplifier output — and this new approach simply makes it obvious that that is what regulated supplies do — did not appeal to him any more than it does to me, and the project went forward with unregulated supplies. And no hum.

In the foregoing argument there is one point that has been oversimplified a little. Making a small-signal clipping circuit is straightforward. Making a clipping circuit that is wholly distortion-free below the clipping point is anything but straightforward. As I described in Chapter 4, it can be done, with some non-obvious circuitry. You will, I hope, forgive me for not revealing it at the moment, but I rather hope that someone might buy the idea off me.

Design Considerations for Power Supplies

A typical unregulated power supply is shown in Figure 26.2. This is wholly conventional in concept, though for optimal hum performance the wiring topology and physical layout need close attention, and this point is rarely made.

In a multichannel amplifier, the power supply will fall into one of three types. In order of increasing cost, and allegedly decreased interaction between channels:

- The transformer, rectifiers and reservoir capacitors are shared between channels.
- Each channel has its own transformer secondary, rectifiers and reservoirs. There is a single transformer but only the core and primary are shared.
- Each channel has its own transformer, rectifiers and reservoirs. Nothing except possibly the mains inlet and mains switch are shared.

In reality, the only interaction experienced with (1) and (2) is a variation in maximum power output depending on how the other channels are loaded. With competent design signal, crosstalk via the power supply should simply not happen.

For amplifiers of moderate power the total reservoir capacitance per rail usually ranges from 4700 to 20,000 µF, though some designs have much more. Ripple current ratings must be taken seriously, for excessive ripple current heats up the capacitors and reduces their lifetime. It is often claimed that large amounts of reservoir capacitance give 'firmer bass', presumably following the same sort of vague thinking that credits regulated power supplies with giving 'firmer bass', but it is untrue for all normal amplifier designs below clipping.

I do not propose to go through the details of designing a simple PSU at this point, because such information can be found in standard textbooks, but I instead offer below some hints and warnings that are either rarely published or are especially relevant to audio amplifier design.

Mains Connectors

Small and medium power amplifiers almost universally use what is commonly just called an 'IEC connector' into which plugs 'an IEC lead'. There is in fact a whole series of IEC connectors of different sizes. The popular one for amplifiers, computer, and most domestic electronics is called the C14 male inlet connector and the matching connector on the cable is called C13. They are normally rated at 10 Amps maximum. There is a high-temperature version of this called C16 which is intended for use on kettles and the like; these are not normally used on amplifiers. The 10 Amp rating naturally limits the current that can be drawn from the mains and hence the output

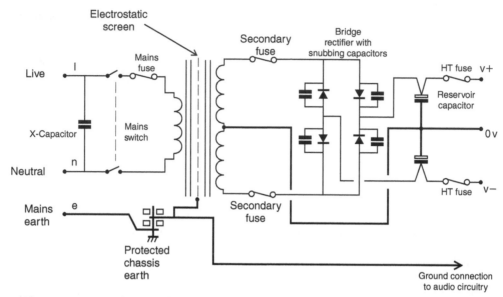

Figure 26.2. A simple unregulated power supply, including rectifier-snubbing and X-capacitor.

power of an amplifier. Most equipment has to run from 110 V mains as well as 230 V and at the lower voltage the maximum power that can be drawn is therefore limited to 1100 W. This is a definite restriction if you are trying to make a reasonably powerful amplifier that can drive 4 Ω or 2 Ω loads.

It is therefore very fortunate that there is a more capable connector in the shape of the IEC C20, which is rated at 16 Amps in Europe or 25 Amps in the USA. (In the UK 16 Amps is more than the official capacity of the 13-Amp mains plug.) The matching connector on the mains cable is called C19. This connector is the largest IEC type. It is considerably bigger physically than the C14, and to my mind has a definite 'means business' look to it, which is no bad thing on a powerful amplifier. These connectors are freely available and can be sourced from China. I have used them many times with every satisfaction — one example is the Cambridge Audio 840 W Class XD amplifier (2007).

The small figure-of-eight section mains connector which is sometime found on low-power equipment is the C8, rated at 2.5 Amps. There are only two pins and the common version is not polarised so it can only be used with double-insulated equipment. The matching connector on the mains cable is called C7.

There are several other IEC connectors with C-numbers but they are rarely encountered.

It is worth mentioning that in the UK the 13 Amp mains plug which goes into the wall socket is unique

in that it contains it own mains fuse, usually of 3, 5, or 13 Amp rating. This is to allow use with ring mains. The plug and wall socket is defined by British Standard BS 1363. It is polarised so you always know which connection is live. Modern plugs have insulated sleeves on the pins which prevent live contacts from being exposed while the plug is being inserted or removed. The plug and wall socket also has earth-pin operated shutters on the live and neutral contacts to prevent the insertion of foreign objects. This does not exhaust the safety features; BS 1363 specifies a high retention force for the contacts, so that the plug is virtually impossible to dislodge by accidental impact or by pulling on the cable, which enters the plug from the bottom to discourage yanking on it. The internal wiring is such that if the cable is pulled out of the plug, the earth wire will break last. The BS 1363 plug/socket is used in the United Kingdom, the Republic of Ireland, Sri Lanka, most Arab states, Cyprus, Malta, Iraq, Gibraltar, Hong Kong, Jordan, Malaysia, Singapore, Botswana, Ghana, Kenya, Uganda, Nigeria, Tanzania, Zambia, and Zimbabwe.

Mains Transformers

The mains transformer will normally be either the traditional E-&-I frame type, or a toroid. The frame type is used where price is more important than compactness or external field, and vice versa. There are various other types of transformer, such as C-core, or R-core, but they do not seem to be able to match the low external field of the toroid, while being significantly more expensive than the frame type.

The procurement of the mains transformer for a given voltage at a given current is simple in principle, but in the field of audio power amplifiers always seems to involve a degree of trial and error. This is because when transformers are used in unregulated power supplies for audio power amplifiers, the on-load voltage has to be accurate; the power output in Watts depends on the square of the rail voltage. Watts do not have a direct relation to subjective volume, but are psychologically an important part of the written spec. An amplifier which on review does not quite meet its published power output gives a poor impression. The subjective difference between 199 W and 200 W is utterly negligible but the two figures look quite different lying there on the paper. It is therefore normal practice to err on the side of higher rather than lower output power; this should not be taken too far as the amplifier will be running hotter than necessary.

The main reason for this is that the voltage actually developed on the reservoir capacitors depends on losses that are not easily predicted, and this is inherent in any rectifier circuit where the current flows only in short sharp peaks at the crest of the AC waveform.

First, the voltage developed depends on the transformer regulation, i.e., the amount the voltage drops as more current is drawn. (The word 'regulation' in this context has nothing to do with negative-feedback voltage control — unfortunate and confusing, but there it is.) Transformer manufacturers are usually reluctant to predict anything more than a very approximate figure for this.

Voltage losses also depend strongly on the peak amplitude of the charging pulses from the rectifier to the reservoir; these peaks cause voltage drops in the AC wiring, transformer winding resistances, and rectifiers that are rather larger than might be expected from just considering the mean DC current. Unfortunately the magnitude of the peak current is poorly defined, being affected by wiring resistance and transformer leakage reactance (a parameter that transformer manufacturers are even more reluctant to predict) and calculations of the extra peak losses are so rough that they are of doubtful value. There may also be uncertainties in the voltage efficiency of the amplifier itself, and there are so many variables that it is only realistic to expect to try two or even three transformer designs before the exact output power required is obtained. I have run projects where the transformer was exactly right the first time, but that was maybe 10% of cases, and I might as well be honest and put them down to good luck.

The power output of an amplifier depends on when it starts clipping — a common criterion is that the rated power is given when the THD due to clipping reaches 1%. Given the usual unregulated power supply, clipping is controlled by the troughs of the ripple waveform rather than its peaks, and the depth of these troughs is a function of the size of the total reservoir capacity. Since large electrolytics have relatively wide tolerances, this introduces another uncertainty into the calculations.

Secondly, the voltage losses in the power amplifier itself are not that easy to predict, some of the clipping mechanisms being quite complicated in detail. The inevitable conclusion is that the fastest way to reach a satisfactory transformer design is to make only approximate calculations, order a prototype as soon possible, and fine-tune the required voltage from there.

Since most amplifiers are intended to reproduce music and speech, with high peak-to-average power ratios, they will operate satisfactorily with transformers rated to supply only 70% of the current required for extended sinewave operation, and in a competitive market the cost savings are significant. Trouble comes when the amplifiers are subjected to sinewave testing, and a transformer so rated will probably fail from internal overheating, though it may take an hour or more for the temperatures to climb high enough. The usual symptom is breakdown of the winding insulation, the resultant shorted turns causing the primary mains fuse to blow. This process is usually undramatic, without visible transformer damage or the evolution of smoke, but it does of course ruin an expensive component.

To prevent such failures when a mains transformer is deliberately underrated, some form of thermal cut-out is essential. Self-resetting cut-outs based on snap-action bimetal discs are physically small enough to be buried in the outer winding layers and work very well. They are usually chosen to open the primary circuit at 100 °C or 110 °C, as transformer materials are usually rated to 120 °C unless special construction is required. Once-only thermal cut-outs can also be specified, but their operation renders the transformer almost as useless as shorted turns do — it is rarely economic to rewind transformers. The point is that they are required for safety reasons; the transformer will fail in a controlled fashion rather than relying on internal shorting and consequent fuseblowing, and they are significantly cheaper than self-resetting cut-outs.

If the primary side of the mains transformer has multiple taps for multi-country operation, remember that some of the primary wiring will carry much greater currents at low voltage tappings; the mains current drawn on 90 V input will be nearly 3 times that at 240 V, for the same power out.

Transformer Mounting

Mounting frame transformers is straightforward; bolts go through holes in the frame and into the chassis. There may be an orientation that minimises the hum induced into the electronics, and this needs to be considered at the mechanical design stage. These transformers are usually, but not always, mounted with their sides parallel to the sides of the chassis for aesthetic reasons, and rotating them to minimise hum is not common practice.

Toroidal transformers introduce some extra considerations. It is well known that toroids can be rotated to minimise induced hum, and it is a very good idea to allow for this by making the transformer lead-out wires long enough.

Toroidal transformers are typically mounted by sandwiching them between two dished plates, or one dished plate and a dished area pressed into a chassis plate. The plates are then held in place by a single large bolt passing through the centre, as shown in Figure 26.3. Neoprene washers are used at top and bottom to prevent the pressure from the plates putting undue pressure on the outer windings. In some cases a large flat washer is used underneath the chassis to spread the loading from the central bolt.

The fixing bolt must be secured with some kind of locking nut or locking washer. The toroid will be the heaviest part of the amplifier, and you really do not want it bouncing around inside the equipment because vibration in transit has loosened the nut. It is important not to over-tighten the bolt and put undue stress on the windings; in a production situation a torque-wrench setting is usually specified.

Very small toroids can be mounted simply by putting a fixing bolt through a central filling of epoxy potting compound. This would not be safe for larger sizes as the potting compound is only adhering to the tape on the inside of the toroid, and any serious vertical force will either tear the tape or rupture the bond between tape and potting. Nevertheless, large toroids very often do have their centre filled with potting compound; this is to deal with side-forces, at which it is good because one side is in compression, and *not* vertical forces. These side-forces are typically produced by the 1-metre drop-test.

It is well known that when a toroidal transformer is mounted, it is essential to avoid creating a shorted-turn through the central bolt. However, the mistake shown in Figure 26.4a does still occur and the result will inevitably be blown primary fuses and profuse profanity. Slightly more subtle is the dangerous situation shown in Figure 26.4b, where a shorted-turn is created when the equipment lid is slightly deformed by placing a heavy item on top of it. The clearance between the top of a toroid mounting bolt and the lid

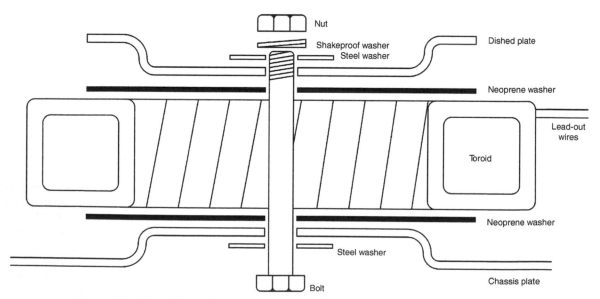

Figure 26.3. Toroidal transformer mounting. For clarity, the fixing bolt is partly withdrawn.

(a)

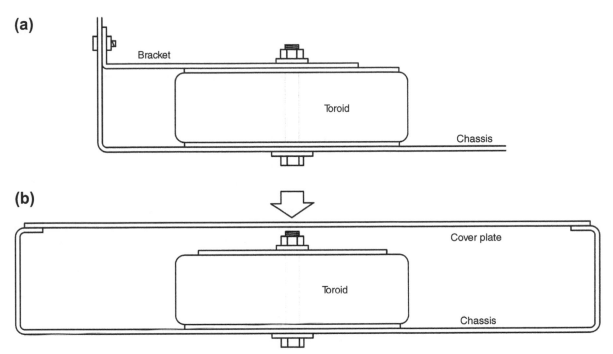

(b)

Figure 26.4. How NOT to mount a toroidal transformer.

must always be checked. If you've got it wrong and you are surrounded by a thousand sets of metalwork, a thin layer of tough insulation on the inside of the lid will get you out of trouble.

Transformer Specifications

A transformer specification needs to be a formal document. There are many factors to nail down and the usual result is an electrical schematic showing the primaries, secondaries, screen, etc., and a mechanical drawing showing maximum dimensions, supplemented by quite a lot of text.

The specification process usually starts with an informal discussion with the manufacturer to determine the approximate physical size of the transformer for the VA required. This must be done before you freeze the mechanical size of your product. A procedure for getting a good first estimate of the secondary voltage required is given in Chapter 6.

The list below gives an idea of what you need to specify when ordering a toroidal transformer:

Electrical Specifications

Having done the basic calculations, you have (you hope) a pretty good idea of what DC voltage you will need on your main supply rails to give the desired power into a given load impedance, and have done your best to translate that into a required AC voltage from the transformer secondary.

Manufacturers are not normally prepared to give exact figures as to how transformer regulation will affect the DC voltage available after rectification, and so when specifying the secondary AC voltage, it is realistic to aim for getting this exactly right under only one loading condition. This is usually the 'rated load' which is almost always 8 Ω. (The choice of what 'rated load' you put on the rear panel can have implications for safety testing, see Chapter 29 on Testing and Safety.)

1. Primary structure, i.e., will there be dual-voltage capability? If so, will there be two primary windings or a single tapped primary? The former makes better use of the copper but voltage-switching is more complicated.
2. Presence or otherwise of electrostatic screen between primary and secondary. Use one if you possibly can because it effectively stops RF getting in and out of the unit.
3. Secondary voltage with no load, i.e., with no external load on the amplifier. This will be greater than the voltage in (2) and you will need to make

sure the reservoir capacitor voltage ratings are high enough.

4. Secondary voltage with rated amplifier load, such as 8 Ω. (This is the 'Design Point' that has to be accurate.)

5. Secondary voltage with heavy amplifier load, such as 2 or 4 Ω. This will obviously be less, due to resistive losses in the windings. If it comes out too low, then it can be pulled up a bit by using thicker wire for windings, but this will increase the overall size.

6. Length, diameter, and colour of all lead-out wires. Length stripped of insulation, and if tinned.

Mechanical Specifications

1. Maximum diameter (the diameter will not be constant going around the periphery — it will typically be greater near the lead-out wires).

2. Maximum height.

3. State if manufacturer is expected to supply mounting hardware, i.e., dished plate, neoprene washers, fixing bolt & nut, locking washer, etc.

4. Specify the central potting, stating the size of the hole through it for the fixing bolt. A crucial thing to specify is how close the potting comes to the top of the toroid; it must be below the dished part of the top mounting plate, or the plate will not sit properly on the top of the transformer.

Safety Issues

The internal safety requirements, such as the thickness of insulation between windings, are usually left to the manufacturer. It is common, however, to specify the safety requirements for the lead-out wires, with phrases such as 'Must be UL-approved'.

Transformer Evaluation

When a sample transformer is ordered, there are several aspects of its performance that need to be looked at. Most are straightforward, e.g., is it the right physical size? does it have the right lead-out colours? and so on, but some aspects need a little more thought.

Unless your transformer manufacturer is hopelessly incompetent, the secondary voltages should be roughly what you asked for, but, for the reasons detailed above are unlikely to be exactly right. This is of no importance in secondaries for powering regulated supplies to run

opamps, etc., but because of its strong effect on the power output figure in Watts, the main HT rail secondary voltage has to be correct.

When checking power output, it is obviously important to have the incoming mains at exactly the right voltage, as errors here will feed directly into erroneous measurements. Once again, this is particularly important since the output in Watts varies as the square of the voltage. The usual practice is to use a variable autotransformer to fine-tune the mains voltage, its output being monitored by a DVM with the usual measure-average-but-call-it-RMS calibration. Another option is to use a ferroresonant constant voltage transformer, but these have several disadvantages; the output waveform is usually more of a square wave than a sine, and there is a fixed output voltage. They are also heavy and expensive.

The ideal solution is to use a mains synthesiser, which can output a good sinewave of variable voltage and variable frequency at a serious power level; the only downside is that it is a very expensive piece of equipment that will only be used relatively infrequently. I have only ever worked with one manufacturer that had one of these to hand They went bust, though ownership of a mains synthesiser was not the reason.

Particular difficulties can arise when you are in a country with 230 V mains, and testing transformers for equipment aimed at the American and Canadian markets (115 V) and Japan (100 V). Now the variable auto-transformer is required to make a major change in voltage rather than a small adjustment, and the distortion of its output waveform will usually be severe. This renders the reading of the aforementioned measure-average-but-call-it-RMS voltmeter inaccurate, as the waveform distortion alters the relationship between average and peak values of the mains, and it is the peak value which determines the voltage produced by the amplifier power supply. The normal result is that the measured amplifier power output is lower than expected at 115 V and 100 V, and this can lead to baffled exchanges with your transformer manufacturer, who knows quite well that what he has supplied is correct.

If you have no plans to invest in a mains synthesiser, the second-best solution is to get a large fixed auto-transformer that reduces the 230 V mains to 115 V, and use that to feed the variable auto-transformer. The latter now only has to make small voltage corrections and the waveform of its output will be much less distorted than if it was doing the whole voltage step-down itself.

Evaluating a transformer sample for safety is somewhat problematical. You can do the standard insulation tests, and you can check that the lead-outs are at least labelled with the right approvals. The internal construction can only be investigated by taking a sample apart, the issue here being proper insulation between windings, especially where the lead-outs from an inner winding pass through an outer winding. If you use a reputable manufacturer you are most unlikely to have trouble in this area — if you don't, you may not find out until you submit the transformer to a test house for safety approval, by which time you're usually a long way down the road to production.

Very often the most critical part of the evaluation is the amount of hum that the transformer induces into the electronic circuitry. This has its own section just below.

Transformers and Hum

All transformers, even high-quality toroidal ones, have a significant hum field, and this can present some really intractable problems if not taken very seriously from the start of the design process; the expedients available for fixing a design with excessive transformer hum are limited in number. In comparison, the fields from AC wiring are much smaller, unless the cabling arrangements are really peculiar. Here are some factors to consider:

1. Make sure that the transformer is as far as possible from any sensitive electronics. This sounds simple — you just put them on opposite sides of the box, no? Unfortunately other practical considerations may get in the way. The electronic PCBs may be so large that however they are mounted, part of their area is near the transformer. It is also not a good plan to put a heavy transformer at the extreme end of the box, as this makes it awkward to pick up and carry; when we approach a solid object like an amplifier case, we expect the centre of gravity to be in the middle. There may also be an aesthetic requirement that the transformer should be in the centre of the box. The visual appeal with the lid off is a significant marketing factor.
2. Use a toroidal transformer. They are more expensive for the same VA, and harder to mount, but the reduction in hum field is significant, and they are used wherever external fields are an issue.
3. If you are using a toroid, make sure it can be rotated to minimise hum. It is not usually economic to

optimise the orientation for each example of a product, but toroids made by reputable manufacturers should not vary much in the shape of their hum field and the orientation can be fixed at the design stage. The limitation of this technique is that if the susceptible electronics is spread out over space; very possibly left and right channels will be on opposite sides of the enclosure, and with dreadful certainty it will be found that the hum minimum for one channel is something like the maximum for the other. However, with suitable layout, rotation can be very effective. One prototype amplifier I have built had a sizeable toroid mounted immediately adjacent to the TO-3 end of the amplifier PCB; however, complete cancellation of magnetic hum (hum and ripple output level below -90 dBu) was possible on rotation of the transformer. Some toroids have single-strand secondary lead-outs, which are too stiff to allow rotation; for experimental use these can be cut short and connected to suitably large flexible wire such as 32/02, with carefully sleeved and insulated joints.

4. If you are using a frame transformer, its external field can be significantly reduced by specifying a hum strap, or 'belly-band' as it is sometimes rather indelicately called. This is a wide strip of copper that forms a closed circuit around the outside of the core and windings, so it does not form a shorted turn in the main transformer flux. Instead it intersects with the leakage flux, partially cancelling it.
5. Use transformer screening. Because of its physical construction, a toroidal transformer cannot use the hum strap method to reduce the external field. The usual approach is to wrap the outside of the toroid in one or more layers of silicon steel, the intention being screening rather than the creation of a shorted turn. The success of this depends on using high-quality silicon steel, or better still GOSS (Grain-Oriented Silicon Steel), and even then the reduction in hum figures from the affected circuitry is not likely to be more than 6 dB. It may sound unlikely, but it is a fact that the method of making GOSS was discovered in 1935 — by a Mr N. P. Goss. Mu-metal, a nickel-iron alloy (75% nickel, 15% iron, plus copper and molybdenum) is the most effective magnetic screening material, but it is expensive and has a disconcerting habit of losing its magical properties if deformed.
6. Go to a manufacturer with a reputation for making low-field transformers. At least one toroid manufacturer specialises in low-field designs for audio

applications, and their products can be 10 dB better than a standard-quality transformer; on the downside, the price will be something like twice as much. Low-field transformers are usually slightly larger than a conventional design.

7. Put the transformer in another box that can be positioned some distance away. This is obviously an expensive approach, and raises interesting questions about running high-current connections between the amplifier box and the transformer box. It is usually inappropriate. It is, however, undeniably effective. There is more on this approach below.

Induced hum varies proportionally with the incoming mains voltage, and this needs to be borne in mind during testing if your mains voltage varies significantly.

External Power Supplies

However much care is taken, it is very difficult to keep all traces of transformer-induced hum out of the signal circuitry. It is highly irritating to find that despite the cunning use of low-noise circuitry, the noise floor is defined by the deficiencies of a component — for the ideal transformer would obviously have no external field — rather than the laws of physics as articulated by Johnson.

The ultimate solution to the problem is to put the mains transformer in a separate box, which can be placed a metre or so away from the amplifier unit, and powering it through an umbilical lead.

Advantages:

- The transformer field hum problem is authoritatively solved.
- Will appeal to some potential customers as a 'serious' approach to high-end audio.

Disadvantages:

- The cost of an extra enclosure plus an extra cable and connectors, indicator lights, etc. The connectors will have to be multi-pole and capable of handling considerable voltages and currents. The transformer box must have fuses or other means of protection in case of short-circuits in the cable.
- A significant proportion of users will, exhortations to the contrary notwithstanding, promptly place the amplifier box directly on top of the transformer box, immediately defeating the whole object. This is particularly likely if the two boxes have the same footprint, and so look as if they *ought* to be stacked together. However, all is not lost in this situation, as

the transformer is still physically further away from the sensitive electronics (though if the transformer has a large field emerging from its ends, things may actually be worse) and there are now two extra layers of steel interposed. Assuming the boxes are made of steel, that is.

- The voltages involved will probably be above the limit set by the Low Voltage Directive, so it will be necessary to ensure that the connector contacts cannot be touched. If the cable has a connector at both ends, then both must be checked for this. A cable that is captive at the power supply end makes this issue simpler and will also save the cost of a mating pair of connectors, which may be considerable.

The most important design issue is the distribution of the power supply components between the two boxes. One approach is to put just the mains transformer in the power supply box. This has the disadvantage that the current in the umbilical cable consists of short charging pulses of large magnitude at a frequency of 100 or 120 Hz, and these will not only experience a greater voltage-drop in the cable resistance than a steady current, but also give rise to much greater I^2R heating. The latter is unlikely to cause problems in the cable itself, but can easily be fatal to the contacts of connectors. Speaking from bitter experience, I can warn that connectors that appear to have a more than adequate safety margin can fail under these conditions, and it is best to keep connectors out of charging pulse circuits.

The alternative is put not just the mains transformer but also the rectifiers and reservoir capacitors in the power supply box. The current in the umbilical cable is now rectified and smoothed DC, and it is much easier to specify connectors to cope with it. The snag is that the reservoir capacitors have two functions: as well as smoothing the rectified DC, they also hold a store of energy which can be drawn on during output peaks. The resistance of the cable between the reservoir capacitors and the power amplifier which will cause unwanted voltage drops when there are sudden demands for load current, which can significantly reduce peak power outputs during toneburst testing. Another worry is that the extra resistance in the supply rails could imperil the stability of the amplifier, though the use of generous local decoupling capacitors should be enough to deal with this problem.

A solution to both problems is the provision of significant amounts of capacitance at both ends; the capacitors in the power supply box deal with the smoothing, while those at the amplifier end provide a ready reserve of electricity. In this case, the current through the cable will still show some charging peaks, the size of which will depend on the proportion of the total capacitance at each end and the cable resistance. This could be artificially increased by adding series resistors of small ohmic value but high wattage, making an RC filter that will reduce the ripple seen by the amplifier. This is a bit of a doubtful remedy as it will reduce the power output on sustained signals, and it is a very poor way to reduce amplifier noise derived from the supply rails, as will be described later.

There you have some of the pros and cons of external amplifier power supplies. It is not quite the expensive-but-foolproof solution it first appears to be, and the design issues require careful thought.

Inrush Currents

When a transformer is abruptly connected to the mains supply, it takes a large current that decays exponentially; this is called the inrush current, (or sometimes the turn-on surge, or even the 'inductive surge') and it is highly inconvenient as it can be much greater than the normal current drawn, even at maximum output into the lowest rate load impedance. This inrush current is not a danger to the transformer, which has a big thermal mass; but it can and will blow primary fuses and trip house circuit breakers. With small and medium-sized transformers the problem is not serious, but it does mean that you have to be very careful in sizing the fuse or fuses in the primary circuit, making sure that they have a high enough rating so their life is not impaired by repeated inrush currents.

With a large transformer (say, bigger than 500 VA for a toroid), the inrush becomes large enough to trigger domestic overload protection. Since most houses now have magnetic circuit breakers rather than wire fuses in the mains distribution panel, this is not as inconvenient as it used to be, but is still thoroughly annoying, and will quickly earn you the enmity of your customers. The inrush issue has to be taken very seriously as it can cause problems which only show up when the unit is out in the field. There is anecdotal evidence that circuit breakers in Germany, while nominally rated the same as those in Britain, actually respond somewhat faster, so a design which has received careful checking in one country may cause serious trouble in another.

Inrush current is most conveniently measured with purpose built-instruments such as the Voltech power analyser range. A cheaper method is to use a current transformer (typically of the 'giant-clothes-peg' type clamped around one of the primary connections, and connected to a digital oscilloscope; this is naturally only cheaper if you already have a digital oscilloscope. It is characteristic of inrush current that its peak value varies widely from one switch-on to the next, as it depends crucially on the point of the mains cycle at which the transformer is connected. If you're unlucky, the transformer core briefly saturates and a big peak current is drawn by the primary. For this reason, repeated tests − possibly up to fifty − have to be done before you are confident you have experienced the worst case. This often has to be spread out over some time to avoid over-taxing inrush suppression components.

Toroidal transformers typically take greater inrush currents than frame types, due to their lower leakage reactance. There is a component of the inrush current that is due to the charging of the power supply reservoir capacitors, but this is usually small compared with the transformer inrush. As a rough guideline, if your transformer is bigger than 500 VA you should consider using inrush suppression. If in doubt, then at least make provision for adding it to the design in the development phase.

The inrush current is controlled by making sure there is enough series resistance in the transformer primary circuit to keep the flood of Amperes down to an acceptable level. The two main ways of doing this are to use an inrush suppressor component, or a relay that switches resistance into circuit for starting.

Inrush Suppression by Thermistor

An inrush suppressor component (sometimes called a surge limiter) is a giant thermistor whose resistance drops to a low value as it is heated by the current passing through it; they are usually of the disc type. The inrush suppressor is inserted in series with the transformer primary. The thermal inertia of its mass causes the resistance to drop relatively slowly, so the inrush current is restricted. Because of their thermistor action, these components run very hot in the low-resistance state (about 200 °C) and must be mounted with caution to ensure they do not melt the plastic of adjacent components. The component leads must be left long enough to avoid thermal degradation of the solder joints with the PCB, and if these leads are insulated, it must be with a high-temperature material such as fibre-glass sleeving. They are also likely to burn the fingers of service personnel − it is only polite to put a HOT warning on the PCB silkscreen.

Inrush suppressors require a cool-down time after power is removed. This cool-down or 'recovery' time allows the resistance of the NTC thermistor to increase sufficiently to provide the required inrush current suppression the next time it is needed. The necessary time varies according to the particular device, the mounting method and the ambient temperature, but a typical cool-down time is about one minute. This is usually specified by the manufacturer as a thermal time-constant with values ranging from 30 to 150 seconds, the longer times being for the larger and more highly rated versions.

Inrush suppressors are available in many different sizes. The quickest design method is to select a few types that can handle the maximum current in the primary circuit, and try them out to see which is the most effective at controlling the peak inrush value.

Inrush Suppression by Relay

In this method a series resistance is placed in series in the transformer primary circuit when mains is first applied. This limits the inrush current, and is then switched out by a relay after a suitable inrush control period, typically about one second. The basic circuit is shown in Figure 26.5a. The inrush resistor has to sustain a very large short-term overload, so a chunky wirewound type is appropriate, and it is vital to ensure

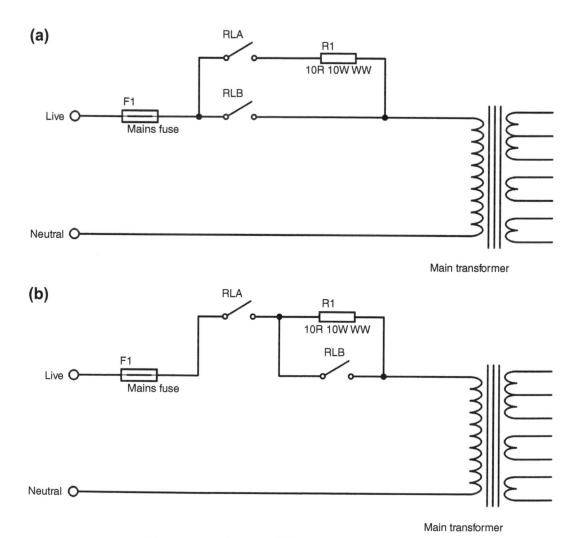

Figure 26.5. Relay-controlled inrush suppression circuits.

that it can cope with this overload many times over the life of the amplifier.

However ... resistor manufacturers are noticeably reluctant to specify how their products will cope with such conditions. It is therefore a good plan to use inrush suppression in its intended final form from the very start of the development process; by the time all other design issues have been addressed, you will almost certainly have put the inrush suppression through enough operating cycles to have confidence in its durability. (Using inrush suppression from the start may well be essential anyway to prevent the workbench circuit-breakers from tripping.)

The inrush current is a complex phenomenon and the resistance value and power rating of the resistor is usually determined by experience rather than protracted mathematical analysis. Here are some typical values that I have used with success:

10 Ω 10 Watts for a 800 VA toroid
10 Ω 20 Watts for a 1300 VA toroid

Wirewound resistors come in a limited number of types for sizes above 10 W, and it is often more convenient to use two 10 W resistors in parallel when a 20 W capability is required. Do not put the resistors in series; with a parallel connection, if one of the resistors fails open-circuit (which is by far the most likely mode), the inrush system will probably keep working satisfactorily.

If the resistor is correctly sized, after a single inrush event, it should be warm rather than hot. Repeated and rapid cycling of the power, as may occur in testing, can cause it to get very hot, and could eventually lead to failure. Fortunately this is not likely to occur in service.

The circuitry must be arranged so that if the power is turned off then immediately turned on again, inrush suppression still operates for the full period. This situation is called a 'hot restart'.

Many amplifiers are not simply switched on and off, but have an on/standby system where the mains switch initially applies power only to a small transformer that energises a small amount of control circuitry. A low-current switch, which can be more cosmetically attractive than something hefty enough to control the full mains power, activates the control circuitry and causes it to close a relay that energises the main supply. When this function is combined with inrush protection, there are usually two identical relays in the primary circuit as shown Figure 26.5a; at switch-on RLA closes and applies power to the transformer through the inrush resistor R1. After a second or so RLB closes and shorts out the resistor; RLA is now doing nothing so it is de-energised after a very short delay to make sure that RLB is fully closed. The alternative arrangement at Figure 26.5b should be avoided as it requires both relays to be energised all the time, which is a pointless waste of perfectly good electricity.

Fusing

You will have noticed that there are three levels of fusing in Figure 26.2. Each has a job to do.

Firstly, there is a mains fuse in the live side of the incoming mains, internal to the equipment. This is nothing to do with protecting the amplifier and everything to do with protecting its owner. If a short-circuit develops between live and a grounded chassis, this fuse will blow and disconnect the live. It is essential that the ground connection has a low resistance so this will blow reliably when required. There is no protection against shorts to ground from the cabling between the mains connector and the mains fuse except for the building circuit-breakers (in the UK such a fault will probably blow the fuse in the mains plug before tripping the breaker), so mains inlet connectors with built-in fuseholders that incorporate this connection internally are an excellent idea. The mains fuse must be rated to withstand the inrush current of the transformer over the long term. These fuses should be slow-blow.

Secondly, there are two secondary fuses between the mains transformer and the bridge rectifier. These have only one role − to protect the transformer in case the bridge rectifier fails short-circuit. This is, it has to be admitted, very unlikely, but if it does happen, the transformer, which is almost certainly the most expensive component in any amplifier, is protected. The mains fuse cannot be guaranteed to blow because it has to be sized to withstand the transformer inrush currents, and the transformer thermal cut-out (if fitted) cannot be relied upon to act quickly enough to save the transformer. These fuses are invariably slow-blow.

Thirdly, there are two HT fuses in the supply rails. These are not intended to protect the amplifier components. As has been remarked on many occasions, the output devices usually sacrifice themselves to protect the rail fuses. They are actually intended rather to minimise the damage if the output devices do fail short-circuit, preventing melting tracks, overheated cabling, and generally unsafe conditions. These fuses may be either fast or slow-blow; they must be sized so they can cope when the amplifier is delivering maximum power (with 10% high mains) into the lowest load impedance.

Fuseholders may heat up when carrying heavy currents and consideration should be given to using heavy-duty types. Keep an eye on the fuses; if the fuse-wire sags at turn-on, or during transients, the fuse will fail after a few dozen hours, and the rated value needs to be increased.

When selecting the value of the mains fuse in the transformer primary circuit, remember that toroidal transformers take a large current surge at switch-on. The fuse will definitely need to be of the slow-blow type.

Rectification

The rectifier (almost always a packaged bridge) must be generously rated to withstand the initial current surge as the reservoirs charge from empty on switch-on. Rectifier heatsinking is definitely required for all but the smallest amplifiers; the voltage drop in a silicon rectifier may be low (1 V per diode is a good approximation for rough calculation), but the current pulses are large and the total dissipation is significant.

Reservoir capacitors must have the incoming wiring from the rectifier going directly to the capacitor terminals; likewise the outgoing wiring to the HT rails must leave from these terminals. In other words, do not run a tee off to the cap, because if you do, its resistance combined with the high-current charging pulses adds narrow extra peaks to the ripple crests on the DC output and may worsen the hum/ripple level on the audio.

The cabling to and from the rectifiers carry charging pulses that have a considerably higher peak value than the DC output current. Conductor heating is therefore much greater due to the higher value of I-squared-R. Heating is likely to be especially severe if connectors are involved.

RF Emissions from Bridge Rectifiers

Bridge rectifiers, even the massive ones intended solely for 100 Hz power rectification, generate surprising quantities of RF. This happens when the bridge diodes turn off; the charge carriers are swept rapidly from the junction and the current flow stops with a sudden jolt that generates harmonics well into the RF bands. The greater the current, the more RF produced, though it is not generally possible to predict how steep this increase will be. The effect can often be heard by placing a transistor radio (long or medium wave) near the amplifier mains cable. It is the only area in a conventional power amplifier likely to give trouble in EMC emissions testing.[3]

Even if the amplifier is built into a solidly grounded metal case, and the mains transformer has a grounded electrostatic screen, RF will be emitted via the live and neutral mains connections. The first line of defence against this is usually four snubbing capacitors of approximately 100 nF across each diode of the bridge, to reduce the abruptness of the turn-off. If these are to do any good, it is vital that they are all as close as possible to the bridge rectifier connections. Such capacitors must be of a type intended to withstand continuous AC stress. Using mains-rated X-capacitors is recommended.

The second line of defence against RF egress is an X-capacitor wired between Live and Neutral, as near to the mains inlet as possible (see Figure 26.1). This is usually only required on larger power amplifiers of 300 W total and above. The capacitor must be of the special type that can withstand direct mains connection. 100 nF is usually effective; some safety standards set a maximum of 470 nF. A drain resistor should be connected across the X-capacitor because if the equipment mains switch is open, and the mains lead is disconnected at the peak of the mains waveform, the X-capacitor can be left with enough charge to give a perceptible shock if the mains plug pins are touched. The resistor value should be low enough so that the X-capacitor is discharged to a safe voltage in a small fraction of a second, without being so low as to point-lessly dissipate heat. The voltage rating of the resistor should be watched; this is not usually a problem for 1/4 W sizes and above.

Relay Supplies

It is very often most economical to power relays from an unregulated supply. This is perfectly practical as most relays have a wide operating voltage range. Hum induced by electrostatic coupling from this supply rail can be sufficient to compromise the noise floor; clearly the likelihood of this depends on the physical layout, but it is inevitable that signal paths and the relay come into proximity at the relay itself. It is therefore desirable to give this line some degree of smoothing, without going to the expense of providing another regulator and heatsink. (There should be no possibility of direct coupling between the signal ground and relay power ground; these must only join right back at the power supply.) This method of relay driving is more power efficient than a regulated supply rail as it does not require a voltage drop across a regulator that must be sufficient to prevent drop-out and consequent rail ripple in low-mains conditions.

Simple RC smoothing is quite adequate for this purpose and there is no need to consider the use of expensive chokes, which would probably cost more than a regulator, take up more space, radiate magnetic fields and generally be a pain in the amp. Because relays draw relatively high currents, a low R and a high capacitance value for C are necessary to minimise voltage losses in R and changes in the rail voltage as different numbers of relays are energised.

Figure 26.6 shows a typical power supply circuit giving a regulated +5 V rail to power a microcontroller, with the addition of an RC smoothed +9 V rail to power relays. The RC smoothing values shown are typical, but are likely to need adjustment depending on how many relays are powered and how much current they draw. The R is low at 2.2 Ω and the C high at 4700 uF.

Note the 10 nF capacitor across the transformer secondary; this part must be an X-capacitor or other type rated for continuous AC stress. This is typical of the extra components required to meet modern EMC standards.

Power Supply-rail Rejection in Amplifiers

The literature on power amplifiers frequently discusses the importance of power-supply rejection in audio amplifiers, particularly in reference to its possible effects on distortion![4]

I have (I hope) shown in earlier chapters that regulated power supplies are just not necessary for an exemplary THD performance. I want to confirm this by examining just how supply-rail disturbances insinuate themselves into an amplifier output, and the ways in which this rail-injection can be effectively eliminated. My aim is not just the production of hum-free amplifiers, but also to show that there is nothing inherently

mysterious in power-supply effects, no matter what Subjectivists may say on the subject.

The effects of inadequate power-supply rejection ratio (PSRR) in a typical Class-B power amplifier with a simple unregulated supply may be two-fold:

1. A proportion of the 100 Hz ripple on the rails will appear at the output, degrading the noise/hum performance. Most people find this much more disturbing than the equivalent amount of distortion,
2. The rails also carry a signal-related component, due to their finite impedance. In a Class-B amplifier this will be in the form of half-wave pulses, as the output current is drawn from the two supply-rails alternately; if this enters the signal path, it will degrade the THD seriously.

The second possibility, the intrusion of distortion by supply-rail injection, can be eliminated in practice, at least in the conventional amplifier architecture so far examined. The most common defect seems to be misconnected rail bypass capacitors, which add copious ripple and distortion into the signal if their return lines share the signal ground; this was denoted No. 5 (Rail Decoupling Distortion) on my list of distortion mechanisms in Chapter 5.

This must not be confused with distortion caused by *inductive* coupling of halfwave supply currents into the signal path − this effect is wholly unrelated and is completely determined by the care put into physical layout; I labelled this Distortion Six (Induction Distortion).

Assuming the rail bypass capacitors are connected correctly, with a separate ground return, ripple and distortion can only enter the amplifier directly through the circuitry. It is my experience that if the amplifier is made ripple-proof under load, then it is proof against distortion-components from the rails as well;

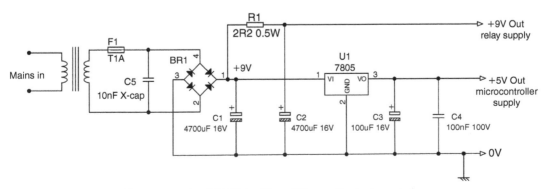

Figure 26.6. A +5 V PSU with an RC smoothed relay supply.

this bold statement does, however, require a couple of qualifications:

Firstly, the output must be ripple-free *under load,* i.e., with a substantial ripple amplitude on the rails. If a Class-B amplifier is measured for ripple output when quiescent, there will be a very low amplitude on the supply-rails and the measurement may be very good; but this gives no assurance that hum will not be added to the signal when the amplifier is operating and drawing significant current from the reservoir capacitors. Spectrum analysis could be used to sort the ripple from the signal under drive, but it is simpler to leave the amplifier undriven and artificially provoke ripple on the HT rails by loading them with a sizeable power resistor; in my work I have standardised on drawing 1 Amp. Thus one rail at a time can be loaded; since the rail rejection mechanisms

are quite different for V+ and V-, this is a great advantage. Drawing 1 Amp from the V- rail of the typical power amplifier in Figure 26.7 degraded the measured ripple output from -88 dBu (mostly noise) to −80 dBu.

Secondly, I assume that any rail filtering arrangements will work with constant or increasing effectiveness as frequency increases; this is clearly true for resistor-capacitor (RC) filtering, but is by no means certain for *electronic* decoupling such as the NFB current-source biasing used in some designs in this book. (These will show declining effectiveness with frequency as internal loop-gains fall.) Thus, if 100 Hz components are below the noise in the THD residual, it can usually be assumed that disturbances at higher frequencies will also be invisible, and not contributing to the total distortion.

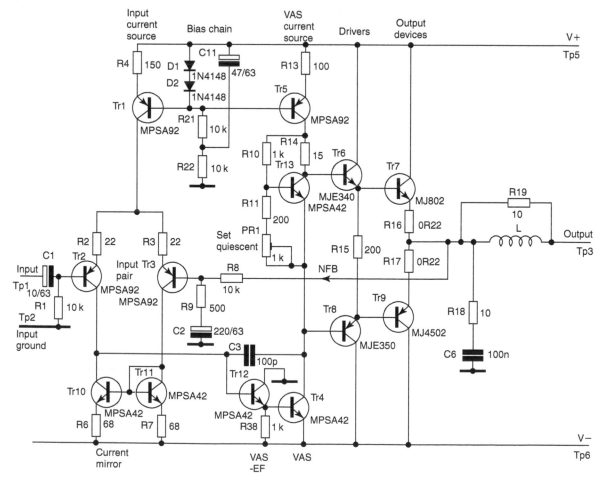

Figure 26.7. Diagram of a generic power amplifier, with diode biasing for input tail and VAS source.

To start with some hard experimental facts, I took a power amplifier — similar to Figure 26.7 — powered by an unregulated supply on the same PCB (the significance of this proximity will become clear in a moment) driving 140 Wrms into 8 Ω at 1 kHz. The PSU was a conventional bridge rectifier feeding 10,000 μF reservoir capacity per rail.

The 100 Hz rail ripple under these conditions was 1 V pk–pk. Superimposed on this were the expected halfwave pulses at signal frequency; measured at the PCB track just before the HT fuse, their amplitude was about 100 mV peak-peak. This doubled to 200 mV on the downstream side of the fuse — the small resistance of a 6.3 A slow-blow fuse is sufficient to double this aspect of the PSRR problem, and so the fine details of PCB layout and PSU wiring could well have a major effect. (The 100 Hz ripple amplitude is of course unchanged by the fuse resistance.)

It is thus clear that improving the *transmitting* end of the problem is likely to be difficult and expensive, requiring extra-heavy wire, etc., to minimise the resistance between the reservoirs and the amplifier. It is much cheaper and easier to attack the *receiving* end, by improving the power-amp's PSRR. The same applies to 100 Hz ripple; the only way to reduce its amplitude is to increase reservoir capacity, and this is expensive.

A Design Philosophy for Supply-rail Rejection

Firstly, ensure there is a negligible ripple component in the noise output of the quiescent amplifier. This should be pretty simple, as the supply ripple will be minimal; any 50 Hz components are probably due to magnetic induction from the transformer, and must be removed first by attention to physical layout. It is assumed that obvious problems, such as rail bypass capacitors putting ripple into the signal, have been avoided.

Secondly, the THD residual is examined under full drive; the ripple components here are obvious as they slide evilly along the distortion waveform (assuming that the scope is synchronised to the test signal). As another general rule, if an amplifier is made visually free of ripple-synchronous artefacts on the THD residual, then it will not suffer detectable distortion from the supply-rails.

PSRR is usually best dealt with by RC filtering in a discrete-component power amplifier. This will, however, be ineffective against the sub-50 Hz VLF signals that result from short-term mains voltage variations being reflected in the HT rails. A design relying wholly on RC filtering might have low AC ripple figures, but would show irregular jumps and twitches of the THD residual; hence the use of constant-current sources in the input tail and VAS to establish operating conditions more firmly.

The standard opamp definition of PSRR is the dB loss between each supply-rail and the effective differential signal at the inputs, giving a figure independent of closed-loop gain. However, here I use the dB loss between rail and output, in the usual non-inverting configuration with a C/L gain of 26.4 dB. This is the gain of the amplifier circuit under consideration, and allows dB figures to be directly related to testgear readings.

Looking at Figure 26.7, we must assume that any connection to either HT rail is a possible entry point for ripple injection. The PSRR behaviour for each rail is quite different, so the two rails are examined separately.

Positive Supply-rail Rejection

The V+ rail injection points that must be eyed warily are the input-pair tail and the VAS collector load. There is little temptation to use a simple resistor tail for the input; the cost saving is negligible and the ripple performance inadequate, even with a decoupled mid-point. A practical value for such a tail-resistor would be 22 k, which in SPICE simulation gives a low-frequency PSRR of −120 dB for an undegenerated differential pair with current-mirror.

Replacing this tail resistor with the usual current source improves this to −164 dB, assuming the source has a clean bias voltage. The improvement of 44 dB is directly attributable to the greater output impedance of a current source compared with a tail resistor; with the values shown this is 4.6 M, and 4.6 M/22 k is 46 dB, which is a very reasonable agreement. Since the rail signal is unlikely to exceed +10 dBu, this would result in a maximum output ripple of −154 dBu.

The measured noise floor of a real amplifier (ripple excluded) was −94.2 dBu(EIN = −121.4 dBu) which is mostly Johnson noise from the emitter degeneration resistors and the global NFB network. The tail ripple contribution would be therefore 60 dB below the noise, where I think it is safe to neglect it.

However, the tail-source bias voltage in reality will not be perfect; it will be developed from V+, with ripple hopefully excluded. The classic method is a pair of silicon diodes; LED biasing provides excellent temperature compensation, but such accuracy in setting DC conditions is probably unnecessary. It may

Table 26.1. How decoupling improves hum rejection

	No decouple (dB)	Decoupled with 47 µF (dB)
2 diodes	−65	−87
LED	−77	−86
NFB low-beta	−74	−86
NFB high-beta	−77	−86

be desirable to bias the VAS collector current-source from the same voltage, which rules out anything above a Volt or two. A 10 V Zener might be appropriate for biasing the input pair tail-source (given suitable precautions against Zener noise) but this would seriously curtail the positive VAS voltage swing.

The negative feedback biasing system used in some designs provides a better basic PSRR than diodes, at the cost of some beta-dependence. It is not quite as good as an LED, but the lower voltage generated is more suitable for biasing a VAS source. These differences become academic if the bias chain mid-point is filtered with 47 µF to V+, as Table 26.1 shows; this is C11 in Figure 26.7.

As another example, the amplifier in Figure 26.7 with diode-biasing and no bias chain filtering gives an output ripple of −74 dBu; with 47 µF filtering this improves to −92 dBu, and 220 µF drops the reading into limbo below the noise floor.

Figure 26.8 shows PSpice simulation of Figure 26.7, with a 0 dB sinewave superimposed on V+ only. A large Cdecouple (such as 100 µF) improves LF PSRR by about 20 dB, which should drop the residual ripple below the noise. However, there remains another frequency-insensitive mechanism at about −70 dB. The study of PSRR greatly resembles the peeling of onions, because there is layer after layer, and even strong men are reduced to tears. There also remains an HF injection route, starting at about 100 kHz in Figure 26.9, which is quite unaffected by the bias-chain decoupling.

Rather than digging deeper into the precise mechanisms of the next layer, it is simplest to RC filter the V+ supply to the input pair only (it makes very little difference if the VAS source is decoupled or not) as a few Volts lost here are of no consequence. Figure 26.9 shows the very beneficial effect of this at middle frequencies, where the ear is most sensitive to ripple components. I have never yet found it necessary to use V+ supply filtering in practical designs.

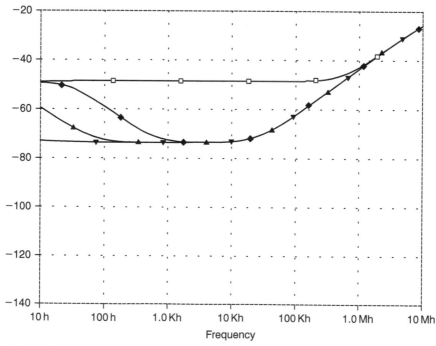

Figure 26.8. Positive-rail rejection; decoupling the tail current-source bias chain R21, R22 with 0, 1, 10 and 100 µF.

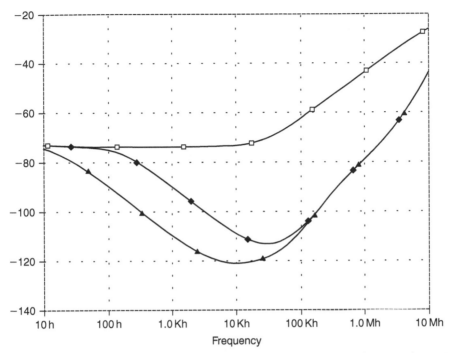

Figure 26.9. Positive-rail rejection; with input-stage supply-rail RC filtered with 100 Ω and 0, 10 and 100 μF. Same scale as Figure 26.8.

To summarise the main mechanisms of V+ PSRR in the LF region:

Imperfect bias generator for the input pair and VAS current-sources	Figure 26.8
Imperfect filtering of supply to bias generator for current-sources	Figure 26.8

Negative Supply-rail Rejection

The V- rail is the major route for injection, and a tough nut to analyse. The well-tried wolf-fence approach is to divide the problem in half, and in this case, the Fence is erected by applying RC filtering to the small-signal section (i.e., input current-mirror and VAS emitter) leaving the unity-gain output stage fully exposed to rail ripple. The output ripple promptly disappears, indicating that our wolf is getting in via the VAS or the bottom of the input pair, or both, and the output stage is effectively immune. We can do no more fencing of this kind, for the mirror has to be at the same DC potential as the VAS. SPICE simulation of the amplifier with

a 1 V (0 dBV) AC signal on V- gives the PSRR curves in Figure 26.10, with Cdom stepped in value. As before there are two regimes, one flat at -50 dB, and one rising at 6 dB per octave, implying at least two separate injection mechanisms. This suspicion is powerfully reinforced because as Cdom is increased, the HF PSRR around 100 kHz improves to a maximum and then degrades again; i.e., there is an optimum value for Cdom at about 100 pF, indicating some sort of cancellation effect. It is just a happy coincidence that this is the usual value for compensation purposes. In the V+ case, the value of Cdom made very little difference.

A primary LF ripple injection mechanism is Early effect in the input-pair transistors, which determines the −50 dB LF floor of Curve 1 in Figure 26.10, for the standard input circuit (as per Figure 26.10 with Cdom = 100 pF).

To remove this effect, a cascode structure can be added to the input stage, as in Figure 26.11. This holds the Vce of the input pair at a constant 5 V, and gives Curve 2 in Figure 26.12. The LF floor is now 30 dB lower, although HF PSRR is slightly worse. The response to Cdom's value is now monotonic; simply a matter of more Cdom, less PSRR. This is a good

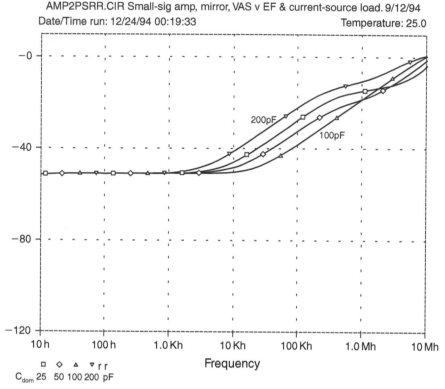

Figure 26.10. Negative-rail rejection varies with Cdom in a complex fashion; 100 pF is the optimal value. This implies some sort of cancellation effect.

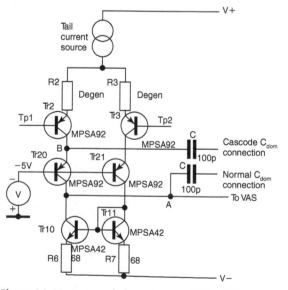

Figure 26.11. A cascoded input stage; Q21, Q22 prevent AC on V- from reaching TR2, TR3 collectors, and improve LF PSRR. B is the alternative Cdom connection point for cascode compensation.

indication that one of two partly cancelling injection mechanisms has been deactivated.

There is a deep subtlety hidden here. It is natural to assume that Early effect in the input pair is changing the signal current fed from the input stage to the VAS, but it is not so; this current is in fact completely unaltered. What *is* changed is the integrity of the feedback subtraction performed by the input pair; modulating the Vce of TR1, TR2 causes the output to alter at LF by global feedback action. Varying the amount of Early effect in TR1, TR2 by modifying VAF (Early intercept voltage) in the PSpice transistor model alters the floor height for Curve 1; the worst injection is with the lowest VAF (i.e., Vce has maximum effect on Ic) which makes sense. Because it is the feedback subtraction which is compromised, increasing the open-loop gain of the amplifier does not improve (or indeed affect at all) the LF negative rail PSRR.

We still have a LF floor, though it is now at -80 rather than −50 dB. Extensive experimentation showed that this is getting in via the collector supply of TR12, the VAS beta-enhancer, modulating Vce and adding a signal to

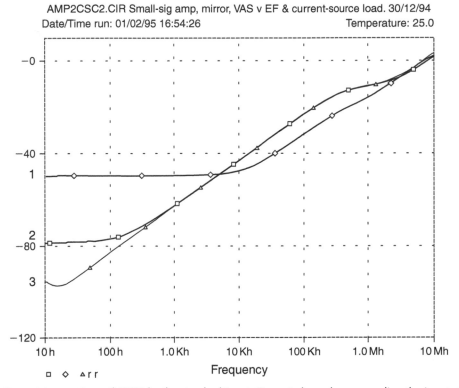

AMP2CSC2.CIR Small-sig amp, mirror, VAS v EF & current-source load. 30/12/94
Date/Time run: 01/02/95 16:54:26   Temperature: 25.0

Figure 26.12. Curve 1 is negative-rail PSRR for the standard input. Curve 2 shows how cascoding the input stage improves rail rejection. Curve 3 shows further improvement by also decoupling TR1 2 collector to V−.

the inner VAS loop by Early effect once more. This is easily squished by decoupling TR12 collector to V−, and the LF floor drops to about -95 dB, where I think we can leave it for the time being (Curve 3 in Figure 26.12).

Having peeled two layers from the LF PSRR onion, something needs to be done about the rising injection with frequency above 100 Hz. Looking again at the amplifier schematic in Figure 26.7, the VAS immediately attracts attention as an entry route. It is often glibly stated that such stages suffer from ripple fed in directly through Cdom, which certainly looks a prime suspect, connected as it is from V- to the VAS collector. However, this bald statement is untrue. In simulation it is possible to insert an ideal unity-gain buffer between the VAS collector and Cdom, without stability problems (A1 in Figure 26.13) and this absolutely prevents direct signal flow from V- to VAS collector through Cdom; the PSRR is completely unchanged.

Cdom has been eliminated as a direct conduit for ripple injection, but the PSRR remains very sensitive to its value. In fact, the NFB factor available is the determining factor in suppressing V- ripple-injection,

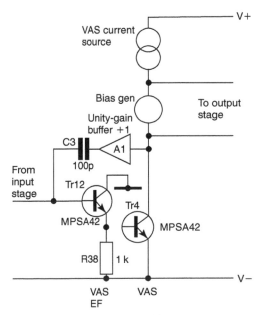

Figure 26.13. Adding a Cdom buffer A1 to prevent any possibility of signal entering directly from the V- rail.

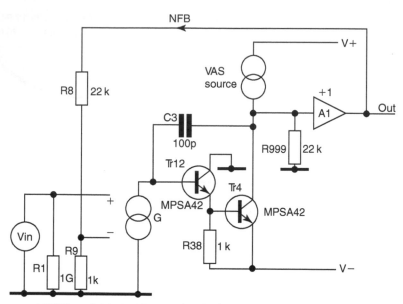

Figure 26.14. A conceptual SPICE model for V- PSRR, with only the VAS made from real components. R999 represents VAS loading.

and the two quantities are often numerically equal across the audio band.

The conventional amplifier architecture we are examining inevitably has the VAS sitting on one supply-rail; full voltage swing would otherwise be impossible. Therefore the VAS input must be referenced to V-, and it is very likely that this change-of-reference from ground to V- is the basic source of injection. At first sight, it is hard to work out just what the VAS collector signal *is* referenced to, since this circuit node consists of two transistor collectors facing each other, with nothing to determine where it sits; the answer is that the global NFB references it to ground.

Consider an amplifier reduced to the conceptual model in Figure 26.14, with a real VAS combined with a perfect transconductance stage G, and unity-gain buffer A1. The VAS beta-enhancer TR12 must be included, as it proves to have a powerful effect on LF PSRR.

To start with, the global NFB is temporarily removed, and a DC input voltage is critically set to keep the amplifier in the active region (an easy trick in simulation). As frequency increases, the local NFB through Cdom becomes steadily more effective, and the impedance at the VAS collector falls. Therefore the VAS collector becomes more and more closely bound to the AC on V-, until at a sufficiently high frequency (typically 10 kHz) the PSRR converges on 0 dB, and everything on the V- rail couples straight through at unity gain, as shown in Figure 26.15.

There is an extra complication here; the TR12/TR4 combination actually shows *gain* from V- to the output at low frequencies; this is due to Early effect, mostly in TR12. If TR12 was omitted, the LF open-loop gain drops to about −6dB.

Reconnecting the global NFB, Figure 26.16 shows a good emulation of the PSRR for the complete amplifier of Figure 26.14. The 10−15 dB open-loop-gain is flattened out by the global NFB, and no trace of it can be seen in Figure 26.16.

Now the NFB attempts to determine the amplifier output via the VAS collector, and if this control was perfect, the PSRR would be flat at the LF value. It is not, because when Cdom takes effect, the NFB factor begins to falls with rising frequency, so PSRR deteriorates at exactly the same rate as the open-loop gain falls. This can be seen on many opamp spec sheets, where the V- PSRR falls off from the dominant-pole frequency, assuming conventional opamp design with a VAS on the V- rail.

The open-loop gain of the amplifier can also be altered by changing the amount of emitter degeneration in the input stage. Increasing these resistors reduces the PSRR in the HF region but does not affect the LF region. The PSRR here is proportional to the square of the open-loop gain.

Clearly a high global NFB factor at LF is vital to keep out V- disturbances. In Chapter 7, I rather tendentiously suggested that apparent open-loop bandwidth

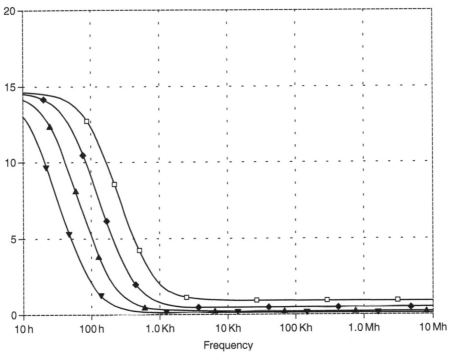

Figure 26.15. Open-loop PSRR from the model in Figure 26.14, with *C*dom value stepped. There is actually gain below 1 kHz.

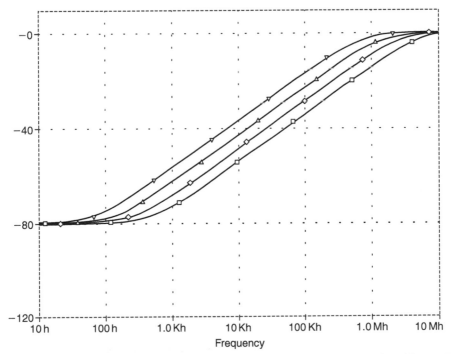

Figure 26.16. Closed-loop PSRR from Figure 26.13, with Cdom stepped to alter the closed-loop NFB factor.

Table 26.2. Effect of Rnfb value on rail ripple rejection

Rnfb	Ripple out (dBu)
None	83.3
470 k	85.0
200 k	80.1
100 k	73.9

could be extended quite remarkably (without changing the amount of NFB at HF where it matters) by reducing LF loop gain; a high-value resistor Rnfb in parallel with Cdom works the trick. What I did not say was that a high global NFB factor at LF is also invaluable for keeping the hum out; a point overlooked by those advocating low NFB factors as a matter of faith rather than reason.

Table 26.2 shows how reducing global NFB by decreasing the value of Rnfb degraded ripple rejection in a real amplifier.

Having got to the bottom of the V- PSRR mechanism, in a just world, our reward would be a new and elegant way of preventing such ripple injection. Such a method indeed exists, though I believe it has never before been applied to power amplifiers. The trick is to change the reference, as far as Cdom is concerned, to ground. Figure 26.11 shows that cascode-compensation can be implemented simply by connecting Cdom to point B rather than the usual VAS base connection at A. Figure 26.17 demonstrates that this is effective, the PSRR at 1 kHz improving by about 20 dB.

I shamelessly borrowed this technique from opamp technology; so far as I can determine it was originated by B. K. Ahuja in 1983[5, 6]. It is also called 'Indirect feedback compensation' in the literature.[7] I introduced this compensation method to the power amplifier business while designing for the late lamented TAG-McLaren Audio company. To the best of my knowledge this was the first time that had been done. I applied it to all the amplifiers I produced while I was there; it proved extremely successful and was used as one of the Unique Selling Propositions in the advertising material.

To summarise the mechanisms of V− PSRR in the LF and HF regions:

LF region

Early effect in the input transistor pair	Figure 26.10
Early effect in the EF transistor of the EF-VAS	Figure 26.12
Not affected by open-loop gain.	

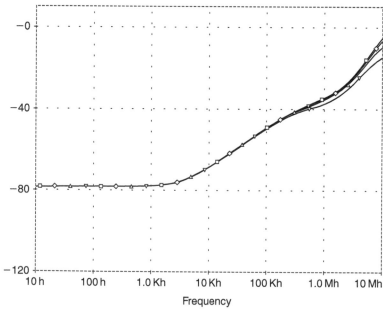

Figure 26.17. Using an input cascode to change the reference for Cdom. The LF PSRR is unchanged, but extends much higher in frequency. (Compare Curve 2 in Figure 26.12.) Note that the Cdom value now has little effect.

HF region

Change of VAS reference	Figures 26.16, 26.17, 26.18

Elegant or not, the simplest way to reduce ripple below the noise floor still seems to be brute-force RC filtering of the V- supply to the input mirror and VAS. This is effective against all the V- mechanisms listed above and they do not need to be thought about in detail. The technique may be crude, but it is highly effective, as shown in Figure 26.18. Good LF PSRR requires a large RC filtering time-constant, and the response at DC is naturally unimproved, but the real snag is that the necessary voltage drop across R directly reduces amplifier output swing, and since the magic number of watts available depends on voltage squared, it can make a surprising difference to the raw commercial numbers (though not, of course, to perceived loudness).

With the circuit values shown, 10 Ω is about the maximum tolerable value; even this, gives a measurable reduction in output. The accompanying C should be at least 220 μF, and a higher value is desirable if every trace of ripple is to be removed.

Negative Sub-rails

A complete solution to this problem, which prevents ripple intruding from the negative supply rail without compromising the output voltage swing is the use of a separate sub-rail to power the small-signal stages. This is arranged to be about 3 Volts below the main heavy-current V- supply rail, being supplied from an extra tap on the mains transformer secondary winding, via an extra rectifier and a reservoir capacitor, as in Figure 26.19 which shows a typical power supply for an amplifier or amplifiers giving about 90 W into 8 Ω.

Since the current required is only a few mA, half-wave rectification and a small reservoir capacitor are

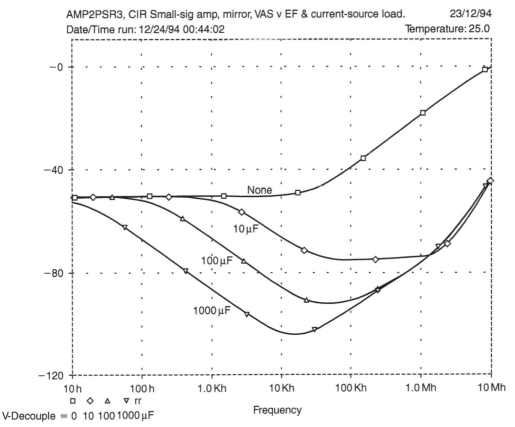

Figure 26.18. RC filtering of the V- rail is effective at medium frequencies, but less good at LF, even with 1000 μF of filtering. R = 10 Ω.

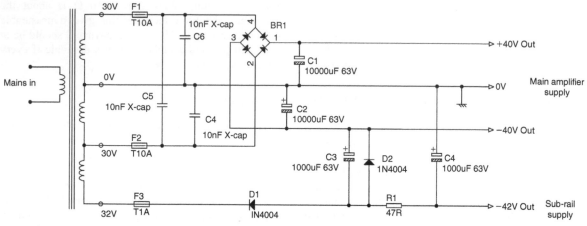

Figure 26.19. Adding a V- sub-rail supply with RC filtering to power the small-signal stages of an amplifier.

all that is required. The sub-rail is given some simple RC smoothing by R1 and C4, and the result is that ripple intrusion from the negative supply rail is below the noise floor. This stratagem may also improve VAS linearity, as the greatest curvature in its characteristic is likely to be at the negative end of its voltage swing.

The timing with which the rails come up and collapse is important, because if the sub-rail is less negative with respect to 0 V than the main V- rail, the negative side driver may be excessively reverse biased. Therefore the sub-rail reservoir C3 is connected to the V- rail, rather than ground; C4 has to be connected to ground if it is to perform its function of filtering the sub-rail. Note the clamping diode D2 which prevents C3 from becoming reverse-biased when the power is turned off, as C4 will probably discharge faster than the main rails.

References

1. Sinclair, R (ed.) *Audio and Hi-Fi Handbook*, 3rd edn, Oxford: Newnes, 2000, p. 266.

2. Linsley-Hood, J, Evolutionary Audio. Part 3, *Electronics World*, Jan. 1990, p. 18.

3. Williams, T, *EMC for Product Designers*, Oxford: Newnes Butterworth-Heinemann, 1992, p. 106.

4. Ball, G, Distorting Power Supplies, *EW+WW*, Dec. 1990, p. 1084.

5. Ribner, D B and Copeland, M A, Design Techniques for Cascoded CMOS Opamps, *IEEE J. Solid-State Circuits*, Dec. 1984, p. 919.

6. Ahuja, B K, Improved Frequency Compensation Technique for CMOS Opamps, *IEEE J. Solid-State Circuits*, Dec. 1983, pp. 629−633.

7. Saxena, V and Baker, R J, Indirect Feedback Compensation of CMOS Op-amps, *IEEE Workshop on Microelectronics and Electron Devices*, conference paper, 2006.

Power Amplifier Input Systems

Most of this book deals with the actual power amplifier itself, and its immediate ancillary circuitry such as power supplies, overload and DC offset protection, control of fan cooling, and so on. It is quite feasible to take one or more such amplifiers and put them in a box with no further complications. Such a product was the much-loved Quad 303 power amplifier, which had no controls at all unless you count the mains voltage selector. However, amplifiers are very often also provided with input circuitry that gives a balanced input, gain control, filtering, level indication, and so on. This chapter deals with these subsystems, defining the amplifier input system as any part of the signal path before the actual power amplifier stage.

Balanced interconnections are seeing increasing use in hi-fi applications, and they have always been used in the world of professional audio. Their importance is that they can render ground loops and other connection imperfections harmless. Since there is no point in making a wonderful power amplifier and then feeding it with a mangled signal, making an effective balanced input is of the first importance and I make no apology for devoting a large part of this chapter to it.

Figure 27.1 shows a typical input system; many variations on this are of course possible. Firstly, RF filtering is applied at the very front end to prevent noise break-through and other EMC problems. The filtering must be done before the incoming signal encounters any semiconductors where RF demodulation could occur, and can be regarded as a 'roofing filter'. At the same time, the bandwidth at the low end is given an early limit by the use of DC-blocking capacitors, and over-voltage spikes are clamped by diodes. The input amplifier presents a reasonably high impedance to the outside world and almost invariably in professional amplifiers, and increasingly in hi-fi amplifiers, it is balanced so noise produced by ground loops and the like can be

rejected. Sometimes the input is connected directly to a line output so that amplifiers can be daisy-chained together; this is much more economical of cable than having multiple fan-out cables each running from the source to one amplifier; the downside is that the failure of one amplifier in the chain can affect the feed to all of them. While the bandwidth of the amplifier has been very roughly circumscribed by the RF filtering and DC-blocking, it is usual to define it more precisely by the use of a subsonic filter, and less commonly, an ultrasonic filter. These very often come next in the signal path so they remove unwanted signals as soon as possible, and can benefit from a low-impedance drive from the input amplifier. After this comes the gain control, if there is one, and whatever buffering arrangements may be needed to drive the actual power amplifier stage.

If only a limited gain range is required, which is often the case, it is sometimes possible to combine it with the input amplifier in an active-gain-control format, which has advantages in minimising noise and maximising headroom in the input system. If a wide gain range is desirable, the gain control is almost always after the balanced input amplifier because making balanced inputs that retain good common-mode rejection when their gain is varied over a wide range is not so simple. More on that later …

External Signal Levels

There are several standards for line signal levels. The −10 dBV standard is used for a lot of semi-professional recording equipment as it gives more head-room with unbalanced connections − the professional levels of +4 dBu and +6 dBu assume a balanced output which inherently gives twice the output level for

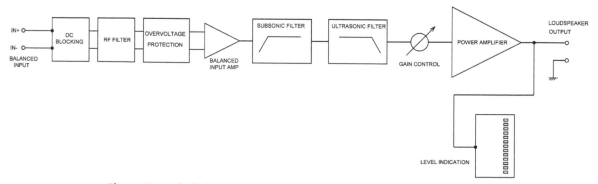

Figure 27.1. Block diagram of a comprehensive power amplifier input system.

Table 27.1. Nominal signal levels

	V rms	dBu	dBV
Semi-professional	0.316	−7.78	−10
Professional	1.228	+4.0	+1.78
German ARD	1.55	+6.0	+3.78

the same supply rails as it is measured between two pins with signals of opposite phase on them. See Table 27.1.

Signal levels in dBu are expressed with reference to 0 dBu = 775 mV rms; the origin of this odd value is that it gives a power of 1 mW in a purely historical 600 Ω load. The unit of dBm refers to the same level but takes the power rather the voltage as the reference − a distinction of little interest nowadays. Signal in dBV (or dBV) are expressed with reference to 0 dB = 1.000 V rms.

These standards are well established, but that does not mean all amplifiers adhere to them and will give full output for a +4 or +6 dBu output. To take just one current example, the Yamaha P7000S requires +8 dBu (1.95 Vrms) to give its full output of 750 W into 8 Ω.

In the hi-fi world there is little consensus on the input sensitivity of hi-fi power amplifiers, The input required for full output can range from to 0 to +10 dBu or more. Many pieces of equipment, including preamplifiers and power amplifiers designed to work together, have both unbalanced and balanced inputs and outputs. The consensus appears to be that if the unbalanced output is, say, 1 Vrms, then the balanced output will be created simply by feeding the in-phase output to the hot output pin, and also to a unity-gain inverting stage, which drives the cold output pin with 1 Vrms phase-inverted. The total balanced output voltage is therefore 2 Vrms, and so the balanced input must have a gain of ½ or −6 dB relative to the unbalanced input to maintain consistent signal levels.

Internal Signal Levels

It is necessary to select a suitable nominal level for the signal passing through the input system. It is always a compromise − the signal level should be high so it suffers the minimum of degradation by the addition of noise as it passes through the circuitry, but not so high that it is likely to suffer clipping before the gain control. It has to be considered that the gain control may be maladjusted by setting it too low and turning up the input level from the source equipment, making input clipping more likely. The levels chosen are usually in the range 388 mV to 775 Vrms (−6 to 0 dBu). Since the maximum

output swing of an opamp is around 9 Vrms, this gives from 27 to 19 dB of headroom before clipping. If there is no gain control, then headroom in the input circuitry is not an issue. The power amplifier will always clip before the input circuitry, even though the latter is running off supply rails of only ±15 V.

In deciding the gain structure of the overall amplifier, it is wise to consider the needs of the power amplifier stage first. Making a linear Class-B power amplifier is always a challenge, and the only sensible way to meet it is to use as much negative feedback as can be safely applied without risking instability. This tends to result in a power amplifier stage gain in the region of 20 to 30 times. Thus a powerful amplifier delivering 500 W into 8 Ω, which is an output voltage of 63 Vrms, will need an input voltage of between 2.1 Vrms (+8.7 dBu) and 3.15 Vrms (+12.2 dBu). These levels are well above any standard for equipment interfacing and at least 6 dB of gain will have to be provided by the input system, the exact amount depending on what you want the sensitivity of your amplifier input to be. This sort of gain can be provided by opamps without adding significant distortion, and this is the best way to do it. The gain of a power amplifier stage can be increased by reducing the negative feedback ratio, and reducing the size of the compensation capacitor in proportion, to keep the feedback factor the same, but this process does seem to degrade the distortion performance by a certain amount. This is probably because the smaller compensation capacitor means there is less local feedback around the VAS.

If the incoming signal does have to be amplified, this should be done as early as possible in the signal path, to get the signal well above the noise floor as quickly as possible. If the gain is implemented in the first stage (i.e., the input amplifier, balanced or otherwise), the signal will be able to pass through later stages, such as filters, at a high level and so their noise contribution will be less significant.

Unbalanced Inputs

The simplest unbalanced input feeds the incoming signal directly to the input of the power amplifier. This is not normal practice as at least some RF filtering will be required for EMC purposes. In addition, the power amplifier input impedance may be deliberately kept low, to 2 kΩ or less, so that offset voltages and noise generated in the feedback network are minimised (this is described in detail in Chapter 6; in this case, a buffer amplifier will be needed). Figure 27.2 shows an unbalanced input amplifier, with the added

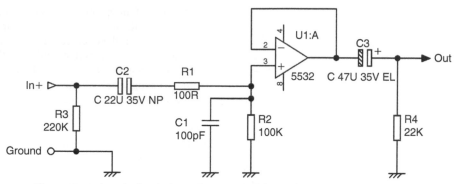

Figure 27.2. A typical unbalanced input amplifier with associated components.

components needed for interfacing to the real world. The opamp acts as a unity-gain buffer or voltage follower; if you need gain just add two feedback resistors. A 5532 bipolar type is used here to keep the noise down; with the relatively low source impedances that are most likely to be encountered here, an FET-input opamp would be 10 dB or more noisier. R1 and C1 make a first-order low pass filter to remove incoming RF before it has a chance to reach the opamp and demodulate into the audio band; once this has occurred, any further attempts at RF filtering are of course useless. R1 and C1 must be as close to the input socket as physically possible to prevent RF from being radiated into the equipment enclosure before it is shunted to ground.

There is of course an inherent problem in selecting component values for input filters of this sort: we don't know what the output impedance of the source equipment is. If the source is an active preamp stage, then the output impedance will probably be around 50 Ω, but it could be as high as 200 Ω or more. If a so-called 'passive preamplifier' is in use, i.e., just an input selector switch and a volume control potentiometer, then the output impedance will almost certainly be a good deal higher. (At least one passive preamplifier uses a transformer with switched taps for volume control, but I think analysing that might be too much of a digression just at the moment.) If you really feel you want to use such a doubtful piece of equipment, then a sensible potentiometer value is 10 kΩ, and its maximum output impedance (when it is set for 6 dB of attenuation) will be 2.5 kΩ, which is very different from 50 Ω. This resistance is in series with R1 and affects the turnover frequency of the RF filter. While it is very desirable to have effective RF filtering, it is also important to avoid a frequency response that sags significantly at 20 kHz.

If we take 2.5 kΩ as a worst-case source impedance and add the 100 Ω of R1, then 2.6 kΩ and 100 pF together give us −3 dB at 612 kHz; this gives a loss at 20 kHz of only 0.0046 dB, so possibly C1 could be usefully increased; for example, if we made it 220 pF, then the 20 kHz loss is still only 0.022 dB. If we stick with C1 at 100 pF and assume an active output with a 50 Ω impedance in the source equipment, then together with the 100 Ω resistance of R1 we have a total of 150 Ω, and 150 Ω with 100 pF gives −3 dB at 10.6 MHz.

This all seems very sensible and satisfactory, until you take a quick look at the sort of potentiometer values that passive preamplifiers really employ. I did a rapid survey, and while 10 kΩ seems to be a popular value, I quickly found one model with a 20 kΩ potentiometer, and another that had a value as high as 100 kΩ. The latter would have a maximum output impedance of 25 kΩ, and would give very different results with a C1 value of 100 pF − the worst-case frequency response would now be −3 dB at 63.4 kHz and −0.41 dB at 20 kHz, which is not good.

To put this into perspective, C1 is almost certainly going to be smaller than the capacitance of the interconnecting cable. Audio cable capacitance is usually in the range 50 to 150 pF/metre, so with a 2.5 kΩ source impedance, you can only permit yourself a rather short run before there are significant effects on the frequency response; with a 25 kΩ source impedance, you can hardly afford to have any cable at all. This is just one reason why 'passive preamplifiers' are not a good idea.

Another important constraint is that the series resistance R1 must be kept as low as practicable to minimise the generation of Johnson noise; but lowering this resistance means increasing the value of shunt capacitor C1, and if it becomes too big, then its impedance at high audio frequencies will become too low. This can have two bad effects: too low a roll-off frequency if the input is connected to a source with a high output impedance, and a possible increase in distortion at high audio frequencies because of excessive loading on the source output stage.

Replacing R1 with a small inductor will give significantly better filtering but at increased cost. This is usually justifiable in professional audio equipment, but it is much less common in hi-fi. If you do use inductors, then it is essential to check the frequency response to make sure the LC circuit is well-damped and not peaking at the turnover frequency.

C2 is a DC-blocking capacitor to prevent voltages from ill-thought-out source equipment from getting into the circuitry. Note that it is a non-polarised type as voltages from the outside world are of unpredictable polarity. It is rated at 35 V so that even if it gets connected to defective equipment with an opamp output jammed against one of the supply rails, no harm will come to our input circuit. R2 is a DC drain resistor that prevents the charge put on C2 by aforesaid doubtful external equipment from remaining there for a long time and causing a thud when the connections are altered; as with all such drain resistors, its value is a compromise between draining the charge off the capacitor reasonably quickly, and keeping the input impedance suitably high. The input impedance of the input circuit is R2 in parallel with R3, i.e., 220 kΩ in parallel with 100 kΩ, which comes to 68 kΩ. This is a nice high value and should work well with just about any source equipment you can find.

Because of the presence of C2, R3 is required to provide biasing for the opamp input; it needs to be quite a high value to keep the input impedance up, and bipolar input opamps draw significant input bias current. The Fairchild 5532 data sheet quotes 200 nA typical, and 800 nA maximum, and these currents would give a voltage drop across R2 of 20 mV and 80 mV respectively. This offset voltage will be faithfully reproduced at the output of the buffer, with the opamp input offset voltage added on; this is only 4 mV maximum and so will not affect the final voltage much, whatever its polarity. The 5532 has NPN input transistors, and so the bias current flows into the input pins, and the voltage at Pin 3, and hence the output will therefore be negative with respect to ground.

Such DC voltages are big enough to generate unpleasant thumps and bumps if the input stage is followed by any sort of switching, so further DC blocking is required in the shape of C3; R4 is another DC drain resistor to keep the output at zero volts. It can be made rather lower in value than the input drain resistor R3 as the only requirement is that it should not significantly

load the opamp output. FET-input opamps have much lower input bias currents, so that the offsets they generate as they flow through biasing resistors are usually negligible, but they still have input offsets of a few milliVolts, so DC blocking is still required if switches or relays downstream are to act silently.

With appropriate use of blocking capacitors, DC offsets should not cause trouble upstream or downstream, but you need to make sure that the offset voltages are not so great that the output voltage swing of the opamp is significantly affected. The effect of our worst-case 80 mV offset here is trivial.

Balanced Interconnections

Balanced inputs on power amplifiers are used to prevent noise and crosstalk from affecting the input signal, especially in applications where long interconnections are used. They are standard on professional amplification equipment, and are steadily becoming more common in the world of hi-fi. A balanced input amplifier is sometimes called a line receiver. The basic principle of balanced interconnection is to get the signal you want by subtraction, using a three-wire connection. In some cases a balanced input is driven by a balanced output, with two anti-phase output signals; one signal wire (the hot or in-phase) sensing the in-phase output of the sending unit, while the other senses the anti-phase output.

In other cases, when a balanced input is driven by an unbalanced output, as shown in Figure 27.3, one signal wire (the hot or in-phase) senses the single output of the sending unit, while the other (the cold or phase-inverted) senses the unit's output-socket ground, and once again the difference between them gives the wanted signal. In either of these two cases, any noise voltages that appear identically on both lines (i.e., common-mode signals) are in theory completely cancelled by the subtraction. In real life the subtraction falls short of perfection, as the gains via the hot and cold inputs will not be precisely the same, and the degree of discrimination actually achieved is called the Common-Mode Rejection Ratio or CMRR. There will be more on this very important quantity later.

It is tedious to keep referring to non-inverting and inverting inputs, and so these are usually abbreviated to 'hot' and 'cold' respectively, though this does not necessarily mean that the hot terminal carries more signal voltage than the cold one. For a true balanced connection, the voltages will be equal. The 'hot' and 'cold' terminals are also often referred to as IN+ and IN−, and this latter convention has been followed in the diagrams here.

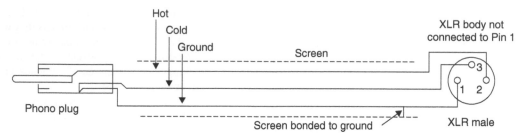

Figure 27.3. Unbalanced output to balanced input interconnection.

The subject of balanced interconnections is a large and subtle one, and a big fat book could be written on this topic alone. A classic paper on the subject is Muncy.[1] To keep it to a reasonable length, this section has to concentrate on the areas most relevant to power amplifier interconnection.

The advantages of balanced interconnects are:

- Balanced interconnections discriminate against noise and crosstalk, whether they result from ground currents, or electrostatic or magnetic coupling to signal conductors.
- Balanced connections make ground loops much less intrusive, and usually inaudible, so people are less tempted to start 'lifting grounds' to break the loop. This tactic is only acceptable if the equipment has a dedicated ground-lift switch that leaves the external metalwork firmly connected to mains safety earth. In the absence of this facility, the optimistic will remove the mains earth (not quite so easy now that moulded mains plugs are standard), and this practice is of course dangerous, as a short-circuit from mains to the equipment chassis will result in live metalwork.
- A balanced interconnection incorporating a true balanced output gives 6 dB more signal level on the line, which should give 6 dB more dynamic range. This is true with respect to external or ground noise, but also with respect to the balanced input amplifier noise. As the section below describes, a standard balanced input is usually much noisier than an unbalanced input.
- Balanced connections are usually made with XLR connectors. These are a professional 3-pin format, and are a much superior connector to the phono (RCA) type normally used for unbalanced connections. More on this below.

The disadvantages of balanced interconnects are:

- Balanced inputs are inherently noisier than unbalanced inputs by a large margin, in terms of the noise generated by the input circuitry itself rather than external noise. This may appear paradoxical but it is all too true, and the reasons will be fully explained in this chapter.

- More hardware means more cost. Small-signal electronics is relatively cheap; unless you are using a sophisticated low-noise input stage, of which more later, most of the extra cost is likely to be in the balanced input connectors.
- Balanced connections may not provide much protection against RF intrusion — both legs of the balanced input would have to demodulate the RF in equal measure for common-mode cancellation to occur. This is not very likely, and it is important to provide the usual input RF filtering to avoid EMC difficulties.
- There are more possibilities for error when wiring up. For example, it is easy to introduce an unwanted phase inversion by confusing hot and cold in a connector, and this can go undiscovered for some time. The same mistake on an unbalanced system interrupts the audio completely.

Balanced Connectors

Balanced connections are most commonly made with XLR connectors. These are a professional 3-pin format, and are a much better connector in every way than the usual phono (RCA) type. Phono connectors have the great disadvantage that if you are connecting them with the system active (inadvisable, but people are always doing inadvisable things), the signal contacts meet before the grounds and thunderous noises result. XLRs are wired with Pin 2 as hot, Pin 3 as cold, and the Pin 1 as ground.

The main alternative to the XLR is the stereo jack plug. These are often used for line level signals in a recording environment, and are frequently found on the rear of professional power amplifiers as an alternative to an adjacent XLR connector. Jack sockets can be obtained with switching contacts that can be used to disable the XLR input to prevent the intrusion of noise. Balanced jacks are wired with the tip as hot, the ring as cold, and the sleeve as ground.

Big sound reinforcement systems often use large multiway connectors that carry dozens of 3-wire balanced connections.

Balanced Inputs: Electronic vs Transformer

Balanced interconnections can be made using either transformer or electronic balancing. Electronic balancing has many advantages, such as low cost, low size and weight, superior frequency and transient response, and no problems with low-frequency linearity. While it is still sometimes regarded as a second-best solution, the performance is more than adequate for most professional applications. Transformer balancing has some advantages of its own, particularly for work in very hostile RF/EMC environments, but serious drawbacks. The advantages are that transformers are electrically bullet-proof, retain their high CMRR performance forever, and consume no power even at high signal levels. Unfortunately they also generate LF distortion, particularly if they have been made as small as possible to save weight and cost. They tend to have HF response problems due to leakage reactance and distributed capacitance, and inevitably they are heavy and expensive compared with any electronic input. The first two objections can be surmounted with extra electronic circuitry, but the last two cannot. Transformer balancing is therefore relatively rare, even in professional audio applications, and most of this chapter deals with electronically-balanced inputs.

Balanced Inputs and their Common-Mode Rejection Ratio

The Common Mode Rejection Ratio or CMRR is a measure of how well a balanced interconnection does its job. It is the difference between the differential and common-mode gain, and is usually measured in dB.

A vital point is that even if you have a balanced input with a superb CMRR when measured alone, the CMRR of the whole balanced interconnection, consisting of output − cable − balanced-input may be significantly worse. This is because the output impedances feeding the interconnection have an important effect.

Figure 27.4 shows a balanced interconnection reduced to its bare essentials; two source resistances and a standard differential amplifier. The balanced output in the source equipment is assumed to have two exactly equal output resistances Rout+, Rout−, and the balanced input in the receiving equipment has two exactly equal input resistances R1, R2. The balanced input amplifier senses the voltage difference between the points marked IN+ (hot) and IN− (cold) and ideally completely ignores common-mode voltages which are present on both. Suppose a differential voltage input between IN+ and IN− gives an output voltage of 0 dB; then reconnect the input so that IN+ and IN− are joined together and the same voltage is applied between the two of them and ground. Ideally the result would be zero output, but in this imperfect world it won't be, and in real life the output could be anywhere between −20 dB (for a bad balanced interconnection) and −140 dB (for a very good one). This figure is the CMRR. In one respect, balanced audio connections have it easy. The common-mode signal is normally well below the level of the unwanted signal, and so the common-mode range of the input is not an issue.

In Figure 27.4, the differential voltage sources Vout+, Vout− which represent the actual balanced output are set to zero, and Vcm, which represents the common-mode voltage drop down the cable ground, is set to 1 Volt to give a convenient result in dBV. The

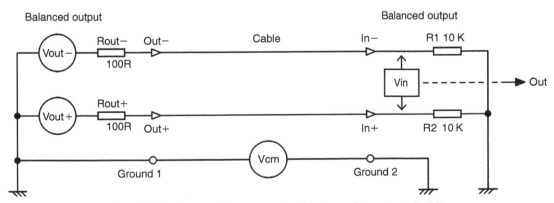

Figure 27.4. Balanced interconnection showing influences on CMRR.

output resulting from the presence of this voltage source is measured by a mathematical subtraction of the voltages at IN+ and IN− so there is no actual input amplifier to confuse the results with its non-ideal CMRR performance.

Let us start with Rout+, Rout− = 100 Ω and R1, R2 = 10 k, which are plausible values and nice round figures. When all four resistances are exactly at their nominal value, the CMRR is infinite, which on my SPICE simulator rather interestingly appears as exactly −400 dB. If one of the output resistors or one of the input resistors is then altered in value by 1%, then the CMRR drops like a stone to −80 dB. If the deviation is 10%, things are predictably worse and the CMRR degrades to −60 dB.

The essence of the problem is that we have two resistive dividers, and we want them to have exactly the same attenuation. If we increase the ratio between the output and input resistors, by reducing the former or increasing the latter, the attenuation gets closer to unity and variations in either resistor have less effect on it. If we increase the input impedance to 100 kΩ, things get ten times better, as the Rin/Rout ratio has improved from 100 to 1000 times. We now get −100 dB for a 1% resistance deviation, and −80 dB for a 10% deviation. An even higher input impedance of 1 MΩ raises Rin/Rout to 10,000 and gives −120 dB for a 1% resistance deviation, and −100 dB for a 10% deviation.

In practical circuits, the combination of 68 Ω output resistors and a 20 kΩ input impedance is often encountered; the 68 Ω resistors are about as low as you want to go with conventional circuitry, to avoid HF instability. The 20 kΩ input impedance is what you get if you make a basic balanced input amplifier with four 10 kΩ resistors. I strongly suspect that that value is chosen because it looks as if it gives standard 10 kΩ input impedances − in fact, it does nothing of the sort, and the common-mode input impedance, which is what matters here, is 20 kΩ on each leg; more on that later. It turns out that 68 Ω output resistors and a 20 kΩ input impedance give a CMRR of −89.5 dB for a 1% deviation; much better than you will see in practice.

The conclusion is simple; we want to have the lowest possible output impedances and the highest possible input impedances to get the maximum common-mode rejection This is highly convenient because low output impedances are already needed to drive multiple amplifier inputs and cable capacitance, and high input impedances are needed to minimise loading and maximise the number of amplifiers that can be driven.

The worst situation occurs when an unbalanced output is driving a balanced input. In this case the cold input may be connected directly to ground, as in Figure 27.3 above. With a 100 Ω output resistance and 10 kΩ input impedance, the CMRR collapses to −40.2 dB; with 68 Ω output resistors and a 20 kΩ input impedance it only falls to −49.4 dB, demonstrating again the importance of keeping a high ratio between the output and input resistors. To prevent this CMRR degradation, the connection to ground should be via the same impedance as that of the unbalanced output. This is called an impedance-balanced output, and is described more fully at the end of this chapter; note that only the impedances are balanced; the signal is unbalanced.

The Basic Balanced Input

Figure 27.5 shows the basic balanced input amplifier using a single opamp. To make it balanced, R1 must

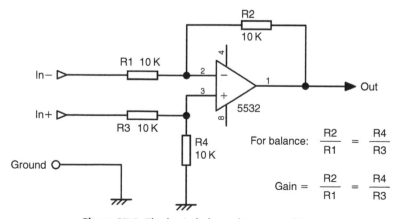

Figure 27.5. The basic balanced input amplifier.

Table 27.2. The input impedances for different input drive conditions

Case	Pins driven	Hot input res	Cold input res
1	Hot only	20k	Grounded
2	Cold only	Grounded	10k
3	Both (balanced)	20k	6.66k
4	Both common-mode	20k	20k
5	Both floating	10k	10k

Table 27.3. How resistor tolerances affect the CMRR with realistic opamp o/l gain

R1	R1 deviation (%)	Gain x	CMRR dB
10k	0	100,000	−94.0
10.0001k	0.001	100,000	−96.5
10.001k	0.01	100,000	−90.6
10.01k	0.1	100,000	−66.5
10.1k	1	100,000	−46.2
11k	10	100,000	−26.6
10k	0	1,000,000	−114.1
10.0001k	0.001	1,000,000	−110.5
10.001k	0.01	1,000,000	−86.5
10.01k	0.1	1,000,000	−66.2
10.1k	1	1,000,000	−46.2
11k	10	1,000,000	−26.6

be equal to R3 and R2 equal to R4. The gain is R2/R1 (= R4/R3). The standard one-opamp balanced input or differential amplifier is a very familiar circuit block, but its operation often appears somewhat mysterious. Its input impedances are *not* equal when it is driven from a balanced output; this has often been commented on,[2] and some confusion has resulted.

The root of the problem is that a simple differential amplifier has interaction between the two inputs, so that the input impedance on the cold input depends strongly on the signal applied to the hot input. Since the only way to measure input impedance is to apply a signal and see how much current flows into the input, it follows that the apparent input impedance on each leg varies according to the way the inputs are driven. If the amplifier is made with four 10 K resistors, then the input impedances Z are as in Table 27.2.

Some of these impedances are not exactly what you would expect; a full explanation is given in my book, *Small Signal Audio Design*.[3]

Practical Common-Mode Rejection

We saw earlier that to get a good CMRR the first thing to do is get the overall plan of the balanced interconnection right by having low and balanced output impedances at the sending end, and high and balanced input impedances at the receiving end. It is now time to look at the CMRR of the balanced input stage itself. There are two main influences on this: the opamp properties, and the accuracy of the resistors within the input amplifier.

Even if perfectly matched resistors are assumed, the CMRR of an opamp balanced input stage will not be infinite, because the mathematics of the subtraction assume that the two opamp inputs are at exactly the same voltage. This is an approximation because it depends on the efficacy of the negative feedback, and hence on the open-loop gain of the opamp, and that is neither infinite, nor flat with frequency up to the

ultra-violet. CMRR will fall off at high frequencies. However, in general, the influence of the opamp properties on the CMRR is usually small. The truth of the matter is that in real life the most important effect on the CMRR is the accuracy of the resistor values in the input amplifier itself.

Table 27.3 shows the results of SPICE simulation for various inaccuracies in *one* of the four resistors. The results are shown for opamp LF open-loop gains of 100,000 and 1,000,000, but the effect of finite opamp bandwidth has not been taken into account. R1 is varied while R2, R3 and R4 are kept at exactly 10 kΩ. The balance output impedance is taken as exactly 100 Ω in each leg.

The results show that our previous calculations, which took only output and input impedances into account, were actually highly optimistic. If a 1% tolerance resistor is used for R1 (and at the time of writing the Sixth Edition, there really is no financial incentive to use anything sloppier), the CMRR is dragged down to −46 dB; the same results apply to varying any other one of the four resistances. If you can run to a 0.1% tolerance for these components, the CMRR is a rather better −66 dB; however, such resistors are not ten times more expensive than 1% types — the factor is more like a hundred. It is clear that opamp gain has no significant effect when you are using resistors of practical accuracy.

In real life — a phrase that keeps creeping in here, and shows how many factors affect a practical balanced interconnection — all four resistors will of course be subject to a tolerance, and a more realistic calculation would produce a statistical distribution of CMRR rather than a single figure. One method is to use the Monte Carlo function in SPICE, which runs multiple simulations with random component variations and collates the results. However you do it, you must know (or assume) how the resistor values are distributed within their tolerance window. Usually you don't know, the manufacturer can't or won't tell you, and finding out by measuring hundreds of resistors is not a task that appeals to all of us.

It is, however, easy to calculate the worst-case CMRR, which occurs when all resistors are at their limit of tolerance in the most unfavourable direction. The CMRR in dB is then:

$$CMRR = 20 \log \left(\frac{1 + R2/R1}{4T/100} \right) \qquad \text{Equation 27.1}$$

Where R1 and R2 are as in Figure 27.4, and T is the tolerance in percentage.

This tells us that 1% resistors give a worst-case CMRR of only 34.0 dB, that 0.5% parts give only 40.0 dB and expensive 0.1% parts yield but 54.0 dB. Things are, however, nothing like as bad in practice, as the chance of all four resistors being as wrong as possible is actually vanishingly small. I have measured the CMRR of more 1% resistor balanced inputs than I care to contemplate, but I do not ever recall that I ever saw one with an LF CMRR worse than 40 dB.

Errors in the resistors around the input amplifier have much more effect on the CMRR than do imbalances in the output resistance/input impedance system that we looked at earlier; this is because in a well-designed interconnection the output resistances and input impedances are of a different order of magnitude, whereas the amplifier resistors are, if not of the same value, then of the same order. If you are designing both ends of a balanced interconnection and you are spending money on precision resistors, you should put them in the input amplifier, not the balanced output.

There are 8-pin SIL packages that offer four resistors that should have good matching; be wary of these as they usually contain thick-film resistive elements that are not perfectly linear. In a test I did, a 10 k resistor with 10 Vrms across it generated 0.0010% distortion. In the search for perfect audio, resistors that do not stick to Ohm's Law are not a good start.

It has to be said at this point that simple balanced input amplifiers with four 1% resistors are used extensively in the audio business, and almost always prove to be up to the job in terms of their CMRR. When a better CMRR is required, one of the resistances is

made trimmable with a preset, as shown in Figure 27.6. This has to be adjusted in manufacture, but it is a quick set-and-forget adjustment that need never be touched again unless one of the four fixed resistors needs replacing, and that is extremely unlikely. CMRRs at LF of more than 70 dB can easily be obtained, but the CMRR at HF degrades due to the opamp gain roll-off and stray capacitances.

Figure 27.7 shows the CMRR measurements for a trimmable balanced input amplifier. The flat line at −50 dB was obtained from a standard non-adjustable balanced input using four 1% 10 kΩ unselected resistors, while the much better (at LF, anyway) trace going down to −85 dB was obtained from Figure 27.7 by using a multi-turn preset for PR1. Note that R4 is an E96 value so a 1 K pre-set can swing the total resistance of that arm both above and below the nominal 10 kΩ.

The CMRR is dramatically improved by more than 30 dB in the region 50−500 Hz where ground noise tends to intrude itself, and is significantly better across almost all the audio spectrum.

The upward sloping part of the trace in Figure 27.7 is partly due to the finite open-loop bandwidth of the opamp, and partly due to unbalanced circuit capacitances. The CMRR is actually worse than 50 dB above 20 kHz, due to the stray capacitances in the multi-turn preset. In practice, the value of PR1 would be smaller, and a one-turn pre-set with much less stray capacitance used. Still, I think you get the point; trimming can be both economic and effective.

The Practical Balanced Input

The simple circuit shown back in Figure 27.5 is not fit to face the world without some additional components. Figure 27.8 shows a more fully dressed version.

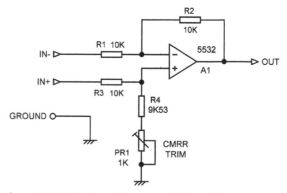

Figure 27.6. A balanced input amplifier, with preset pot to trim for best LF CMRR.

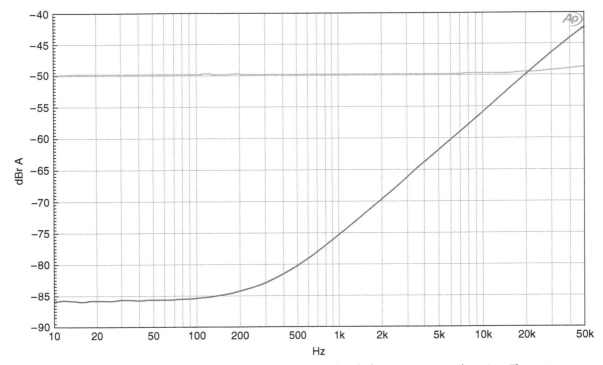

Figure 27.7. The CMRR of a standard balanced amplifier compared with the preset- trimmed version. The opamp was a 5532 and all resistors were 1%. The trimmed version gives better than 80 dB CMRR up to 500 Hz.

Firstly, and most important, C5 has been added across the feedback resistor R2; this prevents stray capacitances from Pin 2 to ground causing extra phase-shifts that lead to HF instability. The value required for stability is small, much less than that which would cause an HF roll-off anywhere near the top of the audio band. The values here of 10 kΩ and 27 pF give −3 dB at 589 kHz, and such a roll-off is

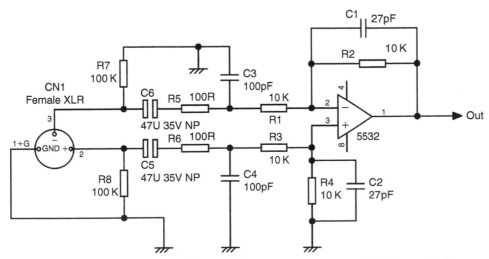

Figure 27.8. Balanced input amplifier with the extra components required for practical use.

only down by 0.005 dB at 20 kHz. C6, of equal value, must be added across R4 to maintain the balance of the amplifier, and hence its CMRR, at high frequencies.

Secondly, RF filtering is added. C5 and C6 are not particularly effective at RF rejection as C5 is not connected to ground, and there is every chance that RF will demodulate at the opamp inputs. An RF filter is therefore added to each input, in the shape of R1, C1 and R2, C2. The capacitors here will shunt incoming RF to ground before it reaches the opamp.

It was explained at some length in the previous section on unbalanced inputs that it is not at all easy to guess what the maximum source impedance will be, given the existence of so-called 'passive preamplifiers'. Such a device clearly cannot give a balanced output (unless it is fitted with a transformer), but there is no reason why it could not be used to feed a balanced input, and so it needs consideration.

As before, circuit resistances must be kept as low as practicable to minimise the generation of Johnson noise. The situation here is, however, different from the unbalanced input as there have to be resistances around the opamp, and they must be kept up to a certain value to give acceptably high input impedances; this is why a simple balanced input like this one is appreciably noisier than an unbalanced input. There is therefore more freedom in the selection of the values of R1 and R2. With the values shown, if we once more assume 50 Ω output impedances in both legs of the source equipment output, then together with the 100 Ω resistances we have a total of 150 Ω, which with 100pF gives −3 dB at 10.6 MHz.

Returning to a possible passive preamplifier with a 10 kΩ potentiometer, its maximum output impedance of 2.5 kΩ plus 100 Ω with 100 pF gives −3 dB at 612 kHz, which remains well clear of the top of the audio band.

As with the unbalanced input, replacing R1 and R2 with small inductors will give much improved RF filtering but at increased component cost. Ideally a common-mode choke (usually two bifilar windings on a small toroid core) should be used, as this improves performance. Once more, check the frequency response to make sure the LC circuits are well-damped and not peaking at the turnover frequency.

C3 and C4 are DC-blocking capacitors. Once again, they are rated at 35 V to protect the input circuit, and are non-polarised types as voltages from outside may be of either polarity. The lowest input impedance that can occur with this circuit when using 10 kΩ resistors, is, as described above, 6.66 kΩ when it is being driven in the balanced mode. The low-frequency rolloff is therefore −3 dB at 0.51 Hz. This may appear to be undesirably low, but the important point is not the roll-off but possible loss of CMRR at low frequencies due to imbalance in the values of C3 and C4; they are electrolytics and will have a significant tolerance. Therefore they should be made large so their impedance is a small part of the total circuit impedance. 47 uF is shown here but 100 uF or 220 uF can be used to advantage if there is the space to fit them in. The low-end frequency response must be defined somewhere in the input system, and the sooner the better, but this is not the place to do it. It should be done with a non-electrolytic capacitor which will have a closer tolerance; preferably it should be a polypropylene type which does not generate distortion.

R3, R4 are DC drain resistors that prevent charges lingering on C3 and C4. These can be made lower in value than for the unbalanced input as the input impedances are lower, and a value of, say, 100 kΩ rather than 220 kΩ makes relatively little difference to the total input impedance.

There now follows a collection of balanced input circuits that offer advantages or extra features over the standard balanced input configuration that has just been described in remorseless detail. To make things clearer, the circuit diagrams mostly omit the stabilising capacitors, input filters, and DC blocking circuitry discussed above. They can be added in a straightforward manner; in particular bear in mind that a stabilising capacitor like C5 is often needed between the opamp output and the negative input to guarantee freedom from high-frequency oscillation.

Combined Unbalanced and Balanced Inputs

Very often both unbalanced and balanced inputs are required, and it is extremely convenient if it can be arranged so that no switching between them is required; switches cost money. A handy way to do this is shown in Figure 27.9, which for clarity omits most of the extra components required for practical use that are referred to above. For balanced use, simply connect to the balanced input and leave the unbalanced input unterminated. For an unbalanced input, simply connect to the unbalanced input and leave the balanced input unterminated. No mode switch is required. These unterminated inputs sound as though they would cause a lot of extra noise, but in fact the circuit works very well and I have used it with success in high-end equipment.

As described above, in the world of hi-fi, balanced signals are at twice the level of the equivalent unbalanced signals, and so the balanced input must have

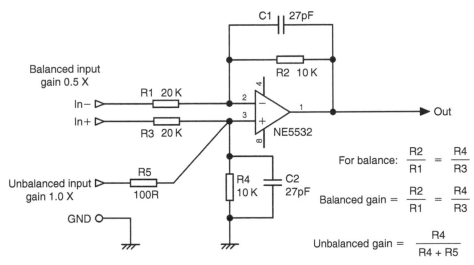

Figure 27.9. Combined balanced and unbalanced input amplifier with no switching required.

a gain of ½ or −6 dB relative to the unbalanced input to keep the same system gain by either path. This is done here by increasing the value of R1 and R3 to 20 kΩ. The balanced gain of this circuit can be made either greater or less than unity, but the gain via the unbalanced input is always unity. The differential gain of the amplifier and the constraints on the component values for balanced operation are shown in Figure 27.5, and are not repeated in the text to save space. This applies to the rest of the balanced inputs in this chapter.

There are a few compromises in this scheme. The noise performance in the unbalanced input mode is worse than for the sort of dedicated unbalanced input circuitry described earlier in this chapter, because R2 remains effectively in the signal path in unbalanced mode. Also, the input impedance of the unbalanced input cannot be very high because it is determined by the value of R4, and if this is raised all the resistances around the opamp must be increased proportionally and the noise performance is markedly worsened. A vital point is that only one input cable should be connected at a time. If an unterminated cable is left connected to an unused input, then the extra cable capacitance to ground will cause frequency response anomalies and can in bad cases cause HF oscillation. A warning on the back panel is a very good idea.

Variable-gain Balanced Inputs

A variable-gain balanced input is advantageous because it gives the opportunity to get the incoming signal up to the desired internal level as soon as possible, exposing it to the minimum contamination from circuit noise. If the input stage can attenuate as well as amplify, it also avoids the possibility of internal clipping that cannot be prevented because the stage doing the clipping is before the gain control. Unfortunately, making variable-gain differential stages is not so easy; the obvious method is to use dual-gang pots to vary two of the resistances, but this is clumsy and will give a CMRR that is both bad and highly variable due to the inevitable mismatches between pot sections. For a stereo input the resulting 4-gang pot is not an attractive proposition.

There is, however, a way to get a variable gain with good CMRR and a single pot section. The principle is essentially the same as for the switched-gain amplifier above: to maintain constant the source impedance driving the feedback arm. The principle is shown in Figure 27.10. To the best of my knowledge I invented this circuit in 1982, but so often you eventually find out that you have re-invented rather than invented; any comments on this point are welcome. The feedback arm R2 is of constant resistance, and is driven by voltage-follower U1:B. This eliminates the variations in source impedance at the pot wiper, which would badly degrade the CMRR. R6 limits the gain range and R5 modifies the gain law to give it a more usable shape. Bear in mind that the centre-detent gain may not be very accurate as it partly depends on the ratio of pot track (often no better than +/−10%, and sometimes worse) to 1% fixed resistors.

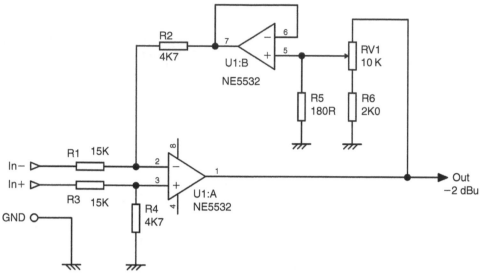

Figure 27.10. Variable gain balanced input amplifier.

This stage is very useful as a general line input with an input sensitivity range of −20 to +10 dBu. For a nominal output of 0 dBu, the gain of Figure 27.10 is +20 to −10 dB, with R5 chosen for 0 dB at the central wiper position. An opamp in a feedback path appears a dubious proposition for stability, because of the extra phase-shift it can introduce, but here it works as a voltage-follower, so its bandwidth is maximised and in practice the circuit is dependably stable.

The Instrumentation Amplifier

Just about every book on balanced or differential inputs includes the three-opamp circuit of Figure 27.11 and praises it as the highest expression of the differential amplifier. It is usually called the instrumentation amplifier configuration because of its undoubted superiority for data-acquisition. (Note that specialised ICs do exist that are sometimes also called instrumentation amplifiers or in-amps; these are designed to give very high CMRR without external resistors. They are expensive and are not in general optimised for audio applications.)

The beauty of this configuration is that the dual input stage buffers the balanced line from the input impedances of the final differential amplifier, so the four resistances around it can be made much lower in value, reducing their Johnson noise and the effect of opamp current noise significantly, and simultaneously reaping the CMRR benefit of presenting high input impedances to the balanced line. This can also be achieved simply by

adding voltage-followers to each input of a standard balanced input stage, as described later in this chapter.

The unique feature of the instrumentation amplifier configuration, which is often emphasised because of its unquestionable elegance, is that the dual input stage can have any value of differential gain, but the common-mode gain is always unity; this is *not* affected by mismatches in R3 and R5. The final amplifier U2:A then does its usual job of common-mode rejection, and the combined result can be a very high CMRR.

This does, however, assume a high gain in the first stage. A data-acquisition application may need a gain of a thousand times, which will allow a stunning CMRR to be achieved without using precision resistors, but the gains that are required in a line input are much lower, and often attenuation rather than gain is required. A notable exception to this is a line input to an active crossover, where typical system design allows internal levels of up to 3 Vrms to be used without loss of overall headroom.[4]

The instrumentation amplifier can also be very effective when gain in the line input stage is required for a power amplifier. Assume we want 200 Wrms into 8 Ω. That is a voltage output of 40 Vrms, and if the power amplifier itself has a gain of 20 times, we will have to feed it with 2 Vrms. Assume the input sensitivity has to be 500 mVrms for full output, and we clearly need a line input stage with a gain of 4 times (+12.0 dB).

If we use the strategy of using unity-gain input buffers followed by a balanced amplifier working at low impedances, as described later, the required gain can be obtained by altering the ratio of the resistors in

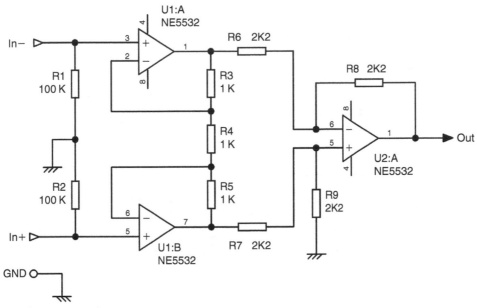

Figure 27.11. The instrumentation amplifier configuration, configured for unity gain.

the balanced amplifier. With suitable resistor values, such as R1 = R3 = 560 Ω, R2 = R4 = 2k2, the output noise from the combined stage is −100.9 dBu, so its Equivalent Input Noise (EIN) is −100.9 dBu − 12.0 = −112.9 dBu, which is reasonably quiet. All the noise measurements here are 22−22 kHz, rms sensing, unweighted, inputs terminated by 50 Ω, and all opamps are 5532. Using the LM4562 would reduce noise somewhat.

Figure 27.12 shows an instrumentation input configured for a gain of 4 times by reducing R4. To be precise, the gain is 3.94 times, as close as you can get with single E24 values. The resistors in the second stage have been reduced to 820 Ω to improve noise performance. I measured it and the theoretical CMRR improvement really is obtained. To take just one set of results, when I built the second stage it gave a CMRR of 56 dB working alone, but when the first stage was added, the CMRR leapt up to −69 dB, an improvement of 13 dB. These one-off figures (13 dB is actually better than the 12 dB improvement you would get if all resistors were exact) obviously depend on the specific resistor examples I used, but you get the general idea. CMRR was flat across the audio band.

I realise that there is an unsettling flavour of something-for-nothing about this, but it really works. The total gain required can be distributed between the two stages in any way, but it should all be concentrated in Stage 1, as shown, to obtain the maximum CMRR benefit.

But have we sacrificed noise performance in improving the CMRR? Certainly not. Fascinatingly, it is considerably improved. The noise output of this instrumentation amp is −105.4 dBu, which is no less than 4.5 dB better than the −100.9 dBu we got from the more conventional buffer−balanced amp configuration. This is because the opamps in Stage 1 of the instrumentation amp are working in better conditions for low noise than those in Stage 2. They have no significant resistance in series with the non-inverting inputs to generate Johnson noise, or turn opamp current noise into voltage noise. As they implement all the gain before the signal reaches Stage 2, the noise contribution of the latter is therefore less significant. The EIN of this instrumentation amp is −117.3 dBu.

This approach can be extended to give even more CMRR improvement if we can configure Stage 1 to have more gain without causing headroom problems. The instrumentation amplifier in Figure 27.13 has had the gain of Stage 1 doubled to 8 times (actually 7.67 times, or +17.7 dB, as close as you can get with single E24 values) by further reducing R4, while the input resistors to Stage 2 have been doubled in value so that stage now attenuates; having a gain of 0.5 times, it keeps the total gain at approx 4 times.

The theoretical CMRR improvement is once more the gain of Stage 1, which is 7.67 times, (+17.7 dB). Stage 2 gave a measured CMRR of 53 dB working alone, but when Stage 1 was added, the CMRR

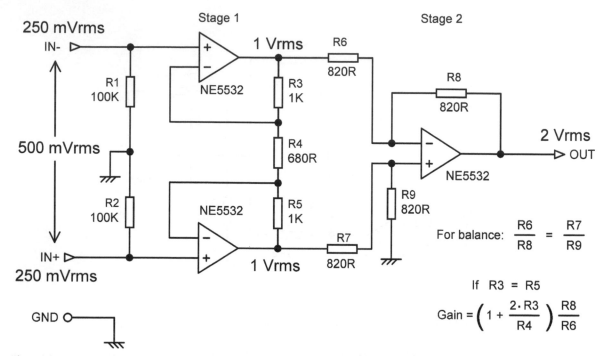

Figure 27.12. Instrumentation amplifier configured for 4 times (+11.8 dB) gain. All the gain is in Stage 1, Stage 2 having unity gain. CMRR is improved by 12 dB without having to improve resistor precision.

increased to −70 dB, an improvement of 17 dB, significantly greater than the 14 dB improvement obtained when Stage 1 had a gain of 4 times. The noise output was −106.2 dBu, which is better than the four-times instrumentation amplifier by 0.8 dB, and better than the more conventional buffer−balanced amp configuration by a convincing 5.3 dB. EIN is −118.1 dBu.

You may well be getting nervous about having a gain as high as 8 times followed by attenuation; have we compromised the headroom? The input stage has a gain of 8 times so a 500 mVrms balanced input will give 2 Vrms at each output of Stage 1. These outputs are in anti-phase, and their difference is 4 Vrms. When they are subtracted in Stage 2, with 0.5 times

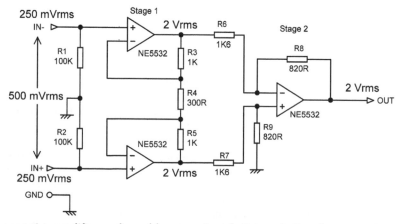

Figure 27.13. Instrumentation amplifier configured for approximately 8 times (+17.7 dB) gain in the first stage. Stage 2 has a gain of 0.5 times, keeping overall gain at approximately 4 times but improving CMRR further.

gain, we get 2 Vrms at the final output. In balanced operation all three opamps will clip simultaneously, so there is no headroom compromise at al.

So far, so good; all the voltages are well below clipping. But we have assumed a balanced input, and unless you are designing an entire system, that is a dangerous assumption. If the input is unbalanced but at the same level, we will have 500 mVrms on one input pin, and nothing on the other. You might fear we would get 4 Vrms on one output of Stage 1 and nothing on the other; that would still be some way below clipping but the safety margin is shrinking fast. In fact, things are rather better than that. The cross-coupling of the feedback network R3, R4, R5 greatly reduces the difference between the two outputs of stage 1 with unbalanced inputs. With 500 mVrms on one input pin, and nothing on the other, the Stage 1 outputs are 2.19 Vrms and 1.68 Vrms; the higher voltage is at the output of the amplifier associated with the driven input. These voltages naturally sum to 4 Vrms, reduced to 2 Vrms by Stage 2. Maximum output with an unbalanced input is slightly reduced at 9.1 Vrms; this slight restriction does not apply to the Figure 27.12 version with only 4 times gain in Stage 1. Since the power amplifier will clip with 2 Vrms in, this restriction has no consequences unless there is a gain control between the instrumentation amplifier and the power amplifier.

2 Vrms is a relatively high input to a power amplifier; if 1 Vrms was all that was required then reducing the output of the instrumentation amplifier with a 6 dB attenuator will also reduce the noise, and we retain the full CMRR improvement. This technique is dealt with in detail in;[4] note that the source impedance seen by the power amplifier must be kept low to prevent the introduction of distortion (see Chapters 5 and 6). In some cases it may be possible to work with even higher gain in Stage 1; increasing it to 21 times gave

me a rather spectacular CMRR of −86 dB, an improvement of no less than 25 dB, using unselected 1% resistors.

Transformer Balanced Inputs

When it is important that there is no galvanic connection (i.e., no electrical conductor) between two pieces of equipment, transformer inputs are indispensable. They are also useful if EMC conditions are severe. Figure 27.14 shows a typical transformer input. The transformer usually has a 1:1 ratio, and is enclosed in a metal shielding can which must be grounded. Good line transformers have an inter-winding shield that must also be grounded or the high-frequency CMRR will be severely compromised. The transformer secondary must see a high impedance as this is reflected to the primary and represents the input impedance; here it is set by R2, and a buffer drives the circuitry downstream. In addition, if the loading is too heavy, there will be increased transformer distortion at low frequencies. Line input transformers are built with small cores and are only intended to deliver very small amounts of power; they should not be used as line output transformers. An ingenious approach to solving the distortion problem by operating the transformer core at near-zero flux was published by Paul Zwicky in 1986.[5]

There is a bit more to loading the transformer secondary correctly. If it is simply loaded with a high-value resistor, there will be peaking of the frequency response due to resonance between the transformer leakage inductance and the winding capacitance. This is shown in Figure 27.15, where a Sowter 3276 line input transformer (a high-quality component) was given a basic resistive loading of 100 kΩ. The result was Trace A, which has a 10 dB peak around 60 kHz.

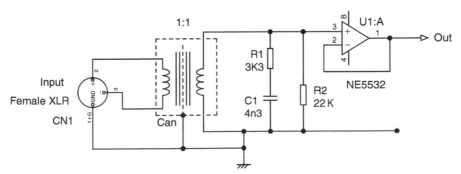

Figure 27.14. A transformer balanced input. R1 and C1 are the Zobel network that damps the secondary resonance.

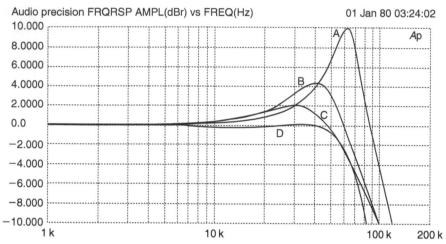

Audio precision FRQRSP AMPL(dBr) vs FREQ(Hz)   01 Jan 80 03:24:02

Figure 27.15. Optimising the frequency response of a transformer balanced input with a Zobel network.

This is bad not only because it accentuates the effect of out-of-band noise, but because it also affects the audio frequency response, giving a lift of 1 dB at 20 kHz. Reducing the resistive load would damp the resonance, but it would also reduce the input impedance. The answer is to add a Zobel network, i.e., a resistor and capacitor in series, across the secondary; this has no effect except high frequencies. The first attempt used R1 = 2k7 and C1 = 1 nF, giving Trace B, where the peaking has been reduced to 4 dB around 40 kHz, but the 20 kHz level is actually slightly worse. R1 = 2k7 and C1 = 2 nF gave Trace C, which is a bit better in that it only has a 2 dB peak. A bit more experimentation ended up with R1 = 3k3 and C1 = 4.3 nF (3n3 + 1nF) and yielded Trace D, which is pretty flat, though there is a small droop around 10 kHz. The Zobel values are fairly critical for the flattest possible response, and must certainly be adjusted if the transformer type is changed.

Input Overvoltage Protection

Input overvoltage protection is not common in hi-fi applications, but is regarded as essential in professional amplifier use. The normal method is to use clamping diodes, as shown in Figure 27.16, that prevent certain points in the input circuitry from moving outside the supply rails.

This is very straightforward, but there are two points to watch.

Firstly, the ability of this circuit to withstand excessive input levels is not without limit. Sustained overvoltage may burn out R5 and R6, or pump unwanted amounts of current into the supply rails; this sort of protection is mainly aimed at transients.

Secondly, diodes have a non-linear junction capacitance when they are reverse biased, so if the source impedance is significant the diodes will cause distortion at high frequencies. To quantify this problem here are a few figures. If the circuit of Figure 27.16 is being fed from the usual kind of line output stage, the impedance at the diodes will be approximately 1 kΩ and the distortion introduced with an 11 Vrms 20 kHz input will be below the noise floor. However, in a test I conducted where the impedance was increased to 10 kΩ with the same input, the THD at 20 kHz was degraded from 0.0030% to 0.0044% by adding the diodes. I have worked out a rather elegant way to eliminate this effect completely, but this is not the place to disclose it. As you might have guessed, I am rather hoping to sell the idea.

Noise and the Input System

In Chapter 6 the sources of noise inside the actual power amplifier were examined. As an example we looked at a real amplifier with a closed-loop gain of +27.2 dB and measured noise at the amplifier output of a pleasingly low −95.6 dBu over the standard 22−22 kHz bandwidth. This is a noise level referred to the amplifier input of only −122.6 dBu. We will take those as standard values for the time being. It is immediately clear that almost anything we connect to the input of this amplifier is going to compromise the noise performance. But by how much?

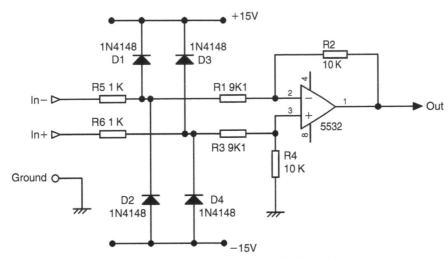

Figure 27.16. Input overvoltage protection for a balanced input amplifier.

As a first example, consider that in some amplifier designs a first-order RC low-pass filter is placed immediately before the power amplifier input, in the pious hope of preventing slew-rate limiting. It was offered as a magical panacea against TIM (Transient Intermodulation Distortion) before it became clear to everybody — and it took an unconscionably long time — that TIM was just another way of referring to slew-limiting, which rarely, if ever, occurred. This was always a dubious expedient if there was no buffering before the RC filter because the source resistance of the external equipment would affect the turnover frequency; the resistor was usually in the range 470 Ω–1 kΩ in the hope that this would be significantly greater than the external source resistance and so minimise the variation. If we take an 820 Ω resistance as typical (for some reason it does seem to have been a particularly popular value), then its Johnson noise is −123.5 dBu. If we RMS sum this with the amplifier EIN of −122.6 dBu, then the result is −120.0 dBu, and we have degraded the noise performance by 2.6 dBu already with this one component, with no active circuitry upstream. (It might be as well to mention here that putting resistances directly in series with a power amplifier input is a bad plan for another reason — it induces input current distortion. This ticklish topic is dealt with in Chapter 6.)

Let us now see what the noise consequences are of putting active circuitry that actually does something useful in front of the power amplifier. I need to say at this point that the opamp noise data quoted here is taken from extensive real-life measurements rather than theoretical calculations, but averaged over a relatively

small number of samples. It is my experience that (providing you stick to reputable manufacturers) the noise performance of bipolar opamps such as the 5532 shows relatively little variation. The aim here is to show the general principles of low-noise design rather than get too picky about the last decimal place.

Firstly, we put a simple unity-gain buffer in front of the power amplifier stage; this might be done to drive an input resistor that has been given a low value to minimise input offset voltages, so we can still present a high input impedance to the outside world, or to prevent input current distortion by driving the amplifier from a very low impedance. (The latter issue is described in Chapter 6.) We will take the impedance seen by this buffer stage as 50 Ω, which is about as low as we might hope for; the Johnson noise from this is only −135.2 dBu. The noise output of a NE5532 unity-gain buffer with these input conditions is −119 dBu; this is a very low value and is obtained only because there are no medium-value feedback resistances in the opamp circuit and all we are seeing is the opamp voltage noise. When this noise level is added to the power amplifier EIN of −122.6 dBu, it gives an RMS total of −117.4 dBu. We have added what at first sight is the quietest possible preceding stage but amplifier noise output has already been increased by 5.2 dB. If we use an LM4562 unity-gain buffer, then its calculated voltage noise is −125.7 dBu, and the total EIN is −120.9 dBu, only 1.7 dB worse, and a more respectable result.

It is much more useful to put a balanced input stage in front of the power amplifier itself, as it gives all the benefits of common-mode rejection. The standard one-opamp unity-gain balanced input is very commonly made with four 10 kΩ resistors; this is a compromise which, as we have seen, gives a respectable if not stunning 20 kΩ common-mode impedance on each input, combined with levels of Johnson noise that are usually considered acceptable; see Figure 27.17a. The noise output of a 4 x 10 kΩ balanced input amp using a NE5532 is −105.1 dBu, which completely swamps the power amplifier EIN, and degrades the overall noise performance by 17.5 dB. The noise at the amplifier output increases from −92 dBu to −75 dBu. This is not a very happy outcome, but is the inevitable result of using conventional balanced input technology. The ideal would be an input stage that has negligible noise

compared with the power amplifier; if we accept that the power amplifier noise should be increased by only 0.1 dB, a few seconds juggling with RMS-summation shows that the input stage noise output would have to be a very low −139 dBu. That is the Johnson noise of a 21 Ω resistor basking in a room temperature of 25°C, and is clearly a pretty tall order. If we reluctantly agree that the power amplifier noise can be worsened by 1.0 dB, which would be hard to detect even in ABX testing, we then need an input stage with a noise output of −128.5 dBu, which is still an ambitious target. We might decide, in a fleeting moment of realism, that we can live with equal noise contributions from the input stage and the power amplifier; in other words, the input stage noise will degrade the overall

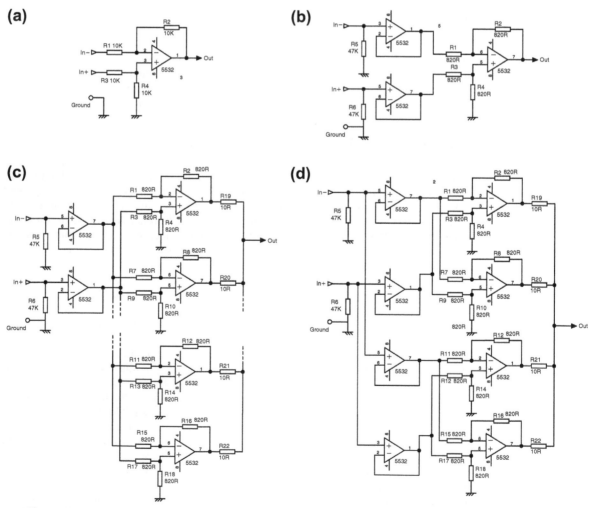

Figure 27.17. Low noise unity-gain balanced inputs using multiple buffers and multiple differential amplifiers.

Table 27.4. How noise from an input stage degrades power amplifier output noise performance

Input stage noise o/p dBu	Input noise summed with power amp EIN dBu	Power amp noise worsened by dB
−139.00	−122.50	0.10
−128.50	−121.61	0.99
−122.60	−119.59	3.01
−105.10	−105.02	17.58

noise output by 3 dB. That will require an input stage noise output that is the same as the power amplifier EIN, which is −122.6 dBu. That might be attainable, with a bit of thought. These noise requirements figures are summarised in Table 27.4, where the bottom line shows the effect of the standard balanced input with four 10 k resistors.

So, we have a target: a balanced input stage with an output noise of −122.6 dBu. Let's see what can be done about it.

Low-noise Balanced Inputs

We have seen that the noise output of the standard one-opamp balanced input with four 10 kΩ resistors in Figure 27.17a is −105.1 dBu; we clearly need to apply some more serious electronics to the noise problem. In all the measurements that follow, the source impedance was 50 Ω to ground on both inputs. We have seen that the instrumentation amplifier configuration is of limited use for audio work as it only gives an improvement in CMRR commensurate with the gain of its first stage. However, if we reduce it to a standard differential amplifier with a unity-gain buffer on each input, we can reduce the value of the four resistors around the final differential amplifier, reducing their Johnson noise, and at the same time increase the input impedance presented to the outside world, and so possibly improve the CMRR. This arrangement is shown in Figure 27.17b. There is a limit to how far the four resistors can be reduced, as the differential stage has to be driven by the input buffers, and it also has to drive its own feedback arm, but against this is the relatively small part of the output swing that is used if the input stage is directly coupled to the power amplifier; the latter will always clip a long time before the opamps get anywhere near their maximum output. If NE5532s are used, a safe value that gives

no measurable deterioration of the distortion performance is about 820 R, and an NE5532 differential stage alone (without the buffers) and 4 x 820 Ω resistors gives a noise output of −111.7 dBu, which is 6.6 dB lower than the standard 4 x 10 kΩ version. Adding the two input buffers degrades this only slightly to −110.2 dBu, because we are adding only the voltage noise component of the two new opamps, which is uncorrelated between the two of them, and we are still 5.1 dB quieter than the 4 x 10 k version. The interesting point here is that we have three opamps in the signal path instead of one, but we get a significantly lower noise level. Overall noise, however, is still degraded by 12.6 dB.

This might appear to be all we can do; it is not possible to reduce the value of the four resistors around the differential amplifier any further without compromising linearity. In fact, there is almost always some way to go further in the great game that is electronics, and here are three possibilities. A step-up transformer could be used to exploit the low source impedance (remember we are still assuming the source impedances are 50 Ω) and it might well work superbly in terms of noise alone, but transformers are always heavy, expensive, susceptible to magnetic fields and of doubtful low-frequency linearity. Alternatively, we could design a discrete-opamp hybrid stage that uses discrete input transistors, which are quieter than those integrated into IC opamps, coupled to an opamp to provide raw loop gain; this can be effective but you need to be very careful about high-frequency stability. Thirdly, we could design our own opamp using all discrete parts; this approach tends to have fewer stability problems but does require rather specialised skills, and the result takes up a lot of PCB area.

If those three expedients are rejected, now what? One of the most useful techniques in low-noise electronics is to use two identical amplifiers so that the gains add arithmetically, but the noise from the two separate amplifiers, being uncorrelated, partially cancels. Thus a 3 dB noise advantage is gained each time the number of amplifiers used is doubled. This technique works very well with multiple opamps; let us apply it and see how far it may be usefully taken.

Since the noise of a single 5532 unity-gain buffer is only −119 dBu, and the noise from the 4 x 820 Ω differential stage (without buffers) is a much higher −111.7 dBu, the differential stage is the place to start work. We will begin by using two identical 4 x 820 Ω differential amplifiers as shown in Figure 27.17c, both driven from the existing pair of input buffers. This will give partial cancellation of both resistor and opamp noise from the two stages when their outputs are summed. The main question is how to sum the two amplifier outputs; any active solution would introduce another opamp, and hence more noise, and we would almost certainly wind up worse off than when we started. The answer is, however, beautifully simple. We just connect the two amplifier outputs together with 10 Ω resistors; the gain does not change but the noise output drops. The signal output of both amplifiers is nominally the same, so no current should flow from one opamp output to the other. In practice, there will be slight gain differences due to resistor tolerances, but with 1% resistors I have never experienced any problems. The combining resistor values are so low at 10 Ω that their Johnson noise contribution is negligible.

We therefore have the arrangement of Figure 27.17c, with single input buffers (i.e., one for each of the two inputs) and two differential amplifiers, and this drops the noise output by 2.3 dB to −112.5 dBu, which is quieter than the original 4 x 10 k version by a hefty 7.4 dB. We do not get the full 3 dB noise improvement because both differential amplifiers are handling the noise from the input buffers; this is correlated and so is not reduced by partial cancellation. Power amplifier output noise is now only worsened by 10.5 dB. The role of the input buffer noise is further emphasised if we take the next step of using four differential amplifiers. (There is nothing special about using amplifiers in powers of two. It is perfectly possible to use three or five differential amplifiers in the array, which will give intermediate amounts of noise reduction.)

So, sticking with single input buffers, we try the effect of four differential amplifiers. These are added on at the dotted lines in Figure 27.17c. We get a further improvement, but only by 1.5 dB this time. The output noise is down to −114.0 dBu, quieter than the original 4 x 10 kΩ version by 8.9 dB, but still making the power amplifier 9.2 dB noisier when connected. You can see that at this point we are proceeding by decreasing steps, as the input buffer noise begins to dominate, and there seems little point in doubling up the differential amplifiers again; the amount of hardware would be getting out of hand, as would the PCB area occupied. The increased loading on the input buffers is also a bit of a worry.

A more fruitful approach is to tackle the noise from the input buffers, by doubling them up as in Figure 27.17d, so that each buffer drives only two of the four differential amplifiers. This means that the buffer noise will also undergo partial cancellation, and will be reduced by 3 dB. There is, however, still the contribution from the differential amplifier noise, and

so the actual improvement on the previous version is 2.2 dB, bringing the output noise down to −116.2 dBu. This is quieter than the original 4 x 10 k version by a thumping 11.1 dB, but still makes the power amplifier noisier by 7.3 dB, and we are some way short of our target of 3 dB. Using this input stage, if we RMS-sum its noise output with the power amplifier EIN, the effective input noise at the power amplifier is −115.3 dB, and the lesson is that the power amplifier noise is no longer negligible; now it increases the total noise at the output by 0.9 dB. Remember that there are two inputs, and 'double buffers' means two buffers per input, giving a total of four in the complete circuit.

Since doubling up the input buffers gave us a useful improvement, it's worth trying again, so we have a structure with quad buffers and four differential amplifiers, as shown in Figure 27.18, where each differential amplifier now has its very own buffer. This improves on the previous version by a rather less satisfying 0.8 dB, giving an output noise level of −117.0 dBu, quieter than the original 4 x 10 kΩ version by 11.9 dB. Connecting this input stage to the power amplifier increases its noise output by 6.7 dB. The small improvement we have gained indicates that the focus of noise reduction needs to be returned to the differential amplifier array, but the next step there would seem to be using eight amplifiers, which is not very appealing. Thoughts about grains of corn on chessboards tend to intrude at this point.

This is a good moment to pause and see what we have achieved. We have built a balanced input stage that is quieter than the standard circuit by 11.9 dB, using standard components of low cost. We have used increasing numbers of them, but the total cost is still small compared with power transistors, heatsinks, and transformers. The power consumption is clearly greater, but trivial compared with the quiescent current of the average Class-B power amplifier. The technology is highly predictable and the noise reduction reliable, in fact, bullet-proof. The linearity is as good as that of a single opamp of the same type, and in the same way there are no HF stability problems. The noise performance of the various circuits described are summarised in Table 27.5.

I don't want you to think that this noise-reduction exercise is simply a voyage off into pure theory. As an example of this technique in action, consider the Cambridge Audio 840 W power amplifier, which, I will modestly mention in passing, won a CES Innovation Award in January 2008. This unit has both unbalanced and balanced inputs, and for the reasons given above, conventional technology would have meant

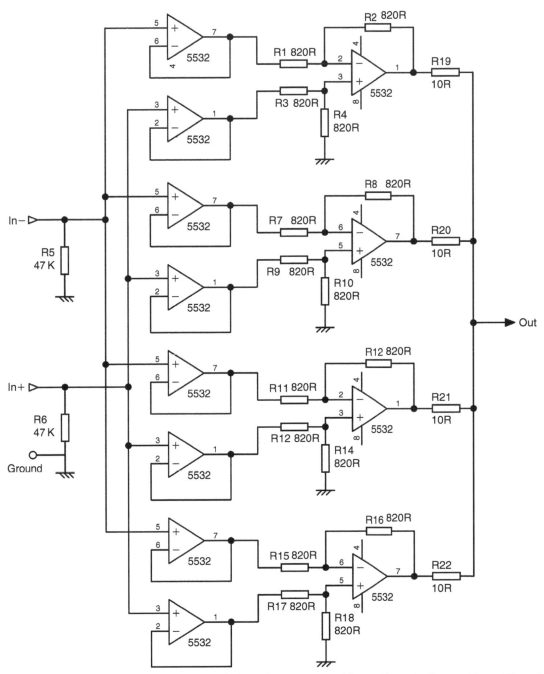

Figure 27.18. The final 5532 low noise unity-gain balanced input stage, with quad input buffers and four differential amplifiers. The noise output is only −117.0 dBu.

that the balanced inputs would have been the noisier of the two. Since the balanced input is the 'premium' input, many people would think that was the wrong way round. We therefore decided that the balanced input was

required to be quieter than the unbalanced input. Using 5532s in a structure similar to those outlined above, this requirement proved quite practical, and the finalised balanced input design was both economical

Table 27.5. A summary of the noise improvements made to the balanced input stage

Buffer	Differential amplifier	Input stage noise output dBu	Improvement on previous version dB	Improvement over 4x10kΩ diff amp dB	Power amp noise degraded by dB
None	Standard diff amp 10K 5532	−105.10		0	17.58
None	Single diff amp 820R 5532	−111.70	6.60	6.6	11.24
Single 5532	Single diff amp 820R 5532	−110.20		5.1	12.64
Single 5532	Dual diff amp 820R 5532	−112.50	2.30	7.4	10.50
Single 5532	Quad diff amp 820R 5532	−114.00	1.50	8.9	9.16
Dual 5532	Quad diff amp 820R 5532	−116.20	2.20	11.1	7.30
Quad 5532	Quad diff amp 820R 5532	−117.00	0.80	11.9	6.66
Quad 5532	Quad diff amp 820R AD797	−119.90	2.90	14.8	4.57
Dual AD797	Quad diff amp 820R AD797	−120.60	0.70	15.5	4.12
	Notional input stage	−122.60		17.5	3.01
	Notional input stage	−128.50		23.4	0.99
	Notional input stage	−139.00		33.9	0.10

and quieter than its unbalanced neighbour by a dependable 0.9 dB.

Two other versions were evaluated that made the balanced input quieter than the unbalanced one by 2.8 dB, and by 4.7 dB, at slightly greater cost and complexity. These were put away for possible future upgrades.

The Choice of Opamps

It is the near-universal choice to make the amplifier input system out of opamps, though there are notable exceptions; a few 'high-end' power amplifiers have quite complex discrete circuitry for implementing balanced inputs.

Until recently, the 5532 would have been the automatic choice for opamps in the signal path, despite its great age (it was introduced in 1978). It has very low distortion and bipolar input devices that are quiet at the kind of impedances met with in most audio circuitry, with an input noise density of 4 nV/√Hz (typical at 1 kHz). The current-noise density is 0.7 pA/√Hz.

The National LM4562 is a bipolar opamp which has finally surpassed the 5532 in performance. It was introduced by National Semiconductor at the beginning of 2007. Initially it was very expensive, but since the Fifth edition of this book was published, the price has fallen to something like six times that of the 5532, making it a viable alternative when achieving the best noise and distortion performance is more important

than cost. The LM4562 is a dual opamp — there is no single or quad version. The input noise voltage is typically 2.7 nV/√Hz, substantially lower than the 4 nV/√Hz of the 5532. For suitable applications with low source impedances, this gives a noise advantage of 3 dB or more. With a medium or high source impedance (such as a moving-magnet cartridge), it is noisier than the 5532, because its current-noise density is higher at 1.6 pA/√Hz. It is not fussy about decoupling, and as with the 5532, 100 nF across the supply rails usually ensures stability. Whether decoupling from rails to ground is required depends on the application. The typical slew rate is ±20 V/μs, but the minimum is a bit lower at ±15 V/μs.

The AD797 has remarkably low voltage noise at 0.9 nV/√Hz (typical at 1 kHz) but a high current noise of 2 pA/√Hz. It remains a specialised and expensive part, costing approx 25 times as much as a 5532 at the time of writing. It is only available in single versions, while the 5532 is a dual, so the cost factor per amplifier is actually 50 times. It has the reputation of being difficult to stabilise at HF, but in my experience it is not too hard.

Using an Internal Balanced Power Amplifier Interface

The use of a balanced input to a power amplifier is a thoroughly good idea, cancelling out ground voltages and rendering ground loops innocuous. This technology is useful even *inside* the power amplifier unit, where ground currents can otherwise put a limit on achievable hum performance; typically it may be necessary when passing the signal at line level from one PCB to another. The obvious approach is to put a differential amplifier in front of the power amplifier, but this rather smacks of overkill to deal with internal grounding problems, and adding an opamp in such a position almost invariably degrades the overall noise performance. At this point the idea may strike — we have an amplifier here already, with a differential-pair input stage: can it be made into a differential power amplifier? The answer is, yes, if you do it with a little care.

Now and again the circuit shown in Figure 27.19a has appeared in the literature. It suggests that all you have to do to make a differential power amplifier is to treat the bottom of the negative feedback network as the cold input. There are several things wrong with this.

Firstly, if the cold input becomes disconnected, say, by someone pulling out the input connector — the closed-loop gain of the amplifier is abruptly reduced to unity. This will almost certainly render it completely unstable, with massive high-frequency oscillation and an excellent chance of damaging both the amplifier and any attached loudspeakers. This problem can be prevented by adding a low value resistor R3 at the bottom of the negative feedback network as shown in Figure 27.19b. In this example the closed-loop gain is 25 times (+28.0 dB) with the cold input connected and 22.8 times if it goes open; a power amplifier with reasonable stability margins should not be worried by this. The gain change can be reduced by making R3 smaller, so long as it remains high in value compared with the ground resistance. For example, making R3 lower at 4R7 gives a normal gain of 25 times, and 23.9 times with cold disconnected, and I have found this value works well in practical use.

Secondly, it is not a proper differential circuit as the gains via the two inputs are not quite the same. The gain via the inverting input is R1/R2, while the gain via the non-inverting input is R1/R2 + 1. The common-mode gain (by simulation) is +1.8 dB, so the CMRR is only 26.2 dB. This can be easily corrected by using the circuit shown in Figure 27.19c, which adds a little attenuation to the non-inverting input in the form of R4 in conjunction with R5. R4 is very often already present as part of an RC input filter, and it is simple to scale the filter values so that R5/R4 = R1/R2. Likewise R5 is usually already there to define the DC level of the input stage. The input gains are now the same and good common-mode rejection is achieved.

The values shown reduce the normal gain very slightly to +27.6 dB, and reduce the common-mode gain sharply to −12.4 dB, so the CMRR improves markedly to 40.0 dB. In a perfectly balanced circuit, and with an infinite open-loop amplifier gain, the CMRR would in theory be infinite, but here the CMRR achievable is determined by the realistic open-loop gain assumed for the amplifier when the simulations were run. Open-loop gain was set to 35,000 with a single pole at 20 Hz. This gives a flat CMRR curve up to 10 kHz.

This can be compensated for by increasing R4 to 110 Ω which improves the CMRR to 44.5 dB, but heading in this direction involves cancelling two quite different effects and is not very respectable. If 40 dB is not enough, then there is something seriously wrong with your grounding system and you need to seek out the root cause of the problem.

Finally, it is not a proper differential circuit in that the two inputs are not interchangeable. If you are trying to correct an absolute phase error by driving the cold input (which is not exactly good practice in itself), you will soon find it is not at all suitable for the task, having much too low an impedance for use as a signal input. This is not a problem if the cold

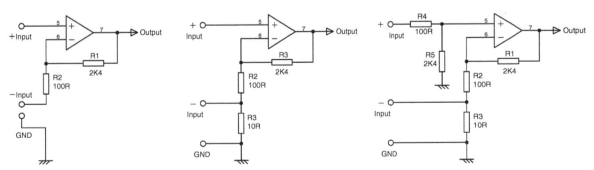

Figure 27.19. Balanced power amplifier interfaces: (a) is dangerous; b) works very nicely; c) also works very nicely but with a better CMRR.

input is used appropriately, to cancel the voltage drop on the ground connection. Since the resistance of any sensible ground is a small fraction of an Ohm, a cold input impedance of 4.7 Ω or 10 Ω will not cause significant errors.

While this approach could be used to interface to the outside world, it is much more useful for making internal amplifier connections between input circuitry and the power amplifier itself. In one instructive case there was a difference in ground potential between the two sections, which were on separate PCBs; both gave an exemplary hum-free performance alone, but when the signal connection between them was made, hum appeared. It was at a very low level but it was uncomfortably visible on the amplified noise output. There was clearly a 50 Hz current flowing through the ground of the signal connection, for when the ground was reinforced with the traditional piece of 32/02 green-and-yellow wire, the hum level dropped significantly. It has to be said that the deep reason for this current remained obstinately obscure, despite intense investigation. The most obvious possible cause was that capacitive coupling between primary and secondary of the mains transformer was causing 50 Hz currents to flow — but the transformer was a high-quality item with a grounded interwinding screen to prevent just that happening, and removing the ground rather disconcertingly made no difference to the problem.

The total noise level at the power amplifier output was −94.1 dBu over the 22−22 kHz bandwidth. When the PCB−PCB interconnection was changed to the balanced system, this dropped to −94.4 dB. That may sound like a pretty trivial improvement, but in fact it meant that the obtrusive 50 Hz component disappeared. To quantify this, a narrowband noise measurement, made with a 50 Hz bandpass filter, was reduced from −100.8 dBu to −109.6 dBu. The coherent 50 Hz waveform disappeared completely, leaving just the random excitation of the bandpass filter by wideband noise. The balanced interface was working beautifully.

References

1. Muncy, N, Noise Susceptibility in Analog and Digital Signal Processing Systems, *JAES*, (6), June 1995, p. 447.

2. Winder, S, What's the Difference? *Electronics World*, Nov. 1995, p. 989.

3. Self, D, *Small Signal Audio Design*, Burlington, MA: Focal Press, 2010, pp. 352−354.

4. Self, D, *Active Crossover Design*, Burlington, MA: Focal Press, 2011, pp. 419−425.

5. Zwicky, P, Low-Distortion Audio Amplifier Circuit Arrangement, US Patent No. 4,567,44, 1986.

Input Processing and Auxiliary Systems

This chapter deals with input processing functions such as gain control and filtering, and other power amplifier features that are not part of the signal path, but perform ancillary functions such as level indication or control of on/stand-by switching.

Ground Lift Switches

If balanced inputs with high CMRR are used, they should deal effectively with the humming and buzzing ground currents that result from the existence of ground loops. However, in an emergency it may be useful to have a means of lifting the ground without creating a shock hazard. It has to be said that it is often not very effective, and is a last resort when the audience is starting the slow hand-clap.

The typical ground lift switch breaks the direct connection between mains earth and the equipment signal earth, but leaves them connected by a resistor, usually in the range 10–100 Ω. This resistor is high enough to prevent significant ground loop current passing, but low enough to prevent the signal earth from floating about. Without it, the signal circuitry may have a 120 V AC voltage on it, due to inter-winding capacitance in the mains transformer, if it is not otherwise grounded. If the mains transformer has an interwinding screen, then this must remain connected to mains earth, and not the signal earth.

The resistor is usually shunted with small capacitor (usually 10 nF to 100 nF) which makes the impedance low at radio frequencies but does not let a significant current at 50 Hz flow.

Phase Reversal Facility

A feature found on a small minority of professional amplifiers is a phase reverse switch that can be used to correct phase reversals elsewhere in the audio system. If only a balanced input is provided, phase reversal is easily implemented by a switch that swaps over the inputs. If there is also an unbalanced input, however, then things are a bit more difficult; it will be necessary to add a unity-gain inverting stage to create the anti-phase signal.

Gain Control

Gain controls on hi-fi power amplifiers are relatively rare as volume is almost always controlled from the preamplifier. They do, however, come in very handy when setting up a system, and this is more true, the

more powerful the amplifier. Alternatively, a switch giving 20 or 30 dB of attenuation would be useful, but unless you are a hi-fi reviewer you are unlikely to be changing the system around often enough to justify the cost.

Professional amplifiers usually have gain controls; there are usually separate controls for each channel as such an amplifier is more likely to be handling two bands of a multi-amped loudspeaker system rather than two stereo channels. If stereo *is* being handled, ganged potentiometers are not very satisfactory as gain controls because all but the most expensive tend to have poor channel balance at large attenuations. For this reason, a large number of professional amplifiers use switched attenuators based on rotary switches and resistor ladders.

At least one manufacturer (Harrison) has produced a range of professional amplifiers with VCA modules that allow remote control of output level by means of a DC voltage.

Subsonic Filtering: High-pass

Subsonic filters are a common feature of professional amplifier input systems. Power amplifiers are not likely to be damaged by subsonic inputs, but the loud-speakers they drive are. It is essential to prevent large and uncontrolled excursions of the loudspeaker bass units. This is particularly important when using reflex (ported box) loudspeakers that have no restraint on cone movement at very low frequencies. In sound rein-forcement applications, subsonic signals can be gener-ated by microphones with insufficient blast protection, by direct-injected bass guitars, and in many other ways. Mixing desks almost always include effective low-frequency filters on microphone input channels, but these are usually configured for a fairly high roll-off such as 100 Hz to deal with microphone placement problems, so it is still good practice to incorporate true subsonic filters in the power amplifiers. In hi-fi applica-tions subsonic filtering is usually done at the preampli-fier end, but there is very often some last-ditch low-end bandwidth limitation in the power amplifier as well. In this case, the most common source of subsonic signals are the warps on vinyl records.

While large subsonic signals are not likely to actually damage a power amplifier, they will consume valuable headroom by pushing the amplifier into clipping that would not happen if it was only reproducing the wanted signals in the audio band, and this is another reason for including an effective subsonic filter.

The high-pass filters used are typically of the second-order or third-order Butterworth (maximally flat) configuration, giving roll-off rates of 12 dB/octave and 18 dB/octave respectively, the latter being preferable. Fourth-order 24 dB/octave filters are sometimes used but are less common, as people get worried (from all the evidence, unnecessarily) about the possible subjective effects of rapid phase changes at the very bottom of the audio spectrum. The Butterworth response (sometimes called the Butterworth alignment) is not the only one possible; the Bessel alignment gives a slower roll-off, but aims for linear phase, i.e., a constant delay versus frequency, and so reproduces the shape of transients better. There are other filter alignments such as the Chebyshev which give faster initial roll-offs than the Butterworth, but they do so at the expense of ripples in the passband or stopband gain so they are not used in this sort of application.

The most popular filter configuration is the well-known Sallen and Key type. The second-order version is very simple to design; the two series capacitors C1, C2 are made equal and R2 is made twice the value of R1. A second-order Butterworth filter with a −3 dB point at 20 Hz is shown in Figure 28.1a. The response is 12.3 dB down at 10 Hz, and 24.0 dB down at 5 Hz, by which time the 12 dB/octave slope is well established. Above the −3 dB roll-off point, the response is −0.78 dB down at 30 Hz, which is intruding a little into frequencies we want to keep. Other roll-off frequencies can be obtained simply by scaling the component values while keeping C1 equal to C2 and R2 twice R1.

Third-order filters are a bit more complicated, and many versions of them use two opamps instead of the one required for a second-order filter. But it can be done with just one, as in Figure 28.1b, which is a third-order Butterworth filter also with a −3 dB point at 20 Hz. The resistor value ratios are now 2.53:1.00:17.55, and the circuit shown uses the nearest E24 values to this − which by happy chance come out as E12 values. These values are not quite mathematically correct, and there is a little gain peaking of 0.001 dB around 80 Hz, according to SPICE simulation, whereas the true Butterworth response has no peaking at all but stays flat until the roll-off. If you can't live with this, then you will have to go to E96 resistors; R1 = 12.1 k, R2 = 4.75 k and R3 = 84.5 k give a very accurate response with no peaking at all.

For the third-order filter the response is 18.6 dB down at 10 Hz, and 36.0 dB down at 5 Hz, which is some pretty serious filtering, but the 30 Hz response is only −0.37 dB down, which demonstrates that a third-order filter is much more effective than a second-order one for this sort of job. As before, other roll-off frequencies can be had by scaling the component values while keeping the resistor ratios the same.

An important consideration with frequency-dependent networks like filters is the input impedance; this can sometimes drop to unexpectedly low values, putting excessive loading on the previous stage. In the high-pass case, the input impedance is high at low frequencies but falls as frequency increases. For the second-order version, it tends to the value of R2. R1 is bootstrapped and has no effect. In the third-order version, it tends to the value of R1 in parallel with R3, which here is 10.58 k. In neither case should the previous stage be embarrassed by the loading.

Because of the large capacitances, the noise generated by the passive elements in a highpass filter of this sort is usually well below the opamp noise. For the values used here, SPICE shows that the resistors in the second-order filter produce −132.4 dBu at the filter

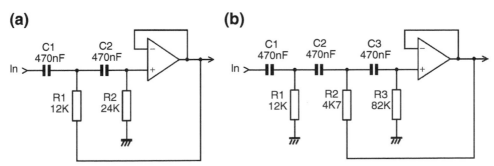

Figure 28.1. Subsonic filtering: (a) second-order and b) third-order Butterworth high-pass filters, both −3 dB at 20Hz.

output, while in the third-order they produce −125.0 dBu at the output (22 kHz bandwidth, 25 °C).

Ultrasonic Filtering: Low-pass

Amplifier input systems are sometimes fitted with an ultrasonic filter (somehow a 'supersonic filter' sounds wrong) to define the upper limits of the working bandwidth, though they are rather less common than subsonic filters. This is *not* a duplication of the input RF filtering which, as described in Chapter 27, has to be designed very cautiously as the source resistance is unknown. If an active low-pass filter is used, driven from a low and known source impedance, the turnover frequency of the filter can be accurately defined, and therefore set much lower in frequency without the fear that it will ever encroach on the wanted audio bandwidth. One of the main uses of an ultrasonic filter is the protection of the power amplifier and loudspeakers against ultrasonic oscillation in the audio system.

The filters used are typically second-order with roll-off rates of 12 dB/octave; third-order 18 dB/octave filters are rather rarer at the high end of the audio spectrum, probably because there seems to be a general feeling that phase changes are more audible at the top end of the audio spectrum than the bottom. As with subsonic filters, either the Butterworth (maximally flat) or the Bessel type can be used. It is unlikely that there is any real audible difference between the two types of filter in this application, as most of the action occurs above 20 kHz, but using the Bessel alignment does require compromises in the effectiveness of the filtering because of its slow roll-off, as I will demonstrate.

Figure 28.2a shows a second-order Butterworth low-pass filter. This time the resistors are equal while the capacitors must have a ratio of two. This creates problems as capacitors are available in a much more limited range of values than resistors − often in the E6 series, which runs 10, 15, 22, 33, 47, 68 − so filter values often have to be made up of two capacitors in parallel. (It is perfectly possible to make Sallen and Key filters where both the Rs and the Cs are equal; you just have to replace the unity-gain buffer with an amplifier with a gain of 1.586 times.[1] However, this gain is often unwanted and inconvenient.)

The Butterworth filter in Figure 28.2a has a −3 dB point set at 50 kHz, and this gives almost exactly 0.0 dB at 20 kHz, so there is no intrusion into the audio band. The response is −12.6 dB at 100 kHz and a useful −24.9 dB at 200 kHz. C1 is made up of two 2n2 capacitors in parallel.

But supposing we are worried about linear phase and we want to use a Bessel filter. The only circuit change is that C1 is 1.335 times as big as C2 instead of twice, but the response is very different. If we design for −3 dB at 50 kHz again, we find that the response is − 0.39 dB at 20 kHz, which is not exactly a stunning figure to put in your spec sheet. If we decide we can live with about −0.1 dB at 20 kHz, then the Bessel filter has to be designed to be −3 dB at 72 kHz, and this is the Bessel filter shown in Figure 28.2b, with C1 made up of two 1 nF capacitors in parallel. Due to this change, and the inherently slower roll-off, the response is only down to −5.8 dB at 100 kHz, and −15.8 dB at 200 kHz; the latter figure is almost 10 dB worse than for the Butterworth filter. Think hard before you decide to go for the Bessel option.

Once again we need to consider the way the filter input impedance loads the previous stage. In this case, the input impedance is high in the pass-band, but above the roll-off point it falls until it reaches the value of R1, which here is 1 kΩ. This is because, at high frequencies, C1 is not bootstrapped, and the input goes through R1 and C1 to the low-impedance opamp output. Fortunately this low impedance only occurs at

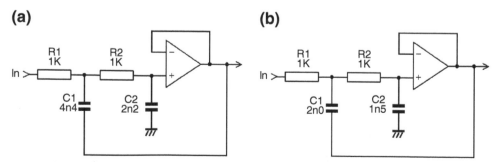

(a)   **(b)**

Figure 28.2. Ultrasonic filtering; (a) Butterworth and (b) Bessel second-order filters.

high frequencies, where one hopes the level of the signals to be filtered out will be low.

Another important consideration with low-pass filters is the balance between the R and C values in terms of noise performance. R1 and R2 are in series with the input and their Johnson noise will be added directly to the signal. Here the two 1 kΩ resistors generate −119.2 dBu of noise (22 kHz bandwidth, 25 °C), while SPICE simulation of the complete filter gives −118.9 dBu, which is pretty close; this ignores the opamp noise, which must be included to give the overall noise performance. The obvious conclusion is that R1 and R2 should be made as low in value as possible without causing excess loading, (and 1 kΩ is not a bad compromise) with C1, C2 proportioned to maintain the same roll-off frequency.

Combined Filters

The subsonic and ultrasonic filters can be combined into one stage for convenient bandwidth definition. This is feasible only because the two turnover frequencies are widely separated. Figure 28.3 shows a second-order Butterworth high-pass filter combined with a second-order Butterworth low-pass filter; the response of the two filters is exactly the same as described above for each separately, with the slight proviso that the mid-band gain is now −0.088 dB rather than precisely unity. Hopefully your overall system design can cope with this. The loss occurs because the series combination of C3 and C4, together with C2, form a capacitive potential divider with this loss figure, and this is one reason why the turnover frequencies need to be widely separated for the combining of the filters to work. If C3, C4 were smaller and C2 bigger, the loss would be greater.

A third-order high-pass filter can be used instead of the second-order version, in exactly the same way.

The only difference is that the mid-band loss is now −0.15 dB because there are three capacitors in the high-pass filter rather than two. Combined filters have the advantage that the signal now has to only pass through one opamp rather than two, and it can also be very useful if you only have one opamp section available, and you would otherwise need another dual package, of which one half would just sit there twiddling its inputs and consuming quiescent current.

This is not the place to go any deeper into the vast subject of active filters. A large number of them are described in my new book, *Active Crossover Design*[1] with simple recipes and design rules. If, however, you want to get into the complex algebra, two good in-depth references are Van Valkenburg[2] and Williams and Taylor[3].

Electronic Crossovers

A few manufacturers (BGW is one example) have produced amplifiers with internal electronic crossovers to split the incoming signal into three or more bands which are applied to separate power amplifiers and separate kinds of loudspeaker which specialise in a given frequency range. The subject of electronic crossover design is a large and complex area, and there is no room to go into it here.

Digital Signal Processing

A minority of professional power amplifiers include Digital Signal Processing facilities. The types of processing offered include filtering, equalisation, delay, level control by limiting, and the implementation of electronic crossover systems. The delay option is particularly useful in compensating for sound delays resulting from speaker placement.

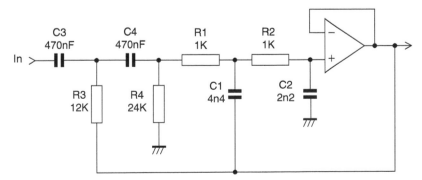

Figure 28.3. Subsonic and ultrasonic second-order Butterworth filters combined.

An LCD screen is usually provided to allow changes in configuration and parameter settings. A number of storable user-defined DSP presets are usually available so that the amplifier can quickly be reconfigured for different uses. DSP facilities are somewhat outside the scope of this book and will not be considered further here.

Signal-present Indication

Professional amplifiers are often fitted with a 'signal-present' indicator that gives reassurance that an amplifier — possibly in a bank of twenty others — is receiving a signal and doing something with it. The level at which it triggers must be well above any likely noise level, but also well below the maximum amplifier output. They are usually provided for each channel of a multi-channel amplifier, and are commonly

set up to illuminate when the channel output level exceeds 2 Vrms, which is equivalent to 1/2 W into an 8 Ω load, or 1 W into a 4 Ω load. A trigger level of 4 Vrms is also used.

A vital design consideration is that the operation of the circuit should not introduce distortion into the signal being monitored; this could easily occur by electrostatic coupling or imperfect grounding if there is a comparator switching on and off at signal frequency. This is not a problem with clipping detectors as when clipping occurs, the signal is already distorted. A typical circuit would comprise just the bottom step of the LED bar-graph meter shown in Figure 28.4 below.

Output Level Indication

Many power amplifiers include some indication of the output level. This may be an LED bar-graph, analogue

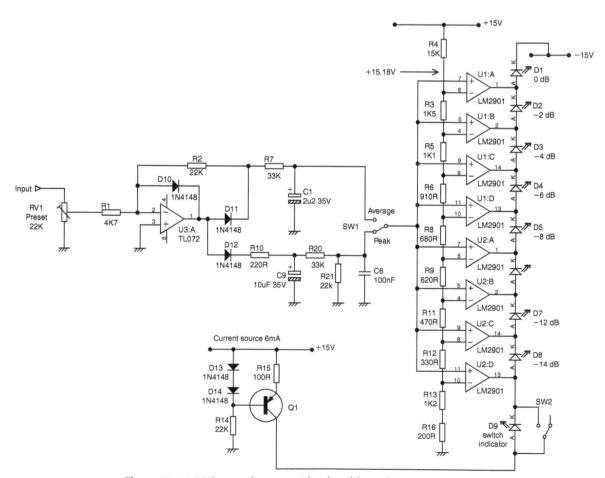

Figure 28.4. LED bar-graph meter with selectable peak/average response.

VU-type meters (though these are used mostly as a fashion statement on hi-fi amplifiers, being too fragile for on-the-road use) or simply a clipping indicator light on its own.

VU meters consist of a relatively low resistance meter winding driven by rectifier diodes, sometimes with a series resistor. It is important to remember that this represents a horribly non-linear load to an external circuit, and it must never be connected across a signal path unless it has near-zero impedance. In practice, a buffer stage is always used between a signal path and the meter to give complete isolation.

Bar-graph meters are commonly made up of an array of LEDs. Vacuum fluorescent displays are sometimes used but require hefty tooling charges if you want a custom display, and their high-voltage operation makes driving them more complicated.

An LED bar-graph meter can be made effectively with an active-rectifier circuit and a resistive divider chain that sets up the trip voltage of an array of comparators; this allows complete freedom in setting the trip level for each LED. A typical circuit which indicates from 0 dB to −14 dB in 2 dB steps with a selectable peak/average characteristic is shown in Figure 28.4 and illustrates some important points in bar-graph design.

U3 is a half-wave precision rectifier of a familiar type, where negative feedback servos out the forward drop of D11, and D10 prevents opamp clipping when D11 is reverse-biased. The rectified signal appears at the cathode of D11, and is smoothed by R7 and C1 to give an average, sort-of-VU response. D12 gives a separate rectified output and drives the peak-storage network R10, C9 which has a fast attack and a slow decay through R21. Either average or peak outputs are selected by SW1, and applied to the non-inverting inputs of an array of comparators. The LM2901 quad voltage comparator is very handy in this application; it has low input offsets and the essential open-collector outputs.

The inverting comparator inputs are connected to a resistor divider chain that sets the trip level for each LED. With no signal input, the comparator outputs are all low and their open-collector outputs shunt the LED chain current from Q1 to −15 V, so all LEDs are out. As the input signal rises in level, the first comparator U2:D switches its output off, and LED D8 illuminates. With more signal, U2:C also switches off and D7 comes on, and so on, until U1:A switches off and D1 illuminates. The important points about the LED chain are that the highest level LED is at the bottom of the chain, as it comes on last, and that the LED current

flows from one supply rail down to the other, and is not passed into a ground. This prevents noise from getting into the audio path. The LED chain is driven with a constant-current source to keep LED brightness constant despite varying numbers of them being in circuit; this uses much less current than giving each LED its own resistor to supply rail, and is universally used in mixing console metering. Make sure you have enough voltage headroom in the LED chain, not forgetting that yellow and green LEDs have a larger forward drop than red ones. The circuit shown has plenty of spare voltage for its LED chain, and so it is possible to put other indicator LEDs in the same constant-current path; for example, D9 can be switched on and off completely independently of the bar-graph LEDs, and can be used to indicate the status of a ground-lift switch or whatever. An important point is that in use the voltage at the top of the LED chain is continually changing in 2-Volt steps, and this part of the circuit should be kept well away from the audio path to prevent horrible crunching noises from crosstalking into it.

This meter can of course be modified to have a different number of steps, and there is no need for the steps to be the same size. It is as accurate in its indications as the use of E24 values in the resistor divider chain allows.

If a lot of LED steps are required, there are some handy ICs which contain multiple open-collector comparators connected to an in-built divider chain. The National LM3914 has 10 comparators and a divider chain with equal steps, so they can be daisy-chained to make big displays, but some law-bending is required if you want a logarithmic output. The National LM3915 also has 10 comparators, but with a logarithmic divider chain covering a 30 dB range in 3 dB steps.

Signal Activation

With increasing attention being paid to economy in the use of energy, it is now quite common for power amplifiers to have a stand-by mode where the main transformer is disconnected from the supply when the unit is not in use, but a small stand-by transformer remains energised to run a housekeeping microcontroller. This is particularly pertinent if large Class-A amplifiers are in use. It is therefore very convenient to have the amplifier wake-up automatically when a signal is applied. It is now only necessary to pop in a CD or whatever, and start it playing; there is no need to push a button on the power amplifier. This is especially useful when large amplifiers are in use that are hidden away out of sight, or when monobloc power amplifiers are used, in which case, two buttons in different locations have to be pushed.

Signal activation operates by detecting when a low-level signal appears at any of the amplifier inputs, low-level in this case meaning a long way, such as −60 dB, below that which gives maximum output but hopefully well clear of the noise floor to avoid false triggering. The idea is that triggering on a low enough level will mean that you don't miss much of the start of the music; but you are always going to miss a fraction of a second. If the amplifier has a wake-up time of several seconds, which is quite common, because of the need for an inrush suppression delay, followed by a period in which muting is maintained while the internal voltages settle, you are going to miss rather more and you will probably find yourself restarting the CD. Signal activation is not without its limitations. In the case of multi-channel amplifiers, seven or more inputs must all be monitored for the onset of a valid audio signal.

The principle of signal activation is very simple, and it appears straightforward to implement — you just have to sum all the amplifier inputs together (you cannot use the amplifier outputs, of course, because the amplifiers are not active yet), put them through a high-gain amplifier, and apply the result to a level detector which in turn signals a microcontroller or otherwise wakes up the system. However, it may not come as a total surprise that in the real world things are a little more complicated than that. The challenges and their solutions are described with reference to the signal activation system shown in Figure 28.5, which is loosely based on one of my commercial designs. It is considered in

some detail below to bring out the various important points. The design interfaces to a microcontroller, but the same design principles can be used to interface with discrete logic.

Firstly, if you have a high-gain amplifier connected to the amplifier inputs, it is going to be clipping hard all the time when normal signal levels are applied. This is not a good thing, as it takes sharp-edged gulps of current from the opamp supply rails and dumps them into ground; even with good grounding practice it is possible for these ugly currents to contaminate a signal that you are trying to reproduce with less than 0.001% THD. In addition, the high-gain amplifier output will consist of a series of sharp edges that may get into the audio path by simple capacitive crosstalk. It is always wise to put the activation circuitry as far away as possible physically from susceptible audio paths, but in modern equipment that is not usually very far. Screening plates would obviously help with this but extra bits of metalwork cost money and are to be regarded as a last resort.

The answer to the problem is to prevent the high-gain amplifier from clipping by clamping its output, applying increased negative feedback when the limits of the desired output excursion are reached. In this way the output of the stage always remains under negative feedback control. There are many ways to do this, but the simplest is adding a couple of Zener diodes to the feedback path. This approach greatly reduces the potential problems, but there is of course still a distorted signal on the high-gain amplifier output, and this must still be kept away from sensitive audio circuitry.

Secondly, it is important to avoid false triggering. If an amplifier switches on and off the supply to the main

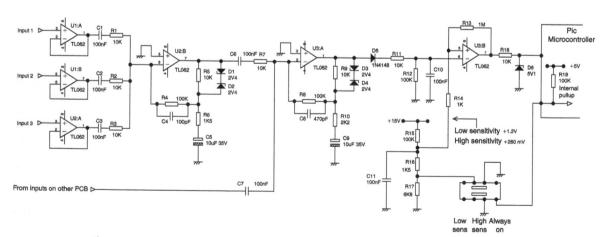

Figure 28.5. Circuit diagram of a signal-activation system for a multi-channel amplifier.

transformer with relays, these will usually be quite hefty and will make very audible clicking noises when switching from stand-by to the active state and *vice versa*. If mains noise, etc. is continuously initiating these cycles of clicking, the paying customer is soon going to get irritated. This is bad enough, but there is also the point that most inrush-protection circuitry will not take kindly to continuous on/off cycling, and ultimately the inrush resistor might overheat and burn out. Not good.

Ideally, the activation system should not trigger even when all the inputs of the amplifier are left open-circuit. This is a rather severe requirement, and not everyone would agree it was necessary, but it can be met by informed design and thorough testing of prototypes. The design shown here includes a switchable high/low sensitivity control so that signal activation is still usable in non-optimal conditions, and also an 'always on' switch position to cater for those who feel that amplifiers have to be powered constantly if they are to give of their best.

The main defence against false triggering is the restriction of the bandwidth the activation system responds to. Curtailing the frequency response at the low end prevents hum from giving a false trigger; likewise, limiting the high-frequency response discriminates against noise and transients. In the case of an open-circuit input, the typical hum waveform from electrical pick-up is a severely distorted travesty of a 50 Hz sine wave, with strong harmonics going up to at least 500 Hz, and so the low-frequency response must be curtailed quite dramatically. When the amplifier is activated, it is important that it remains on for some time after the signal inputs cease, say, 30 seconds or longer. This will prevent the amplifier going to stand-by between tracks of music.

Thirdly, since all the inputs meet at the signal-activation system, it is important that this is not a source of crosstalk between them.

The signal activation system of Figure 28.5 was designed for a seven-channel home theatre amplifier which has happily proved to have a long life in the market-place. The electronics were divided into two big PCBs, one carrying three channels of power amplifier and the other four. Here is the description.

The seven inputs are of the usual unbalanced type with phono (RCA) connectors. The activation circuitry is on the PCB carrying the three power amplifiers, and each input is connected to a unity-gain buffer (U1:A, U1:B, and U2:A), which eliminates the possibility of crosstalk between them. The three buffered outputs are DC-blocked by C1, C2, and C3 and then fed through

R1, R2 and R3 into the virtual-earth summing point of amplifier U2:B. The small capacitor value of 100 nF together with a 10 k input resistance gives an LF roll-off of −3 dB at 160 Hz to discriminate against hum. It might be objected that the buffers are superfluous as the virtual-earth will stop signals coming in one input and then sidling out of the other, but in real life a virtual-earth is not a perfect earth, and carries enough residual signal to compromise the crosstalk performance. The other point is that R1, R2 and R3 are low enough in value to excessively load the amplifier inputs. They have to be relatively low to give high gain with a feedback network that uses practicable resistor values.

It should be explained here that one of the reasons why the high-gain amplifier is divided into two stages is that the large amount of closed-loop gain required (about +70 dB) would mean a shortage of open-loop gain at high audio frequencies if only one opamp was used.

The first amplifier stage U2:B has a mid-band gain of +38 dB, and uses shunt feedback in the usual way to generate a virtual-earth at Pin 6. The feedback network R4, R5, R6 is in the form of a T-network C4 across R4 provides an HF roll-off of −3 dB at 16 kHz to discriminate against HF noise. C5 reduces the gain to unity at DC, and gives an LF roll-off at 11 Hz. D1 and D2 are 2V4 Zener diodes that provide output clamping by increasing the negative feedback when the output exceeds about 3 V peak in either direction.

The second amplifier stage U3:A is a similar shunt-feedback stage with a virtual-earth at Pin 2. It is fed from the first stage just described via capacitor C6 and input resistor R7. Once again the capacitor value of 100 nF together with R7 gives an LF roll-off of −3 dB at 160 Hz. The other PCB, which carries the other four power amplifiers, also has four more identical unity-gain buffers feeding another first amplifier stage. On this PCB there also reside DC-blocking and input components equivalent in function to C6 and R7, and there is an important point relating to why they are there and not on the same PCB as the second amplifier stage. At first sight it is risky to send a signal from one PCB to another in current-mode (i.e., at virtual-earth) because such signals are vulnerable to capacitive crosstalk unless they are screened. For cost reasons, no screening was used here; inter-PCB signals were carried by a ribbon cable and adding a screened wiring assembly to this was not a tempting proposition. However, the first amplifier stages raise the signal level sufficiently so that there is no possibility of crosstalk from other signals causing false triggering of the activation system.

The real benefit of this philosophy is that the signal, being in current mode and at a negligible voltage, cannot itself crosstalk to parts of the main audio paths. This approach works very well.

The second gain stage U3:A has a mid-band gain of +35 dB, using shunt feedback through T-network R8, R9, R10 to generate the virtual-earth at Pin 2. C8 across R8 provides an HF roll-off of −3 dB at 3.4 kHz which discriminates heavily against HF noise. C9 reduces the stage gain to unity at DC, rolling off at 7.2 Hz, and D3 and D4 are once more 2V4 Zener diodes which clamp the output to about 3 V peak.

The output of the second gain stage drives a simple peak detector made up of D5 and C10. This only gives half-wave rectification but seems to be perfectly satisfactory in practice and the extra expense of full-wave rectification appears to be pointless. R11 gives a slow attack time to further discriminate against isolated noise pulses, and R12 defines the decay time. The stored voltage on C10 is applied to comparator U3:B, which has a threshold voltage generated by the divider R15, R16, R17 and switched by SW1 to give the high and low sensitivity settings. For the low sensitivity mode, one section of SW1 removes the short across R17 and increases the threshold voltage from +260 mV to +1.2 V, reducing sensitivity by 13 dB.

R13 and R14 provide a small amount of positive feedback to introduce a little hysteresis and give clean comparator switching. The comparator output is clamped to +5 V by R18 and Zener D6, and applied to the input port of a microcontroller, in this case one of the PIC family. Another input port is used in conjunction with the remaining half of SW1 to sense when sensitivity switch SW1 is in the 'always on' position; the internal pull-up facility of the PIC is used to simplify this bit of the circuit. Note that TL062 opamps, rather than the more familiar TL072, are used because of their lower input offset voltage. This is particularly important in the comparator stage where the voltages are low.

The long turn-off delay before the unit returns itself to stand-by is implemented in software in the microcontroller, as it is inconveniently long to be done in hardware. It is also possible to incorporate further discrimination against false-triggering in the software, for example, by disregarding single input pulses that are not followed by further input signals within a specified time interval.

The circuitry described above fully met the demanding requirement that it should not false-trigger when all the amplifier inputs were left open-circuit. In other applications this immunity will depend on many details of the design, such as the input impedances, the type of input connectors used, their proximity to mains wiring, and so on.

12 V Trigger Activation

Another method of activating a power amplifier from stand-by is the use of a 12 V trigger. Typically a preamplifier (which will be at hand for access to the controls) is connected to a remote power amplifier by a cable with 3.5 mm jack plugs at each end. When the preamplifier is turned on, it sends out a +12 V DC signal that tells the power amplifier to come out of stand-by and become active. A vital point is that the connection must be opto-isolated to prevent the formation of a ground loop; this is done at the input (power amplifier) end. A typical +12 V trigger input and output system is shown in Figure 28.6.

It is important to remember that a 12 V trigger line might be connected to almost anything, as it is not an audio in/out and there are a lot of people out there who are pretty vague about how it works. Having once seen a cable with a 13-Amp mains plug on one end and a phono connector on the other, I believe anything is possible. (I also saw the result of plugging this lead into an expensive Revox reel-to-reel recorder; the owner, who was lucky to be alive, did not seem to like the sound of 'Beyond economic repair, guv.')

The 12 V trigger output must therefore be protected against short-circuits and against being connected to reverse voltages. In Figure 28.6 the PIC microcontroller switches on Q3, which switches on Q2, which in turn applies +15 V to the 100 mA 78L12 regulator U2. This not only provides a regulated +12 V output, but is also current-limiting. D3 protects against intrusive reverse polarities.

The trigger input must be protected against excessive input voltages and reverse-polarity; it must also be designed to reject quite high levels of electrical noise, at both low and high frequencies. R1 limits the current drawn from the transmitting unit, C1 filters out noise and D2 protects against reverse polarity. The 3V3 Zener D1 ensures that incoming voltages have to exceed a threshold of about 5 V before the opto-isolator is activated. The opto output turns on Q1, which sends a low to the PIC when an incoming trigger occurs; the values of R2 and R3 depend on the opto characteristics.

Infra-red Remote Control

An infra-red remote control facility is now very common on preamplifiers for source selection, volume

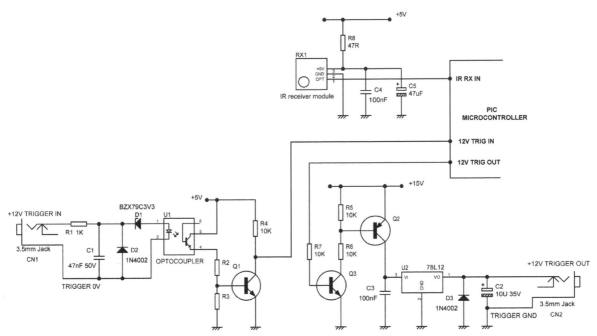

Figure 28.6. Typical 12V trigger in/out system and IR receiver facility for a power amplifier.

and balance control, muting, and on/stand-by control. Very often there are other control functions as well. The application of IR control to power amplifiers is rather rarer, but it is sometimes used for on/stand-by switching. Commands are transmitted using the Philips/Sony RC5 code modulated onto a carrier, typically at 37 kHz; this in turn is modulated on the IR emitted by the hand controller. The receiving circuitry required is very simple to arrange, most of the complexity being contained in a small transistor-sized component such as the Toshiba TSOP348XX series. The IR sensor, amplifier, AGC loop, bandpass filter, and demodulator circuitry are all integrated; carrier frequencies between 30 and 56 kHz are available. The only real precaution required with these devices is to make sure you have effective supply rail decoupling close to the module. This is carried out by R8 and C4, C5 in Figure 28.6. After demodulation from the carrier the RC5 decoding is carried out in software by the microcontroller.

Other Amplifier Facilities

There are several other facilities that may appear on a professional amplifier, but are much less likely to be found in the hi-fi world.

- Temperature indication: some amplifiers go beyond a simple 'over-temperature' indicator, and have a bar-graph display that reads the heatsink temperature. This can be useful as a rise in temperature due to obstructed ventilation or whatever can be detected before it puts the amplifier into shutdown. It does, of course, assume that someone has the time to keep an eye on a dozen or more temperature displays.
- Fan-running indicator: An LED that illuminates when a thermostatic fan-control system turns the fan on. This gives confidence that the cooling system is working, and can also give advanced warning of imminent overheating before shutdown becomes necessary.
- Fuse indicators: A few amplifiers are fitted with LEDs that indicate when internal fuses have blown.

References

1. Self, D, *Active Crossover Design*, Burlington, MA: Focal Press, 2011, p. 196.

2. Van Valkenburg, M E, *Analog Filter Design*, New York: Holt-Saunders International Editions, 1982.

3. Williams, A B and Taylor, F J, *Electronic Filter Design Handbook*, 4th edn, Maidenhead: McGraw-Hill, 2006.

Chapter **29**

Testing and Safety

Out of this nettle, danger, we pluck this flower, safety.

William Shakespeare, *Henry IV.1*

Simulating Amplifiers

The use of SPICE simulation is invaluable in amplifier design, cutting out the need for most of the mathematics and allowing purely conceptual models to be evaluated. I use it all the time. There are, however, some limitations which should be appreciated:

Firstly, SPICE should not be used to in attempts to determine the distortion performance of complete amplifiers. This is due to limitations in the semiconductor models. SPICE uses approximate mathematical expressions to define how beta changes with collector current, how Early effect influences collector current (it is modelled as linear, but this is a great oversimplification), and how the non-linear base-collector capacitance Cbc changes with voltage. These limitations are of minor importance in most SPICE applications, but have a large effect on simulated distortion results.

Secondly, it is necessary to be very cautious when mixing virtual components like ideal current sources with simulated real components. I stumbled into this one when doing some investigations into how slew-rate was affected by VAS loading. To save time, the input stage was represented as a differential voltage controlled current source (VCIS). The slew-rates obtained were impressive but not at all in line with reality, and the reason was that the VCIS was happily driving circuit nodes in the amplifier to kilovolt levels in order to force current through the circuitry. It was rapidly replaced by a transistor differential pair and the results became realistic. None of the versions of SPICE that I have used give explicit warning when component capabilities are exceeded.

The simulations in this book were run on B2SPICE and PSPICE as appropriate.

Prototyping Amplifiers

There are two main ways to prototype an audio amplifier. If you are confident that your circuitry is correct, or at any rate as close as you can get without testing it, the best way is to go straight to PCB. The first version may not be completely correct, but since it is very unusual to get from prototype to production without at least one board iteration (though I have done it), you will have at least one chance to put

things right. Going straight to PCB also allows all the mechanical dimensions to be validated.

Alternatively, if you are trying out new circuitry that you are far from sure of, or just researching ideas, the small-signal parts of the amplifier can be built by plugging components and links into a prototype board (also known as a breadboard, protoboard, or plugboard).This may sound like an iffy business, but it is entirely practical, and many of the measurements in this book were taken from amplifiers so constructed. There are two limitations; firstly, prototype boards are obviously not meant to handle large currents, and so the driver and output transistors are mounted on an adjacent heatsink, and the power supply is a separate item. The second point is that prototype boards have a non-negligible capacitance between the contact rows, typically 5 pF. In a three-stage amplifier this is usually not important, as there are no high-impedance circuit nodes, and the worst that can happen is that your compensation Miller capacitance is 5 pF more than you thought it was. That problem can be avoided simply by leaving a contact row between the collector and base of the VAS transistor, and grounding it. Other amplifier configurations such as the four-stage are likely to be more vulnerable to the inter-contact capacitance.

Testing and Fault-finding

Testing power amplifiers for correct operation is relatively easy; faultfinding them when something is wrong is not. I have been professionally engaged with power amplifiers for a long time, and I must admit I still sometimes find it to be a difficult and frustrating business.

There are several reasons for this. First, almost all small-signal audio stages are IC-based, so the only part of the circuit likely to fail can be swiftly replaced, so long as the IC is socketed. A power amplifier is the only place where you are likely to encounter a large number of components all in one big negative feedback loop. The failure of any components may (if you are lucky) simply jam the amplifier output hard against one of the rails, or (if you are not) cause simultaneous failure of all the output devices, possibly with a domino-theory trail of destruction winding through the small-signal section. A certain make of high-power amplifier in the mid-1970s was a notorious example of the domino-effect, and when it failed (which was often), the standard procedure was to replace *all* of the semiconductors, back to and including the bridge rectifier in the power supply.

The advice given here is aimed primarily at the power amplifier designs included in this book, but is general enough to apply to most semiconductor power amplifiers.

By far the most important step to successful operation is a careful visual inspection before switch-on. As in all power amplifier designs, a wrongly installed component may easily cause the immediate failure of several others, making fault-finding difficult, and the whole experience generally less than satisfactory. It is therefore most advisable to meticulously check:

- That the supply and ground wiring is correct.
- That all transistors are installed in the correct positions.
- That the drivers and TO3 output devices are not shorted to their respective heatsinks through faulty insulating washers.
- That the circuitry around the bias generator transistor in particular is correctly built. An error here that leaves this transistor turned off will cause large currents to flow through the output devices and may damage them before the rail fuses can act.
- That the bias adjustment is set to minimum.

For the Trimodal amplifier in Chapter 17, I recommend that the initial testing is done in Class-B mode. There is the minimum amount of circuitry to debug (the Class-A current-controller can be left disconnected, or not built at all until later) and at the same time the Class-B bias generator can be checked for its operation as a safety-circuit on Class-A/AB mode.

The second step is to obtain a good sinewave output with no load connected. As explained in the next section, it is strongly recommended that the supply voltage is increased slowly from zero. A good power amplifier design will give a visibly correct sinewave with only a few volts on the supply rails, and the risk of damage is minimised.

A fault will often cause the output to sit hard up against either rail; this should not in itself cause any damage to components. Since a power amplifier consists of one big feedback loop, localising a problem can be difficult. The best approach is to take a copy of the circuit diagram and mark on it the DC voltage present at every major point. It should then be straightforward to find the place where two voltages fail to agree, e.g., a transistor installed backwards usually turns fully on, so the feedback loop will try to correct the output voltage by removing all drive from the base. The clash between 'full-on' and 'no base-drive' signals the error.

When checking voltages in circuit, bear in mind that in my designs the feedback network capacitor is protected against reverse voltage in both directions by diodes which will conduct if the amplifier saturates in either direction.

This DC-based approach can fail if the amplifier is subject to high-frequency oscillation, as this tends to cause apparently anomalous DC voltages. In this situation the use of an oscilloscope is really essential. An expensive oscilloscope is not necessary, but a bandwidth of at least 50 MHz is essential to avoid missing some kinds of parasitic oscillation. (Though they will certainly make their presence felt by their effect on the THD residual.) A digital scope is at a serious disadvantage here, because HF oscillation is likely to be aliased into nonsense and be hard to interpret.

The third step is to obtain a good sinewave into a suitable high-wattage load resistor. It is possible for faults to become evident under load that are not shown up in Step 2 above.

Setting the quiescent conditions for any Class-B amplifier can only be done accurately by using a distortion analyser. If you do not have access to one, the best compromise is to set the quiescent voltage-drop across both emitter resistors to 10 mV when the amplifier is at working temperature; disconnect the output load to prevent DC offsets causing misleading current flow. This should be close to the correct value, and the inherent distortion of the designs is so low that minor deviations are not likely to be very significant. This implies a quiescent current of approximately 50 mA.

It may simplify faultfinding if the diodes in the collectors of the protection transistors are not installed until the basic amplifier is working correctly, as errors in the SOAR protection cannot then confuse the issue. This demands some care in testing, as there is then no short-circuit protection.

I insert here a few precautions learnt the hard way; as Benjamin Franklin put it, experience keeps a dear school, but fools will learn in no other.

- Make sure the reservoir capacitors are *fully* discharged before applying your soldering iron to the circuitry. Use a 10 Ω wirewound resistor to do this, not a screwdriver.
- Remove the earth clip from the oscilloscope probe unless you are really using it (not likely). The little croc clip will sooner or later (in my world, sooner) drape itself across some circuitry and things will end badly.
- If you have one of those handy trimming tools for bias adjustment, take off the metal clip and throw it away *before* you drop the tool into a 200 W amplifier.

- If the THD is too high, check that the output connections are tight, if they are in the form of binding posts.

Powering up for the First Time

Testing an amplifier by hitting it with its full operating voltage is a risky business. Slowly winding up a variable voltage transformer from zero is usually much safer. But not always — some amplifiers with complicated compensation schemes are not unconditionally stable (in the true sense of the word) and are in fact very unstable when operated from supply rails that are much below normal. I am thinking here particularly of the four-stage Otala architecture described in Chapter 4. This is never a problem with three-stage amplifiers using straightforward dominant pole Miller compensation.

It is clearly desirable for an amplifier to start working, even if imperfectly, as soon as possible when the supply rails are being wound up from zero. The sooner you can find out if something is amiss, the better chance you have of avoiding damage. Taking power transistors off heatsinks and replacing them is a time-consuming business, and it doesn't take a lot of that sort of thing to knock a hole in your profit margins.

How low a voltage can a power amplifier be expected to work from, and give a clean-looking sinewave which is very good evidence that all is well? The measurements in Table 29.1 were made on a three-stage power amplifier with a very similar circuit to the Blameless amplifiers, designed to work from ±65 V rails. It first began to show signs of life at a rail voltage of ±3.8 V, though the maximum unclipped output level was very restricted at 128 mV.

Even so, a visually good sinewave was obtained with no load; a sinewave with 0.51% THD is visually perfect, and even the 5% distortion in the loaded condition, which is of course a more searching test that all is

well, is not that easy to see as it is all crossover distortion due to the bias being set to minimum. As the supply rail is raised further, the maximum output increases rapidly and the distortion, both unloaded and loaded, falls rapidly. By ±5 V you can be pretty sure that all is as it should be, especially if you have a note of what output and distortion to expect under these conditions. Continuous checking of device temperatures with a questing finger is a wise precaution; if something is getting disquietingly hot at less than a tenth of the intended supply voltage, some prompt investigation is called for. After that, a quick check at perhaps half-rail voltage, and you can then apply full volts with some confidence.

The very great desirability of this gradual start-up procedure has implications for the design of the power amplifier. Don't go for current-source biasing schemes that need a lot of volts to work properly. And *don't* put a resistor in the tail of the input stage pair, as I have done in the past. In past editions both the Load Invariant amplifier (2k2), and the Trimodal amplifier (1 kΩ) had these resistors, I put my hands up; it was a mistake. The notional function of the resistor in the tail was to minimise the damage if the tail current-source transistor failed short-circuit; this is actually very unlikely, and I have yet to come across a case of it. The unwanted result was that these amplifiers would not work on very low rail voltages because of the voltage drop across the tail resistor caused by the 6 mA tail current.

This cautious start-up procedure can protect you from even the most bone-headed mistakes. On one notable occasion a Blameless amplifier was powered up in this way with the supply rails reversed. The negative supply rail fuse blew but the amplifier survived completely undamaged in all other respects.

Power Supplies for Testing

The previous section assumes that the amplifier is being powered from a simple unregulated supply (which is what I strongly recommend, see Chapter 26). In many cases, initial tests can be done with a dual-rail bench supply. My Blameless amplifiers are quite happy when powered this way, though the output power is limited by the supply current capacity, and it does allow a cautious start-up without the expense of buying a big variable transformer. My amplifiers all have quite size-able rail decoupling capacitors on the amplifier PCB, (usually 220 uF) and this is adequate to ensure stability. Other amplifier designs may not take so well to being

Table 29.1. Amplifier distortion levels at very low rail voltages

Supply rail Volts	Output level RMS	THD No Load (%)	THD into 8 Ω (%)
± 5.4V	640 mV	0.038	0.75
± 4.1V	256 mV	0.17	1.6
± 3.8V	128 mV	0.51	5

powered in this way and you should keep a careful lookout for HF instability.

Safety when Working on Equipment

This section considers the safety of the designer and the service technician. The recommendations here are advisory only. Regulations bearing on the safety of the user are backed by law; they are considered in the next section.

There are some specific points that should be considered:

1. An amplifier may have supply-rails of relatively low voltage, but the reservoir capacitors will still store a significant amount of energy. If they are shorted out by a metal finger-ring, then a nasty burn is likely. If your bodily adornment is metallic, then it should be removed before diving into an amplifier.

2. Any amplifier containing a mains power supply is potentially lethal. The risks involved in working for some time on the powered-up chassis must be considered. The metal chassis *must* be securely earthed to prevent it becoming live if a mains connection falls off, but this presents the snag that if one of your hands touches live, there is a good chance that the other is leaning on chassis ground, so your well-insulated rubber-soled shoes will not save you. All mains connections (Neutral as well as Live, in case of mis-wired mains) must therefore be properly insulated so they cannot be accidentally touched by finger or screwdriver. My own preference is for double insulation; for example, the mains inlet connector not only has its push-on terminals sleeved, but there is also an overall plastic boot fitted over the rear of the connector, and secured with a tie-wrap. Note that this is a more severe requirement than BS415 which only requires that mains should be inaccessible until you remove the cover. This assumes a tool is required to remove the cover, rather than it being instantly removable. In this context a coin counts as a tool if it is used to undo giant screwheads. If you are working on equipment with exposed mains voltages, taking the time to improvise some temporary insulation with plastic sheet and tape might just save your life.

3. Switch-mode supplies are even more dangerous, as they contain capacitors charged to 400 V DC by rectification of the incoming mains. DC supplies are proverbially dangerous as the contraction of the muscles may mean you cannot let go. Be VERY careful with these things. Set up the equipment so it is impossible to touch the 400 V section.

4. Be very wary about leaning over equipment you are not sure of, and never do it when you are switching on for the first time. I was once involved with the use of an outside consultant to design a switch-mode supply, and I retain vivid memories of the first time he switched one of the prototypes on. There was a violent explosion directed upwards, followed by an almost perfect scale reproduction of the Bikini Atoll mushroom cloud. 'Ah,' said the consultant. 'This is an opportunity to refresh the design!' The moral of this story is not that you should never employ consultants, but that you should never lean over unproven equipment.

5. A Class-A amplifier runs *hot* and the heatsinks may well rise above 70°C. This is not likely to cause serious burns, but it is painful to touch. You might consider this point when arranging the mechanical design. Safety standards on permissible temperature rise of external parts will be the dominant factor.

6. Readers of hi-fi magazines are frequently advised to leave amplifiers permanently powered for optimal performance. Unless your equipment is afflicted with truly doubtful control over its own internal workings, this is quite unnecessary. (And if it *is* so afflicted, personally I would turn it off Right Now!) While there should be no real safety risk in leaving a soundly constructed power amplifier powered permanently, I see no point and some potential risk in leaving unattended equipment powered; if you prefer big Class-A amplifiers, there may be a hefty impact on your electricity bill.

7. The Dress Code for working on power amplifiers comes down to jeans and T-shirt, for practicality rather than as a style statement. Cotton is resistant to molten solder, while fabrics based on polyester and the like melt instantly and allow the hot metal straight through to your quivering flesh. Put on your steampunk goggles if you are likely to be spattering molten solder about, or there is a possibility that something will explode. Shoes with plastic soles provide some protection against electric shock, but all too often one hand or elbow will be resting on something grounded. Safety boots with reinforced toe-caps are a good plan if you are likely to be dropping 1 kVA toroids on your feet.

Warning

This section of the book is intended to provide a starting-point in considering safety issues. Its main purpose is to alert you to the various areas that must be considered. For reasons of space it cannot be a comprehensive manual that guarantees equipment compliance; it cannot give a full and complete account of the various safety requirements that a piece of electronic equipment must meet before it can be legally sold. If you plan to manufacture amplifiers and sell them, then it is your responsibility to inform yourself of the regulations. The regulations change, always in the direction of greater safety and hence greater severity, and you must keep up to date. All the information here is given in good faith and is believed to be correct at the time of writing, but I accept no responsibility for its use.

Safety Regulations

The overall safety record of audio equipment is very good. I am only aware of one British case in which an amplifier set fire to a house; I can assure you it was not made by a company I have ever worked for. There is, however, no room at all for complacency. The price of safety, like that of liberty, is eternal vigilance. Safety regulations are not in general hard to meet so long as they are taken into account at the start of the mechanical design phase, and enough time is set aside for checking component approvals and safety testing.

European safety standards are defined in a document known as BS EN 60065:2002 'Audio, Video and Similar Electronic Apparatus: Safety Requirements'. The BS EN classification means it is a European standard (EN) having the force of a British Standard (BS). The latest edition was published in May 2002. It is produced by CENELEC, the European Committee for Electrotechnical Standardisation.

In the USA, the safety requirements are set by the Underwriters Laboratories, commonly known simply as 'UL'. The relevant standards document is UL6500, 'Audio/Video and Musical Instrument Apparatus for Household, Commercial, and Similar General Use', ISBN 0-7629-0412-7. The name 'Underwriters Laboratories' indicates that this institution had its start in the insurance business, allegedly because American houses tend to be wood-framed and are therefore more combustible than their brick counterparts.

The requirements for Asian countries are essentially the same, but it is essential to decide at the start which countries your product will be sold in, so that all the necessary approvals can be obtained at the same time. Changing your mind on this, so things have to be re-tested, is very expensive.

Electrical Safety

This is safety against electrical shocks. There must be no 'hazardous live' parts accessible on the outside of the unit, and precautions must be taken in the internal construction so that parts do not become live due to a fault.

A part is defined as 'hazardous live' if under normal operating conditions it is at 35 V AC peak or 60 V DC with respect to earth. Under fault conditions 70 V AC peak or 120 V DC is permitted. Professional equipment, defined as that not sold to the general public, is permitted 120 V rms; there are also special provisions for audio signals. You are strongly advised to consult page 50 of BS EN 60065:2002 for more detailed information.

To determine if a 'hazardous live' part is accessible, the jointed test finger called 'Test Probe B' (to IEC 61032) is used; it is pushed against the enclosure or inserted through any openings. This is repeated with 'small finger probes' (see IEC 61032 again). Openings are also tested by inserting a 4 mm diameter metal test pin by up to 100 mm.

Connectors are also tested by inserting a 1 mm diameter metal test pin by up to 100 mm. No hazardous voltages may be touched by it. This may be a serious problem with phono (RCA) connectors and some female XLR connectors, which allow the test pin to pass right through the rear of the connector and into the equipment. The force used on the 1 mm test pin is 20 Newtons, the force exerted by a 2.04 kg weight. This rules out blocking test probes with cunningly placed vertical electrolytics.

Mains connections are always well insulated and protected where they enter the unit, normally by IEC socket or a captive lead, so the likeliest place where such voltages may appear is on the loudspeaker terminals of a power amplifier. An amplifier capable of 80 W into 8 Ω will have 35 V AC peak on the output terminals when at full power.

This seems like a tricky situation, but the current interpretation seems to be that the contacts of loudspeaker terminals, if they are inaccessible when they are fastened down, may be 'hazardous live', provided they are marked with the lightning symbol on the adjacent panel. Strictly speaking, one should consider the operation of connecting speaker cables to be 'by hand' and therefore the contacts should be inaccessible at all times, i.e., closed or open. However, the general view seems to be that the connection of speaker terminals is a rare event and that adequate user instructions will be sufficient for the 'by hand' clause to be disregarded.

The instructions would be of the form: 'Hazardous live voltages may be present on the contacts of the loudspeaker terminals … before connecting speaker cables disconnect the amplifier from the mains supply … if in doubt consult a qualified electrician.' I would remind readers at this point that such an interpretation appears to be the current status quo, but things can change and it is their responsibility to ensure that their equipment complies with the regulations.

Loudspeaker terminals that can accept a 4 mm banana socket from the front have been outlawed for some time. Existing parts can be legally used if an insulating bung is used to block the hole; indeed, this is the preferable way, because a lot of people (myself included) like to use banana plugs. It is therefore wise to configure the bung so it can be removed by the customer on his own responsibility without too much of a struggle.

Unless the equipment is double-insulated, also known as Class II, an essential safety requirement is a solid connection between mains ground and chassis, to ensure that the mains fuse blows if Live contacts the metalwork. The differences between Class I (grounded) and Class II (double-insulated) are described in Chapter 25. British Standards on safety require the mains earth to chassis connection to be a *Protected Earth,* clearly labelled and with its own separate fixing. A typical implementation has a welded ground stud onto which the mains-earth ring-terminal is held by a nut and locking washer; all other internal grounds are installed on top of this and secured with a second nut/washer combination. This discourages service personnel from removing the chassis ground in the unlikely event of other grounds requiring disconnection for servicing. A label warning against 'lifting the ground' should be clearly displayed.

The ground wire from the chassis to the rear of the IEC socket, if it is soldered, must be wrapped around the terminal to make a sound mechanical joint before soldering, and *not* just poked through the hole and soldered. If push-on connectors instead of solder are used, no further restraint of the ground wire is required, but adding a cable-tie close to the IEC connector to keep Live, Neutral, and Ground together is a sensible extra precaution.

In the internal construction, two of the most important requirements to be observed are known as 'Creepage and Clearance'. Creepage is the distance between two conductors along the surface of an insulating material. This is set to provide protection against surface contamination which might be sufficiently conductive to create a hazard. While the provisions of BS EN 60065:2002 are

Table 29.2. Creepage distances between conductors

Conductors	Creepage Distance (mm)
Live to Earth	3
Neutral to Earth	3
Live to Neutral	6
Live to low-voltage circuitry	6

complex, taking into account the degree of atmospheric pollution and the insulating material involved, the usual distances employed in domestic equipment are shown in Table 29.2.

Different sorts of Live tracks (e.g., before and after a mains switch, or a fuse) must have a minimum creepage distance of 2.5 mm between them for standard 230 V mains. More information can be found on p. 74 of BS EN 60065:2002.

Clearance is the air gap between two conductors, set to prevent any possibility of arcing; obviously the spacing between live conductors and earthed metalwork is the most important. The minimum air spacing is 2 mm. More information can be found on p. 70 of BS EN 60065:2002.

Live cables must be fixed so that they cannot become disconnected, and then move about creating a hazard. This is important where cables are connected directly into a PCB. If the solder joint to the PCB breaks, they must still be restrained. The two most common ways are:

1. Fixing the cable to an adjacent cable with a cable tie or similar restraint. The tie must be close enough to the PCB to prevent the detached cable moving far enough to cause a hazard. Obviously there is an assumption here that two solder joints will not fail at the same time. See Figure 29.1.

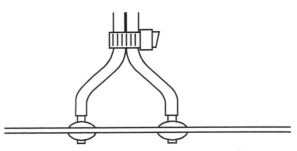

Figure 29.1. Cable restraint by fixing it to an adjacent cable.

2. Passing the cable through a plain hole in the PCB, and then bending it round through 180° to meet the pad and solder joint, as shown in Figure 29.2. This is often called 'hooking' or 'looping'.

Shocks from the Mains Plug

The need for EMC approval has resulted in X-capacitors being connected between Live and Neutral, and if the equipment mains switch is open, these can hold a charge when the equipment is unplugged, depending on the mains voltage at the instant of breaking contact. To prevent shocks from mains plug pins, a drain resistor should be connected across the X-capacitor. An ordinary 270 kΩ 1/4 W resistor can be used. A 220 kΩ resistor is often seen in this position, but at 230 V it dissipates 240 mW, which is not exactly a large safety margin; 270 k only dissipates 196 mW and is to be preferred. 270 kΩ with an X-cap of 470 nF gives a time constant of 0.12 seconds, so in the worst case the voltage on the plug pins will have dropped from 230 V to 32 V in a quarter of a second, and it is going to be difficult to get your fingers on the pins faster than that when you're unplugging something.

This is one time when the voltage rating of resistors matters. 1/4 W parts are usually rated at 500 V or 700 V and this is fine for 230 V mains; 1/8 W resistors are often only rated at 200 V and are not suitable for this application, even if their value is such that they could withstand the power dissipation.

Touch Current

As mentioned in Chapter 25 when describing Class I and Class II equipment, the amount of current that can flow to ground via a human being when they touch the casework is an important issue. A Class I (grounded) piece of equipment in normal use should have no touch current at all, as even a tenuous metallic connection to ground (and hopefully it is not tenuous) will have a negligible resistance compared with the body and no

current will flow. For this reason Class I equipment is tested for touch current with the protective earthing connection disconnected.

Class II equipment has no ground connection, and the primary-to-secondary capacitance of the mains transformer can allow enough current to flow through to the casework for it to be perceptible in normal use. Clearly, if the current was large enough, it would be hazardous.

Touch current is measured using a special network that connects the equipment to ground via resistors and capacitors, and expressed in terms of the voltage that results; this is then compared with the voltages that make a part 'hazardous live'. The special network is defined in Annexe C of BS EN 60065:2002.

Here are a few more miscellaneous safety requirements, not necessarily enshrined in BS EN 60065:2002.

- Mains fuse ratings must be permanently marked, and a legend of the form 'WARNING: replace with rated fuse only' must be marked on the PCB.
- Internal wiring does not have to be colour-coded (e.g., brown for live, blue for neutral) except for ground wiring, which must be green with a yellow trace.
- Crimp terminals on mains switches do not require colour-coding of their plastic shrouds.
- It is essential to keep an eye on mains transformer construction. With increasing globalisation, transformers are now being made in parts of the world which do not have a long history of technological manufacturing, and mistakes are sometimes made, for example, not using adequate insulation between primary and secondary.

Case Openings

As remarked elsewhere, in the section on mechanical design, case openings are subject to strict dimensional limits. A width of 3 mm is the maximum permitted. The old 'gold-chain' test has been removed from the latest edition of the standard, and is replaced by a narrow rigid test probe.

Equipment Temperature and Safety

There are limits on the permissible temperature rise of electronic apparatus, with the simple motivation of preventing people from burning themselves on their cherished hi-fi equipment. The temperature allowed is quoted as a rise above ambient temperature under specified test conditions. These conditions are detailed

Figure 29.2. Cable restraint by hooking the cable through the PCB.

below. There are two regimes of ambient considered: 'Moderate Climate' where the maximum ambient temperature does not exceed 35°C and 'Tropical Climate' where the maximum ambient temperature does not exceed 45°C. In the Tropical regime, the permitted temperature rises are reduced by 10°C. The temperature rise regulations are specified on pages 37–41 of BS EN 60065:2002 (Section 7).

The permitted temperature rise also depends on the material of which the relevant part is made. This is because metal at a high temperature causes much more severe burns than non-metallic or insulating material, as its higher thermal conductivity allows more heat to flow into the tissue of the questing finger.

The external parts of a piece of equipment are divided into three categories:

1. Accessible, and likely to be touched often. This includes parts which are specifically intended to be touched, such as control knobs and lifting handles.

Metallic, normal operation:	Temperature rise 30°C above ambient
Metallic, fault condition:	Temperature rise 65°C above ambient
Non-metallic, normal operation:	Temperature rise 50°C above ambient
Non-metallic, fault condition:	Temperature rise 65°C above ambient

This is usually an easy condition to meet, as knobs and switches are only connected to the internals of the amplifier via a shaft and a component such as a potentiometer or rotary switch that does not have good heat-conducting paths. Handles can be more difficult as they are likely to be secured to the front panel through a substantial area of metal, in order to have the requisite strength.

2. Accessible, and unlikely to be touched often. This embraces the front, top and sides of the equipment enclosure.

Metallic, normal operation:	Temperature rise 40°C above ambient
Metallic, fault condition:	Temperature rise 65°C above ambient
Non-metallic, normal operation:	Temperature rise 60°C above ambient
Non-metallic, fault condition:	Temperature rise 65°C above ambient

This is the part of the temperature regulations that usually causes the most grief. To work effectively, internal heatsinks have vents in the top panel above them, allowing convective heat flow. The escaping air heats the top panel and this can get very hot. Some amplifier designs have a plastic grille over the heatsink. This has several advantages. Since plastic is more economical to form than metal, the grille can have a structure that is more open and gives a larger exit area, while still complying with the 3 mm width limit for apertures. The grille itself is also allowed to get 20°C hotter because it is non-metallic, and for the same reason it conducts less heat to the surrounding metal top panel.

3. Not likely to be touched. This includes rear and bottom panels, unless they carry switches or other controls which are likely to be touched in normal use, external heatsinks and heatsink covers, and any parts of the top enclosure surface that are more than 30 mm below the general level.

Normal operating conditions:	Temperature rise 65°C above ambient

The permitted temperature under fault conditions is not specified, but it is probably safe to assume that a rise of 65°C is applicable.

4. The bottom panel is not likely to get very hot unless heatsinks are directly mounted on it, as it gets the full benefit of the incoming cool air. The rear panel can be a problem as its upper section will be heated by convection, and is typically at much the same temperature as the top of the unit; it also often carries a mains switch, which takes it out of this category.

The test conditions under which these temperatures are measured are as follows:

- One-eighth of the rated output power into the rated load.
- All channels driven and with rated load attached.
- The signal source is pink noise which is passed through an IEC filter to define the bandwidth to about 30 Hz–20 kHz. The details of the filter are given in Annexe C of BS EN 60065:2002.
- The mains voltage applied is 10% above the nominal mains voltage, so in Europe it is 230 V + 23 V = 253 V.

More information on the test conditions can be found on pp. 24–27 of BS EN 60065:2002.

The introduction of temperature rise regulations caused external heatsinks to become a rarity, despite the recognition that heatsinks are rarely going to be touched in normal operation. It is usually much more cost-effective to have the heatsinks completely enclosed by the casework, with suitable vents at top and bottom to allow convection. The heatsinks can then be run much hotter, so they can be smaller, cheaper and lighter, obviously assuming that the semiconductor temperature limits are observed; the limit for power transistors is usually 150°C, and for rectifiers 200°C. This usually allows the heatsinks to be safely run at 90°C or more, depending on the details of transistor mounting and the amount of power dissipated by each device. Hot heatsinks are more effective at dissipating heat by convection, but on the downside the restriction caused by the top and bottom vents, which must be of limited width, impairs the rate of airflow.

An exception to this is the use of massive heatsinks to form part of the case, to make an aesthetic statement. In this case the heatsinks are likely to be much larger for structural reasons than required for heat dissipation, and meeting the temperature-rise requirements is easy. Since aluminium extrusions are relatively expensive, this approach is restricted to 'high-end' equipment.

The importance of determining the temperature rise of the amplifier is such that it must be done as soon as possible in the product development process. If there are problems with EMC approval, they can usually be fixed by relatively minor internal modifications. Heat problems are much more intractable, because fixing them may entail major mechanical redesign and rethinking of the aesthetics, or as a last resort reducing the amplifier power rating. It is therefore essential to test as early as possible, even if you haven't procured all the parts yet. Get a mains transformer that is roughly right, and use a variable transformer to get it to the exact voltage required. Mechanical parts not ready? Block up the holes with cardboard and tape. It is important, however, to use the correct heatsink and get the ventilation arrangements as accurate as possible. The height off the test bench must be correct as it has a strong effect on air flow through vents in the bottom of the chassis.

Touching Hot Parts

It was described above how heatsinks are often mounted internally, with air circulation through protective grilles. The air holes in these grilles must be small enough to prevent parts that exceed the temperature rise regulations from being touched. The holes permitted are somewhat larger than those allowed near electrically hazardous parts. Testing is done with two probes. The 'toddler-finger probe', also known as Test-probe 19, simulates the finger of a child of 36 months or younger. It has a diameter of 5.6 mm and is articulated, with a hemispherical end. The other probe, Test-probe 18 covers persons from 36 months to 14 years; the diameter is 8.6 mm, and it is also articulated with a hemispherical end. The test force for both finger-type probes is 20 Newtons. On the Earth's surface a mass of 2.04 kg exerts a force of 20 Newtons on its support. It is unlikely you will be doing the testing anywhere else.

Instruction Manuals

The instruction manual is very often written in a hurry at the end of a design project. However, it must not be overlooked that it is part of the product package, and must be submitted for examination when the equipment itself is submitted for safety testing. There are rules about its contents; certain safety instructions are compulsory, such as warnings about keeping water away from the equipment.

Chapter 30

A Brief History of Solid-state Power Amplifiers

Those who cannot remember the past are condemned to repeat it.

George Santayana

This chapter attempts a very brief history of solid-state audio power amplifier technology. It may appear a touch Anglo-centric as I am much more familiar with English publications than American ones. In Britain, innovative designs were, until recently, usually published in *Wireless World*. I would be glad to hear of suggestions for amplifiers that deserve to be included in this chapter in later editions.

First Beginnings: 1953

One of the earliest transistor amplifier circuits was that shown in Figure 30.1; it was revealed by Mr G. Sziklai of RCA in a lecture in London in 1953.[1] While the familiar transistor symbol has yet to be invented, the design looks startlingly modern, directly coupled to the loudspeaker and with a fully complementary output stage.

According to Mr Sziklai, the power output was 500 mW at 2% distortion, with a gain of 28 dB. It is not clear if that was voltage or power gain.

Transformer-coupled Transistor Power Amplifiers: 1960s

In the early history of transistor amplifiers, transformers were often used for coupling between stages and matching to a loudspeaker impedance. They were, as now, relatively heavy and expensive components with dubious low-frequency linearity, but at the time they allowed the best use to be made of transistors that were both expensive and limited in performance.

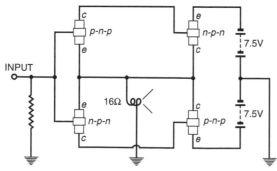

Figure 30.1. A Class-B push-pull transistor amplifier from RCA: 1953.

Transformers can exhibit overshoot and ringing if not properly terminated, and many early transformer-coupled amplifiers showed these defects even into a purely resistive load.

Figure 30.2 shows a typical low-power Class-B power amplifier using both driver and output transformers. Such circuitry was very widely used in transistor radios and other battery-powered equipment in the early and mid-1960s. The power output was typically 500 mW or less.

The driver stage Q1 works in Class-A, and care is needed that its standing collector current does not push the driver transformer towards saturation. There is no negative feedback; in more sophisticated versions, feedback was taken either directly from the output transformer secondary or via a tertiary winding. Typical impedance ratios for the transformers are shown. The output transformer was there for three reasons: firstly, the output transistors then available were happier with medium currents and voltages rather than high currents and low voltages. Secondly, if you are making a small loudspeaker, I suspect it is easier to make one with an impedance of 3 Ohms because there are fewer voice-coil turns, and such a low impedance is not a good fit with transistors even now. The impedance transformation is 500/3 = 167 times, giving a turns ratio of about 12:1. The final and most important function of the transformer was to add together the two halves of the Class-B output from the output devices. This can also be done by vertically stacking the output devices, as we shall see shortly.

The transformers were typically tiny and the performance mediocre at best. There was no thermal compensation of the quiescent current, which was set as low as possible to extend battery life. A pre-set is shown but the bias was often set by fixed resistors. As battery voltage declined, the crossover distortion became severe. Battery replacement was usually prompted when the amplifier would reproduce only the peaks of the audio.

A less common amplifier configuration had an output transformer, for all the reasons given above, but instead of the driver transformer, there was a transistor phase-splitter stage that gave push-pull drive to the output devices, illustrated in Figure 30.3. While this seems highly desirable to modern eyes, it never caught on at the time — this design goes all the way back to 1955. You must remember that transistors then were *expensive*, probably costing more than the small driver transformer. There was also the snag that the phase-splitter stage was inherently limited to a gain of one, as one of the outputs effectively came from an emitter-follower. This was not a cost-effective use of a transistor.

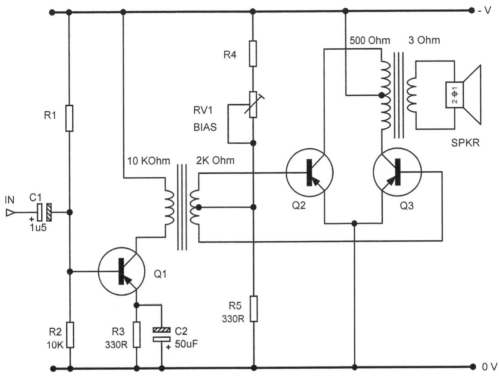

Figure 30.2. A generic low-power transistor amplifier with both driver and output transformers. Component values are omitted where they depend on supply voltage. Note PNP transistors and negative rail at the top.

The power output was only 50 mW; no distortion figures are available.

An alternative approach is to use a transformer only between the drivers and output devices, a choice driven by the availability of only one polarity of output devices; PNP in germanium. The transformer gave DC isolation between the two separate windings driving the output devices, and allowed two output transistors of the same polarity to be stacked on top of each other, so avoiding the need for a heavy and expensive output transformer to sum the two push-pull outputs and present the combined signal to the load. Several such designs were published in *Wireless World*.

Butler published a design for a transformer-coupled transistor power amplifier in 1958.[2] This had an initial phase-splitter (outside of any negative feedback loop) and all the rest of the amplifier worked in push-pull. It appears to be a quite sophisticated design with semi-local feedback loops that included the coupling transformer between the driver and output stages.

In 1961, R. C. Bowes published a transformer-coupled transistor power amplifier,[3] giving up to 10 watts into 15 Ω, which was then a relatively common

impedance for loudspeakers. The circuit had an initial phase-splitter, this time inside a global feedback loop, and two push-pull driver transistors in a long-tailed pair, driving two primary windings of the coupling transformer. There were separate quiescent current adjustment for each of the two output devices.

Arthur Bailey brought out another transformer-coupled amplifier giving 20 Watts into 16 Ω in 1966,[4] which appears to have been no great advance on the earlier designs. He felt obliged to admit in the article that the use of a coupling transformer might be considered old-fashioned.

The Rogers Ravensbourne incorporated what may have been the last transformer-coupled power amplifier to be introduced (in 1968). The architecture of this amplifier, shown in Figure 30.4, resembles that of a traditional valve amplifier, with a single-transistor input amplifier and feedback subtractor T6, driving a 'concertina' phase splitter T7 which provides in-phase and anti-phase signals to the two drivers T8, T9. These devices work in Class-A and are coupled to the push-pull output emitter followers T10, T11 by the transformer. This gives DC isolation and allows the

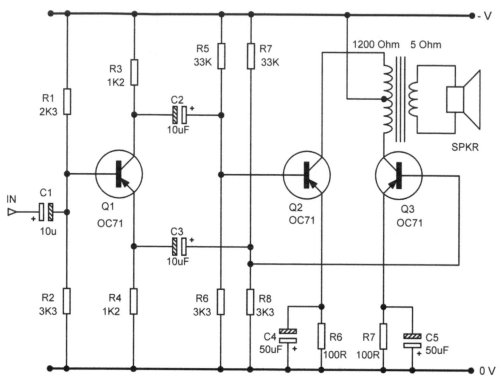

Figure 30.3. Low-power (50 mW) transistor amplifier from Philips in 1955 with output transformer only, the push-pull output devices being driven by a phase splitter. Note PNP transistors and negative rail at the top.

two emitter-followers to be stacked on top of each other. There may well be some impedance transforming going on, but that currently remains obscure as I'm afraid I have no details of the transformer, beyond the fact that it was quadrafilar wound to minimise leakage inductance. Unlike a valve amplifier, there is no output transformer to match the output stage to the load. The series feedback connection via R71 appears to set the stage gain to a modest 5.7 times; (+15.1 dB); there is no overall DC feedback because of the transformer in the forward path.

The compensation scheme appears to be a pole-zero RC network in the collector of T6; the zero introduced by R56 may be intended to cancel a pole somewhere else.

The preset control RV6 controls the DC conditions in the first two stages of the amplifier, presumably with the intention of setting T7 emitter and collector to about 1/4 and 3/4 of the +33 V rail respectively, to get the maximum signal swing. I would have thought that DC negative feedback through R54 would have been able to handle that if the circuit was suitably configured. There are separate pre-sets for the quiescent current in

the two output emitter-followers; I would have thought that adjustment of these would need to be carefully co-ordinated to get the output voltage at the right quiescent level, in the absence of any overall DC feedback. The only overload protection is the two fast-blow fuses in the supply rails, and there is no DC-offset protection. Despite this, the Ravensbourne had a reputation for reliability. Table 30.1 shows the official specifications.

Some strange figures there. It seems very wrong that the power into 15 or 8 Ohms is exactly the same, and that the power into 4 Ohms is half that into 8 Ohms. The NFB figure was 'calculated' from the cut-off frequency of the output devices and looks highly dubious.

The Ravensbourne was reviewed by *Gramophone* in 1968.[5] They liked it.

The Lin 6 W Amplifier: 1956

Figure 30.5 shows the famous Lin circuit,[6] generally considered to be the starting point for the design of

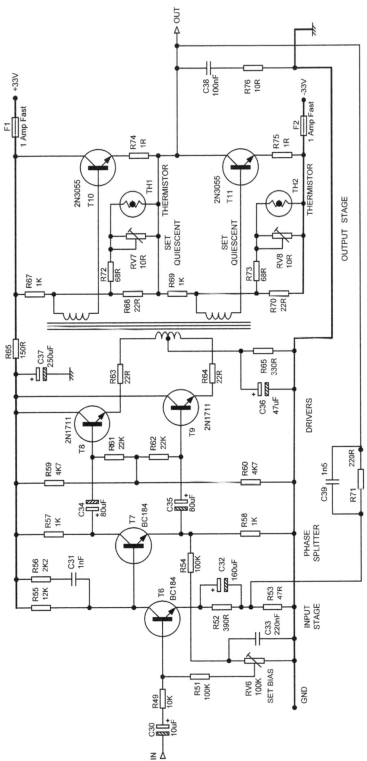

Figure 30.4. Schematic of the Rogers Ravensbourne transformer-coupled power amplifier: 1968. The original component numbering is preserved.

Table 30.1. Ravensbourne specifications

Power output:	25 + 25 Watts RMS into 15 or 8 ohms 13 + 13 Watts RMS into 4 ohms, 25 to 25 kHz
Harmonic distortion	Better than 0.06%, up to 15 Watts into 15 ohms at 1 kHz (0.1% average at 25 Watts)
IM distortion	Better than 0.25% up to 15 Watts into 15 ohms at 1 kHz (0.5% average at 25 Watts)
NFB	36 dB at 15 ohms, Stability margin 16 dB. Resistive or Loudspeaker load

modern power amplifiers, i.e., those without coupling transformers. H. C. Lin was a research engineer at RCA when he developed it. The amplifier was of course built wholly with germanium transistors,

silicon devices lying in the future. The Lin amplifier is a two-stage design, with a feedback-subtractor/ VAS Q1 followed by a unity-gain quasi-complementary output stage. There are two feedback paths: a DC path via R2, which in conjunction with R1 establishes the DC conditions, and AC feedback via R4, which you will note is taken from outside the output coupling capacitor C5. The use of shunt feedback made the input impedance rather low and in the complete published circuit, which included a preamplifier with volume and bass and treble controls, it was driven from the collector of a preceding stage.

The collector of Q1 is bootstrapped by C4, to increase the feedback factor, and thermal compensation of the output-stage biasing is done in a very familiar way by thermistor TH1. What is rather startling is the absence of emitter resistors in the output stage to make the biasing less critical. Given the much greater

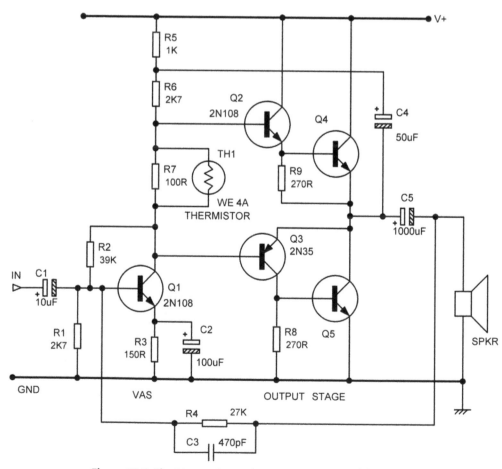

Figure 30.5. The Lin quasi-complementary power amplifier: 1956.

propensity of germanium transistors to go into thermal runaway, this seems puzzling.

The original article claimed it gave 6 Watts into 16 Ω at 1% THD, 'at mid frequencies' but a graph in the article shows that THD was much increased at 2 kHz, not a very high frequency in audio terms, reaching 2.2 % at 6 Watts out.

Despite the publication of the Lin configuration, power amplifiers with coupling transformers continued to appear for at least another 12 years, for example, the Rogers Ravensbourne described above. Nonetheless the Lin two-stage configuration was very influential as when transistors became freely available, it was the most economical way to make a power amplifier of reasonable quality. To take one example from many, the Metrosound ST20 integrated amplifier was a British design which flourished in the early 1970s. The power amplifier looked very much like the Lin, the main differences being the addition of 0.5 Ω emitter resistors and the replacement of the thermistor with a Vbe-multiplier

with preset adjustment. THD was about 1% at 10 Watts into 8 Ω.

The Tobey & Dinsdale Amplifier: 1961

Another milestone in power amplifier development was a design by Tobey & Dinsdale who published it in *Wireless World* in November 1961.[7] One version is shown in Figure 30.6; it has a three-stage architecture, unlike the two-stage arrangement of the Lin. There is a single input transistor to perform the feedback subtraction, a bootstrapped VAS, and a quasi-complementary output stage employing germanium PNP power transistors. There is only one temperature-compensation diode, when three would be expected to compensate the Vbe's in the output stage. It was AC-coupled to the load via a series capacitor of ungenerous size. The negative rail is at the top of the drawing, following the original, to make the configuration more familiar. This is necessary because the majority of the transistors are PNP.

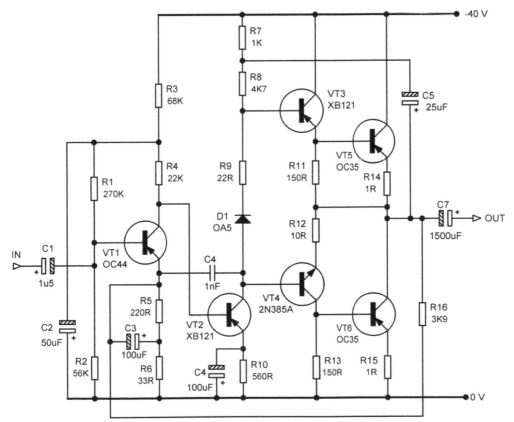

Figure 30.6. The Tobey & Dinsdale quasi-complementary power amplifier: 1961.

Table 30.2. Published distortion performance of the Tobey & Dinsdale amplifier (10W, 400 Hz)

Harmonic	Percentage
Second	0.1
Third	0.2
Fourth	0.05
Fifth	0.04
Sixth	0.02
Seventh and above	Less than 0.01

The distortion performance was much better than that of the Lin circuit, no doubt largely due to the extra open-loop gain provided by a three-stage architecture, but it was still very poor by modern standards. THD was quoted as 0.25 %, with plenty of high harmonics (presumably measured with a wave-analyser) as shown in Table 30.2.

Dinsdale alone published a preamp and power amp combination in January 1965,[8] the power amplifier being almost unchanged.

The Bailey 30 W Amplifier: 1968

This amplifier design was highly regarded when it was published in *Wireless World*.[9] It broke new ground in having a fully complementary output stage using silicon transistors and VI-sensing overload protection.

The schematic is shown in Figure 30.7. The input stage is a single transistor which inevitably generated second-harmonic distortion; it lacks the DC balance of an input pair and so a pre-set was required to set the output voltage. Clearly Arthur Bailey did not put all his faith in this arrangement, as you will note there is a 2000 uF non-polarised capacitor in series with the output, made up of two 4000 uF electrolytics back-to-back. Possibly it was intended to protect loudspeakers from DC-offset faults, as there was no output relay; its presence is discreetly ignored in the original article. The VAS collector load is bootstrapped, as in the earlier Tobey & Dinsdale amplifier. Distortion was quoted as 0.05% at 30 W, 8 Ω, 1 kHz.

The original article is notable in that it contained the first suggestion that Early effect and non-linear Cbc could seriously degrade VAS linearity.

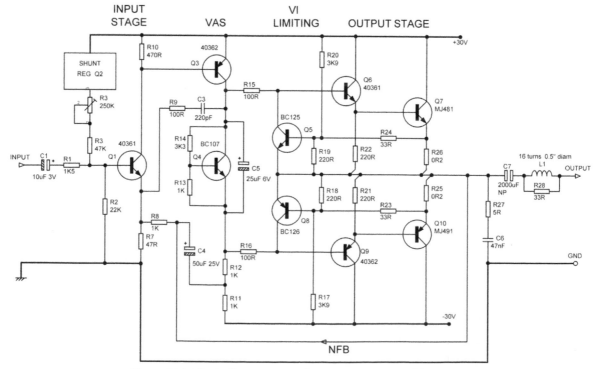

Figure 30.7. The Bailey 30W complementary power amplifier: 1968.

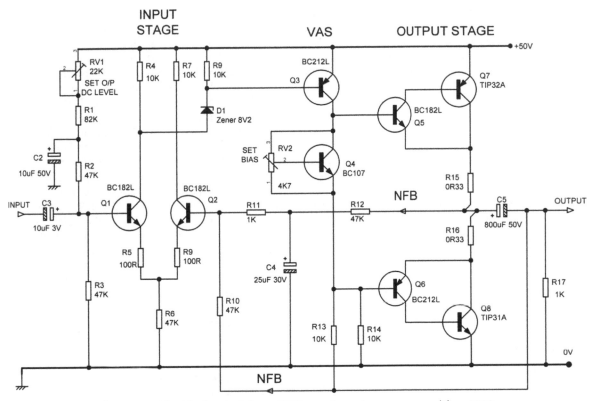

Figure 30.8. The Hardcastle & Lane 15W complementary power amplifier: 1969.

Hardcastle & Lane 15 W Amplifier: 1969

This is one of the earliest published designs to make use of a differential pair input stage.[10] However, the DC balance of this configuration was not exploited and the amplifier uses a single rail and an output coupling capacitor. Figure 30.8 shows the schematic, note the output devices are complementary silicon TIP types. Our dynamic duo in fact seem to have been unclear on exactly how their circuit worked, for the level-shifting Zener diode is quite unnecessary and might well degrade the noise performance. If the Zener voltage is 8.2 V, as the text implies, the input stage collector currents make no sense so I suspect there is a typo somewhere, probably in the value of R6.

The collector of the VAS is bootstrapped by connecting R13 to the loudspeaker. This crude method puts a DC current through the voice-coil, and R17 is needed to stop the amplifier going mad if the loudspeaker is disconnected. There are separate AC and DC feedback paths, AC via R10, and DC via R12. The authors give no reason why this was adopted, but one wonders if taking the AC feedback from after the pitifully under-sized output capacitor C5 was intended to reduce the distortion it undoubtedly created. The compensation is a bit of mystery; there is no Miller capacitor, but the input pair are degenerated with 100 Ω and R14 appears to be loading the VAS, which, if it is required for stability, makes it a poor substitute for a Miller capacitor. There are no base-emitter resistors on the silicon output transistors and this must have impaired turn-off at high frequencies.

The original design was fitted with simple current-limiting circuitry. This is not shown for clarity. Distortion was quoted as 0.035% at 30 W, 8 Ω, 1 kHz, the '15 W' in the amplifier name referring to the power into a 15 Ω load. Alarmingly, the THD was quoted as 0.135% at 15 W, 15 Ω, 1 kHz, when it should have been lower than the 8 Ω case.

The History of VAS Improvements

It is often difficult to pin down the first time that a particular circuit feature was used. For example, in Chapters 7 and 8 it is explained how adding an emitter-follower inside the Miller compensation loop, to create an EF-VAS, transforms the distortion performance. My first encounter with this very effective technique was in 1975 in the Cambridge Audio P60. An EF-VAS was also used in the Armstrong 621 power amplifier in the 1970s.

Another (less effective) way of enhancing a VAS is to cascode it. The valve cascode dates back to 1939. My research has not yet revealed when a cascode VAS was first used in an audio power amplifier, but the technique was certainly well known in the mid-1970s.

The first use of double input stages to drive a push-pull VAS that I know of was by Dan Meyer in 1973 in an amplifier design known as the Tigersaurus.[11] It was also employed in Jim Bongiorno's Ampzilla design in 1974;[12] he has said the idea was conceived some time before that.

If anyone has information on the earlier use of these technical features, I would be very glad to hear about it.

The History of other Technical Features

Information on the history of other amplifier technical features can be found in the relevant chapters of this book, thus:

Non-switching amplifiers	Chapter 4
Error correction	Chapter 4
Non-linear Cbc VAS distortion	Chapter 7
The EF-VAS	Chapter 7
The cascode VAS	Chapter 7
Output-inclusive compensation	Chapter 13
Two-pole compensation	Chapter 13
Class-G	Chapter 19
Class-D	Chapter 20

Transistors and FETs

The early solid-state power amplifiers used germanium power transistors because that was all that was available. Silicon power transistors were much tougher and could handle greater power, so they were used pretty much as soon as they became freely available. Their fast adoption was partly because no new design techniques were

required – just adjustments to the bias generators to allow for the higher Vbe. A milestone in power transistor technology was the arrival of the 2N3773, which could handle 16 Amps and 140 Volts, and dissipate 150 W via its TO-3 package. This allowed powerful amplifiers such as the Phase Linear and Crown series to be built. The 2N3773 was (and indeed is) an NPN device. At the time there were no comparable PNP transistors, so output stages were inevitably quasi-complementary. (Interestingly, it appears that now there is now a PNP complement to the 2N3773, called the 2N6609.)

The first silicon complementary power transistors had a limited power capability. The most common pair was the 2N3055 (NPN) and 2N2955 (PNP) rated at 100 V. Better devices such as the MJ802/MJ4502 came along (still in TO-3 packages), and after that sustained-beta devices such as the 2SC3281/2SA1302 pair made it easier to design amplifiers for low distortion into sub-8 Ω loads. This trend has continued with modern devices such as the Sanken 2SC3264 and 2SA1295 (in MT-200 packages) showing even flatter beta characteristics with collector current.

Hitachi lateral MOSFETs were launched in 1977.

Dead Ends of Amplifier Technology 1: Ultrasonic Biasing

In 1967 there was general agreement that crossover distortion was a major problem – perhaps the major problem – in transistor power amplifiers. Ferranti came up with their own solution, at the London Audio Fair in that year.[13] They announced a 30 Watt amplifier with no bias generator and therefore no quiescent current as such. Instead the crossover region was smoothed over by applying a 100 kHz signal derived from a multivibrator to the output stage. This arrangement was claimed to be very stable, and to dispense with a preset adjustment, but not surprisingly the idea was never heard of again.

Dead Ends of Amplifier Technology 2: Sliding-bias Amplifiers

The sliding-bias amplifier was one of many attempts to combine the low distortion of Class-A with the high efficiency of Class-B; the idea had some popularity in the late 1950s and early 1960s, stimulated by a Mullard design. The idea was that the amplifier would operate in push-pull Class-A at low levels, using a modest quiescent current. As the output level increased, the currents in the output devices were sensed and the bias reduced so that when full output was reached, operation was in Class-B. The current-sensing system relied on heavy resistor-capacitor filtering to keep the audio out of its path, and there lies the problem. When a sudden large transient comes along, how quickly can the bias controller respond? It seems certain that there would be periods of crunchingly under-biased Class-B operation during level changes, in what one might call an 'oops, sorry' mode. People who take measurements seriously are sometimes accused of 'designing for constant-amplitude sinewaves'; this really does seem to be a case of doing just that. A rather crude version using an output transformer was described by Thomas Roddam in *Wireless World* in 1962.[14]

A special case of the sliding bias principle was the 'Pi-mode' amplifier, also publicised by Mullard; the 'Pi' comes from the mathematics of Class-B power dissipation. This had the transition between Class-A and Class-B set at 40% of full power, the result of this being that the current drawn from the supply rail (only one) remained constant no matter what output power or Class was in use at a given moment. This was considered to be an advantage as it eliminated supply voltage variations with current drawn, and allowed simple RC smoothing to be effective, though that would clearly have been inefficient due to the voltage drop in the resistor. Increasing the PSRR of the amplifier itself would have been a sounder approach. A 10 Watt Pi-mode amplifier based on the Mullard design was published in *Wireless World* in 1963,[15] and that appeared to be the end of the matter.

However, in the audio business it is noticeable that very few technologies actually die completely, even though they should (ultrasonic biasing does, however, seem to be one example of complete extinction). The sliding-bias amplifier was the subject of a lively discussion on the DIYaudio forum as recently as 2007.[16] There have also been at least two manufacturers who have adopted what they called 'adaptive biasing' to reduce Class-A dissipation. One assumes that the bias control circuitry must be much more sophisticated and rapid in action than the historical sliding-bias amplifiers.

References

1. Anon, Transistor Circuits and Applications, *Wireless World*, Aug. 1953, p. 369.

2. Butler, F, Transistor Audio Amplifier, *Wireless World*, Nov. 1958, p. 529.

3. Bowes, R C, Transistor Audio Amplifier, *Wireless World*, July 1961, p. 342.

4. Bailey, A, High-Performance Transistor Amplifier, *Wireless World*, Nov. 1966, p.542.

5. *Gramophone* June 1968, p. 110.

6. Lin H C, Quasi-Complementary Transistor Amplifier, *Electronics,* Sept. 1956, p. 173ff.

7. Tobey, R and Dinsdale, J, Transistor Audio Power Amplifier, *Wireless World*, Nov. 1961, p. 565.

8. Dinsdale, J, 'Transistor High-Quality Audio Amplifier, *Wireless World*, Jan. 1965, p. 2.

9. Bailey, A, 30 W High Fidelity Amplifier, *Wireless World*, May 1968, p. 94.

10. Hardcastle, I and Lane, B, Low-Cost 15 Watt Amplifier, *Wireless World*, Oct. 1969, p. 456.

11. Meyer, D, Tigersaurus: Build This 250-Watt Hifi Amplifier, *Radio-Electronics,* Dec. 1973, pp. 43–47.

12. Bongiorno, J, http://www.ampzilla2000.com/Amp_History.html (accessed Mar. 2012).

13. Anon, London Audio Fair, *Wireless World*, May 1967, p. 249.

14. Roddam, T, Sliding-Bias Amplifiers, *Wireless World*, May 1962, p. 241.

15. Osborne, R and Tharma, P, Transistor High-quality Amplifiers, *Wireless World*, June 1963, p. 300.

16. http://www.diyaudio.com/forums/solid-state/100736-mullard-ss-power-amplifier.html (accessed Feb. 2012).

Index

T

Technics amplifiers: SU-V2 78, 79; SU-V2X 614; SU-V303 614; SU-V5 188; SU-V505 614; SU-V7A 614; Su-V707 614; SU-V505 97; SU-V96 514; SE-A100 85; SE-A3 K 97; SU-Z65 97; V1X 97; V2 97; V2X 97; V3 97; V4 97; V4X 97; V5 97; V6 97; V6X 97; V7 97; V8 97; V8X 97; V9 97

Technics New Class-A 97

Technics non-switching system 97

temperature changes, ambient 531—2

temperature coefficient (tempco): creating higher 531; creating lower 532; variable-tempco bias generator 530—5

temperature indication 683

temperature rise and safety 692—4

temperature sensors: integral 527—30; position 522—3

testing: DC offset protection 581; and fault-finding 686—8; overload protection 567; power supplies 688—9; powering up 688; and safety 685—94; servo 552; total harmonic distortion (THD) 14—16

thermal capacity 514

thermal compensation/thermal dynamics 505—42; accuracy required 506—9; basic 509—10; bias error assessment 510; CFP output stage 522; CFP output stage modelling 518, 519; EF output stage 520—2; EF output stage modelling 511—18; experiment 535—8; integral temperature sensors 527—30; integrated absolute error criterion 518—19; junction-temperature estimator 523—7; sensor location 522—3; simulation 510—11; variable-tempco bias generators 530—5

thermal coupler 506, 512, 514-5, 614, 616

thermal cycling (failure mode) 556

thermal distortion 120, 281—3

thermal protection 581—5

thermal resistance 196, 420, 444, 474, 510-8, 610, 614

thermal simulation 510—11, 527

thermal switch 584

thermal washers 610

ThermalTrak transistors 538—42

thermistors 582—3

three-stage amplifiers 75—6

thyratrons 486

Tigersaurus amplifier 704

time: amplitude distribution with 4—5; dead-time 492; relay drop-out 588

TO-3 package 615, 616—17

TO-3P package 253, 615, 616

TO-220 package 615, 616

TO-225AA package 615, 616

TO-264 package 615

Tobey & Dinsdale amplifier (1961) 701—2

tone-controls 19, 101

total harmonic distortion (THD) *see* harmonic distortion/THD

touch current 692

touching hot parts 694

transconductance 4, 58, 71, 75, 114, 126-9, 134-5, 138, 146, 162, 165, 172, 178, 182, 191, 212, 217, 224, 270, 331, 384, 432

transdiode, in output stage 244, 246

transformer balanced inputs 653, 663—4

transformer-coupled transistor power amplifiers (1960s) 696—8

transformers: evaluation 628—9; hum 604—5, 629—30; mains 599; mounting 626—7; safety 628; specifications 627—8; toroidal 626—7, 629, 631, 634

transient intermodulation distortion (TID) 77, 127, 665

transistor equation 244

transistors, historical development 704

transistor sockets 616

translinear loop 100

trigger, 12V 682, 683

trimodal amplifier 432—4, 438, 444—5; adaptive 446

Trio-Kenwood non-switching system 97

triple-based output 245—50, 262—3

tropical climates 693

two-pole compensation 341—9, 478—81

two-stage amplifiers 76

two-way speaker loads 380

U

ultrasonic biasing 704

ultrasonic filters 676—7

unbalanced inputs 649—51, 658—9

unbalanced outputs 649, 651

unconditional stability 329

undervoltage protection 491

Underwriter's Laboratories (UL6500 safety requirements) 690

unity-gain buffer 59, 288

V

VAF (SPICE parameter) 166—7, 178—9, 470, 640

valve sound 18

variable-gain balanced inputs 659—60

variable-tempco bias generator 530—5

VAS current-limiting 196, 221, 589

The Signal Transfer Company

The Signal Transfer Company is the only source for PCBs guaranteed to comply with the preamp and power amplifier design philosophies pioneered by Douglas Self

Shown above is the Class-G power amplifier, combining improved efficiency with first-class performance. The design is described in detail by Doug Self in Chapter 19.
The following PCBs are available:

- The Compact Class B power amplifier - Blameless performance in a small space
- The Load-Invariant power amplifier - very low distortion into heavy loads
- The Trimodal power amplifier - ultra-low distortion Class-A, switchable to Class-B
- The Class-G power amplifier - higher efficiency than Class-B, with mode indicator LED
- Power amplifier power supply - includes dual-rail opamp supply
- Amplifier and speaker protection card - DC offset and overtemp protection, with startup delay

- The Precision Preamplifier - active gain-control and variable frequency tone controls
- RIAA phono preamps - MM and MC inputs, superb RIAA accuracy. With unbalanced or balanced outputs
- Ultra-low noise balanced line input - with CMRR enhancement

Our PCBs have been designed with meticulous care at every point. The power amplifier board layouts are precisely the same as those approved by Douglas Self when his famous series of articles on power amplifier distortion appeared in Electronics World. You can therefore be confident that proper operation is built-in.

We supply the finest quality double-sided fibreglass PCBs, with a full solder mask, gold-plated pads, and a comprehensive silk-screen component overlay. Each PCB is supplied with extensive constructional notes, previously unpublished information about the design, and a detailed parts list to make ordering components simple.

Kits of parts to build the above PCBs are also available. These contain all PCB-mounted parts, including machined heatsink coupling plates for the power amplifiers, as shown in the illustration above. All products are also available fully built and tested.

For prices and more information go to http://www.signaltransfer.freeuk.com or contact:

The Signal Transfer Company,
Unit 9A Topland Country Business Park,
Cragg Road,
Mytholmroyd,
West Yorkshire,
HX7 5RW,
England.
Tel: 01422 885 196

Milton Keynes UK
Ingram Content Group UK Ltd.
UKHW052016071024
449327UK00027B/2290

9 780240 526133